Trace Element Metabolism in Animals–2

Proceedings of the Second International Symposium on Trace Element Metabolism in Animals, held in Madison, Wisconsin

Sponsored jointly by The Steenbock Symposium Fund, Department of Biochemistry, University of Wisconsin-Madison
and
The United States Department of Agriculture

Trace Element Metabolism in Animals–2

Edited by
W. G. Hoekstra, Ph.D., The University of Wisconsin
J. W. Suttie, Ph.D., The University of Wisconsin
H. E. Ganther, Ph.D., The University of Wisconsin
Walter Mertz, M.D., United States Department of Agriculture

University Park Press
Baltimore

Butterworths
London

University Park Press
International Publishers in Science and Medicine
Chamber of Commerce Building
Baltimore, Maryland 21202

Typeset in the United States of America by The Composing Room of Michigan
Printed in the United States of America by Universal Lithographers, Inc.

Library of Congress Cataloging in Publication Data

International Symposium on Trace Element Metabolism in
Animals, 2d, University of Wisconsin–Madison, 1973.
Trace element metabolism in animals–2.

Sponsored jointly by the Steenbock Symposium Fund;
Dept. of Biochemistry, University of Wisconsin–Madison;
and the U. S. Dept. of Agriculture.
1. Trace element metabolism–Congresses.
I. Hoekstra, W. G., ed. II. Steenbock Symposium Fund.
III. Wisconsin. University–Madison. Dept. of Biochemistry. IV. United States. Dept. of Agriculture.
V. Title.[DNLM: 1. Trace elements–Metabolism–Congresses. W3 IN924R 1973 / QU130 I634 1973t]
QP 531.I57 1973 591.1'33 74-11167

Published jointly by
Butterworth & Co. (Publishers) Ltd., London
and
University Park Press, Baltimore
ISBN 0-408-70647-3 (Butterworths)
ISBN 0-8391-0696-3 (University Park Press)

Contents

SHORT COMMUNICATIONS

Preface

This book is the published proceedings of the Second International Symposium on Trace Element Metabolism in Animals held at the University of Wisconsin-Madison, Wisconsin, U. S. A. on June 18–22, 1973. The first of these symposia (TEMA–1) was held in Aberdeen, Scotland in July of 1969, and the proceedings, *Trace Element Metabolism in Animals,* edited by C. F. Mills, were published by E. & S. Livingstone, Edinburgh and London, 1970.

The 1973 symposium was sponsored jointly by the Harry Steenbock Symposium Fund, Department of Biochemistry, University of Wisconsin-Madison, and by the United States Department of Agriculture. This symposium was also the Third Harry Steenbock Symposium. The proceedings of the first and second Steenbock Symposia have been published: H. F. DeLuca and J. W. Suttie (eds.), *The Fat-Soluble Vitamins,* The University of Wisconsin Press, Madison, Wisconsin, 1969; and R. D. Wells and R. B. Inman (eds.), *DNA Synthesis in Vitro,* University Park Press, Baltimore, Maryland, 1973. The fourth Steenbock Symposium, *Structure and Conformation of Nucleic Acids and Protein-Nucleic Acid Interactions,* will be held at the University of Wisconsin-Madison in June, 1974, coordinated by M. Sundaralingam and S. T. Rao.

The editors express their sincere appreciation to Mrs. Harry Steenbock for the endowment which supports the Harry Steenbock Symposia. Professor Harry Steenbock (1886–1967) was a distinguished Professor of Biochemistry at the University of Wisconsin from 1916 until 1956 and later an Emeritus Professor renowned for his many contributions, especially for his discovery that ultraviolet irradiation of foods produces vitamin D. While Steenbock's work centered on fat-soluble vitamins, he was no stranger to the trace element field, and co-discovered the essentiality of copper in preventing milk-induced anemia in the laboratory rat.

We wish also to acknowledge with appreciation the aid of the United States Department of Agriculture in providing support for the symposium and for the publication of these proceedings. Programs of the United States Department of Agriculture have contributed extensively to our knowledge of the role of trace elements in animal and human nutrition and metabolism. The desire of the USDA to disseminate such knowledge for use throughout the world is an admirable and important goal, and we are pleased to have had this opportunity to help in this effort.

It should be apparent from this publication that knowledge in the field of trace elements in animals and man is advancing rapidly. We hope that we have covered most of the high points in this field since the 1969 TEMA symposium and have eliminated gaps which existed previously. Undoubtedly, some important investigators and investigations have been missed for a variety of reasons, and for this, we apologize. Of special satisfaction in regard to this symposium was the merging of basic and applied knowledge of trace element metabolism in farm and laboratory animals with that of human subjects, and particularly, the merging of scientists in these various disciplines. The designation TEMA clearly implies trace element metabolism in man *as well as* other animals. We thank sincerely the participants of the symposium, and especially the session chairmen, for their contributions.

We look forward to the Third TEMA Symposium to be held in Munich, West Germany in July, 1977, organized under the guidance of Professor Dr. Manfred Kirchgessner, Institut für Tierernährung, 8050 Friesing-Weihenstephan, Munich, West Germany.

The editors wish to acknowledge with thanks the guidance of other members of the TEMA International Guiding Committee, Dr. C. F. Mills, chairman (Aberdeen, Scotland), Dr. M. Kirchgessner (Munich, Germany), Dr. V. V. Kovalsky (Moscow, U.S.S.R.), and Dr. E. J. Underwood (Wembley, Western Australia).

Finally, we wish to thank Mrs. Mary Parker for her excellent secretarial and managerial assistance, the wives of the local (Madison) organizers, Mrs. Leone Suttie, Mrs. Marion Ganther, and Mrs. JoAnn Hoekstra for their generous help during the symposium, and the many graduate students, technicians, secretaries, and others who contributed to the success of the symposium. From the editors' point of view, the hard work necessary in organizing and publishing such a symposium has been well rewarded, and we feel grateful for having had this opportunity.

W. G. Hoekstra
J. W. Suttie
H. E. Ganther
Walter Mertz

Registered Participants

Olav Alvares, Dept. of Oral Pathology, University of Illinois, 808 South Wood Street, Chicago, Illinois 60612

C. B. Ammerman, Dept. of Animal Science, Nutrition Laboratory, University of Florida, Gainesville, Florida 32601

Ernest D. Andrews, Wallaceville Animal Research Centre, Ministry of Agriculture and Fisheries, Private Bag, Upper Hutt, New Zealand

Jean Apgar, U.S. Plant, Soil and Nutrition Lab, Soil and Water Conservation Research, ARS, USDA, Tower Road, Ithaca, New York 14850

W. D. Armstrong, Dept. of Biochemistry, 227 Millard Hall, University of Minnesota, Minneapolis, Minnesota 55455

E. Aughey, Dept. of Veterinary Histology, University of Glasgow, Glasgow, Scotland

W. R. Beisel, U.S. Army Medical Research Institute for Infectious Diseases, Frederick, Maryland 21701

Marvin C. Bell, AEC Ag. Research Laboratory, University of Tennessee, 1299 Bethel Valley Road, Oak Ridge, Tennessee 37830

Duane Benton, Ross Laboratory, 625 Cleveland Avenue, Columbus, Ohio 43216

Eleanor Berman, Dept. of Biochemistry, Cook County Hospital, 1825 West Harrison Street, Chicago, Illinois 60657

W. T. Binnerts, Laboratorium voor Fysiologie der Dieren, Landbouwhogeschool, Agricultural University, Wageningen, The Netherlands

Linda C. Bloomer, Dept. of Internal Medicine, Hematology Division, University of Utah Medical Center, Salt Lake City, Utah 84112

Ian Bremner, Rowett Research Institute, Bucksburn, Aberdeen, AB2 9SB, United Kingdom

E. J. Briskey, Campbell Institute for Food Research, Camden, New Jersey 08109

R. Brown, Accu-Labs Research, Inc., 9170 West 44th Avenue, Wheat Ridge, Colorado 80033

Robert E. Burch, VA Hospital, 4101 Woolworth Avenue, Omaha, Nebraska 68105

F. J. Burger, Dept. of Chemical Pathology, Universiteit van die Orange-Vrystaat, Bloemfontein South Africa

Raymond F. Burk, U.S. Army Medical Research and Nutrition Laboratory, Fitzsimons General Hospital, Denver, Colorado 80240

Edith M. Carlisle, Dept of Environmental and Nutritional Sciences, University of California-Los Angeles, Los Angeles, California 90024

Earle E. Cary, U.S. Plant, Soil and Nutrition Lab, Tower Road, Ithaca, New York 14850

J. K. Chesters, Nutritional Biochemistry Dept., Rowett Research Institute, Bucksburn, Aberdeen, AB2 9SB, United Kingdom

Anthony V. Colucci, Bioenvironmental Laboratory Branch, Environmental Protection Agency, National Environmental Research Center, Research Triangle Park, North Carolina 27711

Robert J. Cousins, Dept. of Animal Sciences, Bartlett Hall, Rutgers University, New Brunswick, New Jersey 08903

Jose-Ma Culebras-Poza, Dept. of Biochemistry, Faculty of Pharmacy, Universidad Complutense, Madrid (3), Spain

Joe Dahmer, Murphy Products Company, Burlington, Wisconsin 53105

Thomas Daniels, Dept. of Biochemistry, University of Minnesota, Minneapolis, Minnesota 55455

George K. Davis, Div. of Sponsored Research, 219 Graduate School & International Studies Building, University of Florida, Gainesville, Florida 32601

A. T. Diplock, Royal Free Hospital Medical School, Dept. of Biochemistry, University of London, 8 Hunter Street, London, WC1N 1BP, England

Vern N. Dodson, Dept. of Preventive Medicine, Mayo Clinic, Rochester, Minnesota

E. A. Doisy, Jr., Dept. of Biochemistry, St. Louis University, 1402 South Grand Boulevard, St. Louis, Missouri 63104

R. J. Doisy, Dept. of Biochemistry, Upstate Medical Center, Syracuse, New York 13210

Delbert J. Eatough, Center for Thermochemical Studies, 191 EDLA, Brigham Young University, Provo, Utah 84601

Margaret E. Elmes, Physiology Dept., Medical Biology Centre, Queen's University, Belfast, BT9 7BL, Northern Ireland

Lawrence C. Erway, Dept. of Biological Sciences, Brodie Science Complex, University of Cincinnati, Cincinnati, Ohio 45221

Gary Evans, Human Nutrition Laboratory, USDA, P.O. Box D, University Station, Grand Forks, North Dakota 58201

Joe L. Evans, Dept. of Animal Science, Bartlett Hall, Box 231, Rutgers University, New Brunswick, New Jersey 08903

Eugene Faltin, Carnation Farms, Carnation, Washington 98014

Martin Ferguson, Dept. of Oral Pathology, University of Illinois, 808 South Wood Street, Chicago, Illinois 60612

Arthur Flynn, Dept. of Surgery, Cleveland Metropolitan General Hospital, 3395 Scranton Road, Cleveland, Ohio 44109

Richard M. Forbes, Dept. of Animal Science, University of Illinois, Urbana, Illinois 61801

W. Forth, Abteilung fur spezielle Pharmakologie, University of the Saarland, D-665 Homburg/Saar, West Germany

Gary Fosmire, Human Nutrition Laboratory, USDA, 2420 Second Avenue North, Grand Forks, North Dakota 58201

M. R. Spivey Fox, Div. of Nutrition BF124, Food and Drug Administration, 200 C Street, S.W., Washington, D.C. 20204

E. Frieden, Dept. of Chemistry, Florida State University, Tallahassee, Florida 32306

Douglas V. Frost, 17 Rosa Road, Schenectady, New York 12308

Howard E. Ganther, Dept. of Nutritional Sciences, 45A Home Economics Building, University of Wisconsin, Madison, Wisconsin 53706

Carmen Garcia-Amo, Dept. of Biochemistry, Faculty of Pharmacy, Universidad Complutense, Madrid 3, Spain

Thomas Gasiewicz, Dept. of Medicine and Dentistry, University of Rochester, Rochester, Minnesota 55901

I. Gedalia, Preventive Dentistry Chemistry, Faculty of Dentistry, Hadassah Medical School, Jerusalem, Israel

C. Goudie, Poultry Science Dept., Meat and Animal Science Building, University of Wisconsin, Madison, Wisconsin 53706

E. Grassmann, Inst. fur Tierernahrung der Univ. Technischen, 8050 Friesing-Weihenstephan, Munchen, West Germany

E. M. Gregory, Dept. of Biochemistry, Duke University Medical Center, Durham, North Carolina 27514

Clark Gubler, Dept. of Chemistry, 659 WIDB, Brigham Young University, Provo, Utah 84602

Betty Hackley, National Marine Fisheries Service, College Park, Maryland 20740

Andrew W. Halverson, Station Biochemistry Department, South Dakota State University, Brookings, South Dakota 57006

K. Michael Hambidge, Dept. of Pediatrics, University of Colorado Medical Center, Denver, Colorado 80220

Sam L. Hansard, Animal Science Dept., University of Tennessee, Knoxville, Tennessee 37916

J. Hartmans, I.B.S., P.O. Box 14, Wageningen, The Netherlands

W. Bernard Healy, Soil Bureau, Dept. of Scientific and Ind. Research, Private Bag, Lower Hutt, New Zealand

Mary Heckman, Ralston Purina Company, 835 South 8th Street, St. Louis, Missouri 63188

R. G. Hemingway, Glasgow University, Veterinary School, Bearsden, Glasgow, Scotland

Delbert D. Hemphill, Environmental Trace Substances Center, University of Missouri, 426 Clark Hall, Columbia, Missouri 65201

Robert I. Henkin, Experimental Therapeutics Branch, National Heart and Lung Institute, National Institutes of Health, Bethesda, Maryland 20014

A. Hennig, Karl-Marx Universitat, Sektion Tierproduktion und Veterinarmedizin, 69 Jena, Dornburger Strabe 24, German Democratic Republic

John L. Herrman, VA Hospital, Zorn Avenue, Louisville, Kentucky 40202

M. Hidiroglou, Canadian Dept. of Agriculture, Animal Research Institute, Central Experiment Farm, Ottawa, Canada

John M. Hill, Dept. of Food Science and Nutrition, Brigham Young University, Provo, Utah 84601

R. Hill, Royal Veterinary College, Boltons Park, Potters Bar, Herts, England

John P. Hitchcock, Dept. of Animal Husbandry, 204 Anthony Hall, Michigan State University, East Lansing, Michigan 48823

W. G. Hoekstra, Dept. of Biochemistry, University of Wisconsin, Madison, Wisconsin 53706

Leon L. Hopkins, Jr., ARS, USDA, P.O. Box E, Ft. Collins, Colorado 80521

Ronald Horst, Dairy Science Dept., University of Wisconsin, Madison, Wisconsin 53706

Donald J. Horvath, Animal Science Dept., West Virginia University, Morgantown, West Virginia 26506

Jeng M. Hsu, Biochemistry Research Lab., VA Hospital, 3900 Loch Raven Boulevard, Baltimore, Maryland 21218

John W. Huckabee, Environmental Sciences Division, Oak Ridge National Laboratory, P.O. Box X, Oak Ridge, Tennessee 37830

E. W. D. Huffman, Jr., Huffman Laboratories, Inc., 3830 High Court, P.O. Box 350, Wheat Ridge, Colorado 80033

Lucille Hurley, Dept. of Nutrition, University of California, Davis, California 95616

R. M. Izatt, Center for Thermochemical Studies, 191 EDLA, Brigham Young University, Provo, Utah 84601

R. M. Jacobs, Food and Drug Administration (BF-124), 200 C Street, S.W., Washington, D.C. 20204

K. J. Jenkins, Trace Element and Pesticide Section, 2044 Neatby Building, Animal Research Institute, Canada Dept. of Agriculture, Ottawa, Ontario, Canada

Herman L. Johnson, Bioenergetics Division, USAMR and NL, Fitzsimons Medical Center, Denver, Colorado 80240

R. L. Johnson, Environmental Sciences Association, Inc., Burlington, Massachusetts 01803

Ryoji Kawashima, Dept. of Animal Science, Faculty of Agriculture, University of Kyoto, Kyoto, Japan

Henry Kayongo-Male, 204 Anthony Hall, Dept. of Animal Husbandry, Michigan State University, East Lansing, Michigan 48823

Eldon W. Kienholz, Dept. of Animal Science, Colorado State University, Fort Collins, Colorado 80521

M. Kirchgessner, Inst. fur Tierernahrung der Univ. Technischen, 8050 Friesing-Weihenstephan, Munchen, West Germany

Les Klevay, Human Nutrition Laboratory, USDA, 2420 Second Avenue, North, Grand Forks, North Dakota 58201

V. V. Kovalsky, Vernadsky Institute of Geochemistry and Analytical Chemistry, USSR Academy of Sciences, Vorobievskoe Shosse, 47a, Moscow, U.S.S.R.

F. H. Kratzer, Dept. of Avian Sciences, University of California, Davis, California 95616

Pao-Kwen Ku, Dept. of Animal Husbandry, Michigan State University, East Lansing, Michigan 48823

Joe Kubota, SCS, U.S. Plant, Soil

and Nutrition Laboratory, Cornell University, Ithaca, New York 14850

H.-J. Lantzsch, Abt. Tierernährung, Universitat Hohenheim, 7 Stuttgart 70, BRD, Germany

James W. Lassiter, Dept. of Animal Science, University of Georgia, Athens, Georgia 30601

Roland M. Leach, Jr., Poultry Science Dept., 205 Animal Ind. Building, Pennsylvania State University, University Park, Pennsylvania 16802

H. J. Lee, C.S.I.R.O., Div. of Nutritional Biochemistry, Kintore Avenue, Adelaide, South Australia

Josef Leibetseder, Institute of Physiology, School of Veterinary Medicine, Vienna, Austria

R. W. Leucke, Dept. of Biochemistry, Michigan State University, East Lansing, Michigan 48823

O. A. Levander, Vitamin and Mineral Nutrition Lab, ARS, USDA, Beltsville, Maryland 20705

Gwyneth Lewis, Ministry of Agriculture, Fisheries and Food, Central Veterinary Laboratory, New Haw, Weybridge, Surrey, England

G. Paul Lynch, Ruminant Nutrition Laboratory, Nutrition Institute, Bldg. 200, ARC-East, Beltsville, Maryland 20705

A. MacPherson, Chemistry Dept., West of Scotland Agricultural College, Auchincruive, Ayr., Scotland, U.K.

Nolan F. Mangelson, Dept. of Chemistry, Brigham Young University, Provo, Utah 84601

John L. Martin, Dept. of Biochemistry, Colorado State University, Ft. Collins, Colorado 80521

Karl E. Mason, Nutrition Program Director, National Institute of Arthritis, Metabolism and Digestive Diseases, National Institutes of Health, Bethesda, Maryland 20014

Edward J. Massaro, Dept. of Biochemistry, State University of New York, G-56 Capen Hall, Buffalo, New York 14214

Gennard Matrone, Biochemistry Dept., North Carolina State University, Raleigh, North Carolina 27607

K. P. McConnell, Dept. of Biochemistry, University of Louisville, Louisville, Kentucky 40202

Walter Mertz, Vitamin and Mineral Nutrition Lab, Human Nutrition Research Division, ARS, USDA, Beltsville, Maryland 20705

H. H. Messer, Dept. of Physiology, University of British Columbia, Vancouver 8, Canada

Alfred H. Methfessel, VA Hospital, Metabolic Research Section, P.O. Box 55, Hines, Illinois 60141

Elwyn R. Miller, Animal Husbandry Dept., Michigan State University, East Lansing, Michigan 48823

James K. Miller, UT-AEC Agricultural Research Lab, University of Tennessee, 1299 Bethel Valley Road, Oak Ridge, Tennessee 37830

Lorraine T. Miller, Dept. of Foods and Nutrition, Oregon State University, Corvallis, Oregon 97331

Samuel T. Miller, Medical Dept., Brookhaven National Laboratory, Upton, L.I., New York 11973

W. J. Miller, Dairy Science Dept., Livestock and Poultry Building, University of Georgia, Athens, Georgia 30601

C. F. Mills, Dept. of Nutritional Biochemistry, Rowett Research Institute, Bucksburn, Aberdeen AB2 9SB, Scotland

David Milne, Dept. of Nutrition, L.A.I.R., The Presideo, San Francisco, California 94129

Milagros Miro, Dept. of Agronomy, University of Puerto Rico, Mayaguez, Puerto Rico 00708

Eugene R. Morris, Nutrition Institute, ARS, USDA, Agricultural Research Center, Beltsville, Maryland, 20705

Forrest Nielsen, Human Nutrition Laboratory, ARS, USDA, Box D, University Station, Grand Forks, North Dakota 58201

Hipolito V. Nino, Nutritional Biochemistry Section, Center for Disease Control, Atlanta, Georgia 30333

Donald Oberleas, Medical Research Service, VA Hospital, Allen Park, Michigan 48101

Claude L. Onkelinx, U-136, University of Connecticut, Storrs, Connecticut 06268

Donald E. Orr, Jr., Dept. of Animal Husbandry, Michigan State University, East Lansing, Michigan 48823

J. Pallauf, Inst. fur Tierernahrung der Univ. Techneschen, 8050 Friesing-Weihenstephan, Munchen, West Germany

Bozidar Panic, Institute for the Application of Nuclear Energy in Agriculture, Veterinary Medicine & Forestry, Zemun, Baranjska 15, Yugoslavia

J. Pařizek, Institute of Physiology, Czechoslovakian Academy of Sciences, Prague 4-KRC, Czechoslovakia

H. Mitchell Perry, Jr., Dept. of Medicine, Washington University School of Medicine, St. Louis, Missouri 63106

David H. Petering, Dept. of Chemistry, University of Wisconsin-Milwaukee, Milwaukee, Wisconsin 53201

Harold Petering, Dept. of Environmental Health, Kettering Laboratories, University of Cincinnati, 3223 Eden Avenue, Cincinnati, Ohio 45219

Henry A. Peters, Dept. of Neurology, University of Wisconsin, Madison, Wisconsin 53704

Cecil Pinkerton, Environmental Protection Agency, NERC, Research Triangle Park, North Carolina 27711

Magnus Piscator, Dept. of Environmental Hygiene, Karolinska Institute, Stockholm, Sweden

D. B. R. Poole, The Irish Agricultural Institute, Dunsinea, Castleknock, Co. Dublin, Ireland

A. L. Pope, Meat and Animal Science Dept., University of Wisconsin, Madison, Wisconsin 53706

Walter J. Pories, Dept. of Surgery, Cleveland Metropolitan General Hospital, 3395 Scranton Road, Cleveland, Ohio 44109

H. Porter, Dept. of Neurology, New England Medical Center Hospital, 171 Harrison Avenue, Boston, Massachusetts 02111

Kumar Prabhala, Indian Veterinary Research Institute, Izatnagar (U.P.), India

J. Quarterman, Rowett Research Institute, Bucksburn, Aberdeen AB2 9SB, Scotland

George H. Reussner, Corporate Research Dept., General Foods Corporation, White Plains, New York 10602

Paul J. Reynolds, U.S.D.A., ARS, Nutrition Institute, Ruminant Nutrition Lab, Bldg. 200, ARC-East, Beltsville, Maryland 20705

Bartolome Ribas-Ozonas, Dept. of Biochemistry, Faculty of Pharmacy, Universidad Complutense, Madrid 3, Spain

Mary E. Richardson, Environmental Protection Agency, Armed Forces Institute of Pathology, Washington, D.C. 20306

Alan Richter, Laboratory Cell Suppliers, 303 East Second Street, Frederick, Maryland 21701

Keith E. Rinehart, Ralston Purina Company, Checkerboard Square, St. Louis, Missouri 63188

P. Roth, Institut fur Tierernährung der Univ. Techneschen, 8050 Friesing-Weihenstephan, Munchen, West Germany

John T. Rotruck, Procter and Gamble, Miami Valley Research Labs, P.O. Box 39175, Cincinnati, Ohio 45239

Paul Saltman, Office of the Vice Chancellor for Academic Affairs, University of California, San Diego, La Jolla, California 92037

Harold H. Sandstead, Human Nutrition Laboratory, 2420 Second Avenue, North, P.O. Box D, University Station, Grand Forks, North Dakota 58201

A. Santos-Ruiz, Dept. of Biochemistry, Faculty of Pharmacy, Universidad Complutense, Madrid 3, Spain

F. J. Schwarz, Inst. fur Tierernahrung der Univ. Techneschen, 8050 Friesing-Weihenstephan, Munchen, West Germany

Klaus Schwarz, Lab ot Exptl. Metabolic Diseases, VA Hospital, 5901 East Seventh Street, Long Beach, California 90801

Bruce A. Scoggins, Howard Florey Institute of Experimental Physiology & Medicine, University of Melbourne, Parkville, Victoria, 3052, Australia

Roy Scott, Department of Urology, Royal Infirmary, Glasgow G4 OSF, Scotland

B. G. Shah, Health Protection Branch, Dept. of National Health and Welfare, Food Research Laboratories, Tunney's Pasture, Ottawa K1A OL2, Ontario, Canada

Raymond J. Shamberger, Dept. of Biochemistry, Cleveland Clinic, Cleveland, Ohio 44106

Thomas R. Shearer, University of Oregon Dental School, 611 S.W. Campus Drive, Portland, Oregon 97201

Samuel I. Shibko, Division of Toxicology, BF150, Food and Drug Administration, 200 C Street, S.W., Washington, D.C. 20204

Ray L. Shirley, Nutrition Laboratory, University of Florida, Gainesville, Florida 32601

Helen Sievers, Laboratory of Medicine, Mayo Clinic, Rochester, Minnesota 55901

Leon Singer, Dept. of Biochemistry, 227 Millard Hall, University of Minnesota, Minneapolis, Minnesota 55455

J. Cecil Smith, Jr., Trace Element Research Lab, V.A. Hospital, Washington, D.C. 20422

John R. J. Sorenson, Dept. of Environmental Health, Kettering Laboratory, University of Cincinnati, Cincinnati, Ohio 45219

A. G. Spais, Lab of Internal Medicine, Faculty of Veterinary Medicine, University of Thessaloniki, Thessaloniki, Greece

Julian E. Spallholz, Dept. of Biochemistry, Colorado State University, Ft. Collins, Colorado 80521

Herta Spencer, Metabolic Section, VA Hospital, Hines, Illinois 60141

Paul E. Stake, Dairy Science Department, University of Georgia, Athens, Georgia 30602

F. H. Steinke, Central Research, Ralston Purina Company, 900 Checkerboard Square, St. Louis, Missouri 63188

Paul Stitt, Wisconsin Whey Research Company, 1318 South 8th Street, Manitowoc, Wisconsin 54220

William H. Strain, Dept. of Surgery, Cleveland Metropolitan General Hospital, 3395 Scranton Road, Cleveland, Ohio 44109

M. L. Sunde, Poultry Science Department, University of Wisconsin, Madison, Wisconsin 53706

F. William Sunderman, Jr., Dept. of Laboratory Medicine, University of Connecticut Health Center, Farmington, Connecticut 06032

John W. Suttie, Department of Biochemistry, University of Wisconsin, Madison, Wisconsin 53706

N. F. Suttle, Biochemistry Department, Moredun Research Institute, Gilmerton, Edinburgh, Scotland

Doris C. Sutton, U.S. Atomic Energy Commission, 376 Hudson Street, New York, New York 10014

Richard Theuer, Mead Johnson Research Center, Nutritional Research Dept., Evansville, Indiana 47711

D. J. Thompson, International Minerals and Chemical Corp., IMC Plaza, Libertyville, Illinois 60048

Iain Thornton, Dept. of Geology, Imperial College, Prince Consort Road, London SW7, England

Edward W. Toepfer, Chief, Vitamins and Minerals Lab, USDA, Beltsville, Maryland 20705

Robert W. Tuman, Dept. of Biochemistry, Upstate Medical Center, Syracuse, New York 13210

Duane E. Ullrey, Animal Husbandry Dept., Anthony Hall, Michigan State University, East Lansing, Michigan 48823

Alejandro Uribe-Peralta, Calle 51A, No. 5-15, Bogota DE, Colombia

B. L. Vallee, Biophysics Research Lab, Peter Bent Brigham Hospital, Boston, Massachusetts 02115

D. R. Van Campen, Plant, Soil and Nutrition Lab, USDA, Tower Road, Ithaca, New York 14850

C. J. A. van den Hamer, Interuniversity Reactor Institute, Berlageweg 15, Delft, The Netherlands

A. I. Venchikov, Ashkhabad 740024, Besmeinskaya Street-4, U.S.S.R.

P. Venkateswarlu, Dept. of Biochemistry, 227 Millard Hall, University of Minnesota, Minneapolis, Minnesota 55455

Charles W. Weber, Poultry Science Dept., 324 Agricultural Sciences Bldg., University of Arizona, Tucson, Arizona 85721

Eugene D. Weinberg, Dept. of Microbiology, Jordan Hall, Indiana University, Bloomington, Indiana 47401

Ross M. Welch, U.S. Plant, Soil and Nutrition Lab, USDA, Tower Road, Ithaca, New York 14850

Ulrich Weser, Physiologisch-Chemisches Institut., Universitat Tubingen, Hoppe-Seyler-Strasse 1, 74 Tubingen, West Germany

Paul H. Weswig, Agricultural Chemistry, Oregon State University, Corvallis, Oregon 97331

P. D. Whanger, Dept. of Agricultural Chemistry, Oregon State University, Corvallis, Oregon 97331

Gerald Wiener, Animal Breeding Research Organization, West Mains Road, Edinburgh EH9 3JQ, Scotland

Robert Willes, Toxicology Division, Food Research Laboratories, Health Protection Branch, Tunney's Pasture, Ottawa, Ontario, K1A OL2, Canada

Wayne R. Wolf, Nutrition Institute, USDA, Room 213, Bldg. 307, ARC-East, Beltsville, Maryland 20705

Additional Participants

C. A. Baumann, Biochemistry Dept., University of Wisconsin, Madison, Wisconsin 53706

Paul Bray, Physiological Chemistry Dept., University of Wisconsin, Madison, Wisconsin 53706

R. H. Burris, Biochemistry Dept., University of Wisconsin, Madison, Wisconsin 53706

Louis W. Chang, Dept. of Pathology, University of Wisconsin, Madison, Wisconsin 53706

H. F. Deutsch, Physiological Chemistry Dept., University of Wisconsin, Madison, Wisconsin 53706

Mahmoud M. El-Begearmi, Poultry Science Dept., University of Wisconsin, Madison, Wisconsin 53706

Merle A. Evenson, Dept. of Medicine, University of Wisconsin, Madison, Wisconsin 53706

David Farb, Physiological Chemistry Dept., University of Wisconsin, Madison, Wisconsin 53706

Regina Fenster, Biochemistry Dept., University of Wisconsin, Madison, Wisconsin 53706

Mary Ruth Horner, Nutritional Sciences Dept., University of Wisconsin, Madison, Wisconsin 53706

Suzanne Jenkins, Nutritional Sciences Dept., University of Wisconsin, Madison, Wisconsin 53706

Nancy E. Johnson, Nutritional Sciences Dept., University of Wisconsin, Madison, Wisconsin 53706

Ed Kasarskis, Biochemistry Dept., University of Wisconsin, Madison, Wisconsin 53706

Hellen Linkswiler, Nutritional Sciences Dept., University of Wisconsin, Madison, Wisconsin 53706

Alexander Martin, Dept. of Anatomy, University of Wisconsin, Madison, Wisconsin 53706

Evelyn McGown, Biochemistry Dept., University of Wisconsin, Madison, Wisconsin 53706

Christine Olson, Nutritional Sciences Dept., University of Wisconsin, Madison, Wisconsin 53706

David Parmeler, Physiological Chemistry Dept., University of Wisconsin, Madison, Wisconsin 53706

Dale Smith, Dept. of Agronomy,

University of Wisconsin, Madison, Wisconsin 53706

Roger A. Sunde, Biochemistry Dept., University of Wisconsin, Madison, Wisconsin 53706

Anne Swanson, Biochemistry Dept., University of Wisconsin, Madison, Wisconsin 53706

Shyyhwa Tao, Biochemistry Dept., University of Wisconsin, Madison, Wisconsin 53706

June B. Tartt, Nutritional Sciences Dept., University of Wisconsin, Madison, Wisconsin 53706

Jane Voichick, Nutritional Sciences Dept., University of Wisconsin, Madison, Wisconsin 53706

Pat Wagner, Nutritional Sciences Dept., University of Wisconsin, Madison, Wisconsin, 53706

Ching Fu Wang, Biochemistry Dept., University of Wisconsin, Madison, Wisconsin 53706

Trace Element Metabolism in Animals – 2

INTRODUCTORY LECTURE

TRACE-ELEMENT RESEARCH: RECENT DEVELOPMENTS AND OUTLOOK

WALTER MERTZ

Nutrition Institute, Agricultural Research Service, United States Department of Agriculture, Beltsville, Maryland

Since Professor Eric Underwood presented the opening address to the 1st International Symposium on Trace Element Metabolism in Animals in Aberdeen four years ago, many significant advances have been made in this field. Not only has our knowledge of new trace elements with essential functions increased by the addition of tin, vanadium, nickel, fluorine, and silicon to the list of essential elements, but new findings have identified certain trace elements as substances of great public health concern in areas with malnutrition problems and in industrialized societies. The work on metabolic and nutritional aspects of iron, zinc, and chromium deficiency has acquired great urgency in view of the demonstration of marginal deficiencies in man. Much progress has been made in our understanding of interactions between trace elements, such as the protective effect of selenium against heavy metals which is an example of the great potential of trace-element research to meet the problems of the future. In these past four years, we have also seen how changing practices of man have, in some areas, caused a redirection of our concerns, for example in the field of iodine nutrition.

All these examples are representative of two underlying developments that may well determine the direction of trace-element research in the future. The first is the man-made redistribution of trace elements in the environment, and the second the changing habits of man, which profoundly influence his nutrition as well as that of his farm animals. Man is redistributing trace elements along a concentration gradient; concentrated deposits which are generally protected and harmless to man are exploited and are eventually dissipated in the environment. This has reached levels which may be detrimental to health on life-long exposure. Redistribution also occurs against a concentration gradient, when essential trace elements are taken out of the soil and gradually concentrated in the food chain until they reach toxic concentrations in sewage sludge. This is an example of redistribution causing two problems: the first is a possible depletion of agricultural soils of available essential trace elements, and the second is the disposal of high, potentially toxic levels in the byproducts of the food chain. The changes in man's practices include the use of new high-yielding varieties of plants. Our knowledge about their trace-element metabolism and its impact on human and animal nutrition is minimal. The changes also include intensive production practices to raise animals. The best-known example is the practice that resulted in the appearance of zinc deficiency in agricultural animals in the United States. In human nutrition the predictable trend toward partitioning of foods and the increasing use of food analogs may lead to serious problems unless trace-element research can make recommendations to safeguard an optimal intake of essential trace elements.

These developments will affect many aspects of the human environment but it is likely that they will express themselves first through changes in the balance of trace elements.

These changes have occurred in the past and will continue in the future. They are slow and subtle changes. The consequences for human and animal health can be expected to be just as subtle. They will not amount to a matter of life or death within our lifetime, and probably not even to a question of serious disease. The main problem of trace-element research in the future may well be the definition of slight deviations of biological functions from optimal to suboptimal owing either to marginal overexposure or to marginal deficiency. It will be of utmost importance for trace-element research to recognize and diagnose these subtle changes and to assess their consequences for health if we want to make recommendations for prevention and remedy. This means that ever more sensitive diagnostic criteria are needed. In some countries we have begun to use neurological and behavioral criteria to determine air-quality standards. We are studying the consequences of a very marginal zinc deficiency for the offspring of experimental animals rather than death or the rate of weight gain as a criterion. We have learned to prevent death from lead poisoning and now we have become interested in the early effects of overexposure of this metal. As we have learned to prevent and to cure severe selenium deficiency in farm animals, our interest is turning to the diagnosis and study of marginal selenium deficiency or overexposure in human and animal nutrition and to the potential connection of such states to a variety of human diseases. It is evident that we can expect to solve these complex problems only if we use everything that the vast field of modern science has to offer. To probe into the entatic sites of metalloenzymes is an essential part of these efforts, as is the study of mechanisms by which plants extract trace elements from the soil. To define the subtle consequence of marginal under- or overexposure is as important as the identification of essential functions for trace elements or as the improvement of analytical methods. It is only through multidisciplinary efforts that we can identify the optimal levels of the trace elements in the environment, and only with this understanding can we hope to create and maintain an optimal environment for animals and man.

THE ENTATIC PROPERTIES OF COBALT CARBOXYPEPTIDASE AND COBALT PROCARBOXYPEPTIDASE[1]

BERT L. VALLEE

Biophysics Research Laboratory, Department of Biological Chemistry, Harvard Medical School and Division of Medical Biology, Peter Bent Brigham Hospital, Boston, Massachusetts

The key to the specificity and catalytic potential of metalloenzymes seems to lie in the diversity and topological arrangement both of their active site residues and of their metal atoms, all of which interact with their substrates. The behavior of metals in metalloenzymes may be compared with that of the functional amino acid side chains of such and of other (nonmetallo) enzymes. In both, amino acid side chains exhibit unusual chemical reactivities toward specific organic reagents, reactivities not observed in simple peptides or denatured proteins. The hyper-reactivities of, for example, functional sulfhydryl, histidyl, and lysyl groups of particular enzymes to organic reagents, specific for these residues under mild conditions (1), may be related to unique chemical properties of these amino acid side chains. These, in turn, may be attributable, perhaps, to environmental factors originating from the three-dimensional environment in which they may find themselves. Questions then arise as to the relation of such chemical hyper-reactivity to biological function.

A number of years ago we became interested in applying the principles of chelate chemistry to metalloenzymes and to comparing the spectrochemical properties of metal complex ions with those of metalloenzymes. We found that their absorption, CD, and EPR spectra were quite different. Without exception, the spectra of the metalloenzymes then known were quite unlike

1. This work was supported by Grant-in-Aid GM-15003 from the National Institutes of Health, Department of Health, Education, and Welfare.

those of model compounds, indicating to us that in the former the metal is bound in low symmetry, reflecting a distorted geometry of the binding sites (2, 3). Since then this deduction has been supported by numerous investigations of other metalloenzymes and confirmed by X-ray crystallographic analyses of the few metalloenzymes which have been completed so far, e.g., rubredoxin, carboxypeptidase A and carbonic anhydrase; in each case, the metal atom is bound in low symmetry and there is significant distortion from regular octahedral or tetrahedral geometries. On the other hand, when these or other enzymes are denatured, and are therefore inactive, the geometry of their coordination sites becomes more conventional, i.e., similar to that of metal complex ions and of metalloproteins which do not exhibit biological activities. Thus, spectroscopic, optical rotatory, and magnetic properties of metalloenzymes (2–6) suggest that characteristics of their metal atoms are also affected by the three-dimensional protein environment of active enzymatic sites. This seems analogous to the considerations outlined for functional amino acid residues. In addition, it has been noted frequently that substrates or inhibitors render the unusual absorption, CD, and EPR spectra of metalloenzymes more nearly similar to those of well-defined model systems. These observations have emphasized the singular nature of metals and their coordination in metalloenzymes, perhaps a reflection of their chemical features which are significant in catalysis.

Closer inspection has revealed that the binding sites of metalloenzymes are polydentate, the metal ion being coordinated by from three to five ligands. Metal complex ions employing similarly polydentate coordination are generally insoluble in nonpolar solvents, raising the question of solvation, of heterogeneity, and phase differences at the active sites of enzymes which often comprise both hydrophylic and hydrophobic residues. Suitable water-soluble model systems that could serve for comparative purposes are not available as yet in adequate numbers.

The foregoing observations led us to advance the hypothesis that the irregular metal binding site of metalloenzymes reflects an energy state which is favorable to catalysis, an *entatic state* (2). The word *entasis,* derived from the Greek, symbolizes a state of strain or stress, thought to be a chemical reflection of biological potential—in this instance for enzymatic catalysis. The entatic state differs from the familiar concept of 'strain' in enzyme catalysis in that 'entasis' does not involve substrate, being an intrinsic property of the enzyme itself. 'Strain' is usually reserved for the thermodynamic consequences of enzyme-substrate interaction.

The entatic state thus refers to energy relationships at the active site of the enzyme *prior to* the formation of an enzyme-substrate complex; it implies a topographic domain of the enzyme poised for catalysis. This operational definition permits studies based on hypotheses concerning the properties of the enzyme which are essential to catalysis, but without making judgments on the merit of the hypotheses or advancing detailed explanations for the basis of the observation. The precepts of operationalism pertinent to this mode of procedure have been discussed (7, 8). The entatic-state hypothesis has served us as a major focus of experimental design, and this discussion will indicate the directions taken.

CARBOXYPEPTIDASE A OF BOVINE PANCREAS

Work performed in our laboratory on carboxypeptidase A, a single-chain enzyme, will be discussed for illustrative purposes.

A number of proteolytic enzymes contain zinc, which is essential for catalysis, but it can be replaced with catalytically active transition metals, such as cobalt whose probe properties are more favorable than those of zinc. Among these peptidases, carboxypeptidase A of pancreas has been studied most extensively (9–11). Some of its properties are summarized in Table 1. The spectra of cobalt carboxypeptidase A are indicative of asymmetry around the cobalt atom involved in catalysis (12).

The visible spectrum of cobalt carboxypeptidase has a shoulder at about 500 nm and maxima at 555 and 572 nm, both with absorptivities ϵ slightly greater than 150 (Fig. 1b). Two infrared bands are centered at 940 nm and 1570 nm ($\epsilon = 20$). The location of the high-energy band at 940 nm is unusual, since the near-infrared absorption bands of divalent cobalt complex ions are

Table 1. Some properties of carboxypeptidase A

Exopeptidase	
Molecular weight: 34,600	
Single polypeptide chain	
1 Zn atom, catalytically active toward peptides and esters	
Amino acid residues essential to activity:	2 tyrosines 1 glutamic acid 1 arginine
Apoenzyme: inactive	
Replacement of Zn with Co^{2+}, Mn^{2+}, Ni^{2+} :	enzymes active as both peptidases and esterases
Replacement of Zn with Cd^{2+}, Hg^{2+}, Pb^{2+}, Rh^{2+} :	active as esterases only
Other metal replacements: inactive	

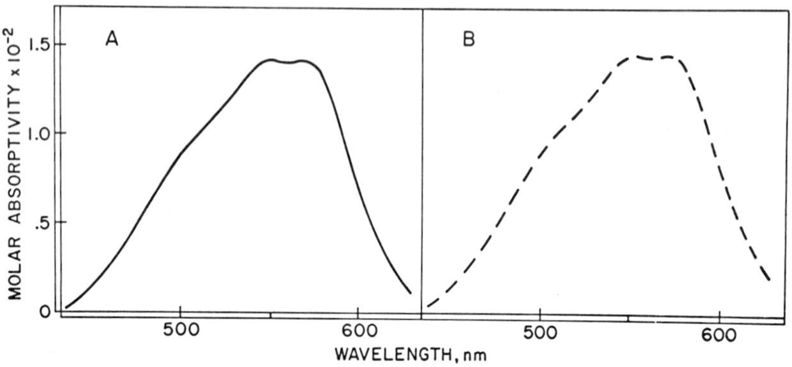

Fig. 1. Absorption spectrum of (A) cobalt procarboxypeptidase ———, and (B) cobalt carboxypeptidase - - -. Conditions: 0.005 M Tris-Cl, 1.0 M NaCl, pH 7.1, 25°C (13).

generally found at longer wavelengths. In cobalt spectra the position of the infrared bands often probe the metal environment more effectively than do the bands in the visible spectra. In the present instance the IR spectrum suggests that the interaction of cobalt with its ligands is stronger than that in cobalt complex ions.

At liquid-helium temperatures, absorption measurements show an increase in spectral resolution. There are two new shoulders: the maximum at 555 nm is shifted to 532 nm, but that at 572 nm is maintained at this wavelength. Absorptivity is not reduced greatly, arguing against participation of vibronic interactions in the generation of the intensity of the spectra. The spectral changes at 4.2°K are reversed by a gradual increase of the temperature to 25°C. A magnetic field of 47,000 G renders the absorption band at 572 nm optically active (12).

The geometry of the metal binding site is reflected in the circular dichroic spectra which may also encompass information on the influence of vicinal factors on the cobalt atom. A negative band at 538 nm and the shoulder near 500 nm probably correspond to the low-wavelength maximum and shoulder of the absorption band at room temperature enhanced in the spectrum observed at 4.2°K. The spectral detail suggests the existence of distinct transitions with different polarization.

Agents which inhibit enzymatic activity also alter the absorption and CD spectra of cobalt carboxypeptidase. Spectral details observed as a function of the structure and concentration of inhibitors added are shown in Table 2. They are characteristic for inhibitors with different functional groups.

The CD spectrum of the cobalt carboxypeptidase glycyl-L-tyrosine complex is particularly interesting. It differs both from that of the enzyme itself and from that of most of the enzyme-inhibitor complexes examined. The extremum of the negative Cotton effect of the cobalt enzyme at 537 nm shifts to 555 nm with an inversion of sign; the molar ellipticity increases from −500 to +2000, a four-fold change in magnitude (Fig. 2b). The striking inversion of sign as well as the marked enhancement suggest major electronic rearrangements around the cobalt atom upon interaction with substrates. This impression is reinforced by the fact that the band at 940 nm splits into two, now located at ~850 and ~1150 nm, respectively, while that at 1570 nm is shifted to 1420 nm.

The absorption and CD spectra of cobalt carboxypeptidase indicate that the enzymatically active metal occupies an asymmetric site and that the geometry of coordination of the cobalt atom is irregular. The spectra, while delineating constraints on the nature of the metal-site geometry, also suggest that the outer cobalt electrons are delocalized, evidence of strong interactions with the protein. These changes could reflect directional forces generated at the active site by the asymmetry around the cobalt atom, magnified and shifted upon enzyme-substrate interaction. Thus, in this instance, cobalt serves simultaneously as an integral part of the catalytic apparatus and as a spectral probe which is capable, perhaps, of providing new insight into enzyme action.

The spectra of cobalt neutral protease and of cobalt thermolysin resemble those of cobalt carboxypeptidase, except that the signs of the CD bands are opposite.

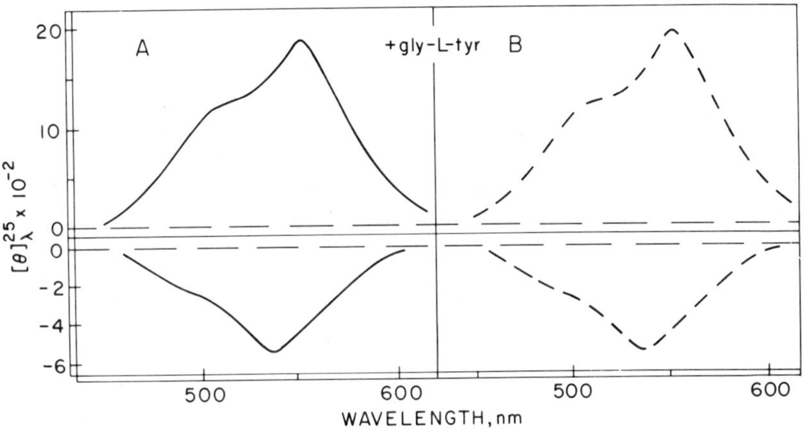

Fig. 2. Circular dichroic spectra of (A) cobalt procarboxypeptidase ——— (lower, CD spectrum; upper, CD spectrum after addition of glycyl-L-tyr, 20 mM) and (B) cobalt carboxypeptidase - - - (lower, CD spectrum; upper, CD spectrum after addition of glycyl-L-tyrosine, 10 mM). Conditions: 0.005 M Tris-Cl, 1.0 M NaCl, pH 7.1, 25°C (13).

Table 2. Inhibitor effects on cobalt carboxypeptidase adsorption and circular dichroic spectra[1,2]

Inhibitor	Spectral K_I [mM] or I (mM)	Absorption Maxima (nm) (ϵ cm^{-1} M^{-1})		CD bands (nm) ([θ], deg cm^2/decimole)
		Visible	Infrared	
β-Phenylpropionate	(0.5)	505(110),555(160),575(155),610(80)	1040	490(−250),543(−650)
	(2)	510(125),555(150),590(120)	1100	490(−450),536(−900), 580(−800)
β-Iodopropionate	(0.5)	505(110),558(178),575(165)	1030	490(−200),551(−650)
	(5)	520(140),555(160),590(130)	1090	495(−375),555(−850), 590(−550)
Phenylacetate	(2)	510(125),555(155),575(165),605(95)	1030	495(−200),539(−400)
	(20)	515(160),575(180)	1090	510(100),536(−275), 600(125)
Indole-3-acetate	(0.8)	510(120),557(180),573(180),610(90)	1010	401(50),555(−175)
	(25)	520(130),565(160),590(130)	1070	492(−300),537(−500)
Butyrate	(1)	505(105),555(165),570(170),605(100)	1010	495(−150),573(−450)
	(20)	515(130),575(150)	1090	495(−150),585(−700)

Acetate	(1000)	515(135),575(155)	1050	510(100),555(-300)
D-Phenylalanine	(0.5)	545(165),580(205),602(180)	1030	510(100),575(100)
L-Phenylalanine	[3]	510(130),555(200),574(205),610(140)	1000	510(200),545(0), 580(100)
Benzoate	[20]	505(105),560(165),575(160),610(90)	1000	485(-50),505(25), 560(-400)
Thiobenzoate	[5]	505(115),555(275),574(300),625(125)[3]	1080	475(-100),510(0) 570(-700),370(-2000)
L-β-phenyllactate	[<0.5]	510(135),560(195),572(190)	970	495(-300),527(-325) 585(100)
L-Lysyl-L-Tyrosylamide	[7]	515(175),555(175)	1030	505(1500),550(1050)

[1] 1 M NaCl, 0.005 M Tris-Cl, pH 7.1, 25°C.

[2] The effects of inhibitor on the cobalt spectra were examined in H_2O. Hence, in the infrared only effects on the band at 940 nm could be examined.

[3] High background has prevented resolution of absorption below 390 nm.

PROCARBOXYPEPTIDASE A

The entatic-site hypothesis would suggest that a metalloprotein, thought to be enzymatically inert but exhibiting an entatic spectrum, actually has catalytic potential which may not have been previously realized or appreciated. The examination of this proposition would constitute an examination of the hypothesis and of its predictive value. The zymogen of carboxypeptidase A, procarboxypeptidase, lends itself particularly well to such experiments. Both the bovine zymogen, molecular weight 87,000 and the enzyme, molecular weight 34,600, contain a single zinc atom which can be replaced with cobalt.

Generation of functional potential is the essence of zymogen activation, a process which has been thought to involve conformational changes that might either affect the catalytic site or reveal, form, or complete the substrate-binding site, or all of these. Our studies have demonstrated that the spectrochemical properties of the cobalt procarboxypeptidase-carboxypeptidase pair are virtually identical and, based on this, the zymogen was shown to exhibit substantial peptidase activity, as great or greater than that of the native enzyme (13).

The visible spectrum of cobalt procarboxypeptidase exhibits maxima near 555 and 572 nm ($\epsilon \sim 145$), and a shoulder near 500 nm. It is almost identical to that of cobalt carboxypeptidase in all respects (Figs. 1A,B), as are the infrared bands.

The coordination of the cobalt atom in both the zymogen and the enzyme is similarly asymmetric as indicated by the striking identity of their CD (Figs. 2A,B) and MCD spectra. Remarkably, the perturbation of these spectra by inhibitors such as β-phenylpropionate and of substrates such as glycyl-L-tyrosine is virtually identical to that of cobalt carboxypeptidase. The CD spectrum of the glycyl-L-tyrosine cobalt procarboxypeptidase complex is nearly identical with that of the corresponding complex with cobalt carboxypeptidase, indicating, first, that glycyl-L-tyrosine binds to the zymogen, and, second, that it influences the cobalt environment of both proteins in a similar fashion.

It has long been known that the zymogen exhibits very low activity toward N-substituted dipeptides such as carbobenzoxyglycyl-L-phenylalanine, contrasting with the respective carboxypeptidases. This difference could be due to the fact that the *binding residues* of the zymogen are inadequate to recognize such substrates, but that the catalytic site does pre-exist prior to activation, contrary to past beliefs. The absorption and CD spectra suggested that in both the zymogen and the enzyme the metal is poised to participate in the catalytic process. Hence, we employed peptide substrates, minimal in size and complexity but with the specificity requirements of the enzyme. Further, the amplification of activity achieved by the substitution of cobalt for zinc was expected to facilitate recognition of catalytic potential.

Haloacylated aromatic amino acids fulfill these conditions and cobalt procarboxypeptidase catalyzes their hydrolysis as readily as does the native enzyme. Among the series of acyl phenylalanine substrates studied, dichloroacetylphenylalanine is hydrolyzed most rapidly, with a turnover number of 980 min^{-1}, followed by chloroacetyl, bromoacetyl, acetyl-glycyl-, iodoacetylphenylalanine, and acetylphenylalanine. The rates of hydrolysis are equal to or greater than those observed for the native zinc enzyme (Table 3). In addition to the acyl phenylalanine substrates listed here, the acyl derivatives of other amino acids have yielded analogous results.

In contrast, the cobalt and zinc zymogens hydrolyze the conventional blocked di- and tri-peptide substrates such as carbobenzoxyglycyl-L-phenylalanine and carbobenzoxyglycylglycyl-L-phenylalanine at about 1/40th the rate of native zinc carboxypeptidase (Table 3).

Both cobalt procarboxypeptidase and the native zinc enzyme exhibit Michaelis-Menten kinetics toward chloro- and dichloroacetylphenylalanine, and their pH otpima, k_{cat} and K_M values are all virtually the same. Further, both β-phenylpropionate and β-phenyllactate are competitive inhibitors with

Table 3. Activities of bovine cobalt procarboxypeptidase A [(PCPD)Co] and of native bovine zinc carboxypeptidase A [(CPD)Zn], toward various substrates (13)

Substrates[1]	[(PCPD)Co] $V'/e(min^{-1})$	[(CPD)Zn] $V/e(min^{-1})$	$\frac{[(PCPD)Co]}{[(CPD)Zn]} \times 100$ $V'/V(\%)$
Ac Phe	3	18	600
I Ac Phe	150	80	190
Cl Ac Phe	900	720	125
Br Ac Phe	520	540	96
Cl_2 Ac Phe	980	2100	47
Ac Gly Phe	340	1640	21
Cbz Gly Phe	180	7350	2.5
Cbz Gly Gly Phe	200	7840	2.5

[1] Ac Phe, acetylphenylalanine; I Ac Phe, iodoacetylphenylalanine; Cl Ac Phe, chloroacetylphenylalanine; Br Ac Phe, bromoacetylphenylalanine; Cl_2 Ac Phe, dichloroacetylphenylalanine; Cbz Gly Phe, carbobenzoxyglycyl-L-phenylalanine; Cbz Gly Gly Phe, carbobenzoxyglycylglycyl-L-phenylalanine.

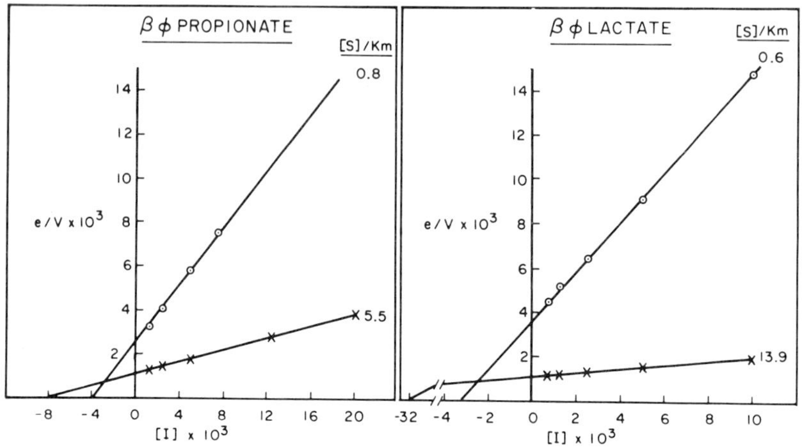

Fig. 3. Inhibition of cobalt procarboxypeptidase catalyzed hydrolysis of chloroacetylphenylalanine by β-phenylpropionate and β-phenyllactate. K_M for chloroacetylphenylalanine is 0.9×10^{-3} M and the ratios of $[S]/K_M$ were chosen both above and below this value. Conditions: 0.05 M Tris-Cl, 1.0 M NaCl, pH 7.5, 25°C (13).

identical apparent K_i values (Fig. 3), and perturbation spectra of the cobalt zymogen-enzyme pair constitute another link between their spectral and catalytic features. We are unaware of previous instances in which metal spectra have served as guides to the existence of activity in a metalloprotein, based on consideration of symmetry and geometry of metal coordination.

Zymogen activation has long been thought to present unusual opportunities to understand further the structural bases of enzyme action. This process has been thought to involve rearrangements or major changes in protein structure, leading to the proper alignment of amino acid side chains from various regions of the molecule so as to produce effective substrate binding sites. The mechanisms by which active center residue(s) acquire enzymatic capacity, and the possible interdependence and interrelationships of these processes, have been the subject of much speculation. Since cobalt procarboxypeptidase is at least as effective a catalyst toward these haloacylamino acids as is the native zinc enzyme, conformational rearrangements need not be postulated to account for their hydrolysis, much as the features of the activation process that lead to enhanced binding and catalysis of conventional substrates remain to be defined.

Metal substitutions (14), inhibitor and substrate binding (14), and organic modifications (15) suggested long ago that the bovine procarboxypeptidase A molecule contains significant aspects of the active center of carboxypeptidase. Thus, bovine procarboxypeptidase A exhibits low (5–6%) esterase activity toward hippuryl-dl-β-phenyllactic acid (HPLA) prior to activation (16). We have suggested earlier that some features of the active center carboxypeptidase, already existent in the zymogen, allow the hydrolysis of some substrates, but that activation ultimately results in the acquisition of catalytic capacity and specificity (14). Thus, the catalytic apparatus apparently exists in procarboxypeptidase prior to activation, but the substrate binding loci are developed incompletely (17).

The carboxypeptidase catalyzed hydrolysis of HPLA, its peptide analog benzoylglycyl-L-phenylalanine (BGP), and carbobenzoxyglycyl-L-phenylalanine (CGP) are characterized by multiple kinetic anomalies (18) which are not apparent in the kinetics of the present acylamino acid derivatives. Detailed kinetics of the bovine procarboxypeptidase A catalyzed hydrolysis of HPLA, BGP, and CGP are not yet available for comparison with those of the native enzyme. However, based largely on results with these substrates (18), we have suggested previously that the anomalies for the native zinc enzyme might relate to their size relative to its binding area, which is designed to accommodate much larger peptides and proteins. Our 'dual site' model postulated distinct but overlapping dipeptide and ester binding sites, as well as productive and nonproductive (i.e., inhibitory binding) modes, for their respective enzyme complexes (19, 20). Whether or not such considerations would be relevant to the zymogen is not known, nor is it possible to judge at present whether or not inadequate development of productive or nonproductive binding sites might pertain equally to dipeptides and esters. The observed activities of this zymogen toward such and similar substrates could be due to the insufficient development, imbalance, or productive and nonproductive binding modes, and preponderance of ester *versus* peptide binding sites or combinations of these. In the absence of systematic kinetic investigations of the bovine cobalt procarboxypeptidase A catalyzed hydrolysis of natural or synthetic substrates of varying chain lengths, such considerations render unprofitable speculations regarding detailed mechanism(s) designed to account for the low rates of hydrolysis of particular substrates by this zymogen.

In the present study the catalytic effectiveness of the cobalt atom at the active site (12, 18) combines with the productive binding of the halogenated acylamino acids to yield rates far above such 'background'. For these substrates they become, in fact, equal to or greater than those of the hydrolytic rates of the native enzyme. The present data suggest that the active metal site is entatic prior to activation of bovine procarboxypeptidase A. We are presently examining conditions requisite for detailed kinetic comparisons between the characteristics of various zymogen and enzyme derivatives acting on different substrates.

The present findings demonstrate the existence of both catalytic and binding sites of the cobalt zymogen for acylamino acids. They suggest, moreover, that some of the sites required for the binding of conventional substrates must pre-exist in the bovine zymogen. Hence, the magnitude of conformational changes at the catalytic site thought to accompany its activation must be very small, no matter what their magnitude might be elsewhere in the molecule. The process of activation may require a less substantial conformational change of the catalytic subunit than might be inferred from past general views of such mechanisms.

In formulating the entatic-state hypothesis we have emphasized the importance of spectra as probes of active sites of metalloenzymes. Such spectra reveal geometric and electronic detail thought to be pertinent to the function of such systems (2, 5, 6). The hypothesis gains significance from experiments designed to evaluate critically inherent assumptions and postulates. The present enzymatic findings bear out the implications of the entatic spectrum of bovine cobalt procarboxypeptidase A. They support the hypothesis and suggest that it may profitably guide the exploration of catalytic potential and substrate specificity of metalloproteins and of metal-protein complexes.

We have previously commented upon the fact that both selective chemical modifications of specific, hyper-reactive amino acid side chains of a large number of (non)metalloenzymes and the physical-chemical properties of metalloenzymes constitute apparent links between chemical and catalytic properties of these proteins (2). Hence, the present discussion of metallozymogen/enzyme systems also directs attention to serylzymogen/enzyme systems which might be examined profitably against the background of the above discussion.

Overall, the examination of metalloenzymes has generated hypotheses and experimental results which, while pertinent to their mechanisms, may likely prove important to those of other enzymes. Even the process of zymogen activation has thereby been elucidated, an unlikely turn of events at first glance but one that might have been anticipated on reflection.

REFERENCES

1. Vallee, B. L., and Riordan, J. F. 1969. Chemical approaches to the properties of active sites of enzymes. *A. Rev. Biochem.* 38: 733.
2. Vallee, B. L., and Williams, R. J. P. 1968. Metalloenzymes: the entatic nature of their active sites. *Proc. Natn. Acad. Sci. U.S.A.* 59: 498.
3. Vallee, B. L., and Williams, R. J. P. 1968. Enzyme action: views derived from metalloenzyme studies. Chem. Br. 4: 397.
4. Vallee, B. L., and Wacker, W. E. C. 1970. Metalloproteins. *In* H. Heurath (ed.), *The Proteins,* Vol. 5, Academic Press, New York.
5. Williams, R. J. P. 1971. The entatic state. *Cold Spring Harb. Symp. Quant. Biol.* 36: 53.

6. Vallee, B. L., Riordan, J. F., Johansen, J. T., and Livingston, D. M. 1971. Spectro-chemical probes for protein conformation and function. *Cold Spring Harb. Symp. Quant. Biol.* 36: 517.
7. Bridgeman, P. W. 1946. *The Logic of Modern Physics,* Macmillan, New York.
8. Ogden, C. K., and Richards, I. A. 1947. *The Meaning of Meaning,* Harcourt-Brace, New York.
9. Neurath, H., Bradshaw, R. A., Petra, P. H., and Walsh, K. A. 1970. Bovine carboxypeptidase A–activation, chemical structure and molecular heterogeneities. *Phil. Trans. R. Soc.* B 257: 159.
10. Lipscomb, W. N., Reeke, G. N., Jr., Hartsuck, J. A., Quiocho, F. A., and Bethge, P. H. 1970. Structure of carboxypeptidase A. VIII. Atomic interpretation at 0.2nm resolution, a new study of the complex of glycyl-L-tyrosine with CPA, and mechanistic deductions. *Phil. Trans. R. Soc.* B 257: 177.
11. Vallee, B. L., Riordan, J. F., Auld, D. S., and Latt, S. A. 1970. Chemical approaches to the mode of action of carboxypeptidase A. *Phil. Trans. R. Soc.* B 257: 215.
12. Latt, S. A., and Vallee, B. L. 1971. Spectral properties of cobalt carboxypeptidase. The effects of substrates and inhibitors. *Biochemistry* 10: 4263.
13. Behnke, W. D., and Vallee, B. L. 1972. The spectrum of cobalt bovine procarboxypeptidase A. *Proc. Natn. Acad. Sci. U.S.A.* 69: 2442.
14. Piras, R., and Vallee, B. L. 1967. Procarboxypeptidase A-carboxypeptidase A interrelationships. Metal and substrate binding. *Biochemistry* 6: 348.
15. Freisheim, J. H., Walsh, K. A., and Neurath, H. 1967. The activation of bovine procarboxypeptidase A. II. Mechanism of activation of the succinylated enzyme precursor. *Biochemistry* 6: 3020.
16. Yamasaki, M., Brown, J. R., Cox, D. J., Greenshields, R. N., Wade, R. D., and Neurath, H. 1963. Procarboxypeptidase A-S6. Further studies of its isolation and properties. *Biochemistry* 2: 859.
17. Neurath, H., and Walsh, K. A. 1970. Evolutionary interrelationships of proteolytic enzymes as a tool for probing internal structure. *Proc. VIIIth International Congress of Biochemistry,* p. 68.
18. Davies, R. C., Riordan, J. F., Auld, D. S., and Vallee, B. L. 1968. Kinetics of carboxypeptidase A. I. Hydrolysis of carbobenzoxyglycyl-L-phenylalanine, benzoylglycyl-L-phenylalanine, and hippuryl-dl-β-phenyllactic acid by metal-substituted and acetylated carboxypeptidases. *Biochemistry* 7: 1090.
19. Vallee, B. L. 1967. Approaches to the mechanism of action of carboxypeptidase A. *Proc. VIIth International Congress of Biochemistry,* p. 149.
20. Vallee, B. L., Riordan, J. F., Bethune, J. L., Coombs, T. L., Auld, D. S., and Sokolovsky, M. 1968. A model for substrate binding and kinetics of carboxypeptidase A. *Biochemistry* 7: 3547.

INVOLVEMENT OF COPPER AND THE CUPREINS IN THE BIOCHEMISTRY OF OXYGEN[1]

ULRICH WESER

Physiologisch-Chemisches Institut der Universität Tübingen, Tübingen, Federal Republic of Germany

Our information regarding both the structure and biochemical action of a considerable number of copper proteins has increased during the last 30 years. The main results have been reviewed or presented in symposia (1–8). In the present survey special attention is placed on the cupreins, which have recently become the most actively studied copper proteins. The cupreins are frequently called hemocuprein, hepatocuprein, cerebrocuprein, cytocuprein, or erythro-cupro-zinc protein. At the moment no final decision can be made concerning the proper naming of this copper protein. Recently the name superoxide dismutase was introduced owing to the accelerated dismutation of superoxide into oxygen and peroxide in the presence of this metalloprotein. Whether or not the above biochemical action can be considered the physiological role of the cupreins, however, requires further study.

The first isolation of cuprein from bovine tissues was achieved in 1939 (9). The isolated protein contained approximately 0.34% copper and the molecular weight was 35,000. In contrast to the greenish-blue copper protein prepared from erythrocytes, a colorless protein of identical molecular weight and copper content was found in liver. However, Mohamed and Greenberg (10) successfully isolated a colored equine hepatocuprein. From the late fifties studies on the cupreins have progressively intensified (11–15). Copper-balance studies revealed that some 50–60% of the erythrocyte copper is present in erythrocuprein (14). A similar copper content was found in a soluble copper protein, designated cerebrocuprein I, in normal human brain (16–18). Already, a striking similarity of all these soluble tissue Cu proteins had become apparent.

1. Supported by DFG grants We 401/4, 401/6, and 401/9, and a personal EMBO long-term fellowship.

During the past eight years comprehensive research on both the human and the bovine cupreins has been conducted (19–43). Deutsch and co-workers (23) proved the identity of erythro-, cerebro-, and hepatocuprein prepared from human tissues. The physicochemical properties, immunochemical behavior, metal content, and amino acid analysis were identical in each of the copper proteins employed.

In 1970 two zinc ions in addition to copper were found in the human cupreins (23). This result was confirmed using the cupreins isolated from bovine tissues (26, 27, 44). Treatment with protein-unfolding agents such as urea and/or sodium dodecylsulfate combined with short heating resulted in the appearance of 16,000 molecular weight subunits (28, 36). Owing to the well-established superoxide dismutase activity of the cupreins, McCord and Fridovich (24) assigned this biochemical activity as the predominant physiological role of these copper proteins. However, detailed enzymic studies (45–48) led to the proposal of another, probably even more important, function of erythrocuprein, namely the scavenging of singlet oxygen in metabolism. Surprisingly, a similar biochemical action was shown using the copper chelates of lysine, tyrosine and histidine (49).

ISOLATION OF THE CUPREINS

Generally two different isolation procedures are used. Treatment with organic solvents such as ethanol/chloroform and acetone (9–11) was the method originally devised and is still successfully used today (24, 26, 27). Substantial modifications were developed later by introducing ion exchange and gel chromatography. In the alternative method only aqueous solutions are used, which are subjected to different chromatographic steps (21, 22). The last method represents a gentler treatment of the protein. However, cuprein was found to be one of the most stable proteins (33) which survived without any measurable damage a short treatment with organic solvents. The first isolation process is advantageous with respect to the economy of time, since, for example, possible deterioration of the protein due to microbial growth or other harmful side reactions will be minimized. The conversion into the apoprotein was most successful using chelator-equilibrated columns (30, 32). This apoprotein proved homogeneous and essentially metal free as examined by gel filtration with and without an applied electrical field, ultracentrifugation, and atomic absorption spectroscopy, respectively.

ANALYTICAL DATA

In many laboratories the copper content of the cupreins was independently measured to be 2 g atom per 32,600 ± 200 g protein (9, 12, 18, 20, 23, 26,

28, 36). The concentration of the second metallic component was 2 g atom zinc per mole protein (23, 26, 27, 44). Unfortunately, we do not know whether or not the 16,000 mol wt subunits contain one copper and one zinc each. The alternative possibility would be a two-copper-containing subunit bound with a two-zinc-containing second subunit. Although there are some preliminary X-ray crystallographic data (55) of the monoclinic and orthorhombic forms of erythrocuprein available, no final decision regarding the metal binding sites can be made at present. In all of the cupreins an unusually low concentration of aromatic amino acid residues can be seen (Table 1). Glycine and aspartate are the most abundant amino acids.

OPTICAL ACTIVITY

The absorption spectra of human and bovine cupreins are depicted in Fig. 1. In the case of human cuprein absorption bands appear at 254, 260, 266, and 326 nm and shoulders at 270, 282, and 289 nm. The low content of aromatic amino acids was confirmed by the unusually low absorption in the 280 nm region (ϵ_{265} = 17,000). A significant difference from human cuprein in both

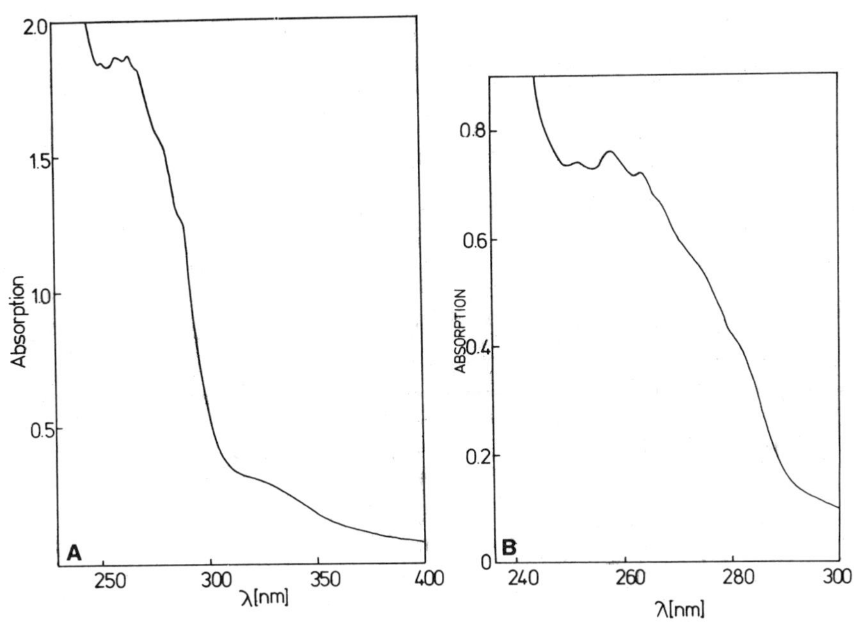

Fig. 1a. Absorption spectrum of human cuprein in the ultraviolet region. Recording with a Unicam SP 1800 using 1 cm light path quartz cells.

Fig. 1b. Absorption spectrum of bovine cuprein in the ultraviolet region. Recording with a Unicam SP 1800 using 1 cm light path quartz cells.

Table 1. Amino acid composition of cupreins purified from various sources

	Human							Bovine				Garden	Neurospora	Saccharomyces	
	Erythrocytes				Liver		Brain	Erythrocytes			Heart	pea	crassa	cetevisiae	
	(12)	(20)	(21)	(22)	(18)[1]	(22)	(22)	(26)	(36)	(51)	(36)	(52)	(53)	(54)	
Lys	20.3	22.9	22.2	23.2	17.7	22.5	22.1	20.9	22	22.8	22	10	12	18	21.3[3]
His	14.5	16.5	15.8	12.8	10.4	13.3	13.7	15.3	16	16.8	16	18	11	11	10.7
Arg	9.1	8.6	6.8	7.8	7.1	8.4	8.1	7.2	10	8.7	9	6	9	7	8.1
Asp	41.3	40.2	36.5	36.2	27.8	36.9	36.6	35.1	35	34.1	36	45	36	32	33.7
Thr	18.3	16.9	15.8	16.2	12.0	16.1	16.4	24.9	26	22.4	25	30	26	18	18.8
Ser	19.7	19.8	19.8	20.3	14.1	20.9	20.7	14.1	20	15.4	21	14	14	20	21.3
Glu	28.2	18.1	26.3	27.4	22.8	28.7	27.6	27.8	24	22.5	27	19	20	25	27.3
Pro	11.1	11.3	10.4	11.3	8.9	10.7	10.0	11.6	14	14.6	15	14	14	20	15.6
Gly	45.6	52.2	50.1	47.1	36.6	48.0	47.5	51.6	50	50.4	54	56	39	40	40.1
Ala	20.8	21.4	19.8	20.6	16.1	21.2	20.9	18.4	21	19.6	22	21	20	24	27.5
Val	23.0	28.4	27.6	29.6	19.3	28.2	29.7	28.3	28	28.4	26	21	22	28	27.2
Met			0.0	0.1	0.6	0.3	0.0	2.0	0	2.1	0	0	0	2	0.0
Ile	11.0	16.0	16.4	16.5	10.7	16.1	16.8	19.0	17	18.1	17	20	13	9	11.4
Leu	17.8	19.7	17.4	19.4	15.0	19.4	18.5	16.8	20	17.5	18	21	11	11	14.4
Tyr	1.9	0.0	0.0	0.1	1.5	0.3	0.0	1.8	2	2.3,2.0[4]	2	0	2	2	2.4
Phe	9.9	8.1	7.8	8.9		8.2	8.4	8.0	10	8.5	10	9	6	10	12.8
Cys[5]	11.3	5.1	6.6	6.7	4.3	7.7	6.3	6.2		5.9		6	3	4	3.4
Cys			0.7					1.0[6]		2.0		0			
Try	0.6							0.4		0[4]				0	
Total	304	305	300[7]	304	225	310	303	309[7]	315	310[7]	320	310	258	281	296

[1] Assuming 33,600 mol. wt.

[2] Assuming 32,700 mol. wt.

[3] Values in this column recalculated using analytical data of (44,50).

[4] From reference (32).

[5] Determined after oxidation.

[6] After treatment with 8M guanidine hydrochloride.

[7] Cysteine excluded.

the absorption profile and the molar absorption in the UV region is apparent using bovine cuprein (Fig. 1b). The molar coefficient of absorption was calculated to be 9840 at 259 nm.

In the visible region the cupreins from either biological source displayed virtually identical absorption spectra. A shoulder was detectable at 430 nm and an absorption maximum was clearly visible in the 680 nm region ($\epsilon \sim$ 300). Recording of the spectra at low temperature (77°K) resulted in a dramatic rise of the absorption (Fig. 2).

An interesting phenomenon was observed when apoerythrocuprein was partially reconstituted with Cu^{2+} and/or Zn^{2+}. The apoprotein and the Zn^{2+} protein displayed no visible absorption. The addition of two Cu^{2+} to apoerythrocuprein resulted in an absorption band at 740 nm which was clearly shifted by over 50 nm to 680 nm when the two zinc ions were also present (Table 2). Furthermore, the shoulder at 430 nm was only detectable using the native or the fully reconstituted cuprein. From this observation and from other measurements including EPR and XPS data it can be concluded that copper and zinc are in close proximity to each other (56).

The CD properties can be seen in Fig. 3. Upon applying a magnetic field of 50 kG the resolution of the bands was much improved. Two (and possibly a third) MCD extrema of the copper d→d transitions at 595 and 640 nm were clearly detectable. The complementary usefulness of MCD and CD for resolv-

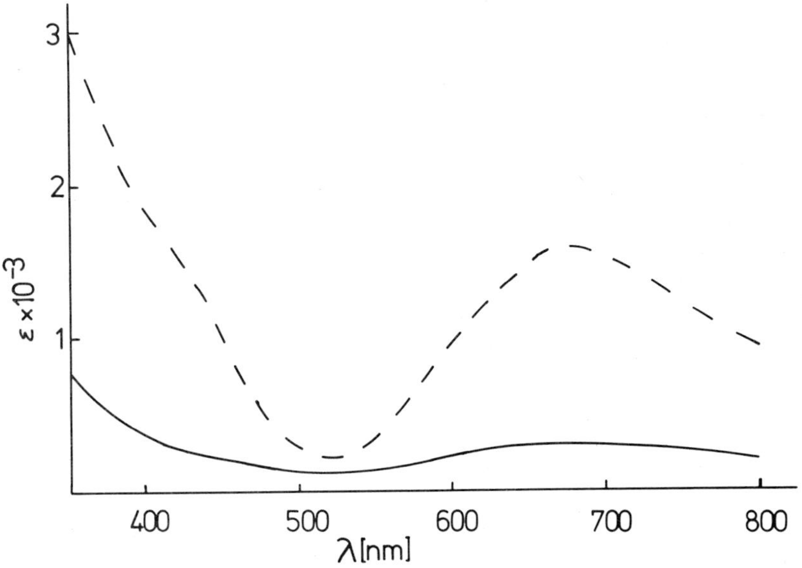

Fig. 2. Absorption spectrum of bovine cuprein in the visible region. The same spectrum was obtained using human cuprein. - - - Spectrum recorded at 77° K (28).

Table 2. Absorption properties of native and partially reconstituted cupreins

	λ_{max} nm	λ_{max} (ϵ value)	Shoulder at 430 nm
Native erythrocuprein	680	(−313)	Distinct
Apoprotein			
2Zn - Apoerythrocuprein			
2Cu - Apoerythrocuprein	740	(−280)	
2Cu - 2Zn Apo erythrocuprein	680	(−330)	Distinct

ing overlapping absorption bands is obvious. The presence of urea or dodecylsulphate had virtually no effect on the CD properties. The question remains as to whether this protein has mainly random coil or folded-sheet structure or a protein conformation which differs from both.

X-RAY PHOTOELECTRON SPECTROSCOPY (XPS)

X-ray photoelectron spectroscopy proved especially convenient for studying the presence of those metals which display no Mössbauer or EPR activity.

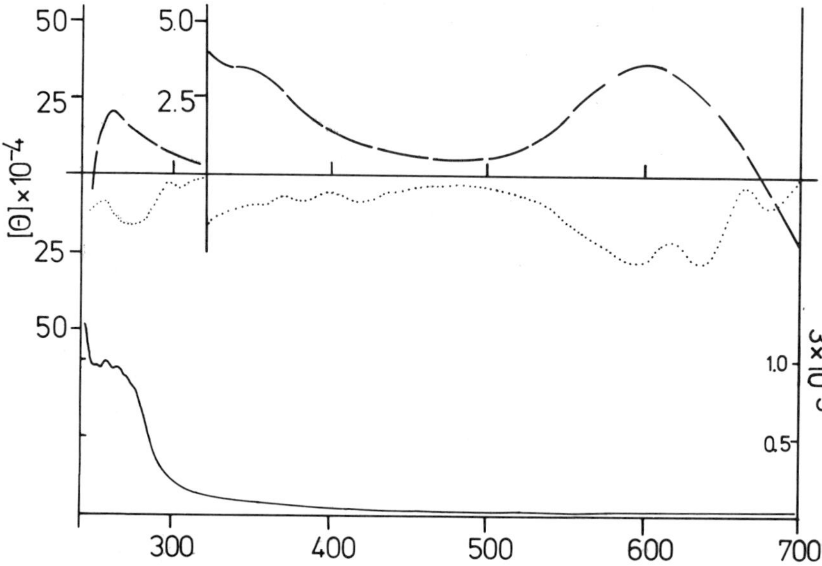

Fig. 3. Absorption ———, circular dichroism - - -, and magnetic circular dichroism . . . . properties of bovine cuprein. For experimental details see (28).

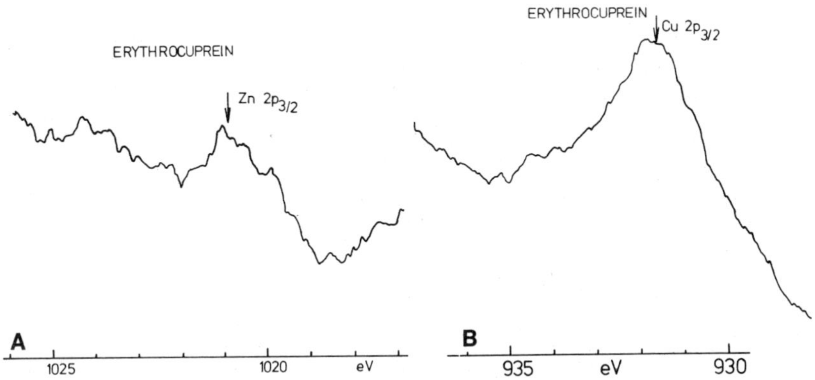

Fig. 4. X-ray photoelectron spectra of native bovine cuprein: (a) the Zn $2p_{3/2}$ signal; (b) the Cu $2p_{3/2}$ signal (41).

The binding energies of the core electrons of Cu $2p_{3/2}$ and Zn $2p_{3/2}$ in bovine cuprein, respectively, were measured and compared with the corresponding values of different Cu/Zn amino acid chelates (41). The binding energy of the core electrons seen in the Zn $2p_{3/2}$ signals were 1021.5 ± 0.2 eV and identical with the electron binding energies of low molecular weight zinc chelates. Upon employing low molecular weight Cu complexes and native cuprein, marked differences of the electron binding energies were observed. The Cu $2p_{3/2}$ value for the Cu amino acid chelates was 934.2 eV; the value for the cuprein Cu $2p_{3/2}$ signal was 931.9 eV (Fig. 4).

BIOCHEMICAL ACTION OF THE CUPREINS

A large number of oxidases are known to contain copper in the active site. In this connection we were interested to examine the following question: how many electrons can be transferred by the cupreins? Dithionite was used as a reducing agent and the course of the reduction was measured employing visible or EPR spectroscopy (Fig. 5a (28)). From the slope of the straight line obtained after plotting the main EPR signal *versus* different dithionite concentrations it was clearly shown that two electrons were transferred (Fig. 5b (28)).

For almost 30 years the cupreins were thought to act exclusively as copper carriers. It was rather attractive to come to this conclusion since, for example, over 50% of the erythrocyte copper is located in erythrocuprein (14), and it was concluded that the physiological role of this protein was to store and/or transport copper inside the cell. In 1969 the purification of a protein displaying superoxide dismutase activity was successful using bovine erythrocytes. This protein was shown to be identical with the long-known

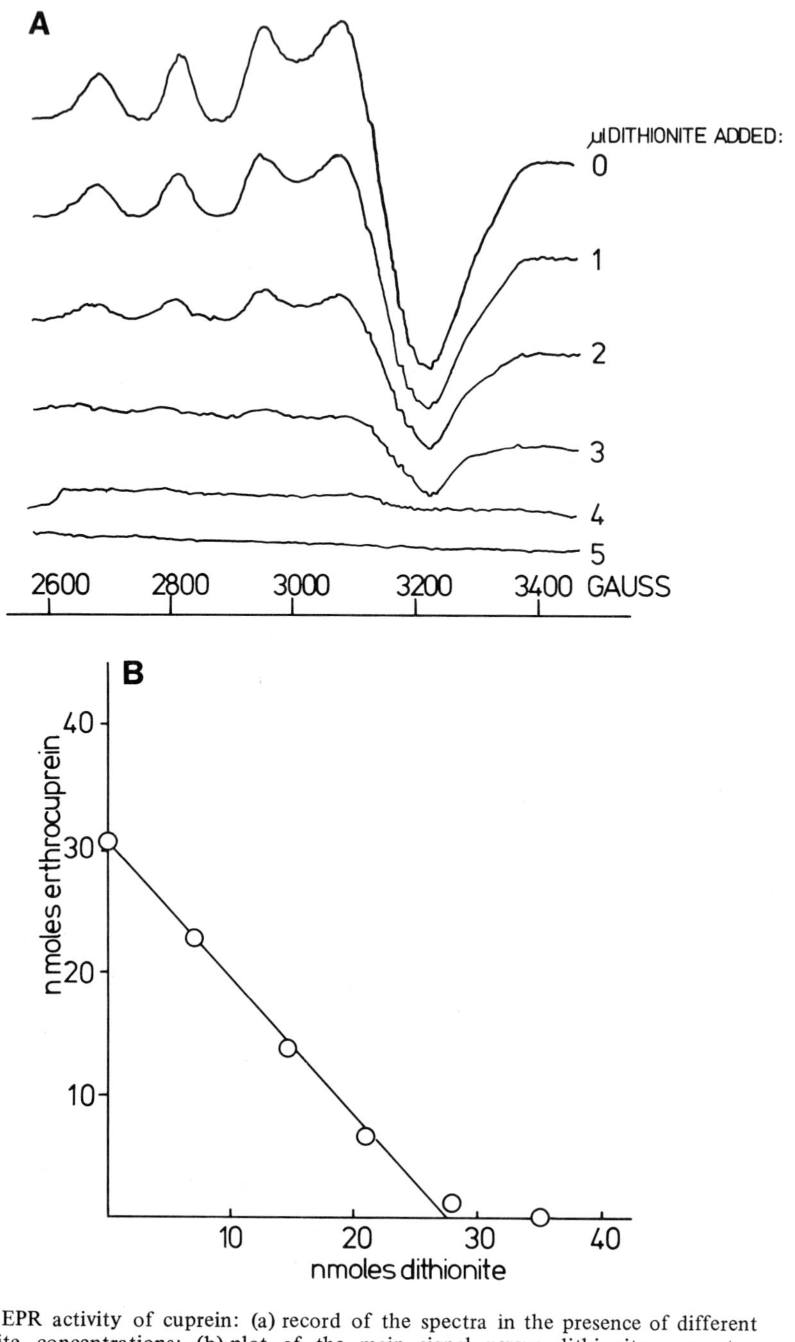

Fig. 5. EPR activity of cuprein: (a) record of the spectra in the presence of different dithionite concentrations; (b) plot of the main signal *versus* dithionite concentrations (28).

erythrocuprein (24). In the presence of erythrocuprein the spontaneous disproportionation of $\dot{O}_2^-$ is accelerated by approximately three orders of magnitude, i.e., the cupreins react directly with the $\dot{O}_2^-$ species. The rate constants of the disproportionation of $\dot{O}_2^-$ were determined at three different pH values (56, 57):

$$\begin{aligned}
&H\dot{O}_2 + H\dot{O}_2 \longrightarrow H_2O_2 + O_2, && k_1 = 7.6 \times 10^5\ M^{-1}\,sec^{-1}\\
&H\dot{O}_2 + \dot{O}_2^- \xrightarrow{H_2O} H_2O_2 + O_2 + OH^-, && k_2 = 8.5 \times 10^7\ M^{-1}\,sec^{-1}\\
&\dot{O}_2^- + \dot{O}_2^- \xrightarrow{2H_2O} H_2O_2 + O_2 + 2OH^-, && k_3 \sim 10^2\ M^{-1}\,sec^{-1}
\end{aligned}$$

In the presence of erythrocuprein a second-order rate constant for the reaction between O_2^- and erythrocuprein of about $2 \times 10^9\ M^{-1}\,sec^{-1}$ (58, 59) was determined. This value is near to the upper limit expected for a diffusion-controlled enzyme-substrate reaction (60). In contrast to the above pH-dependent spontaneous disproportionation, no such dependency was observed. In fact the same numerical value was obtained between pH 5 and pH 9.

The determination of the rate constant was possible using pulse radiolysis. With this method the O_2^- concentration was sufficiently high to measure its absorption at 250 nm. However, in biochemical systems the O_2^- concentration is far too low to be measured directly. Indirect measurements using the oxidative or reductive function of the O_2^- upon colored substances of high absorption coefficients are commonly employed (8). Owing to the EPR activity of O_2^- its quantitation during enzymic reactions and in the presence or absence of cuprein should be possible (61, 62). This last method would be the second direct method for the evaluation of O_2^-.

SUBSTRATE SPECIFICITY

The O_2 molecule can also be taken as a stable biradical. Unlike the N_2 molecule two π^*-antibonding molecular orbitals of the oxygen are occupied according to Hund's law, each with one electron. In the ground state both electrons have the same spin direction. The binding situation of this electronic triplet ground state is normally expressed as

$$^3\Sigma_g^-[\pi,\ y(4)\pi_x^*(\uparrow)\pi_y^*(\uparrow)]\,.$$

After excitation of this triplet ground state two metastable singlet states $^1\Sigma_g^+$ and $^1\Delta_g$ are possible. (The higher excited states will not be discussed here.)

$$^1\Sigma_g^+[\pi_{xy}(4)\pi_x^*(\uparrow)\pi_y^*(\downarrow)]$$

and

$$^1\Delta_g\,[\pi_{xy}(4)\pi_x^*(\uparrow\downarrow)\pi_y^*(O)]\,.$$

The required energies to obtain these excited states are 22 kcal (8000 cm^{-1}) for $^1\Delta_gO_2$ and 37 kcal (13,000 cm^{-1}) for $^1\Sigma_g^+O_2$. The $^1\Delta_gO_2$ has both antibonding electrons located with antiparallel spin in one π^*-antibonding molecular orbital, leaving the second unoccupied. Regarding the $^1\Sigma_g^+$ oxygen, both electrons of the triplet ground state remain in the two π^*-antibonding orbitals. However, these electrons have antiparallel spin. Principally, the conversion of $^1\Sigma_g^+$ oxygen into $^1\Delta_g$ oxygen is possible. According to Khan (63) peroxide is a convenient source for $^1\Delta_g$ oxygen while $^1\Sigma_g^+$ oxygen, apparently, is preferentially formed using superoxide (45).

Singlet oxygen can be conveniently prepared using potassium superoxide (63). Unfortunately, we do not know the exact mode of singlet-oxygen formation. Nevertheless, the generation of singlet-type oxygen in aqueous media, for example, during the enzymic catalysed oxidation of xanthine, was taken into consideration. Arneson (45) concluded from his experiments using xanthine oxidase and aldehyde oxidase that singlet oxygen was being formed from superoxide which was monitored by chemiluminescence. Singlet oxygen may even be formed in the absence of O_2^- (46, 64). During the reaction of lipoxidase on its substrate linoleic acid a 'singlet-oxygen-like' intermediate was observed which was quenched rather specifically by erythrocuprein. The usual cytochrome C reductase assay was not applicable in this system (46). Thus we in our laboratory (47–49) and Rotilio (46) propose that the main physiological activity of the cupreins is the quenching of highly reactive singlet-oxygen species or some intermediates, rather than accelerating the O_2^- disproportionation as shown using *in vitro* systems.

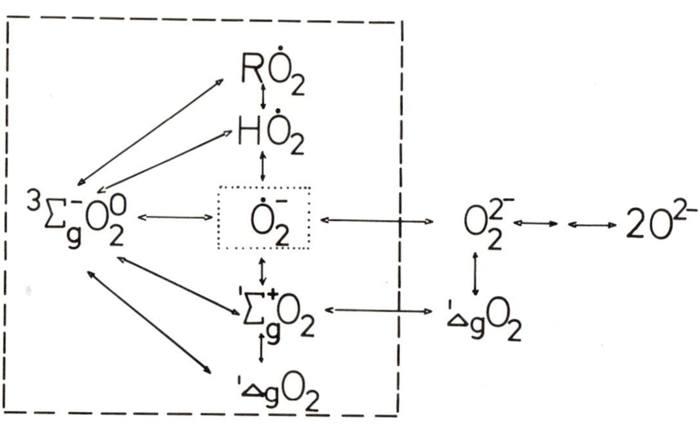

Fig. 6. Substrate specificity of the cupreins.

A critical examination of the substrate specificity for the cupreins is summarized in Fig. 6. Considering the fact that the cupreins react extremely fast with O_2^- ($\sim 2 \times 10^9 M^{-1} sec^{-1}$), it seems likely that this Cu-Zn-protein could well react with some intermediates of O_2^-.

BIOCHEMICAL ACTION OF LOW AND HIGH MOLECULAR WEIGHT COPPER CHELATES

At present our knowledge of the structure of almost all copper proteins including the cupreins is rather limited. Thus, for the time being great caution is required in the interpretation of physical and chemical properties of these metalloproteins (5, 6). A considerable number of studies using Cu-model complexes of low molecular weight were carried out as an approach to understanding these structural and functional properties. The biochemical action of Cu in the different Cu proteins could formally be summarized as:

$$\left(\begin{bmatrix}Cu^+\\Cu^+\end{bmatrix} {}^1\Delta_g O_2 \leftrightarrow \begin{bmatrix}Cu^+\\Cu^+\end{bmatrix} {}^1\Sigma_g^+ O_2\right) \leftrightarrow \underset{\text{I}}{\begin{bmatrix}Cu^+\\Cu^+\end{bmatrix}} \ldots {}^3\Sigma_g^- O_2 \leftrightarrow \underset{\text{II}}{\begin{bmatrix}Cu^{2+}\\Cu^+\end{bmatrix}} \ldots O_2^-$$

$$\leftrightarrow \underset{\text{III}}{\begin{bmatrix}Cu^{2+}\\Cu^{2+}\end{bmatrix}} \ldots O_2^{2-} \quad \underset{\text{IV}}{\begin{bmatrix}Cu^{2+}\\Cu^{2+}\end{bmatrix}} \ldots O^{2-}.$$

In Cu chelates of high covalence no precise decision can be made regarding the oxidation state of the copper. Thus, the oxidation states depicted in I–IV are merely of formal nature. The redox mesomerism which implies the different biochemical action of the chelated metal ion is highly interesting. Type I would represent the reversible oxygenation as found in hemocyanin, type II would be the superoxide dismutation, provided O_2^- really is the substrate, and types III and IV are represented by the catalytic and oxidative action displayed by a considerable number of copper proteins (polyphenol oxidases, amine oxidases, etc.). The biochemical specificity of each chelated copper is more or less dictated by the macromolecular ligands.

An extensive study on the catalatic and peroxidatic activity of different Cu complexes was performed by Sigel (66). It was shown that at least one coordination site of the copper had to be vacant to allow the binding of the substrate [Fig. 7 (65)].

The type III complex remained essentially inactive. Proof of an intermediary Cu^{2+}-peroxi-complex was shown by an absorption shoulder at 360 nm (5). This biochemical reactivity was induced not only by ethylene diamine and its derivatives, but also by using monomeric and polymeric amino acids and nucleotides, respectively.

We wanted to examine the question as to what degree copper chelates using both the free amino acids or some low molecular weight peptides would be able to display superoxide dismutase activity. Of all the amino acids employed which were present in bovine cuprein, only the copper complexes of lysine, tyrosine, and histidine were able to show enzymic activities similar to those reported for the cupreins, i.e., inhibition of the cytochrome C reductase activity and the quenching of singlet-type oxygen (Fig. 8, Table 3). In the presence of the metal-free ligands and Cu^{2+} alone no such biochemical activity was detected. It has to be emphasized that the employed Cu-chelate concentrations were in the micromolar range, i.e., their concentration was almost three orders of magnitude lower compared with the Cu^{2+} concentrations employed by Sigel (6). Owing to the extremely low copper amino acid concentrations required for superoxide dismutase activity the assignment of the exact number of coordinated amino acids with Cu^{2+} in solution cannot be stated.

The Cu chelates employing EDTA or serum albumin as ligands display no superoxide dismutase activity. It is very attractive to conclude that, as in the case of the catalatic action of Cu^{2+}, one or more Cu coordination sites are fully accessible to water or the substrate. A residence lifetime of 4×10^{-6}–10^{-8} sec (67) on each of the two Cu in the cupreins is not very different compared with the second-order rate constant for the $\dot{O}_2^-$-cuprein reaction (58, 59, 68). In other words, the coordinated water can be most rapidly

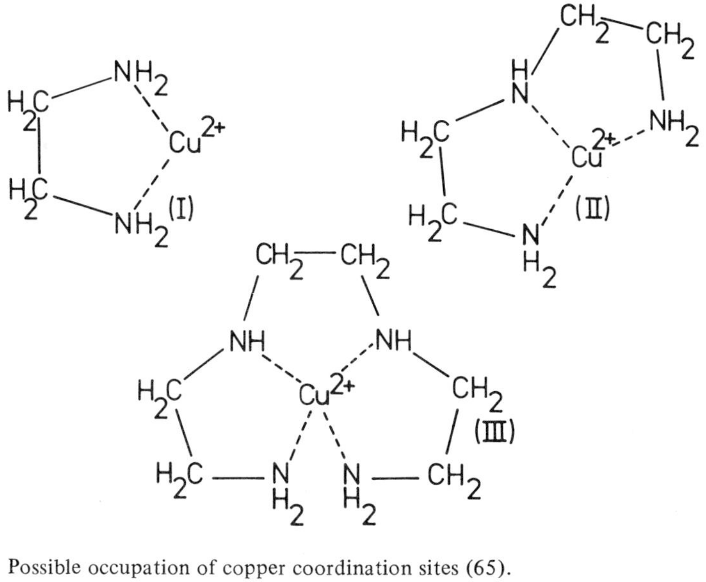

Fig. 7. Possible occupation of copper coordination sites (65).

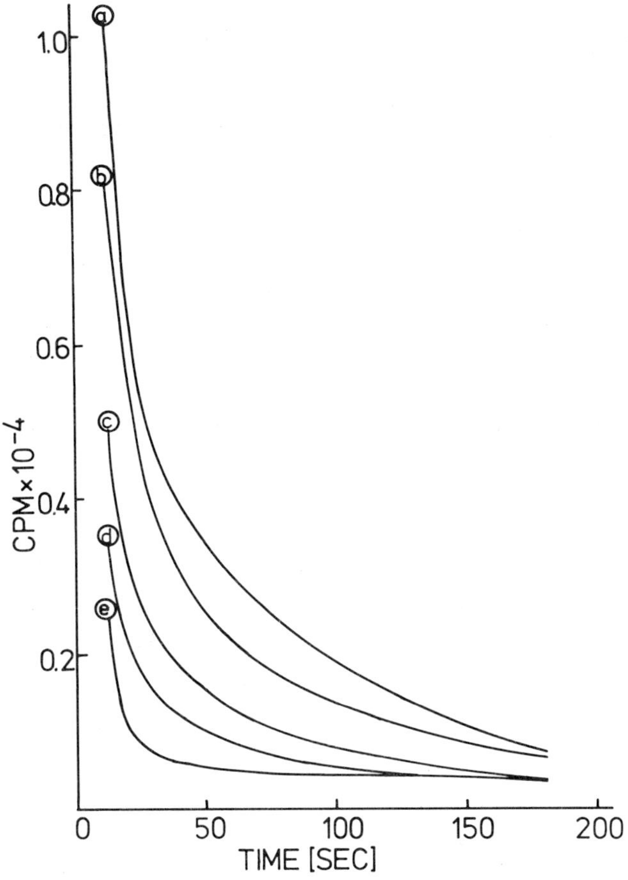

Fig. 8. Chemiluminescence assay of bovine cuprein and different Cu^{2+} amino acid chelates at pH 7.8: (a) none; (b) $Cu(Lys)_2$, 50 nM; (c) $Cu(His)_2$, 100 nm; (d) Cu-Tyr, 145 nM; (e) bovine erythrocuprein, 8 nM. The enzymic catalysed oxidation of xanthine by xanthine oxidase in the presence of luminol was employed. In the case of the Cu^{2+} amino acid chelates, equimolar concentrations of EDTA were added to avoid unspecific inhibition of the xanthine oxidase. For further details see (49).

exchanged by O_2^- and/or singlet-type oxygen. However, this fast exchange is only possible provided the copper is easily accessible to the substrate. From XPS measurements it could be concluded that Cu^{2+} might be located in the outer sphere of the protein molecule, which would support the findings of Fee (67), Rotilio (58, 68), and others (59).

At the moment no final decision can be made as to whether or not erythrocuprein is really an essential enzyme in metabolism. In order to assign

Table 3. Superoxide dismutase activity of different Cu^{2+} L-amino acid chelates[1]

Cu^{2+} chelate	Required equivalent of chelated Cu^{2+} to yield $[\text{Cyt } C_{red}] \times [\text{Cyt } C_{ox}]^{-1} = 1$ (μM^{-1})
Native bovine erythrocuprein	16.7
1-Cu-apoerythrocuprein	1.0
$Cu(Lys)_2$	0.9
Cu Tyr	0.6
$Cu(His)_2$	0.5
$Cu(\text{His-methylester})_2$	0.2
Cu-Leu-Tyr	0.5
Cu-Lys-Ala	0.2
Cu-His-Leu-Gly	0.6
Cu-Gly-His-Leu	0
Cu-His-Leu-Leu	0.3
Cu-Leu-His-Leu	0.04
Cu-Leu-Leu-His	0
Cu^{2+} .aq	0
Cu-EDTA	0
Cu-bovine serum albumin	0

[1]For experimental details see Fig. 8 and (49).

a specific role, two essential criteria must be fulfilled. Firstly appropriate ligands bound with the Cu^{2+} must be able to induce a 'superoxide dismutase' activity, and secondly the protein molecule has to be in a position to protect the free coordination site of the copper ion from unspecific and undesired binding with a large number of naturally occurring ligands.

Whether the scavenging of O_2^-, some intermediary superoxide complexes, or singlet-type oxygen is the important function of cupreins is a most interesting question which will be a challenging task for further studies. All these excited oxygen species except the superoxide anion itself would be extremely reactive and 'burn' anything in the living cell. Nevertheless, the ubiquity of copper and the cupreins in a great number, if not all, aerobic cells and their reactivity in the biochemistry of oxygen suggests a most important and possible essential function of this copper protein.

REFERENCES

1. Dawson, C. R., and Malatte, M. F. 1945. The copper proteins. *Adv. Protein Chem.* 2: 179–248.

2. Hamilton, G. A. 1969. Mechanisms of two- and four-electron oxidations catalysed by some metalloenzymes. *Adv. Enzymol.* 32: 55–96.
3. Malkin, R., and Malmström, B. G. 1970. The state and function of copper in biological systems. *Adv. Enzymol.* 33: 177–244.
4. Freeman, H. C. 1967. Crystal structures of metal-peptide complexes. *Adv. Protein Chem.* 22: 257–424.
5. Peisach, J., Aisen, P., and Blumberg, W. E. (eds.). 1966. *The Biochemistry of Copper,* Academic Press, New York.
6. Giretti, F. (ed.) 1968. *Physiology and Biochemistry of Haemocyanins,* Academic Press, New York.
7. Fridovich, I. 1972. Superoxide radical and superoxide dismutase. *Abstract Communications, Metalloenzymes Conference, Oxford,* pp. 22–24.
8. Weser, U. 1973. Structural aspects and biochemical function of erythrocyprein. *Structure and Bonding,* Vol. 17. Springer Verlag, New York, pp. 1–65.
9. Mann, T., and Keilin, D. 1939. Haemocuprein and hepatocuprein, copper-protein compounds of blood and liver in mammals. *Proc. R. Soc.* B 126: 303–315.
10. Mohamed, M. S., and Greenberg, D. M. 1954. Isolation of purified copper protein from horse liver. *J. Gen. Physiol.* 37: 433–439.
11. Markowitz, H., Cartwright, G. E., and Wintrobe, M. M. 1959. Studies on copper metabolism. XXVII The isolation and properties of an erythrocyte cuproprotein (erythrocuprein). *J. Biol. Chem.* 234: 40–45.
12. Kimmel, J. R., Markowitz, H., and Brown, D. M. 1959. Some chemical and physical properties of erythrocuprein. *J. Biol. Chem.* 234: 46–50.
13. Nyman, P. O. 1960. A modified method for the purification of erythrocuprein. *Biochim. Biophys. Acta* 45: 387–389.
14. Shields, G. S., Markowitz, H., Klassen, W. H., Cartwright, G. E., and Wintrobe, M. M. 1961. Copper metabolism. XXX. Erythrocyte copper. *J. Clin. Invest.* 40: 2007–2015.
15. Malmström, B. G., and Vänngård, T. 1960. Electron spin resonance of copper proteins and some model complexes. *J. Molec. Biol.* 2: 118–124.
16. Porter, H., Folch, P. J., Beggrovs, J., and Ainsworth, S. 1957. Cerebrocuprein I. A copper-containing protein isolated from brain. *J. Neurochem.* 1: 260–71.
17. Porter, H., and Ainsworth, S. 1959. The isolation of the Cu containing protein cerebrocuprein I from normal human brain. *J. Neurochem.* 5: 91–98.
18. Porter, H., Sweeny, M., and Porter, E. 1964. Human Hepatocuprein. Isolation of a copper protein from the subcellular soluble fraction of adult human liver. *Archs Biochem. Biophys.* 105: 319–325.
19. Stansell, M. J., and Deutsch, H. F. 1965. Preparation of crystalline erythrocuprein and catalase from human erythrocytes. *J. Biol. Chem.* 240: 4299–4305.
20. Stansell, M. J., and Deutsch, H. F. 1965. Physciochemical studies of crystalline human erythrocuprein. *J. Biol. Chem.* 240: 4306–4311.
21. Hartz, J. W., and Deutsch, H. F. 1969. Preparation and physicochemical

properties of human erythrocuprein. *J. Biol. Chem.* 244: 4565–4572.

22. Carrico, R. J., and Deutsch, H. F. 1969. Isolation of human hepatocuprein and cerebrocuprein. *J. Biol. Chem.* 244: 6084–6093.
23. Carrico, R. J., and Deutsch, H. F. 1970. The presence of zinc in human cytocuprein and some properties of the apoprotein. *J. Biol. Chem.* 245: 723–727.
24. McCord, J. M., and Fridovich, I. 1969. Superoxide Dismutase. An enzymic function for erythrocuprein (Hemocuprein). *J. Biol. Chem.* 244: 6049–6055.
25. McCord, J. M., and Fridovich, I. 1969. The utility of superoxide dismutase in studying free radical reactions. *J. Biol. Chem.* 244: 6056–6063.
26. Bannister, J., Bannister, W., and Wood, E. 1971. Bovine erythrocyte cuprozinc protein. 1. Isolation and general characterisation. *Eur. J. Biochem.* 18: 178–186.
27. Wood, E., Dalgleish, D., and Bannister, W. 1971. Bovine erythrocyte cuprozinc protein. 2. Physicochemical properties and circular dichroism. *Eur. J. Biochem.* 18: 187–193.
28. Weser, U., Bunnenberg, E., Cammack, R., Djerassi, C., Flohé, L., Thomas, G., and Voelter, W. 1971. A study on purified bovine erythrocuprein. *Biochim. Biophys. Acta* 243: 203–213.
29. Weser, U. 1971. Erythrocuprein. *Angew. Chem.* 83: 939 ff.
30. Weser, U., and Hartmann, H. J. 1971. Preparation of pure bovine apoerythrocuprein by gel filtration. *FEBS Lett.* 17: 78–80.
31. Weser, U., and Voelcker, G. 1972. Evidence for accelerated H_2O_2 formation by erythrocuprein. *FEBS Lett.* 22: 15–18.
32. Weser, U., Barth, G., Djerassi, C., Hartmann, H. J., Krauss, P., Voelcker, G., Voelter, W., and Voetsch, W. 1972. A study on purified apoerythrocuprein. *Biochim. Biophys. Acta* 278: 28–44.
33. Weser, U., Bohnenkamp, W., Cammack, R., Hartmann, H. J., and Voelcker, G. 1972. An enzymic study on bovine erythrocuprein. *Z. Physiol. Chem.* 353: 1059–1068.
34. Bohnenkamp, W., and Weser, U. 1972. Untereinheiten des Erythrocupreins. *Z. Physiol. Chem.* 353: 695–696.
35. Rotilio, G., Finazzi-Agro, A., Calabrese, L., Bossa, F., Guerrieri, P., and Mondovi, B. 1971. Studies of the metal sites of copper proteins, ligands of copper in hemocuprein. *Biochemistry* 10: 616–621.
36. Keele, B. B., McCord, J. M., and Fridovich, I. 1971. Further characterisation of bovine superoxide dismutase and its isolation from bovine heart. *J. Biol. Chem.* 246: 2875–2880.
37. Fee, J. A., and Gaber, B. P. 1972. Anion binding to bovine erythrocyte superoxide dismutase. *J. Biol. Chem.* 247: 60–65.
38. Fee, J. A. 1973. Studies on the reconstitution of bovine erythrocyte superoxide dismutase I. *Biochim. Biophys. Acta.* 295: 87–95.
39. Fee, J. A. 1973. Studies on the reconstitution of bovine erythrocyte superoxide dismutase II. *Biochim. Biophys. Acta* 295: 96–106.
40. Fee, J. A. 1973. Studies on the reconstitution of bovine erythrocyte superoxide dismutase III. *Biochim. Biophys. Acta* 295: 107–116.

41. Jung, G., Ottnad, M., Bohnenkamp, W., Bremser, W., and Weser, U. 1973. X-ray photoelectron spectroscopic studies of copper- and zinc-amino acid complexes and superoxide dismutase. *Biochim. Biophys. Acta* 295: 77–86.
42. Rotilio, G., Calabrese, L., Bossa, F., Barra, D., Finazzi-Agro, A., and Mondovi, B. 1972. Properties of the apoprotein and role of copper and zinc in protein conformation and enzyme activity of bovine superoxide dismutase. *Biochemistry* 11: 2182–2186.
43. Rotilio, G., Morpurgo, L., Giovagnoli, C., Calabrese, L., and Mondovi, B. 1972. Studies of the metal sites of copper proteins. Symmetry of copper in bovine superoxide dismutase and its functional significance. *Biochemistry* 11: 2187–2192.
44. Weser, U., Prinz, R., Schallies, A., Fretzdorff, A. M., Krauss, P., Voelter, W., and Voetsch, W. 1972. Microbial and hepatic cuprein (superoxide dismutase) from saccharomyces cerevisiae and bovine liver. *Z. Physiol. Chem.* 353: 1821–1831.
45. Arneson, R. M. 1970. Substrate induced chemiluminescence of xanthine oxidase and aldehyde oxidase. *Archs Biochem. Biophys.* 136: 352–360.
46. Finazzi-Agro, A., Giovagnoli, C., Del Sole, P., Calabrese, L., Rotilio, G., and Mondovi, B. 1972. Erythrocuprein and singlet oxygen. *FEBS Lett.* 21: 183–185.
47. Zimmermann, R., Flohé, L., Weser, U., and Hartmann, H. J. 1973. Inhibition of lipid peroxidation in isolated inner membrane of rat liver mitochondria by superoxide dismutase. *FEBS Lett.* 29: 117–120.
48. Weser, U., and Paschen, W. 1972. Mode of singlet oxygen induced chemiluminescence in the presence of erythrocuprein. *FEBS Lett.* 27: 248–250.
49. Joester, K. E., Jung, G., Weber, U., and Weser, U. 1972. Superoxide dismutase activity of Cu^{2+}-amino acid chelates. *FEBS Lett.* 25: 25–28.
50. Weser, U., Fretzdorff, A., and Prinz, R. 1972. Superoxide dismutase in baker's yeast. *FEBS Lett.* 27: 267–269.
51. Bohnenkamp, W. 1973. Untereinheiten von Erythrocuprein. Ph.D. Thesis, University of Tübingen, Germany.
52. Sawada, Y., Ohyama, T., and Yamazaki, I. 1972. Preparation and physico-chemical properties of green pea superoxide dismutase. *Biochim. Biophys. Acta* 268: 305–312.
53. Misra, H. P., and Fridovich, I. 1972. The purification and properties of superoxide dismutase from neurospora crassa. *J. Biol. Chem.* 247: 3410–3414.
54. Goscin, S. A., and Fridovich, I. 1972. The purification and properties of superoxide dismutase from saccharomyces cerevisiae. *Biochim. Biophys. Acta* 289: 276–283.
55. Richardson, D. C., Bier, C. J., and Richardson, J. S. 1972. Two crystal forms of bovine superoxide dismutase. *J. Biol. Chem.* 247: 6386–6369.
56. Rabani, J., and Nielsen, S. O. 1969. Absorption spectrum and decay

kinetics by O_2^- and HO_2 in aqueous solutions by pulse radiolysis. *J. Phys. Chem.* 73: 3736–3744.
57. Behar, D., Czapski, G., Rabani, J., Dorfman, L. M., and Schwarz, H. A. 1970. Acid dissociation constant and decay kinetics of the perhydroxyl radical. *J. Phys. Chem.* 74: 3209–3213.
58. Rotilio, G., Bray, R. C., and Fielden, M. E. 1972. A pulse radiolysis study of superoxide dismutase. *Biochim. Biophys. Acta* 268: 605–609.
59. Klug, D., Rabani, J., and Fridovich, I. 1972. A direct demonstration of the catalytic action of superoxide dismutase through the use of pulse radiolysis. *J. Biol. Chem.* 247: 4839–4842.
60. Eigen, M., and Hammes, G. G. 1963. Elementary steps in enzyme reactions (as studied by relaxation spectrometry). *Adv. Enzymol.* 25: 1–38.
61. Knowles, P. F., Gibson, J. F., Pick, F. M., and Bray, R. C. 1969. Electron spin resonance evidence for enzymic reduction of oxygen to a free radical, the superoxide ion. *Biochem. J.* 111: 53–58.
62. Massey, V., Strickland, S., Mayhew, S. G., Howell, L. G., Engel, P. C., Matthews, R. G., Schuman, M., and Sullivan, P. A. 1969. The production of superoxide anion radicals in the reaction of reduced flavins and flaoproteins with molecular oxygen. *Biochem. Biophys. Res. Commun.* 36: 891–857.
63. Khan, A. K. 1970. Singlet molecular oxygen from superoxide anion and sensitized fluorescens of organic molecules. *Science* 168: 476–477.
64. Chan, H. W. S. 1971. Singlet oxygen analogs in biological systems. Coupled oxygenation of 1,3-dienes by soybean lipoxides. *J. Am. Chem. Soc.* 93: 2357–2358.
65. Sigel, H., and Müller, U. 1966. Über Struktur und Aktivität der den H_2O_2 Zerfall katalysierenden Cu^{2+}-Komplexe. II. Einfluss der Zahl der freien Koordinationsstellen. *Helv. Chim. Acta* 49: 671–681.
66. Sigel, H. 1969. Zur katalatischen und peroxidatischen Aktivität von Cu^{2+}-Komplexen. *Angew. Chem.* 81: 161–194.
67. Gaber, B. P., Brown, R. D., Koenig, S. H., and Fee, J. A. 1972. Nuclear magnetic relaxation disperson in protein solutions. V. Bovine erythrocyte superoxide dismutase. *Biochim. Biophys. Acta* 271: 1–5.
68. Bannister, J. V., Bannister, W. H., Bray, R. C., Fielden, E. M., Roberts, P. B., and Rotilio, G. 1973. The superoxide dismutase activity of human erythrocuprein. *FEBS Lett.* 32: 303–306.

DISCUSSION

Schwarz (Long Beach). I suggested earlier that there may be a normal peroxide pathway in metabolism and that lipid peroxides may be normal intermediates. Could you point out where cuprein, for instance, would fit into such a peroxide pathway?

Weser. We have considered whether the cupreins are active in preventing lipid peroxidation. Our experiments, conducted in cooperation with Flohé, suggested that glutathione peroxidase was not as active as cupreins in protecting the inner membrane of mitochondria from the reactions of oxygen with a reducing system such as the unsaturated fatty acids. This complex situation needs further elucidation because we don't know whether the damaging species is singlet oxygen, superoxide, or peroxide.

Schwarz (Long Beach). Perhaps it's not just a question of detoxification. There must be a compartmentalization of some sort and there must be steps in the normal utilization of the various intermediates in this peroxide pathway. Chance has shown, for example, that 80% of the liver catalase is saturated with hydrogen peroxide and, from the turnover rate of that enzyme and the total amount of enzyme, he calculated that 5–15% of all the oxygen which is consumed by the liver is going through the peroxide pathway.

Weser. Other workers have also found that catalase is not sufficient to protect the erythrocyte membrane; you also need the cupreins. Glutathione in the presence of oxygen and an electron transferring system will produce superoxide and the related oxygen species; perhaps this relates to the requirement for cupreins.

Frieden (Tallahassee). Don't you think that there is a catalytic advantage to having the copper bound, perhaps in entatic form, in this specific protein?

Weser. Yes, most metalloenzymes are in a kind of entatic state and I agree that model complexes are always a bad example for showing activity. However, the model complexes at concentrations in the micromolar region had about 10% of the activity of the native protein which we considered surprisingly high reactivity, but of course the native protein is still better.

BIOCHEMICAL FUNCTIONS OF ZINC WITH EMPHASIS ON NUCLEIC ACID METABOLISM AND CELL DIVISION

J. K. CHESTERS

Rowett Research Institute, Bucksburn, Aberdeen, U.K.

Todd, Elvehjem and Hart (1) showed that zinc deficiency reduced the rate of growth of rats. Growth can be divided into two aspects, increase in cell size and in cell number, and the rates of synthesis of protein and DNA within a tissue have been used as approximate indicators of these two types of growth. A number of groups have investigated the effects of Zn deficiency on these synthetic rates in the rat (2–7). Some of these investigations suffered from problems associated with providing adequate controls for animals whose food intake was depressed by the deficiency, but the general concensus which emerged was that Zn deficiency caused a more marked inhibition of DNA synthesis than of protein synthesis. This may indicate that increase in cell number was more severely affected by Zn deficiency than was increase in cell size. Zn has also been implicated in the processes of wound healing (8–11) in which it may have been required for the phase of cell proliferation (12, 13). Studies of carrageenin granulomas in Zn-deficient rats showed that multiplication of fibroblasts continued to occur after the time at which those of control rats had ceased to divide and had differentiated into fibrocytes.[1] The mitotic index of the basal layer of rat epidermis was reduced by Zn deficiency, but in the oesophagus and the acinar cells of the pancreas Zn deficiency increased the mitotic index and in liver parenchymal cells it was unaffected by the deficiency (14). Incorporation of ^{3}H-thymidine into 12 d rat embryos was decreased by Zn deficiency, but the mitotic index of the neuroepithelium of the deficient embryos was higher than that in the control embryos (15). The

1. Personal communication from Dr. B. F. Fell, Rowett Research Institute, Bucksburn, Aberdeen.

relationship between the Zn status of animals and the growth of malignant tissues is unresolved. Dosing with Zn inhibited the incidence of tumors in hamsters treated with a carcinogen (16) but stimulated the development of tumors in rats (17). Zn deficiency in rats and mice inhibited the growth of a range of tumors (18), but it has been postulated as a possible cause of a high incidence of esophageal cancer in man (19). Despite the range and sometimes apparently contradictory nature of the results cited most of them indicate a role of Zn in cellular multiplication or differentiation.

Zn availability at the specific tissue or organ under investigation is difficult to control in experiments with whole animals. Fujioka and Lieberman (2) used an infusion of EDTA to reduce trace-metal availability in partially hepatectomised rats. They found that DNA but not protein synthesis was inhibited and that only Zn reversed the inhibition of DNA synthesis. They also found that the degree of incorporation of ^{3}H-thymidine obtained with varying Zn concentrations was proportional to the number of cells labeled with thymidine, suggesting that Zn was required for the initiation of DNA synthesis within a cell rather than for synthesis per se.

TISSUE-CULTURE EXPERIMENTS

Investigations using tissue-culture techniques allow a more precise definition of the environment of the cells than do studies with whole animals. Present technology, however, is generally inadequate to provide tissue-culture media sufficiently low in Zn to limit growth, and chelators have been used to restrict trace-metal availability in *in vitro* studies of Zn metabolism.

Kidney Cortex Cells

Lieberman and co-workers investigated the effects of EDTA on primary cultures of rabbit kidney cells (20, 21). Synthesis of DNA was almost totally inhibited by the chelator and was restored only by the addition of Zn. The effects of EDTA and Zn on RNA and protein synthesis were much less marked. The inhibition of DNA synthesis was accompanied by reductions in the activities of both thymidine kinase (EC 2.7.1.21) and DNA polymerase (EC 2.7.7.7).

Lymphocytes Stimulated with Phytohemagglutinin (PHA)

When stimulated by PHA, lymphocytes undergo a series of biochemical and morphological changes leading to cell division after approximately 72 hr. Studies by Alford (22) and Chesters (23) have shown that EDTA added to the cultures inhibited subsequent DNA synthesis and cell division even when the ratio of calcium to EDTA was greater than 2:1. Of the trace metals tested only Zn caused full recovery of activity in the EDTA-inhibited cultures, although further addition of Fe increased the rate of thymidine incorporation above that in control cultures treated with PHA alone. In subsequent studies EDTA was always added in combination with an optimal amount of Fe.

Figure 1 shows the time courses of RNA and protein synthesis in stimulated lymphocytes in the presence and absence of EDTA and Zn. In all instances the time course for stimulation with PHA alone (not shown) was virtually identical to that with PHA + EDTA + Zn. The initial stimulation of both RNA and protein synthesis by PHA was not affected by EDTA, but from about 8 hr absence of Zn in EDTA-treated cultures prevented further increase in the rate of protein synthesis and caused a marked reduction in the rate of increase in RNA synthesis.

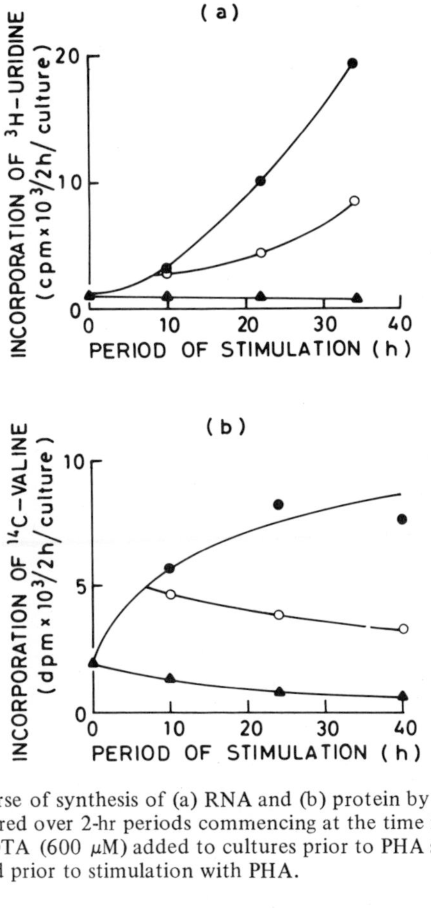

Fig. 1. Time course of synthesis of (a) RNA and (b) protein by lymphocytes. Incorporations were measured over 2-hr periods commencing at the time indicated. ▲ No additions to cultures; ○ EDTA (600 μM) added to cultures prior to PHA stimulation; ● EDTA and Zn (50 μM) added prior to stimulation with PHA.

The effects of EDTA and Zn on the type of RNA synthesised 20 hr after stimulation with PHA were investigated. Addition of ^{3}H-uridine to the cultures was carried out and incubation continued for 30 min or 20 hr before the RNA was extracted and analysed by sucrose gradient centrifugation (Fig. 2). After 30 min, the incorporation of uridine in the EDTA-treated cultures was only 54% of that in cultures treated with EDTA + Zn, but by 20 hr the incorporation in the cultures lacking Zn was 96% of that in the controls. The patterns of labeling suggested a slower but relatively normal synthesis of RNA in the EDTA-inhibited cultures with no gross change in the types of RNA synthesized. However, the method of analysis was not adequate to detect minor changes in the species of RNA synthesized.

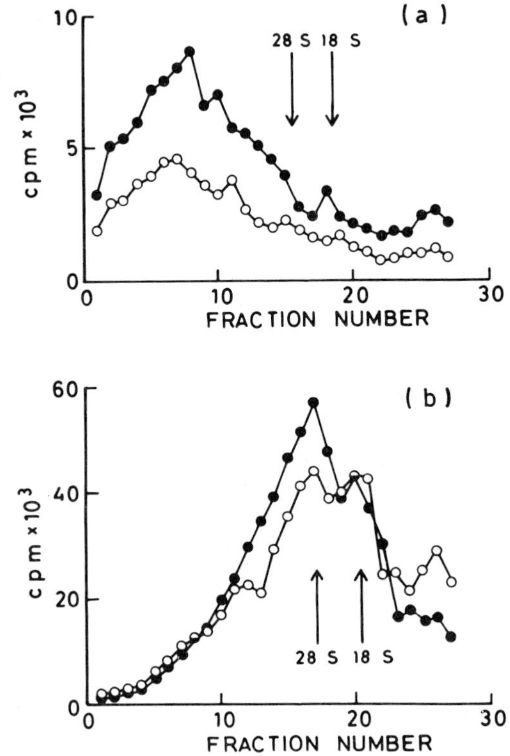

Fig. 2. Incorporation of ^{3}H-uridine into RNA of PHA-stimulated lymphocytes. The cells were incubated for 20 hr following stimulation, ^{3}H-uridine was added, and the incorporation was investigated after a further (a) 30-min, (b) 20-hr incubation. RNA was extracted from pooled cultures and the distribution of radioactivity determined after centrifugation through a 5–20% sucrose gradient. ○ Cultures received EDTA and PHA at zero time; ● EDTA, Zn, and PHA added at zero time.

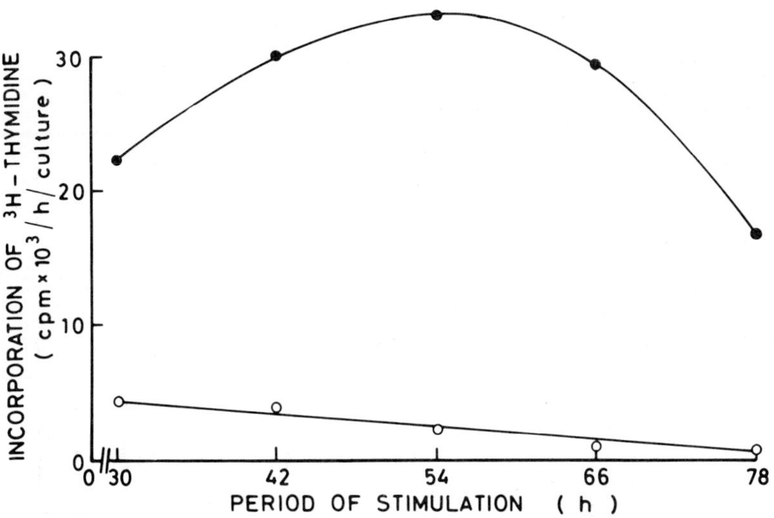

Fig. 3. Time course of DNA synthesis by stimulated lymphocytes. DNA synthesis was estimated by incorporation of ^{3}H-thymidine in a 1-hr period starting at the time indicated. ○ Cultures received only EDTA and PHA; ● EDTA, Zn, and PHA added at zero time.

The reduced rate of DNA synthesis in EDTA-inhibited cultures could have been caused by a reduction in the number of cells synthesizing DNA, or it could have resulted from lower concentrations of the necessary enzymes in each cell causing a reduced rate of synthesis. In the latter case, one would have expected an extension of the period of synthesis since the total amount of DNA to be replicated would be the same in cultures with and without adequate Zn. The time course of synthesis of DNA in the presence of EDTA or EDTA + Zn (Fig. 3) suggests the former explanation. In addition, a combination of autoradiography and liquid-scintillation counting of cells labeled with ^{3}H-thymidine in cultures subject to varying degrees of Zn limitation showed that incorporation was proportional to the number of cells synthesizing DNA. Reduced availability of Zn appears therefore to have limited the number of cells which were stimulated rather than the rate of synthesis of DNA per stimulated cell. The timing of the requirement for Zn was investigated by delaying the addition of EDTA or Zn (Fig. 4). There was a delay of 8–16 hr between the requirement for Zn and the synthesis of DNA. This time course of inhibition by EDTA corresponds closely with the sensitivity of DNA synthesis to low doses of actinomycin D (24).

PHA-stimulated lymphocytes show an early increase in K^+ uptake which is associated with changes in the permeability of the lymphocyte mem-

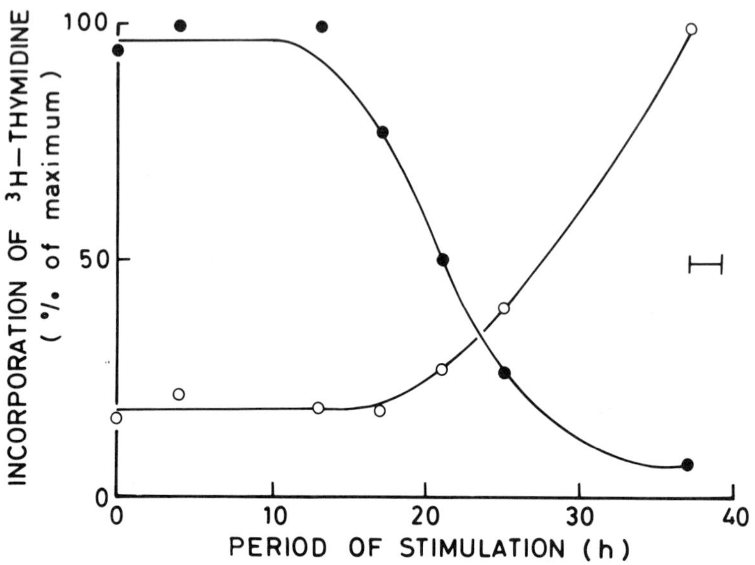

Fig. 4. Requirement for Zn for DNA synthesis by stimulated lymphocytes. ⊢⊣ Period of incorporation of ^{3}H-thymidine; ○ EDTA added to PHA-stimulated cultures at the times indicated; ● Zn added at the time indicated to cultures which had received EDTA and PHA at zero time.

brane (25). Recent studies in our laboratory have shown that EDTA did not affect the stimulation of ^{42}K uptake during the first 6 hr after PHA addition.

Chick Embryo Cells

Rubin (26) showed that EDTA inhibited the synthesis of DNA by chick embryo cells in tissue culture and that the inhibition could only be reversed by Zn. There was a delay of approximately 9 hr between addition of EDTA and half-maximal inhibition of DNA synthesis; the reversal of EDTA inhibition by Zn occurred after a similar delay. The decrease in thymidine incorporation caused by EDTA was shown to result from a reduction in the number of cells synthesizing DNA rather than the rate of synthesis by individual cells. Thymidine incorporation by cultures whose growth was limited by population density was unaffected by EDTA, although the rapid increase in DNA synthesis following a change of medium was inhibited. Protein synthesis by these cultures was much less inhibited by EDTA than was DNA synthesis.

Recently Rubin (27) found that glucose uptake by chick embryo cells was not affected by EDTA.

Malignant and Virus-transformed Cells

EDTA did not significantly inhibit DNA synthesis in HeLa and L cells (20), ascites tumor cells (28), or in chick embryo cells transformed by Rous sarcoma virus (26).

DISCUSSION

With the exception of malignant and virus-transformed cells, low availability of Zn resulted in an inhibition of DNA synthesis in all the mammalian systems investigated. The fundamental differences between these two types of cells are still not fully understood, but seem to be associated with the control processes which regulate cell replication. In both types of cell protein synthetic activity is less dependent on cell replication than is the synthesis of DNA because of protein synthesis associated with cell enlargement and protein turnover. The much greater sensitivity of DNA synthesis to EDTA in normal than in malignant cells (where cell multiplication is less subject to controls) and the relative insensitivity of protein synthesis to EDTA suggest a role of Zn in the processes which control cell replication. In each experimental system Zn was required for DNA synthesis at a time which corresponded closely with that at which the enzymes involved in DNA synthesis first appear. Furthermore, in cultured kidney cells lack of Zn resulted in lower activities of at least two of these enzymes (19). Since Zn has been shown to be a constituent of DNA polymerase (EC 2.7.7.7) (29), it is possible that Zn was required solely for the synthesis of Zn-containing enzymes involved in DNA synthesis. However, EDTA-treated cultures also had a requirement for Zn for the synthesis of RNA and protein (20, 23). In lymphocytes the time at which RNA synthesis was first affected by lack of Zn corresponded with the time at which the increase in activity of RNA polymerase (EC 2.7.7.6) becomes sensitive to inhibition by puromycin (30). The inhibition of synthesis of lymphocyte DNA and protein by EDTA was the same in both magnitude and timing as that caused by the addition of low concentrations of actinomycin D (24). A similar parallelism was found between the effects of EDTA and actinomycin D on DNA synthesis by cultured kidney cells (20, 21, 31). The combined evidence suggests that Zn was required for the processes causing a change in the pattern of enzymes synthesized by the cells in addition to being necessary as a constituent of some of these enzymes.

Changes in the genetic expression of cells are generally associated with early alterations in both nuclear metabolism and membrane function (32). With lymphocytes, increase in K^+ uptake caused by PHA stimulation has been shown to be unaffected by EDTA, and glucose transport by chick embryo cells was similarly unaltered by EDTA. It seems unlikely, therefore, that the primary site of action of Zn was associated with the changes in membrane permeability.

The similarity of the effects of EDTA and actinomycin D, an inhibitor of RNA synthesis, on the synthesis of DNA and protein suggested that Zn was primarily involved in nuclear metabolism, possibly in the synthesis of one or more types of RNA. However, with stimulated lymphocytes, the types of RNA synthesized in cultures treated with EDTA were similar in general pattern to those in control cultures. There was no evidence of a selective inhibition of the synthesis of mRNA, tRNA, or rRNA, although the method of analysis, sucrose gradient centrifugation, was not sufficiently sensitive to detect minor changes in the range of species of which these basic types were composed.

Changed expression of the genetic potential of a cell is dependent on the unmasking of portions of the genetic information not previously transcribed or an alteration in the relative rates of transcription of previously transcribed regions of the DNA. Each of the systems studied showed a close correlation between the rate of DNA synthesis and the number of cells synthesizing DNA under conditions of varying Zn availability. Considered along with the fact that two of the enzymes for DNA synthesis showed parallel changes in activity in EDTA-treated cultures (20), this suggests that Zn was required for the unmasking of the genetic information to synthesize a group of enzymes. Recent investigations in our laboratory have shown that low concentrations of actinomycin D, which were insufficient to inhibit the synthesis of DNA in control cultures, significantly stimulated its synthesis in EDTA-treated cultures (Fig. 5). It seems possible that actinomycin D, by binding to the DNA, inhibited the unmasking and transcription of some of the rRNA genome and liberated Zn from this portion of the DNA. Since ribosome availability was unlikely to limit protein synthesis at very low concentrations of actinomycin D (24), increased Zn availability within the nucleus could have led to increased mRNA synthesis and consequently to an increase in the enzymes available for DNA synthesis. Such an explanation again suggests a role for Zn in the processes leading to, or involved in, the transcription of DNA into RNA.

The hypothesis is presented that Zn is required for the alteration of expression of the genetic potential of a cell which occurs during transformation of a "resting" (G_0) cell into one in the phase of preparation for replication of its DNA (G_1). Similar changes may also occur when cells have completed division and re-enter a G_0 phase. Most of the effects of Zn cited earlier could possibly be explained by a requirement for Zn in the changes of

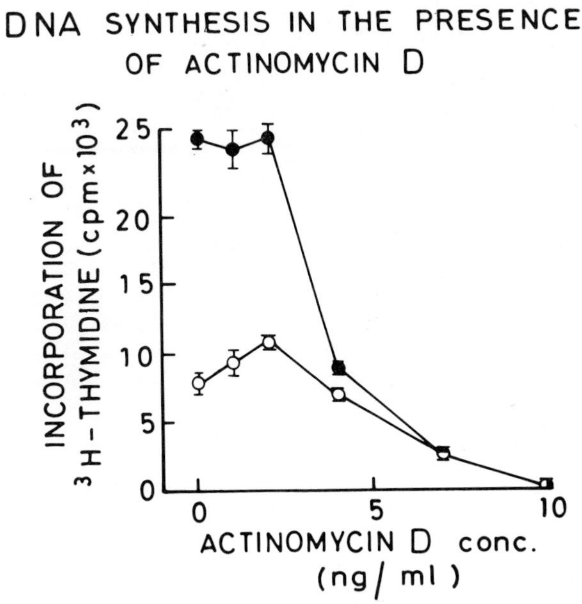

Fig. 5. Effect of actinomycin D on DNA synthesis by lymphocytes stimulated with PHA. DNA synthesis was estimated by ^{3}H-thymidine incorporation between 40 and 41 hr after PHA addition. ○ Cultures received EDTA and PHA at zero time; ● EDTA, Zn, and PHA added at zero time.

genetic expression involved in these transitions to and from the "resting" or G_0 phase of the cell cycle. Cell proliferation in many tissues involves a series of G_0 to G_1 transitions and malignancy might be described as the failure of cells to return to a G_0 phase after cell division or the excessive transition of G_0 cells into the G_1 phase. The normal differentiation of daughter cells derived from the basal epithelium of the esophagus and small intestine involves entry of previously dividing cells into a G_0 phase. Therefore changes in the availability of Zn might be expected to affect growth, wound healing, differentiation of the cells of the G.I. tract, and tumor development. Although we are far from a full understanding of the role of Zn in cell replication it is hoped that this hypothesis will provide a basis for further investigation.

REFERENCES

1. Todd, W. R., Elvehjem, C. A., and Hart, E. B. 1934. Zinc in the nutrition of the rat. *Am. J. Physiol.* 107: 146.
2. Fujioka, M., and Lieberman, I. 1964. A Zn^{2+} requirement for the

synthesis of deoxyribonucleic acid by rat liver. *J. Biol. Chem.* 239: 1164.

3. Williams, R. B., Mills, C. F., Quarterman, J., and Dalgarno, A. C. 1965. The effect of zinc deficiency on the *in vivo* incorporation of ^{32}P into rat liver nucleotides. *Biochem. J.* 95: 29P.
4. Buchanan, P. J., and Hsu, J. M. 1968. Zinc deficiency and DNA synthesis in liver. *Fed. Proc.* 27: 483.
5. Weser, U., Seeber, S., and Warnecke, P. 1969. Reactivity of Zn^{2+} on nuclear DNA and RNA biosynthesis of regenerating rat liver. *Biochim. Biophys. Acta* 179: 422.
6. Standstead, H. H., and Rinaldi, R. A. 1969. Impairment of deoxyribonucleic acid synthesis by dietary zinc deficiency in the rat. *J. Cell Physiol.* 73: 81.
7. Williams, R. B., and Chesters, J. K. 1970. The effects of early zinc deficiency on DNA and protein synthesis in the rat. *Br. J. Nutr.* 24: 1053.
8. Pories, W. J., Henzel, J. H., Rob, C. G., and Strain, W. H. 1967. Acceleration of wound healing in man with zinc sulphate given by mouth. *Lancet* 1: 121.
9. Standstead, H. H., and Shepard, G. H. 1968. The tensile strength of healing skin incisions in zinc deficiency. *Fed. Proc.* 27: 484.
10. Greaves, M. W. 1972. Zinc in cutaneous ulceration due to vascular insufficiency. *Am. Heart J.* 83: 716.
11. Lavy, U. I. 1972. The effect of oral supplementation of zinc sulphate on primary wound healing in rats. *Br. J. Surg.* 59: 194.
12. Hsu, T. H. S., and Hsu, J. M. 1972. Zinc deficiency and epithelial wound repair: an autoadiographic study of ^{3}H-thymidine incorporation. *Proc. Soc. Exp. Biol. Med.* 140: 157.
13. Rahmat, A. 1972. Studies on the effect of zinc on wound healing. Ph.D. Thesis, University of Aberdeen, U.K.
14. Fell, B. F., Leigh, L. C., and Williams, R. B. 1973. The cytology of various organs in zinc-deficient rats with particular reference to the frequency of cell division. *Res. Vet. Sci.* 14: 317.
15. Swenerton, H., Schrader, R., and Hurley, L. S. 1969. Zinc-deficient embryos: reduced thymidine incorporation. *Science* 166: 1014.
16. Poswillo, D. E., and Cohen, B. 1971. Inhibition of carcinogenesis by dietary zinc. *Nature, Lond.* 231: 441.
17. Dvizhkov, P. P., and Potapova, I. N. 1970. Blastomogenic properties of metallic zinc and zinc chloride. *Vitalstoffe* 15: 43.
18. De Wys, W., and Pories, W. 1972. Inhibition of a spectrum of animal tumors by dietary zinc deficiency. *J. Natn. Cancer Inst.* 48: 375.
19. Kmet, J., and Mahboubi, E. 1972. Oesophageal cancer in the Caspian littoral of Iran: initial studies. *Science* 175: 846.
20. Lieberman, I., and Ove, P. 1962. Deoxyribonucleic acid synthesis and its inhibition in mammalian cells cultured from the animal. *J. Biol. Chem.* 237: 1634.
21. Lieberman, I., Abrams, R., Hunt, N., and Ove, P. 1963. Levels of enzyme activity and deoxyribonucleic acid synthesis in mammalian cells cultured from the animal. *J. Biol. Chem.* 238: 3955.

22. Alford, R. H. 1970. Metal ion requirements for phytohaemagglutinin-induced transformation of human periferal blood lymphocytes. *J. Immun.* 104: 698.
23. Chesters, J. K. 1972. The role of zinc in the transformation of lymphocytes by phytohaemagglutinin. *Biochem. J.* 130: 133.
24. Kay, J. E., Leventhal, B. G., and Cooper, H. L. 1969. Effects of inhibition of rRNA synthesis on the stimulation of lymphocytes by PHA. *Exp. Cell Res.* 54: 94.
25. Quastel, M. R., and Kaplan, J. G. 1970. Early stimulation of potassium uptake in lymphocytes treated with PHA. *Exp. Cell Res.* 63: 230.
26. Rubin, H. 1972. Inhibition of DNA synthesis in animal cells by EDTA and its reversal by zinc. *Proc. Natn. Acad. Sci. U.S.A.* 69: 712.
27. Rubin, H., and Koide, T. 1973. Inhibition of DNA synthesis in chick embryo cultures by deprivation of either serum or zinc. *J. Cell Biol.* 56: 777.
28. Weser, U., Seeber, S., and Warnecke, P. 1969. Zur Wirkung von Zn^{2+} auf die Nuklein-Säurebiosynthese von Ascites-Tumorzellen. *Experientia* 25: 489.
29. Slater, J. P., Mildvan, A. S., and Loeb, L. A. 1971. Zinc in DNA polymerase. *Biochem. Biophys. Res. Commun.* 44: 37.
30. Pogo, B. G. T. 1972. Early events in lymphocyte transformation by PHA. *J. Cell Biol.* 53: 635.
31. Lieberman, I., Abrams, R., and Ove, P. 1963. Changes in the metabolism of RNA proceeding the synthesis of DNA in mammalian cells cultured from the animal. *J. Biol. Chem.* 238: 2141.
32. Cooper, H. L. 1971. Biochemical alterations accompanying initiation of growth in resting cells. *In* R. Baseiga (ed.), *The Cell Cycle and Cancer,* Marcel Dekker, New York, p. 197.

DISCUSSION

Hurley (Davis). Our findings with respect to the appearance of enzymes in zinc-deficient embryos may be relevant to the suggested role of zinc in the expression of the genome. In searching for a basic mechanism which would lead to the development of congenital malformations in zinc deficiency, we found that, if a certain developmental type of cell appeared, the enzymes that were normally found in that cell were also present. However, if the particular cell types were not there, then the enzymes were not there either. We never found the presence of a particular cell type with the absence of any of its normal enzymic components. This suggests that zinc is involved in the expression of genome and might release certain blocks of enzymes at the same time.

Mills (Aberdeen). In relation to his widespread interest in zinc and DNA, I should like to ask Dr. Vallee a question. Your early work demonstrated the presence of transition elements in isolated RNA and DNA. Are we any nearer an expression of what metals are normally present *in vivo* and what roles these metals serve in regard to polynucleotide structure? I find it difficult to give more emphasis to zinc in RNA or DNA than perhaps copper or nickel. Have we progressed at all in the last 10 years?

Vallee (Boston). Yes, we found a large number of metals in nucleic acids, the most tenacious including manganese, nickel, and chromium. In the case of zinc, I think evidence has been accumulating for a role in nucleic acid metabolism. Zinc occurs in at least one of the several DNA polymerases of *Escherichia coli* and in RNA polymerase. Dr. Sandstead's experiments suggest effects of zinc deficiency on RNA polymerase in animals. Data on *Euglena gracilis*, which we published about 10 years ago, indicated in zinc deficiency a doubling of DNA content, cessation of RNA synthesis and protein synthesis, and accumulation of nucleotides and amino acids. These cells doubled in size and seemed to do it simultaneously. We have resumed this work lately, and there is no question that at least in that system a deficiency of zinc has a very profound effect primarily on nucleic acid metabolism. I think that Dr. Chesters has pointed out some very interesting approaches to the problem.

BIOCHEMICAL ROLE OF MANGANESE[1]

R. M. LEACH, JR.

Pennsylvania State University, University Park, Pennsylvania

Manganese was shown to be essential in 1931 when this element was found to be necessary for growth and reproduction in rats and mice (1, 2). A few years later it was demonstrated that manganese prevented a skeletal abnormality in chickens called 'perosis' (3). Since that time manganese has been shown to be essential for many species of animals. Recently, the possible occurrence of manganese deficiency in man has been reported (4). The symptoms included impaired blood clotting and lowered serum cholesterol.

BIOCHEMICAL AND PHYSIOLOGICAL CHANGES ASSOCIATED WITH MANGANESE DEFICIENCY

Most of the early studies with manganese deficiency were concerned with attempts to determine the role of manganese in bone formation. The negative findings of many of these studies (5, 6, 7) combined with results of pathological studies (8) led to the hypothesis that chondrogenesis rather than osteogenesis was involved in the skeletal defects associated with manganese deficiency (9). This hypothesis was supported by the finding that the epiphyseal cartilage of manganese-deficient chicks had a severely reduced mucopolysaccharide content (10). Since mucopolysaccharides are important components of cartilage, these observations provided a possible explanation for the skeletal defects associated with manganese deficiency.

The effect of manganese on tissue mucopolysaccharide content may also explain some of the deficiency defects observed with laying hens (11, 12). These defects include reduced egg production, poor shell formation, and an embryonic abnormality called chondrodystrophy. It has been found that

1. Authorized for publication as paper no. 4470 in the journal series of the Pennsylvania Agricultural Experiment Station.

there is a reduction in hexosamine content of the shell matrix associated with the poor shell formation observed in laying hens (13, 14). Although manganese-deficient embryos have not been studied, several types of genetic chondrodystrophy have been found to be characterized by a reduced limb mucopolysaccharide content (15, 16).

Similar observations have been made with young guinea pigs (17) and calves (18) which were born from dams receiving inadequate amounts of manganese. In guinea pigs the concentration of mucopolysaccharides was reduced in both rib and epiphyseal cartilage. As in the chick studies, chondroitin sulfate was the polysaccharide most severely affected by manganese deficiency. Hyaluronic acid and heparin were also reduced. The effect of manganese upon tissue mucopolysaccharide content could also account for the ataxia or postural defects associated with deficiencies of this element (19, 20, 21). It has been reported that deficient animals do not exhibit proper otolith development in the utricular and saccular maculae. Both histochemical staining and ^{35}S metabolism suggested that abnormal otolith development was related to mucopolysaccharide synthesis (22, 23).

Mice having the *pallid* mutant gene also exhibit congenital ataxia that is indistinguishable from that observed with manganese deficiency. The otoliths from *pallid* animals also show evidence of impaired mucopolysaccharide metabolism (22). Of particular interest is the finding that these *pallid* mice show an impairment of manganese metabolism. This was demonstrated by the fact that the feeding of large amounts of manganese to mothers during pregnancy will prevent congenital ataxia in offspring (22). Furthermore, transportation of ^{54}Mn is slower in mice with the *pallid* gene (24).

The mitochondria of manganese-deficient and *pallid* mice have also been investigated (25). Liver mitochondria isolated from deficient mice exhibit normal P:O ratio but a reduced oxygen uptake. Examination of the ultrastructure revealed abnormalities which include elongation and reorientation of cristae. No biochemical or ultrastructural abnormalities were found in the mitochondria from *pallid* mice.

Other defects in carbohydrate metabolism have been observed with manganese deficiency. Many of the guinea pigs with a congenital deficiency have short survival times and exhibit aplasia or hypoplasia of pancreatic tissue. Those that survive to adult age exhibit abnormal tolerance to intravenously administered glucose (26, 27). Newborn guinea pigs also show a lower urinary myoinositol content. It was speculated that this might be related to the impaired carbohydrate metabolism associated with skeletal defects (28).

A new dimension on the symptoms of manganese deficiency has been recently reported by Doisy (4). Presumptive evidence was presented for coincident deficiency of vitamin K and manganese in man. The patient exhibited an inability to elevate clotting proteins in response to vitamin K and, in addition, hypocholesterolemia. The effect on blood clotting was confirmed using the chick as an experimental animal. Chicks deficient in both manganese and vitamin K exhibited a reduced ability to respond to a challenge by vitamin K. These observations were attributed to a probable need for manganese in the synthesis of glycoproteins (prothrombin) and sterols. In this regard it has been found that manganese deficiency does not influence the cholesterol content of the hen's egg (14).

ENZYMATIC ROLE OF MANGANESE

As with most of the other essential transition elements, the relationship between manganese and enzymes can be classified into two categories: (a) metalloenzymes, and (b) metal-enzyme complexes. The various attributes of these categories have been adequately reviewed elsewhere (29, 30). Unlike other transition elements, such as iron, copper and zinc, the number of manganese metalloenzymes is very limited (see Table 1). On the other hand, the enzymes that can be activated by manganese are numerous. These enzymes include hydrolases, kinases, decarboxylases, and transferases (31). The problem with metal activation is that it is usually nonspecific. Thus, it is difficult to correlate physiological defects with biochemical role when the metal has this type of relationship to enzyme action.

MANGANESE AND GLYCOSYLTRANSFERASES

Attempts have been made to relate the impairment in mucopolysaccharide synthesis associated with manganese deficiency to the widespread activation of glycosyltransferases by this element (32). These enzymes are important in

Table 1. Some characteristics of manganese-containing metalloproteins

Protein	M.W.[1]	Mn/mole	Source
Pyruvate carboxylase	500,000	4(II)	Avian liver (37)
Superoxide dismutase	39,500	2(III)	*Escherichia coli* (41)
Superoxide dismutase	40,000	2(III)	*Streptococcus mutans* (42)
Avimanganin	89,000	1(III)	Avian liver (40)
Manganin	56,000	1	Peanuts (43)
Concanavalin A	190,000	1(II)	Jack bean (44)

[1]M.W., molecular weight.

polysaccharide and glycoprotein synthesis. The general reaction may be depicted as follows:

$$\text{UDP-sugar-nucleotide + acceptor} \overset{ME^{2+}}{\rightleftharpoons} \text{product + UDP.}$$

The list of glycosyltransferases which require manganese or some other metal ion for activity is extensive (32). In fact it would be easier to list the transferases which do not show this requirement.

There is indirect evidence linking the activation of glycosyltransferases by manganese to the symptoms observed with a deficiency of this element. First of all the products of glycosyltransferases are reduced in manganese deficiency, e.g., mucopolysaccharides and glycoproteins (prothrombin). Secondly, the glycosyltransferases involved in mucopolysaccharide synthesis require metal ions for activity. Manganese is usually the most effective ion. Lastly, enzyme preparations from deficient tissues incorporate more radioactive substrate into product.

The latter observation requires further explanation. The glycosyltransferases involved in chondroitin sulfate synthesis are found in the 105,000 × *g* sediment obtained by tissue fractionation. These particles contain both the enzyme and acceptor. When incubated with substrates in the presence of manganese, the preparations from the deficient tissues incorporate more substrate than do comparable preparations from control chicks. This is interpreted as an indication of more acceptor sites in the deficient tissues reflecting suboptimum *in vivo* synthetic activity. Carbohydrate analysis of the particulate preparation supports such a hypothesis (33).

However, these findings can only be taken as indirect evidence at best. Certainly other interpretations of the data are possible. Attempts to study these reactions in more detail have been hampered by the fact that the enzyme and acceptor are found in the same particulate preparation. The respective activities are not easily separated.

An alternative approach would be to study the mechanism of action of manganese with a soluble glycosyltransferase. This is what has been done with the galactosyltransferase found in bovine milk (34, 35). The enzyme catalyzes the following reactions:

$$\text{(1) UDP-galactose + N-acetyl-glucosamine} \overset{Mn^{2+}}{\rightleftharpoons} \text{N-acetyl-lactosamine + UDP}$$

$$\text{(2) UDP-galactose + glucose} \overset{Mn^{2+}}{\rightleftharpoons} \text{lactose + UDP.}$$

Which reaction takes place depends upon the presence of lactalbumin as an enzyme modifier as well as substrate and acceptor concentrations. Kinetic studies of these two reactions led to the conclusion that the reaction had an ordered mechanism with the reactants adding in the order Mn^{2+}, UDP-sugar, and acceptor. A further conclusion was that Mn^{2+} reacts with the free enzyme under conditions of thermodynamic equilibrium and does not dissociate after each catalytic cycle.

Barker and associates (36) used affinity chromatography to purify milk galactosyltransferase. The affinity of the enzyme for the absorbant was enhanced by the presence of manganous ion. The type of enhancement observed agreed with the kinetic interactions observed by Morrison and Ebner (34, 35).

It remains to be seen if the above reaction mechanism is typical of other glycosyltransferase enzymes. Although there is evidence linking the need for manganese by glycosyltransferases to the signs of manganese deficiency, conclusive proof will await the results of further investigation.

MANGANESE METALLOENZYMES

Scrutton, Utter, and Mildvan (37) showed that pyruvate carboxylase purified from chick-liver mitochondria contained tightly bound manganese in a stoichiometry of 4 mole manganese (II) per mole enzyme. This enzyme also contains 4 mole biotin and catalyzes the following overall reaction:

$$\text{(1) pyruvate} + \text{ATP} + \text{HCO}_3^- \xrightleftharpoons{\text{acetyl-CoA, Mg}^{2+}} \text{oxalacetate} + \text{ADP} + \text{P}_i.$$

The divalent cation required for activation can be easily removed by dialysis and is not related to the tightly bound metal component of the enzyme. The reaction catalyzed by this enzyme can be further broken down into two steps:

$$\text{(2) enzyme-biotin} + \text{ATP} + \text{HCO}_3^- \xrightleftharpoons{\text{acetyl-CoA, Mg}^{2+}} \text{enzyme-biotin-CO}_2 + \text{ADP} + \text{P}_i.$$

$$\text{(3) enzyme-biotin} + \text{CO}_2 + \text{pyruvate} \rightleftharpoons \text{enzyme-biotin} + \text{oxalacetate}.$$

Studies on the role of the tightly bound manganese resulted in a proposed role for the metal in reaction (3) (38). The electrophilic character of the bound metal facilitates the proton departure from the methyl group of pyruvate and possibly carboxyl transfer from the carboxybiotin residue to pyruvate.

Of particular interest to this discussion is the recent study of the relationship between dietary manganese and the bound-metal content of the enzyme (39). Magnesium was found to replace manganese as the bound metal in pyruvate carboxylase isolated from manganese-deficient chicks. The relative content of the two metals was related to the severity of manganese deficiency. Substitution of magnesium for manganese caused only minor alterations in catalytic properties of the enzyme. Pyruvate carboxylase isolated from mammalian liver was found to contain both magnesium and manganese.

The above findings are in contrast with the observations with avimanganin (40). This is a protein of unknown function which has been isolated from avian liver. Avimanganin contains one mole of manganese (III) per 89,000 mol. wt. Manganese deficiency causes a depletion of manganese from avimanganin as well as a reduction in avimanganin content of liver. Thus two manganese metalloproteins are affected differently by manganese deficiency. In one case (pyruvate carboxylase) magnesium is substituted for manganese under conditions of dietary deficiency, while in the other case (avimanganin) the production of the metalloprotein is limited by dietary deficiency.

SUMMARY

Manganese deficiency in animals and man results in a wide variety of defects which include impairments in growth, reproduction, glucose tolerance, egg shell formation, and blood clotting, as well as skeletal deformities and ataxia. Many, but not all, of these defects may be accounted for by the need for manganese in glycosyltransferase activity. Evidence for this conclusion is based on the fact that many of the deficiency defects can be directly related to reduced products of glycosyltransferase reactions, e.g., mucopolysaccharides and glycoproteins. Secondly, the manganous ion is the most effective activator or metal cofactor for most of the glycosyltransferase enzymes that have been studied.

In contrast, the activity of the manganese metalloenzyme pyruvate carboxylase is not affected by manganese deficiency. Under deficiency conditions magnesium substitutes for the bound manganese, thus maintaining the activity of this enzyme at normal levels. The physiological significance of decreases in avimanganin in manganese deficiency remain to be elucidated.

REFERENCES

1. Kemmerer, A. R., Elvehjem, C. A., and Hart, E. B. 1931. Studies on the relation of manganese to the nutrition of the mouse. *J. Biol. Chem.* 92: 623.
2. Orent, E. R., and McCollum, E. V. 1931. Effects of deprivation of manganese in the rat. *J. Biol. Chem.* 92: 651.
3. Wilgus, H. S., Norris, L. C., and Heuser, G. F. 1936. The role of certain

inorganic elements in the cause and prevention of perosis. *Science* 84: 252.
4. Doisy, E. A., Jr. 1972. Micronutrient controls on biosynthesis of clotting proteins and cholesterol. *In* D. Hemphill (ed.), *Trace Substances in Environmental Health-VI, Proceedings of University of Missouri's 6th Annual Conference on Trace Substances in Environmental Health,* University of Missouri, Columbia, Mo., p. 193.
5. Gallup, W. D., and Norris, L. C. 1938. The essentialness of manganese for the normal development of bone. *Science* 87: 18.
6. Caskey, C. D., Gallup, W. D., and Norris, L. C. 1939. The need for manganese in the bone development of the chick. *J. Nutr.* 17: 407.
7. Parker, H. E., Andrews, F. N., Carrick, C. W., Creek, R. D., and Hauge, S. M. 1955. Effect of manganese on bone formation studied with radioactive isotopes. *Poult. Sci.* 34: 1154.
8. Wolbach, S. B., and Hegsted, D. M. 1953. Perosis: epiphyseal cartilage in choline and manganese deficiencies in the chick. *A.M.A. Archs Path.* 56: 437.
9. Leach, R. M., Jr. 1960. The effect of manganese, zinc, choline, and folic acid deficiencies on the composition of epiphyseal cartilage. Ph.D. Thesis, Cornell University.
10. Leach, R. M., Jr., and Muenster, A. M. 1962. Studies on the role of manganese in bone formation. I. Effect upon mucopolysaccharide content of chick bone. *J. Nutr.* 78: 51.
11. Lyons, M., and Insko, W. M., Jr. 1937. Chondrodystrophy in the chick embryo produced by manganese deficiency in the diet of the hen. *Ky. Agric. Exp. Stn Bull.* No. 371: 63.
12. Lyons, M. 1939. Some effects of manganese on eggshell quality. *Ark. Agric. Expr. Stn Bull.* No. 374: 1.
13. Longstaff, M., and Hill, R. 1972. The hexosamine and uronic acid contents of the matrix of shells of eggs from pullets fed on diets of different manganese content. *Br. Poult. Sci.* 13: 377.
14. Leach, R. M., Jr. 1972. Unpublished data, Pennsylvania State University.
15. Mathews, M. B. 1967. Chondroitin sulfate and collagen in inherited skeletal defects of chickens. *Nature, Lond.* 213: 1255.
16. Gaffney, L., Leach, R. M., Jr., and Buss, E. G. 1972. Unpublished data, Pennsylvania State University.
17. Tsai, H. C. C., and Everson, G. J. 1967. Effect of manganese deficiency on the acid mucopolysaccharides in cartilage of guinea pigs. *J. Nutr.* 91: 447.
18. Rojas, M. A., Dyer, I. A., and Cassatt, W. A. 1965. Manganese deficiency in the bovine. *J. Anim. Sci.* 24: 664.
19. Norris, L. C., and Caskey, C. D. 1939. A chronic congenital ataxia and osteodystrophy in chicks due to manganese deficiency. *J. Nutr. (suppl.)* 17: 16.
20. Hill, R. M., Holtkamp, D. E., Buchanan, A. R., and Rutledge, E. K. 1950. Manganese deficiency in rats with relation to ataxia and loss of equilibrium. *J. Nutr.* 41: 359.
21. Hurley, L. S., Everson, G. J., and Geiger, J. F. 1958. Manganese deficiency in rats: congenital nature of ataxia. *J. Nutr.* 66: 309.

22. Hurley, L. S. 1968. Genetic-nutritional interactions concerning manganese. *In* D. Hemphill (ed.), *Trace Substances in Environmental Health-II, Proceedings of University of Missouri's 2nd Annual Conference on Trace Substances in Environmental Health,* University of Missouri, Columbia, Mo., p. 41.
23. Shrader, R. E., Erway, L. C., and Hurley, L. S. 1973. Mucopolyssacharide synthesis in the developing inner ear of manganese-deficient and pallid mutant mice. *Teratology,* in press.
24. Cotzias, G. T., Tang, L. C., Miller, S. T., Sladic-Simic, D., and Hurley, L. S. 1972. A mutation influencing the transportation of manganese, L-dopa, and L-tryptophan. *Science* 176: 410.
25. Hurley, L. S., Theriault, L. L., and Dreosti, I. E. 1970. Liver mitochondria from manganese-deficient and pallid mice: Function and ultrastructure. *Science* 170: 1316.
26. Shrader, R. E., and Everson, G. J. 1968. Pancreatic pathology in manganese-deficient guinea pigs. *J. Nutr.* 94: 269.
27. Everson, G. J., and Shrader, R. E. 1968. Abnormal glucose tolerance in manganese-deficient guinea pigs. *J. Nutr.* 94: 89.
28. Everson, G. J. 1968. Preliminary study of carbohydrates in the urine of manganese-deficient guinea pigs at birth. *J. Nutr.* 96: 283.
29. Vallee, B. L. 1955. Zinc and metalloenzymes. *In* M. L. Anson, K. Bailey and J. T. Edsall (eds), *Advances in Protein Chemistry,* Vol. 10, Academic Press, New York, p. 317.
30. Malmström, B. G., and Rosenberg, A. 1959. Mechanism of metal ion activation of enzymes. *In* F. F. Nord (ed.), *Advances in Enzymology,* Vol. 21, Interscience, New York, p. 131.
31. Vallee, B. L., and Coleman, J. E. 1964. Metal coordination and enzyme action. *In* M. Florkin and E. H. Stotz (eds), *Comprehensive Biochemistry*, Vol. 12, Elsevier, Amsterdam, p. 165.
32. Leach, R. M., Jr. 1971. Role of manganese in mucopolysaccharide metabolism. *Fed. Proc.* 30: 991.
33. Leach, R. M., Jr., Muenster, A. M., and Wein, E. M. 1969. Studies on the role of manganese in bone formation. II. Effect upon chondroitin sulfate synthesis in chick epiphyseal cartilage. *Archs Biochem. Biophys.* 133: 22.
34. Morrison, J. F., and Ebner, K. E. 1971. Studies on galactosyltransferase: kinetic investigations with n-acetylglucosamine as the galactosyl group acceptor. *J. Biol. Chem.* 246(12): 3977.
35. Morrison, J. F., and Ebner, K. E. 1971. Studies on galactosyltransferase: kinetic investigations with glucose as the galactosyl group acceptor. *J. Biol. Chem.* 246(12): 3985.
36. Barker, R., Olsen, K. W., Shaper, J. H., and Hill, R. L. 1972. Agarose derivatives of uridine diphosphate and n-acetylglucosamine for the purification of a galactosyltransferase. *J. Biol. Chem.* 247: 7135.
37. Scrutton, M. C., Utter, M. F., and Mildvan, A. S. 1966. Pyruvate carboxylase. VI. The presence of tightly bound manganese. *J. Biol. Chem.* 241(15): 3480.
38. Mildvan, A. S., Scrutton, M. C., and Utter, M. F. 1966. Pyruvate carboxylase. VII. A possible role for tightly bound manganese. *J. Biol. Chem.* 241: 3488.

39. Scrutton, M. C., Griminger, P., and Wallace, J. C. 1972. Pyruvate carboxylase: bound metal content of the vertebrate liver enzyme as a function of diet. *J. Biol. Chem.* 247: 3305.
40. Scrutton, M. C. 1971. Purification and some properties of a protein containing bound manganese (avimanganin). *Biochemistry* 10: 3897.
41. Keele, B. B., McCord, J. M., and Fridovich, I. 1970. Superoxide dismutase from *Escherichia coli* B. *J. Biol. Chem.* 245: 6176.
42. Vance, P. G., Keele, B. B., Jr., and Rajogopalan, K. V. 1972. Superoxide dismutase from *Streptococcus mutans. J. Biol. Chem.* 247: 4782.
43. Dickert, J. W., and Rozacky, E. 1969. Isolation and partial characterization of manganin, a new manganoprotein from peanut seeds. *Archs Biochem. Biophys.* 134: 473.
44. Agrawal, B. B. L., and Goldstein, I. J. 1968. Protein carbohydrate interaction. VII. Physical and chemical studies on concanavalin A, the hemogglutinin of the Jack bean. *Archs Biochem. Biophys.* 124: 218.

DISCUSSION

Frieden (Tallahassee). It seems to me that the concentrations of managanous ion required for the variety of transferases which you listed were prohibitively high, that is in the 3–15 mM range.

Leach. I agree, they are very high. I don't know the explanation other than that some of the soluble transferases don't have that high a requirement. The bound transferases may have an artificially high requirement for Mn *in vitro.*

Schwarz (Long Beach). In regard to Dr. Neilsen's findings of cartilage anomalies in nickel deficiency, have you tested nickel in those enzymes which are not manganese responsive, for example the xylosyl transferase system?

Leach. No. We did *in vivo* and *in vitro* work with nickel on the manganese responsive enzymes, but not the xylosyl transferase.

Hurley (Davis). I should like to add to the list of various biochemical roles of manganese the maintenance of normal biochemical function and structure of mitochondria, as evidenced by alterations in mitochondria during manganese deficiency.

BIOCHEMICAL ROLE OF SELENIUM[1]

WILLIAM G. HOEKSTRA

Department of Biochemistry and Department of Nutritional Sciences, University of Wisconsin-Madison, Madison, Wisconsin

The effects of selenium intake on the animal body can be dramatic and have spelled the difference between life or death. Attention was first focused on this element because of its toxicity. Since the discovery of the essentiality of Se for rats by Schwarz and Foltz (1) and its confirmation for chickens by Patterson, Milstrey, and Stokstad (2), and with the recognition of Se deficiency as a practical problem in farm animals (3, 4), the biochemical function of Se as an essential nutrient has attracted major interest. Any idea as to the biochemical role of Se has had to take into account the profound interrelationship of Se to vitamin E and to the sulfur-containing amino acids as first established by Schwarz and Foltz (1). Under appropriate dietary conditions, the condition of liver necrosis in rats can be prevented by either Se or vitamin E and largely prevented, or greatly delayed, by the sulfur-containing amino acids methionine or cystine (5). Although Schwarz (6, 7) has continued to maintain that the metabolic function of Se is distinct from that of vitamin E, some investigators have postulated essentially overlapping functions of these nutrients, such as the idea that both act as rather nonspecific 'antioxidants' (8). That Se could not be completely replaced by vitamin E was shown conclusively by Thompson and Scott (9, 10), and by McCoy and Weswig (11). These observations proved that the function of Se and vitamin E could not be one and the same, although clearly they must intertwine closely to account for the marked mutual sparing effects of the two nutrients. My purpose in this symposium is to review the recent contributions of my collaborators and myself at the University of Wisconsin which establish an

1. Research supported by the College of Agricultural and Life Sciences, University of Wisconsin, Madison, and by United States Public Health Service, Program Grant No. AM-14881.

essential role for Se as a component of the enzyme glutathione peroxidase (glutathione: H_2O_2 oxidoreductase, E.C. 1.11.1.9), which in this paper will be designated GSH-Px. This role does not exclude other possible roles for this element in animal tissues but must be regarded as the most definitive function for Se at this time. In particular, it allows logical speculation on the mechanism of the relation of Se to vitamin E, unsaturated lipids, and the sulfur-containing amino acids, which will receive brief comment. The question will be posed whether decreases in tissue GSH-Px can explain the well-known Se-deficiency defects. Other recently suggested roles of Se will be briefly reviewed.

HOW THE IDEA OF A ROLE FOR SELENIUM IN GLUTATHIONE PEROXIDASE DEVELOPED

Our premise in initiating this research was that both Se and vitamin E have an important role in preventing membrane damage as had been postulated by Tappel (8). However, we were not convinced that such protection resulted from nonspecific 'antioxidant' effects. We selected the red blood cell for study *in vitro* because it represented a rather simple, physiologically meaningful, model system for the study of membrane damage. Increased sensitivity of the erythrocyte to hemolysis *in vitro* was a well-documented effect of vitamin-E deficiency, but erythrocyte hemolysis was considered by many workers to be nonresponsive to dietary Se (12). If the membrane-damage theory was to explain the relationship between vitamin E and Se, it was difficult to understand why the erythrocyte membrane system did not respond to Se. We therefore set about restudying this response and were struck by the observation that most studies of erythrocyte hemolysis involving these nutrients were conducted with the erythrocytes suspended in simple buffered media containing no added energy substrates for the cells. In particular, the omission of glucose was considered of possible importance because of the well-documented relationship of glucose to the problem of 'drug-induced hemolysis' in human subjects (13, 14). In this condition defects in glucose metabolism cause a slow rate of generation of reduced glutathione (GSH). GSH is important in maintaining membrane integrity particularly in the face of an 'oxidant stress' imposed by various drugs such as primaquine, sulfanilimide, etc. GSH is believed to protect the cell by its role in peroxide decomposition and possibly by maintaining protein-SH groups through SH-SS interchange. Some of these relationships are shown in the following reaction sequence:

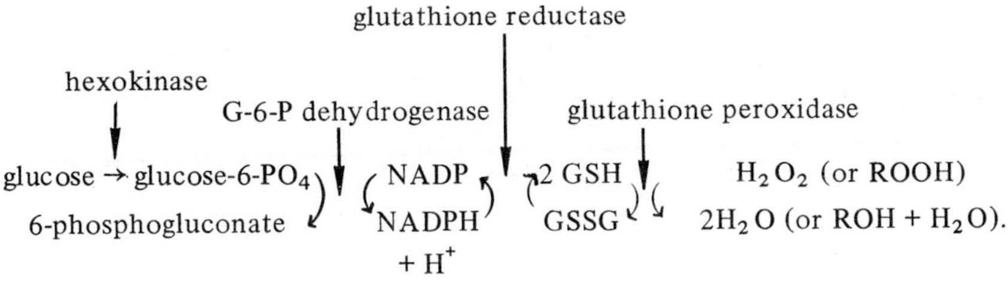

Our rationale was that Se may have a role in this reaction sequence, the consequences of which would be observed only if substrate were provided to fuel the reaction. This seemed an attractive idea because it might relate Se to H_2O_2 and fatty-acid-hydroperoxide decomposition (14, 15) or to SH-SS interchange (16).

Our published results (17–19) fully confirmed this postulate. Only when glucose was included in the incubation medium did dietary Se intake have a marked influence in protecting the erythrocytes against hemolysis. Parenthetically, the initial experiments (17) failed to demonstrate this effect conclusively because some of the rats were succumbing to liver necrosis before the erythrocytes were sufficiently affected, a situation which was remedied by delaying (or preventing?) liver necrosis by supplying additional dietary methionine. Fortunately, we had a sufficient trend in the first experiment to pursue our idea further or we would probably not have discovered this important function of Se.

While vitamin E protected erythrocytes against hemolysis either in the presence or absence of glucose, it had little or no effect in decreasing oxidative damage to hemoglobin during *in vitro* incubation, as assessed by conversion of hemoglobin to methemoglobin, choleglobin, and Heinz bodies (17, 18). However, dietary Se protected both the cell membrane and its contents (hemoglobin) against oxidative damage, which may be an important distinction between the effects of vitamin E and Se. These early studies demonstrated clearly that, contrary to some literature reports, dietary Se did exert a dramatic protective effect against oxidative damage to the cell membrane and other cell components; this effect depended on the presence of glucose and was clearly distinguishable from the effects of vitamin E.

LOCATING THE METABOLIC SITE OF SELENIUM ACTION

The glucose-Se interaction in protecting against oxidative damage pointed to a role for Se in the GSH reaction sequence discussed previously. However, many possibilities existed with respect to the specific site affected by Se. We first attempted to ascertain whether Se was involved in the regeneration of

GSH, the process which is defective in the 'drug-induced' hemolytic anemias, by measuring GSH concentrations in erythrocytes during incubation. Dietary Se deficiency was found to increase the concentration of erythrocyte GSH *in vivo,* rather than decrease it, and had little or no effect on the rate of loss of cell GSH during incubation (18). Therefore the role of Se did not appear to be in the replenishing of GSH but in the utilization of GSH in protecting the cell (i.e., to the right, rather than the left, of GSH in the reaction sequence). This concept was tested directly on erythrocyte hemolysates by assessing the effect of dietary Se on the ability of added GSH to protect hemoglobin from oxidation in the presence of ascorbate or H_2O_2. Only in the hemolysates from rats fed Se did GSH protect hemoglobin against oxidation, thus supporting the idea that Se was involved in the utilization of GSH in protecting the cell (19). Further experiments showed that, in the presence of H_2O_2, GSH disappeared much more rapidly from hemolysates from Se-fed rats than from those from Se-deficient rats (19). This suggested strongly that the rate of the reaction between GSH and H_2O_2 in hemolysates was dependent on Se in the diet. In deficient rats, this reaction rate was little more than the spontaneous chemical reaction rate between GSH and H_2O_2. While this observation implied a dependence of erythrocyte glutathione peroxidase (GSH-Px) on dietary Se, our first postulate was that some low molecular weight form of Se reacted with GSH to form a compound such as GSSeH which was a much better reductant than GSH and could be reused, thereby acting in a catalytic manner. Our reason was that such a compound had been identified (20) and the literature implied that there was probably not sufficient Se in cells to be stoichiometric with most enzymes. However, we could not restore this catalytic activity to Se-deficient hemolysates by incubation with selenite *in vitro* and the activity was found to be heat labile. We therefore decided to test whether Se was a necessary and rather tightly bound component of the enzyme GSH-Px, even though no enzymes containing stoichiometric quantities of Se had yet been isolated. Rats were injected with selenite containing ^{75}Se and 2–4 weeks later glutathione peroxidase was partially purified from their erythrocytes (19). Most of the erythrocyte ^{75}Se accompanied GSH-Px during purification, suggesting that Se is an integral and necessary component of this enzyme. Although we have not pursued purification of the rat enzyme to homogeneity, GSH-Px from ovine erythrocytes was purified to this state of purity.

SELENIUM AS A COMPONENT OF PURIFIED OVINE GLUTATHIONE PEROXIDASE

In order to obtain a larger amount of purified GSH-Px than was easily provided from rat blood, we selected the blood of sheep for further study.

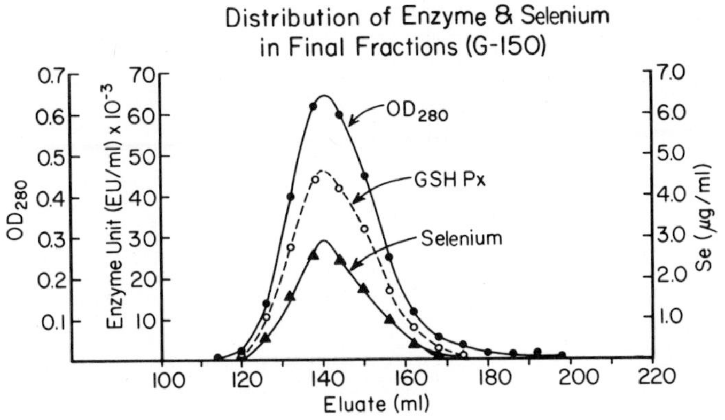

Fig. 1. Elution profile of purified ovine erythrocyte glutathione peroxidase chromatographed on Sephadex-G150 showing coincidence of protein concentration (OD_{280}), glutathione peroxidase activity (GSH Px), and selenium content (measured fluorometrically). No other peaks were observed.

Selenium content during purification of ovine erythrocyte GSH-Px and of the final preparation was followed by fluorometric analysis, thereby allowing calculation of the stoichiometry of Se to protein in the isolated enzyme. A 10-step procedure[2] was developed which provided about a 20% yield of purified GSH-Px. The final preparation was judged to be essentially homogeneous as evidenced by a single, but somewhat diffuse, band on polyacrylamide gel electrophoresis and by the highly uniform elution of the single peak on rechromatography of the final preparation on Sephadex-G150 (Fig. 1). During this 4000-fold purification of ovine GSH-Px, the Se concentration per gram of protein increased by about 3000-fold to a concentration of 0.34% Se in the final preparation, and the Se concentration paralleled roughly the specific enzymatic activity during purification (Fig. 2). By assuming a molecular weight of 85,000, which was estimated from gel chromatography and is close to the molecular weight reported for pure bovine GSH-Px by Flohe's group (21), the final preparation contained 3.7 mole Se per mole of protein. Allowing for some small degree of impurity and inaccuracy in methodology, we considered this to represent a stoichiometry of 4 mole Se per mole GSH-Px (22). This stoichiometry has recently been confirmed by Flohe, Gunzler, and Schock (23) for crystalline bovine erythrocyte GSH-Px analyzed for Se by neutron activation analysis. Flohe *et al.* (21, 23) have demonstrated that bovine GSH-Px has four subunits, making it likely that there is one Se per subunit; however, this has not yet been proven.

2. S. H. Oh, H. E. Ganther, and W. G. Hoekstra, *Biochemistry* (in press).

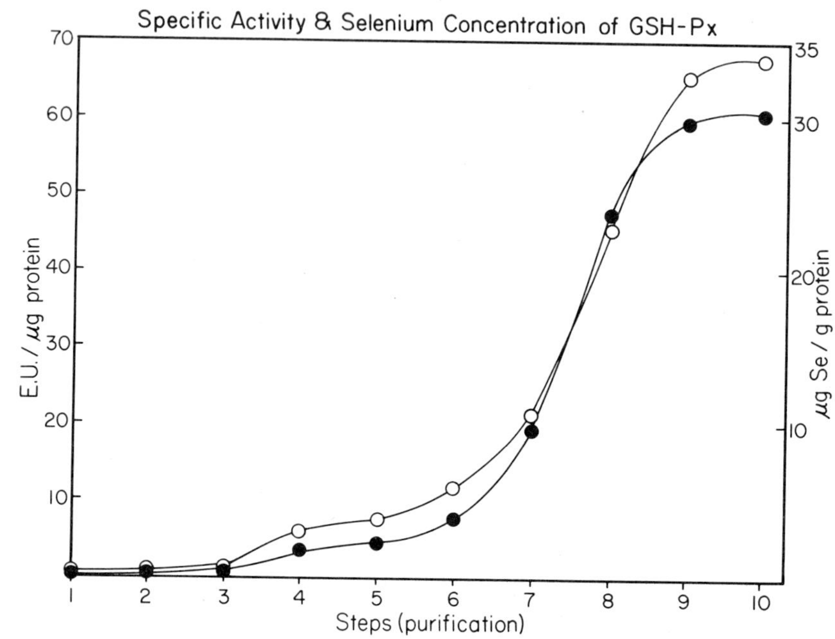

Fig. 2. Specific enzymatic activity (•) and selenium concentration (○) of the active protein fractions following each of 10 purification steps[2] used in the isolation of ovine erythrocyte glutathione peroxidase.

The nature of the Se moiety of GSH-Px remains to be elucidated. We have some evidence that it does not exist as selenite or selenate, but probably as part of an organic Se-containing moiety which is not one of the Se-containing amino acids, selenomethionine or selenocysteine. However, further research is needed before definitive statements can be made as to the nature of the Se component of the enzyme. It is tempting to speculate that the Se moiety may bear a relationship to the organic 'factor-3', which was studied extensively by Schwarz (24) but was not characterized as to its structure. Likewise, the specific role of Se in the enzyme remains to be discovered. It is postulated to have a redox, or electron-transferring, role.

EFFECT OF DIETARY SELENIUM ON THE TISSUE LEVELS OF GSH-Px

Even though our experiments up to this point indicated a decreased activity of GSH-Px in erythrocytes from Se-deficient rats, it was considered important to quantitate this relationship better and extend the studies to additional tissues. A relationship of Se intake to the level of GSH-Px in tissues was of interest because of its possible usefulness in diagnosing Se deficiency, establishing Se requirements, and explaining the defects in various tissues which result from Se deficiency. Our results on the male rat to date will be summarized.[3] The data are most extensive for erythrocytes, because of their ready access in any diagnostic procedure, and liver, because of the relationship of Se to liver necrosis in the rat. Fig. 3 shows the effect of dietary Se levels from < 0.01 ppm (basal diet) to 5 ppm, a chronically toxic level, during several weeks of feeding to male rats after weaning. A profound effect of dietary intake of Se on GSH-Px in erythrocytes is apparent. With the lowest intake of Se GSH-Px decreased to about 18% of the weaning level, while the highest intake of Se raised the GSH-Px activity by about 60%. It is of interest that the enzyme activity did not plateau when the Se requirement was met at about 0.1 ppm

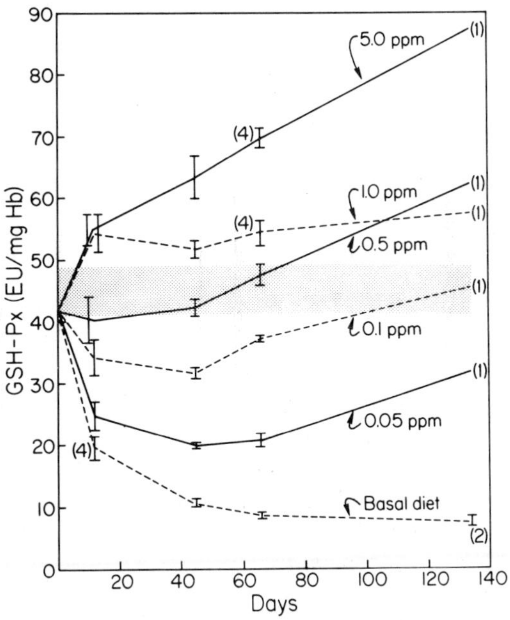

Fig. 3. Effect of dietary Se supplementation from 0 to 5 ppm Se, added as sodium selenite to a torula-yeast-based diet (18), on glutathione peroxidase activity of erythrocytes of weanling male rats of the Holtzman strain. The basal diet contained slightly less than 0.01 ppm Se. Bars represent the mean ± s.e. There were five rats per group except where indicated in parentheses. The shaded area represents the mean ± s.e. for a group of rats fed a practical stock diet adequate in Se. GSH-Px is expressed per milligram of hemoglobin (Hb) since no change in hemoglobin concentration or hematocrit occurred during the experiment. Methods are described in a separate publication.[3]

3. D. Hafeman, R. A. Sunde, and W. G. Hoekstra, *J. Nutr.* (in press).

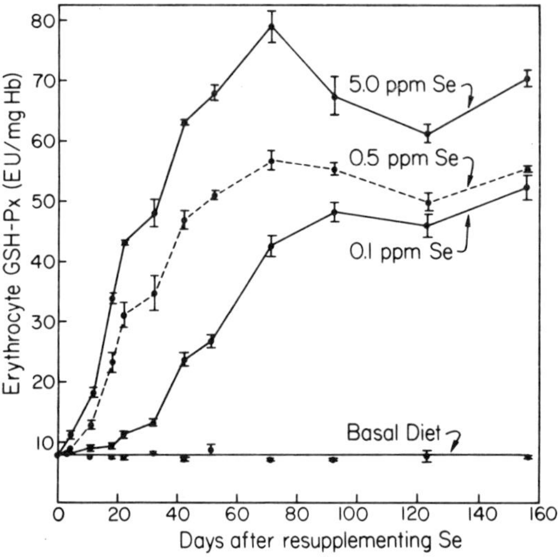

Fig. 4. Restoration of erythrocyte glutathione peroxidase activity by feeding supplemental Se to male Holtzman rats previously depleted of Se by feeding the basal diet for 9 months. There were four rats per group. For further details see legend to Fig. 3.

Se or slightly above, but continued to increase above 'normal' as chronically toxic doses of Se were fed. This additional increase may represent an adaptive response to the oxidant stress imposed by excess selenite. The effectiveness of dietary Se in restoring erythrocyte GSH-Px to Se-depleted rats is illustrated in Fig. 4. The response was more rapid with the higher levels of Se, but in all cases the peak of GSH-Px was reached about 65 days after the initiation of response, which corresponds closely to the life span of rat erythrocytes. This suggests that the enzyme is incorporated into erythrocytes only during erythropoeisis. Again, the ability of high doses of Se to raise the enzyme in erythrocytes to a 'super normal' level is apparent.

The effect of dietary Se level on liver GSH-Px of male rats is shown in Fig. 5. The livers were perfused to eliminate residual blood before assay. Liver GSH-Px was particularly sensitive to dietary-Se intake, dropping to undetectable activity after 25 days of feeding the basal diet (point not shown in Fig. 5). Maximal liver GSH-Px was found with the feeding of 1.0 ppm Se, but 0.5 and 0.1 ppm Se elicited nearly as high an enzyme activity. A level of 0.05 ppm Se was clearly inadequate for maximal, or near-maximal, activity of liver GSH-Px. In contrast to its effect on erythrocytes, 5.0 ppm Se did not result in a 'super normal' level of liver GSH-Px, probably because this level of dietary Se caused noticeable liver damage. When liver GSH-Px was expressed relative to another liver enzyme, histidase, which did not decrease from Se deficiency, this ratio was highest at the 5.0 ppm Se intake; this also suggests that 'inactive' or damaged liver tissue was responsible for the somewhat lower level of GSH-Px when 5.0 ppm Se was fed. Based on liver GSH-Px, the Se requirement of the growing male rat appears to be about 0.1 ppm Se; however, a growth rate comparable with that resulting from the higher levels of Se was achieved at 0.05 ppm supplemental Se. For the rat, liver GSH-Px appears to be a sensitive index of Se adequacy, but further studies are needed to assess the effects of other variables such as sex, vitamin E, unsaturated fatty acids, and oxidant stressors on the amount of this enzyme in liver and its response to Se. In addition, one cannot extrapolate to other species, because, for example, much less GSH-Px is found in the liver of lambs and cows. Studies under way in our laboratory show that feeding of silver or tri-orthocresylphosphate markedly decreases rat-liver GSH-Px.[4]

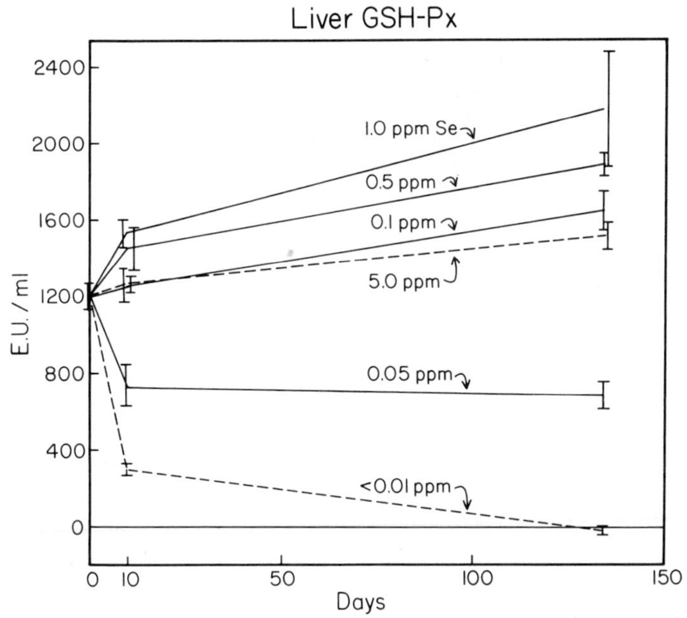

Fig. 5. Effect of dietary Se supplementation from 0 (designated as < 0.01 ppm Se in basal diet) to 5.0 ppm Se on glutathione peroxidase activities of livers of weanling male Holtzman rats. Conditions were as described in legend to Fig. 3. Enzyme activity is reported per milliliter of 20% (w/w) liver homogenate; however, no differences in protein content of homogenates of the various groups occurred so that expression per unit of protein showed a comparable effect.

4. Unpublished work of P. Wagner, A. B. Swanson, H. E. Ganther, and W. G. Hoekstra.

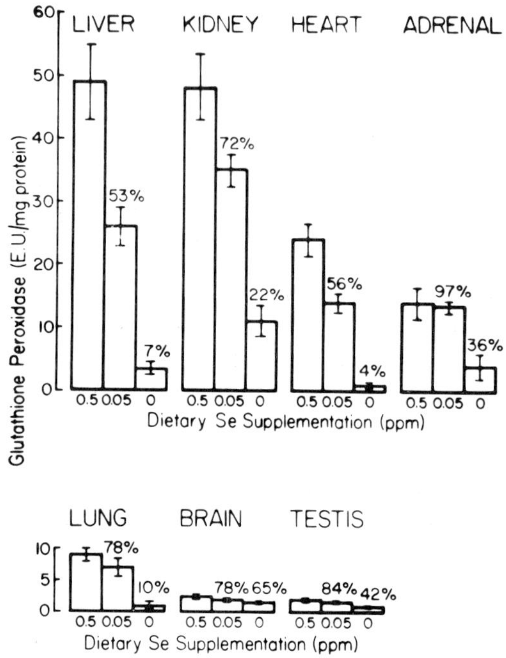

Fig. 6. Effect of three levels of dietary Se supplementation on glutathione peroxidase activities of several organs of male Holtzman rats. Measurements were made during 13–28 weeks of feeding the respective diets. Percentage figures on top of bars represent the enzymatic activity relative to rats fed 0.5 ppm Se. For further details see legend to Fig. 3.

Effects of prolonged Se deficiency (13–28 weeks), both severe (0 ppm supplemental Se to basal diet containing < 0.01 ppm Se) and marginal (0.05 ppm Se), on seven different organs of male rats including liver (unperfused) are presented in Fig. 6.[5] Large differences in amount of enzyme and its response to dietary Se are apparent, with brain showing the least response, and liver, heart, and lung showing the greatest response. Preliminary studies[6] on additional tissues of rats and lambs (data not presented) have shown a marked dependency of tissue GSH-Px to dietary Se for skeletal muscle (lamb), eye lens (rat), white blood cells (rat), lung macrophages (rat), pancreas (lamb), and blood plasma (lamb).

5. G. L. Schwartz and W. G. Hoekstra, unpublished. Data from B.S. Thesis of G.L.S., 1973.

6. Unpublished work of S. H. Oh, R. A. Lawrence, R. E. Serfass, G. L. Schwartz, H. E. Ganther, A. L. Pope, and W. G. Hoekstra.

DOES A LACK OF TISSUE GSH-Px EXPLAIN THE Se DEFICIENCY SYNDROME?

This question cannot be answered at the present time. There are impressive correlations, such as the dramatic responsiveness of liver GSH-Px to Se intake in relation to the protective effect of Se against liver necrosis in the rat. Moreover, it is easy to postulate a role for GSH-Px in preventing liver necrosis and other degenerative conditions, but experiments which prove this point remain to be conducted. Scott and Noguchi (25) have recently demonstrated an impressive relationship of low blood-plasma levels of GSH-Px to exudative diathesis in chicks. However, pancreatic degeneration in the chick was believed to be unrelated to this enzyme because selenite was more effective in eliciting plasma GSH-Px than was selenomethionine, while selenomethionine was more effective than selenite in preventing pancreatic degeneration (26). However, this must be studied further, because the pancreas may simply be better able to utilize the Se of selenomethionine for ultimate synthesis of GSH-Px. At present a role in GSH-Px should be considered as the best-documented function of Se but not necessarily the *only* function.

RELATION OF SELENIUM TO VITAMIN E

The discovery of Se as a component of GSH-Px allows a convenient postulate to explain its relation to vitamin E and to unsaturated lipids as well as the sulfur-containing amino acids. Fig. 7 depicts this postulated relationship schematically. Stated briefly, the postulate is that Se (as a component of

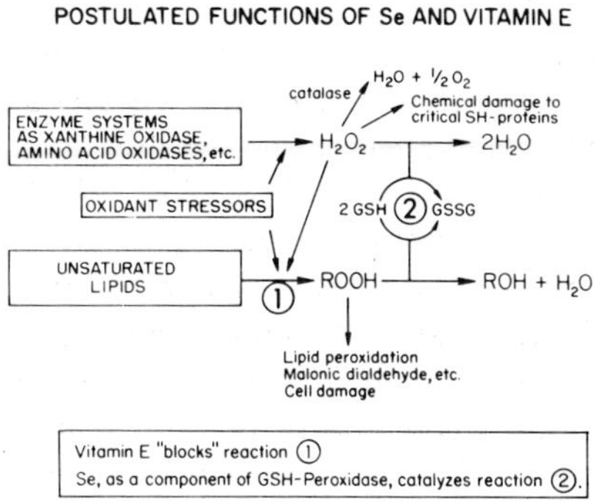

Fig. 7. Schematic representation of the postulated functions of selenium and vitamin E and mechanism of their interrelationship. See text for discussion.

GSH-Px) *catalyzes the destruction* of H_2O_2, lipid hydroperoxides (ROOH), and other hydroperoxides, while vitamin E serves to *decrease the formation* of hydroperoxides, particularly the lipid hydroperoxides. (This is not to imply, necessarily, that vitamin E is simply a lipid antioxidant; it may restrict ROOH formation, instead, by favoring alternate routes of metabolism for hydroperoxide precursors.) The breakdown of hydroperoxides may be the critically important function of GSH-Px, because an alternate system, that catalyzed by catalase, exists for decomposing H_2O_2. Numerous additional suggestions can stem from this postulated scheme. Among these are the following.

1. If the body is deficient in both Se and vitamin E, the tissues of high H_2O_2 production, particularly those also prone to substantial ROOH production because of fatty acid unsaturation, are especially susceptible to degeneration (e.g., the liver of the rat). Either nutrient will correct the problem, vitamin E by decreasing ROOH production (including that promoted by H_2O_2) and Se by increasing the breakdown of ROOH and H_2O_2.
2. If the body has plenty of Se, but inadequate vitamin E, and there is high unsaturation of cell lipids, the capacity to 'destroy' ROOH is exceeded in those organs naturally low in GSH peroxidase (e.g., chick brain, rat placenta?), but not in organs high in GSH peroxidase (e.g., rat liver). Thus, there occurs brain degeneration, resorption of fetuses, etc., which *cannot* be prevented by Se, but can be prevented by vitamin E.
3. If the body has plenty of vitamin E, but inadequate Se, production of ROOH and membrane damage is prevented, but some organs do not have sufficient capacity to destroy the H_2O_2 produced (e.g., pancreas?) because they, perhaps, are low in catalase. Thus, the H_2O_2 will damage some critical SH-proteins (nonmembrane?) and lead to cell degeneration.
4. Species with higher liver H_2O_2 production are particularly susceptible to liver degeneration from the double deficiency (e.g., the rat). For example, it is known that species differ in liver xanthine oxidase, which generates H_2O_2 (the rat is high while birds, which do not develop liver necrosis, are low in this liver enzyme).
5. If ROOH is converted to ROH, malonaldehyde is not produced as substantial lipid peroxidation does not occur. This is the situation *in vivo* in liver if sufficient Se is fed but vitamin E is lacking. Under *in vitro* conditions without glucose the tissue rather quickly loses its ability to regenerate sufficient GSH, and therefore the GSH peroxidase protection is lost and peroxidation results. If GSH is added to such *in vitro* preparations, lipid peroxidation will be prevented. If both Se and vitamin E are lacking, malonaldehyde should be formed *in vivo*, but it may rather

quickly react with $-NH_2$ groups and cross link proteins and phospholipids through Schiff-base-type linkages (age pigment).

6. The mechanism readily explains the sparing effect of vitamin E on the Se requirement. If vitamin E is in rather short supply, more ROOH is produced which requires a substantial quantity of GSH peroxidase, thus requiring more Se. Alternatively, if vitamin E is ample, there will be less ROOH and presumably less need for GSH peroxidase and therefore a lower Se requirement.
7. The sparing effect of Se on the vitamin-E requirement may be explained by the fact that in a relative absence of Se, GSH peroxidase will be lower and the ROOH not adequately handled. Under these conditions an amount of vitamin E would be required largely to stop ROOH production. Alternatively, at high Se and therefore high GSH peroxidase activity, less vitamin E would be required because small to moderate amounts of ROOH could be 'disposed of' without creating tissue damage.
8. The sparing effect of S-amino acids on vitamin E and Se requirements may be explained by the fact that cysteine is a precursor of GSH and this represents a source of GSH other than that produced from reduction of GSSG. This could be important when Se or vitamin E are borderline. If S-amino acids are in short supply, it is likely that some tissues will not contain the maximal amount of GSH + GSSG, and this may limit the GSH peroxidase step.

I must re-emphasize that the above points are hypotheses which must be put to further experimental testing before they can be accepted or rejected.

ADDITIONAL BIOCHEMICAL ROLES OF SELENIUM

It would, indeed, be surprising if GSH-Px were the only functioning form of Se. In fact, additional functions have already been demonstrated or suggested. Two microbial enzymes have recently been implicated to be selenoproteins. These are a 12,000 molecular weight component of the glycine reductase system of *Clostridium sticklandii,* as demonstrated by Turner and Stadtman (27), and almost certainly formic dehydrogenase of *Escherichia coli* (28). Additional selenoproteins in animal tissues have also become evident, such as the muscle protein studied by Whanger, Pederson, and Weswig (29), which appears to have the properties of a cytochrome, and the unidentified proteins of plasma and various tissues demonstrated by the work of Millar (30) and of Burk (31). A role for Se in the microsomal system of liver has been proposed by Diplock (32) and a role in catalyzing cytochrome C reduction by glutathione has been suggested by the work of Levander, Morris, and Higgs (33).

Certain of the selenoproteins may be transport forms or intermediates in the synthesis of GSH-Px. Further work is required to characterize these other possible functioning forms of Se and assess their importance in the Se-deficiency syndrome.

SUMMARY

The discovery of a specific function of Se as a component of glutathione peroxidase in animals has been reviewed. The isolated ovine and bovine erythrocyte enzyme contains 4 mole Se per mole protein. Dietary-Se intake profoundly affects the content of glutathione peroxidase in numerous tissues; however, some organs such as rat liver are particularly susceptible to depletion. Little is yet known of the nature of Se in the enzyme or its specific role in enzymatic activity. Decreases in glutathione peroxidase appear to explain at least some of the degenerative diseases induced by Se deficiency. Knowledge of this role of Se has allowed satisfying postulations for its metabolic and nutritional relationships with vitamin E, unsaturated lipids, and the sulfur-containing amino acids, which were presented and discussed. While this role of Se in glutathione peroxidase must be considered as the most definitive function of Se in animals to date, it is not likely to be the only functioning form of this trace element. Other recently implicated functions were briefly reviewed.

REFERENCES

1. Schwarz, K., and Foltz, C. M. 1957. Selenium as an integral part of factor 3 against dietary necrotic liver degeneration. *J. Am. Chem. Soc.* 79: 3292.
2. Patterson, E. L., Milstrey, R., and Stokstad, E. L. R. 1957. Effect of selenium in preventing exudative diathesis in chicks. *Proc. Soc. Exp. Biol. Med.* 95: 617.
3. Muth, O. H., Oldfield, J. E., Remmert, L. F., and Schubert, J. R. 1958. Effects of selenium and vitamin E on white muscle disease. *Science* 128: 1090.
4. Hartley, W. J., and Grant, A. B. 1961. A review of selenium responsive diseases of New Zealand Livestock. *Fed. Proc.* 20: 679.
5. Schwarz, K. 1965. Role of vitamin E, selenium, and related factors in experimental nutritional liver disease. *Fed. Proc.* 24: 58.
6. Schwarz, K. 1958. Dietary necrotic liver degeneration, an approach to the concept of the biochemical lesion. *In Symposium on Liver Function,* Publ. 4, American Institute of Biological Sciences, Washington, D.C., p. 509.
7. Schwarz, K. 1965. Malnutrition and the liver. *In Proceedings of International Symposium on Comparative Medicine.* Eaton Laboratories, New York, p. 60.

8. Tappel, A. L. 1965. Free radical lipid peroxidation damage and its inhibition by vitamin E and selenium. *Fed. Proc.* 24: 73.
9. Thompson, J. N., and Scott, M. L. 1969. Role of selenium in the nutrition of the chick. *J. Nutr.* 97: 335.
10. Thompson, J. N., and Scott, M. L. 1970. Impaired lipid and vitamin E absorption related to atrophy of the pancreas in selenium-deficient chicks. *J. Nutr.* 100: 797.
11. McCoy, K. E. M., and Weswig, P. H. 1969. Some selenium responses in the rat not related to vitamin E. *J. Nutr.* 98: 383.
12. Sondergaard, E. 1967. Selenium and vitamin E interrelationships. *In* O. H. Muth (ed.), *Selenium in Biomedicine,* Avi Publishing Co., Westport, Conn., p. 365.
13. Beutler, E. 1969. Drug-induced hemolytic anemia. *Pharmac. Rev.* 21: 73.
14. Cohen, G., and Hochstein, P. 1963. Glutathione peroxidase: The primary agent for the elimination of hydrogen peroxide in erythrocytes. *Biochemistry* 2: 1420.
15. O'Brien, P. J., and Little, C. 1969. Intracellular mechanisms for the decomposition of a lipid peroxide. II. Decomposition of a lipid peroxide by subcellular fractions. *Can. J. Biochem.* 47: 493.
16. Dickson, R. C., and Tappel, A. L. 1969. Effects of selenocystine and selenomethionine on activation of sulfhydryl enzymes. *Archs Biochem. Biophys.* 131: 100.
17. Rotruck, J. T., Hoekstra, W. G., and Pope, A. L. 1971. Glucose-dependent protection by dietary selenium against haemolysis of rat erythrocytes *in vitro. Nature New Biol.* 231: 223.
18. Rotruck, J. T., Pope, A. L., Ganther, H. E., and Hoekstra, W. G. 1972. Prevention of oxidative damage to rat erythrocytes by dietary selenium. *J. Nutr.* 102: 689.
19. Rotruck, J. T., Pope, A. L., Ganther, H. E., Swanson, A. B., Hafeman, D. G., and Hoekstra, W. G. 1973. Selenium: Biochemical role as a component of glutathione peroxidase. *Science* 179: 588.
20. Ganther, H. E. 1971. Reduction of the selenotrisulfide derivative of glutathione to a persulfide analog by glutathione reductase. *Biochemistry* 10: 4089.
21. Flohe, L. 1971. Die Glutathionperoxidase: Enzymologie und biologische Aspekte. *Klin. Wschr.* 49: 669. (This review contains references to the original papers of this group to 1971.)
22. Hoekstra, W. G., Hafeman, D., Oh, S. H., Sunde, R. A., and Ganther, H. E. 1973. Effect of dietary selenium on liver and erythrocyte glutathione peroxidase in the rat. *Fed. Proc.* 32: 885 (abstr.).
23. Flohe, L., Gunzler, W. A., and Schock, H. H. 1973. Glutathione peroxidase: a selenoenzyme. *FEBS Lett.* 32: 132.
24. Schwarz, K. 1961. Development and status of experimental work on factor 3-selenium. *Fed. Proc.* 20: 666.
25. Scott, M. L., and Noguchi, T. 1973. Metabolic function of selenium in prevention of exudative diathesis in chicks. *Fed. Proc.* 32: 885 (abstr.).
26. Cantor, A. H., Langevin, M. L., Noguchi, T., and Scott, M. L. 1973. Differing efficacies of selenite and selenomethionine against exuda-

tive diathesis and pancreatic fibrosis in chicks. *Fed. Proc.* 32: 885 (abstr.).
27. Turner, D. C., and Stadtman, T. C. 1973. Purification of protein components of the Clostridial glycine reductase system and characterization of protein A as a selenoprotein. *Archs Biochem. Biophys.* 154: 366.
28. Shum, A. C., and Murphy, J. C. 1972. Effects of selenium compounds on formate metabolism and coincidence of selenium-75 incorporation and formic dehydrogenase activity in cell-free preparations of *Escherichia coli*. *J. Bact.* 110: 447.
29. Whanger, P. D., Pederson, N. D., and Weswig, P. H. 1974. Characteristics of selenium-binding proteins from lamb muscle. This symposium.
30. Millar, K. R. 1972. Distribution of Se^{75} in liver, kidney and blood proteins of rats after intravenous injection of sodium selenite. *N.Z. J. Agric. Res.* 15: 547.
31. Burk, R. F. 1973. Effect of dietary selenium level on ^{75}Se binding to rat plasma proteins. *Proc. Soc. Exp. Biol. Med.* 143: 719.
32. Diplock, A. T. 1974. A possible role of selenium, and vitamin E, in the electron transfer system of rat liver microsomes. This symposium.
33. Levander, O. A., Morris, V. C., and Higgs, D. J. 1974. Relationship between the selenium-catalyzed swelling of rat liver mitochondria induced by glutathione (GSH) and the selenium-catalyzed reduction of cytochrome c by GSH. This symposium.

DISCUSSION

Schwarz (Long Beach). Based on our experimentation from the very beginning, I have been opposed to considering vitamin E or selenium simply as antioxidants. It appears possible that there is a normal peroxide pathway in metabolism and I have suggested before that lipid peroxides may be normal intermediates. What do you consider to be the normal peroxide substrate for glutathione peroxidase?

Hoekstra. It remains controversial as to whether catalase or glutathione peroxidase is more important in destroying hydrogen peroxide. I would agree with Flohe's concept that they're both important and represent a dual system of protection. In terms of substrate specificity, glutathione peroxidase is very nonspecific for the peroxide or hydroperoxide, but highly specific for the hydrogen donor, glutathione. Fatty acid hydroperoxides do not serve as substrates for catalase but should be excellent substrates for glutathione peroxidase if they form *in vivo*. Steroid hydroperoxides, thymine hydroperoxide, and other hydroperoxides may also serve as normal substrates for glutathione peroxidase, which could be important in hydroxylation reactions.

Deutsch (Madison). Patients with acatalasia suffer no overt disease and this is interpreted as meaning that catalase is not very important in hydrogen peroxide metabolism. The K_m for glutathione peroxidase is much more appropriate for catabolizing hydrogen peroxide, while catalase is never saturated and plays a minor role. It's important to look at the K_m values of these systems in assessing their role in hydrogen peroxide decomposition.

Hoekstra. The literature is confusing in terms of K_m values, but Flohe, after careful consideration of K_m values, etc., has calculated that the two enzymes are about equally effective in H_2O_2 breakdown. If catalase is missing, you would not necessarily see overt damage because the other system can handle the H_2O_2, but the hydroperoxides are not acted upon by catalase, so glutathione peroxidase is apparently the more critical of the two.

Pařízek (Prague). Have you any data on the selenium content of glutathione peroxidase in lens. It has been shown by Dr. Pirie and others that this pathway might be of particular interest in relation to the lens.

Hoekstra. We're very interested in that now. We have not isolated the enzyme from lens to assess its selenium content, but we have preliminary data on the glutathione peroxidase activity of the lens of the eye in Se deficiency. Early in the deficiency, up to about eight weeks of deficiency in the lamb, for example, it does not decrease. However, we find very much lower levels of this enzyme in the eye lens of the rat after very long-term Se deficiency such as in the second generation. Oregon State investigators have reported a cataract-like syndrome in second-generation selenium-deficient rats. Therefore, we are very interested in whether there is any significance of selenium deficiency to the cataract problem.

Whanger (Corvallis). Have you done any spectral studies on glutathione peroxidase?

Hoekstra. Other than the simple UV spectrum, we have in cooperation with Dr. Orme-Johnson been doing some preliminary EPR spectral work. When we add glutathione and then H_2O_2 to the purified enzyme and allow it to stand about 2 min before freezing prior to determining the EPR spectrum, we find a signal that could represent a Se free radical. It is also possible, however, that it could represent an oxygen or sulfur free-radical signal, so we must still prove that it is a Se signal and is related to catalysis. This preliminary evidence indicates that Se may be involved in a single-electron transferring process. Apart from that, we haven't done any other spectral work.

Binnerts (Wageningen). We found that during the development of copper deficiency in sheep, the glutathione peroxidase level in the erythrocyte increased very rapidly. This was correlated with the increase of serum aspartate transferase, which is an index of liver damage. Perhaps, during liver damage, the glutathione peroxidase from the liver enters the blood and is attached to the membrane of the erythrocyte, and in this way protects it from further copper deficiency. Can you comment on this hypothesis?

Hoekstra. It seems possible, but I have no idea whether the enzyme binds to the surface of the red cell. Possibly this is an example of an adaptive change in this enzyme, but it would not be expected to occur in mature erythrocytes.

TRACE-ELEMENT INTERACTIONS: EFFECTS OF DIETARY COMPOSITION ON THE DEVELOPMENT OF IMBALANCE AND TOXICITY

C. F. MILLS

Rowett Research Institute, Bucksburn, Aberdeen, U.K.

In a recent discussion of the problems encountered in studies to determine the trace-element requirements of animals (1) it was illustrated that, under practical circumstances, the availability of the 10 essential elements then known could be modified by the concentrations of at least 15 other dietary components. A further 21 interactions influencing trace-element availability have been demonstrated under experimental conditions.

Complex though this situation may appear, it is probable that existing knowledge of the nature of dietary components which influence availability and thus trace-element status is far from complete. Evidence suggesting this is forthcoming from our present inability to predict the development of some zinc-responsive disorders by dietary analysis, by the obscure etiology of endemic goiter in some populations, and by our inadequate understanding of the many situations under which copper deficiency in ruminant animals can arise in situations in which the dietary intake of copper is not abnormally low and the molybdenum intake not excessive.

The objectives of this paper are, firstly, to illustrate how existing knowledge of antagonistic interactions between individual trace elements can contribute to the control or understanding of situations in which risks to health exist from the accumulation of potentially toxic elements, and, secondly, to examine metabolic relationships between zinc, copper, and cadmium in greater detail and to assess the present practicability of developing a working hypothesis to explain some aspects of their mutually antagonistic effects.

CONTROL OF LIVER COPPER ACCUMULATION IN RUMINANTS

The rapid accumulation of copper by the liver of the sheep and young calf and its subsequent release to provoke a hemolytic crisis in death is a frequent cause of economic loss under intensive systems of animal management (2). The causes are several and include failure to monitor the copper content of protein supplements and the contamination of mineral supplements during mixing after the previous processing of high copper rations intended for more tolerant species. The addition of molybdenum to the rations of animals at risk has been suggested as a possible prophylactic measure, but, until there is a more comprehensive definition of quantitative aspects of the molybdenum/copper antagonism, there is understandable reluctance to exploit this possibility.

Basing the approach upon earlier observations that high dietary concentrations of zinc protected the pig against copper intoxication (3), a recent study at the Rowett Research Institute has examined the effects of dietary zinc on copper retention in the livers of weanling sheep offered a semisynthetic diet (4) supplemented with high concentrations of copper. Dorset cross lambs, 5 weeks of age, were used; the experiment was continued for 145 days and the experimental treatments were as given in Table 1. A diet of 100 μg Zn/g, i.e., one-fifth of the tolerable dietary concentration found in work with sheep and calves (5), had a highly significant effect in reducing liver copper storage and markedly inhibited the rise in serum aspartate aminotransferase activity in animals receiving 30 or 60 μg Cu/g diet and starved for 24 hr (Table 2), suggesting that zinc exerted a protective effect against tissue damage in this situation. The last finding is of particular interest in relation to suggestions that copper increases the fragility of lysosomal membranes, an

Table 1. The concentration of Cu[1] in livers of lambs offered Cu- and Zn-supplemented semisynthetic diets for 145 days

		Dietary Cu concentration (μg/g DM)		
		7.5	15	30 (60)[2]
Dietary Zn concentration (μg/g DM)	18	285 ± 131	402 ± 48	888 ± 239
	100	162 ± 57	342 ± 139	690 ± 132

[1] μg/g dry tissue.
[2] Dietary Cu increased to 60 μg/g for last 20 days.
Overall treatment effects due to Zn significant $P < 0.001$.

Table 2. Influence of dietary Zn concentration on increase in serum aspartate aminotransferase activity in sheep subjected to 48-hr starvation after receiving high Cu diets

		mU Activity/100 ml serum	
		Initial	Final
Dietary Cu 30 μg/g			
Dietary Zn (μg/g)	18	77 ± 21	125 ± 11
	100	68 ± 31	73 ± 34
Dietary Cu 60 μg/g			
Dietary Zn (μg/g)	18	66 ± 20	196 ±20
	100	57 ± 18	119 ± 40

effect which is reversible by zinc (6, 7). With this suggestion that zinc may directly reverse a metabolic lesion induced by copper and the more practical point that ruminants exhibit a relatively high tolerance to zinc, it is probable that control of the effects of copper accumulation by the use of zinc could provide a more acceptable alternative to the use of molybdenum in such situations.

EFFECTS OF CADMIUM ON COPPER AND ZINC METABOLISM IN RUMINANTS AND RATS

The work of Pařízek (8) first indicated the existence of antagonistic effects of cadmium upon zinc metabolism. Later work, and particularly that of Hill and Matrone, supported this finding and further illustrated the adverse effects of cadmium upon copper metabolism (9, 10). Among the more significant aspects arising from such studies are implications that tolerance to cadmium may be related to the copper status of the animal. Most studies on the Cd/Cu interaction have been carried out using dietary cadmium concentrations within the range of 20–600 μg Cd/g diet. In order to assess the possible significance of this interaction in situations where animals or man are subject to chronic low-level exposure to cadmium, it is important to determine whether effects can be demonstrated at substantially lower cadmium concentrations which more realistically reflect those encountered as a consequence of environmental contamination. Accordingly, recent work has concentrated on the effects of a 1–12 μg Cd/g diet.

The results of one such study (11) on ewes during pregnancy and on their lambs up to 2 months of age are summarized in Table 3. Although substantially lower dietary concentrations were used in this work the results

Table 3. Effects of elevated dietary contents upon liver Cd, Cu, and Zn in lambs[1]

Liver (μg/g DM)		Cd content of diet (μg/g DM) 0.7	3.5	7.7	12.3	P
Cadmium	(1)	<0.1	<0.1	<0.1	<0.1	
	(2)	<0.1	1.6	3.5	11.0	***
Copper	(1)	183	213	115	73	
	(2)	101	22	62	13	**
Zinc	(1)	178	266	137	148	
	(2)	135	96	103	94	**

[1](1) Lambs at birth where ewes received treatments during last 7 weeks of pregnancy; (2) lambs 2 months after birth

support previous conclusions (12, 13) that while cadmium does not traverse the placenta, it has a small effect in suppressing copper and zinc accumulation by the fetal liver. Of much greater significance were the effects of low-level exposure of the new-born animal to cadmium, where highly significant declines in the liver, copper, and zinc content occurred. Consideration of all the data from this experiment suggests that a dietary content of 3.5 μg Cd/g and a Cd:Cu ratio of 0.7 or above could provoke a marked depletion of tissue copper reserves, but that the adverse effects on zinc metabolism and tissue zinc were of doubtful biological significance.

Recent studies with rats[1] have examined the effects of diets providing as little as 1 μg Cd/g when the copper intake is 'marginal' (2.8 μg Cu/g) and when zinc intake meets requirement or is high. The basal diet (14) provided 0.16 μg Cd/g. Exposure to 1 μg Cd and 30 μg Zn/g diet for 10 weeks promoted a 58% decline in ceruloplasmin activity, a reduction of cortical bone thickness in the femur, and a decline in liver and blood copper content. These effects were exacerbated when the diet contained 5 or 15 μg Cd/g and, particularly so, when the zinc intake was also high (300 or 1000 μg/g).

While such studies do not take into account the possible adverse effects from lifetime exposure to low cadmium concentrations, they clearly suggest that even relatively short-term exposure can have significant metabolic consequences when the copper status of subjects is low. Considerable alarm has been generated as to the biological consequences of the existence of cadmium

1. J. Campbell and C. F. Mills, to be published.

as an environmental pollutant. Much of this attitude arises from the imprudent extrapolation of experimental situations using high exposure levels. Recent progress in the study of the Cd/Cu relationship does, however, suggest that in specific situations a high risk may arise, as, for example, in ruminant animals already exposed to high intakes of molybdenum or in human infants maintained for long periods on diets based predominantly upon dairy and cereal products low in copper.

SPECULATION ON THE SITES OF INTERACTION OF COPPER, ZINC, AND CADMIUM

In a series of studies of the structural requirements for the activity of the carboxypeptidases, Vallee (15) illustrated that the zinc of this enzyme could readily be replaced by a wide range of other metals. Substitution either modified or abolished the activity of the enzyme as a peptidase or esterase. Metal substitution achieved by chemical manipulation *in vitro* has since been achieved with many other metalloenzymes, often with equally striking effects upon activity. To the writer's knowledge there are, regrettably, no instances where the substitution of a metal antagonist at the active center of an enzyme and a subsequent inhibition or modification of its role has been unequivocally demonstrated *in vivo,* thus providing a simple explanation of the metabolic consequences of a competitive antagonism between metals.

Changes in tissue element content or patterns of subcellular metal distribution usually provide the first evidence that such antagonistic interactions are operating. Where there are indications that an antagonist has depleted the tissue content of the target element, it must be presumed that the mode of action of the antagonist is through a restriction of absorption, an inhibition of storage within the tissue matrix, or, alternatively, an enhancement of excretion. That this is an insufficiently detailed concept to account for the mechanism of action of several practically important situations of trace-element antagonism is clearly shown by those circumstances in which the antagonist enhances tissue deposition of the target element, even though there are clearly clinical or biochemical implications that the incorporation of the target element into its functional sites is impaired. Thus, many studies of the effects of molybdenum ingestion by the rat have illustrated the development of clinical syndromes at least partially responsive to copper, even though the storage of copper in the liver is enhanced (16, 17). Similarly the development of clinical signs of zinc deficiency in copper intoxication of the pig is accompanied by an increase in liver zinc content.

Most studies of the Cu/Zn/Cd relationship have strongly emphasized effects of these interactions upon liver composition and have largely neglected effects upon other tissues. In consequence it is often difficult to assess whether the effects of an antagonistic interaction are merely upon the distribution of elements between tissues or whether the primary effects are upon absorption or excretion.

Although the extensive study of compositional changes in the single organ such as the liver has its limitations, it can, however, provide some insight as to the magnitude of the metabolic changes induced by trace-element imbalance. Figure 1 has been prepared from 68 sets of observations by different workers on the effects of dietary imbalance upon liver composition. Each point is defined by the fraction (target-element content in livers of group receiving antagonist over target-element content in appropriate control group); thus numbers less than 1 indicate that the antagonist has reduced the concentration of the target element in the liver and vice versa. As the only

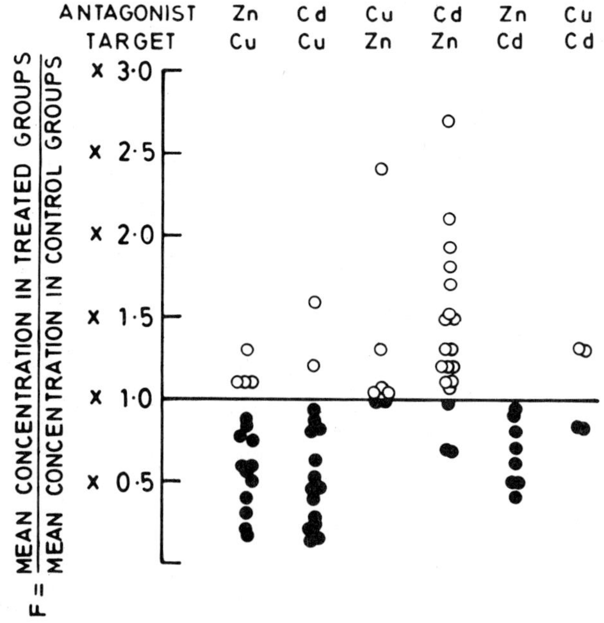

Fig. 1. The influence of dietary imbalances of Zn, Cu, or Cd upon the concentration of these elements in liver tissue. Each column of points indicates the effect of the antagonist included in the diet upon the concentration of the target element. Individual points represent the fraction

$$\frac{\text{mean concentration in livers of groups receiving antagonist}}{\text{mean concentration in livers of appropriate control group}},$$

The plot was constructed from data from the following sources (3, 5, 11, 24, 25, 26, 27, 28, 29, 30, 31, 32).

species difference emerging from this survey was the possibility that cadmium may decrease the liver zinc concentration of the ruminant in contrast to its effect on other species, the data for all species are plotted together and for simplicity no distinction is drawn between different concentrations of antagonistic metal. Despite these limitations the following conclusions can be drawn.

1. Elevated concentrations of cadmium increase the concentration of zinc in the liver.
2. Cadmium strongly depresses the concentration of copper.
3. Zinc depresses the retention of cadmium and also that of copper.
4. No conclusions can be drawn upon the effects of high dietary concentrations of copper upon cadmium.

The key to the understanding of this complex pattern of changes, the greater number of which have been shown to be associated with the development of clinical or metabolic lesions, must lie in the study of those processes which respond to imbalance by the synthesis of components which have a sufficiently high affinity for target elements to influence materially their availability to essential sites on enzymes, or else substantially modify their rate of removal from an organ or from the whole animal. This postulate is similar to that invoked to provide a possible function for the low molecular weight protein metallothionein formed in response to a challenge by cadmium or mercury with the possible object of sequestering these toxic elements from sensitive sites. However, recent evidence suggests that such proteins may have a wider role and are not merely formed in response to mercury or cadmium intoxication. Such evidence is as follows.

1. Proteins of the metallothionein type are present in the livers of ruminant species and the female rat even in the absence of the stimulus from a toxic metal (18, 19, 20).
2. An important determinant of their concentration in the liver appears to be the zinc concentration of that organ, and stimuli which promote zinc redistribution (e.g., restriction of food intake) are also reflected by changes in 'metallothionein' concentration (20).
3. Substantial increases in metallothionein concentration arise following the administration of zinc or copper as well as of cadmium and mercury, and a variable proportion of the total metal-binding groups of this protein bear copper or zinc.

For metalloproteins of this type to be involved in the interactions illustrated in Fig. 1 evidence is demanded that the metal component influences biological half-life and, to account for those situations in which tissue depletion of a target element occurs, that the metalloprotein or its degradation products may be excreted while still retaining a strong affinity for the metal component. There is clear evidence that the cadmium derivative of metallothionein has a long biological half-life (19). Preliminary work at the Rowett Research Institute suggests that the zinc protein may turn over more rapidly (21), while the copper protein is difficult to handle because of instability during preparative procedures. The apparently rapid degradation of mercury metallothionein in the kidney has been commented upon elsewhere (22).

The multiplicity of effects of antagonistic interactions seen in Fig. 1 may be largely explicable if (1) the 'metallothionein-type' proteins can simultaneously bind cadmium, zinc, and copper in varying proportions, (2) the dominant metal influences stability to proteolytic attack, biological half-life, and thus overall distribution of the above metals between cellular components, and (3) the degradation products retain a strong affinity for some or all of the original metal components and retain these metals during transport or excretion. There is already plausible evidence for the existence of 'mixed-metal' metallothioneins (18, 23). The possible influence of the nature of the metal components upon metabolic stability undoubtedly requires investigation, but the concept that stability may differ as a consequence of modifications to secondary structure introduced by differing metal components appears reasonable, particularly in view of the marked differences in tissue retention observed among the metallothioneins ostensibly carrying a single metallic element. The nature of the biological degradation products of such metalloproteins is largely unknown, except for indications that the catabolism of Hg metallothionein in kidney homogenates is accompanied by an increase in lower molecular weight species that still retain a strong affinity for this metal (22). With cysteinyl residues accounting for between 25% and 30% of the component amino acids, such properties can probably be expected from the diffusible oligo-peptides arising by degradation of the parent metalloprotein.

A postulate that 'metallothionein-type' proteins may provide one locus for the interplay of several mutually antagonistic transition elements, and that subsequent effects on tissue retention or loss of these elements may be explicable in terms of differences in turnover rate of the resulting product and the affinity of possible degradation products for these metals, does not presuppose that such interactions occur only in the liver. Proteins of this nature, characterized to varying degrees, have now been detected in liver, kidney, intestinal mucosa, pancreas, spleen, and possibly the testis, so that the possible sites at which transition-element metabolism could be so modified are many. Of particular interest are the possibilities that such interactions could arise in the kidney and in the duodenal mucosa, and so directly influence absorptive processes and endogenous losses. Of equal interest are the possibilities that the entry of cadmium into a predominantly zinc-containing protein confers stability and increases the apparent retention of zinc in that tissue (and vice versa), and that, following induction by zinc or cadmium, the incorporation of mercury or possibly copper may promote instability and the ultimate appearance of readily diffusible, metal-bearing oligo-peptides which could again have a marked influence on tissue distribution and excretory losses of these elements.

From the foregoing arguments it is apparent that much further study is required before a unifying hypothesis can be developed to account for the effects of imbalances among the elements copper, zinc, and cadmium, but it is probable that more detailed investigation of the synthesis and metabolic fate of the 'mixed-metal' metallothioneins may provide a useful model. In more practical spheres a more detailed knowledge of how such interactions influence availability and toxicity of the transition elements will be of increasing value, both in the development of new prophylactic measures and in the realistic appraisal of situations in which a substantial risk to human and animal health may arise.

REFERENCES

1. Mills, C. F., and Williams, R. B. 1971. Problems in the determination of the trace element requirements of animals. *Proc. Nutr. Soc.* 30: 83.
2. Todd, J. R. 1969. Chronic copper toxicity of ruminants. *Proc. Nutr. Soc.* 28: 189.
3. Suttle, N. F., and Mills, C. F. 1966. Studies of the toxicity of copper to pigs. *Br. J. Nutr.* 20: 135.
4. Mills, C. F., Dalgarno, A. C., Williams, R. B., and Quarterman, J. 1967. Zinc deficiency and the zinc requirements of calves and lambs. *Br. J. Nutr.* 21: 751.
5. Ott, E. A., Smith, W. H., Harrington, R. B., and Beeson, W. M. 1966. Zinc toxicity in ruminants I and II. *J. Anim. Sci.* 25: 414, 419.
6. Chvapil, M., Ryan, J. N., and Zukoski, C. F. 1972. The effect of zinc and other metals on the stability of lysosomes. *Proc. Soc. Exp. Biol. Med.* 140: 642.
7. Zukoski, C. F., Ryan, J. N., and Elias, S. L. 1972. *In vivo* and *in vitro* stabilization of lysosomes by zinc. *Fed. Proc.* 3(2): 4050.
8. Pařízek, J. 1957. The destructive effect of cadmium ion on testicular tissue and its prevention by zinc. *J. Endocr.* 15: 56.
9. Hill, C. H., Matrone, G., Payne, W. L., and Barber, C. W. 1963. *In vivo* interactions of cadmium with copper, zinc and iron. *J. Nutr.* 80: 227.
10. Hill, C. H., and Matrone, G. 1970. Chemical parameters in the study of *in vivo* and *in vitro* interactions of transition elements. *Fed. Proc.* 29: 1474.
11. Mills, C. F., and Dalgarno, A. C. 1972. Cu and Zn status of ewes and

lambs receiving increased dietary concentrations of Cd. *Nature, Lond.* 239: 171.

12. Lucis, O. J., Lucis, R. and Aterman, K. 1971. Transfer of ^{109}Cd and ^{65}Zn from the mother to the newborn rat. *Fed. Proc.* 30: 238.
13. Anke, M., Hennig, A., Schneider, H. J., Ludke, H., von Gegern, W., and Schlegel, H. 1970. Interrelationships between Cd, Zn, Cu and Fe in the metabolism of hens, ruminants and man. *In* C. F. Mills (ed.), *Trace Element Metabolism in Animals,* Livingstone, Edinburgh, p. 317.
14. Williams, R. B., and Mills, C. F. 1970. The experimental production of zinc deficiency in the rat. *Br. J. Nutr.* 24: 989.
15. Vallee, B. L., and Wacker, W. E. C. 1970. Metalloproteins. *In* H. Neurath (ed.) *The proteins: Composition, Structure and Function,* Vol. 5, Academic Press, New York.
16. Miller, R. F., Price, N. O., and Engel, R. W. 1956. Added dietary inorganic sulphate and its effect upon rats fed molybdenum. *J. Nutr.* 60: 539.
17. Mills, C. F., and Mitchell, R. L. 1971. Copper and molybdenum in subcellular fractions of rat liver. *Br. J. Nutr.* 26: 117.
18. Bremner, I. 1972. Cu and Zn proteins in ruminant liver. This symposium.
19. Webb, M. 1972. Binding of Cd ions by rat liver and kidney. *Biochem. Pharma.* 21: 2751.
20. Bremner, I., Davies, N. T., and Mills, C. F. 1973. Effect of zinc deficiency and food restriction on hepatic Zn proteins in the rat. *Trans. Biochem. Soc.*, in press.
21. Davies, N. T., Bremner, I., and Mills, C. F. 1973. Studies on the induction of a low-molecular-weight zinc binding protein in rat liver. *Trans. Biochem. Soc.*, in press.
22. Jakubowski, M., Piotrowski, J., and Trojanowska, B. 1970. Binding of mercury in the rat: Studies using $^{203}HgCl_2$ and gel filtration. *Toxic. Appl. Pharmac.* 16: 743.
23. Nordberg, G. F., Nordberg, M., Piscator, M., and Vesterberg, O. 1972. Separation of two forms of rabbit metallothionein by isoelectric focusing. *Biochem. J.* 126: 491.
24. Lee, D., and Matrone, G. 1969. Fe and Cu effects on serum ceruloplasmin activity of rats with Zn-induced Cu deficiency. *Proc. Soc. Exp. Biol. Med.* 130: 1190.
25. Cox, D. H., and Harris, D. L. 1960. Effect of excess dietary Zn on Fe and Cu in the rat. *J. Nutr.* 70: 514.
26. Bunn, C. R., and Matrone, G. 1966. *In vivo* interactions of Cd, Cu, Zn, and Fe in the mouse and the rat. *J. Nutr.* 90: 395.
27. Grant-Frost, D. B., and Underwood, E. J. 1958. Zn toxicity in the rat and its interrelation with Cu. *Aust. J. Exp. Biol. Med. Sci.* 36: 339.
28. Campbell, J., and Mills, C. F. 1973. To be published.
29. Anke, M., Hennig, A., Groppel, B., and Ludke, H. 1971. Der einfluss des Kadmiums auf das Wachstum die Fortpflanzungsleistung und den Eisen, Zink und Kupferstoffwechsel. *Arch. Exp. VetMed.* 25: 799.
30. Banis, R. J., Pond, W. G., Walker, E. F., and O'Connor, J. R. 1969. Dietary Cd, Fe and Zn interactions in the growing rat. *Proc. Soc. Exp. Biol. Med.* 130: 802.

31. Powell, G. W., Miller, W. J., Morton, J. D., and Clifton, C. M. 1964. Influence of dietary Cd level and supplementary Zn on Cd toxicity in the bovine. *J. Nutr.* 84: 205.
32. Vohra, P., and Heil, J. R. 1969. Dietary interactions between Zn, Mn and Cu for turkey poults. *Poult. Sci.* 48: 1686.
33. Webb, M. 1972. Protection by Zn against Cd toxicity. *Biochem. Pharmac.* 21: 2767.

DISCUSSION

Piscator (Stockholm). You said that you think that when metallothionein contains cadmium, the thionein has a very long half-life, but it has a very short one when it's bound to zinc. Aren't you really expressing the half-life of the metals rather than the protein? It would be extremely difficult to imagine a protein in a human being having a half-life, for example, of 15 years according to what you said.

Mills. Yes, I think this is true, but I propose that the half-life of the metal and the half-life of the metalloprotein are interrelated; one can affect the other. The point on which we require experimental observations is whether the metallothioneins can exist in a multi-element form as mixed derivatives, because it's upon this that the whole concept depends. If we envisage separate cadmium, zinc and copper metallothioneins only, we cannot explain the upward shift in zinc half-life when we give cadmium. Your work on rabbit metallothioneins with isoelectric focusing has suggested two different species of metallothionein, one with cadmium and one with cadmium and zinc, but we must establish whether there is a continuous profile of possible compositions which materially influences the fate of the metal.

Schwarz (Long Beach). Even though kidney cadmium would appear to represent a very stable type of situation, we must realize that all these things are in a state of equilibrium. There is a continuous chelate exchange even of cadmium in kidney. One must look at the question of half-life of cadmium and of metallothionein in kidney as a dynamic situation, not as an expression of concentration over a lengthy period of time.

Mills. I feel that we may be neglecting the point that when a protein as metal acceptor is synthesized, secondary conformational changes take place which, until that protein becomes further modified, can remove that metal from the classical equilibrium exchange. Also, such effects of metals as well as anions on protein structure can modify these equilibria. I am simply making a plea that we don't neglect this extremely important aspect of these metal interactions, namely those modifying the secondary structure of a protein and thereby its resistance or susceptibility to metabolic attack, which could have very big effects on both protein and metal turnover.

Gedalia (Jerusalem). How do you explain the very low placental transfer of cadmium? We had a similar problem with fluoride and found that there was a combination of calcium with fluoride such that they were precipitating in the placenta.

Mills. I have no explanation, but it's a very effective block.

Shah (Ottawa). You mentioned that cadmium caused a reduction in the cortical thickness of bone. How is this effect brought about?

Mills. Two possibilities come to mind. One is that in high cadmium exposure there is renal damage and consequently a failure in the conversion of vitamin D to its active metabolites. The second possibility is that there is an antagonism to copper and thus a failure of collagen maturation in bone or defective matrix formation. The second is more tempting to me but we must retain an open mind.

Poole (Dublin). In Ireland we find a low copper status in young beef cattle associated with a moderate excess of molybdenum in the pasture. On some farms clinical symptoms are evident which are rectified by giving parenteral copper and growth rate is significantly increased. On other farms the animals show the same low copper status but appear normal and show no response to parenteral copper. Could we be dealing with a low-grade cadmium toxicity which would precipitate the symptoms of copper deficiency? In the area there have been recent discoveries of ore deposits which include cadmium, and it may be possible that sufficient cadmium is present at the soil surface to cause this disorder.

Mills. This is certainly a possibility. Until one has monitored cadmium contents of tissues, soil, and diet, one cannot know. The difficulty of assessing the clinical status from tissue composition may relate to changes in which metabolic availability is influenced but not necessarily by depressing tissue concentration. Cadmium, zinc, and copper may interact in this way. This is one of the aspects that's making field work so difficult.

CHEMICAL PARAMETERS IN TRACE-ELEMENT ANTAGONISMS[1,2]

GENNARD MATRONE

Department of Biochemistry, North Carolina State University, Raleigh, North Carolina

It is generally recognized that the biologically active forms of trace elements are probably metal chelates. If this premise is true, it follows that many of the biological antagonisms between and among trace elements might well be the result of similarity in chemical properties. Over 15 years ago a search was begun to determine, if possible, the minimum number of chemical properties necessary to predict which trace elements would interact biologically. It was postulated that, in warm-blooded animals, two simplifying assumptions could be made which automatically might eliminate some of the chemical properties from consideration in the living organism; i.e., (a) the temperature for the metal-ligand reaction would be relatively constant, and (b) the pH would also be relatively constant.

The coordination chemistry approach immediately brought d orbitals into consideration since the bulk of the essential cationic trace elements are found in the transition-element groups of the periodic table. Finally, further considerations made it apparent that ligand properties and specifications were also important. In the development of concepts presented herein, the author has drawn freely from the valency bond, molecular orbital, and ligand field theories to explain the properties of coordination compounds. The choice of

1. Supported in part by research grants from the Herman Frasch Foundation and National Institutes of Health (No. AM 13055-04). Contribution from the Department of Biochemistry, School of Agriculture and Life Sciences, and School of Physical and Mathematical Sciences, North Carolina State University, Raleigh, North Carolina. Paper No. 4093 of the Journal Series of the North Carolina State University Agricultural Experiment Station, Raleigh, North Carolina.

2. A part of this material was presented at the Proceedings of the Eighth International Congress on Nutrition held in Prague, Czechoslovakia, and published in Excerpta Medica International Congress Series No. 213, p. 171.

which theory to use in describing a particular property was based on the simplicity of its visualization rather than on its chemical and mathematical rigor. The thought behind this approach was to develop as simple a hypothesis as possible initially, and, if the approximation appeared viable, refinements could be added later.

At this point a simple definition of the type of trace-element interactions under consideration would seem to be in order. Interactions between metal ions in the living organism can be classified into at least two categories: (a) direct interaction, i.e., substitutions of a 'native' metal ion in a biologically active coordination compound by another metal ion; (b) indirect interaction, i.e., interaction of one metal ion with either a precursor in the chelate-formation pathway of another metal ion or interaction of one metal ion with a precursor in one or more of the steps involved in metabolic processes of another such as absorption, transport, or physiological function. Although the biological consequences of either type of interaction could be very similar, the mechanisms of antagonism would be different. Our efforts in this paper will be to attempt to concentrate on the direct type of interaction since *a priori* it seems to be the one with the fewest complications.

PROPERTIES OF TRANSITION ELEMENTS

The majority of the essential trace elements fall among those classified as transition elements in the periodic table. As shown in Table 1, there are three sets of transition elements. Elements which have been classified as essential for animals include eight of the ten in set 1, V, Cr, Mn, Fe, Co, Ni, Cu, and Zn, and one in set 2, Mo. Transition elements within these three sets possess several unique properties including paramagnetism, variable valency, and the ability to form colored ions. Furthermore, as one proceeds from left to right

Table 1. Transition elements

Set 1	Set 2	Set 3
Sc	Y	La
Ti	Zr	Hf
V	Nb	Ta
Cr	Mo	W
Mn	Tc	Re
Fe	Rh	Os
Co	Rn	Ir
Ni	Pd	Pt
Cu	Ag	Au
Zn	Cd	Hg

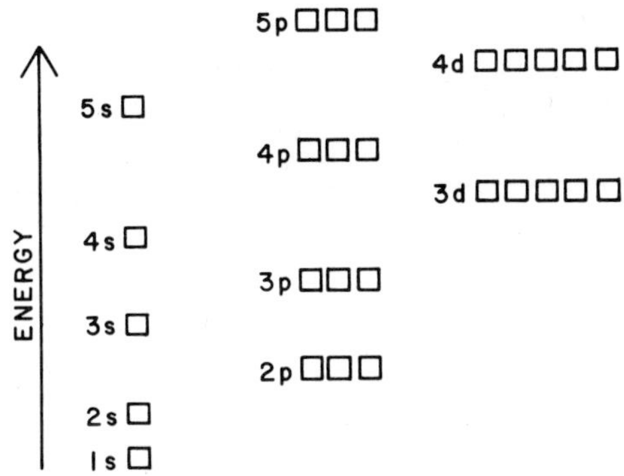

Fig. 1. The approximate relative energies of atomic orbitals.

from set 1 to set 3 in Table 1, each row or triad of elements has many similar properties, e.g., a similar number of oxidation states. Leach and Norris (1) utilized the similarity of chemical properties of the molybdenum triad when they produced a tungsten-induced molybdenum deficiency in chicks. The special properties of the transition elements have been attributed to the approximate relative energy levels of the atomic orbitals (2), as shown in Fig. 1. If one starts with the 1s orbital, tracing the elements from the beginning of the periodic table, the addition of successive electrons into the atomic orbitals is carried out in a regular manner in the order 1s, 2s, 2p, 3s, and 3p. However, at this point, as one can observe in a table of the electronic configuration of the elements in a handbook of chemistry, the 4s orbital is filled before the 3d orbitals. This is due to the fact that the 3d orbitals lie at a slightly higher energy level than the 4s orbital. This situation is repeated three times in the periodic table at the 3d–4s, 5s–4d, and 6s–5d levels, giving rise to the three transition sets shown in Table 1. The five d orbitals are degenerate, i.e., occupy the same energy level. An element such as iron with an atomic number of 26 would have all orbitals through the 3p filled with 18 electrons, leaving eight electrons to be accounted for; two would go into the 4s orbital for reasons stated above, leaving six electrons to be distributed among five d orbitals. How would the six electrons be arranged in the five d orbitals? According to Hund's rule (3), the most stable distribution of electrons in the ground state would be one per orbital with their electron spins all in the same direction. Since there are six electrons, two would have to pair up in one of the five d orbitals, and four d orbitals would have one electron per orbital all spinning in the same direction. Atoms or ions with one or more unpaired electrons exhibit paramagnetism. As will be shown later, the coordination chemistry of the transition elements directly involves the interaction of electrons in the five d orbitals with the ligand field.

COORDINATION NUMBERS AND CONFIGURATION

In general terms the metal ion in a chelate is the acceptor of electrons, and the ligand is the donor of electrons. The number of metal-ligand bonds in a metal chelate or complex is the coordination number (CN). The specification for a chelate is that a ligand must have more than one bond to the metal ion. A ligand which is bonded twice to a metal ion is termed a bidentate; one with three bonds to the metal ion is termed a tridentate, and one with four, a quadridentate, e.g., heme.

The CN under consideration is the favored coordination number, as it is recognized that a metal ion can display more than one CN. A simple approach to the prediction of the probable CN of a metal ion is to use the effective atomic number (EAN) system (4). For example, in this system a metal ion in the first transition group attains an electronic configuration (when complexed or chelated) similar to the next higher inert gas, krypton. The CN is then determined by the number of unfilled orbitals of the metal ion. This is illustrated in Fig. 2. The 36 electrons of krypton completely fill all the orbitals (2 electrons per orbital with opposite spins) from the 1s through the three 4p orbitals. The Zn^{2+} ion with 28 electrons to be distributed has four empty orbitals short of krypton. The predicted coordination number for Zn^{2+} is 4, utilizing one s orbital and three p orbitals. The sp^3 configuration has a tetrahedral shape (Fig. 3). Cuprous copper, which is isoelectronic with Zn^{2+}, also has a CN of 4 and an sp^3 configuration, i.e., tetrahedral shape. Similarity in these chemical properties might be predicted to lead to biological antagonisms between Zn^{2+} and Cu^{+}. The 27 electrons of Cu^{2+} are distributed as shown in Fig. 2 providing one d, one s, and two p orbitals for bonding. Again, the CN is 4, but the dsp^2 configuration has a square planar configuration (Fig. 3). As shown in Fig. 2, the distribution of the 24 electrons of divalent

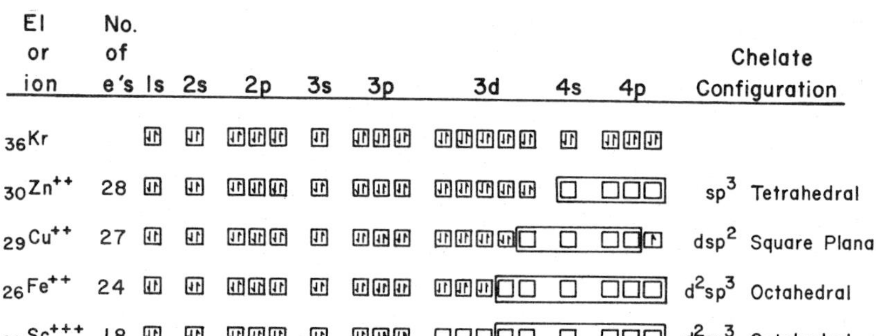

Fig. 2. Common configurations of metal ions and their chelates.

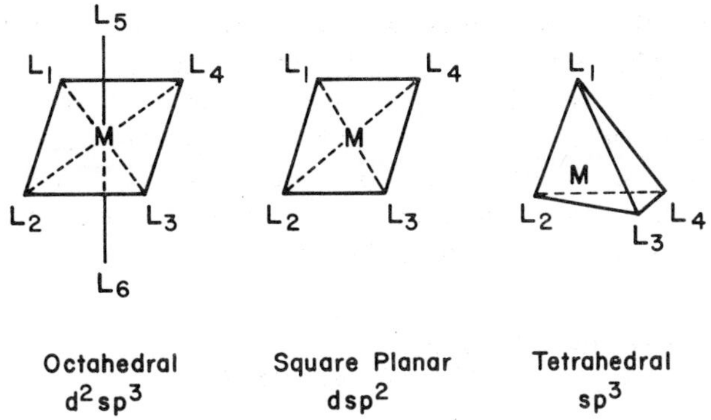

Fig. 3. Configurations of metal complexes.

iron leaves six empty orbitals to be filled by electrons from the ligand(s) forming the iron chelate. The CN is 6 and the configuration is d^2sp^3, an octahedral structure (Fig. 3). Finally, in Sc^{3+}, with 18 electrons distributed from the 1s through the three 3p orbitals and no electrons in the five 3d orbitals, the CN will still be 6 since three of the five orbitals, d_{xy}, d_{xz}, and d_{yz} (Fig. 4), are nonbonding in octahedral complexes (5, 6). It becomes apparent from this simplified approach to CN and configuration of the cations of the transition elements, chelates, or complexes that these properties are highly correlated with the number of d electrons. Both Zn^{2+} and Cu^{+}

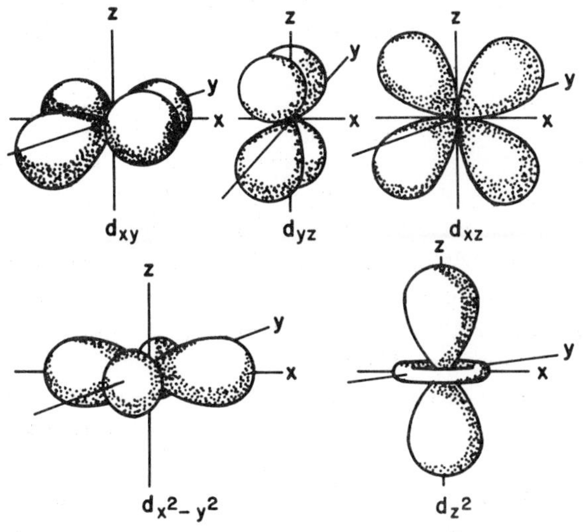

Fig. 4. The d orbitals.

are d^{10} ions which thus form chelates with similar CN and configuration. Both Fe^{3+} and Mn^{2+} are d^5 ions and, therefore, would be expected to form chelates of similar CN and configuration. As a matter of fact, antagonistic effects between manganese and iron have been reported for plants (7) and for animals (8). Further evidence will be presented indicating that when the ions of two transition elements have the same number of d electrons they are prime suspects for acting antagonistically biologically.

FACTORS AFFECTING STABILITY CONSTANTS AND LIGAND ATOM PREFERENCE

The field effects exerted by ligands on the disposition of the d electrons vary with the ligand and the number of d electrons. This arises from the fact that the five d orbitals are not all alike (9). As shown in Fig. 4 (10), three of these orbitals, d_{xy}, d_{xz}, and d_{yz}, resemble one another in producing regions of electron density in the three planes of cartesian axes, but directed between these axes. The other two, $d_{x^2-y^2}$ and d_{z^2}, differ from these three and from each other but are similar in that both produce regions of electron density along the axes rather than between the axes. The d_{xy}, d_{xz}, and d_{yz} orbitals together form a set with spherical symmetry, and the $d_{x^2-y^2}$ and d_{z^2} together form another set with spherical symmetry, but these two sets of orbitals are affected differently by their environment. In an octahedral complex ligands lie along the x, y, and z axes. The electrostatic repulsion between these ligands and the electrons of the central metal ion will be much greater if these electrons are in the $d_{x^2-y^2}$ or d_{z^2} orbitals than if they are in the d_{xy}, d_{xz}, or d_{yz} orbitals. Therefore, the five d orbitals which are all of equal energy (degenerate) in the free ion are split into two groups upon chelation, the energy of the $d_{x^2-y^2}$ and the d_{z^2} orbitals being increased far more than that of the other three by the presence of ligands lying along the axes on which these orbitals produce their regions of maximum electron density. The splitting energy Δ_0 (Fig. 5) caused by the ligand field is termed the crystal field stabilization energy (CFSE) and is represented by E in Fig. 5.

There are several consequences arising from these considerations. First, CFSE increases the stability of the metal complex. Second, metal ions such as Sc^{3+} with no d electrons (d^0) or Cu^+ and Zn^{2+} with the five d orbitals completely filled (d^{10}) have zero CFSE. Third, the splitting energy Δ_0 varies with the type of ligand. When the splitting energy Δ_0 is larger than Π, the energy required to overcome the repulsive interaction of two electrons occupying the same orbital (Hund's rule, i.e., normal tendency for electrons to remain unpaired), the electrons in the d orbitals of the metal ion will be paired to the maximum extent. The resulting CFSE will give greater stability

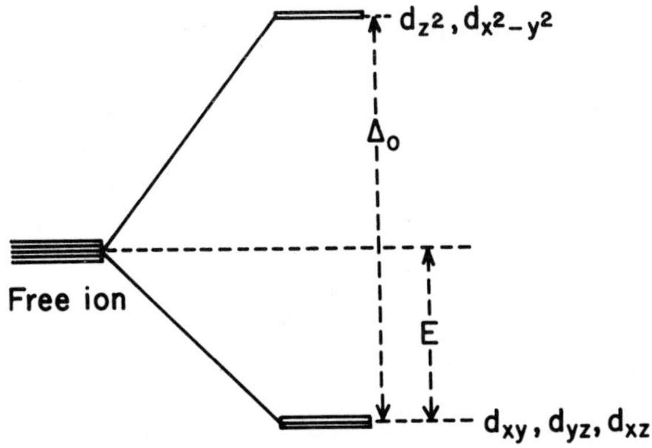

Fig. 5. Energy-level diagram of an octahedral complex.

to the metal chelate. If $\Pi > \Delta_0$, the electrons in the d orbital will not pair, and a 'weaker' complex is obtained. This point is illustrated in Fig. 6. In the Fe^{2+} ion the six d electrons show maximum parallel spins. In the presence of a strong ligand field the electrons are coupled, and the bonding in the metal complex formed is referred to as an inner-bonding (11) or low-spin complex. In the presence of a weak ligand field, $\Pi > \Delta_0$, the distribution of electrons remains as in the free Fe^{2+} ion, resulting in what is termed an outer-bonding (11) or high-spin complex. An inner-bonding complex of a metal ion has a greater stability than an outer-bonding complex. Accordingly, ligands can be arranged in order of increasing ligand field. The list of ligands arranged in order of increasing ligand field produced is known as the spectral chemical series (12), and a shortened version is as follows:

$$I^- < Br^- < Cl^- < H_2O < C_2O_4^{2-} < NH_3 < \text{ethylene diamine} < NO_2^- < CN^-.$$

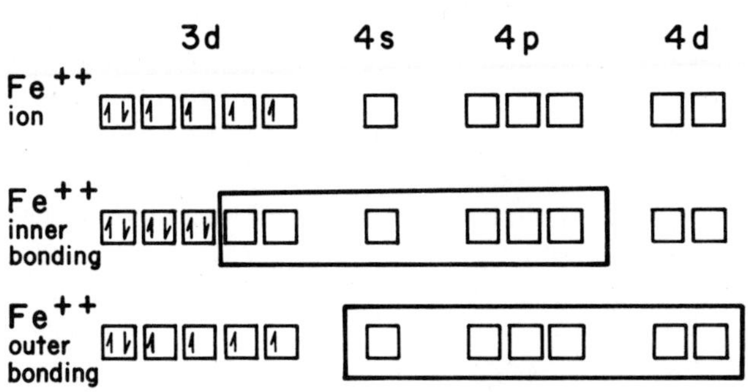

Fig. 6. Inner and outer bonding of ferrous iron as affected by ligand field.

Ions with five d electrons (d^5) such as Mn^{2+} and Fe^{3+} resist coupling or pairing and, therefore, are prone to form high-spin complexes (13). For example, of the ligand series given above, only CN^- is of sufficient ligand-field strength to bring about a low-spin complex of Fe^{3+} (14). These considerations are useful for making other predictions. Metal ions which are capable of having considerable CFSE will react preferentially with ligands effecting strong ligand fields. Thus, whereas most transition-metal ions will bond preferentially to N compared with O, Fe^{3+} and Mn^{2+} have a greater tendency to bond to O than to N (15).

Three other parameters are worth mentioning in connection with factors affecting stability constants. One is the 'ionic potential' (charge/radius) which is a measure of the tendency of a metal ion to associate with electron pairs (from the ligand); another is the 'polarizability' of a ligand which is related to its ability to donate electron pairs (10). The order of polarizability from high to low for ligand atoms encountered in biology is $P > S > N > O$. The third factor, which is related to the charge/radius parameter, is the ionic radius. Generally, if the valence of the metal ions is the same, the stability constant of the complex will increase as the ionic radius decreases. The Irving-Williams series is an example; i.e., for the complexes of almost all ligands, the stability and ionic radii vary in the same way, $Mn^{2+} < Fe^{2+} < Co^{2+} < Ni^{2+} < Cu^{2+} > Zn^{2+}$ (16). The rare exceptions to this rule are usually associated with spin pairing which leads to an extra stabilization of the low-spin configuration. Since the solvent system in living organisms is water, the radius of the hydrated metal ions would be more applicable than the crystal ion radii of the metals which are prevalent in the literature. Unfortunately, measurement of the effective radii of hydrated metal ions is difficult, and, therefore, relatively few values are available in the literature.

CHEMICAL-PARAMETER CONCEPT

In the context of broad generalizations and first approximations, it seems that the most pertinent chemical parameters of the transition elements in biology might be valency, isoelectronicity of the valency shell, coordination number, complex configuration, and radius of the metal ion. It appears, moreover, that most of these properties are reflections of a single chemical parameter, the number of d electrons.

My colleague, Dr. C. H. Hill, and I set out to test these notions over a decade ago. We centered our studies around two essential trace elements, copper and zinc. Our long-range plan was to select a series of elements which one could predict *a priori* would or would not interfere with copper metabolism, since we presumed that, whatever the individual functions of copper

Table 2. Chemical parameters

Ion	Orbital	Configuration	CN
Cu^+	d^{10}	sp^3	4
Zn^{2+}	d^{10}	sp^3	4
Cd^{2+}	d^{10}	sp^3	4
Hg^{2+}	d^{10}	sp	2
Cu^{2+}	d^9	dsp^2	4
Ag^{2+}	d^9	dsp^2	4

were in the animal, one function was that of a catalyst in oxidation-reduction reactions. Moreover, at one time or another, copper would likely be present in both ionic forms, Cu^+ (d^{10}) and Cu^{2+} (d^9). Accordingly, metal ions with either a tetrahedral (sp^3) or a square planar (dsp^2) complex configuration might be expected to be antagonistic to copper *in vivo.* The elements chosen and their pertinent chemical parameters are shown in Table 2. The experiment were carried out with chicks and rats (17–21). In order to minimize nonspecific effects arising from the use of high dietary levels of heavy metals, our basal diet was a copper-deficient diet containing less than 1 ppm of copper. This permitted us to observe effects with minimal dietary levels of the antagonistic metal ion. The evaluation criteria were mortality, hemoglobin level, body weight, and elastin content of the aorta. Each of these factors is adversely affected in copper deficiency. The basic experimental design was to add the metal ions under test to the low-copper basal diet and to the basal diet supplemented with an adequate amount of copper. If the copper-sufficient diet overcame all the adverse effects induced by the challenging metal ion, it was assumed that the test metal ion had interfered primarily with copper metabolism.

From the parameters shown in Table 3, it was predicted that Zn^{2+} would interfere with copper metabolism because it has the same chemical parameters as Cu^+; divalent cadmium would interfere with both copper and zinc metabolism because it had the same chemical parameters as Zn^{2+} and Cu^+; divalent mercury should not interfere with either copper or zinc metabolism since it has a CN of 2 and forms an sp complex configuration; Ag^{2+} [Ag^+ prefers to form linear 2:coordination complexes (12)] should interfere with copper but not with zinc metabolism. The experimental results (17, 19, 20) bore out these predictions. That reverse antagonisms between copper and zinc may occur is suggested by reports that extra zinc must be supplied in the diet when swine are fed high-level copper (22). Although proof that the metal-ion antagonisms reported herein are direct (i.e., the antagonizing metal ion

Table 3. Chemical parameters of oxy-anions

Anions	Number of πd bonds	Configuration
PO_4^{3-}	1	sp^3
VO_4^{3-}	1	sp^3
AsO_4^{3-}	1	sp^3
MoO_4^{2-}	2	sp^3
CrO_4^{2-}	2	sp^3
SeO_4^{2-}	2	sp^3
SO_4^{2-}	2	sp^3

replaces the native metal ion of the biologically active chelate) is not supplied by these experiments, the results are encouraging.

More recently the 'chemical-parameter concept' has been applied to oxy-anions, particularly those of the transition elements (18). The oxy-anions examined experimentally are shown in Table 3. The most important parameters appear to be anion orbital configuration and the number of πd bonds. The orbital diagramming of SO_4^{2-}, MoO_4^{2-}, and VO_4^{3-} shown in Fig. 7 follows

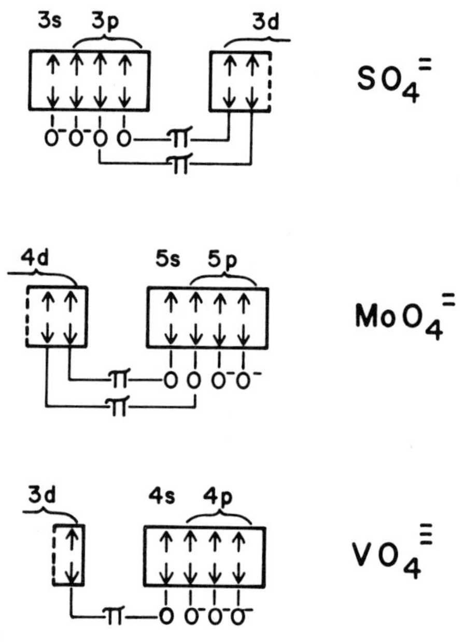

Fig. 7. Orbital diagrams of SO_4^{2-}, MoO_4^{2-} and VO_4^{3-}.

that depicted by Durrant and Durrant (23). The orbitals in the larger square of each diagram determine the configuration, and those in the smaller square the number of πd bonds. It can be seen in Fig. 7 that all three anions have an sp^3 or tetrahedral configuration, but sulfate and molybdate have two πd bonds whereas vanadate has only one πd bond.

As shown in Table 3 all the anions are sp^3 or tetrahedral configuration (24). In studies on the uptake of phosphate by respiring mitochondria, arsenate and vanadate, which have an sp^3 configuration and one πd just as the phosphate anion, were found to be inhibitory, whereas the chromate and the selenate anions, which also have an sp^3 configuration but two πd bonds, did not inhibit phosphate uptake (18). Another example is the finding that molybdate inhibits the reduction of sulfate by ruminal microorganisms (25), and that molybdate absorption was inhibited by sulfate in *in situ* intestinal loop studies with chicks (26). Once again, as shown in Table 3, both the sulfate and molybdate anions have sp^3 configurations and two πd bonds. Thus, these studies indicate that the oxy-anions of similar configuration and the same number of πd bonds also appear to interact antagonistically.

The attempt in this treatment of trace-element antagonisms has been to present a simplified systematic approach for the study of interactions among metal ions, which at the stage of development presented herein is, at best, a crude first approximation rather than a rigorous, highly refined theory. The evidence obtained to date, however, indicates that this approach has validity. It would seem that the chemical-parameter approach could be helpful, not only in studies concerned with the mechanism of trace-element antagonisms but also in studies concerned with mechanisms for absorption, transport, and function of the essential trace elements.

REFERENCES

1. Leach, R. M., and Norris, L. C. 1957. Studies on factors affecting the response of chicks to molybdenum. *Poult. Sci.* 36: 1136.
2. Wells, A. F. 1955. *Structural Inorganic Chemistry,* 3rd edn, Clarendon Press, Oxford, p. 17.
3. Orgel, L. E. 1963. *An Introduction to Transition-Metal Chemistry: Ligand-Field Theory,* Butler and Tanner, London, p. 41.
4. Basolo, F., and Johnson, R. C. 1964. *Coordination Chemistry: The Chemistry of Metal Complexes,* W. A. Benjamin, New York, p. 26.
5. Cartmell, E., and Fowles, G. W. A. 1966. *Valency and Molecular Structure,* 3rd edn, Van Nostrand, Princeton, N. J., and Butterworth, London, p. 227.
6. Larsen, E. M. 1965. *Transition Elements,* W. A. Benjamin, New York, p. 132.
7. Somers, I. I., and Shive, J. M. 1942. Fe-Mn relation in plant metabolism. *Plant Physiol.* 17: 582.

8. Matrone, G., Hartman, R. H., and Clawson, A. J. 1959. Studies of a manganese–iron antagonism in the nutrition of rabbits and baby pigs. *J. Nutr.* 67: 309.
9. Graddon, D. P. 1968. *An Introduction to Co-ordination Chemistry,* 2nd edn, Pergamon Press, Oxford, p. 39.
10. Murmann, R. K. 1964. *Inorganic Complex Compounds,* Reinhold, New York, pp. 72, 40.
11. Cartmell, E., and Fowles, G. W. A. 1956. *Valency and Molecular Structure,* Butterworth, London, p. 206.
12. Graddon, D. P. 1968. *An Introduction to Co-ordination Chemistry,* 2nd edn, Pergamon Press, Oxford, p. 47.
13. Orgel, L. E. 1963. *An Introduction to Transition-Metal Chemistry: Ligand-Field Theory,* Butler and Tanner, London, p. 42.
14. Graddon, D. P. 1968. *An Introduction to Co-ordination Chemistry,* 2nd edn, Pergamon Press, Oxford, p. 49.
15. Basolo, F., and Pearson, R. G. 1963. *Mechanisms of Inorganic Reactions.* John Wiley, New York, p. 60.
16. Orgel, L. E. 1966. *An Introduction to Transition-Metal Chemistry,* 2nd edn, Butler and Tanner, London, p. 87.
17. Hill, C. H., and Matrone, G. 1962. A study of copper and zinc interrelationships. *Proceedings of XII World's Poultry Congress,* p. 219.
18. Hill, C. H., and Matrone, G. 1969. Usefulness of chemical parameters in the study of *in vivo* and *in vitro* interactions of transition elements. *Fed. Proc.* 29: 1474.
19. Hill, C. H., Matrone, G., Payne, W. L., and Barber, C. W. 1963. *In vivo* interactions of cadmium with copper, zinc and iron. *J. Nutr.* 80: 227.
20. Hill, C. H., Starcher, B., and Matrone, G. 1964. Mercury and silver interrelationships with copper. *J. Nutr.* 83: 107.
21. Bunn, C. R., and Matrone, G. 1966. *In vivo* interactions of cadmium, copper, zinc and iron in the mouse and rat. *J. Nutr.* 90: 395.
22. Shuttle, N. F., and Mills, C. F. 1966. Studies of the toxicity of copper to pigs. I. Effects of oral supplements of zinc and iron salts on the development of copper toxicosis. *Br. J. Nutr.* 20: 135.
23. Durrant, P. J., and Durrant, B. 1970. *Introduction to Advanced Inorganic Chemistry,* 2nd edn, John Wiley, New York, p. 187.
24. Durrant, P. J., and Durrant, B. 1970. *Introduction to Advanced Inorganic Chemistry,* 2nd edn, John Wiley, New York, p. 816.
25. Huisingh, J., and Matrone, G. 1972. Copper-molybdenum interactions with the sulfate-reducing system in rumen microorganisms. *Proc. Soc. Exp. Biol. Med.* 139: 518.
26. Huisingh, J., Gomez, G. G., and Matrone, G. 1973. Interactions of copper, molybdenum, and sulfate in ruminant nutrition. *Fed. Proc.* 32: 1921.

DISCUSSION

Grassmann (Munich). You explained the copper-silver antagonism by the same electron configuration of these elements when silver is in the 2^+ valence state. The potential difference of the oxidation of silver to the 2^+ state is very high. Do you think that energy in metabolism is great enough to make this oxidation possible?

Matrone. I don't know. We put in univalent silver and our prediction is based on the assumption that it would be converted to divalent silver. We don't know if, indeed, it was; however, the results agreed with our prediction. The apparent anomaly in terms of inorganic chemistry and what's happening inside the animal body might be interesting to resolve.

Mertz (Beltsville). I am impressed by your hypothesis, but we must realize that the ligand field and molecular orbital theories have been arrived at by calculation or by model compounds. Dr. Vallee told us that in the catalytically active biological substances there is an entasis in which the coordination sphere can be highly distorted and irregular, and perhaps even have some unusual coordination numbers.

Matrone. This should not interfere with the concept which I am trying to promote here. All I am saying is that, if the chemical parameters are the same, then whatever one element is undergoing in the animal body the other element tends to undergo, irrespective of whether an entatic state is involved.

Frieden (Tallahassee). It seems to me that the entatic state is nothing that will defy crystal field theory or molecular orbital theory and it's the active site that we've all pictured. Dr. Vallee suggests that the specific three-dimensional conformation and uniqueness of protein structure probably produces a structure around the metal that we don't adequately recognize yet and can't deal with until we do, but this structure is responsible for the unusual catalytic or other properties of the metalloenzyme.

Matrone. Another point is that an element's metabolism isn't all related to its enzymatic function. There is absorption, transport, etc., at which similarity in chemical properties could cause interactions.

Shirley (Gainesville). About 20 years ago, Dr. Davis and I fed high levels of molybdenum to steers and found that we increased the excretion of phosphorus in the urine about three-fold. Would you have predicted this outcome by your new thinking on electron structure of these elements?

Matrone. No, I would not have predicted it. The electronic and bonding properties of molybdate and phosphate are different, so I would anticipate that you had an indirect, rather than a direct, interaction.

THE BIOCHEMICAL EVOLUTION OF THE IRON AND COPPER PROTEINS[1]

EARL FRIEDEN

Department of Chemistry, Florida State University, Tallahassee, Florida

Iron and copper have become so firmly entrenched in modern biochemical thinking that it is difficult to imagine life in a cell without these two metals. It is quite certain that their essentiality to every form of life from plants to man arises from their role as prosthetic groups in a number of important enzymes and proteins (1). These two metals have become pre-eminent because of their direct involvement in the adaptation to aerobicity and their role as terminal oxidases and respiratory proteins (1–3). If the biochemical evolution of these two metals is considered, a number of important relationships and correlations arise which have not been described previously (Fig. 1).

It is widely accepted that life on earth first evolved in the sea or in saline tidal pools. This has led to an indelible imprint of the composition of the oceans on the chemistry of all cells. Iron and copper compounds are a part of this imprint, since, along with zinc, molybdenum, and manganese ions and with the aid of suitable concentrating mechanisms, the oceans provided adequate amounts of these metal ions for use in evolving biological systems. Apparently no eukaryotes and only a few prokaryotes, e.g., some species of lactic acid bacteria, are capable of surviving without iron or copper. Iron, in particular, has played a major role in the biogeochemical history of the earth. Iron and nickel make up the core of the earth and, since iron occurs most commonly in meteorites, it is probably universally distributed throughout the solar system. The whole earth is believed to be 35% iron and its crust 5% iron. In the reducing atmosphere of the primitive earth, there was a need for the development of ligating and transporting mechanisms for the very insoluble

1. Supported by Grant HL 08344, from the National Institute of Heart and Lung Disease. Paper No. 46 in a series on the Biochemistry of Copper and Iron Metalloproteins.

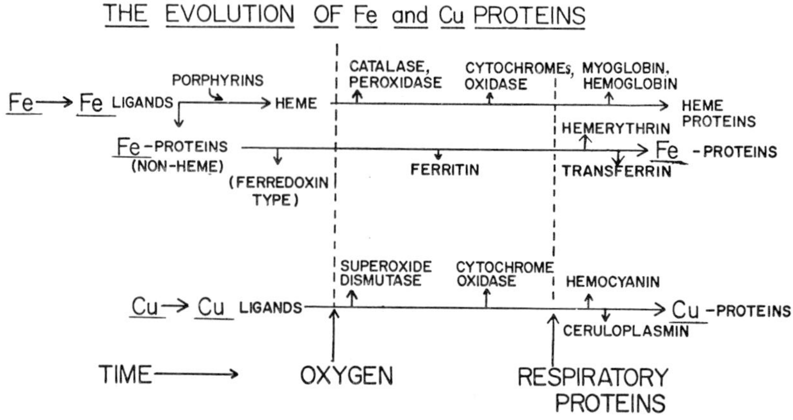

Fig. 1. Diagram depicting the evolutionary sequence and the development of the hemeproteins, the nonheme iron proteins and the copper proteins. The horizontal axis is essentially a time line with the advent of the oxygen era and the respiratory proteins indicated.

ferric hydroxide compounds so that iron could play its vital role in electron transfer and oxidation to pave the way for the aerobic revolution that must have started about 10^9 years ago. Neilands (4) has examined the possibility that the prokaryotes with the most primitive iron metabolism might well be the direct descendents of the primordial cell since only certain anaerobic and lactic acid bacteria do not contain heme compounds. He cites the ferredoxins and the rubredoxins, iron-sulfur compounds found in anaerobic bacteria, as representing the most primordial iron-complexing system. For transport and storage the iron ligands were selectively modified by stepwise transition to a mixed sulfur-nitrogen and sulfur-oxygen binding and, finally, to an all-oxygen type of coordination, characteristic of ferritin, transferrin, and the low molecular weight nonprotein iron-binding siderochromes. For electron transport, oxygenation, and hydroxylation the evolution of iron compounds proceeded from sulfur to nitrogen ligands, culminating in the heme proteins and highly specialized nonheme iron proteins we know today. The porphyrins may have been an especially early iron ligand development. Gaffron (5) has suggested that porphyrins formed geochemically before life began, and that primitive anaerobes may have used these 'exogenous' porphyrins even before they developed enzymic mechanisms for their synthesis.

A corresponding development of copper ligands may have been overwhelmed by the ubiquity of iron in the oceans and on the earth. Copper has trailed iron as the prime metal of civilization. The only copper-porphyrin equivalent to the heme group is found as a brilliant red pigment (turacin) in the wings of the rare African bird, the touraco. The fact that uniquely useful copper proteins developed at all may be attributed to the superior chelating properties of the cupric ion. The heme biosynthetic machinery, however, was too entrenched to permit competition with iron for the protoporphyrin structure. The magnesium porphyrin derivatives in chlorophyll and the cobaltous porphyrin structures in the corrinoids (vitamin B_{12}) represent highly specialized adaptations for which iron may have been chemically unsuitable.

OXYGEN AS AN EVOLUTIONARY FORCE

The major evolutionary adaptation to which iron and copper compounds contributed so handsomely arises from changes in the atmosphere leading to an accumulation of, and a dependence on, oxygen. Gaffron (5) has summarized the major stages of biogenesis in terms of the transition from a highly reducing atmosphere to our present oxygen-rich world. There was first an anaerobic era of excess hydrogen with ammonia, methane, water vapor, and an accumulation of highly reduced organic substances. This was followed by a hydrogen-poor period with the beginnings of inorganic or organic catalysis and traces of oxygen. In the third period, the anaerobic organisms depleted their useful reservoir of organic substances, forcing on the organisms of the earth an essential dependence on photoreduction as the major initial source of energy. This led to the present era of photosynthesis and ascendancy of cells dependent on oxidative energy, eliminating or driving the obligate anaerobes 'underground'. Finally, Bloch (6) has pointed out that oxygen also serves as an obligatory oxidant for a variety of biosynthetic energy-consuming processes. Oxygen itself, then, becomes a major evolutionary force. As we shall see later, the essentiality of iron and copper arises from their ability to promote the effective utilization of oxygen.

How environmental circumstances may determine whether an organism will survive such drastic changes as the step from anaerobic to aerobic conditions was shown in an actual laboratory model. In 1954, Jensen and Thofern (7) found a micrococcus strain which grew well heterotrophically under anaerobic conditions. On contact with oxygen the cells die because they have neither an efficient mechanism for destroying the hydrogen peroxide which might be formed, nor a respiratory system. But, when the nutrient medium contained a simple ferric porphyrin, ordinary heme, these bacteria survived in air. The heme compound was taken up into the cells and there combined with the right kind of protein to provide the bacteria with catalase and respiratory enzymes. The presence or absence of heme in the surroundings at the time oxygen reached the cells determined their fate.

The high degree of adaptability of metals and highly stable organic structures like the porphyrins has been used by Calvin (8) and others to trace the development of metal-ion catalysis. Calvin used catalase activity, the ability to catalyze the decomposition of hydrogen peroxide into water and oxygen, as a model system. He compared the hydrated ferric ion, ferric ion complexed with protoporphyrin, as in heme, and the iron-containing heme group associated with a specific protein, as in the enzyme catalase. The relative catalase activity increased from 1 to 10^3 to 10^{10} from ferric ion to heme to the heme enzyme. This type of change in the catalytic form of ferric ion probably occurred in many more than these three steps, but illustrates the improvement in catalytic activity that must have developed during thousands of years of evolution to achieve the highly efficient metalloprotein catalysts of today. The importance of both iron and copper are reflected in the broad spectrum of useful metalloproteins and metalloenzymes, many of which are listed in Tables 1 and 2. These metalloproteins have evolved to utilize most effectively the versatility of these two metal ions as electron and oxygen carriers and for oxygenation, hydroxylation, and other crucial metabolic reactions, including protection against deleterious oxygen byproducts. Yet, as suggested in Fig. 1, these two metal ions have remained remarkably associated despite the increasing complexity of living organisms.

PROTECTION FROM SUPEROXIDE AND PEROXIDE

A most fundamental relation between copper and iron metalloenzymes has been suggested by recent observations of Fridovich, McCord and their co-workers (9). They have found that a well-known ubiquitous group of copper proteins (2 g atom Cu per 34,000 g) including erythrocuprein, hepatocuprein, etc., possess a unique enzymic activity, superoxide dismutase. The reaction

Table 1. Major types of iron enzymes and proteins

Heme proteins	Nonheme proteins
Hemoglobin: O_2 carrier	Hemerythrin: O_2 carrier (invert.)
Myoglobin: muscle O_2 carrier	Transferrin: Fe transport
Hydroperoxidases: catalase, peroxidase	Ferritin: iron storage
Cytochromes: electron transport	Ferredoxin: electron transport
Cytochrome oxidase: terminal oxidase	Nitrogenase: N_2 fixation
Tryptophan oxygenase: TRP oxidation	Succinic dehydrogenase: dehydrogenation
	Aconitase: Krebs cycle
	Xanthine oxidase: purine metabolism

Table 2. Some examples of copper proteins

Protein; enzyme	Source; function
Hemocyanin	O_2 carrier in invertebrates
Azurin	Electron carrier in bacteria
Plastocyanin	Electron carrier in plants
Tyrosinase	Pigmentation; sclerotization
Ceruloplasmin	Fe mobilization; ferroxidase
Superoxide dismutase	Protection against superoxide
Lysine oxidase	Cross linking of collagen and elastin
Galactose oxidase	Sugar metabolism; contains pyridoxal
Ascorbate oxidase	Terminal oxidase in some plants
Cytochrome oxidase	Terminal oxidase in most cells
Dopamine-β-hydroxylase	Epinephrine biosynthesis; hydroxylation

catalyzed is the dismutation of the superoxide ion to molecular oxygen and hydrogen peroxide:

$$2H^+ + 2O_2^- \xrightarrow[\text{dismutase}]{\text{superoxide}} H_2O_2 + O_2.$$

This furnishes a convenient method of disposing of a highly reactive intermediate, the superoxide ion O_2^-. The superoxide dismutase system is backed up by heme (iron protoporphyrin) enzymes that decompose hydrogen peroxide either to oxygen and water as in catalase, or, with the aid of a hydrogen donor molecule (AH_2), to water and a dehydrogenated product (A) as with the peroxidases:

$$H_2O_2 \xrightarrow{\text{catalase}} H_2O + \tfrac{1}{2}O_2$$

$$AH_2 + H_2O_2 \xrightarrow{\text{peroxidases}} 2H_2O + A.$$

These two types of enzymes provide a disposal route for the two principal toxic byproducts of oxygen reduction, superoxide ion and peroxide. There is no dearth of biochemical reactions which produce hydrogen peroxide. The reactions which produce the superoxide ion have been identified only recently, and the expanding list includes enzymic reactions such as milk xanthine oxidase, rabbit liver aldehyde oxidase, rabbit liver dihydroorotate dehydrogenase, pig liver diamine oxidase, the autooxidation of hemoglobin and myoglobin, and probably many other one-electron transfers (9). It turns out that this protection is a prerequisite for the adaptation of living cells to the utilization of oxygen. It represents an indispensible link between the two groups of metalloproteins, a copper enzyme with superoxide dismutase activity and a heme enzyme with catalase or peroxidase activity. Most of the superoxide dismutases contain zinc, which appears to be important in maintaining the stability of these enzymes but is not directly involved in their catalytic function. A few bacterial and mitochondrial dismutases contain manganese instead of copper and lack zinc (9). Fridovich has recently suggested that the cupro-zinc superoxide dismutases and the mangano-superoxide dismutases evolved independently in protoeukaryotes and prokaryotes respectively. (This conforms to proposals that mitochondria evolved out of an endosymbiotic relationship between these protoeukaryotes and an aerobic prokaryote.)

The presence of these enzymes and the development of life in oxygen has been correlated by McCord, Fridovich and co-workers (9). The earth's atmosphere has evolved from a highly reducing environment to its current oxygen-rich status. It was first recognized by Pasteur that certain bacteria grow best in an oxygen-free atmosphere and will not survive in oxygen. It has now been found that these strict anaerobes show no superoxide dismutase and, generally, no catalase activity. Virtually all aerobic organisms, particularly those containing cytochrome systems, were found to contain both types of enzymes, superoxide dismutase and catalase and/or peroxidase. The evolution of this team of copper and iron enzymes appears to have been a prerequisite for the development of life in oxygen as we find it today. There remains a special group of aerotolerant anaerobes which survive exposure to air and metabolize oxygen to a limited extent but do not contain cytochrome systems. These microorganisms possess superoxide dismutase activity but have no catalase activity.

OXYGEN-CARRYING PROTEINS

Blood has always been associated with animal life and, if for no other reason, iron and copper metalloproteins would be forever linked in nature by their occurrence in blood as the oxygen-carrying chromoproteins. The color of these pigments, red for the iron proteins and blue for the copper proteins, attracted early attention, and they are the oldest recognized metalloproteins. The most prevalent respiratory pigments are of the hemoglobin type, a protein attached to a heme group which consists of a ferrous protoporphyrin complex. Hemoglobin is found in plants, in many invertebrates, and in all vertebrates. Myoglobin is a simpler oxygen-carrying hemoprotein which exists exclusively in muscle. Two small invertebrate phyla, Sipunculida and Brachiopoda, contain hemerythrin, a nonheme iron containing respiratory protein. A larger group of invertebrates, in the four classes Cephalopoda, Crustacea,

Gastropoda, and Xiphosura, have as their oxygen-carrying pigment a blue copper protein, hemocyanin. Thus the ability to serve as respiratory proteins in nature appears to be the exclusive property of the copper and iron proteins (10).

Why has no other essential metal ion or other type of respiratory protein developed to satisfy this important function, principally for the animal kingdom? For a possible answer to this question, we should examine what a protein needs to succeed in this activity. It needs to be able to form a stable dissociable complex with the highly reactive molecule O_2. Transition metals excel in this capacity; few other chemical groups can do this. In fact, all efforts to devise other physiologically compatible model oxygen carriers have failed to date. The compounds that come closest to emulating the oxygen-binding properties of hemoglobin, hemerythrin, and hemocyanin contain cobalt and other transition-metal-ion derivatives. This limits our consideration to metals such as Co, Mn, Mo, and Zn. None of these metal ions match Cu(II) in their ability to form stable chelates with amino acids, peptides, or proteins. The well-known Irving-Williams series on the relative ability of divalent metal ions of the first transition series to interact with nitrogen donor ligands is Cu(II) > Zn(II) > Ni(II) > Co(II) > Fe(II) > Mn(II). In seawater cobalt is less than one-tenth as concentrated as copper and is therefore much less likely to have been utilized in the primitive oceans. Manganous ions are less stable against further oxidation and are much poorer chelators. Though plentiful, zinc does not have some of the key chemical characteristics of the other transition metals, e.g., zinc has a complete set of 3d electrons.

Probably the most compelling reason favoring iron and copper mechanisms, however, is related to our previous discussion about the development of copper and iron enzymes as protectants against toxic oxygen byproducts. It seems a certainty that the development of superoxide dismutase and catalase-peroxidase systems for the survival of aerobes preceded the later evolution of the respiratory pigments and enzymes (Fig. 1). Therefore these cells already had the necessary machinery to utilize copper and iron ions and to convert iron into heme for the biosynthesis of whatever metalloprotein was useful. For iron, this ultimately led to the development of specialized storage and transport proteins, the ferritins, the transferrins, and the mucosal iron-binding proteins. For copper, there was a corresponding development of the serum copper-transport and iron-mobilizing protein ceruloplasmin (ferroxidase) and copper-binding proteins in the liver and other tissues. It was obviously most convenient to use these two metal ions for any new task for which they were capable, including the formation of the oxygen-carrying proteins hemoglobin, hemocyanin, and hemerythrin. It should be emphasized that we do not impose any common genetic origin on the protein moieties of these metalloproteins, although this is not excluded. The genetic history of the two pairs of polypeptide chains which comprise the vertebrate hemoglobins has been described by Ingram (11).

CUPRO-FERRO (HEME) PROTEINS

If a biochemist is asked to identify the *one* enzyme which is most vital to all forms of life, he would probably name cytochrome C oxidase. This is the enzyme, found in all aerobic cells, which is mainly responsible for the introduction of oxygen into the oxidative machinery that produces the energy which we need for physical activity and biochemical synthesis (Fig. 2). When cytochrome oxidase action is blocked, as in cyanide poisoning, all cellular activity grinds to a quick halt, indeed in only a few minutes. This enzyme may be regarded as the ultimate in the integration of the function of iron and copper in biological systems. Here, in a single molecule, we combine the talents of the heme group and copper ions to bind oxygen, to reduce it with electrons from other cytochromes in the hydrogen electron-transport chain, and, finally, to convert the reduced oxygen to water. Yet it is a frustrating fact that this key enzyme has proved to be extremely difficult to isolate, purify, and otherwise successfully study. We do know that each molecule weighs about 140,000 daltons and has two copper ions and two heme groups which are required for its function (12).

A chemically similar enzyme, tryptophan-2,3-dioxygenase, catalyzes the insertion of molecular oxygen into the pyrrole ring of tryptophan yielding N-formylkynurenine. It also is widely distributed and has been purified to homogeneity from rat liver and the microorganism *Pseudomonas acidovorans*

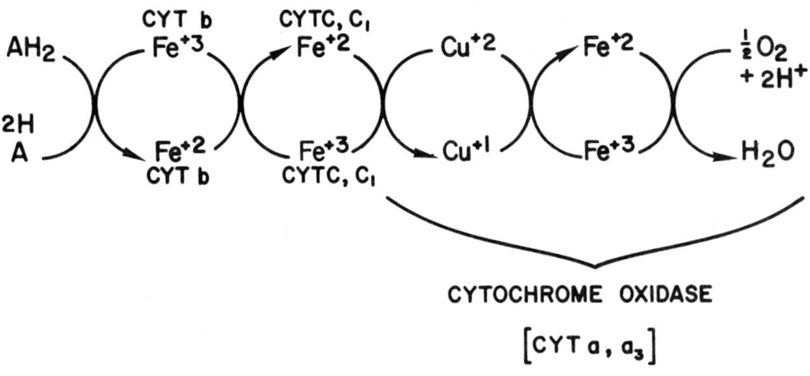

Fig. 2. Diagram illustrating a portion of the cytochrome electron-transport system of mitochondria. The arrows show the direction of the flow of electrons. The sequence Cu → Fe → O_2 in cytochrome oxidase is probable but not certain.

(13). In a molecule weighing 167,000 daltons, it has been found to have two atoms of Cu(I) and two molecules of heme.

Finally, a recently discovered enzyme, cysteamine oxygenase, has been reported to contain one atom each of copper, iron, and zinc ions in a molecule weighing 100,000 daltons (14).

PLANT ENZYMES

Plants have not escaped the influence of the copper-iron relationship. All plants have a small but significant oxidative respiration which utilizes the cytochrome-cytochrome oxidase system. This can be regarded as a vestige from the same development as the primitive aerobic cell. However, iron and copper electron carriers are also involved in the photoinduced electron flow in photosynthetic tissues (15). It is now believed that photosynthesis in green plants proceeds by the cooperation of two photochemical systems, each catalyzed by light absorbed through a specific pigment system. These two photosystems are interrelated by their own cooperative electron-transport system which directs the sequential transfer of electrons derived from water to a reduction of $NADP^+$. While all the components of this system have not been identified, at least three cytochromes (heme) and one ferredoxin (nonheme iron) are involved. Plastocyanin, a small copper protein, is localized in the chloroplasts and functions as an intermediate electron carrier on the reducing side of cytochrome f. The initial transfer of electrons from water requires manganous ions, and magnesium is an essential component of the chlorophyll molecule located in both pigment matrices.

BIOSYNTHESIS OF CONNECTIVE TISSUE PROTEINS

Somewhere in the development of complex organisms from their single-cell precursors, it became advantageous for cells to aggregate to form tissues and, ultimately, organs. The development of extracellular attachments was greatly enhanced by the presence of hardy macromolecules of the connective tissue, particularly collagen and elastin. Recent work on the biosynthesis of these connective-tissue components has revealed a direct role for iron and copper enzymes and a probable role for manganous and zinc ions as well.

Collagen is synthesized by a series of sequential steps involving assembly on ribosomes of a proline- and lysine-rich polypeptide precursor called "protocollagen" (16). The hydroxylation of appropriate proline and lysine residues in protocollagen to hydroxyproline and hydroxylysine occurs before the collagen is extruded into the extracellular matrix. These hydroxylation steps require iron as Fe(II) with a maximum effect at 10 μM; no other metal ion will replace iron. The synthesis of complete collagen can be blocked by an iron chelator, 1 mM a,a'-dipyridyl. The hydroxyamino acids confer additional stability to the collagen molecule by increased prospects for intermolecular hydrogen bonding and for cross linking involving sugar groups.

Since copper deficiency has been shown to lead to aneurisms and soft bones in experimental animals, a role of copper in promoting the tensile strength of the fibrous proteins, collagen and elastin, has long been suspected. It has now been established that a copper enzyme, lysine oxidase, is the determining factor in providing chemical cross linking of polypeptide chains to stabilize these structures (17). The most important cross linkage occurs between four lysyl residues, which form a pyridinium ring, joining neighboring peptide chains. These cross links were sufficiently stable to be isolated after the hydrolysis, yielding the cyclic isomers desmosine (Fig. 3) and isodesmosine. Without lysine oxidase and its induced cross linkages, soluble collagen and elastin molecules with less tensile strength are formed.

It is of interest that two other essential metals are known to be involved in connective-tissue metabolism: manganous ions in mucopolysaccharide biosynthesis, particularly for galactosyl and glycosyltransferase (18), and zinc in as yet unidentified repair and healing processes (19).

CERULOPLASMIN (FERROXIDASE) AND IRON METABOLISM

Finally we consider the relation between copper compounds, particularly the serum copper protein ceruloplasmin, and iron metabolism. The essentiality of copper in the mammal was first recognized in 1928 by Hart *et al.* (20), who reported that the copper-deficient experimental animal became seriously anemic. This anemia has now been related to a low plasma copper and a reduced level of the catalytic serum copper protein ceruloplasmin, now regarded as a molecular link between copper and iron metabolism (21). Ceruloplasmin appears to control the rate of iron uptake by transferrin, the iron-transport protein in serum, which delivers iron to the tissues which need it, e.g., the reticulocyte which is making hemoglobin. We have been able to relate this iron-mobilizing property to the ferroxidase activity of ceruloplasmin, i.e., its ability to catalyze the oxidation of ferrous to ferric ion so that transferrin can bind the ferric ion and transport it to iron-receptor sites wherever they may be (21).

The importance of this interconversion of ferrous to ferric ion in iron metabolism can be appreciated best by a consideration of the important ferrous-to-ferric cycles which seem to be involved in iron metabolism (Fig. 4) (22). These cycles have already been emphasized in regard to their function in the vital hemoproteins such as the cytochromes (Fig. 3) or the nonheme iron enzymes. At the focal point of iron metabolism is the ferroxidase reaction in

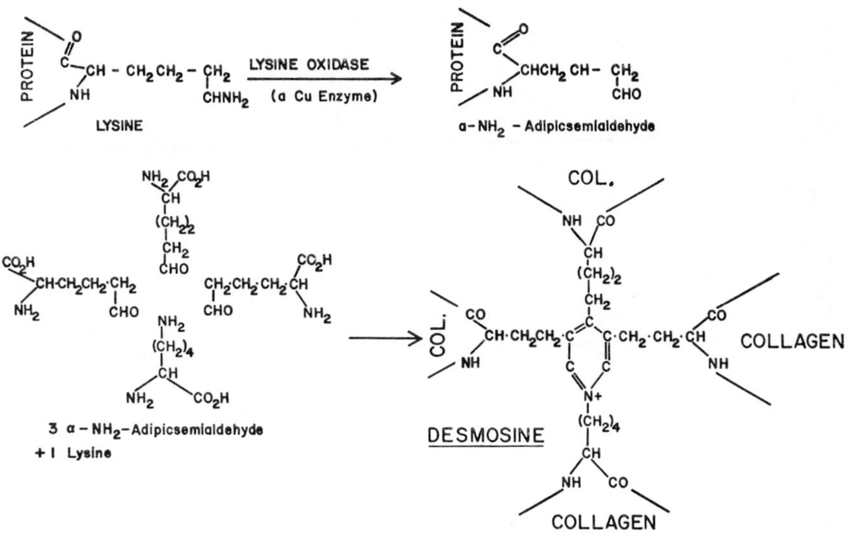

Fig. 3. Formation of desmosine cross links in collagen initiated by the copper enzyme lysine oxidase.

which the oxidation of ferrous ion to ferric ion is catalyzed by ceruloplasmin (ferroxidase I) and by at least one other protein with ferroxidase activity. Iron-utilizing systems have evolved in which iron must go through the ferrous state to be mobilized or integrated into the most prevalent iron compound hemoglobin. For storage as ferritin or for transport as transferrin, however,

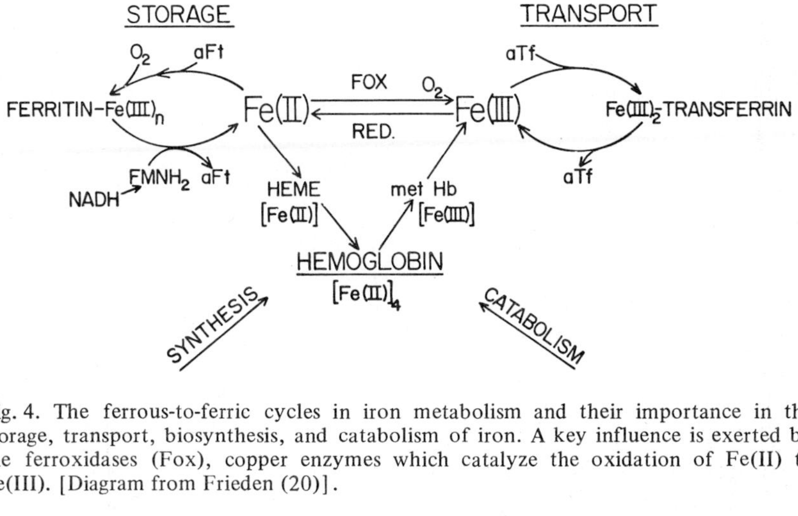

Fig. 4. The ferrous-to-ferric cycles in iron metabolism and their importance in the storage, transport, biosynthesis, and catabolism of iron. A key influence is exerted by the ferroxidases (Fox), copper enzymes which catalyze the oxidation of Fe(II) to Fe(III). [Diagram from Frieden (20)].

iron must be converted to the ferric state. At virtually every known step of iron metabolism, including storage, transport, biosynthesis, and degradation, the ferrous-to-ferric cycles play a significant role in the multimetabolic pathways of the most important metal constituent in all living systems. By virtue of its ferroxidase activity the serum copper enzyme ceruloplasmin exerts its influence on the metabolism of iron by facilitating the ferrous-to-ferric ion cycle. Ceruloplasmin and iron mobilization represent one of the most recent examples of the evolution of copper-iron relationships in the vertebrates.

In mammals the functional relationship between ceruloplasmin (ferroxidase-I), transferrin, and hemoglobin is now established. One example is the dramatic increase (5–10 X) in serum iron in the rooster stimulated by the injection of estrogens (e.g., diethylstilbesterol, β-estradiol), shown to be preceded by a several-fold increase in serum ferroxidase activity but with no change in the transferrin level (23). Further evidence for the interconnection of these metalloproteins has also been found in lower vertebrates, even in the tadpole during metamorphosis (24). In the transition of the tadpole to the bullfrog there is a complete switch in the several globin chains accompanying the total resynthesis of hemoglobins. The appearance of these new hemoglobins is preceded by a 3–4-fold increase in apotransferrin and by a several-fold increase in ferroxidase activity. In this developmental process, the iron mobilization machinery prepares the hematopoietic system for hemoglobin biosynthesis.

SUMMARY

The evolutionary relationships between the iron and copper proteins have been reviewed. It is proposed that these two metals first became associated because of their occurrence in seawater and in their adaptation to oxygen, particularly in providing protection from the highly toxic oxygen byproducts, superoxide ion and peroxide. This took the form of the copper enzyme, superoxide dismutase, and the heme-protein enzymes, catalase and peroxidase. The success of the aerobes was accompanied by the steady development of more sophisticated iron and copper enzymes including the cytochromes, the cupro-ferro (heme) enzyme cytochrome oxidase, and electron-transfer reactions in plants. With the increasing complexity of organisms, the cellular machinery utilizing iron and copper was adapted for the production of the oxygen-carrying proteins, hemoglobins, hemerythrins, and hemocyanins. These later stages of evolution were accompanied by a parallel development in biosynthetic enzymes associated with connective tissue and other more specific processes. Perhaps the most recent example of the continuing association of iron and copper in vertebrates is the role of the serum copper protein ceruloplasmin and its ferroxidase activity in the mobilization of iron for the biosynthesis of hemoglobin and other iron proteins.

REFERENCES

1. Vallee, B. L., and Wacker, W. E. C. 1970. Metalloproteins. *In* H. Heurath (ed.), *The Proteins,* vol. 5, Academic Press, New York.
2. Frieden, E., Osaki, S., and Kobayashi, H. 1965. Copper proteins and oxygen: Correlations in structure and function of the copper oxidases. *J. Gen. Physiol.* 49: 213.
3. Malkin, R., and Malmstrom, B. G. 1970. The state and function of copper in biological systems. *Adv. Enzymol.* 33: 177.
4. Neilands, J. B. 1972. Evolution of biological iron binding centers. *In Structure and Bonding,* Vol. 11, Springer-Verlag, New York, p. 145.
5. Gaffron, H. 1960. The origin of life. *Perspect. Biol. Med.* 3: 163.
6. Bloch, K. 1962. Oxygen and biosynthetic patterns. *Fed. Proc.* 21: 1058.
7. von Jensen, J., and Thofern, E. 1954. Ferriporphyrinchlorid als Bakterienwuchsstoff. IV. Zum Cytochromsystem von *M. pyog.* var. *Aureus. Zeit. Naturf.* 9b, 596.
8. Calvin, M. 1969. *Chemical Evolution.* Oxford University Press, New York.
9. Fridovich, I. 1972. Superoxide radical and superoxide dismutase. *Accts Chem. Res.* 5: 321.
10. Prosser, C. L., and Brown, F. A., Jr. 1961. *In Comparative Animal Physiology,* W. B. Saunders, Philadelphia, Pa., pp. 198–237.
11. Ingram, V. M. 1961. Gene evolution and the haemoglobins. *Nature, Lond.* 189: 704.
12. Chance, B. Yonetani, T., and Mildvan, A. S. (eds). 1971. *Probes of Structure and Function of Macromolecules and Membranes,* Vol. 2, Academic Press, New York, pp. 575–618.
13. Brady, F., Monaco, M. E., Forman, H. J., Schutz, G., and Feigelson, P. 1972. Role of copper in activation of and catalysis by tryptophan-2,3-dioxygenase. *J. Biol. Chem.* 247: 7915.
14. Cavallini, D., Supre, S., Scandurra, R., Graziani, M. T., and Cotta-Ramusino, F. 1968. Metal content of cysteamine oxygenase. *Eur. J. Biochem.* 4: 209.
15. Bishop, N. I. 1971. Photosynthesis: The electron transport system of green plants. *A. Rev. Biochem.* 40: 199.
16. Prockop, D. J. 1971. Role of iron in synthesis of collagen in connective tissue. *Fed. Proc.* 30: 984.
17. Chou, W. S., Savage, J. E., and O'Dell, B. L. 1969. Role of Cu in biosynthesis of intramolecular cross-links in chick tendon collagen. *J. Biol. Chem.* 244: 5785.
18. Leach, R. M., Jr. 1971. Role of manganese in mucopolysaccharide metabolism. *Fed. Proc.* 30: 991.
19. Westmoreland, N. 1971. Connective tissue alterations in zinc deficiency. *Fed. Proc.* 30: 1001.

20. Hart, E. B., Steenbock, H. B., Waddell, J., and Elvehjem, C. A. 1928. Iron in nutrition. VII. Cu as a supplement to iron for hemoglobin building in the rat. *J. Biol. Chem.*, 77: 797.
21. Frieden, E. 1971. Ceruloplasmin, a link between iron and copper metabolism. *In* R. F. Gould (ed.), *Bioinorganic Chemistry, Advances in Chemistry Series,* American Chemical Society, Washington, D.C., pp. 292–321.
22. Frieden, E. 1973. The ferrous to ferric cycles in iron metabolism. *Nutr. Rev.* 31: 41.
23. Planas, J., and Frieden, E. 1973. Serum iron and ferroxidase activity in normal, Cu deficient and estrogenized roosters. *Am. J. Physiol.* 225: 423.
24. Frieden, E., Eaton, D. N., Tripp, M. J., and Eaton, J. E. 1972. Functional relations of iron and copper metalloproteins in development. *Fed. Proc.* 31: 487.

INTERACTION OF SELENIUM WITH MERCURY, CADMIUM, AND OTHER TOXIC METALS

J. PAŘÍZEK, J. KALOUSKOVÁ, A. BABICKÝ, J. BENEŠ and L. PAVLÍK

Institute of Physiology and Isotope Laboratories of the Institutes for Biological Research, Czechoslovak Academy of Sciences, Prague, Czechoslovakia

Research on toxicological interrelations between compounds of selenium and those of certain heavy metals has now a history of about fourteen years. In spite of the growing interest and increasing number of laboratories giving attention to this problem, it seems evident that this time has been too short for obtaining definite answers to many basic questions, including those connected with the mechanisms involved. However, owing to research done so far, it has been possible to learn basic facts, and to recognize their possible implications in relation to the biological effects of selenium and to environmental problems, particularly those connected with cadmium or mercury. On the basis of work done so far it is now possible to formulate clear-cut questions, which can be studied by more appropriate methods.

SELENIUM AND TOXICITY OF CERTAIN HEAVY METALS

In the relatively short history of studies on interrelations between selenium and compounds of certain metals we can trace three lines of research which started and developed, at least for some time, quite separately. We shall discuss one of these lines, namely that concerned with cadmium and mercury, in more detail. However, it should be stated that almost simultaneously with Kar's first studies (1) on cadmium and selenium, Holló and Zlatarov (2) reported on the protective effect of selenate injection against thallium toxicity. As was shown later by Rusiecki and Brzeziński (3), the lethal effects of thallium compounds were prevented in a similar way by peroral administration of selenate, which was accompanied by an increased thallium content in the liver, kidneys, and bones. The studies of Levander and Argrett (4) later revealed that subcutaneous injections of thallium acetate increased the retention of selenium in the liver and kidneys, diminished pulmonary excretion of volatile selenium compounds, and also decreased urinary selenium. This decreased excretion was similar to that observed after the injection of cadmium (5) or mercuric compounds (*vide infra*). Out of all the compounds which decrease pulmonary excretion of selenium, only arsenite enhanced biliary selenium excretion (4). It is apparent, therefore, that the interrelation between thallium and selenium in several respects resembles that between selenium and mercury or cadmium, which will be discussed later.

Interactions of selenium compounds with those of silver, studied by Diplock and his co-workers (6), were discussed at the TEMA 1 meeting (7). It is interesting to note that in the case of silver the toxicity cannot only be prevented by selenium compounds but also by vitamin E. In contrast to mercury, cadmium, and thallium, where at least the first basic facts were obtained from studies on acute toxicity and parenteral administration, results on selenium and silver were obtained in chronic nutritional experiments. From the point of view of the underlying mechanisms and for comparison with the effect of the other metals mentioned, it would be interesting to learn if compounds of silver could have similar effects on selenium metabolism and excretion as those observed in acute experiments with cadmium or mercury.

Present knowledge and apparent lack of a comparative systematic study does not make it possible, so far, to prepare a list of all metals and their compounds the toxicity and metabolism of which could be affected by selenium. On the other hand, with growing recognition of environmental problems connected with cadmium and mercury, it is understandable that the main attention has been focused on the interrelation of selenium with compounds of these two elements.

INTERACTIONS BETWEEN CADMIUM AND SELENIUM

Shortly after the recognition of the fact that parenteral administration of cadmium salts results in testicular necrosis (8) and after the first reports revealing the protective effect of zinc against cadmium toxicity (9), Kar (1) and later Mason (10, 11) called attention to the highly remarkable protective effect of small amounts of selenite in the same situation. In spite of the apparently specific relation of selenium to the spermatogenic pathway (12–16) it should be emphasized here that this protective action of selenite (or selenomethionine) is not confined to testicular effects of cadmium. Administration of selenium compounds also affords full protection against all the other known specific effects of cadmium related to reproduction, including the selective damage produced by cadmium in the nonovulating ovaries (17, 18), in the placenta (19, 20), or the selective toxicity specific for cadmium during the last period of pregnancy (20, 21). Selenium compounds were also shown to prevent the teratogenic effects of cadmium (22). Administration of selenium compounds is known to decrease the mortality of experimental animals given doses of cadmium lethal in the controls (20, 23). Selenium compounds should thus be considered highly effective in decreasing cadmium toxicity in general, and this conclusion seems to deserve further attention in studies on possible adverse biomedical effects of cadmium, including those related to hypertension.

However, results on cadmium-selenium interrelations were also highly inspiring in another respect. Ganther and Baumann (5) discovered several years ago that parenteral administration of cadmium salts markedly influenced selenium metabolism by decreasing respiratory excretion of volatile selenium compounds. From the aspect of mechanisms responsible for the specific effects of cadmium, including those related to reproduction, it seemed to be of interest to learn how far this effect on selenium metabolism was specific for cadmium only.

INTERACTIONS OF COMPOUNDS OF SELENIUM AND MERCURY

The fact that administration of mercuric salts had a similar effect on selenium metabolism (4, 24, 25) as cadmium, was of interest not only in relation to cadmium, but opened up a quite new aspect.

(a) Selenium compounds were highly effective against the toxicity of cadmium salts.
(b) Cadmium and mercuric salts had a similar effect on the selenium metabolism.
(c) It was possible therefore to envisage that selenium compounds could also protect against the toxicity of mercuric compounds.

Further experiments (24) have confirmed this prediction in full. The pathological changes typical for intoxication by inorganic mercuric salts did not appear in animals given selenium compounds (Fig. 1). Animals given a high dose of mercuric compounds lethal to controls survived when treated with compounds of selenium. Selenium compounds protecting experimental animals did not increase excretion, but, on the contrary, increased retention of mercury in the organism and at the same time changed its distribution within the body (27).

We have had the opportunity of reviewing work on these problems on several previous occasions (26–28), discussing at the same time possible mechanisms, including those involving SH groups, selenium, and corresponding metals (26, 27). Even if the underlying chemical reactions responsible

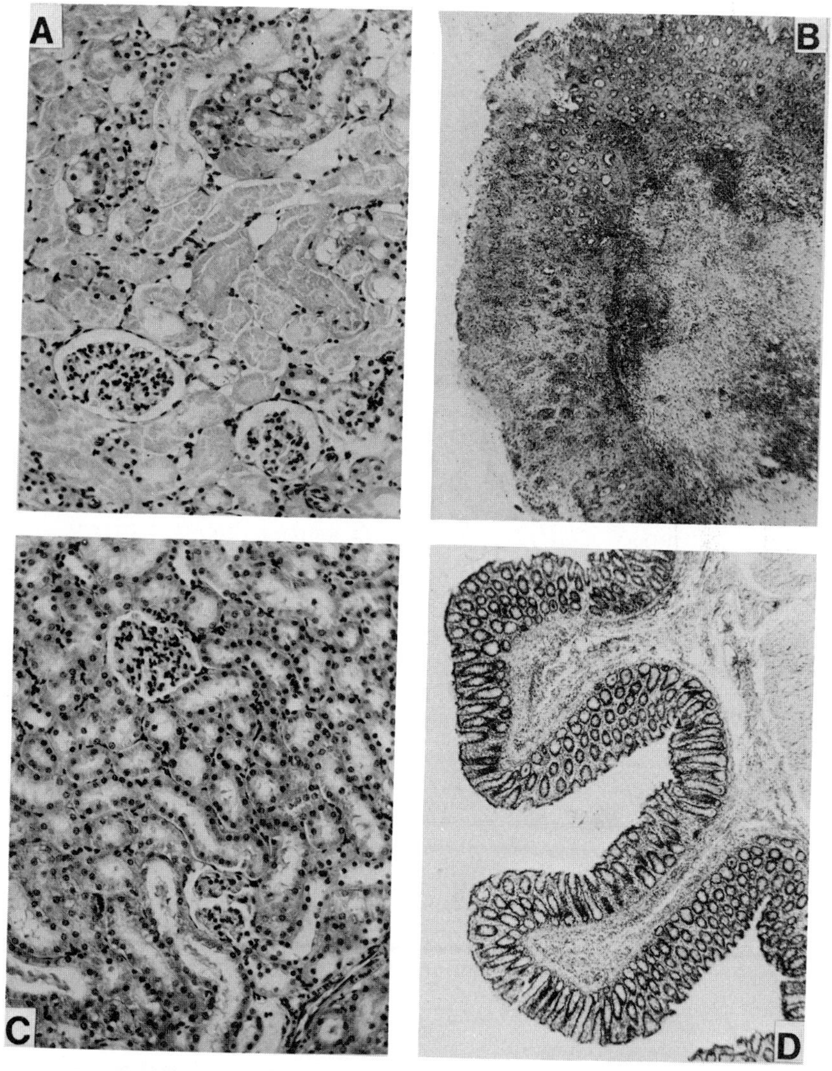

Fig. 1. Selenium and toxicity of mercuric compounds. Kidneys (A, C) and intestine (B, D) of adult female rats given mercuric chloride (A, B) or mercuric chloride and sodium selenite (C, D). Situation 27 hr after subcutaneous injections of 20 μmole/kg body wt. mercuric chloride (26 hr after the same dose of selenite). Experimental conditions otherwise analogous to those in the original publication (24). Formol, hematoxyline eosin stain. Magnification. A, C: 150 ×; B, D: 30 ×.

have not yet been elucidated, changes in binding of mercury and selenium (or cadmium and selenium) could well explain both their altered distribution and chemical reactivity. In this way we could understand not only the detoxifying effects mentioned above, but also the decreased passage of mercury into fetuses and into milk (26, 29) from mothers given selenium compounds and the decreased biological availability of selenium after administration of mercuric compounds (30).

In animals given cadmium or mercuric compounds, selenite or selenomethionine administration can increase the level of these metals in the blood and blood plasma by several orders of magnitude. This characteristic effect is dependent on the dose of selenium, and can be observed after administration of relatively small amounts of selenite and even after peroral administration (27).

The retention of radioactive cadmium in blood plasma of experimental animals, studied within the first few hours after parenteral administration of radioactive cadmium salts, can serve as a sensitive and quantitative method providing valuable information on the effect of selenium compounds. A significant increase in the retention of cadmium in blood plasma can be detected in this way in rats given 0.1 μmole of selenite per kilogram body weight intraperitoneally (Table 1). Four days exposure to drinking water with addition of 1.5 ppm selenite had a very pronounced effect (Table 2).

These amounts of selenium, of course, exceed those in which selenium operates as an essential element. Further research, using more elaborate techniques including dialysis, fractionation of blood-plasma proteins, incuba-

Table 1. The effect of selenite injection on cadmium retention in blood plasma[1]

Dose of sodium selenite (μmole/kg body wt)	No. of rats[2]	Percentage injected dose ^{115m}Cd in 1 ml blood plasma[3]
0	9	0.051 ± 0.008
0.1	9	0.088 ± 0.005
0.2	8	0.174 ± 0.010
0.5	8	0.434 ± 0.029
1.0	8	1.077 ± 0.048
5.0	8	3.176 ± 0.125

[1] $^{115m}CdCl_2$ (10 μmole/kg body wt) given subcutaneously, simultaneously with sodium selenite intraperitoneal injection.

[2] Adult female rats (fasted 24 hr before the injection).

[3] Values are mean ± s.e.

Table 2. The effect of selenite in drinking water on cadmium retention in blood plasma[1]

Drinking water	Percentage injected dose ^{115m}Cd in 1 ml blood plasma[2]
Without addition of selenite	0.043 ± 0.006
With addition of selenite (0.25ppm Se)	0.052 ± 0.005
With addition of selenite (1.5ppm Se)	0.360 ± 0.053

[1]Adult female rats were given water for four days with added sodium selenite. $^{115m}CdCl_2$ (10 μmole/kg body wt) was given subcutaneously 2 hr before decapitation and blood-plasma collection. Eight animals in each group.
[2]Values are mean ± s.e.

tion *in vitro,* etc., and also a selenium-deficient diet, should show the lowest limits of selenium intake which could affect blood-plasma cadmium. However, it is interesting to note that the level of selenium intake shown to have a significant effect on blood-plasma cadmium in these experiments fits in well with the doses of selenium intake shown recently to protect against chronic methylmercury toxicity (31).

All the experiments mentioned so far were acute, and inorganic salts of metals and selenium compounds were given in most cases by parenteral injection. From the point of view of environmental mercury pollution it is highly important that selenite was shown recently by Ganther and his co-workers to affect also chronic toxicity of dietary methylmercury (32). These highly remarkable results have been confirmed by several other laboratories very recently (33, 34).

It can be expected that some laboratories, with facilities for working with these highly neurotoxic, volatile compounds of mercury, already have the answers to some of the following related questions of basic importance. It would be highly desirable to learn if the administration of selenium compounds can affect mercury concentration, distribution, or chemical form in the brain of animals exposed to methylmercury, and if selenium compounds can prevent or alleviate more permanently the characteristic pathological changes resulting from neurotoxicity of this organic mercury compound. We also do not know if mercury in this form has a similar effect on selenium metabolism in the body as inorganic mercuric salts, and if selenium compounds can decrease the passage of mercury into fetuses and milk when mercury is given in the form of its methyl compound. Further work should show if the mechanism by which selenium protects against the toxicity of inorganic salts of mercury and methylmercury are similar.

The practical importance of these results and questions is underlined by recent reports indicating the existence of ecologically significant interrelations between mercury and selenium, existing also under normal conditions in nature. The paper by Ganther *et al.* (32), already mentioned above, revealed that the high mercury levels in tuna were accompanied by a high selenium content and that mercury in this case appeared to be less toxic. Another recent paper from the Netherlands reported that high selenium levels were also found to accompany high mercury levels in liver samples obtained from marine mammals; mercury and selenium concentrations in the tissues of whales and seals showed significant positive correlation (35).

SOME IMPLICATIONS OF THE INTER-RELATION BETWEEN SELENIUM AND MERCURY OR CADMIUM

All these results confirm the validity of certain conclusions discussed previously (26, 28); the fact that selenium compounds can prevent deleterious effects of compounds of mercury or cadmium, increasing simultaneously their retention in the body and their concentration in certain tissues, calls for selenium determinations when comparing the mercury or cadmium content in body tissues with its possible biomedical effects.

In addition to this apparently clear-cut conclusion, some other possible implications seem to deserve further attention. For example, decreased biological availability of selenium in the presence of these metals could be of interest in the light of the growing problems of mineral pollution; on the other hand, it could be questioned how far nutritional selenium deficiency could unmask the toxicity of certain metals in the organism. It would be interesting to study the question of optimal selenium intake and symptoms of selenium deficiency under conditions of controlled environment from the point of view of contamination with trace amounts of heavy metals.

Compounds of such metals as cadmium and mercury can serve as useful experimental tools in studies on selenium metabolism and function; cadmium has been known for many years as a factor inducing respiratory decline (36), and dietary selenite, but not vitamin E, may decrease mitochondrial swelling induced by cadmium or mercuric compounds *in vitro* (37). More recent research has revealed that cadmium or mercuric compounds are potent inhibitors of mitochondrial swelling caused by a combination of glutathione plus selenite.[1]

1. O. A. Levander, V. C. Morris and D. J. Higgs. 1973. Relationship between the selenium-catalyzed swelling of rat-liver mitochondria induced by glutathione (GSH) and the selenium-catalyzed reduction of cytochrome C by GSH. This symposium.

THE EFFECT OF MERCURIC COMPOUNDS ON THE TOXICITY OF METHYLATED SELENIUM METABOLITES

On previous occasions (27, 28), summarizing our knowledge and the resulting implications concerning the interrelations between mercury and selenium, we have had the opportunity of calling attention to the rather complex character of interactions between compounds of mercury and selenium. At the Grand Forks Symposium three years ago (27) we reported that in contrast to the good survival of females receiving selenite a few hours after the administration of mercuric salts, the administration of these compounds in reversed order resulted in high mortality of experimental animals, particularly of adult males. Further research has confirmed the suspicion that simultaneous exposure to mercuric compounds very markedly increases the toxicity of dimethylselenide and, in a similar way, the toxicity of trimethylselenonium ions. As was mentioned previously, the formation of these metabolites is inhibited by pretreatment with mercuric compounds. It is evident that the result is dependent on the form in which selenium is present in the organism.

Furthermore, the toxicity of these methylated selenium metabolites in our experiments was markedly higher in adult males than in females. Both factors, exposure to mercuric compounds as an environmental factor and male sex as a biological factor, were additive in character. As a result of this concomitant potentiation, dimethylselenide in the amounts corresponding to approximately 10 μg selenium per rat can induce lethal, characteristic symptoms in adult male rats exposed to mercuric compounds (Table 3). The dose inducing characteristic toxic effects is more than 10,000 times smaller than the LD_{50} reported for dimethylselenide by McConnell and Portman (38).

Table 3. The effect of mercuric salts on the toxicity of dimethylselenide in adult male rats[1]

	Dimethylselenide (μmole/kg body wt i.p.)	Mortality (percentage of 24 hr)
Mercuric chloride (1 μmole/kg body wt s.c.)	0	0
	0.5	30
	1	60
	5	100
Controls	1	0
	5	40

[1]Adult male rats (63 days postnatal age; 10 animals in each group).

This dose is still quite high in comparison with the amounts of methylated selenium metabolites, which can be present in the organism without excessive selenium intake. However, from the point of view of the sex-linked sensitivity to the acute toxicity of methylated selenium metabolites, it is not without interest that Schroeder (39) observed a higher mortality of male rats in comparison with females in chronic experiments with high dietary selenium.

When considering the interrelation between mercury and different compounds of selenium we should realize, of course, that the biological effect could be quite different when compounds of mercury and selenium interact in food or in the gastrointestinal content, or when they are present simultaneously within the organism. A decreased toxicity of selenite was found by Hill when given perorally with large amounts of mercuric salts (40).

Cadmium or zinc salts (27) did not potentiate dimethylselenide toxicity when injected in amounts equimolar to those of mercuric compounds which were sufficient to have a clear-cut effect. Further research should show if this difference is qualitative or only quantitative in character. It would be highly desirable to learn if organic compounds of mercury, including methylmercury, could have a similar effect on the toxicity of methylated selenium metabolites as shown previously for inorganic mercuric compounds.

SOME OTHER FACTORS AFFECTING TOXICITY AND METABOLISM OF METHYLATED SELENIUM COMPOUNDS

Exposure to mercuric salts or the presence of some quality connected with male sex are not the only factors which can increase the sensitivity to the toxic effects of methylated selenium metabolites. As reported recently elsewhere (28, 41), remarkable sensitivity to the toxic effects of dimethylselenide develops in the maternal organism during lactation. The sensitivity of lactating mothers to the toxicity of dimethylselenide can be abolished completely by artificial weaning, where it disappears gradually within 12–24 hr after removal of sucklings.

As reported previously, adult male rats are not only more susceptible than females to the toxicity of methylated selenium metabolites, but they also retain a higher proportion of ^{75}Se-selenium after the administration of ^{75}Se-dimethylselenide or ^{75}Se-trimethylselenonium salts. More recent experiments (41, 42) have provided an explanation for this sex-linked difference in the retention of methylated selenium metabolites in the organism. Renal excretion of trimethylselenonium ions (into which dimethylselenide can be converted in the organism) is under hormonal control of androgens (or anabolic steroids), which remarkably increase the retention of trimethylselenonium ions in the organism. Further research in progress should explain the underlying mechanisms responsible for this hormonal effect on selenium metabolites. It is interesting to note that it is the anabolic steroids which increase the retention of a trace-element metabolite; further research should show if this could be of physiological significance from the aspect of the biological action of selenium and the hormones involved.

However, it is already clear that the increased retention of methylated selenium metabolites cannot explain by itself the sex-linked difference in their toxicity. Castration of males abolished the sex-linked difference in their retention but not the difference in toxicity; on the other hand, sex-linked difference in the retention of selenium metabolites mentioned above is fully developed in 1-month-old rats, i.e., before the development of sex-linked differences in the toxicity of these selenium compounds. Lactation or exposure to mercuric salts, which increase the sensitivity to the toxicity of methylated selenium metabolites, are not connected with increased retention of these compounds in the organism. It thus seems that some other mechanisms must be operating here. It is evident that a better understanding of factors which could modify the toxicity of selenium metabolites in the organism would be of importance from the aspect of possible therapeutic or prophylactic exploitation of the remarkable potency of selenium compounds to detoxify compounds of mercury in the organism.

REFERENCES

1. Kar, A. B., Das, R. P., and Mukerji, B. 1960. Prevention of cadmium induced changes in the gonads of rat by zinc and selenium—a study in antagonism between metals in the biological system. *Proc. Natn Inst. Sci. India* 26 B(suppl.): 40.
2. Holló, Z. M., and Zlatarov, Sz. 1960. The prevention of thallium death by sodium selenate. *Naturwissenschaften* 47: 87.
3. Rusiecki, W., and Brzeziński, J. 1966. Influence of sodium selenate on acute thallium poisonings. *Acta Polon. Pharmac.* 23: 74.
4. Levander, O. A., and Argrett, L. C. 1969. Effect of arsenic, mercury, thallium, and lead on selenium metabolism in rats. *Toxic. Appl. Pharmac.* 14: 308.
5. Ganther, H. E., and Baumann, C. A. 1962. Selenium metabolism. I. Effects of diet, arsenic and cadmium. *J. Nutr.* 77: 210.
6. Diplock, A. T., Green, J., Bunyan, J., McHale, D., and Muthy, I. R. 1967. Vitamin E and stress. 3. The metabolism of D α-tocopherol in the rat under dietary stress with silver. *Br. J. Nutr.* 21: 115.
7. Diplock, A. T. 1970. Recent studies on the interactions between vitamin E and selenium. *In* C. F. Mills (ed.), *Trace Element Metabolism in Animals,* Livingstone, London, p. 190.
8. Pařízek, J., and Záhoř, Z. 1956. Effect of cadmium salts on testicular tissue. *Nature, Lond.* 177: 1036.

9. Pařízek, J. 1957. The destructive effect of cadmium ion upon the testicular tissue and its prevention by zinc. *J. Endocr.* 15: 56.
10. Mason, K. E., Brown, J. A., Young, J. O., and Nesbit, R. R. 1964. Cadmium–induced injury of the rat testis. *Anat. Rec.* 149: 135.
11. Mason, K. E., and Young, J. O. 1967. Effectiveness of selenium and zinc in protecting against cadmium-induced injury of the rat testis. *In* O. H. Muth (ed.), *Selenium in Biomedicine,* Avi Publishing Co., Westport, Conn., p. 383.
12. Gunn, S. A., Gould, T. C., and Anderson, W. A. D. 1967. Incorporation of selenium into spermatogenic pathway in mice. *Proc. Soc. Exp. Biol. Med.* 124: 1260.
13. Gunn, S. A., and Gould, T. C. 1967. Specificity of response in relation to cadmium, zinc and selenium. *In* O. H. Muth (ed.), *Selenium in Biomedicine,* Avi Publishing Co., Westport, Conn., p. 395.
14. McCoy, K. E. M., and Weswig, P. H. 1969. Some selenium responses in the rat not related to vitamin E. *J. Nutr.* 98: 383.
15. Burk, R. F., Brown, D. G., Seely, R. J., and Scaief, C. C. 1972. Influence of dietary and injected selenium on whole-body retention, route of excretion, and tissue retention of $^{75}SeO_3{}^{2-}$ in the rat. *J. Nutr.* 102: 1049.
16. Brown, D. G., and Burk, R. F. 1973. Selenium retention in tissues and sperm of rats fed a torula yeast diet. *J. Nutr.* 103: 102.
17. Kar, A. B., Das, R. P., and Karkun, J. N. 1959. Ovarian changes in prepubertal rats after treatment with cadmium chloride. *Acta Biol. Med. Germ.* 3: 372.
18. Pařízek, J., Oštádalová, I., Beneš, I., and Piťha, J. 1968. The effect of subcutaneous injection of cadmium salts on the ovaries of adult rats in persistent oestrus. *J. Reprod. Fert.* 17: 559.
19. Pařízek, J. 1964. Vascular changes at sites of oestrogen biosynthesis produced by parenteral injection of cadmium salts: the destruction of placenta by cadmium salts. *J. Reprod. Fert.* 7: 263.
20. Pařízek, J., Oštádalová, I., Beneš, I., and Babický, A. 1968. Pregnancy and trace elements: the protective effect of compounds of an essential trace element–selenium–against the peculiar toxic effects of cadmium during pregnancy. *J. Reprod. Fert.* 16: 507.
21. Pařízek, J. 1965. The peculiar toxicity of cadmium during pregnancy–an experimental "toxaemia of pregnancy" induced by cadmium salts. *J. Reprod. Fert.* 9: 111.
22. Holmberg, R. E., and Ferm, V. H. 1969. Interrelationships of selenium, cadmium, and arsenic in mammalian teratogenesis. *Archs Envir. Hlth.* 18: 873.
23. Gunn, S. A., Gould, T. C., and Anderson, W. A. D. 1968. Specificity in protection against lethality and testicular toxicity from cadmium. *Proc. Soc. Exp. Biol. Med.* 128: 591.
24. Pařízek, J., and Oštádalová, I. 1967. The protective effect of small amounts of selenite in sublimate intoxication. *Experientia* 23: 142.
25. Pařízek, J., Beneš, I., Babický, A., Beneš, J., Procházková, V., and Lener, J. 1969. Metabolic interrelations of trace elements. The effect of mercury and cadmium on the respiratory excretion of volatile selenium compounds. *Physiologia Bohemoslov.* 18: 105.

26. Pařízek, J., Beneš, I., Oštádalová, I., Babický, A., Beneš, J., and Pitha, J. 1969. The effect of selenium on the toxicity and metabolism of cadmium and some other metals. *In* D. Barltrop and W. L. Burland (eds), *Mineral Metabolism in Paediatrics,* Blackwell, Edinburgh, p. 117.
27. Pařízek, J., Oštádalová, I., Kalousková, J., Babický, A., and Beneš, J. 1971. The detoxifying effects of selenium. Interrelations between compounds of selenium and certain metals. *In* W. Mertz and W. E. Cornatzer (eds), *Newer Trace Elements in Nutrition,* Marcel Dekker, New York, p. 85.
28. Pařízek, J. 1972. Toxicological studies involving trace elements. A survey paper. *In Nuclear Activation Techniques in the Life Sciences* 1972, I.A.E.A., Vienna, p. 177.
29. Pařízek, J., Babický, A., Oštádalová, I., Kalousková, J., and Pavlík, L. 1969. The effect of selenium compounds on the cross-placental passage of ^{203}Hg. *In* M. R. Sikov and D. D. Mahlum (eds), *Radiation Biology of the Fetal and Juvenile Mammal,* U.S.A.E.C., Oak Ridge, Tenn., p. 137.
30. Pařízek, J., Oštádalová, I., Kalousková, J., Babický, A., Pavlík, L., and Bíbr, B. 1971. Effect of mercuric compounds on the maternal transmission of selenium in the pregnant and lactating rat. *J. Reprod. Fert.* 25: 157.
31. El-Begearmi, M. M., Goudie, C., Ganther, H. E., and Sunde, M. L. 1973. Attempts to quantitate the protective effect of selenium against mercury toxicity using Japanese quail. *Fed. Proc.* 32: 886.
32. Ganther, H. E., Goudie, C., Sunde, M. L., Kopecky, M. J., Wagner, P., Sang Hwan Oh, and Hoekstra, W. G. 1972. Selenium: relation to decreased toxicity of methylmercury added to diets containing tuna. *Science* 175: 1122.
33. Stillings, B., Lagally, H., Soares, J., and Miller, D. 1972. Effect of cystine and selenium on the toxicological effects of methyl mercury in rats. *Abstracts of Short Communications, 9th International Congress of Nutrition,* Mexico, D.F., p. 206.
34. Potter, S. D., and Matrone, G. 1973. Effect of selenite on the toxicity and retention of dietary methyl mercury and mercuric chloride. *Fed. Proc.* 32: 929.
35. Koeman, J. H., Peeters, W. H. M., Smit, C. J., Tjioe, P. S., and Goeij, J. J. M. 1972. Persistent chemicals in marine mammals. *TNO-nieuws* 27: 570.
36. Schwarz, K. 1965. Role of vitamin E, selenium, and related factors in experimental nutritional liver disease. *Fed. Proc.* 24: 58.
37. Morris, V. C., and Levander, O. A. 1972. Effect of dietary selenium, vitamin E and crude diet on spontaneous, heavy metal induced, or energy-linked swelling of rat liver mitochondria. *Fed. Proc.* 31: 691.
38. McConnell, K. P., and Portman, O. W. 1952. Toxicity of dimethyl-selenide in the rat and mouse. *Proc. Soc. Exp. Biol. Med.* 79: 230.
39. Schroeder, H. A. 1967. Effects of selenate, selenite and tellurite on the growth and early survival of mice and rats. *J. Nutr.* 92: 334.
40. Hill, C. H. 1972. Interactions of mercury and selenium in chicks. *Fed. Proc.* 31: 692.

41. Pařízek, J., Kalousková, J., Beneš, J., Babický, A., Pavlík, L., and Kopoldová, J. 1974. Some hormonal and environmental factors influencing selenium metabolism and action. *In* S. Basta and A. Chavez (eds), *Proceedings of the 9th International Congress of Nutrition,* Karger, Basel, in press.
42. Pařízek, J., Kalousková, J., Beneš, J., Pavlík, L., Babický, A., and Kopoldová, J. 1972. Sex-linked differences in the metabolism and biological effects of certain selenium metabolites. *Physiologia Bohemoslov.* 21: 424.

DISCUSSION

Poole (Dublin). With regard to my previous question to Dr. Mills, do you have anything to add to the possibility that borderline cadmium toxicity may be precipitating clinical copper deficiency in Ireland?

Pařízek. We have not worked on the antagonism between cadmium and copper. Our experiments with zinc and cadmium were mostly done by injection. Perhaps the interaction between cadmium and copper is more important at the site of absorption from the intestine. When we bypass the intestine we get a clear-cut interaction between cadmium and zinc. We were unable to protect the testes against the effects of injected cadmium by injecting copper as we could for zinc.

IRON METABOLISM AS A FUNCTION OF CHELATION[1]

JOHN P. CHRISTOPHER, JOHN C. HEGENAUER, and PAUL D. SALTMAN

Department of Biology, Revelle College, University of California, San Diego, La Jolla, California

CHEMICAL CONSIDERATIONS

The role of chelation in iron metabolism must begin with an appreciation of certain important aspects of the solution chemistry of iron. Foremost is recognition that iron in its two most prevalent oxidation states, ferrous and ferric, rarely, if ever, exists as the free ions under physiological conditions. Ferrous ions are readily oxidized to ferric by dissolved oxygen at a rate which is proportional to the square of the hydroxyl-ion concentration (1). Thus, Fe(III) is the preferred state under all but acidic and anaerobic conditions and in some heme complexes. The situation is further complicated by the low solubility at neutral pH of ferric ions (2), 10^{-17}M, which is 17 orders of magnitude less than that of ferrous ions (3), 10^{-1}M. Consequently, at other than very acid pH (less than 2), ferric ion hydrolyzes to form insoluble ferric hydroxide. Yet, it is possible to produce soluble, polynuclear, spherical particles (of specific composition) if the hydrolysis of ferric ions is conducted in the presence of stabilizing ligands such as citrate (4) or fructose (5). Strong chelators of iron such as nitrilotriacetic acid (NTA) or ethylenediaminetetraacetic acid (EDTA) are capable not only of depolymerizing ferric oxyhydroxide particles but also of preventing their initial formation. The resulting complexes are generally stable under physiological conditions and enhance the availability of the metal.

1. The research support from the National Institutes of Health, Grant No. AM12386-07 is gratefully acknowledged.

CHELATES AND IRON UTILIZATION IN ORGANISMS

All living organisms require iron and are faced with the problem of accumulating a sufficient amount of this metal from the environment. To meet this challenge, many biological systems evolved mechanisms for secreting chelates to solubilize the metal from insoluble minerals and mobilize the iron to the surface of the cell.

In the bacterial world, it has been shown that several species are capable of exuding low molecular weight, organic chelating agents into the environment when these organisms are placed under conditions of iron stress. For example, *Escherichia coli* appears to have evolved just such a system to insure an adequate supply of iron. Under conditions where the iron concentration is less than 22×10^{-19} M the following events apparently occur. There is induction of synthesis of 2, 3-dihydroxy-N-benzoyl-L-serine synthetase which initiates production of the serine derivative (6). Three molecules of the latter are then converted enzymatically to a compound called enterochelin. This is then secreted by the organism into the environment where it avidly binds iron as Fe(III). Once in the cell, a specific esterase hydrolyzes the complex, releasing the iron for utilization by the cell (7). Another example is provided by *Ustilago* which synthesizes and excretes a complex cyclic polypeptide containing hydroxamate-substituted amino acid side chains called ferrioxamines and ferrichromes (8). These are such strong chelating agents that they are capable of attacking the stainless-steel containers in which the organisms are grown. These and other examples indicate the importance of chelation with respect to iron absorption by the bacteria.

Similar behavior is exhibited by higher plants which grow in alkaline soils. Consideration of the aforementioned aspects of iron chemistry shows that the solubility of iron in these soils would be so low that very little of the metal could be mobilized by the roots. To counter this problem, the roots excrete organic acids, such as malonic and citric, into the soil (9). These acids serve to depolymerize and solubilize the ferric hydroxide so that the iron can be utilized by the plant. A striking example of such regulation is exhibited by the soy bean. Particular strains of soy bean, differing by a single allele from the normal, can excrete increased amounts of organic acids, allowing the plant to survive and grow well on alkaline soils. Soil humus also contains a variety of organic molecules which mobilize the iron and permit enhanced plant growth (10). Thus, as was true with bacteria, chelation is also of importance for iron absorption among plants. Enormous amounts of synthetic chelates of Fe(III) such as EDTA and related compounds are used as foliar sprays or soil additive to ensure optimal growth in agricultural crops.

Higher organisms are faced with similar problems of iron absorption. In this respect, it is necessary to consider not only mechanisms for absorption but also the varied interactions between iron and other dietary components. The latter aspect complicates the detailed description of iron absorption in animals. Several constituents of the normal diet are capable of sequestering iron in such a fashion as to render it nutritionally unavailable to the organism. Typical examples of this type of interaction are exhibited by phosvitin of eggs (11, 12), caseins of milk (13), and phytate salts of vegetables (14). Where the diet already contains a low level of iron, the net effect of the presence of such constituents is to place the animal in a condition of iron stress. Humans who consume clay become anemic because the ingested clay is quite efficient in binding the iron in a nutritionally unavailable complex.

MUCOSAL MODELS

Several mechanisms have been proposed for the regulation of iron absorption in animals. One of the first proposals, the 'mucosal-block' hypothesis (15, 16), stressed the role of apoferritin as a membrane-bound facilitating carrier in the mucosal cell. After initial reduction to Fe(II) in the lumen, the iron was reoxidized and bound by apoferritin to form ferritin which then facilitated transport from mucosa to serosa. Release by reduction was proposed at the serosal membrane where Fe(II) passed into the plasma and again became oxidized, then bound by transferrin. The apoferritin was regenerated and recycled to bind more Fe(III) at the mucosal membrane and to carry out the transport. However, many experimental facts are not consistent with such a proposal (17). It was not possible to show saturation of iron absorption over a 3000-fold range in iron dose. Studies under conditions of maximal ferritin concentrations in the mucosa of the small intestine revealed that iron was still absorbed. Also absorption has been shown to increase in patients with hemochromatosis even though there are high iron levels in the mucosa.

An active transport mechanism was proposed by Dowdle, Schachter, and Schenker (18) who utilized everted segments of the rat duodenum from the proximal region. Their *in vitro* experiments showed that the intestinal segments were capable of transporting radioiron from the mucosal to serosal surface against a concentration gradient. The system showed saturation kinetics in its capacity to transport iron, and inhibitors of reactions of oxidative metabolism eliminated transport. Jacobs, Bothwell, and Charlton (19) and Terato, Hiramatsu, and Yoshino (20) arrived at similar conclusions. The former used an isolated duodenal loop and monitored radioiron in the venous effluent. The absorption was decreased by anoxia and metabolic poisons. The latter employed ligated intestinal loops and showed that cycloheximide inhibited the mucosal uptake of iron.

However, Brown and Justus (21) and Pearson and Reich (22), both using everted gut loops, and Helbock and Saltman (23), employing short-circuited intestinal membrane preparations, were unable to show active transport of iron from mucosal to serosal surface. The inhibition observed is not significant enough to account for total iron uptake.

A third proposal (23) for the regulation and facilitation of iron absorption emphasizes the role of low and high molecular weight complexes of iron in maintaining the iron in a soluble form. In this model the iron must be maintained in soluble form, preferably of low molecular weight. The size and reactivity of the chelates play significant roles in both the rate of uptake from the intestine and the utilization within the animal.

EVIDENCE FOR THE CHELATION MODEL

As one aspect of a study on iron transport by rat intestine, Helbock and Saltman (23) reported the fraction of absorbed iron which had accumulated in a specific site as a function of various iron chelates. In the experiments rats were anesthetized and a segment of the intestine, 25 cm distal to the pyloric valve, was isolated by ligatures. Preparations of radio-iron complexes of ethylenediamine tetraacetic acid (EDTA), nitrilotriacetic acid (NTA), and citric acid were injected into the ligated segments. In a parallel experiment the chelates were introduced intravenously in a dose which was equivalent to that absorbed by the intestinal segment. After a 2-hr period the animals were sacrificed, and the tissues and organs removed. The radioactivity of each was determined with a whole-body counter. Negligible amounts of radioiron were detected in the remaining carcass. The results are shown in Table 1, expressed as percentage of total iron in all tissues counted.

Table 1. Organ distribution of ^{59}Fe as a function of chelate[1]

	Intestinal			Intravenous	
Organ	NTA	EDTA	Citrate	NTA	EDTA
Blood	57.0	37.4	76.7	58.4	11.2
Bladder	1.6	41.3	0.4	25.0	75.6
Liver	35.9	5.1	17.7	6.3	1.1
Kidney	4.7	15.8	3.4	8.6	11.7
Spleen	0.8	0.4	1.8	0.8	0.4

[1]All values reported as percentage of total iron in whole animal.

Marked differences are exhibited by the various chelates. Although Fe-EDTA was rapidly absorbed, it was also readily excreted and hence unavailable for utilization by the rat. However, after absorption of both Fe-NTA and ferric citrate, ^{59}Fe quickly found its way to the liver and spleen. Very little was eliminated by way of the kidney and bladder in these cases. The results obtained from the intravenous injection of iron showed essentially the same pattern of accumulation. This is consistent with the idea that these iron chelates may be transported intact across the mucosal cells.

Bates *et al.* (24) have also presented evidence for the facilitation of iron absorption by chelates of Fe(III). Preparations of several radioiron complexes were administered orally at two levels of iron using normal guinea pigs. Retention of ^{59}Fe was measured with a whole-body counter (Fig. 1). The fraction of iron retained after the plateau region had been reached, i.e., stabilization of iron loss, is shown in Table 2.

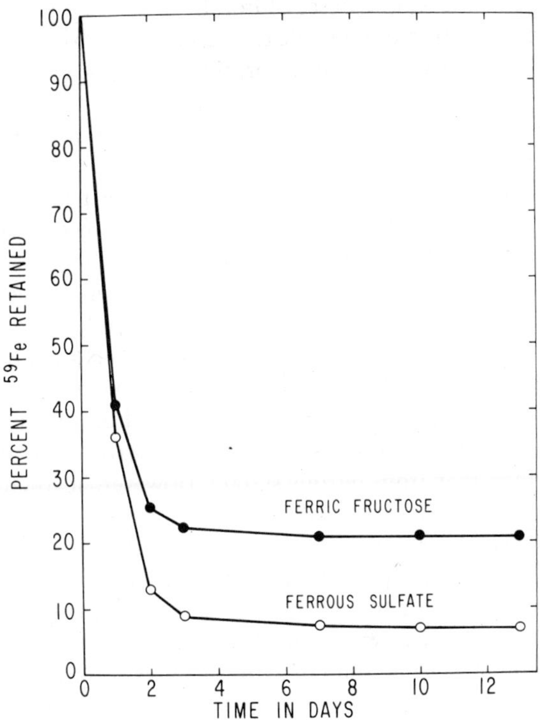

Fig. 1. Retention of radioiron from a single oral dose of ferrous sulfate or ferric fructose as a function of time. The animals were given 1.0 ml of a solution containing either 250 μM ferrous sulfate (14 μg Fe) or 500 μM iron-1M fructose (28 μg Fe). Each datum point represents the average of six animals.

Table 2. Uptake of iron as a function of chelating agent

Complex	Final percentage retention	
	Low iron[1]	High iron[2]
Ferric fructose	21 ± 5	9 ± 2
Ferric citrate	11 ± 3	6 ± 1
Ferric NTA	8 ± 6	6 ± 1
Ferrous sulfate	7 ± 3	6 ± 2
Ferric EDTA	7 ± 2	4 ± 1
Ferric gluconate	N. T.[3]	6 ± 2

[1]Low-iron oral dose, 28 μg Fe in 1.0 ml.
[2]High-iron oral dose, 280 μg in 0.5 ml.
[3]Not tested

At both the high and low doses of iron the fraction of dose retained was greatest for the ferric fructose complex. The only other chelate that resulted in significantly higher levels of iron absorption was ferric citrate at a low-iron dose. The remaining chelates were of similar efficacy in facilitating uptake.

The ability of ferric fructose chelate to enhance iron uptake in man was demonstrated in a study by Pollycove *et al.* (25). These experiments employed the double-labeling technique to compare the relative utilization of ferric fructose and ferrous sulfate within a single subject. Three different preparations of the fructose chelate were used in which the fructose:Fe(III) ratio was 2.5:1, 20:1, and 100:1. Subjects were given 50 mg of iron three times daily, 1 hr before each meal. Ferric fructose and ferrous sulfate were alternated daily throughout a 10-day period. One was labeled with ^{55}Fe, the other with ^{59}Fe. The latter was measured with a whole-body counter. Ratios of ^{55}Fe to ^{59}Fe were determined in the red cells. Iron utilization from the chelate which was prepared at a fructose to Fe(III) ratio of 2.5:1 was three-fold less than that from ferrous sulfate. However, at fructose to Fe(III) ratios of 20:1 and 100:1, iron absorption was essentially the same as and about two-fold greater than ferrous sulfate, respectively.

Avol *et al.* (26) have reported on the utilization of iron from the ferric citrate polymer by guinea pigs and mice. In both cases, the ability to induce ferritin synthesis in the liver was used as the criterion for evaluating absorption from the polymer. One group of guinea pigs was offered ferric citrate *ad libitum* in the drinking water; a second group was given a single injection of iron-dextran intraperitoneally; a third group received neither oral nor parenteral iron supplementation. The animals were followed for two weeks, and the results are shown in Fig. 2. After one week values for liver ferritin iron were similar for guinea pigs given iron either orally or parenterally. However, at the end of the two-week period, the group given ferric citrate exhibited a 40% increase in ferritin iron compared with iron-dextran and a seven-fold greater level than the controls.

Similar results were obtained with mice fed an unsupplemented low-iron diet and given Fe(III) citrate-sucrose solutions of varying iron content. The ferritin-iron levels were followed over a five-week period. It was observed that over a wide range of iron intake, ferritin iron continued to increase exponentially with no indication of reaching saturation in any of the groups. These studies suggest the use of ferritin-iron levels as a criterion for evaluating iron utilization in test animals. Hemoglobin values are no indication of utilization other than that adequate iron is present for hemoglobin synthesis. On the other hand, ferritin-iron levels are a sensitive indicator of the state of body Fe stores.

Experiments have been performed on the facilitation of iron absorption in baby pigs by ferric citrate.[2] Pigs were offered ferric citrate or ferrous sulfate as either a dry or liquid supplement. In addition, a single intramuscular injection of iron-dextran was administered at two days of age to a third group. Both the hemoglobin levels and the liver and spleen ferritin-iron

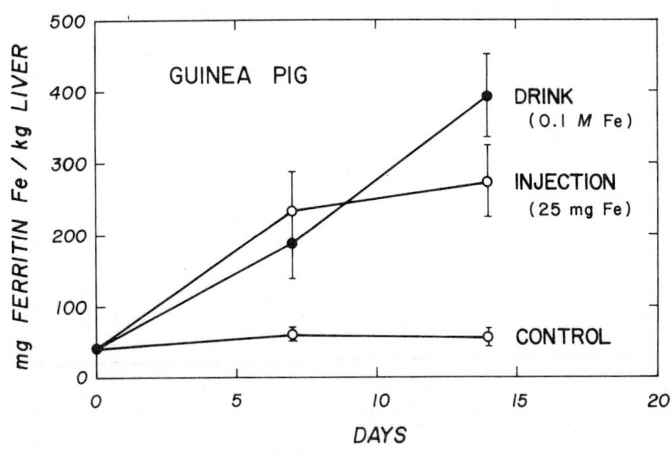

Fig. 2. Accumulation of iron into guinea pig liver ferritin following administration of oral iron *ad libitum* as the hydroxy-iron(III) citrate polymer, or of parenteral iron as a single injection of about 25 mg Fe as the iron-dextran complex. Each datum point represents the average of analyses of ferritin iron in duplicate samples from seven guinea pig livers; error bars are ± s.d.

2. We thank Dr. Duane Ullrey and his colleagues at Michigan State University for their valuable collaboration.

Table 3. Effect of oral iron supplementation for three weeks on iron storage in baby pigs

Treatment	No. of pigs	Ferritin iron				Total iron			
		Liver		Spleen		Liver		Spleen	
		(μg/g)	(mg)	(μg/g)	(mg)	(μg/g)	(mg)	(μg/g)	(mg)
Liquid supplement[1]									
Iron dextran I.M.	4	11.6	1.97	17.1	0.19	19.1	3.02	54.8	0.62
Ferric citrate	4	79.8	10.03	30.3	0.46	83.1	10.41	84.9	1.36
Ferrous sulfate	4	36.5	4.42	11.3	0.14	43.2	5.32	41.5	0.51
Dry supplement[2]									
Iron dextran I.M.	4	22.9	1.66	31.8	0.18	64.8	4.66	56.6	0.42
Ferric citrate	4	26.7	3.54	11.4	0.14	31.6	4.21	36.9	0.48
Ferrous sulfate	4	11.0	1.14	8.7	0.10	25.3	2.69	32.9	0.38

[1]Offered 100 ml of a 10% glucose solution providing 200 mg Fe (as ferric citrate or ferrous sulfate) daily per pig, except for 'iron dextran I.M.' pigs which received only a single intramuscular injection of 100 mg of iron from iron-dextran administered at two days of age.

[2]Offered 17.5 g of Michigan State University Pig Starter daily per pig containing 200 mg Fe (as powdered ferric citrate or ferrous sulfate), except to 'iron dextran I.M.' pigs which received only a single intramuscular injection of 100 mg of iron from iron-dextran at two days of age.

levels were used as criteria for evaluating the utilization of absorbed iron. The results based on the ferritin-iron values are summarized in Table 3.

As a liquid supplement, ferric citrate was better utilized than either ferrous sulfate or intramuscularly administered iron-dextran. Liver ferritin-iron levels for ferric citrate were about two-fold higher than those for ferrous sulfate and seven-fold greater than for iron-dextran. Also, ferric citrate resulted in a spleen ferritin-iron level which was higher than those obtained with the other two treatments.

When offered as a dry supplement, ferric citrate was still more efficiently utilized than ferrous sulfate. However, although there was no difference in liver ferritin iron between ferric citrate and iron-dextran, the latter resulted in about three-fold higher level of spleen ferritin iron. Similar experiments based on hemoglobin levels indicated essentially no difference between ferric citrate and ferrous sulfate, irrespective of the method of offering.

DISCUSSION

The ability of iron chelates to facilitate iron absorption has been studied in a number of experimental animals, including man. In general, all of these Fe(III) chelates used were as good as, if not better than, ferrous sulfate, the accepted standard. The results with human subjects agree with other studies on the effective uptake of ferric fructose (27, 28). In fact, this complex was consistently better utilized than ferrous sulfate. Other results on the relative effectiveness of Fe-EDTA and ferrous sulfate were in agreement with published data (29, 30). In addition, it should be realized that the data were obtained using several different criteria for evaluating iron absorption.

An important consideration in these studies is that the iron was maintained in a soluble form by the chelates. All of the chelates were prepared with Fe(III) and none with Fe(II), the form of iron that is considered to be most readily absorbed. If it were not for the presence of the chelate, Fe(III) would have existed as an insoluble precipitate under experimental conditions and thus have been relatively unavailable for absorption.

The two low molecular weight chelates, Fe-EDTA and Fe-NTA, exist predominantly as monomers, although dimers have been detected (31). As the pattern of site accumulation of absorbed iron suggests, these two chelates are probably transported intact across the intestinal mucosa. Once in the plasma, the step which determines the ultimate utilization by the rat is the rate of transfer of iron from the chelate to transferrin. Since the rate of iron transfer from the NTA-chelate is rapid (32), very little of this iron finds its way to the kidney and bladder. However, transfer from Fe-EDTA is such a slow process (33) that most of this chelate is eliminated before its iron can be donated to transferrin.

The other two complexes, ferric citrate and ferric fructose, present a somewhat more complicated situation. The nature of these two complexes is dependent on the ratio of ligand to Fe(III). At equimolar concentrations of Fe(III) and citrate, a polymer is formed at pH 8–9. This polynuclear complex has been shown to be spherical with a diameter of 70 Å and a calculated molecular weight of 2.1×10^5. The spheres apparently consist of an iron hydroxide core which is coated with bound citrate ions. Under conditions where excess citrate is present, ~30:1, polymer formation is effectively prevented. Instead, an anionic chelate is produced which probably consists of only two citrate ions per Fe(III) (34, 35). A similar situation exists with the ferric fructose system. At low fructose to Fe(III) ratios, a polymer of molecular weight 65,000 can be isolated. However, in the presence of excess fructose a chelate is formed having a maximum molecular weight of 2000 (36). These polymers are capable of existing in equilibrium with low molecular weight species, and hence the rate-determining step for absorption of the polymeric species is probably depolymerization.

These studies also focus attention on the feasibility of employing chelates to supplement the normal diet with iron. Iron-deficiency anemia is a problem of enormous proportions in the world today (37). Although nutritional anemia is more prevalent in underdeveloped countries, iron-deficiency anemia is still a serious public-health problem in the U.S.A. (38). An effective agent which does not harm the flavor or storage properties of foods is required to supplement diets with iron. The results obtained with ferric fructose and ferric citrate deserve consideration. These complexes have been shown not only to facilitate iron absorption but also to be more effective than ferrous sulfate. They are also to be recommended because of their low toxicity and the ability of the body to metabolize the complexing agent.

We believe that the understanding of the solution chemistry of iron, its interaction with low and high molecular weight ligands, and its mechanisms of transport and utilization within the body offers promise for amelioration of iron nutrition in both commercial animal husbandry and prophylactic and pharmaceutical treatments in humans.

REFERENCES

1. Goto, K., Tamura, H., and Nagayama, M. 1970. The mechanism of oxygenation of ferrous ion in neutral solution. *Inorg. Chem.* 9: 963.
2. Biedermann, G., and Schindler, P. 1957. On the solubility product of precipitated iron(III) hydroxide. *Acta Chem. Scand.* 11: 731.
3. Leussing, D. L., and Kolthoff, I. M. 1953. The solubility product of ferrous hydroxide and the ionization of the aquo-ferrous ion. *J. Am. Chem. Soc.* 75: 2476.

4. Spiro, T. G., Pape, L., and Saltman, P. 1967. The hydrolytic polymerization of ferric citrate. I. The chemistry of the polymer. *J. Am. Chem. Soc.* 89: 5555.
5. Bates, G., Hegenauer, J., Renner, J., Saltman, P., and Spiro, T. G. 1973. Complex formation, polymerization, and autoreduction in the ferric fructose system. *Bioinorg. Chem.*, in press.
6. Bryce, G. F., and Brot, N. 1971. Iron transport in *Escherichia coli* and its relation to the repression of 2, 3-dihydroxy-N-benzoyl-L-serine synthetase. *Archs Biochem. Biophys.* 142: 399.
7. O'Brien, I. G., Cox, G. B., and Gibson, F. 1971. Enterochelin hydrolysis and iron metabolism in *Escherichia coli*. *Biochim. Biophys. Acta* 237: 537.
8. Neilands, J. B. 1967. Hydroxamic acids in nature. *Science* 156: 1443.
9. Brown, J. C., and Tiffen, L. D. 1965. Iron stress as related to the iron and citrate occurring in stem oxidate. *Plant Physiol.* 40: 395.
10. Stevenson, F. J. 1972. Role and function of humus in soil with emphasis on absorption of herbicides and chelation of micronutrients. *Biol. Sci.* 22: 643.
11. Morris, E. R., and Greene, F. E. 1972. Utilization of iron of egg yolk for hemoglobin formation by the growing rat. *J. Nutr.* 102: 901.
12. Callender, S. T., Marney, S. R., Jr., and Warner, G. T. 1970. Eggs and iron absorption. *Br. J. Hemat.* 19: 657.
13. Peters, T., Jr., Apt, L., and Ross, J. F. 1971. Effect of phosphate upon iron absorption studied in normal human subjects and in an experimental model using dialysis. *Gastroenterology* 61: 315.
14. Sharpe, L. M., Peacock, W. C., Cooke, R., and Harris, R. S. 1950. Effect of phytate and other food factors on iron absorption. *J. Nutr.* 41: 433.
15. Hahn, P. F., Bale, W. F., Ross, J. F., Balfour, W. M., and Whipple, G. H. 1943. Radioactive iron absorption by gastrointestinal tract. Influence of anemia, anoxia, and antecedent feeding distribution in growing dogs. *J. Exp. Med.* 78: 169.
16. Granick, S. 1954. Iron metabolism. *Bull. N.Y. Acad. Med.* 30: 81.
17. Moore, C. V. 1961. Iron metabolism and nutrition. *Harvey Lect.* 55: 67.
18. Dowdle, G. B., Schachter, D., and Schenker, H. 1960. Active transport of ^{59}Fe by everted segments of rat duodenum. *Am. J. Physiol.* 198: 609.
19. Jacobs, P., Bothwell, T. H., and Charlton, R. W. 1966. Intestinal Fe transport using a loop of gut with an artificial circulation. *Am. J. Physiol.* 210: 694.
20. Terato, K., Hiramatsu, Y., and Yoshino, Y. 1973. Studies on iron absorption. II. Transport mechanism of low molecular iron chelate in rat intestine. *Am. J. Digest. Dis.* 18: 129.
21. Brown, E. B., and Justus, B. W. 1958. *In vitro* absorption of radioiron by everted pouches of rat intestine. *Am. J. Physiol.* 194: 319.
22. Pearson, W. N., and Reich, M. 1965. *In vitro* studies of ^{59}Fe absorption by everted intestinal sacs of the rat. *J. Nutr.* 87: 117.
23. Helbock, H. J., and Saltman, P. 1967. The transport of iron by the rat intestine. *Biochim. Biophys. Acta* 135: 979.

24. Bates, G. W., Boyer, J., Hegenauer, J. C., and Saltman, P. 1972. Facilitation of iron absorption by ferric fructose. *Am. J. Clin. Nutr.* 25: 983.
25. Pollycove, M., Saltman, P., Fish, M., Newman, R., and Tono, M. 1972. Ferric fructose absorption in man. *Abstr. Western Soc. Clin. Res.* Carmel, California.
26. Avol, E., Carmichael, D., Hegenauer, J., and Saltman, P. 1973. Rapid induction of ferritin in laboratory animals prior to its isolation. *Prep. Biochem.*, in press.
27. Brodan, V., Brodanová, M., Kuhar, E., Kordič, V., and Válek, J. 1967. Influence of fructose on iron absorption from the digestive system of healthy subjects. *Nutr. Dieta* 9: 263.
28. Davis, P. S., and Deller, D. J. 1967. Effect of orally administered chelating agents EDTA, DTPA, and fructose on radioiron absorption in man. *Aust. Ann. Med.* 16: 70.
29. Larsen, B. A., Bidwell, R. G. S., and Hawkins, W. W. 1960. The effect of ingestion of disodium ethylenediamine-tetraacetate on the absorption and metabolism of radioactive iron by the rat. *Can. J. Biochem. Physiol.* 38: 51.
30. Hodgkinson, R. 1961. A comparative study of iron absorption and utilization following ferrous sulfate and sodium ironedetate ("Sytron"). *Med. J. Austral.* 48: 809.
31. Gustafson, R. L., and Martell, A. E. 1963. Hydrolytic tendency of ferric chelates. *J. Phys. Chem.* 67: 576.
32. Bates, G. W., Billups, C., and Saltman, P. 1967. The kinetics and mechanism of iron(III) exchange between chelates and transferrin. I. The complexes of citrate and nitrilotriacetic acid. *J. Biol. Chem.* 242: 2810.
33. Bates, G. W., Billups, C., and Saltman, P. 1967. The kinetics and mechanism of iron(III) exchange between chelates and transferrin. II. The presentation and removal with ethylenediaminetetraacetate. *J. Biol. Chem.* 242: 2816.
34. Spiro, T. G., Pape, L., and Saltman, P. 1967. The hydrolytic polymerization of ferric citrate. I. The chemistry of the polymer. *J. Am. Chem. Soc.* 89: 5555.
35. Spiro, T. G., Bates, G., and Saltman, P. 1967. The hydrolytic polymerization of ferric citrate. II. The influence of excess citrate. *J. Am. Chem. Soc.* 89: 5559.
36. Spiro, T. G., and Saltman, P. 1967. Polynuclear complexes of iron and their biological implications. *Struct. Bonding* 6: 116.
37. WHO Scientific Group. 1968. Nutritional anemias. *World Hlth. Org. Techn. Rep. Sers.* 405: 5.
38. Ten-State Nutrition Survey 1968–1970. 1972. U.S. Dept. of Health, Education, and Welfare; Health Sciences and Mental Health Administration; Center for Disease Control; Atlanta, Ga. 30333 DHEW Publ. Nos. (HSM) 72: 8129.

DISCUSSION

Forth (Homburg). You will see that there are only minor differences between your scheme and the scheme which I will present later in this symposium. When the term 'mucosal block' was originally conceived, the discoverers wanted to express nothing but the fact that, as we would say today, iron absorption has saturable kinetics. Use of the term since that time, however, has often been misleading. We have spent or wasted nearly 20 years showing that, in fact, iron absorption occurs after a preceding dose. We now recognize that there is a mediated transport. There is also a diffusion of iron or iron chelates into the organism, and we know that a rather remarkable number of children are toxicated every year by iron doses. Therefore let us use this misleading term 'mucosal block' no longer.

Saltman. That is a very important comment. There are two important points that I'd like to comment on. Iron toxicity in small children, at least in America, results from dosing with ferrous sulphate; what is commonly unrecognized is that the use of certain chelating agents to administer iron to people will completely eliminate such toxicity. I drank a jug of 1 M chelated iron solution without noticeable effect. Secondly, there are endogenous ligands, which are elaborated in the gastric juices, which are very important. We have preliminary evidence that individuals with genetically related hemochromatosis absorb too much iron because of elaboration of low molecular weight chelators which combine with and solubilize iron.

A POSSIBLE ROLE FOR TRACE AMOUNTS OF SELENIUM AND VITAMIN E IN THE ELECTRON-TRANSFER SYSTEM OF RAT-LIVER MICROSOMES[1]

A. T. DIPLOCK

Department of Biochemistry, Royal Free Hospital School of Medicine, University of London, London, U.K.

At the first TEMA symposium at Aberdeen, Scotland, in 1969 I presented some preliminary results (1) which indicated that trace amounts of selenium could be shown to exist in rat-liver tissue in more than one oxidation state, and that the extent of reduction of the selenium was dependent on the presence in the diet of the animals of adequate amounts of vitamin E. The technique employed in these early experiments, which has since been published in detail (2), is shown schematically in Table 1. Rats were given three consecutive oral doses of $Na_2{}^{75}SeO_3$ and their livers subsequently fractionated by the method of Hogeboom (3). Liver subcellular fractions were then treated as shown in Table 1, and the tentative assignment of the designations selenide and selenite was given to the acid-labile and the Zn- and acid-labile selenium respectively. Typical results of an experiment with vitamin-E-deficient and vitamin-E-supplemented rats are given in Table 2. The reduced, acid-labile selenium, which has recently been identified with some certainty as selenide (4), is seen to be present in appreciable amounts only when vitamin E was present in the diet of the animals and the solutions, in which the liver subcellular organelles were isolated, were saturated with antioxi-

1. The collaboration of the following is gratefully acknowledged: J. A. Lucy, C. P. J. Caygill, E. H. Jeffery, A. S. M. GiasUddin, I. Cartwright and S. Silver. I am indebted to the following for financial support: The Medical Research Council, The Royal Society, The University of London, and Beecham Research Laboratories Ltd.

Table 1. Method of fractionation and of measuring different oxidation states of selenium

1. Rats given $Na_2{}^{75}SeO_3$ orally for 3 or 5 days
2. Liver homogenate (10%) prepared in 0.25M sucrose containing α-tocopherol (100 μg/ml) and mercaptoethanol (5mM)
3. Fractionated by method of Hogeboom (3)
4. Mitochondrial and microsomal fractions (1 ml) counted for ^{75}Se radioactivity following resuspension in 0.25 M sucrose containing the antioxidants
5. 1 ml conc. HCl added *rapidly* during rapid passage of O_2-free N_2 for 10 min (4 drops octanol present to prevent foaming)
6. ^{75}Se recounted and loss of counts attributed to *selenide*
7. 200 mg Zn dust added, N_2 passed for a further 10 min
8. ^{75}Se recounted and loss of counts attributed to *selenite*

dants. It was concluded from these results that the selenide was an important, potentially biologically active form of the trace element. The suggestion was made (1) that the selenide might form a part of the active center of a hitherto unidentified class of nonheme iron proteins, the stability of which might be dependent upon the presence, in the associated membrane structure, of α-tocopherol. The present account describes work that has been done in the intervening four years to follow up these early observations.

SUBCELLULAR FRACTIONATION STUDIES

The liver-cell fractionation used in the earlier experiments was simple, and little reliance could be placed on the homogeneity of the fractions. A carefully defined fractionation procedure was therefore devised using the M.S.E. BXIV titanium centrifuge rotor, and this is described in outline in Fig. 1. Using this technique the liver subcellular distribution of ^{75}Se derived from administered doses of $Na_2{}^{75}SeO_3$ was examined in rats given for 3–4 months a normal diet, a vitamin-E-deficient diet followed by a 5-day period during which 3 mg/day of α-tocopherol was given, and a vitamin-E- and Se-deficient diet (5); typical results for the vitamin-E-deficient rats and the deficient rats that were refed vitamin E are given in Figs. 2 and 3.

Fig. 2 shows the results that were obtained in the mitochondrial/lysosomal region of the gradient. The result (Fig. 2a) for the rats that were refed with vitamin E approximates that obtained with normal rats [see (5)] ; the mitochondrial marker enzyme succinate 2-p-iodophenyl-3-nitrophenyl-5-phenyl tetrazolium reductase is well separated from the lysosomal marker acid phosphatase, a fact that was confirmed by electron microscopy (5), and the distribution of mitochondrial marker suggests the presence of two popula-

Table 2. Apparent oxidation state of selenium in rat liver[1]

	Liver homogenized with antioxidants[2]				Liver homogenized without antioxidants[2]			
	Vit. E-deficient rats		Vit. E-supplemented rats		Vit. E-deficient rats		Vit. E-supplemented rats	
	Selenide	Selenite	Selenide	Selenite	Selenide	Selenite	Selenide	Selenite
Mitochondrial fraction (4)	25.5 ± 1.6[ac]	25.2 ± 1.5	36.1 ± 5.1[bc]	20.0 ± 13.7	10.7 ± 1.8[d]	25.2 ± 4.9	12.9 ± 1.1[d]	34.6 ± 6.7
Microsomal fraction (4)	30.2 ± 1.7[ca]	16.1 ± 2.3	43.1 ± 1.7[bc]	3.6 ± 2.1	16.8 ± 5.5[d]	21.5 ± 2.2	22.5 ± 5.8[d]	17.3 ± 3.9

[1]Twenty-four 2-month old vitamin-E-deficient male rats were given a vitamin-E- and selenium-deficient diet for two weeks and then were given a total of 68 μCi of ^{75}Se (36 μg Se) on three consecutive days, during which half the rats received a daily oral dose of 10 mg α-tocopherol. All rats were killed on the fourth day and the livers of lots of three rats were either homogenized and fractionated in sucrose solutions containing 100 μg α-tocopherol per ml and 5 mM mercaptoethanol, or without antioxidants. The particulate fractions were resuspended in 0.25 M sucrose solutions containing antioxidants, and the ^{75}Se radioactivity and the different oxidation states of selenium measured by the method given in Table 1.

[2]Results are expressed as percentage of the total ^{75}Se in the fraction (mean ± s.d.). Numbers in parentheses are numbers of observations. Values marked [a] differ significantly from those marked [b] ($P < 0.001$). Values marked [c] differ significantly from those marked [d] ($P < 0.01$).

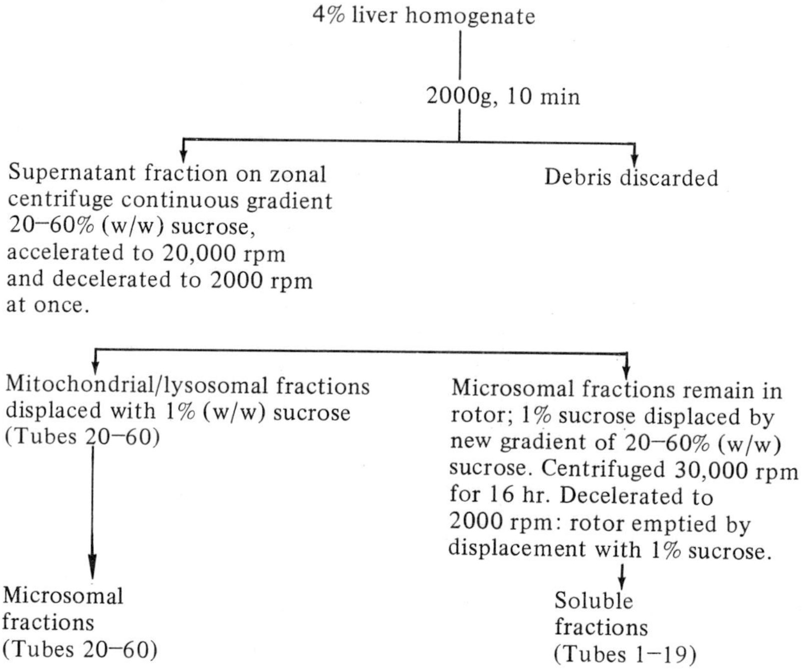

Fig. 1. Methiod of subcellular fractionation of rat liver using a zonal-centrifuge technique. [From Caygill, Lucy, and Diplock (5).]

tions of subcellular organelles of different density. The selenium and the selenide activity are largely located in the lighter of the two populations of mitochondria, and the selenide shows little association with the lysosomal region. In the vitamin-E-deficient rats (Fig. 2b) the mitochondria appear to have increased in density and the association with them of selenium and selenide is lost. The increase in density of the mitochondria appears to offer biochemical confirmation of the electron-microscopic observation of enlarged mitochondria in vitamin-E-deficient rats (5, 6), and the association of selenium (and particularly of selenide) with the mitochondria in normal rats appears to depend on the presence of vitamin E in the diet of the animals.

Fig. 3 shows the results for the microsomal region of the gradient. A technique involving labeling ribosomes with [^{14}C]orotate and the membranes with [^{3}H]choline was employed to establish that the contents of the tubes 32–44 were largely elements derived from the smooth endoplasmic reticulum, and that tubes 48–58 contained largely rough endoplasmic reticulum; this observation was confirmed by electron microscopy. Selenium and

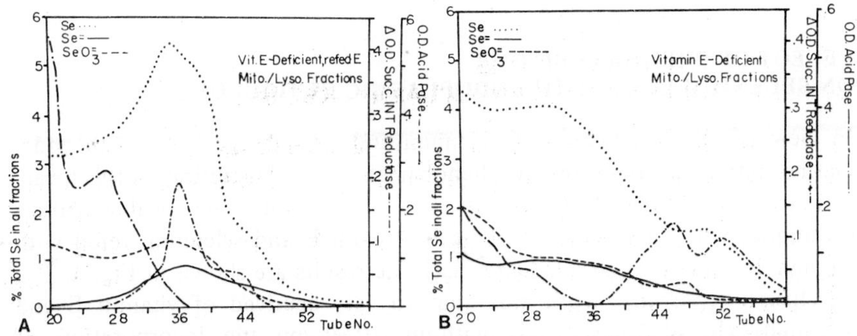

Fig. 2. Distribution of selenium, selenite, selenide, acid phosphatase activity, and succinate-2-*p*-iodophenyl-3-nitrophenyl-5-phenyl tetrazolium reductase activity in the mitochondrial and lysosomal fractions of rat liver. Rats were given the following dietary treatments: (a) vitamin-E deficient, refed vitamin E (3 mg/day for terminal five days), and (b) vitamin-E deficient. On the three days before they were killed all rats received 50 μCi (approximately 8 μg Se) of $Na_2{}^{75}SeO_3$ intraperitoneally. [From Caygill, Lucy and Diplock (5).]

selenide were found to be largely located in the smooth endoplasmic reticulum, and to a lesser extent in the rough reticulum. In vitamin-E-deficient animals the specificity of the localization of selenium and selenide was lost (Fig. 3b). Full details of this work can be seen in our publication (5), in which the conclusion was reached that selenide was particularly located in liver mitochondria and smooth endoplasmic reticulum and that its presence was dependent on the presence of vitamin E in the diet of the animals.

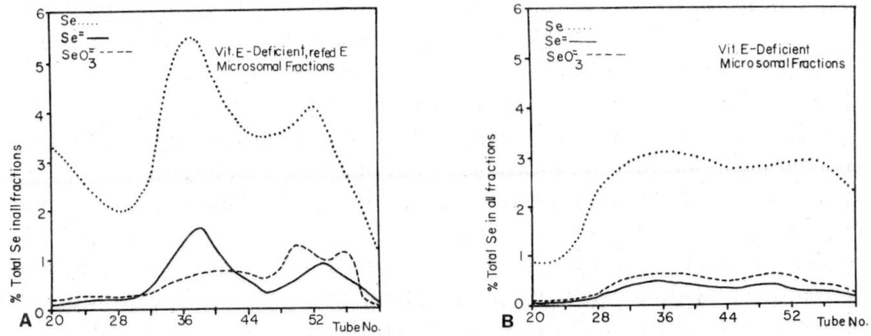

Fig. 3. Distribution of selenium, selenite, and selenide in the microsomal fractions of rat liver. Rats were given the following dietary treatments: (a) vitamin-E deficient, refed vitamin E (3 mg/day for terminal five days), and (b) vitamin-E deficient. On the three days before they were killed all rats received 50 μCi (approximately 8 μg Se) of $Na_2{}^{75}SeO_3$ intraperitoneally. [From Caygill, Lucy and Diplock (5).]

EFFECT OF PHENOBARBITONE ON SELENIUM IN SMOOTH ENDOPLASMIC RETICULUM

In view of the localization of selenium and selenide in the smooth endoplasmic reticulum, the effect of phenobarbitone administration on this localization was examined (7). Experiments were done, using the zonal-centrifuge technique, with rats deprived of both vitamin E and selenium, refed with vitamin E, and given an adequate diet. The results are shown in Fig. 4. The conclusion from this work was that in rats deprived of vitamin E and selenium, phenobarbitone was without effect on the incorporation of $Na_2{}^{75}SeO_3$ into the smooth endoplasmic reticulum, or on its conversion to $^{75}Se^{2-}$. When, however, vitamin E was given at the same time as the phenobarbitone and $Na_2{}^{75}SeO_3$, there was a very large increase in both the total ^{75}Se in the smooth endoplasmic reticulum and on the proportion of the ^{75}Se present as selenide.

MICROSOMAL NONHEME IRON

It was concluded from the above experiments, and others not reported here, that there may be a specific vitamin-E-dependent role for selenium and selenide in the liver smooth endoplasmic reticulum. This was thought to confirm our earlier suggestion (1, 2) that the selenide might form part of the active centre of a nonheme iron protein that might be functional in microsomal electron transfer. Experiments were therefore undertaken (8) to attempt to measure microsomal nonheme iron in rats given various diets. It was expected that, in view of the lability of the selenide, a protein containing selenium in this form would be exceptionally susceptible to oxidation *in vitro.* In each experiment, therefore, the livers were divided into two portions; one portion was used to measure nonheme iron by the conventional method (9), and the other was used for measurement of nonheme iron in solutions that at all times contained large amounts of antioxidants. The details of these experiments, and the results, are given in Table 3. It is evident that in the livers of normal rats little microsomal nonheme iron was measured by the conventional method; when, however, antioxidants were added, a quantity of nonheme iron similar to the amount of cytochrome P_{450} in livers of similar rats [(7) and Table 4], was observed. Phenobarbitone administration was without effect. In vitamin-E-deficient rats, and in vitamin-E- and Se-deficient rats, the nonheme iron was no longer oxidatively destroyed *in vitro.* Refeeding both vitamin E and selenium for three days resulted in a return to the normal situation; however, when vitamin E *alone* was refed, the nonheme

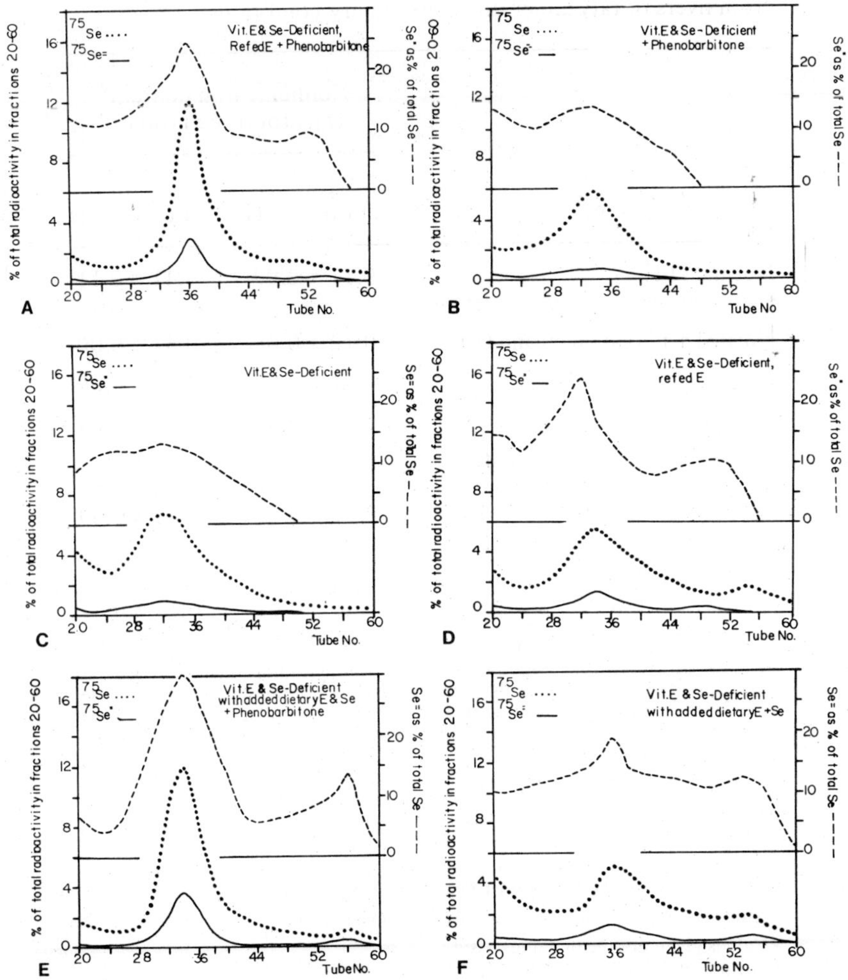

Fig. 4. The effect of phenobarbitone on the distribution of selenium, selenide, and selenide expressed as a percentage of the total selenium in the microsomal fractions of rat liver. Rats were given the following dietary treatments: (a), (c) vitamin-E and selenium-deficient diet, refed with vitamin E (3 mg/day for the terminal five days); (b), (d) vitamin-E- and selenium-deficient diet; (e), (f) vitamin-E- and selenium-deficient diet with added dietary vitamin E and selenium. Rats given treatments (a), (b), and (e) were also given phenobarbitone (1 mg/ml) in their drinking water for the terminal five days. On the terminal five days all rats received 50 μCi (approximately 8 μg Se) of $Na_2\ ^{75}SeO_3$ intraperitoneally. [From Caygill, Lucy and Diplock (5).]

Table 3. Nonheme iron content of microsomal fractions prepared from livers of rats under different dietary conditions[1]

Diet[2]	Nonheme iron content[3] (ng atoms/mg protein) Without antioxidants	With antioxidants
Normal (11)	0.11 ± 0.09	0.41 ± 0.11
Normal + phenobarbitone (6)	0.12 ± 0.08	0.52 ± 0.00
Vitamin-E-deficient (6)	0.51 ± 0.08	0.60 ± 0.04
Vitamin-E- and Se-deficient (6)	0.48 ± 0.00	0.49 ± 0.04
Vitamin-E- and Se-deficient, following 3 days refeeding with E and Se (6)	0.12 ± 0.01	0.42 ± 0.09
Vitamin-E- and Se-deficient, following 3 days refeeding with E *only* (6)	0.48 ± 0.01	0.49 ± 0.01

[1]Paired rat livers were divided into halves; a 10% homogenate was prepared from one half in 0.1 M phosphate buffer pH 7.4 containing 0.25 M sucrose, and 5mM mercaptoethanol, and 100 μg/ml α-tocopherol, and the other half was homogenized in the medium containing no antioxidants. Homogenates were centrifuged at 12,000 rpm for 12 min and 20 ml of supernatant fluid was layered to remove hemoglobin on a gradient consisting of 0.4 M sucrose (4 ml) and 0.3 M sucrose containing 0.23 M NaCl (10 ml). After centrifugation for 1½ hours, at 30,000 rpm, the microsomal pellet was resuspended in 0.01 M phosphate buffer pH 7.4, the protein content estimated, and the suspension diluted to 35 mg protein/ml. Nonheme iron was measured by the method of Doeg and Ziegler (9).

[2]Numbers in parentheses are numbers of observations.

[3]Values given are mean values ± s.d.

[From Caygill and Diplock (8)]

iron remained insensitive to oxidation. These results are indicative of the presence of hitherto undetected nonheme iron in rat-liver microsomes and appear to confirm our hypothesis. The presence of oxidant-insensitive nonheme iron in the deficient rat-liver microsomes is thought to be due to replacement by sulfide of the selenide in the nonheme iron-containing protein. Such a suggestion is not without experimental validation since it has been shown that sulfide and selenide are interchangable artificially in a nonheme iron protein which, in the native state, contains sulfide (9a).

Further work, employing electron spin resonance measurements of the non-heme iron, is in progress.

HYPOTHESIS FOR THE MODE OF ACTION OF VITAMIN E AND SELENIUM

The observations described above have been incorporated into a general hypothesis to attempt to explain the interactions between vitamin E and selenium as a trace element (10). On the basis of molecular model-building studies, it was suggested that the dietary interrelationship of vitamin E and polyunsaturated fatty acids may be due to the occurrence within normal membranes of specific complexes between the vitamin and some of the molecules of polyunsaturated phospholipids (Fig. 5). With the hydrophobic part of the membrane phospholipids interacting in this way with the phytyl side chain of tocopherol, the hydrophylic ring structure would be orientated with the more polar regions of the membrane so that the antioxidant function of the molecule is able to exert a protective role toward selenide-containing (and also possibly sulfide-containing) nonheme iron proteins. Synthetic antioxidants would be able to perform a similar function.

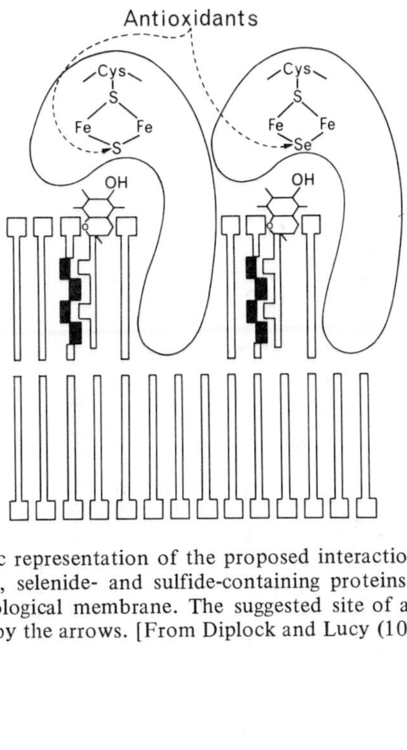

Fig. 5. A diagrammatic representation of the proposed interactions between vitamin E, synthetic antioxidants, selenide- and sulfide-containing proteins, and polyunsaturated phospholipids in a biological membrane. The suggested site of action of the synthetic antioxidants is shown by the arrows. [From Diplock and Lucy (10).]

Table 4. Level of cytochrome P_{450} and cytochrome b_5, and extent of reduction of cytochrome P_{450} by NADH and NADPH, in liver microsomes from normal and vitamin-E-deficient rats, during six days treatment with phenobarbitone[1]

No. of days treatment with phenobarbitone[2]	Cytochrome b_5 (nM/mg microsomal protein)	Cytochrome P_{450}			
		Level (nM/mg microsomal protein)	Reduction by dithionite (%)	Reduction by NADH (%)	Further reduction by NADPH (%)
Vitamin-E-supplemented rats					
1 (8)	0.57 ± 0.26	0.61 ± 0.07	100	8.7 ± 2.1	34.2 ± 5.5
2					
3 (7)	0.88 ± 0.21	1.52 ± 0.27	100	8.7 ± 3.1	41.3 ± 5.7
4 (8)	0.86 ± 0.13	1.58 ± 0.40	100	9.1 ± 2.2	43.8 ± 9.7
5					
6 (7)	0.72 ± 0.02	1.55 ± 0.32	100	7.6 ± 1.4	40.3 ± 4.3
Vitamin-E-deficient rats					
1 (8)	0.45 ± 0.04	0.73 ± 0.04	100	5.8 ± 1.3	39.7 ± 8.6
2 (8)	0.43 ± 0.05	1.09 ± 0.08	100	5.9 ± 2.4	49.0 ± 9.5
3 (7)	0.70 ± 0.02	1.54 ± 0.23	100	7.4 ± 3.4	41.5 ± 7.9
4 (8)	0.71 ± 0.13	1.63 ± 0.39	100	5.2 ± 0.3	43.8 ± 6.6
5 (8)	0.85 ± 0.03	1.50 ± 0.06	100	7.1 ± 1.7	48.3 ± 5.0
6 (7)	0.63 ± 0.03	1.89 ± 0.10	100	5.2 ± 2.8	38.8 ± 10.0

[1]Cytochrome P_{450} and b_5 were measured by the methods of Omura and Sato (11, 12). Reduction by the reduced pyridine nucleotides was measured by the method of Caygill, Diplock, and Jeffery (7).

[2]Numbers in parentheses are numbers of observations.

[From Caygill, Diplock and Jeffery (7)]

FUNCTIONAL STUDIES ON LIVER MICROSOMES

Some experiments have been carried out in an attempt to uncover a metabolic lesion in microsomes derived from the livers of vitamin-E-deficient rats, since it would be expected that a defect in a selenide-containing nonheme iron protein would have functional consequences. The experiments summarized in Table 4 show that the levels of cytochrome P_{450} and cytochrome b_5 are normal in vitamin-E-deficient rats and that their induction with phenobarbitone is unimpaired. The reduction of cytochrome P_{450} by NADPH and NADH was also found to be unaffected. However, it must be pointed out that these figures merely show the complete reduction of the cytochrome by the pyridine nucleotides, and give no indication of the rate of the reaction nor of the behavior of the system when drug metabolism is proceeding. Carpenter (15) has reported defects in drug metabolism in vitamin-E-deficient rats. We have examined the demethylation of aminopyrine and codeine by microsomes of vitamin-E-deficient rats and some results are given in Table 5. The apparent K_m of aminopyrine demethylation is greatly

Table 5. Demethylation of aminopyrine and codeine by liver microsomes from normal and vitamin-E-deficient rats[1]

	Aminopyrine		Codeine	
	K_m ($[M] \times 10^4$)	V_{max} (μmole HCHO/ g liver/hr)	K_m ($[M] \times 10^4$)	V_{max} (μmole HCHO/ g liver/hr)
Normal	2.56 ± 0.03^a	4.9 ± 0.6	1.25 ± 0.01	12.2 ± 0.7^a
Vitamin-E deficient	8.00 ± 0.16^b	4.4 ± 0.7	1.35 ± 0.03	7.3 ± 0.4^b
Vitamin-E and Se deficient	8.35 ± 0.15^b	3.6 ± 0.6		

[1]Liver microsomes were made from normal and vitamin-E-deficient rats and diluted to give a final concentration of 2 mg protein/ml in Tris-chloride buffer, pH 7.4. Aminopyrine and codeine demethylase activity was estimated by the method of Gilbert and Golberg (13), and the formaldehyde produced was measured by the method of Nash (14). Limiting substrate concentrations of 360 nmoles/ml to 20 μmoles/ml aminopyrine (12 observations of each) and 10 to 100 μmoles/ml codeine (8 observations of each) were used, and the apparent K_m and V_{max} of the system were obtained from reversed reciprocal (Lineweaver-Burke) plots of the values obtained and of the substate concentrations.

[2]Values marked [a] differ significantly from those marked [b] ($P < 0.01$).

altered in vitamin-E deficiency and the V_{max} remains unaffected; in a converse manner, the K_m for codeine demethylation is unaffected and the V_{max} is substantially altered. These results indicate the presence of a lesion in microsomal drug metabolism in vitamin-E deficiency, but it is not possible at present to identify it.

CONCLUSION

The results presented here establish that there is a vitamin-E-dependent role for selenium, and particularly for selenide, in two liver subcellular organelles that have an electron-transfer function, viz., mitochondria and smooth endoplasmic reticulum. Studies on the smooth endoplasmic reticulum showed that the uptake of selenium and its conversion to selenide was greatly increased when phenobarbitone was also administered. The suggestion that selenide forms a part of a microsomal nonheme iron protein finds considerable support in the results of the experiments in which nonheme iron was measured, and the defective drug demethylation observed in liver microsomes from vitamin-E-deficient rats may be a manifestation of a defect in a selenide-containing nonheme iron protein.

The scheme for microsomal electron transfer shown by Estabrook and Cohen [(16) and Fig. 6)] has been elaborated by Estabrook *et al.* (17). The oxidation-reduction cycle of cytochrome P_{450} during substrate hydroxylation is a two-electron process, probably involving the transient formation of an oxygenated intermediate of reduced cytochrome P_{450}. In this scheme (17) the first electron for the reduction of the [substrate-$P_{450}{}^{3+}$] complex is thought to be derived from NADPH via a dehydrogenase flavoprotein and 'X', which we suggest may be a selenide- and nonheme-iron-containing protein. Following binding of O_2 to the reduced complex, the second electron is derived from NADH via a second dehydrogenase flavoprotein and cytochrome b_5 to give a [substrate-$P_{450}{}^{2+}$–$O_2{}^-$] complex. Transfer of electrons from the NADPH dehydrogenase flavoprotein and cytochrome b_5 may also occur, so that 'X' is bypassed, and this situation may pertain in the livers of vitamin-E- and selenium-deficient rats where the function of 'X' would be expected to be impaired. The scheme of Estabrook *et al.* (17) is, however, only tentative and the assignment of the position 'X' to the hypothetical

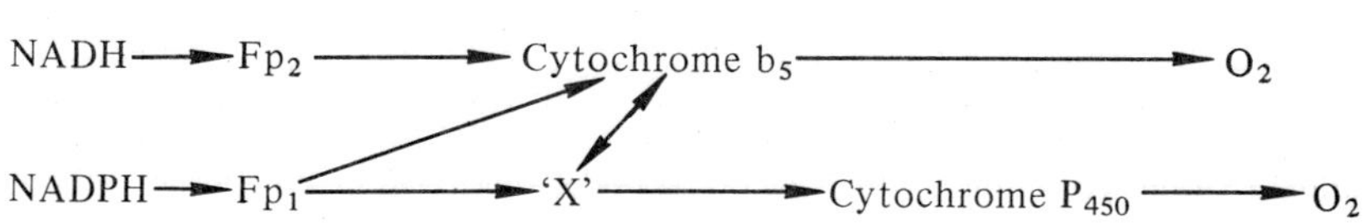

Fig. 6. Possible pathways of electron transfer in rat-liver microsomes. [From Estabrook and Cohen (16).]

iron-selenide protein may need reconsideration. Perhaps 'X' may prove to be a mobile carrier of electrons that functions between the cytochrome b_5 and the cytochrome P_{450} electron-transfer chains. The problem can only be resolved by detailed kinetic studies on microsomal electron transfer in microsome derived from vitamin-E-deficient and vitamin E- and selenium-deficient animals.

REFERENCES

1. Diplock, A. T. 1970. Recent studies on the interaction between vitamin E and selenium. *In* C. F. Mills (ed.) *Trace Element Metabolism in Animals,* Livingstone, Edinburgh, p. 190.
2. Diplock, A. T., Baum, H., and Lucy, J. A. 1971. The effect of vitamin E on the oxidation state of selenium in rat liver. *Biochem. J.* 123: 721.
3. Hogeboom, G. H. 1955. Fractionation of cell components of animal tissue. *In* S. P. Colowick and N. O. Kaplan (eds), *Methods in Enzymology,* Vol. 1, Academic Press, New York, p. 16.
4. Diplock, A. T., Caygill, C. P. J., Jeffery, E. H., and Thomas, C. 1973. The nature of the acid-volatile selenium in the liver of the male rat. *Biochem. J.* 134: 283.
5. Caygill, C. P. J., Lucy, J. A., and Diplock, A. T. 1971. The effect of vitamin E on the intracellular distribution of the different oxidation states of selenium in rat liver. *Biochem. J.* 125: 407.
6. Djaczenko, W., Grabska, J., Urbanowicz, M., and Pezzi, R. 1969. Peculiar mitochondrial forms in the liver parenchymal cells of rats kept on a vitamin E-deficient diet. *J. Microscopie* 8: 139.
7. Caygill, C. P. J., Diplock, A. T., and Jeffery, E. H. 1973. Studies on selenium incorporation into, and electron transfer function of, liver microsomes from normal and vitamin E-deficient rats given phenobarbitone. *Biochem. J.,* 126: 851.
8. Caygill, C. P. J., and Diplock, A. T. 1973. The dependence on dietary selenium and vitamin E of oxidant-labile liver microsomal non-haem iron. *FEBS Lett.,* 33: 172.
9. Doeg, K. A., and Zeigler, D. M. 1962. Simplified methods for the estimation of iron in mitochondria and submitochondrial fractions. *Archs Biochem. Biophys.* 97: 37.

9a. Tsibris, J. C. M., Namtredt, M. J., and Gunsalus, I. C. 1968. Selenium as an acid-labile sulphur replacement in putidaredoxin. *Biochem. Biophys. Res. Commun.* 30: 323.

10. Diplock, A. T., and Lucy, J. A. 1973. The biochemical modes of action of vitamin E and selenium: a hypothesis. *FEBS Lett.* 29: 205.
11. Omura, T., and Sato, R. 1964. The carbon monoxide-binding pigment of liver microsomes. I. Evidence for its haemoprotein nature. *J. Biol. Chem.* 239: 2370.
12. Omura, T., and Sato, R. 1964. The carbon monoxide-binding pigment of liver microsomes. II. Solubilization, purification and properties. *J. Biol. Chem.* 239: 2379.
13. Gilbert, D., and Golberg, L. 1965. Liver response tests. III. Liver enlarge-

ment and stimulation of microsomal processing enzyme activity. *Food Cosmet. Toxic.* 3: 417.
14. Nash, T. 1953. The colorimetric estimation of formaldehyde by means of the Hantsch reaction. *Biochem. J.* 55: 416.
15. Carpenter, M. P. 1972. Vitamin E and microsomal drug hydroxylations. *A. N.Y. Acad. Sci.* 203: 81.
16. Estabrook, R. W., and Cohen, B. 1969. Organisation of the microsomal electron transport system. *In* J. R. Gillette, A. H. Conney, G. J. Cosmides, R. W. Estabrook, J. R. Fouts, and G. J. Mannering (eds.), *Microsomes and Drug Oxidations,* Academic Press, New York, p. 95.
17. Estabrook, R. W., Hildebrandt, A. G., Baron, J., Netler, K. J., and Liebman, K. 1971. A new spectral intermediate associated with cytochrome P-450 function in liver microsomes. *Biochim. Biophys. Res. Commun.* 42: 132.

DISCUSSION

Petering (Milwaukee). How did you measure nonheme iron?

Diplock. The method depends on ethanol extraction of dithionite-reduced iron followed by reaction with bathophenanthroline.

Petering (Milwaukee). Instead of measuring iron, a typical way of doing this is to look for the labile sulfide, which I think would be more specific for nonheme iron proteins than measurement of the iron. Also, what was the average *g* value in your EPR spectrum, because this is also very diagnostic of the type of nonheme iron that you're talking about.

Diplock. In our preliminary work we have a *g* value which is about 2.03 and is rather different from the classical 1.94 of sulfide-containing nonheme iron proteins.

Petering (Milwaukee). A *g* value below 2 which is indicative of the nonheme iron proteins is also found when Se is substituted for S in these proteins, although there is some shift in the EPR signal.

Cary (Ithaca). Do you have any confirmatory data that the rats were actually Se deficient? It's very easy to deplete a rat of vitamin E but much harder to deplete it of Se.

Diplock. If you will accept classical respiratory decline as an indication of Se and vitamin-E deficiency, then yes we do have that confirmation; but if you mean did we measure Se in the tissues, then the answer is no, we did not.

Schwarz (Long Beach). It isn't really that difficult to deplete animals of Se. In our hands, under carefully controlled conditions and feeding a torula yeast diet, we find an almost complete disappearance of glutathione peroxidase from liver within 14 days in weanling rats, which is an indication that selenium has disappeared. Upon resupplementation with very little Se to the animal, the glutathione peroxidase returns to normal within two days.

ADAPTIVE CHANGES OF THE MILK XANTHINE OXIDASE AND ITS ISOENZYMES DURING MOLYBDENUM AND COPPER ACTION

V. V. KOVALSKY, I. E. VOROTNITSKAYA, and G. G. TSOI

Biogeochemical Laboratory of the V. I. Vernadsky Institute of Geochemistry and Analytical Chemistry, USSR Academy of Sciences, Moscow, USSR

The adaptation of organisms to alterations in environmental conditions may be accompanied by substantial changes in the properties of their enzyme systems which were conditioned by the processes of adaptation. In the Biogeochemical Laboratory of the Institute of Geochemistry of the USSR Academy of Sciences, Kovalsky and his collaborators have carried out systematic investigations of the regulation of purine-metabolizing enzymes. This work was started in 1955 by a study of the Ankavan molybdenic province of Armenia, where the correlation between the relatively high molybdenum content in the fodder of sheep and the activity of xanthine oxidase in the blood, liver, kidney, and milk of animals and in human blood was first established (1). Under the conditions present in the high molybdenum province, purine metabolism is increased and uric acid is accumulated in tissues and blood. In animals this excess may be metabolized by urate oxidase. In man this enzyme is probably lacking, and uric acid may be deposited in some tissues to produce signs of a podagric disease which is called endemic molybdenic podagra (2). When the molybdenum content in the diet is high, copper additions may slightly lower the action of molybdenum. However, in model experiments on white rats the presence of two enzyme-activity maxima have been established, one of which corresponded to an elevated molybdenum level in the diet at a normal copper content (Cu/Mo = 1.6) and the other to an elevated copper content at a normal molybdenum content (Cu/Mo = 16). Thus, during a study of the dependence of xanthine oxidase synthesis in liver and kidney tissue of rats on the molybdenum and copper ratio of the diet, we have observed the appearance of an adaptive form of a copper-containing xanthine oxidase (3).

In order to prove that a copper xanthine oxidase was being induced in the animals, it was necessary to determine the changes in copper and molybdenum content of the purified enzyme during the dietary influence of these metals. To solve this question, the xanthine oxidase of cow milk was examined. This allowed us to observe, within the same animal, changes in the properties of xanthine oxidase which were caused by changes in dietary copper and molybdenum content.

METHODS

The first purification of the enzyme was carried out by the method of Gilbert and Bergel (4) as modified by Kozachenko, Vartanyan, and Gonikberg (5). The final purification was achieved by gel filtration on Sephadex G-200. The enzyme purity was determined by the ratio between absorption at 280 nm and absorption at 450 nm, which for pure preparations was equal to 5.0–5.2. Xanthine oxidase activity was determined on SF-4 by measuring the rate of uric acid formation at 295 nm by the method of Avis, Bergel, and Bray (6). FAD in the purified enzyme was determined by the fluorometric method (7), molybdenum by Zn-dithiol (8), copper by diphenylcarbazon (9), and iron by diphenylphenanthraline (10).

RESULTS OF RESEARCH AND DISCUSSION

The composition of xanthine oxidase isolated from the milk of cows being maintained in the Moscow environment on a normal diet is indicated in Table 1. It can be seen from Table 1 that the content of FAD and metals in the prosthetic group of xanthine oxidase depends on the season of the year (physiological state of the animal and composition of the diet). The FAD:Fe:Mo ratio, expressed as molecule:atom:atom is the same in winter and summer at a ratio of 2:8:4, while in the spring this ratio is 2:8:2. The increase in FAD and decrease in molybdenum in the xanthine oxidase molecule during the spring period causes an increase of the FAD/Mo ratio. When the animals were kept on pasture feed (summer period) we have found copper in the xanthine oxidase of milk. Hart *et al.* (11) have also observed seasonal variability of the FAD/Mo ratio in milk xanthine oxidase, but in their experiments this ratio decreased in spring. They did not investigate the copper content in the xanthine oxidase. The presence of copper in the prosthetic group of purified milk xanthine oxidase is not widely recognized, and has been reported only by Roussos and Morrow (12). Comparison of our

Table 1. Seasonal changes in the content and ratio of FAD metals in xanthine oxidase isolated from cow milk[1]

Season	Cow no.	Specific activity (U/mg)	Protein (μg/mg)				Weight ratio				Mole ratio			
			FAD	Fe	Mo	Cu[2]	FAD	Fe	Mo	Cu	FAD	Fe	Mo	Cu
Winter	1	3.9	0.7	0.27	0.2	–	3	1	1	–	2	8	4	–
	2	4.8	0.6	0.27	0.2	–	3	1	1	–	2	8	4	–
	3	8.0	1.2	0.4	0.4	–	3	1	1	–	2	8	4	–
Spring	4	5.2	1.2	0.5	0.16	–	2.5	1	0.5	–	2	8	2	–
	5	9.8	1.1	0.4	0.25	–	3	1	0.5	–	2	8	2	–
	6	7.5	1.0	0.7	0.2	–	2	1	0.5	–	2	8	2	–
Summer	7	9.4	1.2	0.4	0.4	0.02	3.3	1	1	0.03	2	8	4	0.25
	8	7.2	1.5	0.5	0.4	0.01	3.0	1	1	0.03	2	8	4	0.25
	9	6.5	1.4	0.4	0.3	0.02	3.4	1	1	0.04	2	8	4	0.25

[1]The amount of Cu and Mo in the diet varied with the season and was as follows: winter (22 Nov. 1971 thru 7 Feb. 1971) 51.8 mg Cu and 12.6 mg Mo per day, Cu/Mo = 4.1; spring (22 March 1971 thru 18 April 1971) 54.7 mg Cu and 12.2 mg Mo per day, Cu/Mo = 4.5; summer (7 June 1971 thru 24 August 1971) 72.0 mg Cu and 13.8 mg Mo per day, Cu/Mo = 5.2

[2]Copper not detected.

previously obtained data with the observations of Roussos and Morrow led us to the assumption that the presence of copper in cow milk xanthine oxidase is an adaptive reaction dependent upon the copper content in the diet (Table 1).

In experiments of long duration, when cows were kept on diets enriched in copper, we have observed the appearance of copper and a decrease of molybdenum content in purified milk xanthine oxidase. When the cows were given additional molybdenum, a xanthine oxidase relatively enriched in molybdenum and not containing copper was obtained. Data on the influence of feeding additional copper and molybdenum on the content of these metals in the purified enzyme are given in Table 2. It should be noted that the experiments on feeding cows with additional molybdenum were carried out under conditions of lowered molybdenum content of milk xanthine oxidase, as this experiment was conducted on the same cows two weeks after the copper feeding was stopped. Therefore, an increase in molybdenum content of the xanthine oxidase molecule was observed only after prolonged feeding of the animals (27 days). In these experiments enzymatic adaptations were shown which were associated with a possible change of metal content in the enzyme, and may even have been associated with the substitution of one metal by another.

The variability of the ratio between iron, molybdenum, and copper in milk xanthine oxidase reported by various authors and found in our experiments is shown in Table 3. In our investigations xanthine oxidase activity was preserved in cases where molybdenum or copper was missing from the enzyme. It has been shown by Bayer and Voelter (16) that the activity was not altered when iron was artificially eliminated from the enzyme. On the basis of these data it may be considered that FAD is the main factor responsible for xanthine oxidase activity. Only its elimination leads to the loss of oxidizing properties by the enzyme and to the transformation of xanthine oxidase to xanthine hydrogenase. The possibility of retaining the properties of xanthine oxidase by FAD substitution for FMN has also been shown (17).

The iron, molybdenum, and copper contents of xanthine oxidase influence its activity. When FAD and iron increase in xanthine oxidase, the enzyme activity rises. This effect was more noticeable when there was also a simultaneous increase in molybdenum content or the appearance of copper in the enzyme (Table 1). When the animals were given excess molybdenum, the enzyme activity was raised as the iron and molybdenum content of the enzyme increased. When the animals were fed copper, and the iron and molybdenum content of the enzyme decreased, the enzyme activity remained at an average level. Possibly the reserve of metals in the tissues also influences

Table 2. Influence of Cu and Mo in the diet upon their content in purified milk xanthine oxidase[1]

Dietary treatment	Cow no.	Specific activity	Protein (μg/mg)			Weight ratio			Mole ratio		
			Fe	Mo	Cu[2]	Fe	Mo	Cu	Fe	Mo	Cu
Control[3]	3	5.8	0.24	0.28	–	1	1		8	4	
153.6 mg Cu/day for 26 days	3	4.5	0.1	0.11	0.01	0.5	0.5	0.05	8	4	1
Control for 14 days	3	3.8		0.1	–						
100 mg Mo for 27 days	3	8.7	0.2	0.2	–	1	1		8	4	
Control	4	3.1	0.2	0.2	–	1	1		8	4	
307.2 mg Cu for 26 days	4	4.3	0.1	0.09	0.03	0.5	0.5	0.25	8	4	2
Control for 14 days	4	2.5		0.06	–						
200 mg Mo for 27 days	4	9.6	0.25	0.2	–	1	1		8	4	

[1]Data obtained in the period 22 Nov. 1971 thru 7 Feb. 1971

[2]Copper not detected.

[3]The control diet contained 51.8 mg Cu and 12.6 mg Mo per day. The Cu and Mo added was in addition to this.

Table 3. Ratio of FAD:Fe:Mo:Cu in purified milk xanthine oxidase

Conditions when enzyme was isolated	Mole Ratio			
	FAD	Fe	Mo	Cu
Winter	2	8	4	–[1]
Ref. (13)	2	8	2	–[2]
Ref. (14)	2	8	2	–[2]
Ref. (15)	2	8	1	–[2]
Spring	2	8	2	–[1]
Pasture	2	8	4	0.25
Summer				
Cu feeding	2	8	4	1
Cu feeding	2	8	2	1
Cu feeding	2	8	4	2
Mo feeding	2	8	4	–[1]

[1]Cu not detected.
[2]Cu not determined.

the xanthine oxidase activity (3). Thus, when the molybdenum or copper content in the diet increased together with an increase of the content of these metals in tissues, the enzyme activity increased.

Along with functional adaptive changes of the properties of milk xanthine oxidase the existence of various forms in the tissues of different animals is possible. Such forms are probably the xanthine oxidase in chicken liver, with the FAD:Fe:Mo ratio being 1:8:1 (18), or in the mucous membrane of calf intestine, where the mole ratio of FAD:Fe:Cu is 1:17.4:4.2 and molybdenum is absent (19). The question about the possible participation of different metals in the regulation of activity and the existence of various forms of xanthine oxidase may be solved by the separation of purified xanthine oxidase into isoenzymes, and the determination of their specific activity and metal content.

We have separated purified xanthine oxidase by the method of disc electrophoresis in polyacrilamidic gel (7.5%) using a tris-glycinic buffer pH 8.2. The current intensity was 2 mA during the first 15 min and 5 mA for the remaining 60 min. The gels were stained with tetrazolic blue. Six isoforms of the milk xanthine oxidase from Kholmogor cows raised in the Moscow environment were revealed by the method of electrophoresis (Fig. 1). Electrophoretic mobility (relative to the FAD mobility) was 0.40 for KO_1, 0.337 for KO_2, 0.10 for KO_{4a}, 0.075 for KO_{4b}, 0.050 for KO_5, and 0.025 for KO_6. We succeeded in isolating KO_2 and KO_{4a} + KO_{4b}, but the other xanthine oxidase isoenzymes have not been isolated because of their small content. To determine copper and molybdenum in isoenzymes corresponding discs of isoenzymes were cut from the stained gels. The minimum amount of protein needed for copper determination is 0.5 mg and for molybdenum 1.0 mg. When the diet contained a normal molybdenum and copper content, KO_2 proved to be a molybdenic isoform of xanthine oxidase with a specific activity equal to 5.0 U/mg protein and KO_{4a} + KO_{4b} were found to be copper isoforms with a total specific activity of about 1.0 U/mg protein. When animals were fed copper there was a decrease or loss of molybdenum by the KO_2 isoenzyme, which did not influence its specific activity, and an increase of copper in the KO_{4a} + KO_{4b} isoenzymes leading to an increase in their specific activity. Molybdenum feeding resulted in the enrichment of KO_2 with molybdenum and in some cases a substantial increase of its specific activity, and also a decrease of copper content or its loss by the KO_{4a} + KO_{4b} isoenzymes (in one case a substitution of copper for molybdenum was

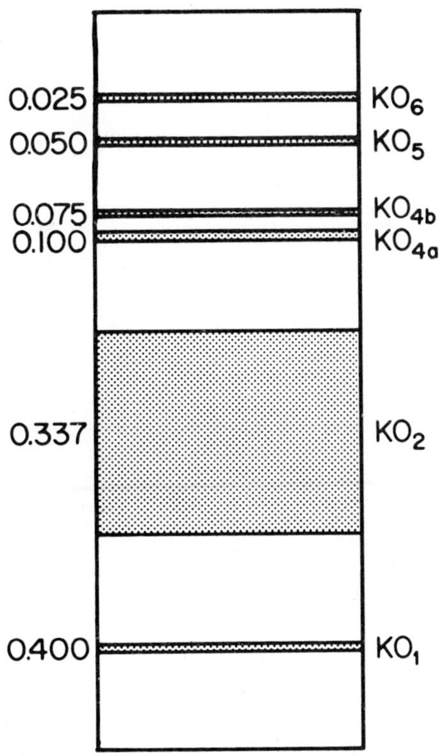

Fig. 1. Isoenzymes of milk xanthine oxidase obtained from Kholmogor cows maintained in Moscow.

Table 4. Cu and Mo content of cow milk xanthine oxidase isoenzymes[1]

Isoenzymes	Cu or Mo content of the diet	Cow no.	Specific activity	Mo[2] (μg/mg)	Cu[3] (μg/mg)
KO_2	Control	1	5.0	0.30	–
	51.8 mg Cu	2	5.1	0.30	–
	12.6 mg Mo	1	1.0	–	0.2
$KO_{4a} + KO_{4b}$		2	1.3	–	0.2
KO_2	154 mg Cu	3	5.0	–	–
	added to the diet	4	5.0	0.05	
$KO_{4a} + KO_{4b}$		3	2.0	–	1.8
		4	1.6	–	0.5
KO_2	307 mg Cu	3	5.7	–	–
	added to the diet	4	5.0	–	–
$KO_{4a} + KO_{4b}$		3	2.0	–	0.5
		4	2.0	–	1.1
KO_2		3	5.0	0.4	–
	100 mg Mo added	4	6.5	0.3	
	to the diet				–
KO_2	200 mg Mo added	3	5.0	0.37	–
	to the diet	4	10.0	0.3	–

[1]Data obtained in the period 22 Nov. 1971 thru 7 Feb. 1972.

[2]–Mo not detected.

[3]–Cu not detected.

noted). The results of the experiments are given in Table 4; the KO_2 (molybdenic) isoenzyme was in all cases more active than the $KO_{4a} + KO_{4b}$ (copper) isoferments.

CONCLUSION

The activity of xanthine oxidase is obviously determined by the ratio of molybdenum-containing or copper-containing isoenzymes which differ in their concentration and metal ratio. The possibility of changing the content of molybdenum and copper in xanthine oxidase and its isoenzymes or of their partial interchangeability under the influence of alterations in dietary content of these metals has been demonstrated for the first time in these studies.

SUMMARY

Changes in the properties of cow milk xanthine oxidase were studied at different levels of copper and molybdenum in the diet. The content of FAD and metals in the prosthetic group of xanthine oxidase which was isolated from the milk of cows maintained on a usual diet is determined by the composition of the diet and the physiological state of the organism. In winter and summer the mole ratio of FAD:Fe:Mo is 2:8:4 and in spring 2:8:2. When the animals are fed on pasture, copper is found in the xanthine oxidase of milk. At a normal copper and molybdenum ratio in the diet we succeeded in dividing the xanthine oxidase of milk into six electrophoretically different isoenzymes; KO_2 contained only molybdenum (specific activity 5.0 U/mg protein), and KO_{4a} + KO_{4b} were isoenzymes which contained copper (specific activity in total about 1.0). Feeding additional copper to the cows led to a decrease of molybdenum, or to its loss from the KO_2 isoenzyme, and to an increase in copper in the sum of the isoenzymes KO_{4a} + KO_{4b}. Feeding molybdenum salts increased the molybdenum content of KO_2 and decreased the copper content of isoenzymes KO_{4a} + KO_{4b}.

REFERENCES

1. Kovalsky, V. V., and Yarovaya, G. A. 1966. Biogeochemical provinces, enriched in molybdenum. *Agrokhimia* No. 8.
2. Kovalsky, V. V., Yarovaya, G. A., and Shmavonyan, D. M. 1961. Changes of purine exchange in man and animals under conditions of biogeochemical molybdenum provinces. *Zh. Obshchei Biologii* 22: No. 3.
3. Kovalsky, V. V., and Vorotnitskaya, I. E. 1969. The role of copper and molybdenum in the regulation of xanthine oxidase and urate oxidase properties. *Dokl. Akad. Nauk SSR* 187: No. 6.
4. Gilbert, D. A., and Bergel, F. 1964. The chemistry of xanthine oxidase. An improved method of preparing the bovine milk enzyme. *Biochem. J.* 90: 350.
5. Kozachenko, A. I., Vartanyan, L. S., and Gonikberg, E. M. 1971. *Biokhimia* 36: No. 1.
6. Avis, P. G., Bergel, F., and Bray, R. C. 1955. The chemistry of xanthine oxidase. I. The preparation of a crystalline xanthine oxidase from cow's milk. *J. Chem. Soc.* Pt II: 1100.
7. Burch, H. B. 1957. Fluorometric assay of FAD, FMN, and riboflavin. *In* S. P. Colowick and N. O. Kaplan (eds.), *Methods in Enzymology*, Vol. 3, Academic Press, New York, p. 960.
8. Marshall, N. J. 1964. Rapid determination of molybdenum in geochemical samples using dithiol. *Econ. Geol.* 59(1): 142.
9. Lapin, L. N., and Rish, M. A. 1957. *Trudy Tadzh. Uchitelsk. Inst.* 4.

10. Brumby, P. E., and Massay, V. 1967. Determination of nonheme iron, total iron and copper. *In* R. W. Estabrook and M. E. Pullman (eds.), *Methods in Enzymology,* Vol. 10, Academic Press, New York, p. 463.
11. Hart, L. I., McGartoll, M. A., Chapman, H. R., and Bray, R. C. 1970. The composition of milk xanthine oxidase. *Biochem. J.* 116: 851.
12. Roussos, G. G., and Morrow, B. H. 1966. Bovine intestinal xanthine oxidase: a metalloflavoprotein containing iron, copper and flavin adenine dinucleotide. *Archs Biochem. Biophys.* 114: 599.
13. Uozumi, M., Hayashikava, R., and Piette, L. H. 1967. ESR and crystallization studies of ironfree xanthine oxidase. *Archs Biochem. Biophys.* 119: 288.
14. Hart, L., and Bray, R. C. 1967. Improved xanthine oxidase purification. *Biochim. Biophys. Acta* 146: 611.
15. Richert, D. A., and Westerfeld, W. W. 1954. The relationship of iron to xanthine oxidase. *J. Biol. Chem.* 209: 179.
16. Bayer, E., and Voelter, W. 1966. Preparation of iron-free active xanthine oxidase. *Biochim. Biophys. Acta* 113: 632.
17. Komai, H., Massey, V., and Palmer, G. 1966. The preparation and properties of deflavo xanthine oxidase. *J. Biol. Chem.* 244: 1692.
18. Remy, C., Richert, D., Doisy, R., Wells, I. C., and Westerfeld, W. W. 1955. Purification and characterization of chicken liver xanthine dehydrogenase. *J. Biol. Chem.* 217: 293.
19. Roussos, G. G. 1967. Nucleic acids, part A. *In* R. W. Estabrook and M. E. Pullman (eds.), *Methods in Enzymology,* Vol. 12, Academic Press, New York, p. 5.

ZINC DEFICIENCY IN CHILDREN[1]

K. M. HAMBIDGE

Department of Pediatrics, University of Colorado Medical Center, Denver, Colorado

Current knowledge of the biochemical and physiological roles of zinc, together with the considerable data now available on the effects of zinc deficiency in animals, suggest strongly that the pediatric subject is of particular importance in any consideration of human zinc deficiency. For example, there is now abundant evidence that zinc is necessary for nucleic acid metabolism, protein synthesis, and cell growth (1); growth retardation is one of the earliest and cardinal features of zinc deficiency in all animal species studied (2). Several of the other major features of zinc deficiency are peculiar to the young growing animal, including delayed sexual maturation (2) and impaired learning ability with abnormal behavior patterns (3). Sandstead has reported that, apart from adequate calories and protein, there is a specific requirement for zinc for optimal brain growth of the rat in the crucial perinatal period (4). The possible implications of this observation and of the impaired learning ability of zinc-deficient growing rats to the development of the human infant have not been explored.

A discussion of zinc deficiency in the growing subject would be incomplete without reference to the fetus. Hurley (5–7) has drawn attention to the large range of major and minor congenital malformations which result from maternal zinc deficiency in the rat, and recently this has also been documented by Warkany and Petering (8). In the pregnant rat an inadequate dietary supply of zinc is not compensated by mobilization of zinc from liver or bone, even under teratogenic conditions of zinc deficiency (9). Thus the

1. This work was supported by Agricultural Research Service, United States Department of Agriculture, Grant No. 12-14-100-9941 (61). Also supported in part by United States Public Health Service Grant No. R01-AM-12432 from NIAMD, and by Grant No. RR-69 from the General Clinical Research Centers Program of the Division of Research Resources, National Institutes of Health.

fetus can be severely damaged by zinc deficiency, though there is little change in the total body content of zinc in the mother. The rat is notably susceptible to the effects of various teratogenic agents and there is no direct evidence at this time that any human congenital malformations are attributable to maternal zinc deficiency. However, Sever has recently quoted some rather scanty epidemiological evidence to suggest a possible teratogenic effect of zinc deficiency in man (10). Moreover, it has been calculated that some pregnant women in the U.S.A. have a marginal to deficient intake of zinc (11). Plasma zinc levels have been reported to be depressed in the third trimester of pregnancy (12–14), but plasma levels are not a reliable index of zinc deficiency during pregnancy as it is known that estrogens lower the plasma zinc without causing a total body depletion of this element (15, 16). However, recent studies on apparently healthy, well-nourished young women in Denver[2] have shown a significant decline in hair zinc in addition to plasma zinc levels between the 17th and 37th weeks of gestation (Table 1). It should also be noted that even at 17 weeks gestation mean plasma zinc levels were significantly lower than normal and 65% of individual subjects had plasma zinc levels more than 2 s.d. below the normal mean; of course, estrogen levels are already elevated at this stage of pregnancy, and it is not known to what extent these low plasma zinc levels may be physiological.

At one time the possibility that the human may be at risk from zinc deficiency was generally discounted, but attitudes have been changing rapidly during the last 10 years. Human zinc deficiency was first recognized by Prasad in 1961 (17); he and his colleagues demonstrated that severe zinc deficiency in adolescents and young adults was not uncommon in rural areas of Egypt and Iran. Dwarfism and absent sexual maturation were outstanding features attributable to this deficiency of zinc. Recent confirmation of the beneficial effect of zinc supplementation on growth in this syndrome has been provided by Halsted and his colleagues with controlled studies in Shiraz, Iran (18). An adequate review of the work of these researchers, though pertinent to the subject of zinc deficiency in children, is outside the scope of this paper, but the magnitude of their contribution should be emphasized. It should be noted that there is evidence that a similar zinc-deficiency syndrome may afflict young children (19); indeed, it appears that such a deficiency may begin in infancy in these countries, particularly in association with protein-calorie malnutrition (20). Typically, these infants after returning home receive a diet containing little animal protein; thus, it is quite possible that zinc deficiency commencing at this time will continue indefinitely. The potential public-health implications of this problem are considerable.

2. K. Michael Hambidge and William Droegemueller (unpublished observations).

Table 1. Comparison of zinc concentrations in plasma and hair of same subjects[1] during early and late pregnancy

	17 weeks gestation	37 weeks gestation	
Plasma zinc (μg/100 ml)	68.2 ± 2.0[2]	56.0 ± 2.1	$p < 0.005$[3]
Hair zinc (ppm)	171 ± 6	156 ± 7	$p < 0.05$

[1]Nineteen subjects were tested.
[2]Mean ± s.e.m.
[3]Paired comparison t test.

The possible significance of this research to the subject of human zinc nutrition in Western countries was not immediately apparent. Major etiological factors in the zinc deficiency observed in the Mid East, especially the large quantities of dietary phytate (21), were not relevant to the United States and elsewhere. Nevertheless, two cases of severe growth retardation and delayed sexual maturation, closely resembling the zinc-deficiency syndrome in the Mid East and responsive to zinc therapy, have recently been reported in the United States, one case complicating a chronic illness (22) and the other a case of regional enteritis and malabsorption (11).

It has been recognized that there is a spectrum of severity of zinc deficiency in adolescents in the Mid East; in less severe cases there is delayed puberty and moderate growth retardation only in otherwise healthy school boys (23), and there is now evidence to suggest that this degree of deficiency is not limited to the Mid East. Certainly, there has been a growing concern recently that at least a marginal degree of zinc deficiency (24), resulting in more subtle but nevertheless important symptomatology, may be relatively common in Western countries. In addition to suboptimal growth, these features may include delayed wound healing (25), impairment of taste acuity (26), poor appetite, and perhaps pica (27). Difficulties in diagnosis present a major problem in defining the incidence and effects of this deficiency. Ideally, this requires an adequately controlled study of zinc supplementation, or perhaps the development of practical methods for monitoring changes in zinc-dependent biochemical functions in the living human subject; in general, these criteria have not been completely fulfilled with studies in this country, and these limitations must be considered in evaluating the present evidence for zinc deficiency in children.

The limitations of plasma zinc determinations alone are illustrated by reference to some of the conditions which have been reported to be asso-

Table 2. Low plasma zinc levels—some associated conditions

Surgery and burns
Decubitus ulcers
Intestinal malabsorption
Parenteral hyperalimentation
Experimental (histidine)
High phytate diet
Cystic fibrosis
Kwashiorkor
Down's syndrome
Nephrosis
Malignant disease
Atherosclerosis
Corticosteroid therapy
Renal failure
Acute infection
Oral contraceptives

ciated with hypozincemia (Table 2) (12). In some instances, for example postoperative surgical patients with poor wound healing and favorable responses to zinc therapy (25), there is good evidence that the hypozincemia reflects an underlying zinc deficiency. In others, for example Down's syndrome (12), the significance of the low plasma zinc is uncertain at this time. In addition, there are certain circumstances in which it is known that the low plasma zinc level does not indicate a body depletion of this element, but rather a change in zinc metabolism and a redistribution of this element within the body; these conditions include acute infection (28) and estrogen therapy.

Despite these limitations, both blood plasma (24) and hair (29–35) zinc levels do reflect dietary zinc intake in animals; moreover, plasma and hair zinc levels were found to be low in zinc-deficient subjects in the Mid East (36–38), and to increase following zinc supplementation. Thus, if sufficient caution is used in interpretation of analytical results, both hair and plasma zinc determinations, considered in association with the clinical features, can be of great value in the detection of zinc deficiency.

A number of investigators have reported evidence of zinc deficiency in association with certain specific conditions and diseases, and some of these are of particular importance in children. Though large-scale studies are lacking, it is known that zinc deficiency may complicate malabsorption syndromes secondary, for example, to regional enteritis (11) and disaccharidase deficiency (39). Children with cystic fibrosis have been reported to have hypozincemia which is limited to those patients who also have growth retardation (12). Mean hair zinc levels are significantly lower than normal in patients with this disease.[3]

Zinc deficiency may also be one etiological factor contributing to failure to thrive in infancy, particularly when anorexia is a major feature. Currently, the evidence for this is based on experience with individual patients and a controlled study is lacking. One recent example was that of a female infant whose appetite started to deteriorate in the first year of life; her weight declined to the third percentile at 6 months. At 18 months of age she became severely anorexic and developed a bizarre form of pica, characterized by eating metallic objects for prolonged periods of time. This persisted until 2 years of age during which interval her height declined from the 25th to the 10th percentile. On the basis of a low hair zinc level (70 ppm) she was then treated with oral zinc supplementation which was followed within a few days by a striking improvement in appetite and cessation of pica; her hair zinc level and growth velocity increased over the next few months (27).

One special area of particular recent concern is that of children receiving either oral or parenteral artificial diets. The therapeutic uses of such diets have increased dramatically in recent years. There has been a recent brief report (40) on symptomatic zinc deficiency in an infant attributed to feeding with a lactose-free commercial formula, which was said to be severely deficient in zinc. No details of zinc levels in the diet or in the infant were given. An increasing number of centers are now routinely adding zinc to the intravenous formula of infants and children receiving prolonged total parenteral hyperalimentation, but this is not yet added deliberately by the companies manufacturing these formulas. Two examples of changes in plasma zinc concentrations that have been monitored in infants receiving total parenteral hyperalimentation are given in Table 3 (41). It will be noted that,

Table 3. Changes in plasma zinc levels in two infants receiving total parenteral hyperalimentation

	Plasma zinc (μg/100ml)		
Day	Infant I	Infant II	Supplemental i.v. zinc (μg/kg/day)
0	64	82	
30	45	40	
			20
40	58	64	
			40
44	114	84	

3. K. Michael Hambidge (unpublished observations).

in the absence of zinc supplementation, plasma zinc levels fell quite rapidly and subsequently increased following the addition of zinc to the intravenous formula.

There is now no doubt that zinc deficiency can occur in children with various disease states and in a number of abnormal circumstances. In combination these conditions are of considerable numerical importance. Furthermore, there is now evidence to suggest that the zinc nutritional status of some normal infants and children in this country is not optimal.

Recent studies (26) designed to determine the value of hair analyses in the assessment of human trace-mineral nutritional status included the collection of samples from 338 apparently normal subjects residing in Denver, Colorado, in order to determine the normal range of hair concentrations of a number of trace metals from birth to maturity. It should be emphasized that it was originally intended that this should be the control group. Therefore, it included only subjects from middle and upper socio-economic families; those living on low income diets, who are at particular risk from a deficient intake of zinc (11), were excluded, as were any subjects who had had a recent or chronic illness or recent surgery. There was no preselection of subjects with respect to height or weight. The frequency distributions for hair zinc concentrations of the 88 normal young adults, and the 132 children over the age of 4 years are shown in Fig. 1. The notable difference between the two groups was the cross-hatched area at the left-hand side of the histogram for the children. This area represents 10 of the 132 children who had hair zinc levels less than 70 ppm. These low levels were unexpected, being more than 3 s.d. below the mean for normal adults, and also being as low or lower than hair zinc levels of Egyptian adolescents with severe zinc deficiency. Accordingly, further information was obtained on these 10 subjects. Eight of the 10 children were found to have heights on or below the 10th percentile, including three with heights and weights below the third percentile. The high incidence of low growth percentiles was not explicable on the basis of parental heights. In addition, seven subjects had poor appetites according to maternal history, though it should be emphasized that these histories were inevitably largely retrospective and must be interpreted with caution. Despite the limitations of these data, the correlation between low hair zinc levels, poor appetite, and poor growth in these children from an apparently normal, unselected population suggested that all three features may be attributable to zinc depletion (poor appetite and retarded growth being two of the earliest and cardinal features of zinc deficiency in young animals).

Henkin had recently demonstrated that abnormalities of taste of unexplained etiology are frequently responsive to zinc therapy (42, 43). Accordingly, detailed investigations of taste acuity were undertaken in six of these children using a technique similar to that described by Henkin (44), and five

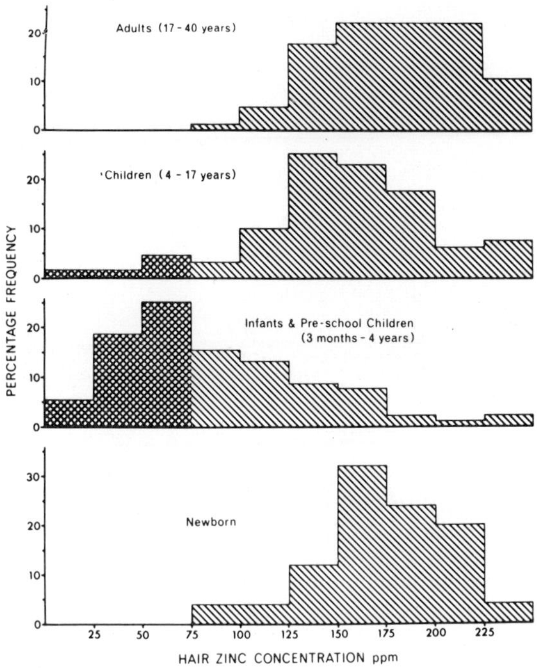

Fig. 1. Percentage frequency distribution of hair zinc concentrations (88 adults, 132 children aged 4–17 years, 93 subjects aged 3 months–4 years, and 25 newborn infants). Checkered area indicates percentage of subjects with hair zinc levels less than 75 ppm. [From (26).]

of the six were found to have evidence of objective hypogeusia (Fig. 2). Only recognition thresholds are shown in Fig. 2, but similar abnormalities were noted for detection thresholds. The shaded area at the bottom of the column for each quality of taste shows the normal range of recognition thresholds obtained by testing 14 control children, ages 5–10 years, whose hair zinc levels were within the normal adult range. Following 1–3 months of dietary zinc supplementation, with as little as 0.2 mg of elemental zinc per kilogram body weight per day, taste acuity had returned to normal in each case. The simultaneous increases in hair zinc concentrations of these subjects ranged from 49 to 107 ppm. Individual children also had encouraging improvement in appetite and in growth velocity, but it will require a larger, adequately controlled investigation to evaluate fully the effects of dietary zinc supplementation on appetite and growth in similar children with low hair zinc levels.

Recently, careful balance studies have been obtained on one boy, aged 7 years, with short stature who had the following evidence of zinc deficiency: a

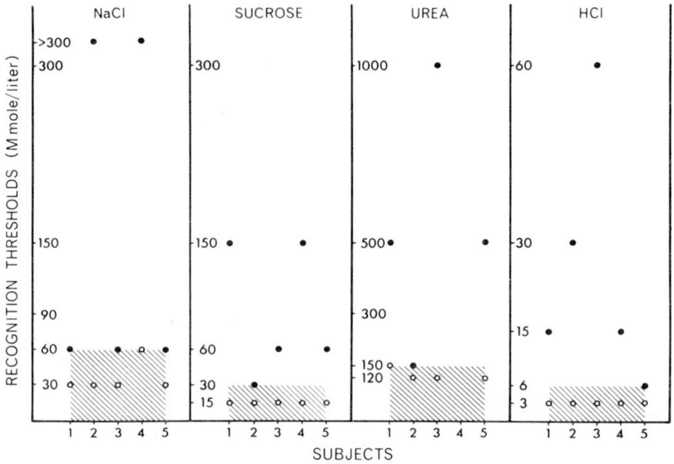

Fig. 2. Recognition thresholds of children with low hair zinc levels before (•) and after (○) dietary zinc supplementation. Shaded area indicates range of recognition thresholds for 14 normal children aged 5–10 years.

hair zinc concentration consistently below 70 ppm over a 9-month period (mean, 37 ppm); a consistently low plasma zinc level (mean, 63 μg/100 ml); a low 24-hr urinary excretion of zinc (mean, 102 μg/24 hr); a low plasma alkaline phosphatase; objective hypogeusia. His total caloric intake, which was low originally (1100 kcal/day), did not change significantly following 4 months of zinc therapy; however, there was a marked change in his selection of food items and, in particular, his average daily consumption of protein derived from animal meats increased from 13 to 34 grams. The characteristics of the initially poor appetite and the changes observed following zinc supplementation in this case tended to confirm the impression obtained from less detailed observations on the children in the original study, though daily caloric intake was more severely restricted in this case.

The original hair zinc studies also included samples from 25 newborn infants and 93 infants and children less than 4 years of age. As shown in Fig. 1, the newborn levels were very similar to those of the adults, but there was a pronounced fall in mean hair zinc levels during the first year of life and levels remained low until 4 years of age; the mean value for this entire group was only 87 ppm. Low hair zinc levels have also been reported by Strain in infants residing in Dayton, Ohio (45), and low plasma zinc levels in infancy have recently been reported by Henkin (46). The significance of these observations is not certain at this time. One possibility is that hair zinc levels are a function of age and that these low levels are physiological during infancy. However, similar low levels of hair zinc have not been found (in apparently normal children) in other countries, including India,[4] Peru (47), and Panama (48), which indicates that the low hair zinc levels are not directly related to age.

Another possible explanation for these observations is that there is an increased physiological requirement for zinc by more vital tissues and organs at this age, which decreases the zinc supply to the hair follicles but does not necessarily entail a significant depletion in the zinc supplied to other tissues. Forty-five percent of the entire group of subjects, aged 3 months to 4 years, had zinc levels less than 70 ppm, but the majority of these were apparently healthy with normal growth rates and no clinical features of zinc deficiency. It is possible that only an exceptionally low hair zinc level or a level that has been consistently low for several months reflects an underlying depletion of zinc that is sufficient to be of practical importance. In the older group of children in this study it is known that seven of the eight children with poor growth and low hair zinc levels had had a hair zinc concentration less than 70 ppm for at least 1 year. Moreover, the decline in growth percentiles of these children commenced before 2 years of age; therefore, if the poor growth was related to zinc deficiency, the latter must have started during infancy. It should also be noted that eight of the younger group (< 4 years) had extremely low hair zinc levels (< 30 ppm), and in this subgroup 75% had evidence of poor growth and poor appetite.

Several etiological factors in zinc deficiency have been identified (49). These include malabsorption of zinc due to dietary factors, especially phytate, or to abnormalities of the intestinal mucosa. Excessive loss of zinc from the body due to blood loss, excessive sweating, or hyperzincuria can also contribute. Many conditions, especially various catabolic states, have now been identified which result in excessive urinary loss of zinc (50), and this may be sufficient to cause a significant body depletion. Whether alterations in zinc metabolism without a reduction in total body zinc can lead to a depletion of any physiologically important pool of zinc is not known.

It has been stated that inadequate intake of dietary zinc is never a cause of human zinc deficiency. This concept requires modification, particularly at times of increased physiological requirements, including times of rapid growth. It is known (2) that young animals have relatively high dietary requirements for zinc compared with mature animals of the same species. Good data for the young human are scanty but it has been calculated (11) that some infants have diets marginal to deficient in zinc in this country.

It seems reasonable to conclude that the zinc nutrition of otherwise normal children with low hair zinc levels, impaired taste acuity, poor appetite, and low growth percentiles may have been at one end of a spectrum of

4. Vijay Kumar and K. Michael Hambidge (unpublished observations).

dietary zinc intake for growing children in this country. The most urgent requirement is for further research, including adequately controlled trials of zinc supplementation, to define the problem accurately and to determine what measures are needed to achieve optimal zinc nutrition in children.

REFERENCES

1. Mills, C. F., Quarterman, J., Chesters, J. K., Williams, R. B., and Dalgarno, A. C. 1969. Metabolic role of zinc. *Am. J. Clin. Nutr.* 22: 1240.
2. Underwood, E. J. 1971. Zinc. *In Trace Elements in Human and Animal Nutrition,* chap. 8, Academic Press, New York, p. 208.
3. Oberleas, D., Caldwell, D. F., and Prasad, A. S. 1971. Behavioral deficit with zinc deficiency. *Psychopharmac. Bull.* 7: 35.
4. Sandstead, H. H., Gillespie, D. D., and Brady, R. N. 1972. Zinc deficiency: effect on brain of the sucking rat. *Pediat. Res.* 6: 119.
5. Hurley, L. S. 1969. Zinc deficiency in the developing rat. *Am. J. Clin. Nutr.* 22: 1332.
6. Hurley, L. S., and Swenerton, H. 1966. Congenital malformations resulting from zinc deficiency in rats. *Proc. Soc. Exp. Biol. Med.* 123: 692.
7. Swenerton, H., Shrader, R., and Hurley, L. S. 1969. Zinc-deficient embryos: reduced thymidine incorporation. *Science* 166: 1014.
8. Warkany, J., and Petering, H. G. 1972. Congenital malformations of the central nervous system in rats produced by maternal zinc deficiency. *Teratology* 5: 319.
9. Hurley, L. S., and Swenerton, H. 1971. Lack of mobilization of bone and liver zinc under teratogenic conditions of zinc deficiency in rats. *J. Nutr.* 101: 597.
10. Sever, L. E., and Emanuel, I. 1973. Is there a connection between maternal zinc deficiency and congenital malformations of the central nervous system in man. *Teratology* 7: 117.
11. Sandstead, H. H. 1973. Zinc nutrition in the U.S.A. *Am. J. Clin. Nutr.* 26: 1251.
12. Halsted, J. A., and Smith, J. C., Jr. 1970. Plasma-zinc in health and disease. *Lancet* 1: 322.
13. Johnson, N. C. 1961. Study of copper and zinc metabolism during pregnancy. *Proc. Soc. Exp. Biol. Med.* 108: 518.
14. Henkin, R. I., Marshall, J. R., and Meret, S. 1971. Maternal-fetal metabolism of copper and zinc at term. *Am. J. Obstet. Gynec.* 110: 131.
15. McBean, L. D., Smith, J. C., Jr., and Halsted, J. A. 1971. Effect of oral contraceptive hormones on zinc metabolism in the rat. *Proc. Soc. Exp. Biol. Med.* 137: 543.
16. Halsted, J. A., Hackley, B. M., and Smith, J. C., Jr. 1968. Plasma-zinc and copper in pregnancy and after oral contraceptives. *Lancet* 2: 278.
17. Prasad, A. S., Halsted, J. A., and Nadimi, M. 1961. Syndrome of iron

deficiency anemia, hepatosplenomegaly, hypogonadism, dwarfism and geophagia. *Am. J. Med.* 31: 532.
18. Halsted, J. A., Ronaghy, H. A., Abadi, P., Haghshenass, M., Amirhakemi, G. H., Barakat, R. M., and Reinhold, J. G. 1972. Zinc deficiency in man. *Am. J. Med.* 53: 277.
19. Eminians, J., Reinhold, J. G., Kfoury, G. A., Amirhakimi, G. H., Sharif, H., and Ziai, M. 1967. Zinc nutrition of children in Fars Province of Iran. *Am. J. Clin. Nutr.* 20: 734.
20. Sandstead, H. H., Shukry, A. S., Prasad, A. S., Gabr, M. K., El Hifney, A., Mokhtar, N., and Darby, W. J. 1965. Kwashiorkor in Egypt: I. Clinical and biochemical studies, with special reference to plasma zinc and serum lactic dehydrogenase. *Am. J. Clin. Nutr.* 17: 15.
21. Reinhold, J. G., Lahimgarzadeh, A., Nasr, K., and Hedayati, H. 1973. Effects of purified phytate and phytate-rich bread upon metabolism of zinc, calcium, phosphorus, and nitrogen in man. *Lancet* 1: 283.
22. Caggiano, V., Schnitzler, R., Strauss, W., Baker, R. K., Carter, A. C., Josephson, A. S., and Wallach, S. 1969. Zinc deficiency in a patient with retarded growth, hypogonadism, hypogammaglobulinemia and chronic infection. *Am. J. Med. Sci.* 257: 305.
23. Ronaghy, H., Spivey Fox, M. R., Garn, S. M., Israel, H., Harp, A., Moe, P. G., and Halsted, J. A. 1969. Controlled zinc supplementation for malnourished school boys: a pilot experiment. *Am. J. Clin. Nutr.* 22: 1279.
24. Zinc in human nutrition. 1970. *Summary of Proceedings of a Workshop,* National Academy of Sciences, Washington, D.C.
25. Pories, W. J., and Henzel, J. H. 1967. Acceleration of wound healing in man with zinc sulphate given by mouth. *Lancet* 1: 121.
26. Hambidge, K. M., Hambidge, C., Jacobs, M., and Baum, J. D. 1972. Low levels of zinc in hair, anorexia, poor growth, and hypogeusia in children. *Pediat. Res.* 6: 868.
27. Hambidge, K. M., and Silverman, A. 1973. Rapid improvement in pica following dietary zinc supplementation. *Archs Dis. Childhood* 48: 567.
28. Beisel, W. R. 1974. The impact of infectious disease on trace element metabolism. This symposium.
29. Reinhold, J. G., Kfoury, G. A., and Thomas, T. A. 1967. Zinc, copper and iron concentrations in hair and other tissues: effects of low zinc and low protein intakes in rats. *J. Nutr.* 92: 173.
30. Miller, W. J., Powell, G. W., and Pitts, W. J. 1965. Factors affecting zinc content of bovine hair. *J. Dairy Sci.* 48: 1091.
31. Lewis, P. K., Hoekstra, W., and Grummer, R. H. 1957. Restricted calcium feeding during zinc supplementation for the control of parakeratosis in swine. *J. Anim. Sci.* 16: 578.
32. Macapinlac, M. P., Barney, G. H., Pearson, W., and Darby, W. J. 1967. Production of zinc deficiency in the squirrel monkey (*Saimiri sciureus*). *J. Nutr.* 93: 499.
33. Ott, E. A., Smith, W. H., Stob, M., and Beeson, W. M. 1964. Zinc deficiency syndrome in the young lamb. *J. Nutr.* 82: 41.
34. Strain, W. H., and Pories, W. J. 1966. Zinc levels of hair as tools in zinc

metabolism. *In* A. S. Prasad (ed.), *Zinc Metabolism,* Charles C Thomas, Springfield, Ill., p. 363.

35. Miller, W. J., Blackmon, D. M., Gentry, R. P., Powell, G. W., and Perkins, H. F. 1966. Influence of zinc deficiency on zinc and dry matter content of ruminant tissues and on excretion of zinc. *J. Dairy Sci.* 49: 1446.
36. Reinhold, J. G., Kfoury, G. A., Ghalambor, M. A., and Bennett, J. C., 1966. Zinc and copper concentrations in hair of Iranian villagers. *Am. J. Clin. Nutr.* 18: 294.
37. Strain, W. H., Steadman, L. T., Lankau, C. A., Berliner, W. P., and Pories, W. J. 1966. Analysis of zinc levels in hair for the diagnosis of zinc deficiency in man. *J. Lab. Clin. Med.* 68: 244.
38. Prasad, A. S., Miale, A., Farid, Z., Sandstead, H. H., Schulert, A. R., and Darby, W. J. 1963. Biochemical studies on dwarfism, hypogonadism, and anemia. *Archs Intern. Med.* 111: 407.
39. McMahon, R. A., Parker, M. L. M., and McKinnon, M. C. 1968. Zinc treatment in malabsorption. *Med. J. Aust.* 2: 210.
40. Moynahan, E. J., and Barnes, P. M. 1973. Zinc deficiency and a synthetic diet for lactose intolerance. *Lancet* 1: 676.
41. Greene, H. L., Merenstein, G., Hambidge, M., Sauberlich, H. E., and Herman, Y. F. 1973. Trace elements and vitamins (V) in total parenteral nutrition (TPN). *Pediat. Res.* 7: 109.
42. Schechter, P. J., Friedewald, W. T., Bronzert, D. A., Raff, M. S., and Henkin, R. I. 1972. Idiopathic hypogeusia: a description of the syndrome and a single-blind study with zinc sulfate. *In* C. C. Pfeiffer and J. R. Smythies (eds.), *International Review of Neurobiology,* Suppl. 1, Academic Press, New York, p. 125.
43. Henkin, R. I., Schechter, P. J., Hoye, R., and Mattern, C. F. T. 1971. Idiopathic hypogeusia with dysgeusia, hyposmia, and dysosmia: a new syndrome. *JAMA* 217: 434.
44. Henkin, R. I., Gill, J. R., and Bartter, F. C. 1963. Studies on taste thresholds in normal man and in patients with adrenal cortical insufficiency: the role of adrenal cortical steroids and of serum sodium concentration. *J. Clin. Invest.* 42: 727.
45. Strain, W. H., Lascari, A., and Pories, W. J. 1966. Zinc deficiency in babies. *Proceedings of VII International Congress of Nutrition,* Vol. 5, Pergamon Press, Oxford, p. 759.
46. Henkin, R. I., Schulman, J. D., Schulman, C. B., and Bronzert, D. A. 1973. Changes in total, nondiffusible, and diffusible plasma zinc and copper during infancy. *J. Pediat.* 82: 831.
47. Bradfield, R. B., Yee, T., and Baertl, J. M. 1969. Hair zinc levels of Andean Indian children during protein-calorie malnutrition. *Am. J. Clin. Nutr.* 22: 1349.
48. Klevay, L. M. 1970. Hair has a biopsy material: I. Assessment of zinc nutriture. *Am. J. Clin. Nutr.* 23: 284.
49. Halsted, J. A. 1970. Human zinc deficiency. *Trans. Am. Clin. Clim. Assn.* 82: 170.
50. Fell, G. S., Cuthbertson, D. P., Morrison, C., Fleck, A., Queen, K., Bessent, R. G., and Husain, S. L. 1973. Urinary zinc levels as an indication of muscle catabolism. *Lancet* 1: 280.

DISCUSSION

Oberleas (Detroit). These studies bring up what I think is an important point. In animal studies, we insist that our control groups be supplemented with the nutrient which we are studying. Yet, in human studies, we take any ambulatory population and consider them normal and use them as a standard. What we need to do is to take a group of people, supplement them with zinc, and then determine hair zinc concentration, plasma zinc concentration, and other parameters in this population and use that as a standard. Even the Iowa growth chart represents an unsupplemented population, and I think if somebody had a couple of thousand 5-year-old children and gave them a zinc supplement, we would find that the Iowa growth chart is not really a good standard.

Mills (Aberdeen). Is there any evidence of changes in the bacterial population of the mouth in any of these subjects? In large experimental animals, one of the early findings has been that subtle changes in the structure of the tongue and buccal mucosa seem to favor bacterial proliferation.

Hambidge. We do not have any data on that at the moment.

Henkin (Bethesda). I'd like to comment on Dr. Mills' question because I think that the changes that occur in the oral environment of any animal that is zinc deficient are very important. There are significant alterations in any number of organs in the mouth in patients who have a zinc deficiency. There are alterations in the anatomy of the taste buds, so that one can actually see anatomical changes. There are anatomical changes in the gland structure both in the mouth and in the nasal mucous membrane, and one can demonstrate these histologically not only by light microscopy but even better by electron microscopy. In addition, there are alterations in glandular secretions so that there are actual alterations in zinc released from these glands, and one can demonstrate specific changes in zinc metabolism in the oral environment.

Quarterman (Aberdeen). We have investigated this aspect of the problem, and found changes in mucus of the saliva and small intestine of both sheep and rats. I do not know to what extent changes in the bacterial flora of the mouth are attributable to these altered mucous secretions.

Hoekstra (Madison). Do you have any data, even retrospective, on actual dietary intakes, and is there anything unusual or atypical about the types of foods that these individuals select?

Hambidge. There is no overall pattern apart from the aversion for animal meats. We have observed a boy who apparently only ate peanut butter, which I believe contains phytate, and drank milk. This seemed an ideal situation to develop zinc deficiency, but as soon as he was supplemented with zinc he wanted to eat everything and had a normal diet. It is difficult to know if any of the dietary abnormalities which were observed in the children we studied were etiological factors in developing the deficiency, or if they were the result of the deficiency.

CHROMIUM AS A DIETARY ESSENTIAL FOR MAN

WALTER MERTZ

Nutrition Institute, Agricultural Research Service, United States Department of Agriculture, Beltsville, Maryland

HISTORY OF NUTRITIONAL RESEARCH ON CHROMIUM

The early interest in the biological role of chromium was oriented toward toxicological, biochemical, and hematological aspects. Like several of the trace elements now recognized as essential, chromium was first known for its toxicity, and the increased incidence of respiratory cancer in workers exposed to high levels of airborne chromate is well established (1). It has also been demonstrated that the addition of chromium *in vitro* can stimulate enzyme systems, as in the case of phosphoglucomutase, succinic cytochrome C dehydrogenase systems, and trypsin. Depending on its concentration, chromium can also inhibit a variety of enzymes *in vitro,* as is true for many other trace elements. The very high chromium concentrations detected in some nucleoprotein fractions and the potential role of chromium in nucleic acid metabolism have received little interest during the past decade, although they present a challenging problem to biochemical research. Chromium in its hexavalent form is a widely used tool for hematological diagnosis, following the discovery that chromates penetrate rapidly and irreversibly the membrane of the red blood cell (2).

The history of nutritional research with chromium can be traced back to the observation that rats developed an impairment of intravenous glucose tolerance when they were fed torula yeast in their diet but not when they were fed brewer's yeast (3). In the absence of any other criteria the agent responsible for this difference was termed glucose tolerance factor (GTF); it was subsequently described as containing trivalent chromium as its active ingredient (4). In the years following this discovery in 1959, a number of studies resulted in a theory of the mode of action of trivalent chromium. It was shown that epididymal adipose tissue from chromium-supplemented rats exhibiting a normal glucose tolerance had a glucose uptake *in vitro* identical with that obtained from chromium-deficient rats. However, significant differences in glucose uptake appeared when the effect of insulin in these two types of tissues was measured. Chromium-deficient tissues responded to insulin *in vitro* with a significantly smaller increase in glucose uptake than did those from chromium-supplemented rats. Furthermore, the poor response to insulin additions in chromium-deficient tissue could be significantly increased by adding submicrogram amounts of chromium in the form of active complexes to the incubation medium. The observation that glucose metabolism was apparently independent of chromium in the absence of insulin, and the significantly greater slope of a dose response curve to insulin as a result of chromium supplementation strongly suggested a close relation between the site of action of these agents. Subsequently, effects of chromium very similar to its effect on glucose uptake were established in a variety of systems, for example, oxidation of glucose, utilization of glucose for lipogenesis, glucose uptake by isolated rat lens, and cell transport of a nonutilizable sugar, D-galactose. This latter observation suggested an insulin-sensitive site on the cell membrane as the primary site of action of chromium. However, there were also experiments that indicated intracellular sites as responsive to chromium deficiency and resupplementation, for example, experiments measuring insulin-induced swelling of liver mitochondria and insulin-induced incorporation of certain amino acids into proteins. Finally, polarographic studies led to the hypothesis that chromium functions by participating in a ternary complex between insulin and membrane receptors, which is involved in the catalysis of the disulfide interchange believed by some to be the initiation of the insulin action (5). During that time there began to appear reports of a more severe chromium deficiency resulting in decreased growth and survival for rats and mice, in a seriously deranged glucose metabolism of rats as evidenced by fasting hyperglycemia and glycosuria, elevated serum cholesterol levels, and increased incidence of aortic plaques (6). Impaired glucose tolerance in squirrel monkeys fed a commercial chow was found to be responsive to chromium supplementation (7). Corneal opacity was described as a symptom of a combined chromium and protein deficiency and the resistance to some forms of stress of chromium-deficient rats was found to be significantly reduced as compared with that of chromium-supplemented animals.

Studies on the effect of supplementation of human subjects with microgram amounts of chromium chloride began in 1962. They demonstrated in a small number of subjects kept under the strict conditions of a metabolic ward that approximately half of those tested showed a significant improvement of their previously impaired glucose tolerance. It was concluded, however, that chromium chloride cannot be considered to be of value in the therapy of diabetes. Several other studies with middle-aged and old human subjects followed, of which one reported no apparent effect whereas the others found various degrees of improvement of glucose tolerance in approximately half of the subjects tested. Of three independent studies conducted with malnourished children, two reported normalization of previously impaired glucose tolerance following one dose of chromium, while the third (performed in Egypt) reported no apparent effect of chromium. However, in this latter study it was determined that the subjects were not chromium deficient. These and other results have been extensively reviewed (5, 6).

PRESENT AREAS OF INTEREST IN CHROMIUM RESEARCH

The results discussed in the first section, which were obtained by independent investigators, form a strong basis for counting chromium among the essential elements for animal species. The demonstration of beneficial effects of chromium supplementation in malnourished children suggests that this element is also essential for man and that deficiencies exist. The following discussion of more recent results does not change the status of chromium as an essential element, but it does show that chromium metabolism is much more complex than was originally believed and that the chromium nutritional state of man is influenced not only by dietary intake but also by a number of additional factors. Our knowledge of chromium metabolism and nutrition is more than ever before in a state of flux, not only with regard to the interpretation of experimental results but even concerning the basic tools of chromium analysis. As our state of knowledge increases, new problems appear and apparently contradictory results ask for reconciliation. Most of these difficulties can be traced to uncertainties about the chemical nature of the most active form of chromium (glucose tolerance factor) and to uncertainties inherent in the present methods of chromium analysis.

Chromium Metabolism: Glucose Tolerance Factor

As has been discussed previously (8), existing data on chromium balance in man are impossible to reconcile unless it is postulated that some food chromium is absorbed to a much higher degree than simple chromium compounds such as chloride. Two independent studies have determined the absorption of orally administered chromium chloride in man as 0.5–0.7% of a given dose. With an average intake of 60 μg/day the amount of chromium absorbed would be sufficient to compensate for only a small fraction of the estimated urinary loss of 7–10 μg/day, and would result in a rapid exhaustion of chromium reserves not compatible with the well-established presence of chromium even in old tissue. It follows that some chromium in foods must be absorbed to a higher degree than the 0.5–0.7% determined for chromium chloride. Furthermore, it has been demonstrated that simple compounds of chromium, regardless of valence state, do not pass into the fetus when orally or intravenously administered to pregnant rats. This observation is in apparent contrast to the high chromium concentrations in the fetus or the newborn mammalian organism. Here again it can be reasoned that while simple forms of chromium do not cross the placenta there must be some chromium complexes present in the maternal food or organism which are transported in a different way into the fetal tissue. This was demonstrated in pregnant rats that were given extracts of brewer's yeast grown in a chromium-containing nutrient solution. Chromium in this form was accumulated in the fetal organs at a higher concentration than in the maternal tissues. It was also observed that, depending on chemical form and valence state, the organ distribution of injected or ingested chromium in experimental animals was different, and that in acute experiments one physiologically important chromium pool was inaccessible to injected chromium chloride but was readily labeled by the injection of chromium in the form of GTF. These findings can be summarized to mean that there are at least two classes of chromium compounds: one category, including chromium salts, which is poorly absorbed and which has no access to physiologically important chromium pools, and another category of organic complexes in which chromium acts like an essential element. This latter category is termed GTF.

GTF is a dietary agent required for normal glucose tolerance in experimental animals and probably also in man. It occurs in brewer's yeast, in foods of animal origins, and in cereal grains. GTF is believed to be a chromium complex with organic ligands, of relatively small size, with an estimated molecular weight of between 300 and 500. It is water soluble and stable against a variety of chemical treatments. Although preparations of high purity, high biological activity, and high chromium content have been obtained through a series of purification procedures, the factor has not yet been isolated or identified. Like inorganic chromium, GTF potentiates the action of insulin *in vitro,* but to a much greater degree than inorganic chromium does.

An evaluation of these findings in a nutritional sense is difficult at the present time. Although many of the facts discussed above suggest that chromium is biologically active only in the form of GTF and that other forms must be converted into the latter before exerting biological activity, there are other data which are difficult to reconcile with this hypothesis. While it is true that adult human subjects do not respond to supplementation with chromium chloride until after a period of several weeks, malnourished children do so immediately. The first example is compatible with the hypothesis that chromium chloride must be converted into GTF chromium, but the second example is difficult to reconcile unless malnourished children have a very high efficiency for this conversion. Rats respond to oral administration of simple chromium complexes overnight and to the intravenous injection of chromium chloride immediately with improvement of glucose tolerance. Furthermore, chromium chloride has a moderate effect even when added *in vitro* to isolated epididymal fat tissue. Thus it appears that for glucose tolerance and intestinal absorption the differences between chromium compounds are quantitative; simple chromium compounds may act as such but the magnitude of their effect is much less than that of GTF chromium. On the other hand, placental transport and, perhaps, the distribution among body compartments appear to be strictly dependent on the chemical form, with simple compounds behaving qualitatively differently from GTF chromium. It is proven that yeast and certain other microorganisms can synthesize GTF from simple chromium salts, but it is not known whether the mammalian organism has this capability. If it has, either by itself or by its intestinal flora, this capability is not very efficient in the majority of adults; it may be the limiting factor of chromium metabolism.

Acute Changes in Plasma Chromium Concentration following Glucose or Insulin Loads

When young healthy subjects are given a test dose of glucose or of insulin there is an acute rise of plasma chromium within 30 to 90 min (9–11). In older subjects, particularly those with impaired glucose tolerance, the plasma chromium concentration does not significantly change. Administration of GTF in rats or supplementation with chromium chloride in man during several weeks has been shown to result in a reappearance of the acute plasma increment, suggesting depletion of a specific chromium pool as the cause for the absence of the acute plasma increment in elderly persons. The measurement of plasma chromium concentrations following a glucose or insulin challenge is therefore believed to be a valid, albeit difficult, tool for assessing the chromium nutrition status. It has been shown and confirmed that part of this chromium increment appearing in the circulation is subsequently lost in the urine (6). This finding has two important implications. It may serve to develop a convenient test to assess the chromium nutritional status by determining the increase of chromium concentrations in urine over the baseline following a glucose load. While there are still many difficulties to be resolved in the determination of chromium in urine, this test would be a great improvement over the repeated determination of plasma chromium. Of even greater importance are the implications of this chromium loss for the chromium balance of man. They strongly suggest that insulin therapy can lead to very considerable losses of chromium from the organism, particularly when fast-acting insulin is administered. These losses in turn increase the chromium requirement, in some cases to a point where it is not met by the normal dietary intake. The significant reduction of chromium concentrations in the hair of juvenile diabetics requiring insulin injections is probably an expression of the excessive urinary loss (12). It is likely that the consumption of substantial amounts of rapidly absorbed sugars results in a similar loss of chromium. As it has been shown that refined sugars do not contain appreciable amounts of chromium (13) an imbalance between intake and excretion can be suspected. The measurement of the acute chromium increment in urine has been found valid in preliminary experiments to determine the chromium nutritional status of persons, for example, during pregnancy. It is believed that after standardization this test can be used to great advantage for the screening of whole population groups.

Here, as in other fields of chromium research, apparently contradictory results have emerged. While our original observation of an acute plasma increment has since been confirmed by three independent investigators, others have reported an acute decrease rather than increase of plasma chromium following glucose and insulin loading in healthy persons (14). Other investigators, using identical techniques, have found increases of circulating chromium after oral administration of glucose and decreases after intravenous administration (15). While it can be stated that the absence of any change in plasma chromium levels following a glucose or insulin challenge is probably indicative of marginal chromium deficiency, it is difficult to reconcile the apparently contradictory findings by various investigators. However, a detailed examination of the plasma chromium response reveals that in many cases an acute increment is preceded by a decline of chromium concentrations from the starting levels (11). Thus, the discrepancy might be time related. On the other hand, it is possible that the two patterns may reflect different nutritional or hormonal states of the subjects examined. In view of the great potential importance of this phenomenon for the diagnosis of chromium deficiency, it appears imperative that these questions be resolved by further research.

Changes of Chromium Metabolism in Diabetes and Pregnancy

In discussing a potential relation between chromium and diabetes mellitus, it must be stated that no evidence is available that would identify chromium deficiency as the primary causative agent in diabetes. The data to be discussed subsequently demonstrate significant correlations between the diabetic state and certain aspects of chromium metabolism, but they do not allow a conclusion as to cause and effect. Insulin-dependent diabetics absorb an oral dose of chromium chloride significantly better than do normal subjects, as shown by measurements of plasma chromium concentration, and they excrete twice as much as do normal subjects in their urine (16). Similarly, the turnover of intravenously injected chromium chloride is significantly faster in insulin-requiring diabetics. These results, obtained in acute experiments with radioactive chromium chloride, are in agreement with the results of chromium analyses in urine of children with diabetes mellitus (10). Two children in particular who were examined before and after insulin therapy showed remarkable differences. In one chromium excretion doubled during insulin treatment; in the other it increased by more than ten-fold. These results suggest that the disturbance in chromium metabolism may be a result rather than a cause of diabetes, and they point out the importance of a careful assessment of chromium nutriture in insulin-requiring diabetics. The chromium content in the livers of subjects with diabetes mellitus was found to be significantly lower than that of normal subjects (17), and hair chromium levels of diabetic, insulin-requiring children were significantly lower than those of normal subjects (12). Experimental diabetes in rats resulted in a significantly different distribution of injected radioactive chromium chloride, with decreases in the kidney and increases in the serum of diabetic rats (18). Diabetes also produced a highly significant redistribution of chromium-51 in the subcellular fractions of rat liver. Significantly more chromium was incorporated into the nuclear and supernatant fractions and significantly less into the mitochondria and microsomes of the diabetic animals.

It has been shown that during the later part of pregnancy the fetus accumulates chromium not only in proportion to the growing body mass but that the chromium concentration in the tissues is greatly increased. It follows that pregnancy might accentuate marginal chromium nutrition in the mother and result in the lowering of body stores and of other indices indicative of chromium deficiency. Hair chromium levels of multiparous women are significantly lower than those of women who have not had children, suggesting that the extra demands of pregnancy are not easily met by the average dietary intake (19). Also, the proportion of women showing no acute change of plasma chromium in response to a glucose challenge increases significantly during the time of gestation and decreased again after parturition (10). Lactation represents another demand for chromium, with human milk containing an average of 12 μg/liter. In order to compensate for the estimated loss of 10 μg in 800 ml of mothers' milk, 1600 μg of chromium would have to be ingested in addition to the ordinary intake if this chromium were absorbed like chromium chloride. Even if the highest ratio of absorption of chromium from GTF (25%) were assumed, however, lactation would result in an additional requirement of 40 μg/day of highly available chromium. This demand is not easily met from average diets. To prove directly that the impairment of glucose metabolism during pregnancy is an expression of marginal chromium deficiency presents one of the most challenging problems of chromium research.

CHROMIUM IN PROTEIN-CALORIE MALNUTRITION

Except for one study, chromium chloride supplementation has been shown to be remarkably effective in malnourished children. In this one study, high plasma chromium concentrations and very high concentrations in the foods consumed were detected, so that the existence of chromium deficiency in the children of this study was extremely unlikely (5). A recent report of unusually high chromium concentrations in common foods of Egypt, where the study was performed, is additional evidence that the children had no chromium deficiency (20). A recent study in Turkey showed normalization of the impaired glucose tolerance of malnourished children in 9 out of 15 cases following chromium supplementation, and no significant effect in control children. As has been mentioned before, the effect of 150 μg chromium given by mouth was almost immediate, in clear contrast to our experience with adult subjects (21). Of the greatest importance is the observation that the children who received the chromium supplement responded to therapy with a significantly greater rate of weight gain than those who were not given chromium. The rate of weight gain in those children who responded to chromium with an improved glucose tolerance was three times greater than that of unsupplemented controls (22). It is not clear whether weight gain is independent of or a result of the improved glucose metabolism. The fact that chromium supplementation can significantly enhance the rate of recovery from malnutrition points out the need to identify those areas of the world in which protein-calorie malnutrition is complicated by chromium deficiency.

RECENT STUDIES WITH GLUCOSE TOLERANCE FACTOR

While it has been established that the addition of GTF to epididymal fat tissue stimulates glucose utilization in the presence of insulin, it remained to be shown whether this effect was truly dependent on the presence of the hormone or merely on an increased glucose flux across the cell membrane. GTF could be part of a membrane carrier system limiting only during increased glucose flux. Therefore, the effect of GTF on glucose oxidation was compared in a system in which cell entry of glucose was stimulated either by increasing glucose concentrations or by insulin. A significant increase of glucose oxidation by GTF was observed only in the presence of insulin (23). These results clearly suggest a direct interaction of GTF with insulin. Further evidence for this direct interaction was obtained from studies investigating the binding of GTF preparations to insulin *in vitro*. When insulin labeled with ^{125}I was reacted with GTF *in vitro* and subsequently chromatographed on Sephadex G-50, the elution pattern of the radioactivity was identical to that of unreacted insulin, indicating that if binding had occurred the molecular weight of GTF must have been small in relation to that of insulin. That binding had indeed occurred was proven by comparing the biological activity of the insulin peaks in the column eluate from unreacted insulin with that of the GTF-reacted insulin. The slope of the dose-response curve of the latter was significantly greater than that of the unreacted insulin peak. Chromatography of GTF on an identical column, but without previous incubation with insulin, yielded a different elution pattern, indicative of a low molecular weight compound (24). It was further postulated on the basis of experiments with acetylated insulin that GTF may bind to alpha or epsilon amino groups of the hormone. The demonstration of a GTF-insulin complex is particularly interesting in the light of the insulin-induced increase of plasma chromium in the intact organism. As the changes of insulin concentrations do not coincide with those of chromium, it is unlikely that a preformed GTF-insulin complex exists in the pancreas. The chromium increment could represent a secretion of GTF with a subsequent complex formation resulting in a tissue-bound insulin of higher activity.

As was mentioned in the preceding sections, the biological activity of chromium compounds as measured by their potentiation of insulin *in vitro* varies greatly. This is also true for compounds administered to the intact animal. This means that the determination of total chromium in foods has little meaning for the assessment of man's intake of available chromium from the diet. Ideally, food-composition studies for chromium should be specific for GTF. Recognizing that prior to the identification of GTF such an undertaking would be difficult, a study was undertaken to correlate the chromium content of different foods with the biological value of the chromium present, as measured by epididymal fat assay. Acid hydrolysates were analyzed for the total chromium of food, and their effect on the potentiation of insulin *in vitro* was measured. A plot of the total chromium concentration *versus* the biological activity *in vitro* established no correlation. On the other hand, when the foods were extracted with 50% ethanol, a highly significant correlation was found between chromium content and biological activity of the extracts (25). Ethanol does not specifically extract GTF from foods, but it leaves behind enough inactive chromium species to allow an assessment of the biological value of foods. On this basis a tentative assessment of biological values for chromium has been calculated, which shows brewer's yeast to be an outstanding source, followed by meats, grain, and some seafoods. Leafy vegetables yielded no biological activity in the ethanol extracts, unless the latter were subjected to acid hydrolysis. These studies demonstrate the difficulty of assessing the intake of available chromium in man on the basis of existing data. Therefore, it is not yet possible to recommend a dietary chromium intake which would safeguard the supply of 7 to 15 μg of available chromium needed to compensate for the daily losses.

ANALYTICAL PROBLEMS

Of all the problem areas of chromium research, analysis is the most difficult and at present the most important one. It must be stated that no one analytical value for chromium in biological materials can be considered absolute. This casts grave doubt on any attempts to compare results obtained by different investigators and on efforts to establish 'normal' chromium concentrations in tissue which could serve as a criterion for an assessment of nutritional status. This reservation does not detract from the validity of comparing relative chromium values of one tissue within one laboratory, for example, measurements of changes in plasma or urine chromium concentration following a glucose stress. However, any nutritional implications, for example, estimates of a human requirement based on analytical values, have to be treated with a great degree of caution.

Historically, the values reported for chromium in blood have shown a tendency to decline, until in the 1960's a plateau appeared to be reached between 20 and 50 ppb. With the use of emission spectrometric techniques and improvements in flame atomic absorption spectrophotometry values of around 10 ppb were reported, and with the advent of flameless atomic absorption spectroscopy the values have declined further to less than 1 ppb. Whereas there is now reasonable agreement among investigators in the USA using the graphite-furnace method, there is still a wide gap between concentrations found by this method and those reported by others using neutron activation analysis. Similar discrepancies exist for chromium in urine. With a supposedly nonvolatile element such as chromium it is usually safe to assume that higher values often reflect contamination and nonspecificity of the method, whereas lower values represent greater specificity and can therefore be judged to be more reliable. This assumption, however, is no longer tenable. It has been shown (13, 20) that an unknown and apparently variable proportion of chromium in biological material is highly volatile and can be lost by such mild procedures as oven drying or sublimation. Neutron activation analysis of Egyptian foods detected losses of up to 90% of the chromium originally present when the samples were dried at 80°C as compared with losses in those dried by lyophilization. That substantial losses of chromium in biological material can occur depending on the method of sample preparation has since been confirmed in our laboratory (W. A. Wolf, unpublished observations). A detailed investigation into the chromium content of different types of sugar and on the effect of ashing procedures thereon revealed considerable differences. For example, the chromium concentration in molasses was 266 ppb with low-temperature ashing, 129 ppb with ashing in a muffle furnace at 450°, and 29 ppb with direct analysis of the sample in a graphite furnace according to the recommended methods (13). Similar differences were detected with unrefined, brown, and refined sugars.

These data cast doubt on any quantitative statement concerning chromium concentrations in biological materials. They make it necessary to re-evaluate critically every step in analysis, from sample collection to the final determination. The complexity of the problem is evident from the fact that the proportion of chromium lost due to sample preparation varies with the individual materials.

CONCLUSION

The essentiality of chromium has been proven in several animal species. Chromium deficiency in humans has been shown to exist in certain areas, associated with protein-calorie malnutrition; marginal chromium deficiency is suspected to exist in the USA during pregnancy and in old age. The biological effect of chromium depends on the chemical nature of the compound in which chromium is present. Chromium occurs in brewer's yeast, animal meats and grains in the form of GTF, a complex of high biological activity. The ability of the mammalian organism to synthesize this factor appears to be limited. A quantitative determination of the human chromium requirement can be made only after GTF has been identified and can be measured directly and after reliable methods of chromium analysis are available.

REFERENCES

1. Baetjer, A. M. 1950. Pulmonary carcinoma in chromate workers. I. A review of literature and report of cases. *Archs Industr. Hyg.* 2: 487.
2. Gray, S. J., and Sterling, K. 1950. The tagging of red cells and plasma proteins with radioactive chromium. *J. Clin. Invest.* 29: 1604.

3. Mertz, W., and Schwarz, K. 1955. Impaired intravenous glucose tolerance as an early sign of dietary necrotic liver degeneration. *Archs Biochem. Biophys.* 58: 504.
4. Schwarz, K., and Mertz, W. 1959. Chromium (III) and the glucose tolerance factor. *Archs Biochem. Biophys.* 85: 292.
5. Mertz, W. 1969. Chromium occurrence and function in biological systems. *Physiol. Rev.* 49: 163.
6. Schroeder, H. A. 1968. The role of chromium in mammalian nutrition. *Am. J. Clin. Nutr.* 21: 230.
7. Davidson, I. W. F., and Blackwell, W. L. 1968. Changes in carbohydrate metabolism of squirrel monkeys with chromium dietary supplementation. *Proc. Soc. Exp. Biol. Med.* 127: 66.
8. Mertz, W., and Roginski, E. E., 1971. Chromium metabolism: The glucose tolerance factor. *In* W. Mertz and W. E. Cornatzer (eds), *Newer Trace Elements in Nutrition,* Marcel Dekker, New York, p. 123.
9. Glinsmann, W. H., Feldman, J. F., and Mertz, W. 1966. Plasma chromium after glucose administration. *Science* 152: 1243.
10. Hambidge, K. M. 1971. Chromium nutrition in the mother and the growing child. *In* W. Mertz and W. E. Cornatzer (eds), *Newer Trace Elements in Nutrition,* Marcel Dekker, New York, p. 169.
11. Behne, D., and Diehl, F. 1972. Relations between carbohydrate and trace-element metabolism investigated by neutron activation analysis. *In Nuclear Activation Techniques in the Life Sciences,* I.A.E.A., Vienna, p. 407.
12. Hambidge, K. M., Rodgerson, D. O., and O'Brien, D. 1968. The concentration of chromium in the hair of normal children and of children with juvenile diabetes mellitus. *Diabetes* 17: 517.
13. Masironi, R., Wolf, W., and Mertz, W. 1973. Chromium in refined and unrefined sugars from various countries, as analyzed by flameless atomic absorption, *In Trace Elements in Relation to Cardiovascular Diseases,* IAEE Tech. Rep. No. 157, p. 77.
14. Davidson, I. W. F., and Burt, R. L. 1973. Physiological changes in plasma chromium of normal and pregnant women: effect of a glucose load, *Am. J. Obstet. Gynecol.* 116: 60.
15. Pekarek, R. S., Hauer, E. C., Wannemacher, R. W., Jr., and Beisel, W. R. 1973. Serum chromium concentrations and glucose utilization in healthy and infected subjects. *Fed. Proc.* 32: 930 (abstr.).
16. Doisy, R. J., Streeten, D. H. P., Souma, M. L., Kalafer, M. E., Rekant, S. I., and Dalakos, T. G. 1971. Metabolism of chromium-51 in human subjects, normal, elderly, and diabetic subjects. *In* W. Mertz and W. E. Cornatzer (eds), *Newer Trace Elements in Nutrition,* Marcel Dekker, New York, p. 155.
17. Morgan, J. M. 1972. Hepatic chromium content in diabetic subjects. *Metabolism* 21: 313.
18. Mathur, R. K., and Doisy, R. J. 1972. Effect of diabetes and diet on the distribution of tracer doses of chromium in rats. *Proc. Soc. Exp. Biol. Med.* 139: 836.
19. Hambidge, K. M., and Rodgerson, D. O. 1969. A comparison of the hair

chromium levels of multiparous and parous women. *Am. J. Obstet. Gynecol.* 103: 320.
20. Maxia, V., Meloni, S., Rollier, M. A., Brandone, A., Patwardhan, V. N., Waslien, C. I., and Shami, S. E. 1972. Selenium and chromium assay in Egyptian foods and in blood of Egyptian children by activation analysis. *In Nuclear Activation Techniques in the Life Sciences,* I.A.E.A., Vienna, p. 527.
21. Gürson, C. T., and Saner, G. 1971. Effect of chromium on glucose utilization in marasmic protein-calorie malnutrition. *Am. J. Clin. Nutr.* 24: 1313.
22. Gürson, C. T., and Saner, G. 1973. Effects of chromium supplementation on growth in marasmic protein-calorie malnutrition, *Am. J. Clin. Nutr.* 26: 988.
23. Roginski, E. E., Toepfer, E. W., Polansky, M. M., and Mertz, W. 1970. Effect of glucose tolerance factor on insulin-stimulated *versus* glucose-stimulated glucose oxidation by rat epididymal fat tissue. *Fed. Proc.* 29: 695 (abstr.).
24. Evans, G. W., Roginski, E. E., and Mertz, W. 1973. Interaction of the glucose tolerance factor (GTF) with insulin. *Biochem. Biophys. Res. Comm.* 50: 718.
25. Toepfer, E. W., Mertz, W., Roginski, E. E., and Polansky, M. M. 1973. Chromium in foods in relation to biological activity. *J. Agric. Food Chem.* 21: 69.

DISCUSSION

Forth (Homburg). What has been found in studies comparing the absorption of cationic and anionic chromium in man?

Mertz. Anionic chromium such as chromate is somewhat better absorbed than cationic chromium. However, we suspect that the chromium in glucose tolerance factor is a cationic form and this is absorbed to an even greater degree than chromate.

Shah (Ottawa). You showed that most of the foods we eat have a low biological value when determined in an *in vitro* assay. On the other hand, when you give chromium chloride to children who are suffering from malnutrition, they show marked improvement in glucose tolerance overnight. It must be that chromium can be converted to glucose tolerance factor in the body, and therefore how relevant is this *in vitro* assay?

Mertz. First of all, I would like to correct a misconception. I do not consider the foods that we generally eat as poor sources. The values were not expressed in percentages, but in arbitrary figures that can be compared with each other. So, except for brewer's yeast which is very active, the members of the second group I showed you, which had values of 2 or 3, are considered, at least by me, as good sources of glucose tolerance factor. As to your second point, we do believe that perhaps children have a very high synthetic capability of putting the inorganic chromium that comes to the gut into a very active form, but we really have no evidence for it.

Gubler (Provo). We have tried labeling the chromium in yeast by giving the yeast radioactive chromium, and then feeding the yeast to rats. We found no better absorption of this radioactively labeled chromium from the yeast than we do from chromium chloride.

Mertz. I don't know how to explain these results other than to suggest that certainly not all of the chromium that goes into the yeast goes into glucose tolerance factor. It is also possible that your methods destroyed this active complex, that it wasn't formed, or that it was lost in some way. We have carried out these absorption studies in quite a few animals and, for us, it works.

Weser (Tubingen). Did you check anemic animals to determine if the plasma chromium content is increased? It is possible for transferrin to bind chromium as a chelator, and in an anemic situation it might bind more chromium because there is less iron to compete with it.

Mertz. This has been investigated by others, but the data are inconclusive and didn't point one way or the other.

Doisy (Syracuse). We had a patient a few years ago with severe hemochromatosis and a very diabetic glucose tolerance. We supplemented him with chromium and after about one month his glucose tolerance was half way back to normal. I think iron overload probably does block binding sites for chromium and, by feeding him extra chromium, we were able to modify his glucose tolerance back toward normal.

IRON ABSORPTION, A MEDIATED TRANSPORT ACROSS THE MUCOSAL EPITHELIUM[1]

W. FORTH

Abteilung für spezielle Pharmakologie, Universität des Saarlandes, Homburg, Saar, Federal Republic of Germany

A particular feature of iron as compared with other heavy metals of biological significance is that, once absorbed, it is released from the body only to a remarkably small extent. In the adult male the daily losses of iron, e.g., in stool and urine or by desquamation of skin, are balanced by the intake with food of 10–15 mg iron. In relation to the total content in the body of 3–5 g, this fraction amounts to 0.2–0.3%. If the demand for iron is increased, e.g., during pregnancy, the amount absorbed is increased. Iron-deficiency results in an increased absorptive utilization of iron in food.

Obviously, absorption plays the determining role for the homoiostasis of iron metabolism. Therefore, the interest of many investigators was focused on the process of iron absorption, i.e., the transport of iron across the mucosal epithelium. The object of this paper is to discuss the properties that enable the mucosal cells to adapt the amount of iron absorbed to the body's needs. The scope of this discussion is limited to this topic, and all factors outside the mucosal cells which influence the absorption of iron additionally are not considered.

'Passive' transport of ions across epithelia is produced either by diffusion or by convection. The energy for 'active' transport of ions is derived from cellular metabolism. In contrast to alkali metal ions, e.g., sodium and potassium, iron shares with other heavy metals the property of not existing in ionic form in biological media. At pH values usually found in a biological environment iron tends, like other heavy metals, to form coordination compounds (1). Therefore, like other heavy metals, iron is virtually nonionized in

1. Investigations in the author's laboratory were supported by a grant of the SFB 38, "Membrane Research," at the University of the Saarland.

plasma but bound to ligands such as transferrin, the iron-binding protein of plasma. Evidently, when investigating the transport of iron across the mucosal epithelium one has to ask whether the mucosal cells also contain ligands to which iron is attached and which may be involved in transport from the brush borders to the basal portion of the mucosal cell where iron is released into blood. This distance is about 100,000–200,000 times the diameter of the Fe^{2+} ion.

In order to characterize the transport system for iron, the following questions will be treated in detail: (1) sites of absorption in the gastro-intestinal tract; (2) kinetics of the transport; (3) anisotropy of the transport system; (4) specificity of the transport system; (5) dependence of the transport on cellular metabolism; (6) present knowledge about the components of the transport system.

SITES OF ABSORPTION

Iron can be absorbed along the entire length of the gastrointestinal tract: in the stomach (2–4), in the ileum (3, 5, 6, 7), and even in the colon (2, 7, 9). It is commonly accepted that the duodenum and the proximal jejunum are quantitatively the most important sections of the small intestine for iron absorption. After oral ingestion iron is absorbed mainly in these sections, as was shown in man by Wheby (10), who used an intestinal tube, and by Hemmati (7) who employed an intestinal capsule with remote control. When estimating the contribution of different sections of the gastrointestinal tract to the total absorption of iron, one also has to take into account differences in the residence time. In the rat the ingesta remain about 20 times longer in the ileum than in the jejunum (11, 12). The prevailing role of the duodenum and jejunum for the absorption of iron is underlined by the observation that in iron deficiency the absorption is increased mainly in these segments (13–15). However, the region of the intestinal tract in which the capacity to absorb iron is increased in iron-deficient rats may also include segments of the proximal ileum (8, 15, 16).

KINETICS OF IRON ABSORPTION

Influence of Concentration

There is much evidence that the amount of iron absorbed does not increase linearly with increasing doses (cf. 46). The most convincing results are given by Wheby, Jones, and Crosby (2) who provided evidence that it is the process of release of iron from the mucosal cells rather than the uptake into the mucosal cells which shows saturation kinetics and, hence, under certain conditions, is limiting for iron absorption. On the basis of this observation the transport system for iron is saturated at a dose range of 5–10 μg in normal rats and 50–100 μg in iron-deficient rats.

Time Course of Absorption

Absorption starts immediately after the exposure of the mucosal epithelium to iron (4, 17–19). Two phases of absorption may be distinguished: a rapid one, lasting up to 2 hr, and, subsequently, a slow phase which in man and animal can operate for 10–12 hr (20–23). The rapid phase appears to prevail in the duodenum and jejunum whereas the slow phase takes place mainly in the ileum (4, 8). In iron-deficient rats 75% of the total amount incorporated was absorbed within the rapid phase (22); this means that a fraction of 25% must be accounted for by absorption in the slow phase. In normal rats the fraction of iron absorbed in the slow phase appears to be greater; it amounts to at least 40% of the total (22, 23).

Preloading of the mucosal epithelium with iron *in vivo* in rats is followed by a diminished transfer of iron in jejunal segments *in vivo* (4) and *in vitro* (15, 24, 25). As measured on intestinal segments of normal rats *in vitro,* the transfer rate returns to normal within 5–6 hr after preloading (4, 15).

ANISOTROPY OF THE TRANSPORT SYSTEM

In jejunal segments *in vitro* of iron-deficient rats an increase could be demonstrated only for the transfer in the direction from the mucosal to the serosal side. In addition, the content of radioiron in the intestinal tissue of iron-deficient rats was increased only if iron was administered at the mucosal side. If supplied from the serosal side, no difference in the iron content of the intestinal tissue could be observed between normal and iron-deficient rats (26, 27). This means that in iron deficiency the transport system has vectorial properties; it is anisotropic. When measuring the specific ^{59}Fe activity in the absorbate and the mucosal tissue, evidence was obtained that in iron deficiency an iron compartment exists in the mucosal cells through which iron is transferred rapidly to the serosal side of the epithelium without an appreciable exchange with other compartments. The specific ^{59}Fe activity in the absorbate of iron-depleted segments *in vitro* increased eight- to nine-fold as compared with that in control segments whereas the specific ^{59}Fe activity in the mucosal tissue increased only three-fold (28).

SPECIFICITY OF THE TRANSPORT SYSTEM FOR IRON

In iron deficiency not only the transport of iron across the mucosal epithelium is increased but also that of heavy metals chemically related to iron, for instance, manganese, cobalt, nickel, chromium, and zinc, but not of copper (cf. Table 1). From these observations the conclusion can be drawn that these heavy metals, with the exception of copper, share, at least in part, the iron-transfer system when passing the mucosal cell from lumen to blood. With regard to the increased capacity of the transfer system in iron deficiency, the specificity of the system is highest for iron (28–32). Mutual inhibition of absorption was shown for iron and cobalt (28–31, 33, 36–40, 43, 44), iron and manganese (28–31, 35), and iron and zinc (28–31). It was demonstrated in rats *in vivo* that cobalt interfering with iron during absorption abolishes the nonlinear portion of the absorption curve for iron when the absorption in a dose range of 0.005–0.5 μM in the presence of 50 μM cobalt was measured (38). Similar results were obtained in perfusion studies on duodenal segments of anesthetized rats (44). From these experiments the conclusion was drawn that acceptor sites for iron, which are involved in the transfer process for iron across the epithelium, are present in the mucosal cells or on their surface. Chemically related metals compete with iron for these acceptor sites. Vice versa, there is experimental evidence that chelating agents administered intraluminally may compete with the cellular acceptor sites for iron. In other words, chelating agents which form stable iron chelates in the gastrointestinal tract may prevent iron being taken up by the acceptor sites of the mucosal cells and, hence, may inhibit the absorption of iron. This holds true for a great variety of chelating agents (cf. 45, 46). Weaker agents, however, which form iron chelates of lower stability, may release their iron to the acceptor sites of the mucosa. The advantage of these chelating agents may be to keep iron available for absorption by preventing the formation of poorly absorbed iron hydroxides and phosphates. The threshold of the thermodynamic stability constants for useful iron chelates administered orally for therapeutic and nutritional purposes amounts to about 10^{12} –10^{13} . Below this value the chelates may release their iron to the acceptor sites of the intestinal mucosa. Above this value the chelator is predominant when competing with the mucosal acceptor sites for iron (1, 45, 46).

DEPENDENCE OF IRON TRANSPORT ON CELLULAR METABOLISM

The view that the transport of iron through the mucosal epithelium is based on a process of simple diffusion was rejected by Hahn *et al.* (47) as early as

Table 1. Absorption of metals other than iron in iron deficiency

Method used	Species	Increased absorption	No change in absorption	References
Jejunal segments *in vitro*	Rat	Co, Ni, Mn, Zn	Cu	(28–32)
Tied-off jejunal segments *in situ*	Rat	Co, Mn, Zn	Cu, Cs, Mg, Hg, Ca	(33)
	Rat	Co, Mn, Zn	Cu	(28–31)
Perfused segments *in situ*	Rat	Co, Mn		(34)
	Rat	Mn		(35)
	Rat	Co, Mn		(36)
In vivo, gastric tube or intragastric injection	Rat	Co, Mn, Zn		(28–31)
	Rat	Co		(37, 38)
	Rat	Co		(39)
	Rat	Cr		(40)
Oral dose	Man	Co		(41)
		Co		(42)
		Mn		(35)

1943: "We cannot believe it is a matter of membrane diffusion but rather a part of the cell metabolism probably involving cell proteins including ferments." Rummel and co-workers (48, 49) obtained the first results which proved that the transfer of iron across the mucosal epithelium is dependent on a chemical process. A diminished mucosal iron content and/or a diminished iron transfer across the mucosal epithelium resulted from addition of metabolic inhibitors or deprivation of oxygen and substrates (14, 24, 50–54). There is one observation which is worthwhile mentioning separately. The increased capacity for the transport of iron in jejunal segments *in vitro* of iron-deficient rats was not restricted by oxygen deprivation, but was abolished by the withdrawal of glucose from the incubation fluid. It can be concluded that anaerobic glycolysis is sufficient to maintain the surplus transfer of iron across the mucosal epithelium (54). In addition to the results mentioned above, which show that the transport of iron across the mucosal epithelium includes energy-requiring steps, there are experiments *in vitro* as well as *in vivo* which demonstrate a dependence of the transport system on the intact protein synthesis of the mucosal cells. Iron absorption is impaired by inhibitors of protein synthesis, e.g., cycloheximide and tetracycline (55–58). Pretreatment of rats with phenobarbital is followed by increased absorption of iron administered by gastric tube (59), but not of glucose, amino acids, or palmitic acid *in vitro* (60). The ability of intestinal iron absorption in mice to adapt to an increased iron demand is genetically controlled (61–69). These data indicate that special cellular proteins are involved, at least as components of the system, in the transport of iron across the mucosal epithelium.

When summarizing these data it can be stated that the capacity to absorb iron, especially to adapt the absorption to the demand of the organism, is a particular property of the proximal section of the small intestine, i.e., the duodenum and jejunum. Obviously, the mucosal epithelium of these sections contains a transport system which performs a directed transfer of iron from the luminal side of the epithelium to the blood side. The capacity of this transport system is limited. It has a high specificity for iron and depends on cellular metabolism. There is evidence that cellular proteins are involved in the transport process.

PRESENT KNOWLEDGE OF THE COMPOSITION OF THE TRANSPORT SYSTEM

In order to characterize the components of the iron-transfer system, the partition of radioiron in different subcellular compartments of mucosal cells of normal and iron-deficient rats was studied during absorption. All fractions obtained, i.e., brush borders and nuclei, mitochondria, microsomes and the particle-free fraction, contained iron. However, next to the brush-border fraction, the highest content of radioiron was found in the cytosol. When measuring radioiron in this fraction during absorption in iron deficiency, it was found that after 10 min exposure iron was taken up rapidly into this fraction and also released rapidly into the organism. This holds true, however, only for iron-depleted intestinal segments of anemic rats. In intestinal segments of normal rats there is only a slow increase of the radioiron content in the particle-free fraction which remains unchanged during the entire experiment lasting three hours (70–75). More than 95% of the iron present in the particle-free fraction of mucosal tissue was bound to high molecular weight compounds. Two iron-binding proteins were isolated and purified from the particle-free fraction and were numbered according to their migration velocity in disc electrophoresis. Protein 1 was heat stable, had a rather high molecular weight (more than 150,000) and could be precipitated, at least partly, with horse-spleen antiferritin (Table 2). In contrast to this ferritin-like protein, protein 2 was heat labile, its molecular weight was about 80,000 and it reacted with rat antitransferrin (74, 75).

There is no doubt that the ferritin-like protein is synthesized within the mucosal cells, whereas the transferrin-like protein could be a contamination of the homogenates with plasma transferrin. After having labeled plasma transferrin with radioiron the contamination of protein 2 in mucosal homogenates

Table 2. Chemical properties of mucosal iron-binding proteins 1 and 2 from homogenates of rat jejunum[1]

	Protein 1	Protein 2
Absorption maximum	–	465 nm
Molecular weight	> 150,000	~ 80,000
Heat stability (80°C, 10 min)	stable	labile
Reaction with antiferritin	+	–
Reaction with antitransferrin	–	+
Isoelectric point (ampholine method)	4.9	5.40[2]
	Liver ferritin: 4.8	Plasma transferrin: 5.20[1]

[1]From 71, 74, and 75.

[2]Value of the second minor band for protein 2 5.12 and plasma transferrin 5.00

by plasma transferrin amounted to less than 0.1% (70, 74, 75). Protein 2 is present mainly in the mucosal tissue of duodenum and jejunum, i.e., the segments distinguished by the highest adsorptive activity for iron. Under the same conditions in ileal mucosal tissue the content of radioiron-labeled protein 1 and 2 amounts to only one-eighth of that obtained in jejunal mucosal tissue (74). In order to decide whether protein 2 is synthesized within the mucosal cells or whether it represents the iron-free fraction of plasma transferrin attached to mucosal cells from the outside, further experiments are needed. When considering the kinetics of radioiron in the particle-free fraction and the partition of radioiron between the iron-binding proteins 1 and 2, there is a striking difference between normal and iron-deficient rats. In iron-deficient animals, uptake and release of radioiron in the particle-free fraction is determined mainly by protein 2. In these experiments protein 1, the ferritin-like protein, takes up iron only slowly and finally it contains about the same amount of radioiron as the ferritin-like protein does in normal animals. In contrast, in normal rats it is protein 2, the transferrin-like protein, that binds only a small amount of radioiron. Here, the time course is determined mainly by the slow uptake of radioiron by protein 1. Thus it can be concluded that the transferrin-like protein 2 is involved in the rapid phase of the transfer process for iron in iron deficiency (71–75). The existence of the two iron-binding proteins in mucosal tissue was confirmed by Worwood and Jacobs (76, 77) and by Sheehan and Frenkel (78).

In 1962 Falconer and Isaacson (86) described a hereditary anemia in litters of mice exposed to X-rays. These animals were called sla-mice, indicating that they suffer from a hereditary sex-linked anemia. Plasma iron concentration and total iron-binding capacity of plasma from sla-mice do not differ from those of normal mice in iron deficiency. The anemia of sla-mice results from insufficient iron absorption (61–69). While the latter does not completely cease, sla-mice have lost the capacity to adapt their iron absorption to a higher demand of the organism. Because of these properties, sla-mice were believed to provide a suitable model for studying the relationship between the mucosal iron-binding proteins and iron absorption. Sla-mice absorbed only one-quarter of that amount which was measured in normal mice (71, 74). Following the feeding of a low-iron diet, the amount of radioiron increased eight-fold in normal mice but only four-fold in sla-mice. However, even under these conditions the amount of iron absorbed in sla-mice did not exceed that measured in normal animals kept on a diet with normal iron content. Most important was the result that, despite the feeding of the low-iron diet for six weeks to sla-mice, the radioiron content in both mucosal proteins remained unchanged. In normal mice the radioiron content in protein 1 increased two-fold and that in protein 2 three-fold under these conditions. The radio-iron content of protein 2 from mucosal homogenates of sla-mice fed a diet with normal iron content was so small that the radioactivity was barely above background. From these experiments the conclusion was drawn that the genetically determined insufficiency of iron absorption in sla-mice is a consequence of the abnormal behaviour of the mucosal iron-binding proteins, especially protein 2. Further experiments are needed to clarify the role of protein 1 during the absorption process in mice.

CONCLUSIONS

The function of the iron-transfer system is to ensure the rapid transfer of the element across the mucosal cells and to prevent reactions of iron with other intracellular ligands and binding sites. This purpose could be achieved by a transferrin-like protein such as protein 2. The fraction of radioiron taken up from the intestinal lumen by the mucosal protein 2 increases with increasing degree of iron deficiency (75). This protein is also able to release iron quickly during the initial rapid phase of iron absorption. This phase is important, especially in iron deficiency. Thus, provided the intracellular origin can be proven experimentally, protein 2 may function as a transcellular carrier for iron during absorption (cf. Fig. 1). In sla-mice suffering from a hereditary defect of iron absorption the synthesis of the transferrin-like protein could be defective or its function could be impaired. The function of an acceptor site for iron has been ascribed to apoferritin in the so-called 'mucosal-block' theory (79, 80). It is now generally accepted that ferritin is not involved in the mechanism of iron absorption in iron deficiency. Although apoferritin is synthesized in the mucosal cells of iron-deficient rats, the rate of synthesis is less than in mucosal cells of normal rats (81). Moreover, in iron-deficient rats iron is taken up by mucosal ferritin to a lesser extent than in normal animals (25, 80–85). Cobalt, the absorption of which is increased in iron deficiency, is not found in ferritin (37). With respect to the rapid phase of absorption, it appears questionable whether ferritin is at all capable of releasing quickly the iron which is stored as ferric hydroxide micelles. Iron deposited into mucosal ferritin, however, is apparently involved in the slow phase of absorption which lasts up to 12 hr. If there is no need for iron in a repleted organism it will be rejected into the intestinal lumen with the extruded mucosal cells at the end of their lifespan (85). Two steps which are important for the transfer of iron across mucosal cells have not yet been taken into account in this discussion: the process of uptake of iron by the luminal membrane, the brush borders of the mucosal cells, and the release of iron across the membranes of

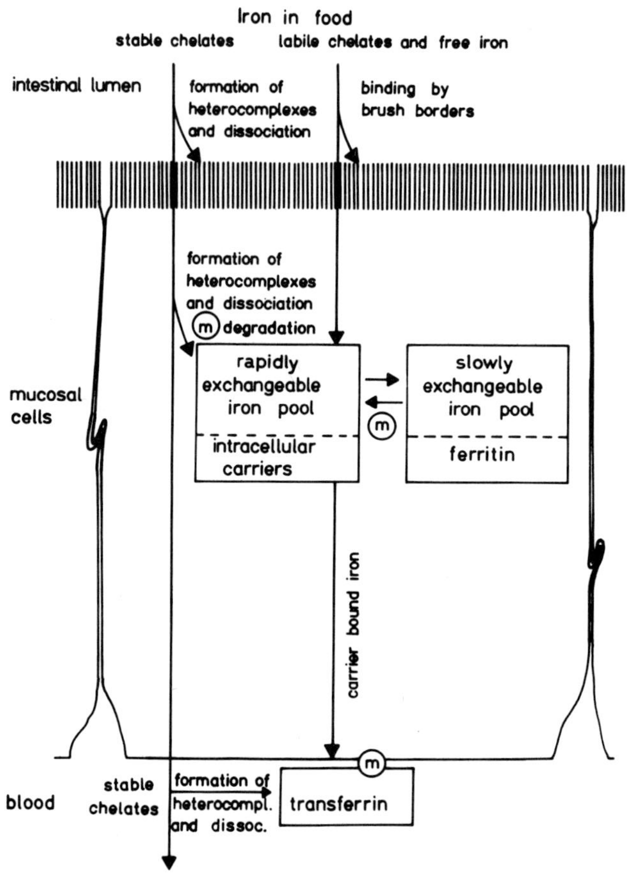

Fig. 1. Scheme of iron transport across mucosal cells of duodenum and jejunum. Chelated iron and free iron ions in food come into contact with binding sites of the brush-border membrane. Iron chelates may form heterocomplexes and dissociate. Simultaneously, iron is taken up by the binding sites of the membrane. Within the cells iron is present mainly in the two pools, the *rapidly exchangeable* one, which is probably identical with the transferrin-like, iron-binding protein 2 (additionally, there may be also low molecular weight ligands for iron in that pool), and the *slowly exchangeable* one, which is probably identical with the ferritin-like iron-binding protein 1. In iron deficiency iron exchanges predominantly with the *rapidly exchangeable* pool. Normally, however, the exchange with the *slowly exchangeable* pool prevails. At the basal pole of the cell 'carrier' bound iron may be released in dependence on a metabolical step, e.g., catalyzed by the hypothetical enzyme 'iron transferase'. Other metabolically dependent steps of iron transport may be the exchange of iron between the two pools and the degradation of iron chelates in the cytoplasma. Stable, membrane-soluble chelates pass the mucosal cells intact. [From Forth and Rummel (46).]

the basal portion of the mucosal cell into the blood. It is known that the capacity of the brush borders to bind iron is increased in iron deficiency, especially in the jejunum (70).

Little information is available concerning the process of iron release across the membranes at the basal pole of the mucosal cell. A hypothesis has been advanced that an enzyme, iron transferase, is involved in this step which is believed to be dependent on cellular metabolism (67).

REFERENCES

1. Forth, W., and Rummel, W. 1974. Gastrointestinal Absorption of Heavy Metals. *In* W. Rummel and W. Forth (eds.), *Pharmacology of Intestinal Absorption, International Encyclopedia of Pharmacology and Therapeutics,* Section 39 B, Pergamon Press, Oxford, in press.
2. Wheby, M. S., Jones, L. G., and Crosby, W. A. 1964. Studies on iron absorption. Intestinal regulatory mechanisms. *J. Clin. Invest.* 43: 1433.
3. Forth, W., Rummel, W., and Pfleger, K. 1968. Der Einfluss von Liganden auf die Retention von Eisen nach oraler Verabfolgung an normale und anämische Ratten. *Naunyn-Schmiedebergs Arch. Exp. Path. Pharmac.* 261: 225.
4. Rhodes, J., Beton, D., and Brown, D. A. 1968. Absorption of iron instilled into the stomach duodenum and jejunum. *Gut* 9: 323.
5. Duthie, H. L. 1964. The relative importance of the duodenum in the intestinal absorption of iron. *Br. J. Haemat.* 10: 59.
6. Chirasiri, L., and Izak, G. 1966. The effect of acute haemorrhage and acute haemolysis on intestinal iron absorption in the rat. *Br. J. Haemat.* 12: 611.
7. Hemmati, A. 1968. Die Bestimmung des Resorptionsortes von Eisen im Intestinalkanal mit einer ferngesteuerten Darmkapsel. *Dt. Med. Wschr.* 93: 1468.
8. Conrad, M. E., Weintraub, L. R., Sears, D. A., and Crosby, W. H. 1966. Absorption of hemoglobin iron. *Am. J. Physiol.* 211: 1123.
9. Chernelch, M., Fawwaz, R., Sargent, T., and Winchell, D. S. 1970. Effect of phlebotomy and pH on iron absorption from the colon. *J. Nucl. Med.* 11: 25.
10. Wheby, M. S. 1966. Site of absorption in man. *Clin. Res.* 14: 50.
11. Dupuis, Y., and Fournier, P. 1963. Lactose and the absorption of calcium and strontium. *In* R. H. Wasserman (ed.) *The Transfer of Calcium and Strontium Across Biological Membranes,* Academic Press, New York, p. 277.
12. Lengemann, F. W. 1963. Overall aspects of calcium and strontium absorption. *In* R. H. Wasserman (ed.) *The Transfer of Calcium and Strontium Across Biological Membranes,* Academic Press, New York, p. 85.
13. Manis, J. G., and Schachter, D. 1962. Active transport of iron by intestine: effects of oral iron and pregnancy. *Am. J. Physiol.* 203: 81.

14. Ruliffson, W. S., and Hopping, J. M. 1963. Maturation, iron deficiency and ligands in enteric radioiron transport *in vitro*. *Am. J. Physiol.* 204: 171.
15. Forth, W., and Rummel, W. 1965. Eisen-Resorption an isolierten Dünndarmpräparaten von normalen und anämischen Ratten. *Naunyn-Schmiedebergs Arch. Exp. Path. Pharmac.* 252: 205.
16. Wack, J. P., and Wyatt, J. P. 1959. Studies on ferrodynamics. I. Gastrointestinal absorption of ^{59}Fe in the rat under differing dietary states. *Pathology* 67: 237.
17. Stewart, W. B., and Gambino, S. R. 1961. Kinetics of iron absorption in normal dogs. *Am. J. Physiol.* 201: 67.
18. Pollack, S. 1968. (Published by W. H. Crosby.) Iron absorption. *In* Ch. Code and W. Heidel (eds), *Handbook of Physiology*, Section 6, Vol. III, American Physiological Society, Washington, D. C., p. 1553.
19. Forth, W. 1968. Eisen- und Kobalt-Resorption am perfundierten Dünndarmsegment. *In* W. Staib and R. Scholz (eds), *Stoffwechsel der Isoliert Perfundierten Leber*, Springer-Verlag, New York, p. 242.
20. Brown, E. B. 1963. The absorption of iron. *Am. J. Clin. Nutr.* 12: 205.
21. Hallberg, L., and Sölvell, L. 1960. Iron absorption during constant in the intragastric infusion of iron in man. *Acta Med. Scand.* 168 (suppl. 358): 43.
22. Wheby, M. S. 1966. The gastrointestinal tract and iron absorption. *Blood*, 22: 416.
23. Forth, W., and Rummel, W. 1966. Abhängigkeit der Eisenresorption von der Eisenbindung durch den Darm. *Med. Pharmac. Exp.* 14: 384.
24. Manis, J. G., and Schachter, D. 1962. Active transport of iron by intestine: features of the two-step mechanism. *Am. J. Physiol.* 203: 73.
25. Manis, J. G., and Schachter, D. 1964. Active transport of iron by intestine: mucosal iron pools. *Am. J. Physiol.* 207: 893.
26. Forth, W., Leopold, G., and Rummel, W. 1968. Eisendurchtritt von der Mucosa zur Serosaseite und umgekehrt an isolierten eisenarmen und normalen Segmenten von Jejunum und Ileum. *Naunyn-Schmiedebergs Arch. Exp. Path. Pharmac.* 261: 434.
27. Forth, W., und Rummel, W. 1969. Beziehungen zwischen dem Durchtritt des Eisens durch die Darmwand und seiner Bindung am Gewebe. *Naunyn-Schmiedebergs Arch. Exp. Path. Pharmac.* 264: 230.
28. Forth, W. 1970. Absorption of iron and chemically related metals *in vitro* and *in vivo;* the specificity of an iron binding system in the intestinal mucosa of the rat. *In* C. F. Mills (ed.), *Trace Element Metabolism in Animals*, Livingstone, Edinburgh, p. 298.
29. Forth, W. 1966. Untersuchungen über die Resorption von Eisen und chemisch verwandten Schwermetallen an Därmen normaler und anämischer Ratten *in vivo* und *in vitro;* Ein Beitrag zur Frage der Spezifität des eisenbindenden Systems in der Mucosa. Ph.D. Thesis, Universität des Saarlandes, Germany.
30. Forth, W. 1971. Resorption von Eisen und chemisch verwandten Metallen *in vitro* und *in vivo;* die Spezifität des eisenbindenden Systems in der Mucosa des Jejunums von Ratten. *In* W. Horst (ed.), *Frontiers of Nuclear Medicine*, Springer-Verlag, New York, p. 83.

31. Forth, W., and Rummel, W. 1971. Absorption of iron and chemically related metals *in vitro* and *in vivo. In* S. C. Skoryna and D. Waldron-Edwards (eds), *Intestinal Absorption of Metal Ions, Trace Elements and Radionuclides*, Pergamon Press, Oxford.
32. Forth, W., Rummel, W., and Becker, P. J. 1966. Die vergleichende Prüfung von Bindung und Durchtritt von Eisen, Kobalt und Kupfer durch isolierte Jejunumsegmente normaler und anämischer Ratten. *Med. Pharmac. Exp.* 15: 179.
33. Pollack, S., George, J. N., Reba, R. C., Kaufman, R., and Crosby, W. H. 1965. The absorption of nonferrous metals in iron deficiency. *J. Clin. Invest.* 44: 1470.
34. Diez-Ewald, M. Weintraub, L. R., and Crosby, W. H. 1968. Interrelationship of iron and manganese metabolism. *Proc. Soc. Exp. Biol. Med.* 129: 448.
35. Thomson, A. B. R., Olatunbosun, D., and Valberg, L. S. 1971. Interrelation of intestinal transport system for manganese and iron. *J. Lab. Clin. Med.* 78: 642.
36. Thomson, A. B. R., and Valberg, L. S. 1972. Intestinal uptake of iron, cobalt and manganese in the iron-deficient rat. *Am. J. Physiol.* 223: 1327.
37. Schade, St. G., Felscher, B. F., Bernier, G. M., and Conrad, M. E. 1970. Interrelationship of cobalt and iron absorption. *J. Lab. Clin. Med.* 75: 435.
38. Schade, St. G., Felsher, B. F., Glader, B. E., and Conrad, M. E. 1970. Effect of cobalt upon iron absorption. *Proc. Soc. Exp. Biol. Med.* 134: 741.
39. Thomson, A. B. R., Shaver, Ch., Lee, D. J., Jones, B. L., and Valberg, L. S. 1971. Effect of varying iron stores on site of intestinal absorption of cobalt and iron. *Am. J. Physiol.* 220: 674.
40. Hopkins, L. L., and Noble, M. A. 1969. Effect of iron deficiency on orally administered Fe (III) upon the gastrointestinal uptake of ^{51}Cr (III) in rats. *Fed. Proc.* 28: 299 (abstr.).
41. Valberg, L. S., and Olatunbosun, D. 1968. Alteration in cobalt absorption in patients with disorders of iron metabolism. *Gastroenterology* 54: 1279.
42. Valberg, L. S., Ludwig, J., and Olatunbosun, D. 1969. Alteration in cobalt absorption in patients with disorders of iron metabolism. *Gastroenterology* 56: 241.
43. Thomson, A. B. R., Valberg, L. S., and Sinclair, D. C. 1971. Competitive nature of the intestinal transport mechanism for cobalt and iron in the rat. *J. Clin. Invest.* 50: 2384.
44. Thomson, A. B. R., and Valberg, L. S. 1971. Kinetics of intestinal iron absorption in the rat: effect of cobalt. *Am. J. Physiol.* 220: 1080.
45. Forth, W., Nell, G., and Rummel, W. 1973. Chelating agents and the transfer of heavy metals across the mucosal epithelium. *In* D. D. Hemphill (ed.) *Trace Substances in Environmental Health,* Vol. VII, University of Missouri, Columbia, Mo.
46. Fourth, W., and Rummel, W. 1973. Iron Absorption. *Physiol. Rev.* 53: 724.
47. Hahn, P. F., Bale, W. F., Ross, J. F., Balfour, W. M., and Whipple, G. H.

1943. Radioactive iron absorption by gastric intestinal tract. *J. Exp. Med.* 78: 169.
48. Jacobi, H., Pfleger, K., and Rummel, W. 1956. Komplexbildner und aktiver Transport durch die Darmwand. *Naunyn-Schmiedebergs Arch. Exp. Path. Pharmac.* 229: 198.
49. Rummel, W., Jacobi, H., and Pfleger, K. 1956. Enterale Resorption kleinster Eisenmengen. *Naunyn-Schmiedebergs Arch. Exp. Path. Pharmac.* 228: 204.
50. Brown, E. B., and Justus, B. W. 1958. *In vitro* absorption of radioiron by everted pouches of rat intestine. *Am. J. Physiol.* 194: 319.
51. Dowdle, E. B., Schachter, D., and Schenker, H. 1960. Active transport of ^{59}Fe by everted segments of rat duodenum. *Am. J. Physiol.* 198: 609.
52. Helbock, H. I., and Saltman, P. 1967. The transport of iron by rat intestine. *Biochim. Biophys. Acta* 135: 979.
53. Pearson, W. N., and Reich, M. 1965. *In vitro* studies of Fe^{59} absorption by everted intestinal sacs of rat. *J. Nutr.* 87: 117.
54. Rummel, W., and Forth, W. 1968. Zur Frage der metabolischen Abhängigkeit von Eisenbindung und Durchtritt durch den isolierten Dünndarm. *Naunyn-Schmiedebergs Arch. Exp. Path. Pharmac.* 260: 50.
55. Greenberger, N. J., and Ruppert, R. D. 1966. Inhibition of protein synthesis: a mechanism for the production of impaired iron absorption. *Clin. Res.* 14: 298.
56. Greenberger, N. J., and Ruppert, R. D. 1966. Tetracycline induced inhibition of iron absorption. *Clin. Res.* 14: 432.
57. Greenberger, N. J., Ruppert, R. D., and Cuppage, F. E. 1967. Inhibition of intestinal iron transport induced by tetracycline. *Gastroenterology* 53: 590.
58. Yeh, S. D. J., and Shils, M. E. 1966. Effect of tetracycline on intestinal absorption of various nutriants by the rat. *Proc. Soc. Exp. Biol. Med.* 123: 367.
59. Teale, F. W. J. 1969. Cleavage of the haem-protein link by acid methylethylketone. *Biol. Biophys. Acta* 35: 543.
60. Thomas, F. B., Baba, N., and Greenberger, N. J. 1972. Effect of phenobarbital on small intestinal structure and function in the rat. *J. Lab. Clin. Med.* 80: 548.
61. Bannerman, R. M., Pinkerton, P. H., and Edwards, J. A. 1968. Hereditary iron-deficiency anemia due to a specific defect of intestinal absorption. *J. Clin. Invest.* 47: 5a.
62. Edwards, J. A., and Bannerman, R. M. 1970. Hereditary defect in intestinal iron transfer in x-linked anemia. *Fed. Proc.* 29: 300.
63. Edwards, J. A., and Bannerman, R. M. 1970. Hereditary defect of intestinal iron transport in mice with sex-linked anemia. *J. Clin. Invest.* 49: 1869.
64. Manis, J. 1970. Active transport of iron by intestine: selective genetic defect in mouse. *Nature, Lond.* 227: 385.
65. Manis, J. 1970. Intestinal iron-transport defect in the mouse with sex-linked anemia. *Am. J. Physiol.* 220: 135.

66. Pinkerton, P. H. 1968. Histobiological evidence of disordered iron transport in the x-linked hypochromic anemia of mice. *J. Path. Bacteriol.* 95: 155.
67. Pinkerton, P. H. 1969. Control of iron absorption by the intestinal epithelial cell. Review and hypothesis. *Am. Int. Med.* 70: 401.
68. Pinkerton, P. H., and Bannerman, R. M. 1967. Hereditary defect in iron absorption in mice. *Nature, Lond.* 216: 482.
69. Pinkerton, P. H., Bannerman, R. M., Doeblin, T. D., Benisch, E. M., and Edwards, J. A. 1970. Iron metabolism and absorption studies in x-linked anemia of mice. *Br. J. Haemat.* 18: 211.
70. Hübers, H., Hübers, E., Forth, W., and Rummel, W. 1971. Binding of iron to a non-ferritin protein in the mucosal cells of normal and iron-deficient rats during absorption. *Life Sci.* 10 (pt. I): 1141.
71. Hübers, H., Hübers, E., Forth, W., and Rummel, W. 1972. Iron absorption and cellular transfer protein in the intestinal mucosa of mice with hereditary anemia. *Naunyn-Schmiedebergs Arch. Exp. Path. Pharmac.* 274: R 56.
72. Hübers, H., Hübers, E., Simon, J., and Forth, W. 1971. A method for preparing stable density gradients and their application for fractionation of intestinal mucosal cells. *Life Sci.* 10 (pt. II): 377.
73. Hübers, H., Hübers, E., Simon, J., Forth, W., and Rummel, W. 1971. The subcellular distribution of ^{59}Fe in mucosal cells of normal and iron deficient rats during absorption. *Naunyn-Schmiedebergs Arch. Exp. Path. Pharmac.* 270: R 65.
74. Forth, W., Hübers, H., Hübers, E., and Rummel, W. 1972. Does a transfer system for iron exist in the mucosal cells of the small intestine? *In* D. D. Hemphill (ed.), *Trace Substances in Environmental Health*, Vol. VI, University of Missouri, Columbia, Mo., p. 121.
75. Hübers, H. 1972. Eine Methode zur Herstellung stabiler Dichtegradienten und ihre Anwendung beim Studium der Eisenresorption. Ph.D. Thesis, Universität des Saarlandes, Germany.
76. Worwood, M., and Jacobs, A. 1971. The subcellular distribution of orally administered ^{59}Fe in rat small intestinal mucosa. *Br. J. Haemat.* 20: 587.
77. Worwood, M., and Jacobs, A. 1971. Absorption of ^{59}Fe in the rat: iron binding substances in the soluble fraction of intestinal mucosa. *Life Sci.* 10 (pt. I): 1363.
78. Sheehan, R. C., and Frenkel, E. P. 1972. The control of iron absorption by the gastrointestinal mucosal cell. *J. Clin. Invest.* 51: 224.
79. Granick, S. 1949. Iron metabolism and hemochromatosis. *Bull. N.Y. Acad. Med.* 25: 403.
80. Granick, S. 1951. Structure and physiological functions of ferritin. *Physiol. Rev.* 31: 489.
81. Brittin, G. M., and Raval, D. 1970. Duodenal ferritin synthesis during iron absorption in the iron-deficient rat. *J. Lab. Clin. Med.* 75: 811.
82. Brown, E. B., and Rother, M. L. 1963. Studies of the mechanism of iron absorption. I. Iron uptake by the normal rat. *J. Lab. Clin. Med.* 62: 357.
83. Brown, E. B., and Rother, M. L. 1963. Studies of the mechanism of iron

absorption. II. Influence of iron-deficiency and other conditions on iron uptake by rats. *J. Lab. Clin. Med.* 62: 804.
84. Charlton, R. W., Jacobs, P., Torrance, J. D., and Bothwell, Th. H. 1965. The role of the intestinal mucosa in iron absorption. *J. Clin. Invest.* 44: 543.
85. Crosby, W. H. 1963. The control of iron balance by the intestinal mucosa. *Blood* 22: 441.
86. Falconer, D. S., and Isaacson, J. H. 1962. The genetics of sex-linked anaemia in the mouse. *Genet. Res.* 3: 248.

DISCUSSION

Saltman (La Jolla). I should like to raise three major issues. I would question the use of ferrous sulfate in these *in vitro* studies. There is no control over what happens to ferrous sulfate in the presence of oxygen, and it would be helpful to know what chelates were used, in what concentrations, and how they were maintained. Secondly, I believe it is important to get away from the use of everted gut loops. I think the sooner that we do that, and start measuring two-way fluxes on open membranes, and monitor those membranes for their viability with respect to sodium transport, the happier we will be. Our experiments really differ remarkably in that over a 50,000-fold concentration range we could not saturate any carrier by measuring two-way flux, and there was no asymmetry except in the anemic animals. We have also been reluctant to treat membranes with metabolic inhibitors which sometimes act like tanning agents on those systems. We have used temperature changes and measured energies of activation, and have found passive rates or passive energies of activation. These are the kinds of concerns I have, and I hope that we shall be able to work out these differences in technique.

Forth. As far as your first point is concerned, we have looked very carefully at what happened to iron with time, and you are right in that we can't inhibit the formation of hydroxides. We have another difficulty and I hope you will agree that we have to chose between two extremes. One extreme is iron sulfate, and the other is any chelate that we add. If we add iron citrate or even EDTA, we can show that we have no precipitation, but under these conditions iron is transferred by a simple diffusion process. This process takes place at the same time as we demonstrate a transport-type phenomena in iron-deficient rats. We can show that, even for EDTA, iron is split in the mucosal tissue of iron-deficient rats and transported to the other side. Therefore we have to decide whether to use a chelate or not, and we have decided not to use one. Secondly, we use *in vitro* segments and I like them very much. In all our experiments we determine the concentration gradient established for glucose or for an amino acid. If these segments were poisoned by withdrawal of oxygen, they failed to establish a glucose gradient. These segments are metabolically alive, and histological inspection indicates that they are in good condition.

Mills (Aberdeen). What is the relationship between your transferrin-like protein component number 2 and serum transferrin? Do you envisage another exchange system operating or do you envisage perhaps a ferroxidase operating? Could you clarify what the relationship of this mechanism is to the sort of changes in iron absorption that occur in a low copper situation? Early work suggested that low copper caused an iron absorptive defect, while the majority of studies would now suggest that perhaps it is an iron mobilization from the liver that is the primary problem.

Forth. I can do nothing but speculate because we have no evidence from our own experiments. We believe that this iron-binding protein 2, which is transferrin like, is really synthesized within the cells, and at the moment I believe that it is kept within the cell. There must, therefore, be another mechanism for the release of iron from the mucosal cell to the blood. We cannot yet really speculate how all of this might tie into the copper-deficient situation, which is at present far from being understood completely as far as the mechanism of iron absorption is concerned.

THE IMPACT OF INFECTIOUS DISEASE ON TRACE-ELEMENT METABOLISM OF THE HOST

WILLIAM R. BEISEL, ROBERT S. PEKAREK, and ROBERT W. WANNEMACHER, JR.

U.S. Army Medical Research Institute of Infectious Diseases, Frederick, Maryland

Highly sophisticated analytical procedures that involve absorption or emission spectroscopy, or neutron activation are now permitting even small alterations in the trace-element content of biological fluids to be documented during a wide variety of pathological conditions. These data are giving rise to new concepts concerning the regulatory mechanisms which influence trace-element metabolism in both health and disease (1).

This review will evaluate current knowledge concerning trace-element responses in the host during diseases of infectious origin. Studies in our laboratories will be summarized and related to the work of others. An attempt will also be made to describe and categorize basic control mechanisms that influence or regulate trace-element metabolism during infection.

The generalized wasting effects of severe illnesses were known to ancient civilizations. Deficiencies in specific elements, however, could not be ascribed to an infectious process until the elements themselves were recognized. Fecal losses of large quantities of "carbonate of soda" and the virtual disappearance of "free alkali" from serum were initially described in acute Asiatic cholera by O'Shaughnessy in 1831 (2). Less than four months later, Latta treated cholera patients with a slightly hypotonic saline and "subcarbonate" mixture in the first recorded instance of intravenous fluid therapy (3). The writings of Heller in 1847 (4), less than 40 years after chlorine was proven to be a chemical element, showed that he was well acquainted with the near absence of chloride from the urine of febrile patients who suffered from various infectious illnesses. He also demonstrated excessive losses of urinary phosphates and sulfates in febrile patients. These findings were confirmed within a decade by Parkes (5) who added nitrogen to the list of elements lost in excess during acute infection.

Of the essential trace metals, copper was first noted by Krebs in 1928 (6) to be increased in its concentration in the serum of patients with various infectious illnesses. This fact was confirmed by Locke, Main, and Rosbach in 1932 (7) who first described low serum iron concentrations in patients with tuberculosis and in horses given diphtheria or tetanus toxins (7). In 1951 Vikbladh (8) reported that serum zinc values were decreased in patients with acute and chronic infections but then returned to normal during recovery. In 1963 Nazarmukhamedova and Nugmanova (9) detected increased serum cobalt and manganese concentrations in patients with infectious hepatitis and, less than three years later, Belozerov (10) described an increase in blood and urinary gallium in patients with hepatitis. Earlier this year Pekarek *et al.* of the U.S. Army Research Institute of Infectious Diseases (11) reported a decrease in postprandial serum chromium concentrations in patients with sandfly fever.

INFECTION-RELATED CHANGES IN HOST IRON METABOLISM

Anemia may develop during infection through one of several mechanisms, including decreased red-cell formation or loss via bleeding or hemolysis. Red cells may become fragile in bacterial infections caused by tubercle bacilli, staphylococci, streptococci, and pneumococci, or certain viruses (12). Direct red-cell destruction by microorganisms occurs in malaria and bartonellosis. Anemia during infection can also be related to a sustained (but reversible) derangement of host iron metabolism. The hypochromic anemia which develops after an infectious process has become chronic is generally termed the 'anemia of infection.'

Iron Metabolism During Infections

With the notable exception of acute infectious hepatitis, virtually all other infectious processes are associated with depressed serum iron concentrations. Low serum iron values are widely documented in bacterial infections (7, 12–28). In sharp contrast single publications describe elevated iron values during bacillary dysentery and typhoparatyphoid diseases (29) and in disseminated candidiasis (30). Low serum iron values are characteristic of experimentally induced bacterial infections in laboratory animals (31–40). Roughly comparable declines in serum iron may be found in viral infections of man (13, 40–43) and experimental animals (44, 45). Hypoferremia occurs in rickettsial illness (40, 41), but is inconsistent during syphilis (7, 46–48). Depressed serum iron values are also characteristic of disseminated parasitic diseases including amebiasis (49), schistosomiasis (50), and fascioliasis (51).

The depression of serum iron and the anemia associated with hookworm infestations are due primarily to chronic blood loss (52, 53).

Serum iron values begin to fall quite early in an infectious process (14, 15, 17), and then rebound promptly with recovery. Our prospective studies in volunteers reveal a fall in serum iron values during the early incubation period of bacterial and viral illnesses, before the onset of fever or symptomatic illness (27, 28, 40–43). Transient hypoferremia may occur in mild infections (15, 27, 42, 43). During generalized infection a factor appears in the serum of patients (28, 39–41) which, if injected parenterally in normal animals, is capable of depressing serum iron and zinc concentrations. This factor, termed leukocytic endogenous mediator (LEM), will be discussed in greater detail.

Low serum iron values cannot be explained by inadequate concentrations of transferrin in the serum. Although total iron-binding capacity may decline somewhat during an infection, serum iron reductions are proportionally far greater (16, 17, 25, 27). Anemia of infection is characterized by hypochromic red blood cells despite adequate stores of hemosiderin iron in tissues (13, 14, 18). There is, typically, a diminished utilization of orally administered iron for red-cell formation during infections of man (14, 54) or experimental animals (33, 34, 55), and even massive intravenous doses of iron fail to correct either anemia or hypoferremia (56). In 1946 Cartwright *et al.* (32) described an accumulation of iron in tissue stores during infection, increased rates of ^{59}Fe clearance from plasma, and an increased intestinal absorption. Increased rates of ^{59}Fe clearance from plasma were also reported by others (19, 57), along with a normal or increased fractional turnover of red cell iron. Reduced uptake of ^{59}Fe by red cells and hepatic nonhemin stores was observed in experimental staphylococcal infection (34, 36). Ferritin iron stores were depleted in the liver and spleen (58, 59), but hemosiderin iron accumulated in the same organs (58, 59). In contrast to these histochemical observations, incorporation of ^{59}Fe into hepatic ferritin was increased during staphylococcal infections of rats (37).

Iron Metabolism During Inflammation

Sterile inflammatory abscesses produced generalized changes in iron metabolism quite similar to those observed during systemic infections (31, 34, 60–62), although ^{59}Fe absorption from the gut seemed to be depressed (63, 64). Cartwright, Gubler, and Wintrobe (60) showed that the decline in serum iron which accompanied a turpentine abscess in dogs could largely be prevented by blocking the reticuloendothelial system with thorotrast. In 1930 Menkin (65, 66) observed that iron given parenterally would accumulate in areas of localized inflammation induced experimentally in animals and in epithelioid and giant cells of tubercular lesions (65, 66).

Thus, changes in host iron metabolism observed during infections and inflammatory states included lowering of serum concentrations, accelerated passage out of the circulation, faster turnover rate of total plasma iron, increased (or decreased) absorption from the gut, and increased storage in the reticuloendothelial system, mainly as hemosiderin; concurrently, incorporation of iron into red blood cells was slowed while that derived from senescent red cells was retained as hemosiderin (62, 67–69). In addition, one of the earliest changes, involving movement of serum iron into hepatic cells, was found to be initiated by an endogenous mediating factor released into the serum by host phagocytic cells, as will be described.

Iron Metabolism in Bacterial Toxemias

Bacterial toxins have been used to elucidate mechanisms by which iron metabolism is regulated during infectious processes. Early studies indicated that serum iron values were depressed by bacterial exotoxins (7, 31). Most investigators have subsequently used killed gram-negative bacilli or purified lipopolysaccharide endotoxins to initiate a depression of serum iron concentrations (16, 38, 70–80). This consistent response is so closely related to the dose of endotoxin administered that lowering of serum iron can be used as a bioassay for endotoxin activity (76). Serum iron values begin to decline within several hours after endotoxin is given, with a nadir within 16–24 hr (70, 72). Total iron-binding capacity remains normal (16) or may decline slowly (75). The fall in serum iron occurs in conjunction with its accumulation in the liver and spleen (74, 77), and a normal red cell incorporation (74). Cortell and Conrad (80) observed an increased clearance of ^{59}Fe from the plasma of animals given endotoxin, and also noted diminished absorption from the gut, which appeared to occur in two distinct stages. Blockade in the transfer of iron from intestinal mucosa cells into the serum was followed by a slowing in the transfer of iron from the lumen of the gut into the mucosal cells (80).

In 1962, Kampschmidt and Upchurch (73) discovered that the acute depression of serum iron was initiated by a heat-labile endogenous factor which appeared in the serum of rats given endotoxin. This observation was confirmed in our laboratories (27, 28, 38–42, 81–88); the mediating factor was also shown to stimulate changes in the metabolism of zinc and copper as well as increased synthesis of 'acute-phase' reactant serum glycoproteins.

Iron Metabolism in Hepatitis

Although of viral origin, acute hepatitis is different from other infectious illnesses in that serum iron values are generally high. This elevation of serum iron concentrations has been reported in considerable detail by many groups during the last three decades (15, 89–108). Elevations in serum iron begin at a considerably later time than the increases in serum bilirubin or the SGOT and SGPT values. Normal, or slightly low, iron values at the onset of illness give way to increases which generally reach a maximum in about 2–3 weeks after the initial appearance of jaundice in uncomplicated infectious hepatitis (15, 91, 95, 96, 103, 105, 106). Total iron-binding capacity does not change appreciably, and, accordingly, iron-binding capacity may become virtually saturated (90, 103, 105).

Speeded rates for the plasma clearance and turnover of ^{59}Fe (109) develop during hepatitis, while the intestinal absorption of orally administered ^{59}Fe may be increased or slightly diminished (102–105). The elevated concentrations of serum iron in hepatitis have been accompanied by an increased excretion of iron in both feces and urine (29, 107). Scuro, Dabrilla, and Innecco (108) attributed high serum iron values during viral hepatitis to the presence of circulating ferritin released from damaged liver cells. Others suggested that the damaged hepatic cells were unable to take up iron from serum at their normal rate (105).

INFECTION-RELATED CHANGES IN HOST COPPER METABOLISM

Increased plasma concentrations of copper and its carrier protein ceruloplasmin appear typical of most infectious illnesses. The initial finding of Krebs (6) of elevated serum concentrations of copper has been confirmed repeatedly in acute and chronic bacterial infections of man (14, 17, 23, 26, 59, 110–115) or laboratory animals (84, 87, 116). Only a single study (20) reported depressed serum copper values in patients with urogenital tuberculosis, and inconsistent changes were noted in syphilis (46, 47). An increase in serum copper has been reported in (117) experimental trypanosomiasis in rats.

Increased plasma copper and ceruloplasmin concentrations have also been found in viral illnesses (40–43, 114, 118), although one report described a depression of copper concentrations in serum and cerebrospinal fluid in patients with tick-borne encephalitis (45). Urinary copper excretion fell below baseline control values in our patients with sandfly fever (40, 41).

Infection-related increases in serum ceruloplasmin concentration have been detected whenever investigated (40, 41, 60, 84, 110, 113–116, 119). The increases in concentration of both serum copper and its binding protein develop concomitantly at the onset of symptomatic illness during an infection (17, 40, 41). These increases occur gradually and often reach their peak only after the period of acute illness has ended. Elevated copper and ceruloplasmin concentrations then slowly revert to normal during convalescence.

Similarly, the synthesis of ceruloplasmin in laboratory animals has been shown to increase slowly during acute experimental infections (84, 116).

In hepatitis the concentrations of copper (90, 97, 99, 101, 114, 120–122) and ceruloplasmin (114, 119, 121–123) were also increased, although the hepatic content of copper remained normal (124). An increased excretion of copper in the urine during hepatitis (122) was ascribed to a postulated increase in the glomerular filtration of microligands that consisted of copper bound to histidine, glutamine, threonine, or cysteine.

Copper and ceruloplasmin concentrations have been observed to increase in the serum of man and animals following the therapeutic (70) or experimental (125) administration of bacterial endotoxins, and in dogs with experimentally induced sterile turpentine abcesses (61). Although changes in copper are slower in onset than the depression of serum zinc and iron, increased copper concentrations can be demonstrated within 24 hr after administration of endotoxin with maximal values within 48 hr (70). These more gradual changes in serum copper values are consistent with the concept that a period of time is required to allow the liver to synthesize and release ceruloplasmin. An increased synthesis of ceruloplasmin is stimulated by LEM (84).

INFECTION-RELATED CHANGES IN HOST ZINC METABOLISM

A depression of plasma zinc values appears to be a characteristic response during infectious diseases. Depressed plasma zinc values have been noted in bacterial infections (8, 27, 28, 40, 41, 113, 126), viral and rickettsial infections (40–43, 86, 88, 118), spirochetal infections (47, 113), parasitic infestations (127, 128), and after the experimental administration of bacterial endotoxin to laboratory animals (129). Prospective studies in our laboratories have shown that a depression of serum zinc values in volunteers begins during the early incubation period of both bacterial (tularemia and typhoid fever) and viral (sandfly fever and attenuated Venezuelan equine encephalitis vaccine) infections, prior to the onset of fever or symptomatic illness (28, 40–43, 83, 130). Urinary zinc excretion falls below normal during the prodromal and early febrile phases of illness, but may increase and become greater than normal in early convalescence.

Following endotoxin administration to rats, a depression in plasma zinc begins within several hours, becomes maximal between 6 and 9 hr postinoculation, and returns to baseline within 24 hr. The magnitude of depression in serum zinc concentration occurs in a linear relationship to the log-dose of endotoxin administered; the change is consistent enough to allow it to be used as a bioassay technique for endotoxin concentration (129) in a manner analogous to that suggested for iron (76). The depression of serum zinc, like that of iron, is mediated by LEM after its release from phagocytic cells (88).

Serum zinc values also tend to be depressed during acute infectious hepatitis. Most investigators have reported appreciable decreases in plasma zinc (122, 126, 131–133). Davis, Musa, and Dormandy (134) saw no changes in plasma zinc during hepatitis; others have reported an increase in serum zinc (135, 136). In the recent studies described by Henkin and Smith (122), the decrease in serum concentrations of total zinc was said to be accompanied by an increased concentration of diffusible zinc in serum and increased urinary excretion. Others also reported this during hepatitis (133, 136), although Ol'gina (135) stated that urinary zinc diminished during hepatitis.

INFECTION-RELATED CHANGES IN THE METABOLISM OF OTHER TRACE ELEMENTS

As noted earlier (9–11), scattered single reports suggest that infectious illnesses may influence the concentration in serum of other trace elements including manganese (9, 101), cobalt (9), gallium (10), and chromium (11). An intracellular role for iodide (137) has been postulated to account, in part, for the bactericidal capabilities of phagocytic cells.

CLASSIFICATION OF PATHOPHYSIOLOGICAL MECHANISMS LEADING TO TRACE-ELEMENT RESPONSES

Trace-element changes during an infectious process can be grouped into broad categories based on their pathophysiological mechanisms. These mechanisms are outlined in Table 1 and will be described in greater detail. The observed changes in the metabolism of a given trace element during infection may be due primarily to one of these mechanisms or to a combination of several that are operating simultaneously.

Table 1. Mechanisms that influence trace-element metabolism during infection

1. Altered intestinal absorption
2. Altered body losses
3. Altered distribution among body tissues
4. Altered carrier-protein concentrations
5. Hormonal interactions
6. Effects of leukocytic endogenous mediator (LEM)
7. Direct interactions with invading microorganisms

Altered Intestinal Absorption

Absorption of a trace element from the intestinal lumen may be increased or decreased during infection. Impairment of intestinal absorption has been postulated to account for some of the nutritional deficits during infection (138). Impaired intestinal absorption is most frequent if diarrhea is an important component of illness. However, relatively little information is available concerning fecal losses of trace elements. Impaired absorption of the cobalt contained in vitamin B_{12} has been reported in patients with hookworm infestation, amebiasis, and in the carriers of the fish tapeworm (139). In the latter instance poor absorption of ^{60}Co-labeled B_{12} has been ascribed to the uptake of this vitamin by tapeworms in the gut (139).

An increase in the intestinal absorption during experimental infections is best documented using ^{59}Fe. Absorption of iron requires an initial uptake of intestinal iron by mucosal cells and its subsequent transfer to the plasma carrier protein transferrin. This process may be speeded up during infection because of hypoferremia and the relatively greater unsaturation of circulating transferrin. Other controlling factors must be considered, however, because the combination of low serum iron and relative unsaturation of transferrin also develops during bacterial endotoxemia, despite the fact that ^{59}Fe absorption is reduced (80).

Excessive Body Losses of Trace Elements

Loss of body stores during infection can occur because of an exaggeration of normal excretory pathways in urine, feces, or sweat (138). Other pathways for trace-element loss during infection can be through sputum or purulent exudates.

Fecal losses are likely to be exaggerated during diarrhea and can include trace elements of endogenous as well as of dietary origin. Total zinc loss via sweat (140) is probably exaggerated during febrile diaphoresis. Losses of a trace element via urine may be increased or decreased during infection. Henkin and Smith (122) suggested that increased urinary excretion of copper and zinc may be brought about through heightened glomerular filtration and loss of the small but variable fractions of serum zinc and copper which are complexed with certain of the amino acids as microligands. Our own studies (83, 85, 86, 88, 130), however, showed that the concentrations of histidine, glutamine, threonine, and cysteine in the serum of patients with bacterial and viral diseases were generally depressed, as were those of most other amino acids. Further, the urinary losses of zinc and copper as well as the urinary losses of the above listed amino acids which form microligands with these elements were generally lower during symptomatic illness than comparable values measured during baseline (pre-infection) control periods in the same volunteers when studies were conducted using careful metabolic-ward collection techniques (130). Additional studies will be needed to define in more precise terms the excretory patterns and renal mechanisms that could account for possible alterations in urinary trace-element losses.

Altered Distribution of Trace Elements Among Body Tissues

Some of the most important changes in host trace-element metabolism involve their redistribution within the body. Such a redistribution may be caused by a pathologic process which leads to cellular injury and death; redistribution may also be initiated and controlled by physiological regulating mechanisms. Any change in the concentration of an element in serum or within a cellular compartment should be viewed as the algebraic summation of all factors that influence its rate of entry into or egress from the body space being measured. Improved techniques and better experimental designs are needed to allow investigators to acquire this type of data.

Escape of a trace element from injured or necrotic cells must be considered a possibility during infection, through mechanisms analogous to the escape of potassium from cells (with subsequent hyperkalemia) such as may follow extensive trauma to soft tissues. Although hyperferremia during hepatitis may imply an escape of iron or ferritin from damaged hepatocytes, functional impairment of hepatic iron uptake serves as an alternative explanation. This latter possibility is supported by the well-documented lack of coincidence in the timing of peak increases in serum iron during hepatitis in relation to the peaking of biochemical indices indicative of acute liver-cell damage.

Accumulation of trace elements in areas of tissue necrosis and injury is another possible cause for pathological alterations in the distribution of trace elements within the body. Calcium becomes localized in areas of tissue injury; there is evidence (65, 66, 69) that iron also becomes localized in areas of inflammation. However, most data suggest that such a process is not the primary mechanism leading to hypoferremia during inflammatory conditions. Rather, hypoferremia appears to be due primarily to a physiologically controlled redistribution of iron leading to its accumulation in hemosiderin stores. Similarly, a physiologically controlled movement of zinc into cells of the liver appears to be responsible for lowered zinc concentrations during infection. Low zinc values in serum cannot be explained by a movement of zinc into areas of tissue injury (141).

The increase in serum copper which typifies virtually all infectious processes undoubtedly accompanies an increased hepatic secretion of ceruloplasmin. As shown by studies in rats (88), the increased synthesis and secretion of ceruloplasmin by the liver leads to depletion of the copper contained in that organ. Similar evidence for hepatic copper depletion has not been obtained in man (124).

Physiologically controlled redistribution of zinc and iron during acute infections has been the subject of many investigations. Much effort has been devoted to the elucidation of the phenomenon that allows for a redistribution of iron in the tissues during infection. It has been difficult to define comprehensively the broad scope of alterations in iron metabolism during infection, because of the closely interlocking mechanisms that control the transportation, accumulation, physiological storage, and release of iron in hepatocytes of the liver and cells of the intestinal mucosa, reticuloendothelial system, and bone marrow. Information is needed at the cellular level to define infection-related influences upon the mechanisms which control the cellular uptake of iron from plasma (especially movement into hepatocytes), its combination with apoferritin to form ferritin, or its more stable fixation as hemosiderin. Experimental evidence generally supports the concept that there is an increased accumulation during infection in the stores of hemosiderin iron in reticuloendothelial cells. Less information is available to document an accumulation of either labile iron stores or hemosiderin in hepatocytes.

The recent studies in our laboratories show that an increased rate of iron flux into the liver is one of the first metabolic changes that can be detected during infection. The earliest redistribution of iron, presumably to hepatocytes but not initially to reticuloendothelial tissues (85), is initiated by LEM released into serum by body cells which are engaged in phagocytic activity. Since LEM is released during endotoxemia and noninfectious inflammatory states as well as during infection, it would appear that the circulation of LEM can account for certain similar derangements of iron metabolism in different pathophysiological conditions that have leukocyte activation and inflammation as common denominators. The same mediator substance is also responsible, it would seem, for the rapid movement of zinc from serum into liver cells (85). Less is known about the intracellular binding and possible hepatic storage of zinc than about the intracellular hepatic localization of iron.

The underlying physiological reasons which necessitate the presence of a fundamental control mechanism that can cause zinc and iron to accumulate within hepatic cells during inflammation have not been identified. However, these metals contribute to the function of enzymes, ribosomal integrity, and other metabolic processes within cells, and their accumulation within hepatocytes during infection and inflammation may be for such purposes.

Altered Concentration of Carrier Proteins

The increase in concentration of ceruloplasmin during infection stands out as the primary example of a change in trace-element metabolism which can be ascribed directly to an alteration in hepatic synthesis and release of its specific binding protein. Although the role of increased ceruloplasmin concentration during infection is not known, this response appears to be stimulated by powerful basic control mechanisms, i.e., increases in ceruloplasmin synthesis occur even in the face of direct damage to hepatic cells (as in hepatitis) or in the presence of severe pre-existing nutritional deficits of amino acids and protein. Ceruloplasmin is one of the many serum glycoproteins whose synthesis is stimulated by LEM (82–84). Although the copper-ceruloplasmin complex is believed to function, in part, as an oxidase that can influence iron transport, low serum iron concentrations in the presence of excess hemosiderin stores during infection do not seem to be reversed by the increased concentrations of ceruloplasmin and copper in the serum.

Iron-binding capacity in the serum is due primarily to the presence of transferrin, the specific carrier protein for iron. As noted above, the concentrations of this protein may decline somewhat during naturally occurring or experimentally induced infections, but the decline is not generally of a magnitude that alone could explain the disappearance of iron from serum. The work of Kampschmidt and Upchurch (75) showed that declines in serum iron could be achieved with doses of endotoxin too small to influence iron-binding capacity. They showed further (75) that, when higher doses of endotoxin were given, the fall in iron-binding capacity occurred too rapidly to be explained by an inhibition of transferrin synthesis, and reasoned that sequestration of transferrin or its adherence to exterior membranes of body cells might explain the transient fall in this binding protein. In any event changes in transferrin do not cause, explain, or even correlate directly with decreases or increases in serum iron concentrations during infection or inflammation.

Although an α_2-macroglobulin is thought to have specific binding affinity for zinc, little is known about the exact relationships between zinc and α_2-macroglobulin during conditions of infection. Attempts to measure changes in this protein have failed to provide an explanation for the relatively consistent decline in serum zinc concentrations during infection (142).

Many trace metals can also be bound nonspecifically to serum albumin. Such binding is less important physiologically than the binding by specific carrier proteins. While concentrations of serum albumin tend to be somewhat lower than normal when an infection becomes chronic, the magnitude of change in total serum albumin concentrations are not generally of a magnitude that could explain alterations of trace elements as observed during infection. Because binding to albumin is relatively weak, albumin may release trace elements for complexing with amino acids (122).

Alterations in the Hormonal Controls of Trace Elements

Metabolism of trace elements and some of their protein carriers can be influenced by hormones. Acute and chronic infections in man are accompanied by many changes in endocrine homeostasis (138). These changes include the following: brief increases in the secretion of glucocorticoids, mineralocorticoids, and adrenal androgens; intermittent increases in growth-hormone secretion; increased rates of thyroid-hormone degradation; hypersecretion of glucagon and insulin; an increased secretion of catecholamines during septic shock (138). Such hormonal changes are generally of short duration and have not been linked directly during an infectious process to the altered metabolism of a trace element. It is possible, however, that the depressed fasting concentrations of serum chromium during infection (11) may be related to the apparent partial resistance by peripheral tissues to the action of insulin, inasmuch as chromium may be required to permit optimal effects of insulin on cell membranes (143).

Leukocytic Endogenous Mediator (LEM)

As referenced earlier a mediating factor initiates a number of changes in host trace-element metabolism during infection. The heat-labile endogenous factor was first found by Kampschmidt and Upchurch (73) to appear in the serum of rats given bacterial endotoxin; serum containing the factor could then be transferred to normal rats to cause a reduction in their serum iron concentrations. Existence of such a hormone-like factor (or factors) has been amply confirmed in our laboratories (27, 28, 38–42, 81–88, 129) and in subsequent studies in Kampschmidt's laboratory (144–147). The factor, which we term LEM, is secreted by phagocytizing cells including neutrophils, macrophages, and monocytes of various species. A heat-labile substance with the properties of LEM has been shown to appear in the serum of human beings who suffer from a wide variety of infectious illnesses, including those caused by viruses or bacteria which contain no endotoxin.

When injected into normal animals, LEM stimulates the onset of an abrupt movement of iron, zinc, and most amino acids into the liver of the recipient animal, and then, after a slight delay, increased RNA formation in hepatic cell nuclei, accumulation of newly synthesized RNA in the membrane-bound hepatic ribosomes, and, finally, an increased hepatic synthesis of various acute-phase plasma glycoproteins, including ceruloplasmin (82, 84,

86, 87). These hormone-like actions of LEM are not mimicked by other hormones, alone or in combination (86, 88).

LEM appears to be a protein of low molecular weight (39, 86), which is released from phagocytic cells that have been stimulated by live or heat-killed microorganisms, bacterial endotoxin, sterile inflammation, chemical irritants, or double-stranded synthetic RNA (148). Kampschmidt, Upchurch, and Eddington (144) suggested that the mediator might be formed by a complex of endotoxin with a protein present in the serum, but arguments against this concept include the stimulation of LEM release from washed leukocytes by organisms and substances that contain no endotoxin (38, 39, 148). LEM is totally inactivated after treatment with proteolytic enzymes (149) and is fully effective in animals that are rendered tolerant to extremely large doses of endotoxin (145–147). Further, the initiation and peak effects of LEM occur sooner than those seen after endotoxin.

Ongoing attempts to isolate, concentrate, and purify LEM show that this factor has many of the properties of the endogenous mediator of fever, endogenous pyrogen (146–147), which is also released by phagocytizing host cells but acts on the hypothalamus instead of the liver. Additional studies will be needed to purify LEM, in order to determine if it is endogenous pyrogen or a separate substance and to define the purpose of its actions on the liver during infection and inflammatory conditions.

Direct Interactions of Trace Elements with Invading Microorganisms

The possibility that invading microorganisms compete with host cells for essential trace elements has been the subject of considerable research (150). Such investigations were designed to determine if the nutritional status of trace elements within host tissues might influence susceptibility to an infection, to learn whether administration of a trace element might benefit the host, or, conversely, to see if such therapy might provide invading microorganisms with essential micronutrients which could allow them to proliferate more luxuriously. These studies are outside the purview of the present review but will be discussed in other papers in this symposium.

SUMMARY

This review describes the infection-related changes in trace-element metabolism of the host, and relates the reported changes in underlying pathophysiologic mechanisms. A depression of serum iron and zinc values is typical of most infections, while serum copper and ceruloplasmin concentrations increase. In infectious hepatitis, serum iron values increase rather than decline.

Loss of body stores of an element may contribute to trace-element changes. However, the earliest changes result from a redistribution of the metals within the body initiated by a hormone-like protein factor which is released from phagocytizing cells. This factor, leukocytic endogenous mediator (LEM), stimulates the liver to take up iron and zinc from serum and to synthesize additional quantities of ceruloplasmin. The actions of LEM can be demonstrated by injecting a normal test animal with sterile serum obtained from an infected subject.

REFERENCES

1. Beisel, W. R., and Pekarek, R. S. 1972. Acute stress and trace element metabolism. *Int. Rev. Neurobiol.* suppl. 1: 53, Academic Press, New York.
2. O'Shaughnessy, W. B. 1831–1832. Experiments on the blood in cholera. *Lancet* 1: 490.
3. Latta, T. 1831–1832. A view of the rationale and results of the treatment of cholera by aqueous and saline injections. *Lancet* 2: 274.
4. Heller, J. F. 1847. Chemische Untersuchung des Harns, der Harnsedimente und Konkretionen am Krankenbette, nebst diagnostischen Beiträgen. *Arch. Physiol. Pathol. Chem. Mikroskop.* IV: 491.
5. Parkes, E. A. 1857. Maculated typhus–sudden termination on the twelfth day. *Med. Times Gazette* 14: 207.
6. Krebs, H. A. 1928. Uber das Kupfer im menschlichen Blutserum. *Klin. Wschr.* 7: 584.
7. Locke, A., Main, E. R., and Rosbach, D. O. 1932. The copper and non-hemoglobinous iron contents of the blood serum in disease. *J. Clin. Invest.* 11: 527.
8. Vikbladh, I. 1951. Studies on zinc in blood. II. *Scand. J. Clin. Lab. Invest.* 3 (suppl. 2): 1.
9. Nazarmukhamedova, M., and Nugamanova, R. N. 1964. A dynamic study of cobalt and manganese in patients with infectious hepatitis. Abstracted in *Ref. Zh. Biol. Khim.* abstr. no. 10F1161.
10. Belozerov, E. S. 1966. Changes of the microelement gallium in patients with epidemic hepatitis. *Stud. Cercet. Inframicrobiol.* 17: 279.
11. Pekarek, R. S., Hauer, E. C., Wannemacher, R. W., Jr., and Beisel, W. R. 1973. Serum chromium concentrations and glucose utilization in healthy and infected subjects. *Fed. Proc.* 32: 930.
12. Smith, C. H. 1966. *Blood Diseases of Infancy and Childhood,* 2nd edn, C. V. Mosby, St. Louis, Mo., p. 207.
13. Heilmeyer, L., and Plotner, K. 1937. *Das Serumeisen und die Eisenmangelkrankheit,* G. Fischer, Jena, p. 50.
14. Schaefer, K. H. 1940. Zur Pathogenese der Infektanämie: insbesondere ihre Beziehungen zum Eisenstoffwechsel des wachsenden Organismus. *Klin. Wschr.* 19: 590.

15. Vahlquist, B. C. 1941. Das Serumeisen. Eine pädiatrischklinische und experimentelle Studie. *Acta Paediat.* 28: 374.
16. Cartwright, G. E., and Wintrobe, M. M. 1949. Chemical, clinical and immunological studies on the products of human plasma fractionation. XXXIX. The anemia of infection. Studies on the iron-binding capacity of serum. *J. Clin. Invest.* 28: 86.
17. Brendstrup, P. 1953. Serum copper, serum iron, and total iron-binding capacity of serum in acute and chronic infections. *Acta Med. Scand.* 145: 315.
18. Adams, E. B., and Mayet, F. G. 1966. Hypochromic anaemia in chronic infections. *S. A. Med. J.* 40: 38.
19. Bothwell, T. H., Callender, S., Mallett, B., and Witts, L. J. 1956. The study of erythropoiesis using tracer quantities of radioactive iron. *Br. J. Haemat.* 2: 1.
20. Gonkzik, M. 1966. The fluctuation of serum iron and copper levels in urogenital tuberculosis. *Z. Urol.* 59: 241.
21. Mukerji, P. K., Khanna, B. K., and Majhur, J. B. 1966. Some observations on serum iron in pulmonary tuberculosis. *Indian J. Chest Dis.* 8: 183.
22. Vannotti, A. 1957. *In* B. N. Halpern (ed.), *Pathophysiology of the Reticuloendothelial System,* Blackwell, Oxford, p. 172.
23. Palukiewicz, J. 1966. Serum iron and copper levels in patients with tuberculosis and impaired liver parenchyma. *Gruzlica* 34: 825.
24. Roberts, P. D., Hoffbrand, A. V., and Mollin, D. L. 1966. Iron and folate metabolism in tuberculosis. *Br. Med. J.* 2: 198.
25. Beard, R. J., and Brooke, B. N. 1967. The effect of postoperative infection on the anaemia of chronic inflammation. *Lancet* 2: 1113.
26. Kalnai, E. H., and Hever, O. 1968. Untersuchungen des Serum-eisen-und Kupferspiegels bei Lungentuberkulose. *Beitr. Klin. Erforsch. Tuberk.* 137: 19.
27. Pekarek, R. S., Bostian, K. A., Bartelloni, P. J., Calia, F. M., and Beisel, W. R. 1969. The effects of *Francisella tularensis* infection on iron metabolism in man. *Am. J. Med. Sci.* 258: 14.
28. Wannemacher, R. W., Jr., DuPont, H. L., Pekarek, R. S., Powanda, M. C., Schwartz, A., Hornick, R. B., and Beisel, W. R. 1972. An endogenous mediator of depression of amino acids and trace metals in serum during typhoid fever. *J. Infect. Dis.* 126: 77.
29. Musabaev, I. K., Amdartsumov, S. M., and Miraliev, A. 1969. Iron metabolism in acute infectious intestinal diseases. *Med. Zh. Uzb.* 1: 11.
30. Caroline, L., Rosner, F., and Kozinn, P. J. 1969. Elevated serum iron, low unbound transferrin and candidiasis in acute leukemia. *Blood* 34: 441.
31. Cartwright, G. E., Lauritsen, M. A., Humphreys, S., Jones, P. J., Merrill, I. M., and Wintrobe, M. M. 1946. The anemia of infection. II. The experimental production of hypoferremia and anemia in dogs. *J. Clin. Invest.* 25: 81.
32. Cartwright, G. E., Lauritsen, M. A., Humphreys, S., Jones, P. J., Merrill,

I. M., and Wintrobe, M. M. 1946. Anemia associated with chronic infection. *Science* 103: 72.
33. Gubler, C. J., Cartwright, G. E., and Wintrobe, M. M. 1950. The anemia of infection. X. The effect of infection on the absorption and storage of iron by the rat. *J. Biol. Chem.* 184: 563.
34. Kawamura, J. 1965. Studies of iron metabolism in experimental infection. *Acta. Paediat. Jap.* 7: 79.
35. Soliman, M. K., Ahmed, A. A. S., El Amrousi, S., and Moustafa, I. H. 1966. Cytological and biochemical studies on the blood constituents of normal and spirochete-infected chickens. *Avian Dis.* 10: 394.
36. Takemoto, M. 1965. Variations in the sideroblast in an experimental infectious disease. *Acta Paediat. Jap.* 69: 579.
37. Urushizaki, I., Tsutsui, H., and Kodama, T. 1965. Iron metabolism in tumor-bearers. 2. Comparison with daikoku rats with *Staphylococcus aureus* infection. *Med. Biol. (Tokyo)* 71: 236.
38. Pekarek, R. S., and Beisel, W. R. 1970. Endogenous mediator of serum iron depression during infection and endotoxemia. *Bact. Proc.* p. 81.
39. Pekarek, R. S., and Beisel, W. R. 1971. Characterization of the endogenous mediator(s) of serum zinc and iron depression during infection and other stresses. *Proc. Soc. Exp. Biol. Med.* 138: 728.
40. Pekarek, R. S., and Beisel, W. R. 1972. The redistribution and sequestering of essential trace elements during acute infection. *Proceedings of International Congress on Nutrition,* p. 201.
41. Pekarek, R. S., and Beisel, W. R. 1971. Metabolic losses of zinc and other trace elements during acute infection. *Program, Western Hemisphere Nutritional Congress* III, p. 43.
42. Beisel, W. R., Wannemacher, R. W., Jr., Pekarek, R. S., and Bartelloni, P. J. 1970. Early changes in individual serum amino acids and trace metals during a benign viral illness of man. *Am. J. Clin. Nutr.* 23: 660.
43. Pekarek, R. S., Burghen, G. A., Bartelloni, P. J., Calia, F. M., Bostian, K. A., and Beisel, W. R. 1970. The effect of live attenuated Venezuelan equine encephalomyelitis virus on serum iron, zinc, and copper concentrations in man. *J. Lab. Clin. Med:* 76: 293.
44. Mirand, E. A., and Grace, J. T., Jr. 1963. Responses of germ-free mice to Friend virus. *Nature, Lond.* 200: 92.
45. Moghilnikov, V. G. 1965. Copper and iron and their quantitative interrelations in the blood serum and cerebrospinal fluid of patients with tick-borne encephalitis. *Zh. Nevropatol. Psikhiatr.* 65: 40.
46. Borisenko, A. M. 1965. Content of copper, zinc and iron in the blood in various stages of syphilis. *Vestn. Dermotol. Venerol.* 39: 52.
47. Tumasheva, N. I., and Borisenko, A. M. 1965. Copper, zinc and iron content in syphilitic patients with lesions of the central nervous system. *Zh. Nevropatol. Psikhiatr.* 65: 37.
48. Olszewska, Z. 1966. Iron metabolism in early symptomatic syphilis. *Pol. Med. J.* 5: 686.
49. Devakul, K., Areekul, S., and Viravan, C. 1967. Vitamin B_{12} absorption test in amoebic liver abscess. *Ann. Trop. Med. Parasit.* 61: 29.
50. Jamra, M., Maspes, V., and Meira, D. A. 1961. Types and mechanisms

of anemia in *Schistosomiasis mansoni. Rev. Inst. Med. Trop., Sao Paulo* 6: 126.
51. Sinclair, K. B. 1965. Iron metabolism in ovine fascioliasis. *Br. Vet. J.* 121: 451.
52. Layrisse, M., Blumenfeld, N., Carbonell, L., Desenne, J., and Roche, M. 1964. Intestinal absorption tests and biopsy of the jejunum in subjects with heavy hookworm infection. *Am. J. Trop. Med.* 13: 297.
53. Miall, W. E., Milner, P. F., and Lovell, H. G. 1967. Haematological investigations of population samples in Jamaica. *Br. J. Prev. Soc. Med.* 21: 45.
54. West, H. D., Jackson, A. H., Elliott, R. R., Hahn, P. F., Patterson, W. A., and Anderson, R. S. 1952. The utilization of ingested iron in disease. *South. Med. J.* 45: 629.
55. Wintrobe, M. M., Greenberg, G. R., Humphreys, S. R., Ashenbrucker, H., Worth, W., and Kramer, R. 1947. The anemia of infection. III. The uptake of radioactive iron in iron-deficient and in pyridoxine-deficient pigs before and after acute inflammation. *J. Clin. Invest.* 26: 103.
56. Kuhns, W. J., Gubler, C. J., Cartwright, G. E., and Wintrobe, M. M. 1950. The anemia of infection. XIV. Responses to massive doses of intravenously administered saccharated oxide of iron. *J. Clin. Invest.* 29: 1505.
57. Bush, J. A., and Cartwright, G. E. 1954. Ferrokinetics in anemia of infection. *J. Clin. Invest.* 33: 921.
58. Wohler, F., and Otte, W. 1961. Zur Pathologie des Speichereisens. II. Mitteilung: Ferritin und Hämosideringehalt menschlicher Organe bei entzündlichen Erkrankungen, Geschwulsten und Krankheiten des blutbildenden Systems. *Acta Haemat., Basel* 26: 81.
59. Tani, P. 1965. Serum iron, copper, iron-binding capacity and marrow hemosiderin in pulmonary tuberculosis. *Ann. Med. Intern. Fenn.* 54 (suppl. 44): 1.
60. Cartwright, G. E., Gubler, C. J., and Wintrobe, M. M. 1950. The anemia of infection. XII. The effect of turpentine and colloidal thorium dioxide on the plasma iron and plasma copper of dogs. *J. Biol. Chem.* 184: 579.
61. Linchevskaya, L. P. 1966. Changes in the arteriovenous difference in the content of iron, copper, and cobalt, as well as in ceruloplasmin and transferrin in aseptic inflammation. *Patol. Fiziol. Eksp. Ter.* 10: 75.
62. Freireich, E. J., Miller, A., Emerson, C. P., and Ross, J. F. 1957. The effect of inflammation on the utilization of erythrocyte and transferrin bound radioiron for red cell production. *Blood* 12: 972.
63. Hahn, P. F., Bale, W. F., and Whipple, G. H. 1946. Effects of inflammation (turpentine abscess) on iron absorption. *Proc. Soc. Exp. Biol. Med.* 61: 405.
64. Gubler, C. J., Cartwright, G. E., and Wintrobe, M. M. 1950. The anemia of infection. XI. The effect of turpentine and cobalt on the absorption of iron by the rat. *J. Biol. Chem.* 184: 575.

65. Menkin, V. 1930. Studies on inflammation. III. Fixation of a metal in inflamed areas. *J. Exp. Med.* 51: 879.
66. Menkin, V., and Menkin, M. F. 1931. The accumulation of iron in tuberculosis areas. *J. Exp. Med.* 53: 919.
67. Cartwright, G. E., Lauritsen, M. A., Jones, P. J., Merrill, I. M., and Wintrobe, M. M. 1946. The anemia of infection. I. Hypoferremia, hypercupremia, and alterations in porphyrin metabolism in patients. *J. Clin. Invest.* 25: 65.
68. Heilmeyer, L., Keiderling, W., and Wöhler, F. 1958. Der Eisenstoffwechsel beim Infekt und die Entgiftungsfunktion des Speichereisens. *Dt. Med. Wschr.* 83: 1965.
69. Heilmeyer, L., and Keiderling, W. 1959. Der Turnover des Hämoglobin- und des Nichthämoglobeneisens beim Gesunden, beim Infekt, und bei malignen Tumoren. *Dt. Med. Wschr.* 84: 724.
70. Brendstrup, P. 1953. Serum copper, serum iron, and total iron-binding capacity of serum during treatment with coli vaccine. *Acta Med. Scand.* 146: 114.
71. Mazur, A., Carleton, A., and Carlsen, A. 1961. Relation of oxidative metabolism to the incorporation of plasma iron into ferritin *in vivo*. *J. Biol. Chem.* 236: 1109.
72. Kampschmidt, R. F., and Schultz, G. A. 1961. Hypoferremia in rats following injection of bacterial endotoxin. *Proc. Soc. Exp. Biol. Med.* 106: 870.
73. Kampschmidt, R. F., and Upchurch, H. F. 1962. Effects of bacterial endotoxin on plasma iron. *Proc. Soc. Exp. Biol. Med.* 110: 191.
74. Kampschmidt, R. F., and Arredondo, M. I. 1963. Some effects of endotoxin upon plasma iron turnover in the rat. *Proc. Soc. Exp. Biol. Med.* 113: 142.
75. Kampschmidt, R. F., and Upchurch, H. F. 1964. Effect of endotoxin on total iron-binding capacity of the serum. *Proc. Soc. Exp. Biol. Med.* 116: 420.
76. Baker, P. J., and Wilson, J. B. 1965. Hypoferremia in mice and its application to the bioassay of endotoxin. *J. Bacteriol.* 90: 903.
77. Kariyone, S., and Miyake, T. 1966. Reticuloendothelial cell in the bone marrow and its relationship to iron metabolism. *Tohoku J. Exp. Med.* 89: 213.
78. Eaves, G. N., and Berry, L. J. 1966. Effect of cortisone on plasma iron concentration of normal and endotoxin-poisoned mice. *Am. J. Physiol.* 211: 800.
79. Fukuda, M., Okada, K., Akikawa, K., Matsuda, M., and Urushizaki, I. 1966. Comparative studies on the biological effect of toxohormone and bacterial lipopolysaccharide. *Gann* 57: 27.
80. Cortell, S., and Conrad, M. E. 1967. Effect of endotoxin on iron absorption. *Am. J. Physiol.* 213: 43.
81. Wannemacher, R. W., Jr., Pekarek, R. S., and Beisel, W. R. 1972. Mediator of hepatic amino acid flux in infected rats. *Proc. Soc. Exp. Biol. Med.* 139: 128.
82. Cockerell, G. L. 1972. Plasma protein and glycoprotein changes in inflammation, infection, and/or starvation. *Fed. Proc.* 31: 710.

83. Wannemacher, R. W., Jr., Pekarek, R. S., and Beisel, W. R. 1972. An endogenous mediator(s) of plasma amino acid flux and trace metal depression during experimentally induced infection in man. *Am. J. Clin. Nutr.* 25: 461.
84. Pekarek, R. S., Powanda, M. C., and Wannemacher, R. W., Jr. 1972. The effect of leukocytic endogenous mediator (LEM) on serum copper and ceruloplasmin concentrations in the rat. *Proc. Soc. Exp. Biol. Med.* 141: 1029.
85. Pekarek, R. S., Wannemacher, R. W., Jr., and Beisel, W. R. 1972. The effect of leukocytic endogenous mediator (LEM) on the tissue distribution of zinc and iron. *Proc. Soc. Exp. Biol. Med.* 140: 685.
86. Wannemacher, W. R., Jr., Pekarek, R. S., and Beisel, W. R. 1973. A hormone-like, leukocytic endogenous mediator (LEM) which can regulate amino acid transport and protein synthesis in liver. *Am. J. Clin. Nutr.* 26: 460.
87. Powanda, M. C., Pekarek, R. S., Cockerell, G. L., Wannemacher, R. W., Jr., and Beisel, W. R. 1973. Mediator of alterations in protein synthesis during infection and inflammation. *Fed. Proc.* 32: 953.
88. Pekarek, R. S., Wannemacher, R. W., Jr., Powanda, M. C., and Beisel, W. R. 1973. Regulation of infection-induced alterations in host metabolism by hormone-like mediator released from polymorphonuclear leukocytes. *Clin. Res.* 21: 608.
89. Brochner-Mortensen, K. 1942. Iron content of the serum in lesions of the liver and bile passages. *Acta Med. Scand.* 112: 277.
90. Brendstrup, P. 1953. Serum iron, total iron-binding capacity of serum, and serum copper in acute hepatitis. *Acta Med. Scand.* 146: 107.
91. Peterson, R. E. 1952. The serum iron in acute hepatitis. *J. Lab. Clin. Med.* 39: 225.
92. Ducci, H., Spoerer, A., and Katz, R. 1952. Serum iron in liver disease. *Gastroenterology* 22: 52.
93. Stone, C. M., Rumball, J. M., and Hassett, C. P. 1955. An evaluation of the serum iron in liver disease. *Ann. Intern. Med.* 43: 229.
94. Shamroth, L., Edelstein, W., Politzer, W. M., and Stevens, N. 1956. Serum iron in the diagnosis of hepatobiliary disease. *Br. J. Med.* 1: 960.
95. Rumball, J. M., Stone, C. M., and Hassett, C. P. 1959. The behavior of serum iron in acute hepatitis. *Gastroenterology* 36: 219.
96. Kaszewska-Jablonska, I. 1963. Serum iron level in epidemic and vaccination-caused hepatitis. *Pol. Arch. Med. Wewn.* 33: 15.
97. Maksimova, L. A., and Strozhe, T. Ya. 1964. Simultaneous determination of copper and iron in blood serum and the clinical significance of determination of these elements in different types of jaundice. Abstracted in *Ref. Zh. Biol. Khim.* abstr. no. 9F1126.
98. Sokolovskaya, Ya. A. 1964. Clinical significance of determination of serum iron level in infectious hepatitis. Abstracted in *Ref. Zh. Biol. Khim.* abstr. no. 2F1176.
99. Yakovlev, A. F. 1967. Copper and iron content and their relation to protein fractions in the blood serum of infectious hepatitis patients. Abstracted in *Chem. Abstr.* 67: abstr. no. 52146e.

100. Frangini, V., Moggi, C., Poggini, G., and Pratesi, C. 1966. Total erythrocytic and serum iron in the course of viral hepatitis. *Riv. Clin. Pediat.* 77: 121.
101. Zhernakova, T. V. 1966. The correlation between iron and copper and between copper and manganese in blood serum in epidemic hepatitis (Botkins disease). *Ter. Arkh.* 38: 59.
102. Turnberg, L. A. 1966. Iron absorption in acute hepatitis. *Am. J. Dig. Dis.* 11: 20.
103. Brodanova, M., Hoenig, V., and Bila, L. 1966. Plasma iron level and total iron binding capacity of the blood in the course of infectious hepatitis in children. *Cesk. Pediat.* 21: 296.
104. Curti, B., Bianco, G., and Doglio, R. 1967. Study of the absorption of iron (ferrous sulfate) in patients with acute viral hepatitis. *Boll. Soc. Ital. Biol. Sper.* 43: 326.
105. Bolin, T., and Davis, A. E. 1968. Iron absorption in infectious hepatitis. *Am. J. Dig. Dis.* 13: 16.
106. Hemmeler, G. 1939. Serumeisen bei Ikterus. *Helv. Med. Acta* 6: 678.
107. Scuro, L. A., and Dobrilla, G. 1967. Siderosis, haemolysis of hepatonecrosis in increasing post-desferrioxamine sideruria in acute viral hepatitis. *Postgrad. Med. J.* 43: 708.
108. Scuro, L. A., Dobrilla, G., and Innecco, A. 1966. Histochemical and functional investigations on the genesis of hypersideremia in acute viral hepatitis. *Policlinico (Med.)* 73: 73.
109. Burger, T., Barna, K., and Keszthelyi, B. 1966. Iron turnover in infectious hepatitis. *Z. Gesamt. Inn. Med.* 21: 368.
110. Markowitz, H., Gubler, C. J., Mahoney, J. P., Cartwright, G. E., and Wintrobe, M. M. 1955. Studies on copper metabolism. XIV. Copper, ceruloplasmin and oxidase activity in sera of normal human subjects, pregnant women, and patients with infection, hepatolenticular degeneration and nephrotic syndrome. *J. Clin. Invest.* 34: 1498.
111. Butt, E. M., and Nusbaum, R. E. 1962. Trace metals and disease. *Ann. Rev. Med.* 13: 471.
112. Pastukhova, G. M., and Raskovalov, M. G., 1964. Copper content in the blood of typhus abdominalis patients. Abstracted in *Ref. Zh. Biol. Khim.* abstr. no. 2F1251.
113. Borisova, M. A. 1966. The indices of ceruloplasmin activity, the blood content of copper and zinc in patients with typhoid and dysentery in levomycetin therapy. *Sov. Med.* 29: 59.
114. Kleinbaum, H. 1968. Ceruloplasmin direkt reagierendes Kupfer und Gesamt Kupfergehalt bei verschiedenen Infektionskrankheiten und entzündlichen Zustandsbildern. *Z. Kinderheilkd.* 102: 84.
115. Chitre, V. S., and Balasubrahmanyan, M. 1969. Changes in serum copper and PPD-Oxidase in different diseases. I. Differential observations in certain types of leprosy and kwashiorkor. *Indian J. Med. Res.* 57: 228.
116. Bozhkov, B. 1967. Study of the synthesis of ceruloplasmin with the use of ^{64}Cu in experimental streptococcal myocarditis. *Patol. Fiziol. Eksp. Ter.* 11: 20.

117. Cicchini, T., and Messeri, E. 1968. Comportamento della cupremia del ratto in corso di infezione sperimentale da *Trypanosoma lewisi. Arch. Ital. Sci. Med. Trop. Parassitol.* 49: 171.
118. Squibb, R. L., Beisel, W. R., and Bostian, K. A. 1971. Effect of Newcastle disease on serum copper, zinc, cholesterol, and carotenoid values in the chick. *Appl. Microbiol.* 22: 1096.
119. Sternlieb, I., and Scheinberg, H. 1961. Ceruloplasmin in health and disease. *Ann. N.Y. Acad. Sci.* 94: 71.
120. Husain, S., and Pohowalla, J. N. 1967. Serum copper levels in jaundice in infancy and childhood. *Indian J. Pediat.* 34: 131.
121. de Jorge, F. B. 1969. Correlation between copper and transaminases in the blood serum in infectious hepatitis during convalescence. *Clin. Chim. Acta* 20: 111.
122. Henkin, R. I., and Smith, F. R. 1972. Zinc and copper metabolism in acute viral hepatitis. *Am. J. Med. Sci.* 264: 401.
123. Trip, J. A. J., Que, G. S., Botterweg-Span, Y., and Mandema, E. 1969. The behaviour of ceruloplasmin fractions in liver diseases. *Acta Med. Scand.* 185: 279.
124. Smallwood, R. A., Williams, H. A., Rosenoer, V. M., and Sherlock, S. 1968. Liver-copper levels in liver disease. Studies using neutron activation analysis. *Lancet* 2: 1310.
125. Roeser, H. P., Lee, G. R., and Cartwright, G. E. 1973. Role of ceruloplasmin (plasma ferroxidase) in hypoferremia associated with chronic inflammation and endotoxin. *Proc. Soc. Exp. Biol. Med.* 142: 1155.
126. Halsted, J. A., and Smith, J. C., Jr. 1970. Plasma-zinc in health and disease. *Lancet* 1: 322.
127. Turk, D. E., and Stephens, J. F. 1967. Upper intestinal tract infection produced by *E. acervulina* and absorption of ^{65}Zn and ^{131}I-labeled oleic acid. *J. Nutr.* 93: 161.
128. Booth, G. H., Jr., and Schulert, A. R. 1968. Zinc metabolism in schistosomes. *Proc. Soc. Exp. Biol. Med.* 127: 700.
129. Pekarek, R. S., and Beisel, W. R. 1969. Effect of endotoxin on serum zinc concentrations in the rat. *Appl. Microbiol.* 18: 482.
130. Wannemacher, R. W., Jr., Pekarek, R. S., Bartelloni, P. J., Vollmer, R. T., and Beisel, W. R. 1972. Changes in individual plasma amino acids following experimentally induced sand fly fever virus infection. *Metabolism* 21: 67.
131. Caviedes, R., Gonzalez, C., and Klinger, J. 1964. Changes in zinc metabolism in hepatic diseases. *Rev. Med. Chil.* 92: 456.
132. Halstead, J. A., Hackley, B., Rudzki, C., and Smith, J. C., Jr. 1968. Plasma zinc concentration in liver diseases. Comparison with normal controls and certain other chronic diseases. *Gastroenterology* 54: 1098.
133. Karlinskii, V. M., and Roomere, P. A. 1965. Changes in the zinc content of the blood serum and urine in Botkin's diseases. *Klin. Med. Mosk.* 43: 78.
134. Davis, I. J. T., Musa, M., and Dormandy, T. L. 1968. Measurements of plasma zinc. I. In health and disease. *J. Clin. Path.* 21: 359.

135. Ol'gina, F. P. 1963. The zinc content of the blood and its elimination from the body in Botkin's disease. *Klin. Med. Mosk.* 41: 23.
136. Kahn, A. M., Helwig, H. L., Redeker, A. G., and Reynolds, T. B. 1965. Urine and serum zinc abnormalities in disease of the liver. *Am. J. Clin. Path.* 44: 426.
137. Klebanoff, S. J. 1967. Iodination of bacteria: a bactericidal mechanism. *J. Exp. Med.* 126: 1063.
138. Beisel, W. R. 1972. Interrelated changes in host metabolism during generalized infectious illness. *Am. J. Clin. Nutr.* 25: 1254.
139. Scudamore, H. H., Thompson, J. H., Jr., and Owen, C. A., Jr. 1961. Absorption of ^{60}Co-labeled vitamin B_{12} in man and uptake by parasites, including *Diphyllobothrium latum. J. Lab. Clin. Med.* 57: 240.
140. Conzolazio, C. F., Matoush, L. O., Nelson, R. A., Harding, R. S., and Canham, J. E. 1963. Excretion of sodium, potassium, magnesium and iron in human sweat and the relation of each to balance and requirements. *J. Nutr.* 79: 407.
141. Lindeman, R. D., Yunice, A. A., Baxter, D. J., Miller, L. R., and Nordquist, J. 1973. Myocardial zinc metabolism in experimental myocardial infarction. *J. Lab. Clin. Med.* 81: 194.
142. McBean, L. D., Smith, J., Cecil, B., Bernard, H., and Halsted, J. A. 1972. Serum zinc and alpha-macroglobulin concentration in patients with various disorders. *Proceedings of Ninth International Congress on Nutrition,* p. 56.
143. Mertz, W. 1969. Chromium occurrence and function in biological systems. *Physiol. Rev.* 49: 163.
144. Kampschmidt, R. F., Upchurch, H. F., and Eddington, C. L. 1969. Hypoferremia produced by plasma from endotoxin-treated rats. *Proc. Soc. Exp. Biol. Med.* 132: 817.
145. Kampschmidt, R. F., and Upchurch, H. 1969. Lowering of plasma iron concentration in the rat with leukocytic extracts. *Am. J. Physiol.* 216: 1287.
146. Kampschmidt, R. F., and Upchurch, H. F. 1970. A comparison of the effects of rabbit endogenous pyrogen on the body temperature of the rabbit and lowering of plasma iron in the rat. *Proc. Soc. Exp. Biol. Med.* 133: 128.
147. Kampschmidt, R. F., and Upchurch, H. F. 1970. The effect of endogenous pyrogen on the plasma zinc concentrations of the rat. *Proc. Soc. Exp. Biol. Med.* 134: 1150.
148. Pekarek, R. S. 1971. Effect of synthetic double-stranded RNA on serum metals in the rat, rabbit, and monkey. *Proc. Soc. Exp. Biol. Med.* 136: 584.
149. Pekarek, R. S., Wannemacher, R. W., Jr., Chapple, F. E., III, Powanda, M. C., and Beisel, W. R. 1972. Further characterization and species specificity of leukocytic endogenous mediator (LEM). *Proc. Soc. Exp. Biol. Med.* 141: 643.
150. Scrimshaw, N. S., Taylor, C. E., and Gordon, J. E. 1968. *Interactions of Nutrition and Infection,* World Health Organization, Geneva, p. 125.

DISCUSSION

Petering (Cincinnati). It has been suggested that lead toxicity is increased during a period of infection. Do you have any comment on this?

Beisel. I cannot shed any light on this. We have never actually looked at lead values in infected individuals.

Spallholz (Fort Collins). Most of the data you presented were in cases of pathogenic disease and fever. Are there changes in zinc and copper metabolism in the noninfectious state, such as the administration of noninfectious antigens?

Beisel. We have seen no such change with noninfectious antigens.

Matrone (Raleigh). Do you have any idea where the copper that's getting into ceruloplasmin is coming from?

Beisel. On studies in rats given repeated doses of LEM we do find enough of a diminution in hepatic copper to be statistically significant at the 5% level. Some individuals have measured copper in autopsy material and did not find much of a change in human liver. My guess would be that the amount of copper needed to saturate the ceruloplasmin being produced is not large in terms of total liver stores, and would be difficult to measure. We think the copper needed for the ceruloplasmin is coming from liver stores, but we have no direct evidence.

Frieden (Tallahassee). I think that LEM is obviously a very exciting factor, but it could be acting in various ways. It could be affecting the release of ferritin iron, and thus affecting the serum level of iron, or even more likely the biosynthesis of liver ferritin, so that actually the iron was being deflected to less useful channels and forming hemosiderin. Also, the ceruloplasmin levels are far in excess of what is needed to mobilize iron in the copper-deficient animal, so that a little more won't make much difference, and there must be some other controlling process which LEM is affecting. Some other agents like estrogens induce a two- to three-fold increase in serum copper and serum ceruloplasmin, and I wondered if estrogens produce differences in other trace metals such as in serum iron and serum zinc?

Beisel. These are questions which are certainly ones we've been considering. The primary effect of LEM in our short-term studies appears to be on the hepatocyte, yet most of the iron stores that have been investigated are in the reticuloendothelial system so that we do not yet know enough about the total processes involved even to approach these points of information. We have not looked at the conversion of apoferritin to ferretin as a factor in the infectious state or as a factor following an endotoxin dose. The total change in the various trace metals is dependent, in a large degree, upon the timing of the experiment and upon the dose responses that one observes. There are certainly differences in various infecting microorganisms and their effect on the total quantities of either ceruloplasmin, iron, or zinc. The subjects that we have studied are individuals who have only one or two days of fever, and don't represent the very sick individual that can be seen in the hospital clinics. Children with pneumococcal meningitis, for instance, may have serum iron levels that are too low to measure by conventional techniques so these changes can be profound. The estrogen effects and other hormonal effects do get involved in infection, but I would say that the changes of hormonal levels in infected individuals are too small and of too brief a duration to account for the changes in trace metals that we see.

Henkin (Bethesda). I should like to comment on the difference between distribution or redistribution of metals in patients with various infections, and what happens to zinc metabolism during hepatitis. There's a difference in these two phenomena in that patients with hepatitis lose zinc from the body. The zinc is not redistributed; it's actually lost in the urine, whereas in Dr. Beisel's example with infection there is a redistribution of zinc.

ROLES OF TEMPERATURE AND TRACE-METAL METABOLISM IN HOST-PATHOGEN INTERACTIONS[1]

EUGENE D. WEINBERG

Department of Microbiology and Program in Medical Sciences, Indiana University, Bloomington, Indiana

In the subtle balance between host and invader, fever may eventually be shown to confer a greater advantage to the defense mechanisms of the host than to the invasive properties of the microorganism . . . it is difficult to believe that this universal response of warm-blooded animals would have survived if it did not, indeed, serve some useful purpose in combating disease. (1)

If fever is often a significant factor in man's resistance to microbial invasion, the fact is not immediately evident from observation of patients with common infectious diseases. (2)

During the past three decades, it has become apparent that specific metabolic reactions of microbial cells may have temperature optima that differ from the temperature at which the generation time is most brief (3, 4). Some of these reactions might be essential for a particular facet of microbial pathogenicity; if so, the amount of virulence at various temperatures could be predicted to be more or less independent of cell number. A corollary prediction is that hosts might be able to suppress virulence of an invading pathogen by altering their body temperatures.

In this paper we shall consider the influence of temperature on two facets of pathogenicity: (1) microbial biosynthesis of toxic secondary metabolites, and (2) ability of microorganisms to acquire growth-essential trace

1. Support for this work was provided in part by research grants from the Department of Animal Health and Nutrition, Commercial Solvents Corporation, and from the National Institute of Child Health and Human Development, USPHS (HD-03038-03).

metals and to tolerate host-defense factors. We shall conclude with a brief survey of host trace-metal metabolic responses to microbial invasion that might be related to temperature suppression of virulence.

ACTION OF TEMPERATURE ON SECONDARY METABOLISM

Microbial cells which have stopped dividing produce synthetases that catalyze transformation of surplus primary metabolites into biochemically bizarre substances termed secondary metabolites (5, 6). The most probable basic function of the latter is that of serving as disposal packages of materials that might otherwise poison the microorganisms; successful completion of the packaging process results in long-term viability, whereas interruption or distortion of the process causes premature microbial death (7). Although the great majority of the many thousands of known secondary metabolites are biologically inert, a small minority are, probably by accident, pharmacologically active against cells either of other microbial groups (i.e., antibiotics), or of plants or animals (i.e., toxins). At least 40 kinds of pharmacologic action have been identified (7), and the molecular mechanisms that cause these actions must far exceed this number.

Despite the existence of an exceedingly diverse array of both inert and toxic products of secondary metabolism, several unifying principles underlying their formation are known. Among these principles is the generalization that secondary metabolism has narrower tolerances for concentrations of trace metals (6) and inorganic phosphate (8), as well as for ranges of pH, redox potential, and temperature (9), than does cell division. Examples of suppression of biosynthesis of secondary metabolites by temperatures that permit normal vegetative growth are contained in Fig. 1 (10–41).

The majority of examples in Fig. 1 are concerned with molecules of high pharmacologic potency; a few cases of less active or inactive materials (e.g., pigments, citric acid) are included to indicate the general nature of the temperature-restriction phenomenon. In each of the 32 systems shown unrestricted vegetative growth of the producer microorganism occurs throughout a range of temperature greater than that covered by the respective arrows. For example, a considerable increase in viable cells of *Serratia marcescens* occurs through a range of more than 30°C, whereas most of the prodigiosin is formed within a range of 16°C (21); likewise, *Staphylococcus aureus* produces most of its toxin at a much narrower range of temperature than that permitted for growth (22) (Fig. 2).

The lengths of the arrows in Fig. 1 indicate the range of temperature through which increasing suppression of yield of metabolite occurs; however, the tips of the arrows do not represent 100% suppression. In the polymyxin system, for instance, 90–95% inhibition is obtained at 37°C (13). In the rubratoxin system the arrow tip at 30°C represents 50% inhibition (40). The entire range of inhibitory temperature examined for V and W antigen formation by nongrowing cells of *Pasteurella pestis* (18) is presented in Fig. 3; the arrow tips for this sytem in Fig. 1 represent approximately 45% suppression of yield.

In most of the 32 systems in Fig. 1 only a single temperature was indicated by the authors as permitting the maximum yield of the particular secondary metabolite. In nine cases, as shown in Fig. 1, narrow ranges of optimum temperature were reported. Probably such ranges exist in each of the other 23 systems, but complete data are not yet available. Each of the 32 systems was studied *in vitro;* reports have not yet been published that provide comparisons of the amounts of toxic secondary metabolites formed in hypothermic, normothermic, and hyperthermic host animals.

Secondary metabolites are formed as a cluster of closely related molecules; for example, there are at least three naturally occurring ochratoxins, four malformins, five mitomycins, eight aflatoxins, ten polymyxins, and more than 20 actinomycins. The proportions of the various molecules in a chemical family, at least for the aflatoxins, is temperature dependent. In one report, for instance, at 72 hr of incubation the combined totals of aflatoxin G_1 and B_1 were 86, 154, and 60 μg/ml at 20, 25, and 30°C, whereas the percentages of G_1 in the mixture at the three temperatures were 74, 30, and 47, respectively (29). This phenomenon could be of critical importance in those host-parasite interactions in which one member of a chemical family of secondary metabolites might be considerably more toxic than other members of that family. Additionally, the amounts and, consequently, the proportions of two different kinds of toxins formed by single microbial cultures can be altered by temperature. For example, in *Aspergillus ochraceus,* 10–20°C favors penicillic acid synthesis, whereas 28°C stimulates ochratoxin A production (42).

Generally, the influence of temperature on secondary metabolism has been observed to occur during some portion of the transitional period between cessation of vegetative growth and the actual synthesis of the secondary substance. This period is comprised of the sequence of events that include (1) repression of that portion of the genome responsible for cell growth, (2) derepression of the portion required for transcription of secondary metabolite synthetases, and (3) translation of the latter.

In some cases, the temperature optima for the transitional period in which synthetase formation occurs differ from that needed for activity of the synthetases. For example, malformin synthetases are optimally formed at 30°C but they function best at 37°C (32). Zearalenone synthetases are

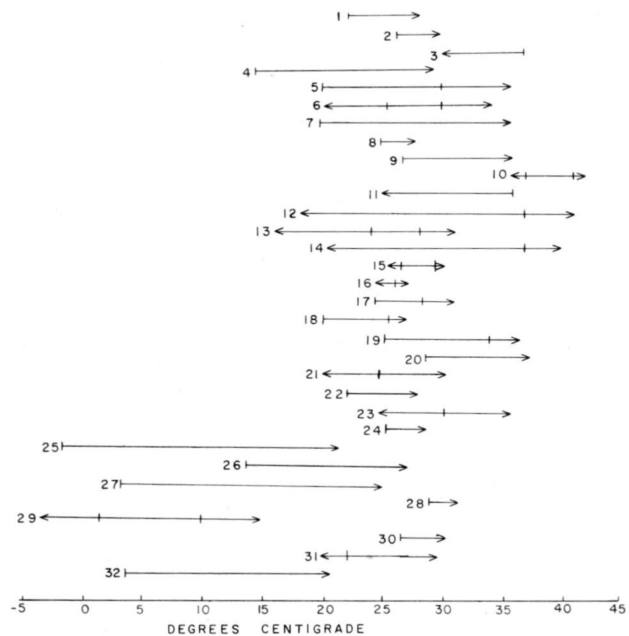

	Arrow no.	Secondary metabolite	Ref. no.
Procaryotes			
1.	*Anabena flos-aquae*	Toxin	(10)
2.	*Aphanizomeron flos-aquae*	Toxin	(10)
3.	*Bacillus cereus*	Toxin	(11)
4.	*Bacillus cereus* var. *alesti*	Pigment	(12)
5.	*Bacillus polymyxa*	Polymyxin	(13)

6.	*Clostridium botulinum* types A,B,E	Toxin	(14,15)
7.	*Listeria monocytogenes*	Hemolysin	(16)
8.	*Microcystis aeruginosa*	Fast-death factor	(10)
9.	*Pasteurella pestis*	Murine toxin	(17)
10.	*Pasteurella pestis*	V & W antigens	(18)
11.	*Pseudomonas aeruginosa*	Pyocyanine	(19)
12.	*Salmonella typhi*	Vi antigen	(20)
13.	*Serratia marcescens*	Prodigiosin	(21)
14.	*Staphylococcus aureus*	Enterotoxin B	(22)
15.	*Streptomyces griseus*	Streptomycin	(23)
16.	*Streptomyces jamaicensis*	Monamycin	(24)
17.	*Streptomyces niveus*	Novobiocin	(25)
18.	*Streptomyces rifamyces*	Rifamycin	(26)
19.	*Vibrio cholerae*	Vascular permeability factor	(27)
Eucaryotes			
20.	*Ashbya gossypii*	Riboflavin	(28)
21.	*Aspergillus flavus*	Aflatoxin	(29,30)
22.	*Aspergillus niger*	Citric acid	(31)
23.	*Aspergillus niger*	Malformin	(32)
24.	*Claviceps purpura*	Ergoline alkaloids	(33)
25.	*Fusarium poae*	Alimentary aleukia toxin	(34)
26.	*Fusarium* sp.	Zearalenone	(35)
27.	*Fusarium tricintum*	Scirpene toxin	(36)
28.	*Gibberella fujikuroi*	Gibberellin	(37)
29.	*Penicillium martensi*	Penicillic acid	(38)
30.	*Penicillium notatum*	Penicillin	(39)
31.	*Penicillium rubrum*	Rubratoxin	(40)
32.	*Penicillium* sp.	Tremortin	(41)

Fig. 1. Suppression of secondary metabolism by temperatures that permit vegetative growth. Vertical bars: temperature at which maximum yield of metabolite is obtained. Arrow tips: temperature at which minimum yield of metabolite is obtained.

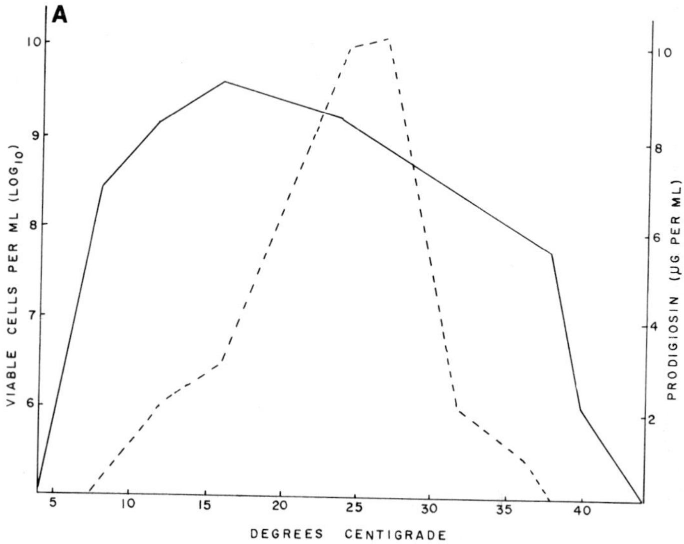

Fig. 2. Effect of temperature on growth and secondary metabolism of (A) *S. marcescens* and (B) *S. aureus.* [Redrawn from Fig. 1 of (21) and from data of Table 2 of (22)]. Broken line: (A) prodigiosin, (B) enterotoxin B.

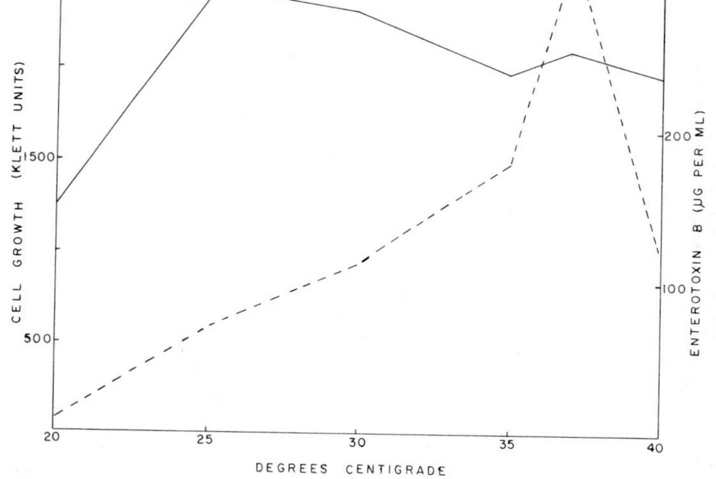

produced at 12–14°C but are maximally active at 27°C (35). Likewise, after the alimentary aleukia toxin synthetases are formed at very low temperatures, a high yield of the toxin is obtained at 22°C (30).

In other secondary-metabolic processes the yield of the metabolites is good at both the temperature optimum for vegetative growth and that best for production of the synthetases. Examples include the formation of polymyxin at either 37 or 30°C (13), of novobiocin at either 32 or 28°C (25), and of *Bacillus cereus* pigment at either 30 or 15°C (12).

There does exist at least one secondary metabolic process in which synthetase activity requires a more restricted temperature range than does vegetative growth. The enzyme that couples the monopyrrole and bipyrrole to form prodigiosin is inactive at temperatures below 10 and above 32°C (43). Even in this system, however, the formation of the prodigiosin synthetases is also inhibited by high temperature. For example, when cells are grown at 37°C for 72 hr and resuspended in a nongrowth medium at 27°C, a lag period of 4–6 hr is required before prodigiosin is first detectable (44). Moreover, 6 μg/ml chloramphenicol added at the start of the lag period completely prevents pigment formation, whereas inhibition of vegetative growth of this strain requires concentrations in excess of 1000 μg/ml. Other

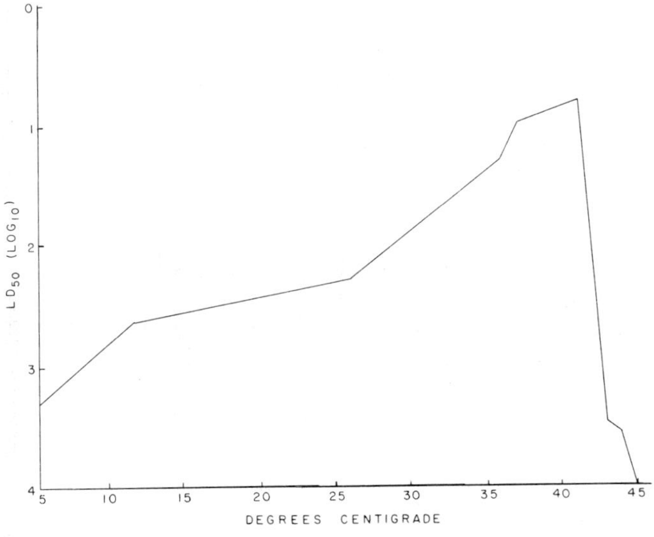

Fig. 3. Effect of temperature on synthesis of V and W antigens by nongrowing cells of *P. pestis* [drawn from data in Table 6 of (18)]. Cells were grown at 5°C and then incubated for 6 hr at the various temperatures indicated on the abscissa. Viable counts remained constant during the 6-hr period except that, at 45°C, a drop of 1 log unit occurred.

inhibitors of macromolecular synthesis are likewise active at the start of the lag period (44).

Suggestions concerning the molecular bases for temperature control of enzyme synthesis have been reviewed (3, 4, 9). A diversity of proposals have been advanced by numerous investigators; thus far, no unitary molecular mechanism is known that can explain the role of temperature in the formation of secondary metabolic synthetases. A reasonable suggestion, however, is that the synthesis of the corresponding repressors is the actual temperature-sensitive reaction (45).

The potential industrial, agricultural, and medical applications of temperature control of secondary metabolism have rarely been exploited. For example, in most industrial fermentations, a single temperature somewhere between the optimum for vegetative growth and the optimum for production of the metabolite is used throughout the entire process (9). Rarely is the temperature purposefully shifted from that needed for prompt growth to that best for synthetase formation and then to that optimal for synthetase activity. When optimal temperature profiles for each of the three phases were followed in one study (39), however, an improvement in yield of only 15% was obtained.

Agricultural scientists are aware that storage of harvested plant foods exposed to moisture permits fungal growth and mycotoxin synthetase activity during daytime warmth, and mycotoxin synthetase production during night-time coolness. Nevertheless, the world-wide problem of toxin synthesis in harvested, stored, plant foodstuffs remains formidable and, in some countries, presents a serious health hazard to humans and domestic animals.

Although the benefit of hyperthermia in the treatment of neurosyphilis and gonorrhoea is well established, its mode of action is unknown (2). A definitive and well-reasoned review (2) on fever as a possible mechanism of resistance to infection contains no known examples of either experimental or natural hyperthermia as an inhibitor of *in vivo* synthesis of secondary metabolic toxins.

ACTION OF TEMPERATURE ON NUTRITIONAL IMMUNITY

An important determinant influencing the outcome of interactions between hosts and invading microorganisms is based on the amount and availability of iron in specific host milieu (46–49). To prevent microbial growth the host employs such powerful iron-binding proteins as transferrin and lactoferrin (47), as well as a mechanism that shifts the metal from the plasma to the liver (50). To capture iron from the host, microorganisms must almost certainly need to synthesize their own iron-sequestering compounds. Zinc is also promptly shifted out of the plasma by the host at the beginning of microbial invasions (46, 50), but it is not yet known if zinc is as critical in the nutrition of pathogens as is iron. In a salient essay on the role of iron in bacterial infections, Kochan (49) has proposed that the term 'nutritional immunity' be designated for those mechanisms whereby the host deprives microbial invaders of essential trace elements.

Although Kochan's initial use of the term refers to inorganic nutrilites, the concept could certainly be extended to embrace organic substances. For example, mammalian species that lack erythritol in their seminal vesicular and placental tissues (51) are considerably more 'nutritionally immune' to brucellosis than are cattle, sheep, goats, and hogs. Likewise, nonrenal tissues that have lower amounts of urea than does the kidney (51) would be somewhat 'nutritionally immune' to those pathogens that produce urease. Earlier studies concerning host-tissue environments that might be deficient in organic nutrilites essential for growth of potential pathogens have been summarized by Garber (52).

To what extent might temperature enhance or depress the efficacy of 'nutritional immunity'? In regard to inorganic nutrilites, such eucaryotes as Ochromonas, Euglena, Saccharomyces, and Candida have elevated trace-metal requirements at elevated temperatures (53, 54). Likewise, such procaryotes as Pseudomonas and Salmonella have increased *in vitro* iron requirements at high temperatures (55, 56). Garibaldi (56) has demonstrated convincingly that high temperature acts by suppressing the ability of the bacteria to form iron-transporting compounds. Normal *in vitro* growth and production of detectable levels of the compounds by *Salmonella typhimurium* occurs at 36.9°C in the presence of 0.5 μM iron; however, at 40.3°C, 50 μM iron is required. Hopefully, this phenomenon will be further examined with a variety of procaryotes and eucaryotes in both *in vitro* and *in vivo* systems.

In regard to organic nutrilites essential for microbial growth, temperature elevation in some cases increases the need, in others decreases the need, and in still others has no effect (57). No information is yet available on possible selective effects of temperature on erythritol utilization by Brucella or on urease formation by either *Corynebacterium renale* or *Proteus mirabilis*.

As mentioned earlier secondary metabolism has a significantly narrower tolerance for trace-metal concentrations than does vegetative cell growth (6). Of the nine trace metals of biological interest (atomic numbers 23–30 and number 42), the concentration of iron is most critical for efficient secondary metabolism in bacteria, as is that of zinc in yeasts and molds. Although at optimum temperatures bacteria grow equally well at concentrations of available iron ranging from 1.0 to 1000 μM, their formation of secondary metabolites generally is stimulated or inhibited by quantities between 3.0 and

100 μM. In view of Garibaldi's observations (56) it can be predicted that temperature control of procaryote secondary metabolism (described earlier in this paper) will be found to be based, at least in part, on the temperature sensitivity of the synthesis of either iron-transport compounds or their corresponding repressors. Of interest in this connection is the demonstration that the ability of trace elements to interfere with citric acid synthesis by *Aspergillus niger* is enhanced at elevated temperatures (58). It is quite probable that, as with procaryotes, trace-element transport systems in eucaryotes will be found to be temperature sensitive.

Since 1928 the level of plasma copper has been known to rise in acute and chronic infections (59). Normal concentrations are approximately 17–20 μM, whereas infected patients have values of 25–38.5 μM. The high levels return to normal in severely ill patients who become 'metabolically exhausted' as well as in cases in which the infection disappears; indeed, the various changes are potentially quite useful in prognoses. Possibly, the hypercupremia might serve to inhibit microbial growth and/or secondary metabolism, and temperature might affect these roles of copper.

Of relevance to this hypothesis is an *in vitro* study of the toxicity of copper to growth of *Escherichia coli* in a synthetic culture medium at various temperatures (60). Briefly, 2 μM permitted immediate growth at 35°C but delayed multiplication for 5, 8, and 12 hr, respectively, at 37, 40, and 42°C. The corresponding delays for 3 μM copper at the four temperatures were 8, 15, 18, and 25 hr. These studies should be extended to include a variety of pathogenic microorganisms grown in culture media with added serum and copper at the appropriate temperatures.

FEVER AS A MECHANISM OF HOST-RESISTANCE

In considering the role of fever in the mechanism of resistance to microbial invasion, some cautionary reminders (2) might be recalled:

> The range of temperature change in febrile animals and man is narrow. A generous scale in man would be 36 to 42°C. For fever to enhance resistance we must have an organism that is pathogenic at 36 to 38°C, the disease must elicit a febrile response, and the elevation of at most 5 to 6 degrees must grievously injure or destroy the parasite, or must activate mechanisms within the host somehow capable of modifying the disease process. (2)

We are further reminded (2) that (1) hosts with high resistance would generally have either no or mild clinical symptoms, and hence have little or no fever, (2) moribund individuals with fulminating infections may have little or no fever, and (3) the antipyretic action of drugs has no obvious influence on the course or outcome of most infections. Nevertheless, fever was proposed (2) to be a possible secondary defense that is called upon when other mechanisms falter and then only in proportion to the body's needs.

In recent years, several observations have been reported on what might first appear as a rather curious association between fever and trace-metal metabolism. For example, children suffering from either fever of infection or a febrile response to immunization have a consistently and profoundly depressed intestinal absorption of iron (61). This fever-associated depression is independent of the host's state of nutrition, growth rate, bone-marrow iron stores, iron saturation of transferrin, plasma iron level, hemoglobin level, and nonfebrile rate of absorption. In a different study (62) using adults an average of 7.2% of an oral dose of radioiron was absorbed into the blood of normal persons, whereas only 1% of the dose was absorbed by febrile patients.

In young adult volunteers following exposure to virulent *Francisella tularensis* plasma iron concentration is lowered in two phases (63). In the first, or exposure, phase a modest decline occurs early in the incubation period and is independent of the development or absence of subsequent illness. In the second, or febrile, phase a greater decline associated with clinical symptoms takes place. It occurs concurrently with the febrile period and, as with fever, it begins at the peak of the neutrophile leucocytosis.

Such host defense cells as neutrophiles, monocytes, and macrophages have been known for a quarter century to produce a pyrogenic substance (which affects hypothalamic control of body temperature) when exposed to endotoxin, bacteria, viruses, or antigen-antibody combinations (1). A decade ago endotoxin was also found to cause hypoferremia (64) and, more recently, to depress intestinal absorption of iron (65). Endogenous pyrogen synthesized by endotoxin-exposed leucocytes produces fever, hypoferremia, and hypozincemia in rats (66, 67). Moreover, either endotoxin, polyriboinosinic polyribocytidylic acids (poly I:C), *Diplococcus pneumoniae,* or *F. tularensis* attenuated vaccine administered intraperitoneally to rats decreases plasma iron and zinc, and a small protein termed endogenous mediator can be obtained from neutrophile leucocytes of peritoneal exudates of the animals (68–70). The mediator, which depresses the plasma level and elevates the liver concentration of each of the two metals in recipient rats, is similar in chemical nature (and may be identical) to that of endogenous pyrogen (50).

In young adult volunteers immunized with live attenuated Venezuelan equine encephalomyelitis virus, plasma iron and zinc concentrations fall early even in those subjects who remain asymptomatic (71). The decrease is intensified just prior to the onset of fever and is maximal at the height of the febrile phase. As the latter wanes plasma iron and zinc levels return to normal. Plasma copper concentrations begin to rise just prior to the onset of fever and remain high until after the fever subsides.

An objection (2) to a possible role of fever in host defense is that, in asymptomatic hosts, fever would not have occurred. This objection can be met by considering the early phase of hypoferremia and hypozincemia to be the initial host-defense mechanism. If this response is not capable of thwarting the invasion, fever might then occur and provide a portion of the fail-safe defense by directly interfering with the ability of the invading procaryotes or eucaryotes to synthesize iron-transport compounds and secondary metabolic toxins. Simultaneously, enhanced declines in plasma iron and zinc levels would be occurring, iron absorption through the intestine would be blocked, and plasma copper levels would be rising.

To what extent does fever, or even normal variation in temperature of body organs, play a critical role in natural resistance to infectious disease? A number of experimental and clinical studies of the role of fever in the outcome of infections caused by procaryotes and one eucaryote have been reviewed (2), but no clear conclusions could be reached. Presumably, in many cases of threatened infectious disease, the activation of the fever fail-safe device is not needed; in still others, it is required and is induced, but fails to protect the host. Unfortunately, the importance of the concentration of available iron (and perhaps also of zinc and copper) in the establishment in the host of pathogenic invaders was not known to earlier investigators, and thus neither monitoring nor manipulation of trace-metal levels was attempted.

In contrast to diseases caused by procaryotes and eucaryotes, the effect of temperature on the interaction between some groups of animal viruses and their hosts has been well explored (2, 72, 73). Small variations in temperature can result in major alterations in the course of virus infections in cell cultures, embryonated eggs, and infected animals. For example, viral replication and clinical symptoms in beagle pups inoculated with canine herpes virus, compared with control-inoculated animals kept at normal body temperature, was greatly enhanced in dogs whose body temperature was maintained at 34.4–36.1°C and greatly decreased in those kept at 38.6–39.4°C (74).

Increased interferon production (72) and increased lysosomal release of ribonucleases (73) at elevated temperatures have been proposed as possible mechanisms for suppression of viral diseases at supraoptimal temperatures. However, in view of the observations (70, 71) that viruses and synthetic interferon inducers cause hypoferremia and hypozincemia in advance of fever, a re-examination of the role of temperature in host-virus interactions with simultaneous monitoring as well as adjustment of trace metals is strongly indicated. Not only should metal-deficient and metal-enriched animal hosts be employed (as has been done recently in diseases caused by procaryotes) (47), but also infected cell cultures should be studied. The minimal requirements of iron and zinc for growth of L mouse cells (75, 76) have been determined; the quantities are essentially similar to those needed for procaryote cell growth. It will be of considerable interest and utility to determine the amounts required by a variety of other mammalian cells derived from normal and tumor tissues, and to compare these with levels needed for virus and interferon synthesis.

In addition to studies on host-virus systems, the interaction of trace-metal metabolism and fever also should be explored in *in vivo* growth of tumor cells. Fever is a common sign in patients with various tumors (1), but to what extent it might aid the host is unknown. The slow sustained fevers characteristic of many neoplastic diseases may result from the production of endogenous pyrogen by the tumor cells themselves or by host monocytes (1). It is well established that plasma iron levels decrease (77, 78) and copper levels increase (79, 80) in hosts with tumors, and in certain kinds of neoplasms plasma zinc decreases (81). Moreover, such agents as bacterial cell walls, endotoxin, BCG, poly I:C, and procaryote and virus infections that cause hypoferremia, hypozincemia, and hypercupremia also retard multiplication of tumor cells and cause remissions in tumor growth (82–85). As with microorganisms it is quite possible that tumor cells require amounts of iron, and perhaps zinc, that are not readily available from hypoferremic and hypozincemic host fluids. Indeed, a general effect of zinc deficiency in rats is that of tumor-growth inhibition, irrespective of cell type, growth rate, or site of growth (86), and, again in parallel with microorganisms, tumor cells might require even more iron and zinc at supraoptimal temperatures than at those to which their hosts are normally adjusted.

SUMMARY

In recent years several effects of supraoptimal temperature on host-pathogen relationships have been identified. Hosts invaded by potential pathogens promptly reduce the quantities of plasma iron and zinc; if this measure fails to halt the progress of infection, endogenous pyrogen is released by host white blood cells. The pyrogen causes an intensified *hypo*ferremia, *hypo*zincemia, *hyper*cupremia, and fever; simultaneously, intestinal absorption of iron is halted. High temperatures which permit normal growth of pathogens usually interfere with formation of microbial iron-transport compounds and of toxins. Fever-induced limitation of iron (and perhaps also of zinc) for microbial growth and toxin synthesis favor survival of the host. Interactions of fever and trace-metal metabolism are apparently not only of benefit to hosts invaded by bacteria and fungi, but also may be important in host defense against viral infections and neoplasms.

REFERENCES

1. Atkins, A., and Bodel, P. 1972. Fever. *New Engl. J. Med.* 286: 27.
2. Bennett, I. L., Jr., and Nicastri, A. 1960. Fever as a mechanism of resistance. *Bacteriol. Rev.* 24: 16.
3. Farrell, J., and Rose, A. H. 1967. Temperature effects on microorganisms. *In* A. H. Rose (ed.), *Thermobiology,* Academic Press, London, p. 147.
4. Farrell, J., and Rose, A. H. 1967. Temperature effects on microorganisms. *A. Rev. Microbiol.* 21: 101.
5. Bu'Lock, J. D. 1961. Intermediary metabolism and antibiotic synthesis. *Adv. Appl. Microbiol.* 3: 293.
6. Weinberg, E. D. 1970. Biosynthesis of secondary metabolites: roles of trace metals. *Adv. Microbial Physiol.* 4: 1.
7. Weinberg, E. D. 1971. Secondary metabolism: raison d'etre. *Persp. Biol. Med.* 14: 565.
8. Gentry, M. J., Smith, D. K., Schnute, S. F., Werber, S. L., and Weinberg, E. D. 1971. *Pseudomonas* culture longevity: control of phosphate. *Microbios* 4: 205.
9. Demain, A. L. 1968. Regulatory mechanisms and the industrial production of microbial metabolites. *Lloydia* 31: 395.
10. Gentile, J. H. 1971. Blue-green and green algal toxins. *In* S. Kadis, A. Ciegler, and S. J. Ajl (eds), *Microbial Toxins,* Vol. 7, Academic Press, New York, p. 27.
11. Bonventre, P. F., and Johnson, C. E. 1970. *Bacillus cereus* toxin. *In* T. C. Montie, S. Kadis, and S. J. Ajl (eds), *Microbial Toxins,* Vol. 3, Academic Press, New York, p. 415.
12. Uffen, R. L., and Canale-Parola, E. 1965. Temperature-dependent pigment production by *Bacillus cereus* var. *alesti. Can. J. Microbiol.* 12: 590.
13. Paulus, H. 1967. Polymyxins. *In* D. Gottlieb and P. D. Shaw (eds), *Antibiotics. II. Biosynthesis,* Springer-Verlag, Berlin, p. 254.
14. Baird-Parker, A. C. 1969. Medical and veterinary significance. *In* G. W. Gould and A. Hurst (eds), *The Bacterial Spore,* Academic Press, London, p. 517.
15. Abrahamsson, K., Gullivar, B., and Molin, N. 1966. The effect of temperature on toxin formation and toxin stability of *Clostridium botulinum* type E in different environments. *Can. J. Microbiol.* 12: 385.
16. Girard, K. F., Sbarra, A. J., and Bardawil, W. A. 1963. Serology of *Listeria monocytogenes.* I. Characteristics of the soluble hemolysin. *J. Bacteriol.* 85: 349.
17. Montie, T. C., and Ajl, S. J. 1970. Nature and synthesis of murine toxins of *Pastuerella pestis. In* T. C. Montie, S. Kadis, and S. J. Ajl (eds), *Microbial Toxins,* Vol. 3, Academic Press, New York, p. 1.

18. Naylor, H. B., Fukui, G. M., and McDuff, C. R. 1961. Effects of temperature on growth and virulence of *Pasteurella pestis*. *J. Bacteriol.* 81: 649.
19. Kurachi, M. 1959. Studies on the biosynthesis of pyocyanine. I. On the cultural conditions for pyocyanin formation. *Bull. Inst. Chem. Res. Kyoto Univ.* 36: 163.
20. Jude, A., and Nicolle, P. 1952. Persistance, à l'etat potentiel, de la capacite d'elaborer l'antigence Vi chez le bacille typhique cultivé en série a basse température. *Acad. Seances C.R. Hebd.* 234: 1718.
21. Williams, R. P., Gott, C. L., Qadri, S. M. H., and Scott, R. H. 1972. Influence of temperature of incubation and type of growth medium on pigmentation in *Serratia marcescens*. *J. Bacteriol.* 106: 438.
22. Dietrich, G. G., Watson, R. J., and Silverman, G. J. 1972. Effect of shaking speed on the secretion of enterotoxin B by *Staphylococcus aureus*. *Appl. Microbiol.* 24: 561.
23. Hockenhull, D. J. D. 1960. The biochemistry of streptomycin production. *Progr. Indust. Microbiol.* 2: 131.
24. Hall, M. J., and Hassall, C. H. 1970. Production of the monamycins, novel depsipeptide antibiotics. *Appl. Microbiol.* 19: 109.
25. Smith, C. G. 1956. Fermentation studies with *Streptomyces niveus*. *Appl. Microbiol.* 4: 232.
26. Sensi, P., and Thiemann, J. E. 1967. Production of rifamycins. *Progr. Indust. Microbiol.* 6: 21.
27. Richardson, S. H. 1969. Factors influencing *in vitro* skin permeability factor production by *Vibrio cholerae*. *J. Bacteriol.* 100: 27.
28. Kaplan, L., and Demain, A. L. 1970. Nutritional studies on riboflavin overproduction by *Ashbya gossypii*. *In* D. S. Ahearn (ed.) *Recent Trends in Yeast Research*, Georgia State University Press, Atlanta, Ga., p. 137.
29. Ciegler, A., Peterson, R. E., Lagoda, A. A., and Hall, H. H. 1966. Aflatoxin production and degradation by *Aspergillus flavus* in 20-liter fermentors. *Appl. Microbiol.* 14: 826.
30. Joffe, A. Z., and Lisker, N. 1969. Effects of light, temperature, and pH value on aflatoxin production *in vitro*. *Appl. Microbiol.* 18: 517.
31. Yamada, K., and Hidaka, H. 1964. Submerged citric acid fermentation of blackstrap molasses. *Agric. Biol. Chem.* 28: 876.
32. Yukioka, M., and Winnick, T. 1966. Biosynthesis of malformin in washed cells of *Aspergillus niger*. *Biochim. Biophys. Acta* 119: 614.
33. Abe, M., and Yamatodani, S. 1964. Preparation of alkaloids by saprophytic culture of ergot fungi. *Progr. Indust. Microbiol.* 5: 204.
34. Joffe, A. Z. 1971. Alimentary toxic aleukia. *In* S. Kadis, A. Ciegler, and S. J. Ajl (eds), *Microbial Toxins*, Vol. 7, Academic Press, New York, p. 139.
35. Mirocha, C. J., Christensen, C. M., and Nelson, G. H. 1971. F-2 (zearalenone) estrogenic mycotoxin from *Fusarium*. *In* S. Kadis, A. Ciegler, and S. J. Ajl (eds), *Microbial Toxins*, Vol. 7, Academic Press, New York, p. 107.
36. Banburg, J. R., Marasas, W. F., Riggs, N. V., Smalley, E. B., and Strong, F. M. 1968. Toxic spiroepoxy compounds from Fusaria and other hyphomycetes. *Biotech. Bioeng.* 10: 445.

37. Jeffreys, E. G. 1970. The gibberellin fermentation. *Adv. Appl. Microbiol.* 13: 283.
38. Kurtzman, C. P., and Ciegler, A. 1970. Mycotoxin from a blue-eye mold of corn. *Appl. Microbiol.* 20: 204.
39. Constantinides, A., Spencer, J. L., and Gaden, E. L., Jr. 1970. Optimization of batch fermentation processes. II. Optimum temperature profiles for batch penicillin fermentations. *Biotech. Bioeng.* 12: 1081.
40. Hayes, A. W., Wyatt, E. P., and King, P. A. 1970. Environmental and nutritional factors affecting the production of rubratoxin B by *Penicillium rubrum* Stoll. *Appl. Microbiol.* 20: 469.
41. Hou, C. T., Ciegler, A., and Hesseltine, C. W. 1971. Tremorgenic toxins from penicillia. III. Tremortin production by Penicillium species on various agricultural commodities. *Appl. Microbiol.* 21: 1101.
42. Ciegler, A. 1972. Bioproduction of ochratoxin A and penicillic acid by members of the *Aspergillus ochraceus* group. *Can. J. Microbiol.* 18: 631.
43. Williams, R. P., Goldschmidt, M. E., and Gott, C. L. 1965. Inhibition by temperature of the terminal step in biosynthesis of prodigiosin. *Biochem. Biophys. Res. Commun.* 19: 177.
44. Qadri, S. M. H., and Williams, R. P. 1972. Biosynthesis of the tripyrrole bacterial pigment, prodigiosin, by nonproliferating cells of *Serratia marcescens*. *Texas Rep. Biol. Med.* 30: 73.
45. Demain, A. L. 1972. Riboflavin oversynthesis. *A. Rev. Microbiol.* 26:369.
46. Weinberg, E. D. 1966. Roles of metallic ions in host-parasite interactions. *Bacteriol. Rev.* 30: 136.
47. Weinberg, E. D. 1971. Roles of iron in host-parasite interactions. *J. Infec. Dis.* 124: 401.
48. Weinberg, E. D. 1972. Systemic salmonellosis: a sequela of sideremia. *Texas Rep. Biol. Med.* 30: 277.
49. Kochan, I. 1972. The role of iron in bacterial infections, with special consideration of host-tubercle bacillus interactions. *Curr. Topics Microbiol. Immunol.* 60: 1.
50. Pekarek, R. S., Wannemacher, R. W., Jr., and Beisel, W. R. 1972. The effect of leucocytic endogenous mediator (LEM) on the tissue distribution of zinc and iron. *Proc. Soc. Expl. Biol. Med.* 140: 685.
51. Keppie, J. 1964. Host and tissue specificity. *Symp. Soc. Gen. Microbiol.* 14: 44.
52. Garber, E. D. 1956. A nutrition-inhibition hypothesis of pathogenicity. *Am. Naturalist* 90: 183.
53. Hutner, S. H., Aaronson, S., Nathan, H. A., Baker, H., Scher, S., and Cury, A. 1958. Trace elements in microorganisms: the temperature factor approach. *In* C. A. Lamb, O. G. Bentley, and J. M. Beattie (eds), *Trace Elements,* Academic Press, New York, p. 47.
54. Roitman, I., Travassos, L. R., Azenedo, H. P., and Cury, A. 1969. Choline, trace elements, and amino acids as factors for growth of an enteric yeast, *Candida slooffii,* at 43°C. *Sabouraudia* 7: 15.
55. Garibaldi, J. A. 1971. Influence of temperature on the iron metabolism of a fluorescent pseudomonad. *J. Bacteriol.* 105: 1036.
56. Garibaldi, J. A. 1972. Influence of temperature on the biosynthesis of

iron transport compounds by *Salmonella typhimurium. J. Bacteriol.* 110: 262.
57. Campbell, L. L., and Pace, B. 1968. Physiology of growth at high temperatures. *J. Appl. Bacteriol.* 31: 24.
58. Kitos, P. A., Campbell, J. J. R., and Tomlinson, N. 1953. Influence of temperature on the trace element requirements for citric acid production by *Aspergillus niger. Appl. Microbiol.* 1: 156.
59. Weinberg, E. D. 1972. Infectious diseases influenced by trace element environment. *Ann. N.Y. Acad. Sci.* 199: 274.
60. Burke, C. M. W., and McVeigh, I. 1967. Toxicity of copper to *Escherichia coli* in relation to incubation temperature and method of sterilization of media. *Can. J. Microbiol.* 13: 1299.
61. Beresford, C. H., Neale, R. J., and Brooks, O. G. 1971. Iron absorption and pyrexia. *Lancet,* March 20: 568.
62. Dubach, R., Callendar, S. T. E., and Moore, C. V. 1948. Studies in iron transportation and metabolism. VI. Absorption of radioactive iron in patients with fevers and with anemias of varied etiology. *Blood* 3: 526.
63. Pekarek, R. S., Bostian, K. A., Bartelloni, P. J., Calia, F. M., and Beisel, W. R. 1969. The effects of *Francisella tularensis* infection on iron metabolism in man. *Am. J. Med. Sci.* 258: 14.
64. Kampschmidt, R. F., and Arredondo, M. I. 1963. Some effects of endotoxin upon plasma iron turnover in the rat. *Proc. Soc. Exp. Biol. Med.* 113: 142.
65. Cortell, S., and Conrad, M. W. 1967. Effect of endotoxin on iron absorption. *Am. J. Physiol.* 213: 43.
66. Kampschmidt, R. F., and Upchurch, H. F. 1970. A comparison of the effects of rabbit endogenous pyrogen on the body temperature of the rabbit and lowering of plasma iron in the rat. *Proc. Soc. Exp. Biol. Med.* 133: 128.
67. Kampschmidt, R. F., and Upchurch, H. F. 1970. The effect of endogenous pyrogen on the plasma zinc concentration of the rat. *Proc. Soc. Exp. Biol. Med.* 134: 1150.
68. Pekarek, R. S., and Beisel, W. R. 1969. Effect of endotoxin on serum zinc concentrations in the rat. *Appl. Microbiol.* 18: 482.
69. Pekarek, R. S., and Beisel, W. R. 1971. Characterization of the endogenous mediator(s) of serum zinc and iron depression during infection and other stresses. *Proc. Soc. Exp. Biol. Med.* 138: 728.
70. Pekarek, R. S. 1971. Effect of synthetic double-stranded RNA on serum metals in the rat, rabbit, and monkey. *Proc. Soc. Exp. Biol. Med.* 136: 584.
71. Pekarek, R. S., Burghen, G. A., Bartelloni, P. J., Calia, F. M., Bostian, K. A., and Beisel, W. R. 1970. The effect of live attenuated Venezuelan equine encephalomyelitis virus vaccine on serum iron, zinc, and copper concentrations in man. *J. Lab. Clin. Med.* 76: 293.
72. Dumbell, K. R. 1967. The effect of temperature on the relation between animal viruses and their hosts. *In* A. H. Rose (ed.), *Thermobiology,* Academic Press, London, p. 219.
73. Lwoff, A. 1969. Death and transfiguration of a problem. *Bacteriol. Rev.* 33: 390.

74. Carmichael, L. E., Barnes, F. D., and Percy, D. H. 1969. Temperature as a factor in resistance of young puppies to canine herpesvirus. *J. Infec. Dis.* 120: 669.
75. Thomas, J. A., and Johnson, M. J. 1967. Trace metal requirements of NCTC clone 929 strain L cells. *J. Natn. Cancer Inst.* 39: 337.
76. Higuchi, K. 1970. An improved chemically defined culture medium for strain L mouse cells based on growth response to graded levels of nutrients including iron and zinc. *J. Cell. Physiol.* 75: 65.
77. Kampschmidt, R. F., Adams, M. E., and McCoy, T. A. 1959. Some systemic effects of toxohormone. *Cancer Res.* 19: 236.
78. Konaka, K., and Matsuoka, T. 1967. The serum iron content in malignant and nonmalignant disease. *Kumamoto Med. J.* 20: 196.
79. Mortazani, S. H., Bari-Hashemi, A., Mozafari, M., and Raffi, A. 1972. Value of serum copper measurement in lymphomas and several other malignancies. *Cancer* 29: 1193.
80. Strain, W. H., Mansour, E. G., Flynn, A., Pories, W. J., Tomaro, A. J., and Hill, O. A., Jr. 1972. Plasma zinc concentration in patients with bronchogenic cancer. *Lancet,* May 6: 1021.
81. Davies, I. J., Musa, M., and Dormandy, T. L. 1968. Measurements of plasma zinc. II. In malignant disease. *J. Clin. Path.* 21: 363.
82. Jordan, R. T., Rasmussen, A. F., and Bierman, H. R. 1958. Effect of group A streptococci on transplantable leukemia of mice. *Cancer Res.* 18: 943.
83. Old, L. J., Benacerraf, B., Clarke, D. A., Carswell, E. A., and Stockert, E. 1961. The role of the reticuloendothelial system in the host reaction to neoplasia. *Cancer Res.* 21: 1281.
84. Fisher, J. C., Grace, W. R., and Mannick, J. A. 1970. The effect of nonspecific immune stimulation with *Corynebacterium parvum* on patterns of tumor growth. *Cancer* 26: 1379.
85. Gelloin, H. V., and Levy, H. B. 1970. Polyinosinic-polycytidylic acid inhibits chemically induced tumorigenesis in mouse skin. *Science* 167: 205.
86. DeWys, W., and Pories, W. 1972. Inhibition of a spectrum of animal tumors by dietary zinc deficiency. *J. Natn. Cancer Inst.* 48: 375.

DISCUSSION

Matrone (Raleigh). You suggested that an increase in plasma copper might be desirable. Would you mind speculating on that?

Weinberg. I am on very shaky ground, but there's one report of an *in vitro* study with no ceruloplasmin present showing that an increase in copper of about 60% is considerably more toxic to bacteria. The toxicity toward the bacteria was markedly increased by going from 35 up to 42°C, and they found that the antibacterial potency was increased much more by small temperature changes than one would have expected. Clearly it is difficult to extrapolate from such simple *in vitro* systems to what happens in body fluids, but I think the time is right to begin such studies.

Gubler (Provo). When we were studying the anemia of infection, the question came up as to whether these changes in iron and copper metabolism were beneficial or detrimental to the progress of the infection. Many people have tried giving cobalt to increase the hemoglobin level, and they have tried giving iron to increase the iron levels, and all these things seemed to make the infection worse instead of better.

Henkin (Bethesda). You have raised many questions, but one which I fully expected you to raise, you didn't. That is, what interrelationship do you see between prostaglandins and these phenomena, particulary since they seem to be so important both in terms of mediating temperature and regulating blood flow? Have you done any experiments relating to that?

Weinberg. I became immersed in the prostaglandin aspect some time ago, but gave up on it. It was very confusing and I just couldn't interpret the possible interactions.

Peters (Madison). I'd like to call attention to a condition in which the iron levels of the body rise precipitously and are directly proportional to the disease process itself, and that's chronic porphyria, a condition which can be reversed by bleeding or by dietary treatment with EDTA which undoubtedly cuts down absorption of iron. Infection is a very prominent feature of the disease, and is probably due to the alterations in iron metabolism. In this condition other chelating mechanisms are brought into play, such as an alteration of tryptophan metabolism in which the metabolites are excellent chelating agents, and the withdrawal of porphyrins themselves have a chelating function.

TRACING AND TREATING MINERAL DISORDERS IN CATTLE UNDER FIELD CONDITIONS

J. HARTMANS

National Council for Agricultural Research (NRLO-TNO), The Hague, Netherlands

One of the major problems to advisors and research workers, when working on field problems, is to indicate which factors are involved when a farmer has problems with his herd such as inadequate feed intake, inadequate growth and condition, complaints about milk production, milk composition (mainly low butter-fat percentage), fertility, weakness and other rearing difficulties, and early 'exhaustion,' e.g., by lameness.

The problems involved are not primarily veterinarian questions, but may be associated with nutrition and possibly with mineral nutrition. In fact management is a complicating factor in almost all cases. It is obvious that these complaints may have many causes.

An approach to attacking these problems on dairy farms in the Netherlands has been worked out by the Committee on Mineral Nutrition, which is a group of research workers and advisors (1–3). In this paper the general procedure will be reviewed, together with some general considerations and the development of criteria for copper and manganese under Dutch conditions.

DIFFERENCES BETWEEN MINERAL EXPERIMENTS AND THE APPROACH TO FIELD PROBLEMS

In animal experiments the supply of the element to be studied is intentionally kept either at a low or a high level. This can be attained directly by a low or high content in the ration, or can be induced by other components of the ration. Such an experiment usually results in the animal reacting by showing clinical signs. These signs may be more or less specific. Particularly in submarginal deficiencies, signs may be unspecific and change from case to case depending on other external conditions. As the experiment proceeds, changes in the chemical composition of body tissues or fluids can be observed, resulting in different levels of the element concerned in the various treatments. On the basis of these levels, criteria for deficient and sufficient supply are developed. In several cases, however, the measured level is also affected by factors other than the supply status of the mineral concerned. Because no general level can be given in such cases to distinguish between deficient and normal supply or between excessive and normal supply, it cannot be used as a consistent criterion for the supply status of the element concerned. This will be illustrated later by the Mn content of hair.

When working on field problems, in fact, the reverse course has to be adopted. Starting from certain clinical signs, the causative factor or factors have to be found. Usually, the general signs mentioned before are so diffuse that they hardly ever allow an indication in a certain direction without extensive analysis of the complaints and checking for possible factors involved.

This indicates the necessity of close cooperation of agronomists and veterinarians in approaching such problems. Neither person is qualified to investigate fully and in depth all the possible problems. In nearly all cases the veterinarian needs the help of the agricultural advisor to check several aspects of farm management and to examine rations and soil, including knowledge of particular soil conditions, etc. Conversely, the agricultural advisor needs the veterinarian for a clinical assessment of the problem, to obtain samples of animal tissue, to interpret the analytical results of these samples, and, finally, to arrive at a differential diagnosis. During the whole procedure there are several crucial occasions at which an exchange of data and insights is necessary, e.g., when making appointments about the task of each, drawing up a sampling plan, interpreting the analytical data, and drawing up advice.

GENERAL PROCEDURE

Appraisal

One of the most important parts of the procedure is an extensive *appraisal.* Before possible faults in the mineral supply are considered, available data should be completed and critically analyzed. In our experience the appraisal should start with a careful checking of the complaints as described by the farmer. The stockman, being interested in production aspects, usually does not mention abnormalities in hair coat, bone deformities, condition, scouring, coughing, etc. Therefore the investigating team should watch the entire herd and note all facts observed with respect to clinical signs, management, fertilization practice, etc.

An analysis of the data follows in order to try to limit the problem to animal age groups (calves, yearlings, heifers or cows), particular periods of lactation and pregnancy, particular seasons of the year or particular weather conditions, specific parts of the farm, and particular sib-groups suggesting a hereditary factor. Some complaints allow a further analysis, e.g., problems involving fertility or milk production.

It is worth mentioning that the usually unspecific syndromes resulting from an analysis of the data may be caused by several mineral imbalances. The syndrome associated with insufficient protein and energy supply may also be deceptively like that of certain mineral imbalances, especially if mild. The same applies to infestations with gastrointestinal parasites. This type of disease is frequently important in young stock, and has a direct effect on development and growth rate and, possibly later, on production and fertility. Therefore, if relevant, checks on parasites and other infectious diseases, on feed intake, analysis of ration components for nutritive value, and testing of drinking-water quality should be carried out. Other management aspects, including rearing practice, should also be examined.

In a large number of cases, the complaints appear to fade away if control measures are taken with respect to the factors mentioned above and closely related to general farm management.

Only after these factors have been corrected and complaints persist is there reason to suspect incorrect mineral nutrition as a causative factor. The sequence of testing (first management and then minerals) is suitable not only because the majority of the complaints appear to be connected with management, but also because the first part can be done without high expense.

Sampling Plan

If in this way it has been established that a problem is probably connected with incorrect mineral supply, a *sampling plan* should be drawn up. The agronomist and the veterinarian therefore have to consider what faults in the mineral supply are likely. The clinical picture, including its differentiation among the different animal groups, and data on farm management may provide hints. However, it is hardly ever justifiable to draw conclusions from such evidence, except from certain typical signs. Suggestions must be confirmed by chemical analysis of substrates from animal, plant, or soil origin. Conversely, certain elements can sometimes be excluded as improbable without chemical data, for instance, cobalt on Dutch clay soils. It is often advisable to discuss the problem with a soil chemist. Certain geological formations and the soils derived from them may be known to have a certain mineral pattern. In the interpretation of these data, fertilization practice and history on the farm should be considered as well.

A sampling plan is drawn up on the basis of elements for which more information is required. Which substrates are the most relevant for checking the different elements? Though the cause of the trouble and its correction may lie in the soil or in herbage composition, it is the animal that is affected, and thus animal substrates are principally the best to show what is wrong. The choice of a relevant substrate is not only dependent on the physiological characteristics of the element but also on the possibilities of easy sampling and eventual storage, a reliable and sufficiently sensitive, and thus inexpensive, chemical method, etc. It may also be important that a criterion gives an early warning, possibly before clinical signs appear. Criteria in ration or soil should only be chosen if analysis of animal substrates is little indicative. Chemical criteria are highly preferred over clinical ones, the latter often being unspecific, especially in marginal deficiencies, and even if specific are sometimes deceptively like those caused by other nutritional or parasitic diseases.

Table 1 indicates which substrates have been chosen in tracing different mineral deficiencies in the Netherlands; a tentative diagram has been developed for checking on excesses. The most useful diagnostic data for a certain element are in bold type; components providing supplementary data are in normal type. Ration and soil testing can be useful for certain elements, once a deficiency has been detected, as a basis for correction and preventive measures. Some criteria have been established after extensive research on the element concerned in the Netherlands; others, being less important under our conditions, have been derived from the literature, e.g., Co and Zn. Se and

Table 1. Substrate samples suggested and components to be estimated in tracing mineral deficiencies

	Substrate						
Mineral	Liver	Blood plasma	Saliva	Milk	Urine	Ration	Soil
Magnesium		**Mg**			**Mg**	Mg+cp+K	
Calcium		Ca				**Ca**	
Phosphorus						**P**	
Sodium			**Na, K**			**Na**	
Chlorine						**Cl**	
Copper	**Cu**	Cu					
Cobalt							**Co**
Iodine				**I**			
Zinc		**Zn**				Zn	
Manganese						**Mn**	

other newer trace elements are not included in our list, since until now there are no indications of abnormalities in their supply.

When a sampling plan has been made on the basis of suspected elements, samples are collected according to guidelines for each substrate and element. These include, e.g., which animals, feedstuffs, or fields should be sampled, the number of samples, and the time of sampling.

Evaluation and Advice

In interpreting analytical results from animal substrates, the average level of particular minerals may be misleading. One or a few values far below or above the average may be much more indicative than the mean of the group. Hence it is important to obtain sufficient samples.

If at any time analysis of soil, plant, or animal substrates indicates a marked deviation from normal, measures should be advised. One should distinguish between two types of measures, namely immediate and the longer-term ones. In the *immediate measures* the cattle are supplied directly with suitable minerals or mineral mixtures to control the deficiency. On the basis of standards for age groups a recommendation can be drawn up. The *longer-term measures* are intended to prevent recurrence. They include fertilization of certain pastures, incorporation of certain mineral salts in concentrates or in roughages, e.g., silage, and changed management to improve intake or absorption of an element by the animal. They are often effective only after some time, but are preferred in practice because they are cheaper in money and labour than continuing the immediate measures.

On farms where mineral deficiency occurs, both immediate and longer-term measures are necessary. Experience has shown that farmers have to be checked to ensure that measures are carried out. Regular and consistent supply of minerals is readily forgotten by pressure of work, especially during the grazing season.

FURTHER DEVELOPMENTS

Evaluation of data obtained in the Netherlands on farms with complaints together with data on farms without complaints, during a period of several years, has shown that affected farms and farms without complaints in the same district have almost identical patterns of mineral status. This not only applies to the mineral status, but also to the incidence of parasitic infestations, deviating ration composition, and other factors connected closely with farm management and soil type. Therefore the conditions on a group of representative farms reflect conditions throughout the area. This indicates that on farms hitherto without complaints the equilibrium may be labile; the occurrence of an additional unfavorable condition may evoke complaints in the herd. This opens the opportunity for advisory officers to concentrate on prevention in districts with the same type of farm and similar soil types.

CONSIDERATIONS IN ADOPTING CRITERIA FOR COPPER AND MANGANESE IN THE NETHERLANDS

Copper

Copper deficiency in cattle is rather common in the Netherlands. The most typical aspect is that the deficiency develops during the pasturing period, and the animals recover on winter rations, especially on hay. Surprisingly, the Cu content in the feed during winter is lower than at pasture. Furthermore, farms with or without Cu deficiency at the end of the grazing season cannot be distinguished on the basis of the Cu content in herbage or soil, while Mo and S, both total and inorganic, are only slightly higher in affected than in unaffected areas. Mo does not exceed 6 ppm at any time of the season, while total S is relatively high, namely 0.3–0.4% in herbage dry matter (4, 5).

Under these conditions a close correlation is found between Cu content in liver and concentration in blood serum (Fig. 1). At a low Cu status, liver and blood both give information on the Cu status of the animal. At a normal blood Cu level, the serum does not indicate the size of the reserve. A normal serum Cu value outside the critical period does not give any indication of what may be expected in the critical period, viz. the autumn. On the other hand, the liver content not only allows evaluation of the actual situation, but also allows estimation of the situation in the most critical period. Conse-

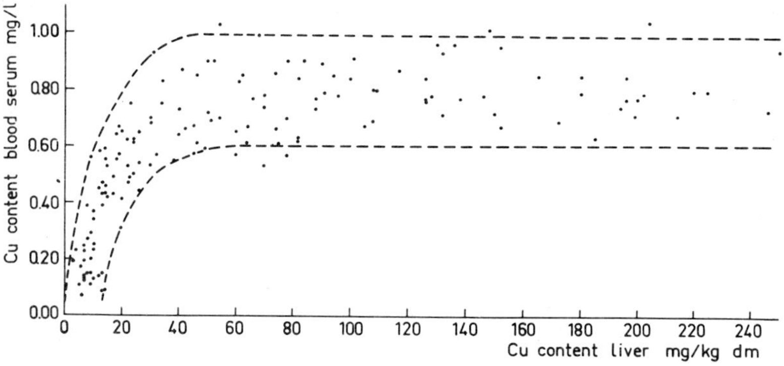

Fig. 1. Relation between the copper contents of liver and blood serum in yearling Friesians.

quently, the Cu content in the liver gives more reliable and detailed information on the Cu status of the animal than the blood value. Interest among Dutch veterinarians in performing liver biopsies is very limited, however, and as a rule blood sampling is carried out, if necessary repeated in autumn. Several animal-health laboratories estimate the oxidase activity of the blood serum as a simplified chemical method. Several authors have demonstrated the close correlation between Cu concentration in blood serum and its oxidase activity (6–8).

Not every decreased serum Cu level is associated with clinical signs, nor with complaints (8, 9). However, the lower the level, the higher the incidence of clear-cut clinical signs and losses in production. Furthermore, experiments have shown that in highly productive dairy cattle the elimination of any decreased serum Cu level by oral supplementation with Cu increases growth rate in young stock and persistency of milk production in adults.[1] Therefore we feel that, costs of supplementation being low, any decreased blood Cu level should be prevented.

From the above it will be clear that, since so many factors are involved in Cu absorption and utilization by the animal (10), the Cu content of the ration can hardly be a criterion in estimating the animal's supply, even if Mo and S contents are considered. Digestibility of the feed and ratio of energy to protein seem to be more important under our conditions (11). It is difficult, however, to quantify their effects.

Manganese

Some earlier publications (12, 13) suggest a Mn requirement for dairy cattle in excess of 100 ppm, owing to high contents of Fe, Ca, P, K, and crude protein in the rations. These reports have led to a series of experiments, most with identical twins, in order to be better informed on the manganese requirement of cattle under Dutch conditions. The experiments showed that much of the syndrome has to be attributed to poor management. Furthermore, in experiments on practical rations, as low in Mn as 16 and 21 ppm in the dry matter of the rations during a period of several years, no clinical signs suggesting Mn deficiency were observed (14).

From Table 2 (A) it appears that the supplementation of a low Mn ration with 109 ppm Mn as manganese sulfate has only a slight effect on the Mn content of the colored hair sampled from the ribs. The effect due to Mn supplementation is far exceeded by the seasonal variation in hair Mn content. We have reason to assume that this effect is due to the type of ration, mainly

1. J. Hartmans, unpublished data.

Table 2. Manganese content in hair and liver of cattle with and without clinical signs of manganese deficiency

Author and experimental treatment	Mn in ration[1] (ppm)	Clinical signs of deficiency[2]	Mn in hair DM[3] At pasture (ppm)	Mn in hair DM[3] On hay (ppm)	Mn in liver DM[4] (ppm)
A. Hartmans (14); identical twins, for 2½ years					
Low Mn	21	–	6.0^a	1.8^b	12.4 ± 0.9
Low Mn + $MnSO_4$	130	–	6.7^a	2.7^c	12.8 ± 1.0
B. Hartmans (14); identical twins, for 3½ years					
Low Mn + 0.4% Ca as $CaCO_3$	16	–	4.4^a	1.7^b	11.3 ± 1.2
Low Mn + 0.4% P as NaH_2PO_4	16	–	4.3^a	1.6^b	10.8 ± 0.5
C. Hartmans (14); calves born from sub A cows					
Dams low Mn	21	–		1.3^a	
Dams high Mn	130	–		1.3^a	
D. Hartmans (14); calves born from sub B cows					
Dams low Mn + Ca	16	–		2.2^a	10.3 ± 1.9
Dams low Mn + P	16	–		2.0^a	12.2 ± 3.5

E. Anke (15)				
Calves, for 20 weeks	23	–	13.9	12.5
Calves, for 20 weeks	5.1	+	4.1	8.6
F. Howes and Dyer (16)				
Newborn calves				
Dams for 12 months	21	–	4.9	8.0 ± 0.5
Dams for 12 months	13	+	4.7	4.5 ± 0.2
Dams for 12 months + 1.5% Ca as $CaCO_3$	14	+	4.8	6.7 ± 0.4
G. Rojas *et al.* (17)				
Newborn calves				
Dams during 12 months	25.1	–		11.8
Dams during 12 months	15.8	+		6.6
Dams during 12 months + 3% $CaHPO_4 \cdot 2H_2O$	16.9	+		7.3

[1] A–E, on dry matter (DM) basis.

[2] – Signs absent; + signs present.

[3] A–D, values of the same experiment with different superscripts ($^{a\ b\ c}$), are significantly different ($P < 0.01$).

[4] A, B, D and F, mean ± standard error.

fresh herbage at pasture and mainly hay during winter, rather than to the season itself. Table 2 further shows the Mn content in hair in our experiments, compared with those of deficient and nondeficient animals in the literature. It is evident that our animals, when fed hay, and in the second experiment also when at pasture, have far lower Mn content in their hair than the animals of Anke (15) and of Howes and Dyer (16) showing clinical signs. We feel, therefore, that the Mn content of the hair is, as yet, not a reliable criterion for estimating the Mn status of animals. Only in comparative experiments can some value be attached to it.

At the same time the Mn content in the liver in our experiments was not different between treatments, nor between seasons. Liver values of about 9 ppm seem to be marginal; at high intakes no increased values occur, except in livers of newborn calves fed high doses of Mn during their first week of life (16).

Though it is likely that the Mn content of liver can be used as a criterion to differentiate between deficient and sufficient supply, we still chose the Mn content of the ration, considering that such a much easier check is justified in view of the absence of a deficiency in the Netherlands hitherto. Thus a content of 25 ppm Mn in the ration dry matter is considered as a marginal level with a certain safety margin.

CONCLUDING REMARKS

1. The above example illustrates that the choice of a criterion may very much depend on the actual situation with respect to the element concerned.

2. The procedure described works under our conditions. We are convinced that the procedure and the criteria can be adapted to local conditions when used elsewhere.

3. We hope that research workers with a practical eye, in cooperation with extension officers and animal health services, will consider our approach to see what it is worth under their conditions, and how it can be improved.

REFERENCES

1. Commissie Onderzoek Minerale Voeding TNO. 1970. Handleiding mineralenonderzoek bij rundvee in de praktijk, 2nd edn, Stichting Landhen Tuinbouwgidsen, Utrecht.
2. Committee on Mineral Nutrition. 1973. Tracing and treating mineral disorders in dairy cattle, Pudoc, Wageningen.
3. Niederländische Kommission zur Untersuchung der Mineralstofffütterung. 1973. Leitfaden zur Beurteilung der Mineralstoffversorgung des Rindes in der Praxis. *Übers. Tierernährg* 1: 89.

4. Hartmans, J. 1969. Copper deficiency in dairy cattle under field conditions. *Agri Digest* 18: 42.
5. Hartmans, J. 1970. The detection of copper deficiency and other trace element deficiencies under field conditions. *In* C. F. Mills (ed.), *Trace Element Metabolism in Animals,* Livingstone, Edinburgh, p. 441.
6. Bosman, M. S. M. 1961. On the relation between copper content and oxidase activity of cattle serum. *Jaarb. Inst. Biol. Scheik. Onderz. LandbGewass.* p. 83.
7. Todd, J. R. 1970. A survey of the copper status of cattle using copper oxidase (caeruloplasmin) activity of blood serum. *In* C. F. Mills (ed.), *Trace Element Metabolism in Animals,* Livingstone, Edinburgh, p. 448.
8. Bingley, J. B., and Anderson, N. 1972. Clinical silent hypocuprosis and the effect of molybdenum loading on beef calves in Gippsland, Victoria. *Aust. J. Agric. Res.* 23: 885.
9. Hartmans, J. 1962. Hypocupraemia and clinical symptoms in cattle. *Jaarb. Inst. Biol. Scheik. Onderz. LandbGewass.* p. 157.
10. Mills, C. F., and Williams, R. B. 1971. Problems in the determination of the trace element requirements of animals. *Proc. Nutr. Soc.* 30: 83.
11. Bosman, M. S. M., and Deijs, W. B. 1969. Biochemical research on copper in roughages and its availability to ruminants. *Jaarb. Inst. Biol. Scheik. Onderz. LandbGewass.* p. 37.
12. Grashuis, J., Lehr, J. J., Beuvery, L. L. E., and Beuvery-Asman, A. 1953. Manganese deficiency in cattle. *Inst. Moderne Veevoeding "De Schothorst", Hoogland,* Med. S 40.
13. Grashuis, J. 1957. Significance of manganese for man and animals. *Landbouwk. Tijdschr.* 69: 642.
14. Hartmans, J. 1972. Manganese experiments with monozygote cattle twins. *Landwirt. Forsch.* 27(II): 1.
15. Anke, M. 1966. Mineral content of hair and several tissues in normal and in iron, copper or manganese deficient calves. *Arch. Tierernährg* 16: 199.
16. Howes, A. D., and Dyer, I. A. 1971. Diet and supplemental mineral effects on manganese metabolism in newborn calves. *J. Anim. Sci.* 32: 141.
17. Rojas, M. A., Dyer, I. A., and Cassatt, W. A. 1965. Manganese deficiency in the bovine. *J. Anim. Sci.* 24: 664.

DISCUSSION

Hidiroglou (Ottawa). You observed more copper deficiency in summer than during the winter. Is this related to the differences in porphyrin content of hay and pasture that you previously postulated?

Hartmans. We did investigate the possibility that porphyrins were involved in the difference between summer and winter rations with respect to copper supply, but we have abandoned this idea because we could not find any copper porphyrins in the rumen contents or feces of cattle. We have been considering the possibility that the availability of copper may be influenced by sulfide formation. We have examined sulfide formation in the rumen as a factor under our conditions, and find that in highly digestible herbage, with a relative excess of protein, there is a considerable formation of sulfides in the rumen. This does not occur in hay and only to a limited extent in silage. We think that much of the data we have is consistent with the idea that sulfide formation in the rumen may be a mechanism of inducing copper deficiency under our conditions rather than molybdenum interactions or other factors.

Thornton (London). Have you any wide-scale programs in The Netherlands to evaluate the economic significance of subclinical disorders under your specific geochemical conditions?

Hartmans. Soil testing is carried out regularly by farmers, and from this we get information on the mineral status of different soil types. We have no special survey in progress, and, although the data from individual farmers may be biased as they are obtained from fields and farms where there are problems, we still have a good usable picture of what the trace-element distribution is in our soils.

Miller (Athens). Is it possible that the manganese content of the hair is not controlled by the diet or by seasonal effects, but that the seasonal effect is perhaps due to factors that are independent of the manganese intake?

Hartmans. It is not a true seasonal effect, but it is an effect caused by the ration. We have shown this in the experiments where we kept animals on hay during the summer. These experiments indicated that manganese from pasture was more available than manganese from hay, and that the manganese content of the liver or blood plasma probably is a better criterion of manganese availability than the hair analysis.

Erway (Cincinnati). In regard to the correlation of manganese in hair with dietary manganese, I'd like to ask what was the pattern of pigmentation in the hair samples used for analysis, and what was the pattern of hair growth?

Hartmans. In our experiments the animals were black or red and white, and there was no difference in manganese content between red and black hairs. We did not see a tendency for the manganese content of the hair to be altered by periods of rapid hair growth. When we kept animals indoors on hay, the content in the hair remained low irrespective of the season.

Mills (Aberdeen). In your experience with these survey situations, how frequently do you come across diarrhea as a consequence of copper deficiency?

Hartmans. Some 10 to 20 years ago we had areas where there was extensive diarrhea in cattle, especially in the fall. These cases have all been cleared up because they gave rise to serious losses and the farmers recognized this. What we deal with now are subclinical losses in which diarrhea is not a serious problem.

Ammerman (Gainesville). Are you measuring total serum copper or ceruloplasmin copper, and do you get a sufficient range in the levels of liver manganese to be useful as a tool in diagnosing the deficiency?

Hartmans. The blood plasma copper values were total plasma copper. We have had the experience that a high dietary supply of manganese does not change the manganese content in liver, but there is data in the literature that would indicate that the manganese content in liver will be decreased if there is insufficient manganese in the diet.

THE SELENIUM-DEFICIENCY PROBLEM IN ANIMAL AGRICULTURE

DUANE E. ULLREY

Michigan State University, East Lansing, Michigan

For years the selenium problem in animal agriculture was considered to be its toxicity. Marco Polo referred to ". . . a poisonous plant . . . which if eaten by [horses] has the effect of causing the hoofs of the animals to drop off" (1). U.S. Army horses at Fort Randall, Nebraska Territory, exhibited signs of selenium toxicity in 1857 which Dr. T. C. Madison called "alkali disease" (2). The relationship between these toxic signs and high intakes of selenium was finally established by researchers of the U.S. Department of Agriculture and the South Dakota and Wyoming Agricultural Experiment Station, and their conclusions were published in 1934 (3). Preoccupation with selenium toxicity, and the inability to detect this element in low concentrations, delayed recognition of its essential metabolic character.

In 1957 Schwarz and Foltz (4) showed that sodium selenite would prevent liver necrosis in rats fed torula yeast diets. Patterson, Milstrey, and Stokstad (5), Scott *et al.* (6), and Stokstad, Patterson, and Milstrey (7) demonstrated in the same year that selenium would prevent the development of exudative diathesis in chicks. Eggert *et al.* (8) in 1957 and Grant and Thafvelin (9) in 1958 established a relationship between selenium deficiency and hepatosis dietetica in swine. In 1958 Muth *et al.* (10) and Hogue (11) showed that selenium would also protect against white muscle disease in young ruminants. These and other significant findings have been reviewed in *Selenium in Nutrition,* published by the U.S. National Academy of Sciences in 1971 (12). It is now apparent that appreciably larger geographical areas in the United States are selenium deficient as compared with those which are selenium toxic (13).

PATHOLOGY OF DEFICIENCY

The selenium-deficiency problem in Michigan is typical of much of the Midwest east of the Mississippi. Cases of selenium-vitamin-E deficiency have been observed in all species of domestic farm animals plus a number of wild species raised in captivity. Michel, Whitehair, and Keahey (14) first diagnosed the deficiency in commercial swine herds in Michigan in 1967. Through June, 1972, 53 farms had submitted pigs to the Michigan State University Diagnostic Laboratory in which lesions of selenium-vitamin-E deficiency were found. Since a considerable differential diagnostic problem is involved (15), it is quite likely that the deficiency was present, but undiagnosed, much earlier than 1967.

Commonly, pigs weighing 20–40 kg die suddenly and a bilateral paleness of skeletal muscles is found. Most prominently affected are the *quadriceps femoris, gracilis, adductor, psoas* and *longissimus dorsi* muscles. Histologically, loss of striations, vacuolization, fragmentation, and mineral deposition in muscle fibers is seen. The most striking sign at necropsy is the change in character of the liver, which is often swollen and pale with focal lesions that give it a roughened appearance. Histological examination reveals lobules that have undergone marked degeneration and necrosis, while adjacent lobules appear normal. Damaged lobules exhibit lysis of the hepatic cells and dilatation of sinusoids with blood, giving the appearance of massive intralobular hemorrhage. Many of the pigs exhibit icterus. Edema is also common in the mesentery of the spiral colon, lungs, subcutaneous tissues, and submucosa of the stomach. Occasional mottling and dystrophy of myocardium is also seen.

Based on controlled research (8, 9, 14, 16–19), it appears that these lesions are the consequence of a primary dietary inadequacy of selenium, or vitamin E, or both. However, the incidence of this disease is increased in response to environmental stress. When a number of litters are weaned and the pigs comingled, the physical exertion of fighting to establish social order precipitates the lesions. If pigs become excessively chilled or overheated in response to very cold or very hot weather, the incidence of the disease increases. On one Michigan farm, when pigs were crowded into an enclosed building without adequate space and ventilation during hot July weather, daily death losses from this disease occurred. On this farm, when pigs from the same litters were housed in an open-fronted building with more space per pig, no pigs died. While the primary problem was a dietary deficiency of selenium, the deficiency was not expressed clinically until a secondary environmental stress was superimposed.

It is not possible to estimate the average percentage mortality or morbidity from the above observations, but we have evidence in the Michigan State University swine herd of the serious economic impact of this nutritional-deficiency disease. When these pigs are fed practical corn-soybean-meal diets, unsupplemented with selenium or vitamin E, a 20% death loss due to this deficiency may be expected. Clinical signs of the deficiency in living pigs are difficult to detect. In one study where all pigs were necropsied whether or not they died before the end of the study, the deficiency was detected clinically in 16%, and these pigs died shortly thereafter. Of all pigs necropsied, 25% had selenium-vitamin-E deficiency lesions. From these limited observations, our experience suggests that in growing pigs mortality is around 15–20%, and morbidity may be 25% or more.

We believe that the deficiency may seriously affect reproductive efficiency as well, but the evidence is less clear. In a Michigan State University comparison of 54 sows receiving an unsupplemented corn-soybean-meal diet with 54 sows receiving supplemental selenium and vitamin E, the latter group weaned 1.1 more pigs per litter, and this difference was statistically significant ($P < 0.05$).

REASONS FOR INCREASED INCIDENCE

Since just eight years ago most nutritionists would have considered a selenium or vitamin-E deficiency in practical swine diets a most unlikely possibility, it is reasonable to ask "Why is this such a problem now?" Basic to our answer is an understanding of the dramatic changes which have taken place in swine management. More and more pigs are being raised in complete confinement, without access to pasture. When this practice includes the breeding herd, the entire swine life cycle is dependent upon the nutrients provided in compounded feeds. Since corn and soybean meal are the primary ingredients in such feeds formulated in the midwestern United States, it is appropriate to consider their selenium and tocopherol content.

With respect to the latter, one must recognize that there are significant differences between naturally occurring tocopherol isomers in their biological activity. If α-tocopherol is given a relative value of 100, then β- and ζ-tocopherols have approximately 33%, and γ-, δ-, ϵ- and η-tocopherols less than 1% of this activity for prevention of fetal resorption in the rat (20). From a practical standpoint, then, the α-tocopherol content of a feed is a much better indicator of its vitamin-E activity than total tocopherol content. α-tocopherol values for corn, soybean meal and pasture plants are presented in Table 1 (21). It is obvious that a swine diet made of corn and soybean meal

Table 1. α-tocopherol concentrations in the dry matter of several feed ingredients[1]

Feed ingredient	α-tocopherol (mg/kg)
Corn, shelled	4.0
Soybean meal, solvent	0.8
Alfalfa, freshly cut	47.0
Perennial rye grass, young freshly cut	69.0

[1]From (21)

will provide much less vitamin-E activity than will pasture. The consequent low vitamin-E intake makes the dietary concentration of available selenium much more critical.

SELENIUM IN SWINE FEEDS

Natural selenium concentrations of corn, sorghum grain, meat meal, soybean meal, and limestone in the United States are presented in Table 2 (22–26). Regional differences for corn are apparent, with relatively higher mean values in South Dakota, Nebraska, and North Dakota, and relatively low values in Michigan, Illinois, Indiana, Ohio, and New York. However, in certain states, such as Iowa, some corn samples were clearly deficient in selenium while others were just as obviously adequate. The overall range for corn was 0.01–2.03 ppm, with the highest state mean of 0.40 in South Dakota.

In a Michigan study (24) one corn hybrid, Michigan 275-2X, was grown at 17 locations and the selenium concentration, soil pH, and yield were compared. The mean selenium concentration was 0.033 ppm, on a dry basis, with a range of 0.013–0.089. Selenium concentrations were positively correlated (r = 0.73) with soil pH ($P < 0.01$) and negatively correlated ($r = -0.48$) with yield ($P < 0.06$). Light, sandy loam soils tended to be associated with lower selenium concentrations in corn as compared with heavier, clay loam soils, but sandy loam soils also tended to have a lower pH.

Groce (24) also studied 20 different corn hybrids grown in a single location. All of these samples were very low in selenium concentration (mean of 0.014, dry basis), reflecting perhaps the low available soil selenium levels and the relatively low soil pH (6.0). The correlation ($r = -0.14$) of selenium levels with yield was not statistically significant. Replicate selenium determinations on each hybrid were highly repeatable and it appeared that, at this location, the three-fold range of selenium values (0.007–0.024 ppm) represented real hybrid differences.

Table 2. Selenium concentration in major feed ingredients used in United States swine feeds

Ingredient[1]	Origin	No. of samples	Range (ppm)	Mean (ppm)	Ref.
Corn	Ill.	31	0.02–0.15	0.05	(22)
Corn	Ind.	17	0.01–0.15	0.04	(22)
Corn	Iowa	25	0.02–0.16	0.05	(22)
Corn	Iowa	1	0.32		(23)
Corn	Kan.	1	0.99		(22)
Corn	Mich.	5	0.03–0.04	0.03	(22)
Corn	Mich.	17	0.01–0.09	0.03	(24)
Corn	Mich.	20	0.01–0.02	0.01	(24)
Corn	Minn.	23	0.02–0.19	0.09	(22)
Corn	Mo.	4	0.02–0.09	0.05	(22)
Corn	Nebr.	6	0.04–0.81	0.35	(22)
Corn	N.Y.	1	0.02		(25)
Corn	N.Dak.	5	0.09–0.26	0.19	(22)
Corn	Ohio & Ind.	5	0.06–0.15	0.09	(23)
Corn	S. Dak.	9	0.11–2.03	0.40	(22)
Corn[2]	Texas	1	0.11		(26)
Corn	Wisc.	5	0.02–0.13	0.04	(22)
Sorg.gr[2]	Texas	1	0.28		(26)
Sorg.gr.[3]	Texas	1	0.07		(26)
Meat meal	Iowa	1	0.84		(23)
Meat meal	Ohio	2	0.13–0.24	0.18	(23)
SBM (44)	Ill.	2	0.20–0.21	0.20	(23)
SBM (44)	Iowa	1	1.04		(23)
SBM (44)	Ohio	4	0.05–0.13	0.10	(23)
SBM (49)	Ind.	1	0.09		(24)
Limestone	Ohio	1	0.01		(23)

[1]Sorg.gr., sorghum grain; SBM, soybean meal.

[2]Dryland, Williamson County.

[3]Irrigated, Baily County.

When grains and supplements in various states were combined to make complete swine feeds, the selenium concentrations shown in Table 3 were found (15, 24, 27, 36, 37, 46). These values ranged from 0.03 to 0.49 ppm, with the same regional differences seen in corn. In the absence of supplemental vitamin E, those diets containing 0.06 ppm selenium or less resulted in signs of deficiency.

Table 3. Selenium concentration in complete swine feeds manufactured in the United States

Feed	Origin	Samples	Range (ppm)	Mean (ppm)	Ref.
Wheat-soy	Ark.	1	0.15		(36)
Corn-soy	Idaho	1	0.09		(36)
Corn-soy	Ill.	1	0.04		(36)
Corn-soy	Ind.	1	0.05		(36)
Corn-soy	Iowa	1	0.24		(36)
Corn-soy	Mich.	1	0.04		(36)
Corn-soy	Mich.	1	0.06		(15)
Corn-soy	Mich.	1	0.05		(27)
Corn-soy	Mich.	2	0.04–0.04	0.04	(46)
Corn-soy	Mich.	2	0.05–0.05	0.05	(24)
Corn-soy	Nebr.	1	0.33		(36)
Corn-soy	N.Y.	1	0.04		(36)
Barley-soy	N.Dak.	1	0.41		(36)
Corn-soy	S.Dak.	1	0.49		(36)
Corn-soy	S.Dak.	2	0.30–0.45	0.38	(37)
Corn-soy	S.Dak.	2	0.24–0.44	0.34	(37)
Corn-soy	Va.	1	0.03		(36)
Grain-supp.	Wisc.	1	0.18		(36)
Grain-supp.	Wyo.	1	0.16		(36)

Table 4. Selenium concentration in *longissimus* muscle of swine receiving only natural dietary selenium[1]

State	Diet Se (ppm)	*Longissimus* muscle Se (ppm)[2]
Ark.	0.15	0.21 (3)
Idaho	0.09	0.11 (3)
Ill.	0.04	0.06 (3)
Ind.	0.05	0.06 (3)
Iowa	0.24	0.28 (4)
Mich.	0.04	0.05 (3)
Nebr.	0.33	0.31 (4)
N.Y.	0.04	0.05 (3)
N. Dak.	0.41	0.39 (2)
S. Dak.	0.49	0.52 (2)
Va.	0.03	0.03 (3)
Wisc.	0.18	0.12 (3)
Wyo.	0.16	0.31 (4)

[1]From (36)

[2]Number of pigs in parentheses.

Table 5. Effect of natural dietary selenium levels and sodium selenite supplementation upon selenium concentration in *longissimus* muscle, liver and kidney[1]

Item	Michigan State University trial[2]				South Dakota State trial[3]		
	MSU basal + 0.4 ppm Se	SDS basal	SDS basal + 0.1 ppm Se (WD)[4]	SDS basal + 0.1 ppm Se (NWD)[4]	SDS basal	SDS basal + 0.1 ppm Se (WD)[4]	SDS basal + 0.1 ppm Se (NWD)[4]
Natural diet Se (ppm)	0.04	0.44	0.44	0.44	0.24	0.24	0.24
Supplemental Se (ppm)	0.40		0.10	0.10		0.10	0.10
Total grower diet Se (ppm)	0.44	0.44	0.54	0.54	0.24	0.34	0.34
Total finisher diet Se (ppm)	0.44	0.44	0.44	0.54	0.24	0.24	0.34
No. pigs	4	4	4	4	4	4	4
Longissimus muscle Se (ppm)	0.12^{a}	0.48^{b}	0.50^{b}	0.45^{b}	0.31^{d}	0.35^{d}	0.35^{d}
Liver Se (ppm)	0.61^{a}	0.84^{b}	0.96^{c}	0.92^{bc}	0.73^{d}	0.82^{de}	0.90^{e}
Kidney Se (ppm)	2.14^{a}	2.17^{a}	2.45^{a}	2.33^{a}	2.43^{d}	2.59^{d}	2.81^{d}

[1]From (37)

[2]Values in MSU trial with different superscripts [abc] are significantly ($P < 0.05$) different.

[3]Values in SDS trial with different superscripts [de] are significantly ($P < 0.05$) different.

[4]WD, Se withdrawn 60 days before slaughter; NWD, Se not withdrawn (fed until slaughter).

SELENIUM IN SWINE TISSUES

Since it has been proposed (28) that supplemental dietary sodium selenite be used to prevent the development of selenium deficiency in swine, it is appropriate to consider the effect of such a supplement on swine-tissue selenium concentrations. Much of the selenium occurring naturally in feedstuffs is in organic form, and thus a comparison of tissue selenium levels resulting from selenium-containing feedstuffs or from sodium selenite would also be useful. Peterson and Butler (29) and Butler and Peterson (30) have produced evidence that much of the selenium in natural feeds is associated with amino acids. Olson *et al.* (31) found that almost half of the selenium in a hydrolysate of crude gluten from seleniferous wheat seeds was in the form of selenomethionine. McConnell and Hoffman (32) demonstrated that rat liver methionyl-tRNA synthetase does not distinguish between [^{14}C]methionine and [^{75}Se]methionine, and found that selenomethionine was incorporated into hepatic polypeptides via the methionine pathway. Other selenium-containing proteins, glutathione peroxidase, and apparently a cytochrome have also been found in animal tissues (33, 34). Ehlig *et al.* (35) have shown that, while selenium from either dietary selenomethionine or sodium selenite may become associated with tissue proteins of the lamb, the association of selenium from selenomethionine was much greater.

Ku *et al.* (36) demonstrated a positive linear relationship between natural selenium levels in swine feeds and the selenium concentration of swine *longissimus dorsi* muscle. These observations are summarized in Table 4. The linear regression relating these two measures was $Y = 0.9602X + 0.0281$, when X is the selenium concentration in the finishing ration and Y is the selenium concentration in the wet *longissimus* muscle. The correlation (r = 0.95) between them was highly significant ($P < 0.01$).

As shown in Table 5, when 0.4 ppm of selenium from sodium selenite was added to a swine diet low in natural selenium (0.04 ppm), selenium concentrations in *longissimus* muscle and liver were significantly ($P < 0.05$) less than those resulting from ingestion of a diet containing the same total selenium level (0.44 ppm) but in a natural form (37). The addition of 0.1 ppm of selenium from sodium selenite to a diet already containing moderate levels (0.24–0.44 ppm) of natural selenium did not increase muscle or kidney selenium levels and had only a small and inconsistent effect on liver selenium levels.

Table 6. Toxicity of selenium administered as sodium selenite in a single oral dose to swine[1]

Pig no.	Initial wt (kg)	Na selenite (g)	Se (mg/kg BW)	Observations	Se in tissues (ppm)			
					Leg muscle	Heart	Liver	Kidney
4282	24	1.20	22.7	Death in 72 hr	T[2]	0.05	2.5	
4754	59	2.28	17.4	Paralyzed; killed after 18 days	0.2	0.1	1.0	0.3
4737	46	1.31	13.2	Sick and recovered; killed at 2½ months		0.5	0.2	0.2
4730	55	1.57	13.2	Sick and recovered; killed at 2½ months		0.1	0.2	0.5
4683	55	1.05	8.8	Sick and recovered; killed at 2½ months				
4703	55	0.52	4.4	Slightly sick; killed at 2½ months		0.05	0.1	0.2
4746	68	0.33	2.2	No effect				

[1]From (38).
[2]Trace.

Table 7. Toxicity of selenium fed continuously in the diet of swine

Total diet Se (ppm)	Suppl. Se (ppm)	Suppl. Se form	No. of pigs	Init. wt. (kg)	Toxicity (% of total)	Se in tissues (ppm) Skeletal muscles	Heart	Liver	Kidney	Ref.
T[1]	0		6	5	0			0.18	0.69	(39)
0.04	0		6	15	0	0.06	0.12	0.17	1.50	(24)
0.09	0.05	Na_2SeO_3	12	15	0	0.08	0.18	0.36	1.88	(24)
0.10	0.10	Na_2SeO_3	2	5	0			0.40	1.23	(39)
0.10	0.10	Se meth[3]	2	4	0			0.22	0.90	(39)
0.14	0.10	Na_2SeO_3	12	15	0	0.09	0.17	0.40	1.82	(24)
0.24	0.20	Na_2SeO_3	12	15	0	0.09	0.19	0.40	1.80	(24)
0.54	0.50	Na_2SeO_3	9	9	0	0.09				(24)
1.04	1.00	Na_2SeO_3	6	45	0	0.13	0.33	0.80	2.10	(24)
5	5	Na_2SeO_3	2	4	0			2.36	2.42	(39)
5	5	Se meth	2	4	0			4.90	6.61	(39)
5	5	Se corn[4]	5	35	0					(41)
7.5[2]	7	Na_2SeO_3	5	15	60					(42)
10	10	Na_2SeO_3	2	4	0			3.62	3.60	(39)
10	10	Se meth	2	5	0			9.92	10.40	(39)
10	10	Se corn	5	35	60					(41)
10.5[2]	10	Na_2SeO_3	8	13	25					(42)
13.5[2]	13	Na_2SeO_3	16	13	31					(43)

20	20	Na_2SeO_3	2	5	50			22.1	12.9	(39)
20	20	Se meth	2	5	0			12.4	18.8	(39)
25	24.5	Na_2SeO_3	2	17	50	0[5]	0[5]	4.0	3.0	(40)
45	45	Na_2SeO_3	2	5	100			7.4	7.3	(39)
45	45	Se meth	2	4	100			18.0	12.0	(39)
49	49	Na_2SeO_3	2	19	100	2.0	5.0	10.0	25.0	(40)
60	60	Na_2SeO_3	2	5	100			11.0	4.8	(39)
60	60	Se meth	2	5	100			18.0	13.6	(39)
100	100	Na_2SeO_3	2	5	100			31.7	27.5	(39)
100	100	Se meth	2	5	100			19.2	16.7	(39)
120	120	Na_2SeO_3	2	4	100			7.6	2.8	(39)
120	120	Se meth	2	5	100			34.5	11.1	(39)
196	196	Na_2SeO_3	2	17	100					(40)
392	392	Na_2SeO_3	2	16	100	0[5]	T[6]	3.0	8.5	(40)
600	600	Na_2SeO_3	2	4	100			9.3	3.0	(39)
600	600	Se meth	2	4	100			27.4	12.8	(39)

[1]Not determined but probably less than 0.02 ppm Se.
[2]Estimated.
[3]Selenomethionine.
[4]Seleniferous corn.
[5]Below undefined limits of detection.
[6]Trace.

SELENIUM TOXICITY

To provide an indication of the margin of safety between oral intakes of selenium adequate to prevent a deficiency and intakes which will produce toxic effects, the data shown in Tables 6 and 7 have been extracted from the literature (24, 38–43). Data on the toxicity of a single oral dose of selenium as sodium selenite are presented in Table 6. Based on these limited data, it would appear that the minimum lethal dose was between 13 and 17 mg Se per kg bodyweight. The minimum no-effect dose appeared to be between 2 and 4 mg Se per kg bodyweight.

Data concerned with the toxicity of selenium fed continuously in the diet of swine are presented in Table 7. Supplemental selenium was presented as sodium selenite, selenomethionine, and seleniferous corn. Dietary selenium levels of 5 ppm produced no effect in any of these three forms. The lowest dietary selenium level at which toxicity was noted was 7.5 ppm, and as much as 45 ppm was needed to produce signs of toxicity in all pigs. Increasing dietary selenium levels resulted in increasing liver and kidney selenium levels up to a dietary selenium concentration of 25 ppm. Higher dietary levels produced variable tissue concentrations dependent largely upon the length of survival.

Various increments of dietary selenium have not been presented to reproducing swine, so the minimum toxic level has not been defined. However, it is apparent from the data in Table 8 that 10 ppm of supplemental selenium from sodium selenite reduces reproductive efficiency of swine (44).

SELENIUM REQUIREMENTS

Groce *et al.* (45) have attempted to define the minimum dietary selenium requirements of growing-finishing swine fed corn-soybean-meal diets. The diets used were low (0.05 ppm) in natural selenium and contained 4.2 mg d-α-tocopherol per kg. Data gathered from these and previous studies (46) concerned the effect of adding 0.05, 0.1, 0.2, 0.5, and 1.0 ppm of selenium from sodium selenite upon development of deficiency signs, tissue selenium concentration (Table 9) and selenium balance (Table 10).

Supplements of 0.1 ppm prevented death losses, gross pathology and histopathological lesions of nutritional muscular dystrophy, and dietary hepatic necrosis. Selenium concentrations in skeletal muscle, myocardium, liver, and kidney were increased by sodium selenite supplementation as compared with values on the basal diet, but tended to reach a maximum at 0.1 ppm supplemental selenium (over a range of 0–0.5 ppm supplemental

Table 8. Effect of selenium on reproduction in swine[1]

Item	Basal	Basal + 10 ppm Se (from Na selenite)
	First farrowing	
No. of gilts	9	10
No. conceived	8	7
Services/conception	1.0	1.3
Gilts farrowing	7	6
Days from weaning to estrus	8	13
Pigs farrowed per litter	8.0	9.8
Live pigs farrowed per litter	8.0	7.7
Birth wt (kg)	1.33	1.19
Pigs weaned per litter	7.3	5.7
Weaning wt (kg)	13.9	10.5
	Second farrowing	
No. of gilts	7	6
Gilts farrowing	6	5
Pigs farrowed per litter	11.2	8.6
Live pigs farrowed per litter	10.5	7.2
Birth wt (kg)	1.31	1.26
Pigs weaned per litter	8.0	5.6
Weaning wt (kg)	13.2	10.7

[1]From (44).

selenium). Even the highest levels of tissue selenium, resulting from feeding 1.0 ppm selenium as sodium selenite, were well below those reported for the same tissues of pigs receiving selenium-adequate diets containing only natural selenium.

Data from the balance trials produced conclusions which parallel those derived from the tissue selenium levels. Supplemental selenite-selenium was absorbed and retained quite well when added to selenium-deficient corn-soybean-meal diets at levels up to 0.1 ppm. Levels of supplemental selenium above 0.1 ppm (0.2 and 0.5 ppm) were absorbed and then excreted in greater proportions in the urine to produce absolute selenium retentions similar to those observed at 0.1 ppm supplemental selenium. Serum selenium concentrations reached a plateau at about the same supplemental selenium levels, indicating the possible existence of tissue and serum thresholds for selenium from sodium selenite in the pig.

From the data available the minimum practical level of selenium supplementation from sodium selenite to prevent deficiencies in confined growing-finishing swine is 0.1 ppm, resulting in a total selenium level in many Midwestern swine diets of about 0.15 ppm. This is 1/50 of the lowest

Table 9. Effect of adding various increments of selenium from sodium selenite to diets low in natural selenium upon tissue selenium concentration[1]

Item	Basal (0.04 ppm Se)	Basal + 0.05 ppm Se	Basal + 0.1 ppm Se	Basal + 0.2 ppm Se	Basal + 0.5 ppm Se	Basal + 1.0 ppm Se
No. of experiments	3	1	2	1	1	1
No. of pigs	18	12	21	12	9	3
Skeletal muscle Se (ppm)	0.05	0.08	0.08	0.09	0.09	0.13
Myocardium Se (ppm)	0.11	0.18	0.17	0.19		0.33
Liver Se (ppm)	0.14	0.36	0.40	0.40		0.80
Kidney Se (ppm)	1.38	1.88	1.82	1.80		2.10

[1]From (24).

Table 10. Effect of adding various increments of selenium from sodium selenite to diets low in natural selenium upon daily selenium balance in the young pig[1]

Item	Supplementary Se (ppm)				
	Basal[2]	0.05	0.1	0.2	0.5
No. of experiments	2	1	2	1	1
No. of pigs	5	3	5	3	2
Se intake (μg/day)	19.9	47.2	65.4	116.1	244.4
Se retention (% of intake)	48.0	53.8	59.7	41.4	15.2
Se retention (μg/day)	9.6	25.4	39.1	48.1	37.1
Se excretion (% of intake)					
Fecal	41.8	37.5	27.4	32.6	23.6
Urinary	10.1	8.9	12.8	26.0	61.2

[1]From (45, 46).
[2]Contained 0.04–0.05 ppm natural Se.

continuously fed dietary selenium level (approximately 7.5 ppm) shown (42) to produce toxicity in swine. Such supplementation would appear to be a safe and scientifically sound nutritional practice.

REFERENCES

1. Polo, M. 1926. *The Travels of Marco Polo* (revised from Marsden's translation and edited with introduction by Manual Komroff), Liveright, New York, p. 81.
2. Madison, T. C. 1860. Sanitary report–Fort Randall. *In* R. H. Coolidge (ed.) *Statistical Report on the Sickness and Mortality in the Army in the United States, Senate Exch. Doc.* 52: 37.
3. Franke, K. W. 1934. A new toxicant occurring naturally in certain samples of plant foodstuffs. I. Results obtained in preliminary feeding trials. *J. Nutr.* 8: 597.
4. Schwarz, K., and Foltz, C. M. 1957. Selenium as an integral part of Factor 3 against dietary necrotic liver degeneration. *J. Am. Chem. Soc.* 79: 3292.
5. Patterson, E. L., Milstrey, R., and Stokstad, E. L. R. 1957. Effect of selenium in preventing exudative diathesis in chicks. *Proc. Soc. Exp. Biol. Med.* 95: 617.
6. Scott, M. L., Bieri, J. G., Briggs, G. M., and Schwarz, K. 1957. Prevention of exudative diathesis by factor 3 in chicks on vitamin E-deficient torula yeast diets. *Poult. Sci.* 36: 1155 (abstr.).
7. Stokstad, E. L. R., Patterson, E. L., and Milstrey, R. 1957. Factors which prevent exudative diathesis in chicks on torula yeast diets. *Poult. Sci.* 36: 1160.

8. Eggert, R. G., Patterson, E., Akers, W. T., and Stokstad, E. L. R. 1957. The role of vitamin E and selenium in the nutrition of the pig. *J. Anim. Sci.* 16: 1032 (abstr.).
9. Grant, C. A., and Thafvelin, B. 1958. Selenium and hepatosis diaetetica of pigs. *Nord. Vet. Med.* 10: 657.
10. Muth, O. H., Oldfield, J. E., Remmert, L. F., and Schubert, J. R. 1958. Effects of selenium and vitamin E on white muscle disease. *Science* 128: 1090.
11. Hogue, D. E. 1958. Vitamin E, selenium and other factors related to nutritional muscular dystrophy in lambs. *In Proceedings of Cornell Nutrition Conference, Feed Manufacturers,* Ithaca, New York, pp. 32–39.
12. Oldfield, J. E., Allaway, W. H., Draper, H. H., Frost, D. V., Jensen, L. S., Scott, M. L., and Wright, P. L. 1971. *Selenium in Nutrition,* National Academy of Sciences, Washington, D.C.
13. Kubota, J., Allaway, W. H., Carter, D. L., Cary, E. E., and Lazar, V. A. 1967. Selenium in crops in the United States in relation to the selenium-responsive diseases of livestock. *J. Agric. Food Chem.* 15: 448.
14. Michel, R. L., Whitehair, C. K., and Keahey, K. K. 1969. Dietary hepatic necrosis associated with selenium-vitamin E deficiency in swine. *J. Am. Vet. Med. Ass.* 155: 50.
15. Trapp, A. L., Keahey, K. K., Whitenack, D. L., and Whitehair, C. K. 1970. Vitamin E-selenium deficiency in swine: differential diagnosis and nature of field problem. *J. Am. Vet. Med. Ass.* 157: 289.
16. Obel, A. L. 1953. Studies on the morphology and etiology of so-called toxic liver dystrophy (hepatosis dietetica) in swine. *Acta Path. Microbiol. Scand.* suppl. 44.
17. Hove, E. L., and Seibold, H. R. 1955. Liver necrosis and altered fat composition in vitamin E deficient swine. *J. Nutr.* 56: 173.
18. Lannek, N., Lindberg, P., Nilsson, G., Nordstrom, G., and Orstadius, K. 1961. Production of vitamin E deficiency and muscular dystrophy in pigs. *Res. Vet. Sci.* 2: 67.
19. Ewan, R. C., Wastell, M. E., Bicknell, E. J., and Speer, V. C. 1969. Performance and deficiency symptoms of young pigs fed diets low in vitamin E and selenium. *J. Anim. Sci.* 29: 912.
20. Lehman, R. W. 1969. Vitamin E—chemistry and analysis. *Proceedings of Eastman Vitamin E Seminary,* Minneapolis, Minn., Eastman Chemical Products Inc., Kingsport, Tenn., pp. 1–23.
21. Ames, S. R. 1956. Roles of vitamin E (α-tocopherol) in poultry nutrition and disease. *Poult. Sci.* 35: 145.
22. Patrias, G., and Olson, O. E. 1969. Selenium contents of samples of corn from midwestern states. *Feedstuffs* 41: 32.
23. Bruins, H. W., Ousterhout, L. E., Scott, M. L., Cary, E. E., and Allaway, W. H. 1966. Is selenium deficiency a practical problem in poultry? *Feedstuffs* 38: 66.
24. Groce, A. W. 1972. Selenium and/or Vitamin E Supplementation of Practical Swine Diets. Ph.D. Thesis, Michigan State University.
25. Pond, W. G., Allaway, W. H., Walker, E. F., Jr., and Krook, L. 1971. Effects of corn selenium content and drying temperature and of

supplemental vitamin E on growth, liver selenium and blood vitamin E content of chicks. *J. Anim. Sci.* 33: 996.
26. Hitchcock, J. P., Tanksley, T. D., and Ullrey, D. E. 1972. Unpublished analyses. Michigan State University.
27. Ullrey, D. E., Groce, A. W., Miller, E. R., Ellis, D. J., and Keahey, K. K. 1970. Vitamin E and/or selenium for swine. *Feedstuffs* 43: 26.
28. Anonymous. 1973. Selenium in animal feed: proposed food additive regulation. *Fed. Reg.* 38: 10458.
29. Peterson, P. J., and Butler, G. W. 1962. The uptake and assimilation of selenite by higher plants. *Aust. J. Biol. Sci.* 15: 126.
30. Butler, G. W., and Peterson, P. J. 1967. Uptake and metabolism of inorganic forms of selenium-75 by *Spirodella oligorrhiza. Aust. J. Biol. Sci.* 20: 77.
31. Olson, O. E., Novacek, E. J., Whitehead, E. I., and Palmer, I. S. 1970. Investigation on selenium in wheat. *Photochemistry* 9: 1181.
32. McConnell, K. P., and Hoffman, J. L. 1972. Methionine–selenomethionine parallels in rat liver polypeptide chain synthesis. *Fed. Proc.* 31: 691 (abstr. 2685).
33. Rotruck, J. T., Pope, A. L., Ganther, H. E., Swanson, A. B., Hafeman, D. G., and Hoekstra, W. G. 1972. Selenium: biochemical role as a component of glutathione peroxidase. *Science* 179: 588.
34. Whanger, P. D., Pedersen, N. D., and Weswig, P. H. 1973. Characteristics of selenium-binding proteins from lamb muscle. This symposium.
35. Ehlig, C. F., Hogue, D. E., Allaway, W. H., and Hamm, D. G. 1967. Fate of selenium from selenite or selenomethionine with or without vitamin E in lambs. *J. Nutr.* 92: 121.
36. Ku, P. K., Ely, W. T., Groce, A. W., and Ullrey, D. E. 1972. Natural dietary selenium, α-tocopherol and effect on tissue selenium. *J. Anim. Sci.* 34: 208.
37. Ku, P. K., Miller, E. R., Wahlstrom, R. C., Groce, A. W., Hitchcock, J. P., and Ullrey, D. E. 1973. Selenium supplementation of naturally high selenium diets for swine. *J. Anim. Sci.* 37: 501.
38. Miller, W. T., and Williams, K. T. 1940. Minimum lethal dose of selenium, as sodium selenite, for horses, mules, cattle, and swine. *J. Agric. Res.* 60: 163.
39. Herigstad, R. R. 1972. Pathology of inorganic and organic selenium toxicosis in young swine. Ph.D. Thesis, Michigan State University.
40. Miller, W. T. 1938. Toxicity of selenium fed to swine in the form of sodium selenite. *J. Agric. Res.* 56: 831.
41. Schoening, H. W. 1936. Production of so-called alkali disease in hogs by feeding corn grown in affected area. *N. Am. Vet.* 17: 22.
42. Wahlstrom, R. C., Kamstra, L. D., and Olson, O. E. 1955. The effect of arsanilic acid and 3-nitro-4-hydroxyphenylarsonic acid on selenium poisoning in the pig. *J. Anim. Sci.* 14: 105.
43. Wahlstrom, R. C., Kamstra, L. D., and Olson, O. E. 1956. The effect of organic arsenicals, chlortetracycline and linseed oil meal on selenium poisoning in swine. *J. Anim. Sci.* 15: 794.
44. Wahlstrom, R. C., and Olson, O. E. 1959. The effect of selenium on reproduction in swine. *J. Anim. Sci.* 18: 141.
45. Groce, A. W., Miller, E. R., Ullrey, D. E., Ku, P. K., Keahey, K. K., and

Ellis, D. J. 1973. Selenium requirements in corn-soy diets for growing-finishing swine. *J. Anim. Sci.* 37: 948.
46. Groce, A. W., Miller, E. R., Keahey, K. K., Ullrey, D. E., and Ellis, D. J. 1971. Selenium supplementation of practical diets for growing-finishing swine. *J. Anim. Sci.* 32: 905.

DISCUSSION

Hidiroglou (Ottawa). Do you observe pancreatic lesions in your selenium-deficient pigs?

Ullrey. Our pathologists have looked at the pancreas, and we haven't been able to find lesions in this species. I am not sure what the difference between the bird and the pig is in this respect.

Hennig (Jena). Have you seen effects on the hair in selenium-deficient pigs, and what are the effects on baby pigs?

Ullrey. We haven't seen any obvious effects on the hair in the selenium-deficient pig. We have seen selenium-deficiency signs in very young pigs, particularly under stressful circumstances where they were chilled within a few days of birth.

Pařízek (Prague). Do you have any information about the influence of sex on the response to high selenium in the diet?

Ullrey. Sex was identified in all the toxicity studies, but I can't recall any sex difference in their sensitivity to high levels.

Shah (Ottawa). Your studies on swine would indicate that selenite was not as well utilized as selenium from naturally occurring selenium sources. Is this true, and what are the natural forms of selenium?

Ullrey. The only measurement we have is the effect on tissue selenium concentration, and no attempt has been made to relate this concentration to enzymatic activity or to some precisely defined biochemical function. It's true that the naturally occurring selenium forms that exist in cereals, as an example, tend to result in higher tissue selenium concentrations than do equal levels of selenium derived from sodium selenite. There are probably various forms of natural selenium, but at least in wheat gluten about 50% of the selenium is in the form of selenomethionine.

Hoekstra (Madison). I might add that it has been very clearly shown that selenomethionine mimics methionine in the protein synthesis system, and that selenomethionine selenium can be metabolized to the active forms, whatever they are. Therefore, whenever you have selenomethionine present, much is nonspecifically incorporated into any methionine-containing protein, and the body keeps increasing in selenium as you increase dietary selenium. The case of selenite is quite a different situation. Except in the ruminant, selenomethionine is not produced from selenite, so that a plateau region is reached where homeostatic regulation prevents the continued increase in tissue selenium.

Andrews (Upper Hutt). We have a number of forms of selenium-responsive sheep diseases in New Zealand: white muscle disease, which everybody knows about, a selenium-responsive unthriftiness, which probably less people know about, and a selenium-responsive infertility. It would appear that the critical selenium content of our pastures is about 0.02 ppm, and we have been unsuccessful in getting any effect from α-tocopherol except a rather transient effect in the case of white muscle disease. It is ineffective against the unthriftiness and the infertility. I note that the lowest selenium concentration that gave a response in swine rations was 0.03 ppm, which we would place in the normal range. Is there any explanation for this apparent discrepancy between the selenium requirement of the sheep and the pig, and do you know of situations where vitamin E does anything that selenium won't do?

Ullrey. The apparent difference between the requirement for sheep and pigs might be related to the forage concentration of α-tocopherol and the tendency for the supply of this very important nutrient to reduce the apparent quantitative need for supplemental selenium. There are indications that the dietary requirement of selenium is similar for sheep and pigs, but it would appear that more work is needed to complete a definition of the requirements. Of course, the requirement needs to be defined in relationship to other dietary components that interact with selenium, and I doubt that there is a single requirement figure that is entirely appropriate.

ZONES OF DISPLAY OF BIOLOGICAL AND PHARMACOTOXICOLOGICAL ACTION OF TRACE ELEMENTS

A. I. VENCHIKOV

Department of Physiology, Medical Institute, Ashkhabad, U.S.S.R.

The fruitful development of ideas about trace elements necessitates a strict separation of trace-element influence as a biological factor from its action upon organs as an exact pharmacotoxicological agent.

In this report facts obtained in earlier studies (1–3) are compared with new data concerning the response of the organism to various trace-element doses. The results of this research indicate the bounds within the limits of which the true properties of trace elements as biological factors are revealed.

In our experiments my collaborators and I used trace elements in approximately the same concentrations usually found in living organisms or in the environment. In such instances trace elements exerted a stimulating effect on the course of a number of physiological processes. Such quantities of trace elements may be regarded as biological levels. If the trace-element concentration was increased, the action was weakened or discontinued. The trace elements became inactive, as it were. A further increase of concentration brought about a resumption of the action in a number of cases. This effect, however, had the nature of the usual pharmacotoxicological action, that is, the trace element exerted an irritating of toxic effect.

As a result of these observations we have established three zones of action of trace elements. We named them (1) biological action zone, (2) inactive zone, and (3) pharmacotoxicological action zone. The scheme of such an action is represented in Fig. 1. The straight line indicates the normal course of physiological processes. From left to right (A to B) increasing of trace-element concentration is indicated. Section BA shows the biological action zone, within the bounds of which a dependence of the effect on the mass of the substance used is observed. The effect declines as the quantity of the substance used is reduced. If the concentration of the trace element is raised

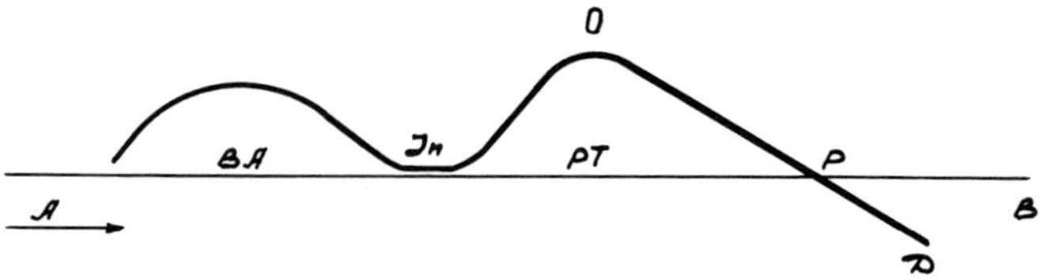

Fig. 1. Action zone of a trace element depending on its concentration. Straight line, value of normal function (initial background); arrow, concentration of active element, gradually rising from A to B; the height of the curve reflects the value of the physiological effect; BA, biological action zone, where the element used enters the internal biochemical structures of living systems as a biological factor; IN, inaction zone of the trace element, which sets in due to its being blocked by the physiological barriers (protective function of an organism); PT, pharmacotoxicological action zone, where the element acts first mainly as a factor irritating the increasing function of living systems (from IN to O) and then, as the concentration is increased, inhibits them (from P to D) (in the limits of this zone the action occurs according to the principle of overcoming the resistance of protective ability of the organism); O, optimal concentration of the irritant which causes the greatest effect; P, transfer of the irritating action to an inhibitive one; D, irreversible reaction of death of the living formations.

above a definite level, the effect is weakened or discontinued and the inaction zone sets in.

This scheme of the trace-element action zones has been deduced on the basis of experiments with guinea pigs, rats, *Bombyx mori,* frogs, and leucocytes of frogs and men. The action of $CuSO_4$, $NiCl_2$, $CrCl_3$, $NaVO_3$, Na_2WO_4, Na_2MoO_4, $ZnCl_2$, $CdCl_2$, $HgCl_2$, KI, and NaI was studied in water solutions in concentrations of 0.003, 0.125, 5, 200 mg%, etc. Daily doses, in terms of the metal, are given below. The examples given below must indicate the presence of zones of trace-element activity described above.

NICKEL EFFECT ON THE INTENSITY OF OXYGEN ABSORPTION BY GUINEA PIGS

In the experiment there were five groups of animals, 13–19 in each group. Except for the control group, they were given daily a water solution of $NiCl_2$ perorally for 40 days, supplying (in terms of the metal per day) one of the following doses: 32 mg/kg, 800 μg/kg, 20 μg/kg, or 0.5 μg/kg. The quantity of oxygen absorbed was measured in a chamber of the closed type (Miropolsky's modification). Before giving nickel the quantity of oxygen absorption was measured for a period of 10 days and then, after defining the resulting baseline, the same measures were repeated when nickel had been given. The changes in average quantities of oxygen absorbed by this group over 10-day periods, expressed as percentages with respect to the baseline, are shown in Fig. 2.

The results indicate the ability of a daily nickel dose of 0.5 μg/kg (curve 5) to increase the intensity of oxygen absorption on the 10th and the 20th days, whereas during the first 20 days a nickel dose of 20 μg/kg (curve 4) had insufficient action and was inactive, and higher doses of 800 μg/kg (curve 3) and 32 mg/kg (curve 2) gave a suppressing action. By the 30th day of nickel introduction, no effect was seen at a dose of 0.5 μg/kg, but the earlier inactive dose of 20 μg/kg began to give a suppressing action, that is, the phenomenon of cumulation had become apparent. With peroral introduction of $NiCl_2$ solution the toxic effect of its large doses did not cause a preliminary increase of gaseous exchange, that is, the function increase was absent.

In experiments with rats a preliminary increase of gaseous exchange occurred (Fig. 3, B, C, group A) when a nickel dose of 800 μg/kg was introduced subcutaneously so that it gave direct action on a tissue. In both types of experiments a decrease of oxygen absorption occurred immediately with a dose of 32 mg/kg. It is necessary to assume that the toxic action of a

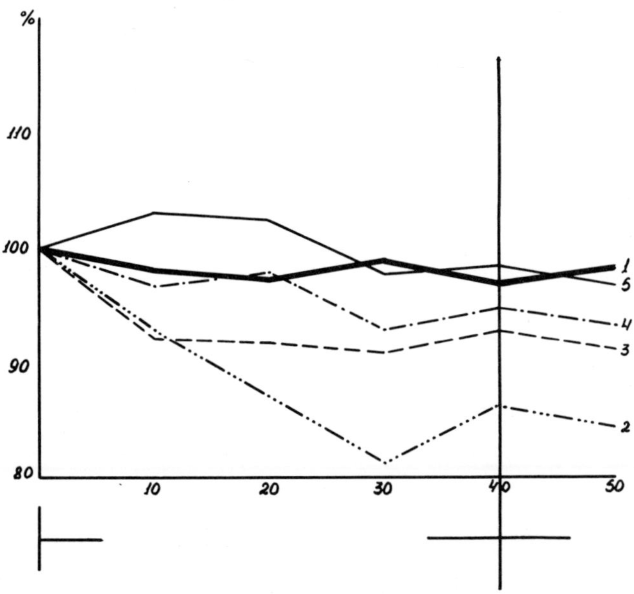

Fig. 2. The influence of peroral administration of various quantities of nickel chloride on oxygen absorption by guinea pigs. The changes are expressed as percentages of the initial quantity of oxygen absorption: curve 1, control group of animals; curve 2, nickel dose 32 mg Ni/kg/day; curve 3, nickel dose 800 μg Ni/kg/day; curve 4, nickel dose 20 μg Ni/kg/day; curve 5, nickel dose 0.5 μg Ni/kg/day. Nickel was given for 40 days. The following 10 days are observations after discontinuing the nickel administration. Abscissa, administration period (days); ordinate, relative oxygen absorption. [L.I. Arutyunov's experiments (4, 5)].

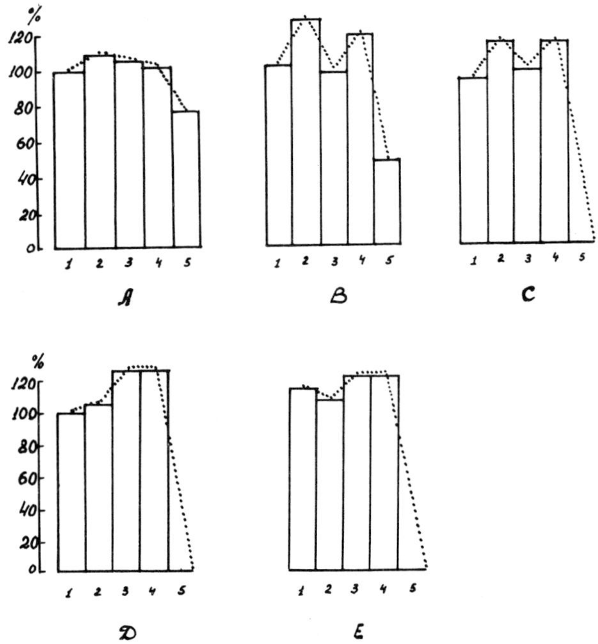

Fig. 3. The influence of daily subcutaneous administration of various quantities of nickel chloride on oxygen absorption by white rats. The changes are expressed as percentages of the initial quantity. Column 1, control group; column 2, 0.5 μg Ni/kg/day; column 3, 20 μg Ni/kg/day; column 4, 800 μg Ni/kg/day; column 5, 32 mg Ni/kg/day. The average data are given for each three-day period of observation. A, 1–3 days; B, 4–6 days; C, 7–9 days; D, 10–12 days; E, 8 days after discontinuing the nickel administration (there were 11–15 animals in the group). Two plateaus of increased oxygen absorption were observed (see B and C). [(4, 5 L.I. Arutyunov's experiments.)]

very high dose suppressed the tissue excitability quickly [L. I. Arutyunov's experiments, refs. (4) and (5)].

CADMIUM EFFECT ON OXYGEN ABSORPTION BY WHITE RATS

In these experiments, generally carried out according to the method described above for nickel, the more pronounced increase of oxygen absorption is apparent using cadmium chloride to supply the metal in a daily 20 μg/kg dose (biological dose) (Fig. 4). At first the 1500 μg/kg dose stimulated oxygen absorption but later it became inactive. Adaptation of the organism to it had occurred. At the 50 mg/kg dose cadmium suppression (toxic action of larger doses) first occurred [Z. N. Nepesova's experiments, refs. (6–9)].

A phenomenon of the same type as seen with $NiCl_2$ and $CdCl_2$ was observed in studies with guinea pigs and rats given $ZnCl_2$ perorally in various quantities (10, 11), $NaVO_3$ (12), $CrCl_3$ (13, 14), and Na_2WO_4 (15, 16). The same effect was observed in frogs placed in NaI solutions of various concentrations for 1 day.

ZINC EFFECT ON OXYGEN ABSORPTION BY *BOMBYX MORI* LARVAE

The described phenomenon of selective action of definite concentrations of trace elements was also observed in experiments with *Bombyx mori* larvae. Using a Tunberg-Winterstein instrument changes in the intensity of oxygen absorption by the larvae effected by daily, in most cases triple, sprinkling of their food by solutions of zinc sulphate or zinc chloride (beginning with 1 ml in the first age and ending with 10 ml in the last age for 50 larvae each time).

It was established that for high concentrations of the solutions ($ZnSO_4$, 10%, 2%, or 0.2%) the degree of oxygen absorption rose as the quantity of

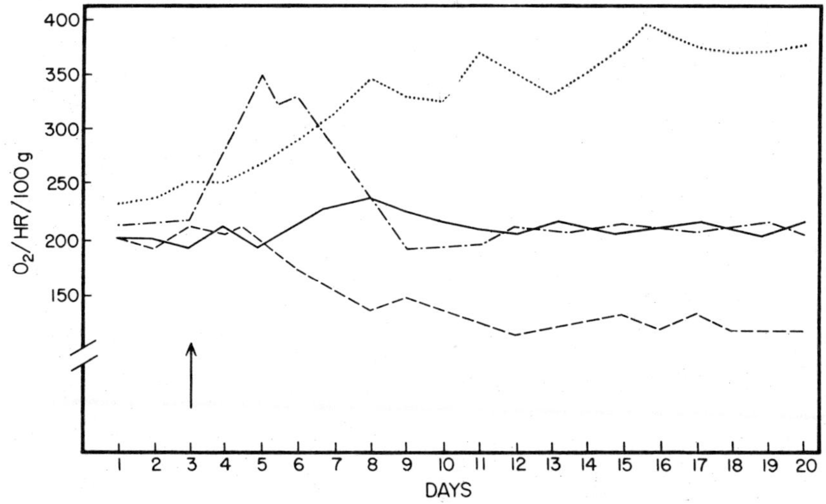

Fig. 4. The influence of daily peroral administration of various quantities of cadmium chloride on oxygen absorption by white rats. The experimental animals received (beginning marked by arrow) a daily dose as follows: ...... 20 μg Cd/kg;-·-·-· 1500 μg Cd/kg; ------ 50 mg Cd/kg; ________ control animals. The quantity of oxygen absorbed per hr per 100 g of the weight is shown, expressed as the average data from 20 rats in each group. A 20 μg/kg dose gives an increasing stimulating action. Note the loss of the initial stimulating action with a 1500 μg/kg dose. A depressing action occurs from the third day of using the 50 mg/kg dose. [Z.N. Nepesova's experiments (6–9)].

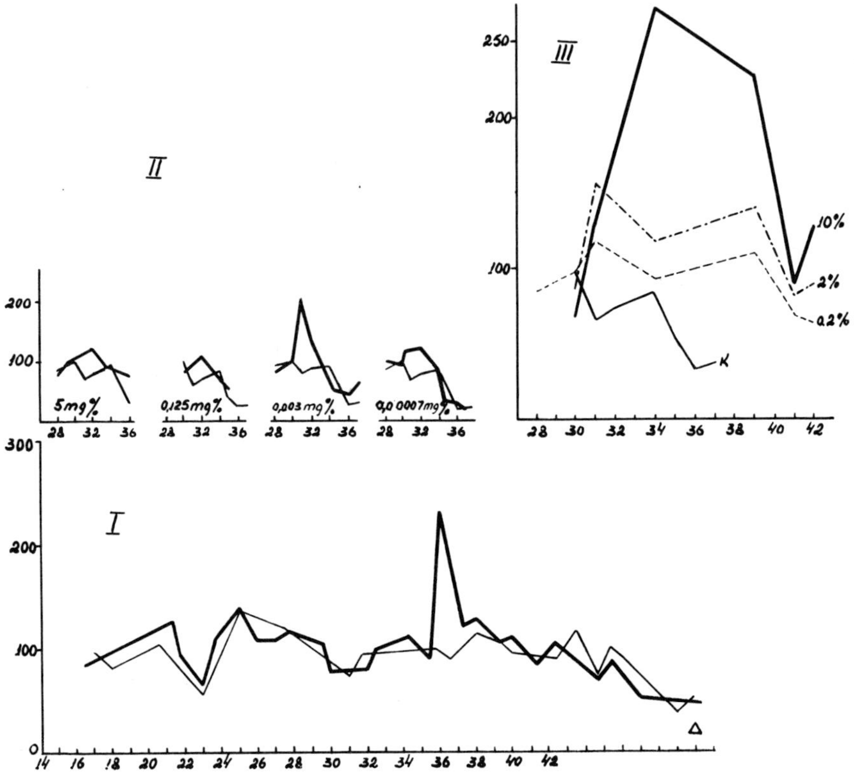

Fig. 5. The influence of solutions of various concentrations of zinc salts on the intensity of oxygen absorption by *Bombyx mori*; abscissa, age of the larvae in days; ordinate, quantity of oxygen absorbed per hr per 100 g larval weight. ——— Control *Bombyx mori* larvae; ——— experimental *Bombyx mori* larvae; △ pupae. I. During the use of small doses a statistically significant action was registered on gaseous exchange only by a solution of zinc chloride in concentrations of 0.003 mg% and only in the fifth age of the larvae (on the 36–38th day). II. The action of other concentrations (5 mg%, 0.125 mg%, or 0.00007 mg%) was not statistically significant. III. During the use of large doses (solutions of zinc sulphate of 0.2%, 2%, or 10%) greater intensity of oxygen absorption was observed in conformity with the increase in the concentration of the solution, but parallel with this the increase in weight declined and the survival of the larvae dropped. (E.V. Cherkasova's experiments.)

the zinc salt used increased ('pharmacotoxicological action zone') (Fig. 5, III). Paralleling the increase in zinc concentration was an increase in its toxic action, expressed as a decrease in the number of larvae putting on weight and an increase in the number of larvae dying. Lowering the concentration of zinc salt to 5–0.125 mg% did not exert any significant effect (actually they were 'inactive') (Fig. 5, II). On the other hand, a 0.003 mg% solution of $ZnCl_2$

caused a statistically significant increase in the intensity of oxygen absorption, but only in the fifth age of the larvae, that is, before the beginning of metamorphosis (Fig. 5, I). The action of definite, very small concentrations of zinc, manifested only in one age, would be regarded as an action of a biological order (E. V. Cherkasova's experiments).

THE INFLUENCE OF SOME TRACE ELEMENTS ON THE ACTIVITY OF SUCCINIC DEHYDROGENASE AND CYTOCHROME OXIDASE

The influence of zinc, cadmium, nickel, chromium, vanadium, and tungsten on the activity of enzymes of tissue respiration was studied under the same experimental conditions as used for guinea pigs and some of the rats. Succinic dehydrogenase activity was measured by the method of Kun and Abood (17), using formazan formed from triphenyltetrazolium chloride, and cytochrome oxidase activity was measured by the method of Vernon (18), using indophenol blue formed with α-naphthol in the oxidation of dimethylparaphenyl amine.

As an example, we give the results of experiments with tungsten. The biological importance of tungsten is unknown, although it is present in plant and animal organisms (19). Examination of its influence on the processes of tissue respiration of guinea pigs showed the ability of small quantities (in daily doses of 13 μg/kg for 9 days) to increase succinic dehydrogenase and cytochrome oxidase activity [Balaeva's experiments, ref. (15) and Fig. 6]. The same type of phenomenon was observed in rats and guinea pigs in experiments with nickel (5), zinc (20), cadmium (8), chromium (14), and vanadium (21). There was a difference between different organs in the rate of response manifested at various doses of the indicated trace elements. In all these experiments a stimulating action by very small quantities (daily doses measured by units and tens of μg/kg) was apparent as a usual phenomenon in the intensity of enzymes of tissue respiration.

In considering the results of these experiments, it is necessary to take into account that the measurement of enzyme activity in organs and tissues was carried out not in a dynamic sense but only after ending trace-element administration. During the whole period of trace-element administration, measured for at least 7–9 days, changes of organism response to trace-element administration could be observed. However, experimental conditions did not allow measurement of these changes. The experiments with cadmium (Fig. 4) and nickel (Fig. 3), for instance, may confirm the possibility of a change of the organism's response depending on the duration of trace-element administration.

The above experiments show the ability of a number of trace elements to increase the level of bioenergetic processes. In connection with this, as our

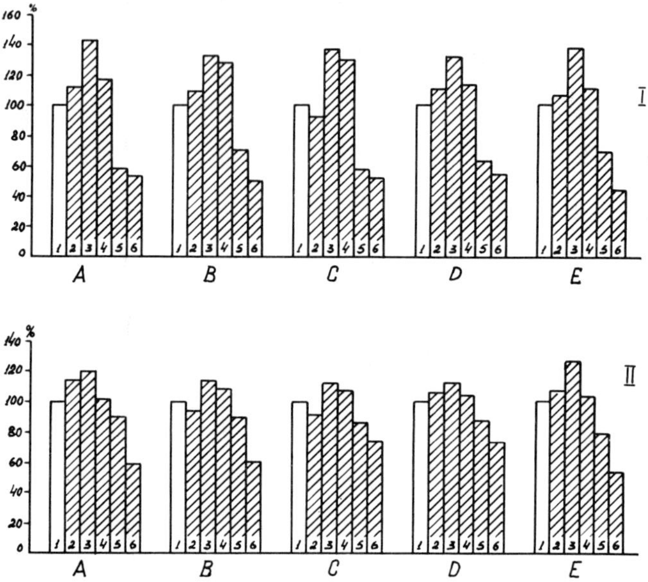

Fig. 6. The influence of various concentrations of tungsten on the activity of succinic dehydrogenase (I) and cytochrome oxidase (II) of guinea-pig organs: A, liver; B, cidney; C, skeletal muscle; D, brain; E, heart. Peroral administration of sodium tungstate into guinea pigs for 9 days in the following daily doses: 1, control; 2, 0.33 μg/kg; 3, 13.36 μg/kg; 4, 534.4 μg/kg; 5, 21.38 mg/kg; 6, 53.4 mg/kg. The changes are expressed as percentages with respect to the control. [S. S. Balaeva's experiments (15).]

experiments showed, there is an increase in the phagocytic activity of leucocytes.

PHAGOCYTIC ACTIVITY OF LEUCOCYTES

Frogs were placed in water containing various concentrations of potassium iodide for periods of 6 hr and 12–15 days. At the end of the period blood was taken from the heart of the frogs and a comparison of the phagocytic activity of the leucocytes made with that of the controls. *Bacterium colli comm.* served as the material for phagocytosis. An emulsion of bacteria in a volume of 0.3 ml was mixed with the blood, diluted 10 times, and kept in a thermostat for 10 min at 20°C. After staining the blood preparation, the number of microbes absorbed by neutrophiles was counted. Then the phagocytic number and phagocytic index were defined. There were 18–20 frogs in each experimental group.

The results of these experiments show the presence of action zones after frogs had been in the KI solution for 6 hr (Fig. 7). After continuous exposure over 12–15 days in solutions of the same KI concentration, a concentration which was previously inactive (5 mg%) began to manifest a toxic action, and a concentration which was previously stimulating (200 mg%) also became toxic. Only the concentrations considered as being in the biological action zone (0.125–0.003 mg%) had a stimulating action (22).

The ability of trace elements in small quantities ('biological') to increase the phagocytic activity of leucocytes was observed by other workers in experiments on frogs with $ZnCl_2$ (23), Na_2MoO_4 (24), $NaVO_3$ (25), and on guinea pigs with $CrCl_3$ (13) and Na_2WO_4 (25).[1]

In most experiments on administration of various doses of a trace element, the quantitative change in content of the element in appropriate organs was measured. With peroral introduction of trace elements into rats and guinea pigs over periods of at least 7–9 days in doses measured by units or tens of μg/kg, we have not observed the true quantitative changes in their

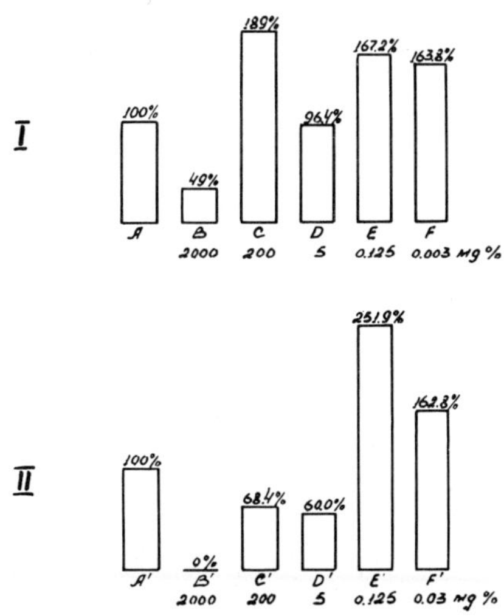

Fig. 7. The influence of the addition of various quantities of potassium iodide to the water surroundings of frogs on the phagocyte activity of their leucocytes. I. Observations for frogs (B, C, D, E, F) after being in the KI solutions for 6 hr. The changes are expressed as a percentage with respect to the control frogs (A), whose strength of phagocytosis is conditionally taken as 100%. II. As in I, but with frogs having been in the KI solutions for 12–15 days. (N.F. Gerasimova's experiments.)

1. Experimental data on the effects of microdoses of copper, cobalt, iron, and iodine on the change of the immunobiological reactivity of an organism, on production of antibodies, and on titre of complement are given in the work of our collaborator L. L. Alkhutova (26). We limit ourselves to mentioning them only, as giving these data would go beyond the bounds of the present paper.

content in organisms and tissues using emission spectral analysis. However, use of radioisotopic methods permitted the observations next reported. Experiments with the radioisotope ^{65}Zn, perorally introduced into rats in various quantities (daily dose of Zn: 0.75, 30, and 1200 μg/kg) showed the following results.

(1) After one day of administration of very small doses (0.75 and 30 μgZn/kg) the degree of its accumulation in different organs (liver, kidneys, lungs, brain, testicles, muscles) was nearly equal. Fluctuations in zinc content were placed in bounds of variability of the phenomenon.

(2) With higher doses (1200 μg/kg), zinc accumulation was obviously increased in kidneys, liver, lungs, and intestines. In contrast, in the brain, testicles, and muscles some tendency for zinc to decrease with a dose increase to 1200 μg/kg was observed. Even allowing for the present large variability of experimental results obtained by radioisotopic methods, the presence of mechanisms preventing excessive zinc entrance to the brain, testicles, and muscles was clearly shown (11, 20).

NORMALIZING AND DESINTOXIC PROPERTIES OF SOME TRACE ELEMENTS

In experiments with vitamin-B_1-deficient pigeons over a period of 20–25 days, administration of cadmium chloride in a daily dose of 20 μgCD/kg lengthened their life by an average of 7 days. In control vitamin-B_1-deficient pigeons the activity of enzymes of tissue respiration (succinic dehydrogenase and cytochrome oxidase) was decreased. Under the influence of the cadmium dose given to the experimental pigeons there was a change in the activity of the enzymes. Cadmium doses of 1500 μg/kg and 50 mg/kg depressed enzyme activity and shortened the life of deficient pigeons (9).

Solutions of copper, zinc, and iodine ($CuSO_4$, $ZnCl_2$, KI), applied in concentrations of 0.125–0.003 mg% in the biological-action zone manifested detoxification properties, as indicated in experiments with Daphnias (genus of freshwater branchiopod crustaceans) poisoned by toxic food which was produced in *Proteus vulgaris* cultures in ground small muscles. Survival of the poisoned Daphnias was increased by the addition of the above-mentioned salt solutions to the surrounding water environment. The same result was observed in experiments with Daphnias in overheated environments (27).

PREPARATION OF EXPERIMENT WITH INDIUM

Biological-action zones were revealed in those trace elements investigated, the biological importance of which was either wholly established or to some degree assumed. For comparison, indium ($InNO_3$) was investigated under the same experimental conditions and with the same indexes of action (oxygen absorption, enzymatic activity of tissue respiration, phagocytic activity of leucocytes). Its content in the animal organisms is 0.016 ppm [Koch and Roesmer, 1962, cited by Bowen (19)]. Its biological importance is wholly unknown. In contrast with the above data, indium was inactive in experiments with rats and guinea pigs (28) when administered in doses close to its content in living organisms and environment.

DISCUSSION

The results of experiments carried out by us between 1942 and 1972 have suggested that mechanisms regulating the entry of trace elements necessary for biological activity to inner systems are present in living organisms. When trace-element concentration is increased above a definite limit its entry into the organism is delayed to some extent. Indications for this were given in our experiments on the permeability of the skin of a frog to various quantities of copper (29) and zinc (30), and of the liver to copper or iron (31). Direct measurement by chemical analysis of the quantity of the substance penetrating physiological barriers showed a considerably greater degree of penetration through barriers when the trace element was used in concentrations close to those in a particular organism (biological concentrations). Upon increasing the concentration the quantity of trace element penetrating the barriers was decreased. A threshold of stimulation of the protective physiological function occurred. However, upon increasing trace-element concentration further, the quantity that penetrated the barriers increased. Action due to "the principle of barrier resistance" occurred. The data given and the results of the following experiments were a foundation for forming a concept of zones of action of trace elements (1, 3, 32). Indexes reflecting the intensity of exchanged processes (mainly tissue respiration) have been important and convincing in the demonstration of this concept. This is understandable if consideration is given to the direct or indirect participation of many trace elements in the exchanged processes of an organism. The presence of trace-element action zones was confirmed in the work of Kulsky (33), Vladimirov (34), and others. Physiological activity of very small quantities of trace elements is recorded in studies with chromium by Mertz and Roginsky (35), and with cobalt by Reshetkina (36) and others. Underwood (37) is in agreement with the concept of the trace-element action zones.

Examining our experimental data it is necessary to take into account that the Arnd-Schultz rule "weak doses excite, strong ones paralyze," long well known to pharmacologists, concerns the pharmacotoxicological action zone.

Trace elements show their action as a biological factor only within definite bounds. In their limits they obey the law of mass action (the effect increasing with increasing mass of the substance to some limit). The high sensitivity of organisms to small quantities of trace elements and the presence of the action zones described above are perhaps inherent in those elements which participate in biological processes.

As indicated above, indium, which is found in animal organisms, had no physiological activity in doses corresponding to its content in the organism, that is, its biological-action zone was not revealed [Bowen's "slightly toxic to plants and to animals" (19)].

The high toxicity of some trace elements, for example, mercury, does not preclude a possible physiological role for them. The experiments of Safarova (38), for example, show that small quantities of mercury (0.22 μg/kg body wt), which are close to its possible content in living organisms, are able to stimulate oxygen absorption. The phenomenon suggests the possibility of a biological role for this highly toxic trace element.

In order to explain the physiological activity of the very small trace-element quantities used by us, it is necessary to proceed from the following observations.

(1) Biologically active substances, participating in exchanged processes and in the manifestations of physiological functions, need very small quantities to produce their effects (enzymes, endocrine gland products, and mediators). The activity of most of them can sometimes occur in concentrations as low as 10^{-9}–10^{-6} M (adrenalin, acetylcholine, thyroxin, etc.). The physiological activity we observed for a number of trace elements in the biological-action zone was also shown for very small quantities.

(2) The entry of a number of trace elements into the structure of protein and other complex organic systems can sharply increase their activity (for example, iron in catalase, cobalt in vitamin B_{12}).

(3) It is also necessary to take into account the large difference in molecular weight between proteins and heavy metals. For example, assuming after Neurath and Bailey (39), that the molecular weight of diphtherial antitoxins is similar to globulins (184,000) and in man its content in blood plasma is 0.1 mg%, calculations show (assuming that each molecule of a bivalent protein requires one atom of a bivalent metal) that the total quantity of this protein substance in man requires only about 1 μg of a metal having an atomic weight of about 60 (for example, copper, zinc, cobalt), and about 3.3 μg is required for a metal with higher atomic weight (for example, mercury). In our calculation we assumed a blood-plasma content in man of 3000 ml.

(4) It is instructive to calculate the quantity of the trace element used as

a percentage of its content in the environment or the organism. For example, in the spring water we used for the experiments on frogs the iodine content was 1.88 μg/liter. Addition of 0.003 mg% KI (calculated in terms of iodine - 23 μg/liter) increases the iodine content by 1122%. This explains why the observed effects on gas exchange in frogs occurred after administration of such small iodine quantities.

(5) The present data about the trace-element content in organs and tissues indicate only its common, gross quantity. A significant part of a trace element is in a bound, deposited form, and only a small part takes an active part in metabolic processes.

(6) The trace elements were used not in food substances, but in water solutions; that is, they entered the biologically active substances in more ionically active forms. Absorption of a trace element from water solution begins in the mouth. By applying, for example, neutron activation methods to the analysis, mineral substances in the form of water solutions which entered the mouth were revealed in blood and different organs within 1 min [(40) and others].

The calculation of the existence of action zones permits the discovery in trace elements of properties that earlier remained unnoticed. The revelation of the ability of biological quantities of a trace element in the animal organism to increase the general level of bioenergetic processes and also its protective responses (phagocytic, detoxifying) suggests a prospective use of trace elements as medical remedies (biotics) and gives a basis to the medical development of the biological principle of treatment of human diseases (3). As remedies that increase the level of bioenergetic processes, trace elements promote the struggle against the impairment of metabolism, particularly those disorders that lead to premature aging (41). These data are considered only briefly; further discussion is beyond the scope of this paper. They are mentioned to show the prospects of investigations of the properties of trace elements used in quantities that relate to their biological-action zone.

CONCLUSION

The investigation of the response of rats, guinea pigs, and frogs to the administration of decreasing quantities of trace elements (water solutions of KI, NaI, $NiCl_2$, $ZnCl_2$, $CdCl_2$, $NaVO_3$, Na_2MoO_4, $CrCl_3$, Na_2WO_4) allowed their action as a pharmacotoxicological agent to be differentiated from their biological action.

(a) The application of aqueous solutions of trace elements in quantities peculiar to the organism (biological) measured, depending on the properties of the trace element, in units or tens of μg/kg body wt per day gives an increase in the level of bioenergetic processes and in the protective response of the organism ('the zone of biological action').

(b) Trace elements in macrodoses give an irritating or depressing effect on the living systems ('the zone of pharmacotoxicological action').

(c) Between zones (a) and (b) there is a transitional zone, within the limits of which the grossly visible effects of the trace element, taken in quantities that are measured by tens or hundreds of μg/kg body wt per day, are absent or considerably weakened ('the inactive zone').

(d) In transition from the inactive zone to the zone of biological action a physiological effect occurs if the quantity of the trace element used is decreased.

REFERENCES

1. Venchikov, A. I. 1947. The physiological properties of concentrations which do not stimulate the barrier functions of the organism. *Collected Reports of 7th All-Union Congress of Physiologists, Biochemists, and Pharmacologists,* Medgiz, Moscow, pp. 5–7.
2. Venchikov, A. I. 1960. About the physiologically active quantities of a trace element and mechanism of manifestation of its action. *Vopr. Pitania, Moscow (Questions Nutr.),* No. 6.
3. Venchikov, A. I. 1962. *Biotics* (*on the theory and practice of the application of trace elements*), Medgiz, Moscow.
4. Arutyunov, L. I. 1968. The influence of nickel on the intensity of tissue respiration. *Izv. Akad. Nauk TSSR* (*News Turkmen Acad. Sci. Ser. Biol. Sci., Ashkhabad*), No. 1.
5. Arutyunov, L. I. 1969. The influence of nickel as a trace element on the activity of cytochromeoxidase and succinicdehydrogenase in the organs of guinea pigs. *Izv. Akad. Nauk TSSR* (*News Turkmen Acad. Sci. Ser. Biol. Sci., Ashkhabad*), No. 1.
6. Nepesova, Z. N. 1964. The influence of various dosages of cadmium on tissue respiration. *Abstracts, Scientific Conference of the State Turkmen Medical Institute, Ashkhabad.*
7. Nepesova, Z. N. 1965. The influence of cadmium as a trace element on the intensity of metabolism. *Collected Abstracts, Conference on Trace Elements in Medicine, Ivano-Francovsk.*
8. Nepesova, Z. N. 1968. The influence of cadmium on the intensity of oxygen absorption by rats. *Izv. Akad. Nauk TSSR* (*News Turkmen Acad. Sci. Ser. Biol. Sci., Ashkhabad,* No. 3.
9. Nepesova, Z. N. 1969. The influence of various dosages of cadmium on the surviving of B-avitaminosed pigeons and on oxidizing-reductive processes in them. *Trace Elements in Medicine, Collected Papers of the 1st All-Union Scientific Conference, Ivano-Francovsk,* p. 231.
10. Cherkasova, E. V. 1969. The influence of various zinc dosages on the degree of its calculation in the organs and tissues of white rats. *Trace*

Elements in Medicine, Collected Papers of the 5th All-Union Scientific Conference, Ivano-Francovsk.
11. Cherkasova, E. V. 1969. The influence of zinc on oxygen absorption by white rats. *Zdrav. Turkmen. (J. Publ. Hlth Turkmen., Ashkhabad)*, No. 7.
12. Kulieva, T. 1970. The influence of various dosages of vanadium on the intensity of oxygen absorption by guinea pigs. *Izv. Akad. Nauk TSSR (News Turkmen Acad. Sci. Ser. Biol. Sci., Ashkhabad)*, No. 4.
13. Ergeshev, I. E. 1972. The action of chrome on the intensity of gas exchange of guinea pigs. *Abstracts, 4th Siberian Conference on Trace Elements in Biosphere and their Application in the Agriculture and Medicine of Siberia and the Far East, Ulan-Ude.*
14. Ergeshev, I. E. 1972. The changes of the activity of some enzymes of tissue respiration under the influence of various chrome quantities. *Zdrav. Turkmen. (J. Publ. Hlth Turkmen., Ashkhabad)*, No. 10.
15. Balaeva, S. S. 1972. The influence of tungsten as a trace element on the activity of some enzymes of tissue respiration. *Abstracts, 5th Conference of Physiologists of the Republic of Middle Asia and Kazakhstan, Ashkhabad.*
16. Balaeva, S. S. 1971. The oxygen absorption by guinea pigs under the influence of tungsten as a trace element. Reported at meeting of the Turkmen Physiological Society, Ashkhabad, 12 November, 1971.
17. Kun, E., and Abood, L. 1949. Colorimetric estimation of succinic dehydrogenase by triphenyltetrazolium chloride. *Science* 109: 144–146.
18. Vernon, N. 1911. The quantitative estimation of the indophenoloxidase of animal tissue. *J. Physiol.* 42: 402–432.
19. Bowen, H. I. 1966. *Trace Elements in Biochemistry*, Academic Press, London.
20. Cherkasova, E. V. 1969. The influence of zinc on oxygen absorption by white rats. *Zdrav. Turkmen. (J. Publ. Hth Turkmen., Ashkhabad)*, No. 7.
21. Kulieva, T. 1971. The influence of vanadium on the activity of cytochrome oxidase and succinic dehydrogenase of the organs of guinea pigs. *Abstracts, Regional Conference of Young Scientists, Ashkhabad.*
22. Gerasimova, N. F. 1966. The influence of iodine as a trace element on the phagocytic activity of leucocytes. *J. Path. Physiol. Exp. Therapy, Moscow*, No. 5.
23. Cherkasova, E. V. 1956. The effect on phagocytosis of additions of zinc chloride solutions to the blood. *Works USSR Soc. Physiol. Pharmac., Moscow* 3: 100–102.
24. Kaprielov, G. M. 1960. The influence of the addition of sodium molybdate to the environment on the phagocytic activity of frog's leucocytes. *Works State Turkmen Med. Inst., Ashkhabad*, Vol. 10.
25. Balaeva, S. S. 1972. The influence of tungsten on the phagocytic activity of leucocytes of guinea pigs. *Zdrav. Turkmen. (J. Publ. Hlth Turkmen., Ashkhabad)*, No. 11: 6–7.
26. Alkhutova, L. M. 1962. The influence of some trace elements on the

indexes of the immunobiological reactivity of the organism. *Zdrav. Turkmen.* (*J. Publ. Hlth Turkmen., Ashkhabad*), No. 6.
27. Gorbatova, V. S. 1958. The role of iodine as a trace element in increasing the resistance of an organism to overheating. *Materials of the 1st Conference of Physiologists and Biochemists of Central Asia and Kazakhstan, Tashkent,* pp. 118–120.
28. Girina, N. Ya. 1972. The influence of indium on the activity of some enzymes of tissue respiration. Reported at the meeting of the Turkmen Physiological Society, Ashkhabad, 19 May, 1972.
29. Cherkasheninova, A. E. 1947. Permeability of the skin to various concentrations of copper. *Bull. Exp. Biol. Med., Moscow* No. 9: 222–224.
30. Cherkasova, E. V. 1955. Permeability of a skin membrane to zinc. *Works State Turkmen Med. Inst., Ashkhabad,* 5–6: 285–291.
31. Venchikov, A. I. 1944. Excitation threshold of the reticulo-endothelial elements of the liver. *Bull. Exp. Biol. Med., Moscow,* 18(9): 16–19.
32 Venchikov, A. I. 1957. On the question of importance of trace elements as factors of mineral nutrition. *Vopr. Pitania (Questions Nutr.), Moscow* No. 3.
33. Kulsky, A. A. 1968. *Silver Water,* Naukova Dumka, Kiev.
34. Vladimirov, V. I. 1969. Dependence of embryonic development and vital steadfastness of a carp for the zinc trace element. *Vopr. Ichtiol.* (*Questions Ichthyol.*) 9:5.
35. Mertz, W., and Roginsky, E. 1963. The effect of trivalent chromium on galactose entry in rat epididymal fat tissue. *J. Biol. Chem.* 238(3).
36. Reshetkina, L. P. 1968. Chronic nutrition disorders in children and trace elements. Thesis, Kiev University, USSR.
37. Underwood, E. J. 1971. *Trace Elements in Human and Animal Nutrition,* 3rd edn, Academic Press, New York.
38. Safarova, R. T. 1971. On the question about the biological role of mercury. *Abstracts, Scientific Conference of the State Turkmen Medical Institute, Ashkhabad.*
39. Neurath, H. A., and Bailey, K. 1954. *The Proteins,* Vol. 2, part B, Academic Press, New York, pp. 740, 801.
40. Robakidze, A. D. 1970. Researches of metabolic mechanism of some elements of mineral waters in the organism with the method of neutron activation. Doctoral Thesis, Tbilisi, USSR.
41. Venchikov, A. I. 1968. Trace elements and the problem of struggle against ageing. Drug therapy in elderly and senile age. *Collected Materials of the All-Union Conference, Kiev.*

THE EFFECT OF CADMIUM AND LEAD ON COPPER AND ZINC METABOLISM[1]

HAROLD G. PETERING

Department of Environmental Health, College of Medicine, University of Cincinnati, Cincinnati, Ohio

Zinc and copper are biologically essential elements for man, animals, and plants (1, 2), and are transported and utilized in living organisms as metalloproteins or metal complexes. Because of the vital nature of these elements and their presence in minute quantities as coordinate covalent complexes, their function is vulnerable to the action of many chemicals and competing transition metals to which an organism may be exposed. The inhibitory effects of these agents may simulate in many ways the deficiency state which can be nutritionally produced for zinc and copper, a situation which permits the quantitation of the inhibitory activity. On the other hand, cadmium and lead are toxic heavy metals which are widely distributed in the environment, the biological effects of which are considered to be among the most insidious of environmental hazards (3–5). It is to explore the extent to which lead and cadmium exhibit some of their toxic effects by virtue of their inhibition of the essential metabolic functions of zinc and copper that this report is made.

Although we shall emphasize the interaction of the toxic heavy metals with zinc and copper, it should be recognized that we must also touch on the effects of these toxic metals on iron metabolism, because of the known interaction of copper and iron in the synthesis of hemoglobin and related heme-iron compounds.

The discussion, therefore, will consist of several sections devoted to considering (a) zinc-copper-iron interactions, (b) cadmium, zinc, and copper

1. The author gratefully acknowledges the assistance of Drs. L. Murthy and D. H. Petering in the preparation of this paper. This work was supported in part by funds from USPHS Grants ES-00159 and OH-00337.

interrelationships, (3) interrelation of lead and copper, and (d) a summary and discussion.

ZINC, COPPER, AND IRON INTERACTIONS

In 1937 Sutton and Nelson (6) reported that feeding rats an excessive amount of zinc carbonate ($> 0.5\%$) reduced their fertility, reduced their normal growth response, and caused a marked reduction in hemoglobin and red blood cells, all of which were related to the level of zinc intake. About a decade later Smith and Larson (7) found that the anemia caused by excessive intake of zinc in rats was alleviated by the concurrent administration of copper but not by iron. This was the first indication of a definite interaction of zinc and copper. Further indications of this interaction came from the report of Duncan, Gray, and Daniel (8), who showed that zinc toxicity not only caused a marked reduction in growth and hemoglobin formation but also a very great reduction in cytochrome C oxidase (cytochrome $a\text{-}a_3$), which is both a heme-iron and a copper enzyme complex. The authors were unable to reverse the growth effect with copper, but the hematologic and enzyme inhibition was reversed by it.

The reverse effect, i.e., the effect of zinc on copper toxicity, has also been found. Thus, Ritchie *et al.* (9) found that manifestations of zinc deficiency were evident in swine receiving an excessive amount of copper, and that these effects were reversed simply by giving more zinc. Van Campen (10) explained this phenomenon as being due primarily to an interference with the intestinal absorption of zinc by the excessive copper, a similar explanation having previously been invoked by Van Campen and Scaife (11) to suggest that the toxicity of zinc was due to an interference with the intestinal absorption of copper.

In a recent paper Lee and Matrone (12) reported that the rapid reduction of ceruloplasmin due to feeding excessive zinc to rats was only partially reversed by feeding more copper, and that to reverse this effect completely it was necessary also to administer iron. Thus we have evidence for the interrelationship of zinc, copper, and iron, when zinc or copper are given in excessive amounts to produce effects which are considered to be manifestations of their toxicity. The real question is whether similar interactions occur when these essential metals are present in the diet in amounts within the physiological range.

In this regard, Petering, Johnson, and Stemmer (13) recently reported that rats fed a zinc-deficient diet responded in growth to zinc administered in the drinking water in a dose-related manner, as did deficient rats given zinc during a repletion period. At the same time that kidney and blood zinc levels were raised with increased zinc intake, liver zinc and copper and kidney copper fell, all in a dose-related manner.

This work led us to develop an experimental model in which weanling or older rats were fed a semipurified diet similar to that described by Petering *et al.* (13) based on dried egg white, corn starch (or sucrose), and corn oil which contained adequate amounts of all vitamins and minerals, with the exception of zinc and copper, which are needed for good growth and reproduction. The missing trace metals were supplied in the drinking water, so that their intake could be carefully controlled. This permitted the study of zinc and copper metabolic interactions in the dose-response regions, i.e., where the intakes of these metals were suboptimal, optimal, and superoptimal. Using this model we have been able to quantitate the effects of cadmium, lead, and other chemical inhibitors of zinc and copper metabolism.

The growth responses of groups of rats receiving levels of copper and zinc in their drinking water above the basal diet in amounts of 0.25, 2.0, and 16.0 μg Cu/ml and 2.5, 10, and 40 μg Zn/ml are presented in Fig. 1. The data, taken from the report of Murthy, O'Flaherty, and Petering (14), show that the growth responses to zinc depend on the levels of copper in the diet, and, conversely, that there is a definite growth response due to copper which is zinc dependent. Furthermore, we found an inverse relationship between serum zinc and serum copper which was highly significant as shown in Fig. 2 (15).

Another interesting physiological parameter which shows interaction of zinc and copper in the normal nutritional range is the effect of these metals on lipid metabolism, which was reported by Murthy *et al.* (14) and which is

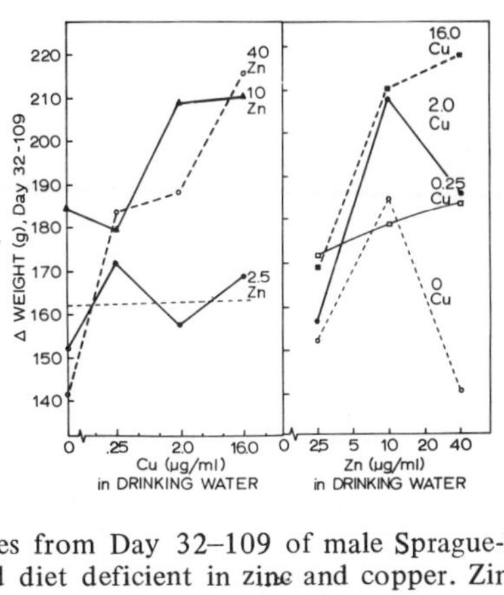

Fig. 1. Growth responses from Day 32–109 of male Sprague-Dawley rats (group averages) fed a semipurified diet deficient in zinc and copper. Zinc acetate was fed at the indicated levels in drinking water from Day 0–109, and copper sulfate was given at the indicated levels from Day 32–109. [Data from (14).]

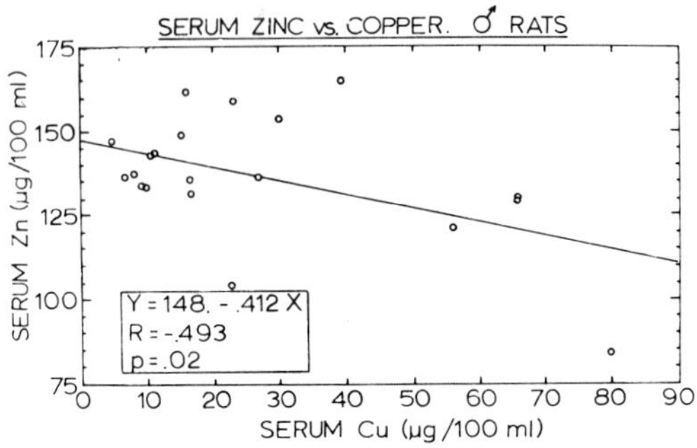

Fig. 2. Serum zinc and copper levels of male Carworth rats fed a semipurified diet deficient in zinc and copper for 60 days; zinc acetate and copper sulfate were given in the drinking water using a factorial design at 2.5, 5.0, 10.0, 20.0, 40.0 μg Zn/ml and 0.25, 0.50, 1.0 and 2.0 μg Cu/ml. [Data from (15).]

illustrated in Figs. 3 and 4. Figure 3 shows that there is an inverse relationship between serum copper and serum cholesterol, the log-linear regression for which is highly significant ($P < 0.01$). Figure 4 shows the effects of copper and zinc on the serum values for cholesterol, and reveals again that, at levels of dietary zinc of 10 μg/ml or greater, there is an inverse relationship between dietary intake of copper and serum lipid values, the patterns being similar in every case. It is interesting to note that at 2.5 μg/ml zinc in the drinking

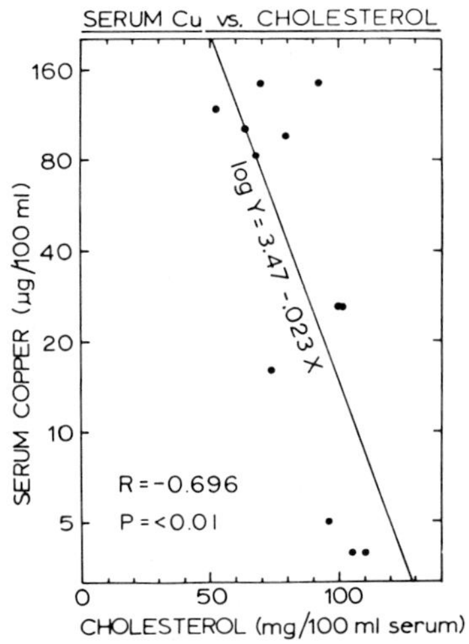

Fig. 3. Serum copper and cholesterol levels of same male Sprague-Dawley rats under the same conditions as described in Fig. 1. [Data from (14).]

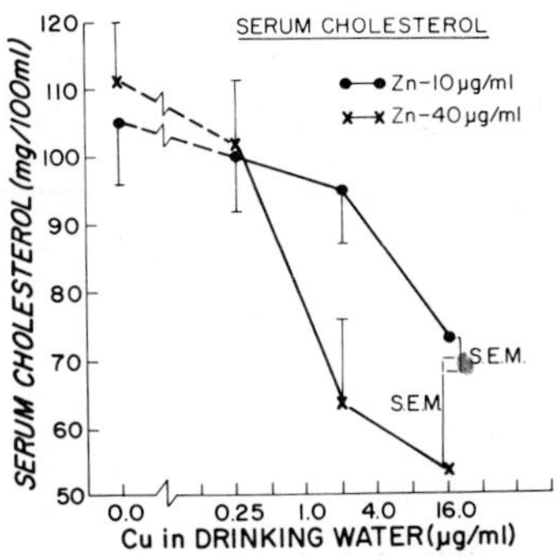

Fig. 4. Serum cholesterol related to dietary intake of zinc and copper of the same male Sprague-Dawley rats under the same conditions as described in Fig. 1. [Data from (14).]

water (a suboptimal level) there was little change in serum cholesterol as Cu was increased. This is important to recognize since in other work, shown in Table 1, we have found that at low levels of copper there is a direct increase in serum cholesterol with increases in dietary zinc (16).

From these reports it is evident that there are definite metabolic interactions of zinc and copper, and both affect lipid metabolism, which points to the need to consider the biologic effects of lead and cadmium in this light and to avoid compartmentalizing their effects.

Table 1. Effect of dietary zinc and copper on serum cholesterol[1]

Copper (µg/ml)	Zinc (µg/ml) 2.5	5.0	10.0
	Cholesterol (mg % ± s.e.m.)		
0.25	79.8 ± 0.8	99.5 ± 5.1	123.8 ± 16.2
0.50	83.5 ± 3.5	103.7 ± 8.7	117.5 ± 12.7
1.00	87.2 ± 12.1	112.8 ± 14.4	109.8 ± 5.7

[1]Fasting levels of serum cholesterol of male Sprague-Dawley (SD) rats fed a semipurified diet deficient in zinc and copper for 67 days after weaning. Zinc acetate and copper sulfate given in drinking water at indicated levels.
Taken from Murthy, O'Flaherty, and Petering (16).

CADMIUM, ZINC, AND COPPER INTERRELATIONSHIPS

Cadmium, being a member of the II B group of the periodic table and coming at the end of the second transition series, has chemical similarities to zinc and mercury. Therefore, it was natural for biochemists to investigate the biological relationship of cadmium to zinc metabolism. The search for interaction of cadmium and copper is of recent origin and stems from finding an interaction between zinc and copper as described above.

The first report that cadmium and zinc have an antagonistic action was that of J. Pařízek (17) in which the necrosis of testes of rats due to subcutaneous administration of cadmium was counteracted by simultaneously injecting zinc acetate at doses 100–200 times that of the cadmium. Supplee (17a) indicated in a later paper that growth inhibition of chicks as well as abnormalities of hocks and feathers due to orally administered cadmium ion could be prevented by the simultaneous dietary administration of zinc ion. His results suggested that the effects were dependent on the dietary ratio of Cd to Zn. Cotzias, Borg, and Selleck (18, 19) also suggested that the toxicity which they found for cadmium given to rabbits could be attributed to a perturbation of zinc metabolism.

Hill *et al.* (20) reported in 1963 that administration of large amounts of cadmium to chicks interfered with the metabolism of copper and iron as well as with zinc. Recently Whanger and Weswig (21) showed that cadmium markedly altered the synthesis of ceruloplasmin in the rat when it was given at very high levels. Finally, Ferm and co-workers (22) found an antagonistic relationship between zinc and cadmium with respect to the teratogenic effects of large subcutaneous doses of cadmium to pregnant hamsters.

Again it is evident that most of the reports indicating that there is interaction between cadmium and zinc or copper metabolism are concerned with experiments in which cadmium was given to the animals at very high dosage levels and so the overt toxicity of cadmium was being studied. Information on the effect of low and moderate doses of cadmium on the zinc and copper metabolism, when the essential metals are being given in the normal dietary ranges, is of great importance in order to evaluate the more subtle aspects of environmental exposure of man to cadmium, which is a present reality in most developed countries. It is also of importance because of the role which cadmium may play as a probe of the mechanisms by which zinc and copper carry out their normal functions in living cells. In our work, therefore, we have focused our attention on the effect of cadmium on the metabolism of zinc and copper in the rat, using the conditions described above for our model system.

Using this model Petering *et al.* (13) found that cadmium was inhibitory of growth and altered the metabolism of zinc and copper when the molar ratio of Zn to Cd was 1:1, but that most of these effects were reversed when the ratio was 4:1. This indicates that an antagonistic action of cadmium could be found in short exposures to cadmium given at low levels, about 3.4 μg Cd/ml (equivalent to 2.0 μg Zn/ml) of drinking water.

The data presented in Tables 2 and 3 are from one experiment with female rats, and show the interaction of zinc and copper and the effect of cadmium on the metabolism of these essential metals. These data, taken from the report of Murthy, Sorenson, and Petering (23), show that zinc intake greatly enhances the effect of dietary copper on serum copper levels, but that copper has no such effect on serum zinc levels. Cadmium depresses serum zinc at all levels of zinc nutriture, and even at the highest intake of zinc the inhibitory action of cadmium is not reversed. On the other hand, the depression of serum copper by cadmium is completely reversed by the highest level of copper and zinc. In this regard zinc enhances the protective effect of dietary copper.

In another study, originating in our laboratory, Book *et al.* (24) found that cadmium altered the tolerance curves of male rats given oral glucose loads when copper was given suboptimally (0.25 μg/ml) and zinc was fed at about the optimal level (10 μg/ml) (Fig. 5). It is noteworthy that cadmium had the same effect as reducing the level of zinc intake from 10 to 5 μg Zn/ml, although the peak of the curve with cadmium was much delayed over

Table 2. Effects of dietary zinc and copper with and without cadmium on serum zinc[1]

		Zinc (μg/ml)		
Cadmium (μg/ml)	Copper (μg/ml)	2.5	10.0	40.0
		Serum zinc (μg/100 ml ± s.e.m.)		
0.0	0.25	108.5 ± 10.9	205.0 ± 19.9	193.5 ± 14.6
	1.00	91.0 ± 11.1	180.8 ± 28.8	211.5 ± 16.7
	4.00	91.5 ± 18.4	172.5 ± 7.8	194.5 ± 20.2
17.2	0.25	84.0 ± 20.8	144.0 ± 18.0	160.0 ± 13.6
	1.00	77.5 ± 10.4	143.5 ± 4.3	163.5 ± 15.2
	4.00	71.5 ± 11.9	151.0 ± 16.5	164.0 ± 8.3

[1]Serum zinc of female Sprague-Dawley (SD) rats fed a semipurified diet deficient in zinc and copper for 73 days after weaning. Cadmium chloride was given during the last 31 days of the experiment in which zinc acetate and copper sulfate were given throughout at the indicated levels, all being in the drinking water (23).

Table 3. Effects of dietary zinc and copper with and without cadmium on serum copper[1]

Cadmium (μg/ml)	Copper (μg/ml)	Zinc (μg/ml) 2.5	10.0	40.0
		Serum copper (μg/100 ml ± s.e.m.)		
0.0	0.25	54.0 ± 22.0	101.5 ± 29.6	99.0 ± 7.0
	1.00	102.5 ± 10.4	146.5 ± 13.2	124.5 ± 10.5
	4.00	118.5 ± 4.0	134.0 ± 11.6	118.5 ± 11.6
17.2	0.25	14.0 ± 6.1	62.5 ± 27.1	25.8 ± 16.8
	1.00	52.0 ± 25.8	85.5 ± 28.5	97.0 ± 20.9
	4.00	101.5 ± 16.8	108.0 ± 4.1	122.0 ± 4.8

[1]Serum copper of the same female Sprague-Dawley (SD) rats as described in Table 2.

those with zinc alone. Cadmium here was given at 17.2 μg/ml of drinking water, which is equimolar with 10 μg Zn/ml, so that the Cd to Zn ratio was 1:1. In other experiments conducted by our group it was found that the oral glucose tolerance of both male and female rats fed a chow diet was abnormal when Cd was given in the drinking water for 10 months at concentrations of 4.3, 8.6, and 17.2 μg/ml, which are equimolar with 2.5, 5.0, and 10.0 μg Zn/ml.

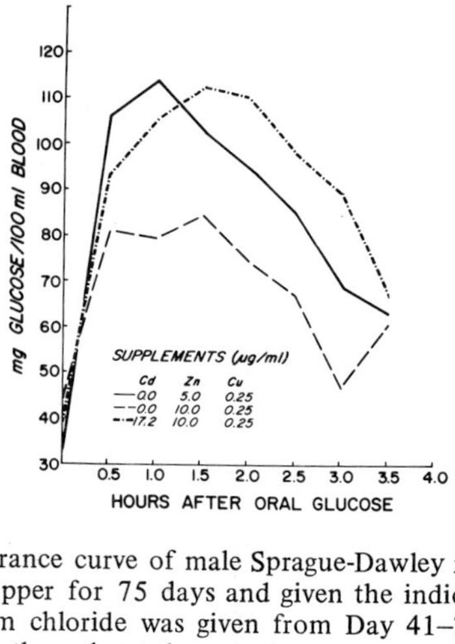

Fig. 5. Oral glucose tolerance curve of male Sprague-Dawley rats fed a semipurified diet deficient in zinc and copper for 75 days and given the indicated supplements in their drinking water. Cadmium chloride was given from Day 41–75, while zinc acetate and copper sulfate were given throughout the experiment.

In their work Book *et al.* (24) found a direct correlation between serum zinc and fasting serum insulin levels, but they were not able to show that the effect of cadmium on glucose tolerance was correlated with serum insulin values, since the toxic metal only depressed serum insulin when zinc was fed at 2.5 μg/ml of drinking water. Since zinc has been assigned a role in the pancreatic storage of insulin (25) it may still be possible that cadmium subtly alters this function of zinc.

These experiments point strongly to an effect of cadmium on the normal physiological metabolism of both zinc and copper, and furthermore suggest that the effects may be focused on a particular aspect of zinc and copper metabolism depending on the dietary balance of the essential metals. These data also indicate that low levels of cadmium may have wide-ranging biologic effects, and that nutritional factors are of importance in preventing the adverse biological effects of this toxic element.

INTERACTION OF LEAD AND COPPER

Lead poisoning involves a number of pathologic conditions depending on the age of the person exposed, and the duration and intensity of the exposure. Among the toxic effects of lead are severe depression of hematopoisis and bone-marrow activity, central nervous system pathology including convulsions and ataxia or paraplegia in the young, and general debility. Until recently there have been no good clues as to the mechanism of the neurological conditions, but the depression of hematologic activity has been ascribed to inhibition of certain steps in the synthesis of heme (3).

In a recent paper Kao and Forbes (26) showed clearly that the earliest effects of oral lead given to rats is a decrease in δ-aminolevulinic acid dehydratase (ALA-D) in red blood cells and kidney, followed by an increase in δ-aminolevulinic acid synthetase (ALA-S) and an increase in urinary ALA. These findings are of considerable importance, since in human beings exposed to lead there is a good negative correlation between blood lead and erythrocyte ALA-D (27). The hematologic depression caused by lead absorption is not only related to porphyrin synthesis but also to iron metabolism, as is shown by the work of Six and Goyer (28). These workers reported that iron deficiency in the rat greatly enhances the toxicity of lead, as shown by elevations of urinary lead and ALA, by increased retention of lead in liver, kidney, and bone, and by the more severe hematologic depression and anemia when compared with rats receiving optimal intake of iron during the lead exposure period.

The report of Six and Goyer was not the first one to indicate a relationship between lead toxicity and the metabolism of an essential transition metal, for in 1958 Rubino *et al.* (29) published data which showed that there was an elevated copper content of red blood cells of patients suffering from lead poisoning, which was correlated with an increase in red blood cell protoporphyrin content. At about the same time Iodice and associates (30, 31) found that ALA-D was depressed in the livers of rats and in the blood of ducklings fed a copper-deficient diet, although a function for copper in ALA-D activity has not been established.

In 1969 Alloway (32), working with Davies at the University of Wales, compiled data on swayback in lambs dropped by ewes pasturing on grasslands which were normal in their copper and molybdenum content but high in lead. These data clearly show that the lambs had sera low in copper, and that swayback was a common occurrence in such lambs, conditions which are usually attributable to a nutritionally induced copper deficiency. They concluded that there was a correlation between the lead in the grass and the apparent copper insufficiency found in the lambs. Following this lead, Klauder in our laboratory began to examine the interaction of dietary copper and lead administered orally to male rats. Using the same model mentioned above, in which the copper content of the diet was controlled, Klauder, Murthy, and Petering (33) found that lead toxicity in rats was increased by a dietary deficiency of copper. Furthermore, they had evidence for a protective role which copper played in reducing lead toxicity, and conversely for a depressive effect which lead had on copper metabolism. These results are presented in Table 4 and Fig. 6 which are taken from the report of Klauder *et al.* (33).

The data given in Table 4 show that ingestion of lead not only caused an anemia in the low-copper rats, which was not evident in the rats receiving normal amounts of copper, but it also reduced serum ceruloplasmin and copper in rats receiving the normal copper where there was no depression of hemoglobin and tended to depress zinc in the rats getting low copper. In addition, it was found that reduction of body growth and retention of lead in kidney and liver were inversely proportional to dietary copper levels. The very pronounced inverse relationship which was evident in all parameters indicated a protective effect of copper at the higher levels. In addition Fig. 6 shows that the amount of lead present in the erythrocytes of the rats, each of which were exposed to 0.5% dietary lead, was inversely related to the amount of copper or ceruloplasmin in the serum. Thus the absorption of lead appears to be markedly reduced by increased dietary intake of copper.

These data offer substantial evidence that copper and lead are interrelated when there is exposure to lead. This fact, taken together with the effect of iron on lead toxicity, offers a basis for understanding many aspects of lead intoxication, especially in the growing young of a given species or of human beings.

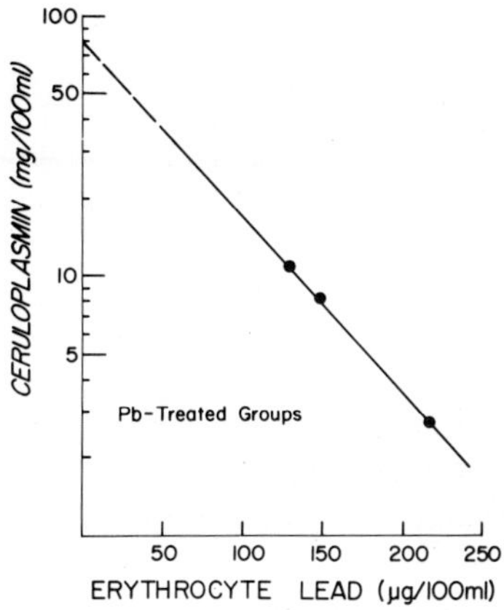

Fig. 6. The effect of increasing ceruloplasm in levels on erythrocyte lead in male Sprague-Dawley rats fed a semipurified diet and 0.5% Pb for 56 days. The groups received 0.5, 1.5, and 2.5 μg Cu/g diet. [Data from (33). Reprinted with permission of University of Missouri Conference on Trace Substances in Environmental Health VI (1973).]

Table 4. Effect of lead on copper metabolism in male rats[1]

	Low Cu^2	Normal Cu^3	Low Cu $+ Pb^4$	Normal Cu $+ Pb^4$
Hematocrit (%)	48.5	51.4	36.2	51.2
Hemoglobin (g %)	12.2	13.2	8.1	12.3
Serum Cu (μg %)	6.0	57.2	7.5	12.4
Serum Zn (μg %)	138	124	98	121
Ceruloplasmin (mg %)	6.7	29.1	2.8	10.7

[1]Parameters affected by lead acetate given in the copper-deficient semi-purified diet to male Sprague-Dawley (SD) rats for 56 days.

[2]Low copper diet contained 0.5 ppm Cu.

[3]Normal copper diet contained 2.5 ppm Cu.

[4]+ Pb diet contained 0.5% Pb.

Data from Klauder *et al.* (33).

SUMMARY AND DISCUSSION

The purpose of this review has been primarily to present the evidence for considering that some of the toxic and environmental health hazards of exposure to cadmium and lead reside in their antagonistic activity with respect to zinc and copper metabolism. Nevertheless, several suggestions as to their mode of action have been made or are obvious, and these will be cited briefly.

Thionein, the low molecular weight protein containing 8.5% sulfur which serves to sequester cadmium (34, 35), has been invoked as a major factor in the biologic effect of cadmium. Thus Evans, Majors, and Cornatzer (36) suggest that the antagonism of copper by zinc and cadmium is due to interference with the binding of cadmium to sites on a thionein-like protein in the intestinal lumen. On the other hand, Webb (37, 38) has shown that both zinc and cadmium cause induction of a thionein-like protein which exists as a protective agent.

Van Campen (10) and Van Campen and Scaife (11) have presented data to show that the oral toxicity of zinc or copper is due in part to interference with absorption of the other element, which of course may be due to interference with binding of one or the other to a protein such as thionein.

Since there is good evidence from the work of Evans and Abraham (39) and Frieden (40) to relate the interaction of copper and iron with the ferroxidase activity of ceruloplasmin, and since we have shown that both cadmium and lead inhibit the formation of ceruloplasmin, one aspect of the biologic effects of these toxic metals may be linked to their effect on the biosynthesis of porphyrins, hemes and cytochromes, and on the mobilization and release of iron.

It would appear obvious from the reports cited here that (a) some aspects of the biologic effects of cadmium and lead are due to inteference with the metabolism of zinc, copper, and probably iron, (b) that nutritional factors may offer significant protection against the adverse effects of cadmium and lead, and (c) that clues as to the mechanism of homeostatic interaction of zinc and copper may follow further study of the biologic action of cadmium and lead.

REFERENCES

1. Underwood, E. J. 1971. *Trace Elements in Human and Animal Nutrition,* 3rd edn, Academic Press, New York (especially chaps 1, 3, 8).
2. Bowen, H. J. M. 1966. *Trace Elements in Biochemistry.* Academic Press, London.

3. National Academy of Science. 1972. Airborne Lead in Perspective. National Academy of Science, Washington, D.C.
4. Friberg, L., Piscator, M., and Nordberg, G. 1971. *Cadmium in the Environment,* CRC Press, Cleveland, Ohio.
5. Fulkerson, W., and Goeller, H. E. (eds.). 1972. *Cadmium, the Dissipated Element.* Rep. No. ORNL NSF-EP-21, Oak Ridge National Laboratory, Tenn.
6. Sutton, W. R., and Nelson, V. E. 1937. Studies on zinc. *Proc. Soc. Exp. Biol. Med.* 36: 211–213.
7. Smith, S. E., and Larson, E. J. 1946. Zinc toxicity in rats. Antagonistic effects of copper and liver. *J. Biol. Chem.* 163: 29–38.
8. Duncan, G. D., Gray, L. F., and Daniel, L. J. 1953. Effects of zinc on cytochrome oxidase activity. *Proc. Soc. Exp. Biol. Med.* 83: 625–627.
9. Ritchie, H. D., Luecke, R. W., Baltzer, B. V., Miller, E. R., Ullrey, D. E., and Hoefer, J. A. 1963. Copper and zinc interrelationships in the pig. *J. Nutr.* 79: 117–123.
10. Van Campen, D. R. 1970. Copper interference with the intestinal absorption of zinc-65 by rats. *J. Nutr.* 97: 104–108.
11. Van Campen, D. R., and Scaife, P. V. 1967. Zinc interference with copper absorption in rats. *J. Nutr.* 91: 473–476.
12. Lee, D., Jr., and Matrone, G. 1969. Iron and copper effects on serum ceruloplasmin activity of rats with zinc-induced copper deficiency. *Proc. Soc. Exp. Biol. Med.* 130: 1190–1194.
13. Petering, H. G., Johnson, M. A., and Stemmer, K. L. 1971. Studies on zinc metabolism in the rat. *Archs Envir. Hlth* 23: 93–101.
14. Murthy, L., O'Flaherty, E., and Petering, H. G. 1972. Effect of dietary levels of copper and zinc on serum lipids in rats. Summary Abstracts, *9th International Congress on Nutrition,* Mexico City, p. 136; Progress Report, Department of Environmental Health, University of Cincinnati (1973).
15. Murthy, L., Klevay, L. M., and Petering, H. G. 1973. Interactions of zinc and copper in the rat at physiological levels of metal nutriture. *J. Nutr.*, in press.
16. Murthy, L., O'Flaherty, E., and Petering, H. G. 1973. Effect of low levels of copper and zinc on lipid metabolism. *A. Rep. Center for the Study of the Human Environment,* Department of Environmental Health, University of Cincinnati, Cincinnati, Ohio, p. 35.
17. Pařízek, J. 1957. The destructive effect of cadmium ion upon the testicular tissue and its prevention by zinc. *J. Endocr.* 15: 56.
17a. Supplee, W. C. 1963. Antagonistic relationship between dietary cadmium and zinc. *Science* 139: 121–122.
18. Cotzias, G. C., Borg, D. C., and Selleck, B. 1961. Specificity of zinc pathway in the rabbit: zinc-cadmium exchange. *Am. J. Physiol.* 201: 63–66.
19. Cotzias, G. C., Borg, D. C., and Selleck, B. 1961. Virtual absence of turnover in cadmium metabolism: Cd^{109} studies in the mouse. *Am. J. Physiol.* 201: 927–930.
20. Hill, C. H., Matrone, G., Payne, W. L., and Barber, C. W. 1963. *In vivo*

interactions of cadmium with copper, zinc, and iron. *J. Nutr.* 80: 227–235.
21. Whanger, P. D., and Weswig, P. H. 1970. Effect of some copper antagonists on induction of ceruloplasmin in the rat. *J. Nutr.* 100: 341–348.
22. Ferm, V. H., and Carpenter, S. J. 1968. The relationship of cadmium and zinc in experimental mammalian teratogenesis. *Lab. Invest.* 18: 429–432.
23. Murthy, L. J. R., Sorenson, J., and Petering, H. G. 1972. *Fed. Proc.* 31: 699; Progress Report, Department of Environmental Health, University of Cincinnati, Cincinnati, Ohio.
24. Book, A. R., Murthy, L. M., Shirley, T., and Srivastava, L. 1973. Effects of cadmium on glucose tolerance and serum insulin, zinc, and copper in male rats. *Fed. Proc.* 32: 468; Progress Report, Department of Environmental Health, University of Cincinnati, Cincinnati, Ohio.
25. Maske, H. 1957. Interaction between insulin and zinc in the islets of langerhans. *Diabetes* 6: 335–341.
26. Kao, R. L. C., and Forbes, R. M. 1973. Effects of lead on heme-synthesizing enzymes and urinary δ-aminolevulinic acid in the rat. *Proc. Soc. Exp. Biol. Med.* 143: 234–237.
27. Hernberg, S., Nikkanen, J., Mellin, G., and Lelius, H. 1970. δ-Aminolevulinic acid dehydrase as a measure of lead exposure. *Archs Envir. Hlth* 21: 140–145.
28. Six, K. M., and Goyer, R. A. 1972. The influence of iron deficiency on tissue content and toxicity of ingested lead in the rat. *J. Lab. Clin. Med.* 79: 128–136.
29. Rubino, G. F., Pagliardi, E., Prato, V., and Giangrandi, E. 1958. Erythrocyte copper and porphyrins in lead poisoning. *J. Haemat.* 4: 103–107.
30. Iodice, A. A., Richert, D. A., and Schulman, M. P. 1958. Copper content of purified δ-aminolevulinic acid dehydrase. *Fed. Proc.* 17: 248.
31. Wilson, M. L., Iodice, A. A., Schulman, M. P., and Richert, D. A. 1959. Studies on liver δ-aminolevulinic acid dehydrase. *Fed. Proc.* 18: 352.
32. Alloway, B. 1969. The soils and vegetation of areas affected by mining for non-ferrous metalliferous ores, with special reference to cadmium, copper, lead and zinc. Ph.D. Thesis, University of Wales, Aberystwyth, chap. 13, pp. 629–683.
33. Klauder, D. S., Murthy, L., and Petering, H. G. 1973. Effect of dietary intake of lead acetate on copper metabolism in male rats. *In* D. D. Hemphill (ed.), *Trace Substances in Environmental Health,* Vol. 6, University of Missouri, Columbia, Mo., pp. 131–135.
34. Kagi, J. H. R., and Vallee, B. L. 1960. Metallothionein: a cadmium- and zinc-containing protein from equine renal cortex. *J. Biol. Chem.* 235: 3460–3465.
35. Kagi, J. H. R., and Vallee, B. L. 1960. Metallothionein: a cadmium- and zinc-containing protein from equine renal cortex. II. Physicochemical properties. *J. Biol. Chem.* 236: 2435–2442.
36. Evans, G. W., Majors, P. F., and Cornatzer, W. E. 1970. Mechanism for cadmium and zinc antagonism of copper metabolism. *Biochem. Biophys. Res. Commun.* 40: 1142–1148.

37. Webb, M. 1972. Binding of cadmium ions by rat liver and kidney. *Biochem. Pharmac.* 21: 2751–2765.
38. Webb, M. 1972. Protection by zinc against cadmium toxicity. *Biochem. Pharmac.* 21: 2767–2771.
39. Evans, J. L., and Abraham, P. A. 1973. Anemia, iron storage and ceruloplasmin in copper nutrition in the growing rat. *J. Nutr.* 103: 196–201.
40. Frieden, E. 1971. Ceruloplasmin, a link between copper and iron metabolism. *Bioinorganic Chemistry,* No. 100, American Chemical Society, Washington, D.C., chap. 4, p. 292.

DISCUSSION

Mertz (Beltsville). Do you have any evidence that trace elements added to the diet act differently than those added to water?

Petering. No, we can find no difference in this respect among the elements that we have been working on.

Horvath (Morgantown). Is your experience in this regard confined to synthetic diets as opposed to natural diets with a more complex mixture of ingredients?

Petering. Yes, I am thinking primarily of the semipurified diet we have been working with.

Erway (Cincinnati). We have taken your analyses of lab chow, 20–50 ppm of zinc, and compared this with the zinc content of our synthetic diet. We have found a fetal inner-ear defect which is less pronounced when we feed a synthetic maternal diet containing 8 ppm zinc than when we feed the commercial lab chow. The dietary form rather than just the total amount of zinc in the diet must be very critical.

Petering. Rat chow varies considerably from most semisynthetic diets in a number of ways, and I would hesitate to ascribe the difference to zinc content until we explored some of the other things.

Sandstead (Grand Forks). Would you comment on the possible effects of selenium on the interactions you have studied.

Petering. I'd rather not speculate on this. We have an average amount of selenium in our diet and have not used either excessive or low concentrations of selenium.

EFFECTS OF FLUORIDE ON THE METABOLISM OF CULTURED CELLS[1]

J. W. SUTTIE, M. P. DRESCHER, D. O. QUISSELL, and K. L. YOUNG

Department of Biochemistry, College of Agricultural and Life Sciences, University of Wisconsin, Madison, Wisconsin

Although much is known about the metabolism and distribution of inorganic fluorides in animals, the gross effects of fluoride ingestion by intact animals, and the effects of fluorides on isolated enzyme systems, the specific sites of fluoride action on cellular metabolism are largely unknown. In an attempt to delineate some of the metabolic alterations which develop in response to fluoride exposure we have begun a series of studies of the effects of fluoride on cultured cells. These studies were originally carried out (1, 2) with suspensions of human epithelial cells (HeLa) but more recently with strain L mouse fibroblasts (3–6). These cells are grown on a complex medium (7) containing 10% bovine serum and glucose as an energy source. They are handled much as a bacterial culture, but have generation times of almost a day rather than a few hours.

FLUORIDE SENSITIVITY

The data shown in the left part of Fig. 1 indicate the effect of fluoride on growth of these cells. Although there have been reports that various strains of cultured cells are sensitive to from 0.1 to 2.0 ppm F in the growth medium (8, 9) these observations have not been confirmed by most workers, and usually little growth inhibition has been seen (1–6, 10–14) until the fluoride

1. Research supported by the College of Agricultural and Life Sciences, University of Wisconsin-Madison, in part by Grant No. AM-15521 from the National Institutes of Health, in part by U. W. Experiment Station project 809, and in part by National Institutes of Health training grant G.M.-00236-BCH.

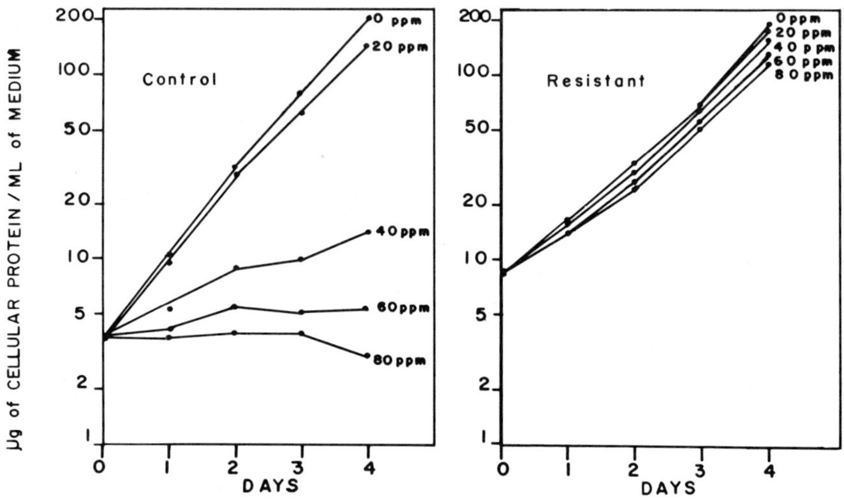

Fig. 1. Effect of fluoride concentration on the growth of the normal strain of L cells (left) and the fluoride-resistant strain of L cells (right). The cells were grown from an initial cell density of 2 × 10^4 cells/ml in 500 ml siliconized florence flasks at 37°C. Growth was measured as an increase in cellular protein.

concentration is raised to 20 ppm F (1.05 mM), and a 50% inhibition of growth is often seen at about 30 ppm F.

Of more interest in assessing the possible metabolic consequences of fluoride exposure is the intracellular fluoride concentration. We have determined this for both L cells and HeLa cells (5) incubated in both 10 and 50 ppm F, and have found the ratio F_{in}/F_{out} is about 0.3–0.4. This ratio was not greatly influenced by either the concentration of fluoride or the duration of incubation.

FLUORIDE-RESISTANT CELLS

Although cultured cells are readily inhibited by high concentrations of fluoride, resistant strains of both HeLa and L cells have been obtained (2, 13) by gradually increasing the concentration of fluoride in the growth media over a period of time. We have recently studied a fluoride-resistant strain of L cells in more detail (4).

The resistant strain of L cells (mouse fibroblasts) was originally developed by progressively increasing the concentration of fluoride in the growth medium until the cells would grow at 70 ppm F (right part, Fig. 1), a concentration of fluoride which was toxic to the original mother strain. It was later shown that these cells are present in the normal population to the extent of roughly 1 in 10 million. The resistant cells had the same energy requirements, the same intracellular Na and K concentrations, and similar cell morphology and size as normal L cells. The basis for the fluoride resistance was shown to be due to a mutational event which resulted in a ratio of intracellular to extracellular fluoride of 0.03 in the resistant cells compared with 0.3 in control cells. The intracellular distribution of Br, I, and Cl was normal in these cells.

When resistant cell cultures were incubated at 0°C, fluoride entered the cell, but it could be removed against a concentration gradient when the temperature was raised to 37°C (Table 1). Fluoride was also found to enter the cells at media fluoride concentrations greater than 100 ppm, and above this level the intracellular distribution of fluoride was similar to that seen in normal cells (Fig. 2) and growth inhibition was seen at similar intracellular fluoride concentrations. On the basis of these observations it has been postulated that the mechanism involved in the resistance was the development of an active transport process which could be saturated by high concentrations of fluoride. It seems most likely that the mutation has been one which altered an existing transport system to give it a greater affinity for fluoride, rather than the development of an entirely new transport system.

METABOLIC CHANGES IN CELLS GROWN IN FLUORIDE

As enolase is commonly cited as an enzyme which is inhibited by fluoride, it is often suggested that fluoride must therefore inhibit cell growth through an enolase-mediated inhibition of glycolysis. It can be shown *in vitro* that enolase is slightly inhibited at 0.5 ppm F, and inhibited about 50% by 10 ppm F. This would correspond to about 1.5 and 30 ppm extracellular fluoride. The intracellular levels of magnesium and phosphate in mammalian

Table 1. Effect of temperature on fluoride distribution in fluoride-resistant L cells[1]

Experimental conditions	^{18}F ratio, in/out
Incubation at 37°C	0.06 ± 0.06
Incubation 1 hr at 0°C	0.19 ± 0.01
Incubation 3 hr at 0°C	0.33 ± 0.01
Incubation 3 hr at 0°C, then 30 min at 37°C	<0.01

[1]Values are mean ± s.e. for four determinations; cells were incubated in 10 ppm F containing ^{18}F. [For details see (4).]

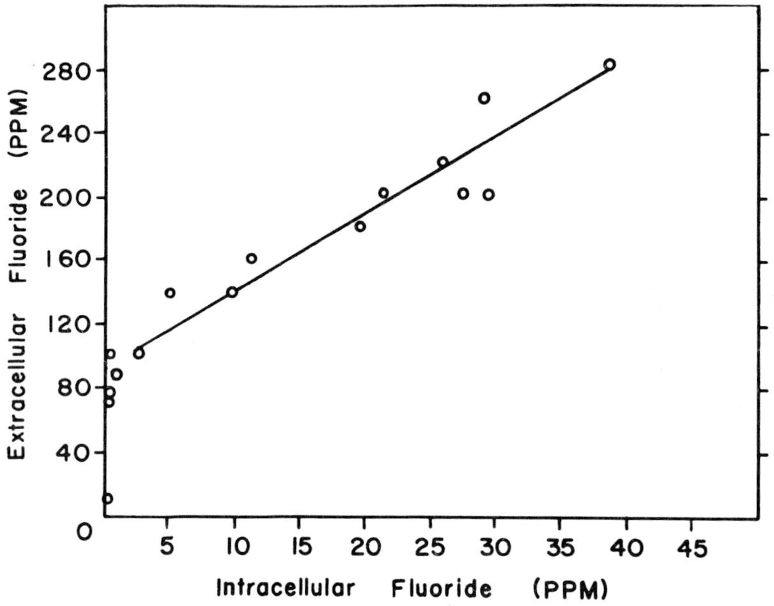

Fig. 2. Effect of fluoride concentration on cellular distribution of fluoride in fluoride-resistant L cells. The cells were incubated in radioactive fluoride and the concentrations calculated from the distribution ratios observed. [For details see (4).]

cells are in the same order of magnitude as those utilized in assaying enolase, so that the question of enolase inhibition seems a reasonable one to pose.

Determining whether or not an enzyme is inhibited *in vivo,* and assessing the significance of this inhibition, is a difficult task, and unambiguous data is difficult to obtain. If the addition of some agent causes the accumulation of the substrate of one enzyme of a metabolic pathway relative to the other metabolites in the pathway, this is usually considered to represent evidence for an inhibition at the site of that enzyme. To determine if enolase was inhibited *in vivo,* the levels of 3-phosphoglyceric acid (3PGA), 2-phosphoglyceric acid (2PGA), phosphoenolypyruvate (PEP), and pyruvate were measured in L cells grown in 5–30 ppm of fluoride. The data in Table 2 indicate that the ratio of 2PGA to PEP (the enolase-catalyzed step) increased as the fluoride level increased. In contrast, although the level of both 2PGA and 3PGA increased, the ratio between them remained relatively constant, and not far removed from the K_{eq} for the reaction.

Even if intracellular enolase was inhibited under these conditions, it may not have had an effect on overall glycolytic flux. It is of interest that fluoride concentrations which had no effect on growth of the cells (5 and 10 ppm F)

Table 2. Effect of fluoride on the intracellular level of glycolytic intermediates in L cells[1]

	Concentration of metabolite					
Treatment	3PGA	2PGA	PEP	Pyruvate	3 PGA/2PGA	2PGA/PEP
Control	0.26	0.07	0.12	0.06	3.9	0.6
5 ppm F	0.53	0.06	0.10	0.06	8.8	0.6
10 ppm F	1.2	0.14	0.15	0.06	8.9	0.9
20 ppm F	3.3	0.36	0.14	0.06	9.2	2.6
30 ppm F	7.8	0.89	0.21	0.04	8.7	4.2

[1]Values are μmole of 3PGA 2PGA, and PEP per gram of cell protein and μmole of pyruvate per equal amounts of cell pellet. The values are the mean of three or four cultures and the s.e. for most measurements was less than 10% of the mean (Drescher and Suttie, unpublished data).

still had an apparent effect on enolase. This is consistent with the usual assumption that, under steady-state conditions, the enolase step is not rate limiting in the regulation of glycolytic flux.

Glycolytic flux in the L-cell cultures was determined by incubating the cells in [^{14}C]glucose and measuring [^{14}C]lactate production. The amount of lactate produced over a 12-hr period from the cultures treated with 30 ppm F (Table 3) was decreased compared with control cultures. This is to some extent a reflection of the fact that by the end of the 12-hr period there are fewer cells in the fluoride-treated cultures. The measurement desired is the amount of the lactate produced per unit time per cell, and equations are available to calculate the instantaneous rate of lactate production at any given time. These values are presented in Table 3, and they indicate that 30 ppm fluoride in the media caused about a 20% reduction in glycolytic rate. Whether this inhibition of glycolysis is related to the decreased rate of cell

Table 3. Effect of fluoride on glycolytic rate of cultured L cells[1]

	Conversion of [^{14}C]glucose to [^{14}C]lactate	
Treatment	Percentage per culture	Percentage per hr/mg protein/ml $\times 10^3$
Control	9.8	10.2
30 ppm F	7.4	8.2

[1]Fluoride and [^{14}C]glucose-UL were added to logarithmically growing cultures and the amount of [^{14}C]lactate was determined after 12 hr and expressed as the percentage of glucose converted to lactate. Similar results were obtained 6 and 24 hr after the addition of fluoride (Drescher and Suttie, unpublished data).

growth is not easily ascertained. The relative inhibition of glycolytic rate is considerably less than the inhibition in rate of new cell-protein accumulation. This does not mean that the effect on glycolysis may not be important, as it is not known what proportion of the total energy produced by glycolysis is used for maintenance of cellular function and what proportion goes to synthesis of new cells.

The conditions under which the cells were grown for the glycolytic flux experiments were somewhat different than those previously used for determining the intracellular concentrations of glycolytic intermediates. For this reason the concentrations of 3PGA, 2PGA, PEP, and pyruvate were measured under conditions identical to those employed for the measurement of glycolytic flux. The results of these studies indicated that the accumulation of 3PGA and 2PGA relative to PEP in fluoride-treated cells the first 6 hr after fluoride was added to these cultures was consistent with the 20–25% inhibition of glycolytic flux noted during this period.

Although the glycolytic rate was apparently inhibited in these cells, they were still able to maintain normal levels of ATP. Over the 48-hr period following the addition of 30 ppm F to L-cell cultures, the level of intracellular ATP fell from roughly 20 μmole/g of cell protein to 10 μmole/g in both control and fluoride-treated cells. It is therefore apparent that the fluoride inhibition of glycolytic flux was not of sufficient magnitude to lower intracellular ATP. These cells do, however, carry out some aerobic metabolism, and the effect of inhibiting glycolysis on ATP levels would depend on the relative contribution of aerobic and anaerobic utilization of glucose.

Another measure of the intracellular inhibition of metabolic pathways would be the possible effect on pyridine nucleotide levels. The effect of fluoride on this parameter appeared to be two-fold (Table 4). Two hours after

Table 4. Effect of fluoride on intracellular levels of DPN and DPNH in L cells[1]

	12-hr incubation		24-hr incubation	
Measurement	Control	30 ppm F	Control	30 ppm F
DPN (μmole/g protein)	3.2	1.4	2.9	1.5
DPNH (μmole/g protein)	0.54	0.41	0.60	0.45
Total DPN + DPNH	3.74	1.81	3.50	1.95
DPNH/DPN	0.17	0.29	0.21	0.30

[1]Fluoride was added to logarithmically growing cultures and the amount of DPN and DPNH determined at 2, 12, 24, and 48 hr. The concentrations measured did not change appreciably at the different times and the data at 2 and 48 hr are similar to those presented. The values are the average of two different cultures which differed by less than 10% (Drescher and Suttie, unpublished data).

Table 5. Effect of fluoride on intracellular levels of sodium and potassium in L cells[1]

Fluoride concentration	Na (mM)	K (mM)
Control	31.8 ± 1.9	157.4 ± 1.3
10 ppm	41.3 ± 2.3	154.4 ± 1.9
30 ppm	35.7 ± 1.3	153.8 ± 0.6
70 ppm	35.1 ± 0.8	138.3 ± 1.2

[1]Cells were incubated in the various concentrations of fluoride for 12 hr. The values are mean ± s.e. of the mean for three or four separate cultures (3).

fluoride was added to growing cultures, the ratio DPNH/DPN was shifted from 0.2 to about 0.3, and the sum of the intracellular DPN and DPNH concentrations started to fall. By 12 hr the total concentration of oxidized and reduced DPN in the fluoride-treated cultures was about 50% of what it was in control cells. The shift in ratio observed by 2 hr remained relatively constant for 48 hr. This shift in ratio occurs early after fluoride addition, and is consistent with what would be expected from the inhibition of enolase in the cells. Both the DPNH/DPN ratio and the concentration of DPN and DPNH in control cells are rather constant values which are not influenced by cell growth. The shift to a new constant value in fluoride-treated cells may therefore be of key importance in the growth-inhibitory effects of fluoride.

There is no doubt that some cellular processes are of more importance to the cell than others, and some indications that inhibitors of energy metabolism may have differential effects on these processes. We have shown that the intracellular concentrations of Na and K are not influenced by fluoride concentrations that inhibit growth by 50%, and they are only slightly altered (Table 5) by sufficient fluoride (70 ppm) to inhibit growth completely.

FACTORS INFLUENCING FLUORIDE TOXICITY

These studies have therefore given us an indication of some of the changes seen in fluoride-treated cells, but have not conclusively indicated the basis for these changes. One other approach which might yield meaningful information on the effects of fluorides on cells would be to determine what factors affect the fluoride sensitivity of cells, and to try to work back from these observations.

The data in Fig. 3 indicate that when various amino acids are added to the growth media in concentrations in excess of those present in the media, some of them do enhance the growth of fluoride-treated cells. Some of the amino acids protect against fluoride, while at the same time decreasing the

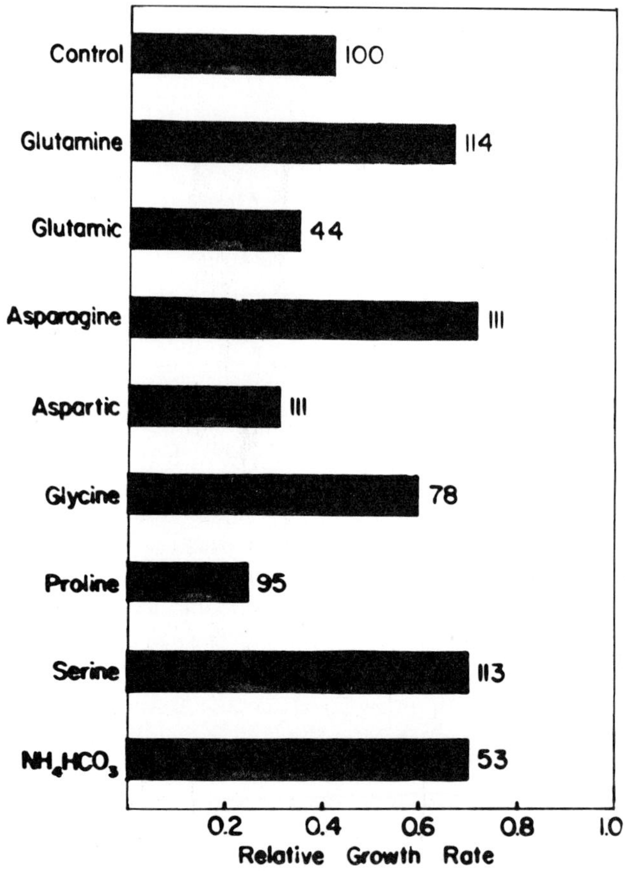

Fig. 3. Effect of the addition of various amino acids on the degree to which fluoride inhibits L-cell growth of the cells over a four-day period in the presence of 30 ppm F compared with the growth rate of the cells growing in the same media without fluoride. The various compounds were added to the media at a concentration of 5mM, and the number beside each bar is the relative growth rate of the cells after this addition compared with the control growth rate (Drescher and Suttie, unpublished data).

rate of growth of control cells, while others decrease the fluoride sensitivity without decreasing the control growth rate. Of these, glutamine has been studied in some detail. The effect is not seen with glutamic acid and little alteration in glutamine metabolism is seen in the presence and absence of fluoride. The basis for the protective effect is not yet clear, but it may involve slight changes in intracellular fluoride concentrations, or some as yet undetermined metabolic alteration.

Cell growth in the presence of fluoride was not altered when the cells were grown in either mannose or galactose, but was slightly impaired when they were grown in the presence of ribose. Fluoride caused little growth inhibition in the presence of glucosamine, but growth of control cells was drastically inhibited. Likewise, the toxicity of fluoride is decreased by anaerobiosis, but the growth rate of the cells is slowed. Because fluoride appeared to be less toxic when cells were growing more slowly, temperature changes were studied as a nonspecific growth inhibitor. Fluoride toxicity decreased as the growth rate of the cells was slowed by incubation at temperatures ranging from 37 to 31°C. When these data are plotted on the same graph (Fig. 4) as the data for growth under anaerobic conditions or in the presence of glucosamine, an interesting relationship is seen. The fluoride inhibition of growth under anaerobic conditions is slightly less than would be expected from the effect of anaerobiosis on growth rate, which indicates that the decrease in toxicity cannot be explained on this basis alone. Toxicity when glucosamine is the substrate is, however, much greater than would be expected on the basis of the generation time involved, which would indicate that glucosamine is in fact detrimental to the cells even in the presence of fluoride and does not offer any protection against the metabolic effects of fluoride.

These studies have not yet provided a definitive explanation of the effects of fluoride on cellular metabolism, but they have, however, opened a number of promising areas for further investigation which might elucidate the

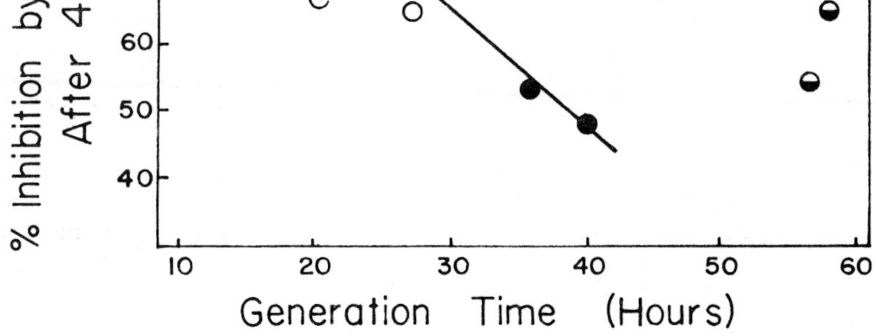

Fig. 4. Effect of temperature, anaerobiosis, and glucosamine addition on generation time of L cells. Cells were incubated for four days at temperatures of 37°, 34.5°, 33°, and 31°C, or they were incubated at 37°C under anaerobic conditions, or in the presence of 1 mg/ml of glucosamine. The cells were grown with or without the addition of 30 ppm F to the media, and the generation time of the cells grown without fluoride is plotted against the percentage inhibition by fluoride (6).

most fluoride-sensitive areas of metabolism. Ultimately such investigations should provide us with an explanation of how this simple inorganic ion can so drastically influence cellular metabolism.

REFERENCES

1. Carlson, J. R., and Suttie, J. W. 1967. Effect of sodium fluoride on HeLa cells I. Growth sensitivity and adaption. *Exp. Cell. Res.* 45: 415–422.
2. Carlson, J. R., and Suttie, J. W. 1967. Effect of sodium fluoride on HeLa cells II. Metabolic alterations associated with growth inhibition. *Exp. Cell. Res.* 45: 423–432.
3. Quissell, D. O., and Suttie, J. W. 1973. Effect of fluoride and other metabolic inhibitors on intracellular sodium and potassium concentrations in L cells. *J. Cell. Physiol.* 82: 59–64.
4. Quissell, D. O., and Suttie, J. W. 1972. Development of a fluoride-resistant strain of L cells: membrane and metabolic characteristics. *Am. J. Physiol.* 223: 596–603.
5. Drescher, M., and Suttie, J. W. 1972. Intracellular fluoride in cultured mammalian cells. *Proc. Soc. Exp. Biol. Med.* 139: 228–230.
6. Young, K. L. 1972. Effects of sodium fluoride on L cell metabolism. M. S. Thesis, University of Wisconsin, Madison.
7. Medappa, K. C., McLean, C., and Rueckert, R. R. 1971. On the structure of Rhinovirus 1A. *Virology* 44: 259-270.
8. Berry, R. J., and Trillwood, W. 1963. NaF and cell growth. *Br. Med. J.* 2: 1064.
9. LeCoultre-Mulder, G. W. A. F., Veldhuizen, C., Bowman, J., and Wise, M. 1969. Influence of the fluoride ion on the growth *in vitro* of human amnion cells, T-(Kidney) cells, and HeLa cells. *Acta Phys. Pharm. Neerland.* 15: 1–19.
10. Armstrong, W. D., Blomquist, C. H., Singer, L., Pollock, M. E., and McLaren, L. C. 1965. Sodium fluoride and cell growth. *Br. Med. J.* 1: 486–488.
11. Albright, J. A. 1964. Inhibitory levels of fluoride on mammalian cells. *Nature, Lond.* 203: 976.
12. Nias, A. H. W. 1965. Sodium fluoride and cell growth. *Br. Med. J.* 1: 1672.
13. Hongslo, J., and Jonsen, J. 1971. Effect of sodium fluoride on cells *in vitro*. *J. Dent. Res.* 50: 717 (abstr.).
14. Pace, D. M., and Elrod, L. M. 1960. Effects of respiratory inhibitors on glucose and protein utilization and growth in strain L cells. *Proc. Soc. Exp. Biol. Med.* 104: 469–472.

DISCUSSION

Singer (Minneapolis). Would you comment on whether the high amounts of fluoride are affecting the magnesium concentration in these cells?

Suttie. We've never measured it. We could measure total intracellular magnesium, but I don't know what relationship that has to the free magnesium concentration within the cell.

Bell (Oak Ridge). Are you getting any indication from your data which would explain the species differences of fluoride tolerance?

Suttie. No, I don't think so. When we talk about difference in resistance, we're talking mainly about the fact that we can feed one species more fluoride than another, and we don't even have data on what the plasma fluoride concentrations are when you maintain different species on the same concentration of fluoride.

Schwarz (Long Beach). You are, of course, dealing with what might be called toxic levels of fluoride in your system. What was the actual fluoride concentration of your control medium, and have you ever seen any positive growth responses to very low levels of fluoride? Also, do you have any evidence that some of the fluoride is converted to organic fluoride derivatives?

Suttie. I think our control fluoride levels are probably around 0.1 ppm or a bit lower, and we have not looked for any growth response at levels slightly above this. We haven't looked for any organic fluoride compounds that might be synthesized, and I am not convinced that toxic organic fluoride compounds are synthesized as readily in some of the other systems as people have claimed. There is no evidence that fluoroacetate or fluorocitrate can be produced in animal cell systems, and the claims that they are formed in a number of common forage plants have not been substantiated by other investigators.

Armstrong (Minneapolis). In your experiment in which you found a reduction of DPN in the cells treated with 30 ppm F, do you know what is happening to the DPN?

Suttie. We don't know. The loss occurs very rapidly and obviously there hasn't been time for it simply to be diluted out by growth. TPNH concentrations are not changed that drastically.

MECHANISMS FOR THE CONVERSION OF SELENITE TO SELENIDES IN MAMMALIAN TISSUES[1]

HOWARD E. GANTHER and H. STEVE HSIEH

Department of Nutritional Sciences, University of Wisconsin, Madison, Wisconsin

The metabolism of selenium in animals shares some similarities with sulfur, as in the case of methionine and its selenomethionine analog, but in general the metabolism of selenium is distinctly different from that of sulfur. In particular, inorganic forms of selenium such as selenate or selenite are readily reduced, even in nonruminants, whereas inorganic sulfate and sulfite are not.

The best-known products of selenium reduction are dimethyl selenide and trimethyl selenide. Both of these are end products of a metabolism that results in the elimination of selenium from the body. Dimethyl selenide is quite volatile and is exhaled through the lungs (1). The trimethylselenonium ion is very polar and is excreted in the urine (2, 3). While methylation of inorganic selenium usually amounts to detoxification (4, 5), in contrast to the case with mercury, Pařízek has demonstrated that the opposite effect can occur when selenite is metabolized to dimethyl selenide, and the animal subsequently injected with mercuric chloride (6). From the work of Diplock (7), we know that animals convert selenite to a form of selenide that can be released from proteins, apparently as hydrogen selenide, upon treatment with acid under anaerobic conditions. Whether this form of selenium represents a biologically active form or just an intermediate in metabolism remains to be established. Clearly, reduction is an important aspect of selenium metabolism in animals. The nature of this process is the subject of this paper.

1. Research supported by the College of Agricultural and Life Sciences, University of Wisconsin, Madison, and by a grant from the National Institutes of Health (AM 14189). The authors express appreciation to Dr. A. T. Diplock for making available the results of his work on selenide formation prior to publication, and to Margaret McNurlan for very capable technical assistance.

BIOSYNTHESIS OF DIMETHYL SELENIDE IN CELL-FREE SYSTEMS

The biosynthesis of dimethyl selenide has been studied for some time in our laboratory as a convenient model system for exploring pathways of selenium reduction (8, 9). The volatile product is stable and readily isolated, and its formation involves a six-electron reduction of selenium, from the 4+ to the 2− state. It was first shown that the process occurred in rat-liver slices, and that the activity of slices *in vitro* could account for the volatilization observed in the whole animal (10). Later, from studies on the reaction in cell-free liver systems (8), further evidence was obtained that the process was enzyme catalyzed and several important characteristics were identified: (1) anaerobic conditions were necessary for optimal activity; (2) there was an absolute requirement for glutathione that could not be met by other thiols; (3) TPNH stimulated the system even in the presence of high levels of glutathione; (4) S-adenosylmethionine was the methyl donor; (5) the system was strongly inhibited by arsenite (50% inhibition at 10^{-6}M) in the presence of a 20,000-fold excess of monothiols.

These properties indicated that the enzymes involved in this pathway would include one or more TPNH-linked reductases, plus at least one methyl transferase (S-adenosylmethionine is not a good alkylating agent per se). The marked inhibition by arsenite at substoichiometric levels relative to selenium and in the presence of a high level of monothiols suggested the probable occurrence of an active-site vicinal dithiol group in some enzyme. In broken-cell preparations the use of anaerobic conditions was essential, suggesting that oxygen-sensitive selenium derivatives were intermediates. It should be mentioned here that in the initial studies with liver slices, a 95% O_2-5% CO_2 atmosphere was routinely employed; such systems are not as active as the homogenate systems but probably function fairly well because of limited diffusion of oxygen into the slice interior. Also, as pointed out by Slater (11), there is a relatively low oxygen tension in the cells caused by the affinity of intracellular oxidases for oxygen. From the preference of the system for glutathione, as compared with other thiols such as 2-mercaptoethanol, it seemed likely that the role of glutathione was not simply to stabilize enzyme sulfhydryl groups, or to serve as a reducing agent. The participation of glutathione derivatives of selenium as intermediates in the pathway was therefore indicated, since the involvement of an enzyme such as glutathione reductase having a high specificity for the glutathione moiety could explain the specific requirement for glutathione. With these clues in mind regarding the nature of the enzymes and intermediates in the overall process, a detailed exploration of the pathway from selenious acid to dimethyl selenide was undertaken.

REACTIONS INVOLVING GLUTATHIONE

Beginning with selenious acid, its reaction with glutathione was investigated. It has long been known that thiols react spontaneously with selenious acid, and a detailed investigation (12) established the general reaction leading to formation of a selenotrisulfide (RSSeSR):

$$4RSH + H_2SeO_3 \rightarrow RSSeSR + RSSR + 3H_2O. \quad (1)$$

The reaction with glutathione proved to be more complex than with other thiols, but conditions for the quantitative preparation of the selenotrisulfide derivative of glutathione and its chromatographic separation from the very similar disulfide were worked out (13). It was then possible to explore in detail the activity of GSSeSG as a substrate for glutathione reductase, a possibility suggested by its close structural similarity to GSSG and supported by preliminary observations with unseparated mixtures of GSSeSG and GSSG (14). Submicrogram amounts of purified yeast glutathione reductase were shown to cause rapid oxidation of TPNH by GSSeSG, with the concomitant appearance of elemental selenium after an initial lag (13). This suggested that the selenopersulfide derivative expected as the initial product of GSSeSG reduction [reaction (2)] was

$$\text{GSSeSG} + \text{TPNH} + \text{H}^+ \xrightarrow{\text{glutathione reductase}} \text{GSSeH} + \text{GSH} + \text{TPN}^+ \quad (2)$$

subsequently decomposing to GSH and elemental selenium. The key to verifying this sequence of events was the use of iodoacetate directly in the glutathione reductase system to trap the labile selenopersulfide as the carboxymethyl derivative. This permitted the stoichiometry of reaction (2) to be verified, and also characterization of the carboxymethyl derivative by thin-layer chromatography and electrophoresis (13). Careful measurements of TPNH oxidation in the absence of iodoacetate showed that more than one mole of TPNH was oxidized per mole of GSSeSG, suggesting the possibility that further reduction of GSSeSG to H_2Se could occur to some extent, either by the direct action of glutathione reductase [reaction (3)], or indirectly (13):

$$\text{GSSeH} + \text{TPNH} + \text{H}^+ \xrightarrow[\text{reductase}]{\text{glutathione}} \text{H}_2\text{Se} + \text{GSH} + \text{TPN}^+. \quad (3)$$

In regard to the direct reduction of GSSeH by glutathione reductase, the compound has a reducible S–Se bond and can be considered as an analogue of GSSG, having one of the glutathione moieties, GS–, replaced by the HSe– group.

Scheme 1, which is a proposed pathway for dimethyl selenide synthesis, incorporates the reactions described above and shows the nonenzymic production of GSSeSG (the first stable intermediate in the pathway) and its reduction to the selenopersulfide (the second intermediate) by glutathione reductase (the first enzymic step). The involvement of glutathione reductase in the latter reaction can explain the requirement for TPNH and the specificity for glutathione in the overall process of dimethyl selenide synthesis. Although glutathione reductase is believed to have an active-site dithiol moiety, which can react with arsenite when the enzyme is in the fully reduced state (15), the states normally existing in its catalytic cycle apparently do not react with arsenite, and neither the yeast nor the liver enzyme is sensitive to low levels of arsenite (16). Thus the marked sensitivity of the dimethyl selenide pathway cannot be explained by the involvement of glutathione reductase, and some other dithiol protein must therefore be involved.

The further reduction of GSSeH to the 2− oxidation state is shown in Scheme 1 as involving the production of H_2Se (or Se^{2-} at physiological pH)

Scheme 1. Pathway for biosynthesis of selenides from selenite

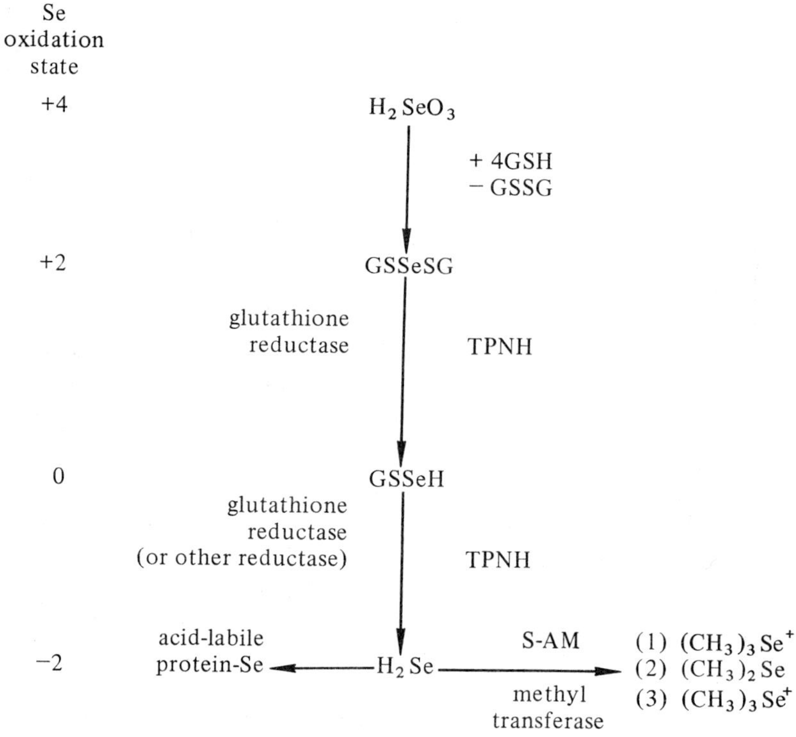

as a formal intermediate prior to methylation. This is not necessarily meant to be an obligatory step, since methylation could occur involving the GSSeH intermediate, forming $GSSeCH_3$, with subsequent reduction to CH_3SeH and final methylation.

FINAL REDUCTION AND METHYLATION

The nature of the final steps of reduction and methylation has been the focus of recent work in this laboratory. These studies have all been conducted by Mr. Hsieh for his dissertation and it is his work which is presented in the remainder of this paper. Our objective was to separate the processes of reduction and methylation and learn more about the two processes by further purification. This would be easier in a completely soluble system and so we turned to kidney. Although not as active as liver, in kidney homogenates the synthesis of dimethyl selenide occurs entirely in the soluble fraction and does not require the microsomal fraction, in contrast to liver. Our expectation was that a reductase of some type was required for the formation of selenide, and a second enzyme was required for methylation. These enzymes should separate at some stage, giving inactive fractions that would be fully active when recombined. Of particular interest was whether glutathione reductase might play a role in the final reduction, or even be the only TPNH-linked enzyme required for reduction, along the lines discussed earlier in connection with reaction (3).

The procedures for preparing homogenates and carrying out the incubations were similar to those described earlier (8) with slight modifications. Preliminary studies had shown that the kidney soluble system was similar to the $9000 \times g$ supernatant fraction of liver with respect to cofactor requirements, pH optimum, and other characteristics. High glutathione concentrations (up to $2 \times 10^{-2}M$) were employed, as with liver, in order to obtain optimal activity. Since it was shown that GSSeH was produced nonenzymically from selenite in the presence of excess GSH (13), selenite was used as the substrate with the expectation that GSSeH would be generated as a substrate for further reduction.

The biosynthesis of dimethyl selenide by the soluble fraction of kidney is shown in Fig. 1. It is somewhat more active than the soluble fraction prepared from the same amount of liver. Both systems show the characteristic lag period (8) and both are strongly inhibited by $10^{-5}M$ arsenite. This completely soluble system was next separated by Sephadex G-75 chromatography into four fractions (Fig. 2). Fraction A consisted of proteins larger than hemoglobin eluting at the void volume, fraction B contained hemoglobin, while fractions C and D contained proteins in the range of 30,000 and 10,000

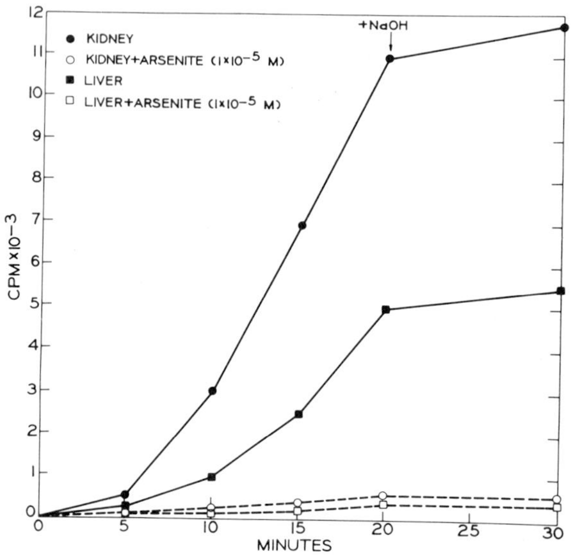

Fig. 1. Dimethyl selenide formation by liver or kidney soluble fraction. Composition of system (mM): $H_2{}^{75}SeO_3$, 0.05; GSH, 20; coenzyme A, 0.2; S-adenosylmethionine, 0.4; TPNH-generating system (isocitrate 4, TPN, 0.13, plus 100 μgm isocitrate dehydrogenase); ATP, 4; MES buffer (2[N-morpholino]ethane sulfonic acid), pH 6.35, 33; enzyme (in this experiment, 0.5 ml of 100,000 × *g* supernatant from 10% liver or kidney homogenate). The system minus selenite was flushed for 6 min at 37°C; then selenite was tipped in to start the reaction. [See (8) for detailed description of incubation procedure.]

molecular weight, respectively, as shown by independent calibration studies with proteins of known molecular weight. The activity of these fractions alone or in various combinations, after concentration by Diaflo ultrafiltration, is summarized in Table 1. Only fraction C had activity by itself. Although C had a higher specific activity than the original soluble fraction, the recovery of activity (units) in C was less than 20% of the activity in the soluble fraction from which it was prepared. The addition of either fraction A or B increased the activity of C, while D had no such effect. A recombination (bottom line, Table 1) of the separated fractions in the proportions found in the soluble fraction gave complete recovery of activity, indicating that no loss of activity occurred through denaturation during the fractionation procedure.

The activity of fraction C could be inhibited by arsenite (Table 2), but higher levels were required to achieve inhibition, compared with the case with kidney soluble fraction. Also, the activity of fraction C preincubated with arsenite was completely restored by dialysis, whereas that of the soluble fraction was not (Table 3). These observations suggest that, in the case of

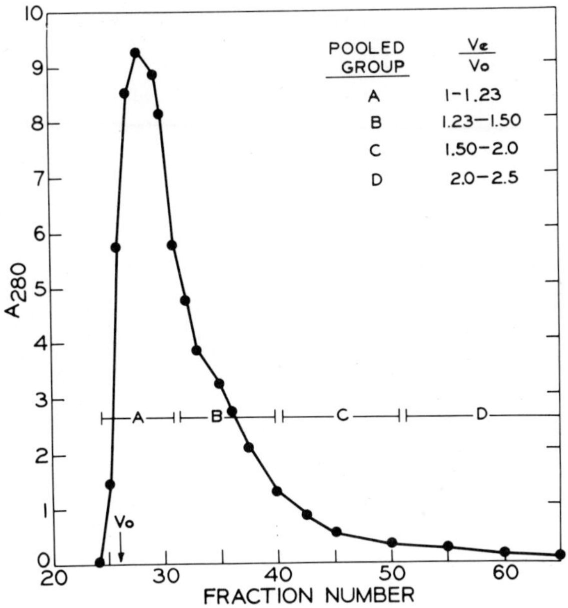

Fig. 2. Separation of kidney soluble fraction on Sephadex G-75. A 2.5 × 94 cm column equilibrated at 4°C with Tris (0.05M, pH 7.8) was used and 5.4 ml fractions were collected.

Table 1. Activity of kidney cytosol fractions obtained by Sephadex G-75 chromatography (see Fig. 2)[1]

Fraction	Protein (mg)	Units[2]	Specific activity[3]
Cytosol	7.35	2.4	32.8
A	5.1	0.01	1.4
B	6.0	0.06	1.8
C	4.8	3.25	67.9
½C	2.4	2.25	94.1
D	3.5	0.07	2.0
1/2 A + 1/2 B	5.55	0.12	2.1
1/2 A + 1/2 C	4.95	3.31	66.6
1/2 B + 1/2 C	5.40	3.42	63.3
1/2 D + 1/2 C	4.15	2.0	60.7
Recombination	5.16	1.94	37.4

[1]See Fig. 1 for incubation procedure.

[2]nmole of dimethyl selenide per min.

[3]Units per mg of protein × 100.

Table 2. Arsenite inhibition of dimethyl selenide formation by kidney fraction C

Arsenite[1] (M)	Specific activity[2]	Percentage inhibition
0	35.4	
10^{-6}	36.1	0
10^{-5}	23.1	35
10^{-4}	14.4	59
10^{-3}	2.0	94

[1]Sodium arsenite was present during the incubation at the concentration indicated. Incubation conditions as described in Fig. 1, with 6.4 mg of fraction C as the enzyme source.

[2]nmole dimethyl selenide/min/mg protein × 100.

fraction C, arsenite might inhibit the system by reacting with selenide intermediates in a stoichiometric fashion, whereas the irreversible inhibition of soluble fraction and inhibition at levels of arsenite below that of the selenium substrate are consistent with binding of arsenite to a protein acting catalytically in the synthesis of dimethyl selenide.

Fraction C was purified further by DEAE cellulose chromatography (not shown) into several protein fractions. One of these, fraction C-III_a, was still

Table 3. Effect of dialyzing arsenite-treated kidney fractions[1]

	Protein (mg)	Specific activity[2]	Percentage inhibition
Soluble fraction			
−As	2.78	30.1	
+As	2.70	16.2	46
Fraction C			
−As	5.18	26.8	
+As	4.85	28.6	0

[1]The fractions were first incubated (as for Fig. 1 but without selenite and S-AM) with or without 1.0×10^{-3}M sodium arsenite for 10 min, then the reaction mixtures were dialyzed 24 hr against three changes of Tris buffer (0.05 M, pH 7.8). The dialyzed fractions were then incubated again as for Fig. 1 to measure dimethyl selenide formation.

[2]nmol dimethyl selenide/min/mg protein × 100.

active by itself, and no evidence was obtained that more than one protein was responsible for the activity of fraction C. It thus appears that the residual activity of fraction C represents an arsenite-insensitive methyl transferase acting on small amounts of hydrogen selenide which can be produced non-enzymically by an excess of glutathione, as will be shown in the following studies. Formal separation of the soluble system into a methylating component (fraction C) and a reducing component (presumably fractions A–B) seems therefore to have been achieved, although the expectation that both of the separated fractions would be completely inactive was not quite correct.

Returning to the reductive process, the formation of hydrogen selenide by both nonenzymic and enzymic processes is shown in Table 4. A small amount of acid-volatilized selenium could be produced by simply mixing GSH and selenious acid at the usual concentrations under nitrogen at pH 7. At this pH no selenium was lost from the flask until acid was added, but in a separate experiment some loss occurred at pH 6.3 without acidification. This latter effect is presumably related to the greater proportion of the more volatile undissociated H_2Se existing at a lower pH. It should be noted here that 2-mercaptoethanol also can reduce sodium selenite to hydrogen selenide, and dithiothreitol is even more effective, as expected from its greater reducing power. In fact, using the apparatus and conditions of Diplock (7), we found that the concentration (5×10^{-3}M) of 2-mercaptoethanol used in his experiments as an 'antioxidant' consistently caused the loss of about 15% of

Table 4. Production of hydrogen selenide by yeast glutathione reductase[1]

Glutathione reductase		Percentage Se volatilized (by difference)	
μgm	units	−HCl	+HCl
0	0	0	7.1
10	1.2	1.5	34.8
40	4.8	2.0	43.2

[1] [^{75}Se] sodium selenite (5×10^{-5}M) was incubated at 37°C with 2×10^{-2}M GSH in phosphate buffer at pH 7 in the presence of a TPNH-generating system plus glutathione reductase (as indicated), and 5 mg of bovine serum albumin, under a stream of nitrogen. After 25 min 12 N HCl (1.5 ml) was added to some of the flasks and incubation continued another 20 min. Volatilization of hydrogen selenide was calculated from the ^{75}Se remaining in the flask. It should be noted that the amount of hydrogen selenide produced with 2×10^{-2}M GSH alone (7%) was less than the normal amount (about 20%); apparently the added serum albumin somehow interfered.

selenite (3×10^{-5}M) as acid-volatile selenium. This suggests that one needs to exercise caution in using 2-mercaptoethanol in experiments such as those of Diplock, even though it was shown that acid-labile selenium was present in tissues that were not treated with the thiol (7).

While Table 4 shows that glutathione can bring about hydrogen selenide production, its formation was stimulated more by the addition of 10–40 μg of purified yeast glutathione reductase (Sigma Chemical Co.). Even 10 μg of enzyme did not give maximum stimulation, whereas submicrogram quantities of the same enzyme are very active in the reduction of GSSeSG to GSSeH (13). This could mean either that GSSeH is a poorer substrate than GSSeSG, or that some other enzyme present as a contaminant in the glutathione reductase is responsible. We favor the former possibility, as a result of other studies in this laboratory with $GSSCH_3$, an analog of GSSeH and GSSG. Using 1–2 μg of glutathione reductase $GSSCH_3$ was found to be a substrate, but $GSSCH_3$ concentrations of 1–10 mM were necessary to achieve a good rate of TPNH oxidation, whereas with GSSG or GSSeSG 0.1 mM or lower concentrations of substrate and submicrogram amounts of enzyme suffice.

It is next shown that yeast glutathione reductase added to fraction C-III_a of kidney stimulated the production of dimethyl selenide (Table 5). The degree of stimulation depended on how much hydrogen selenide was being produced nonenzymically by glutathione. With physiological levels of glutathione (2×10^{-3}M) the activity was more than doubled by the addition of 10 μg of glutathione reductase, and stimulated still further by 50 μg. The maximum activity attained at either level of glutathione was similar, suggesting the common effect with both the enzyme and the GSH was to increase hydrogen selenide production.

Returning to the liver system, the microsomes seem to be the functional counterpart of fraction C in kidney, i.e., the methylating system. Microsomes are active by themselves if high levels of glutathione are provided, but show

Table 5. Stimulation of kidney fraction C-III_a by yeast glutathione reductase[1]

		2×10^{-3}M GSH		2×10^{-2}M GSH	
C-III_a (mg)	Glutathione reductase (μg)	Specific activity	Percentage stimulation	Specific activity	Percentage stimulation
1.31	0	72		120	
1.31	10	160	122	167	38
1.31	50	173	148	162	34

[1]nmole/ dimethyl selenide/min/mg protein $\times$ 100.

Table 6. TPN-dependent stimulation of liver microsomes by liver A_1 soluble fraction

GSH	Components[2]	TPN	Volatilization (by difference) (%)
2×10^{-3}M	Microsomes	+	10
	Microsomes + fraction A[3]	+	59
	Microsomes + fraction A	−	18
2×10^{-2}M	Microsomes	+	39
	Microsomes + fraction A	+	54
	Microsomes + fraction A	−	38

[1]Incubation as for Fig. 1 except as indicated.
[2]5 mg each.
[3]The protein fraction eluting at the void volume, before hemoglobin, when the liver soluble fraction was chromatographed on Sephadex G-75 as described in Fig. 2.

no dependence on TPN (8), which suggests that they rely entirely on selenide produced nonenzymically. Similarly, the synthesis of dimethyl selenide by liver microsomes is stimulated by the liver soluble fraction (8). Further, upon separation of the liver soluble fraction by Sephadex G-75 chromatography, as for the kidney soluble fraction, a fraction A was obtained which stimulated the microsomes (Table 6). The effect of fraction A was largely TPN dependent and was more apparent when the GSH concentration was lowered to a physiological level.

Many experiments have been carried out to determine if yeast glutathione reductase can stimulate dimethyl selenide synthesis by liver microsomes, as it does with kidney fraction C. The results, inexplicably, have tended to be quite variable, but the data in Fig. 3 are believed to be representative. Glutathione reductase did bring about a moderate degree of stimulation, approaching an apparent maximum within the range of 1–10 units. The maximum stimulation obtained was considerably less than that obtained with liver soluble fraction A, however, even though the amount of glutathione reductase activity present in the added fraction A was found by assay to be about 1 unit, an amount which barely stimulated when added as yeast glutathione reductase. One might conclude from this experiment that glutathione reductase was probably not the only protein in the liver soluble fraction or fraction A that was responsible for its stimulatory effect when added to microsomes. The dependence on TPN indicates a TPNH-linked reductase is present, but there might be other TPNH-linked reductases having a molecular weight larger than 75,000 that would be eluted in fraction A, in

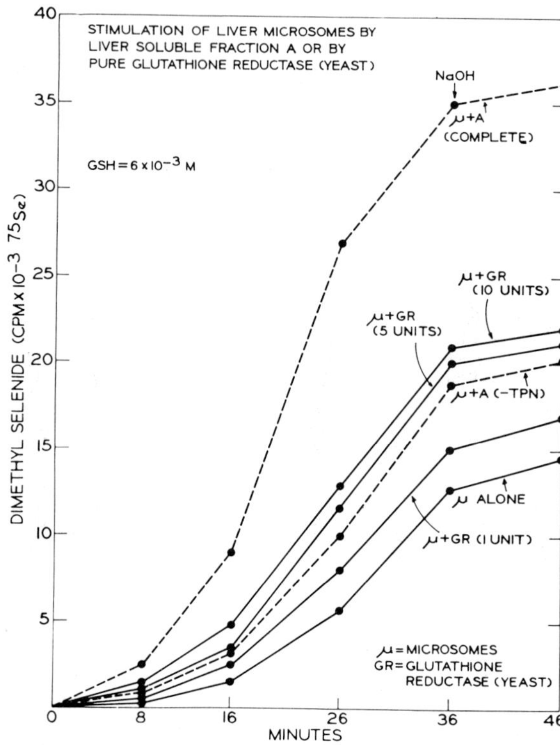

Fig. 3. Stimulation of dimethyl selenide production in liver microsomes by addition of liver soluble fraction A or purified yeast glutathione reductase. Conditions are the same as for Fig. 1 except GSH concentration was 6 mM, and the enzyme source was microsomes (7.05 mg of protein), with or without fraction A (4.28 mg protein), or glutathione reductase (specific activity 165 units/mg).

the void volume for Sephadex G-75, along with glutathione reductase. There are other possible explanations of why yeast glutathione reductase was not equivalent to liver soluble fraction A under the conditions of this experiment, however, and glutathione reductase might still be the major enzyme involved. Direct purification of the active protein(s) in liver soluble fraction A with special reference to glutathione reductase and other TPN-linked reductases present in that fraction are in progress and hopefully will provide further insight into the nature of the reductive process.

In earlier studies, not described here, the interesting and quite surprising finding was made that the microsome system is exceedingly sensitive to arsenite, even in micromolar concentration range. This suggests that the methyl transferase activity of the microsomes involves a dithiol protein. Thus the arsenite sensitivity of the pathway probably is largely expressed at this step, rather than through arsenite inhibition of a reductase. The low sensitivity of glutathione reductase to arsenite certainly fits this interpretation, although we have not yet excluded the possibility of an additional reductase more sensitive to arsenite that might be required in addition to glutathione reductase.

CONCLUSIONS

From the results of the investigations reviewed here, we can say that the pathway for synthesis of dimethyl selenide from selenite in animals is becoming understood in terms of the cofactors, intermediates, and enzymes involved. Both nonenzymic and enzymic reactions are involved. Apparently there are two methyl transferases capable of synthesizing dimethyl selenide, one present in the soluble fraction that is not sensitive to arsenite, and one in the microsomal fraction of liver that is very sensitive to arsenite. These act on selenide produced either nonenzymically or by means of TPNH-linked reductases. The amount of selenide produced nonenzymically by the reducing action of glutathione is relatively small, even when amounts well in excess of physiological levels are employed. The requirement for TPNH indicates that enzymic reduction is important. The involvement of glutathione reductase in one or more of the reductive steps can explain the TPNH-dependence and the specific requirement of the system for glutathione.

The ability of systems in animal tissues to generate highly reactive selenide intermediates opens many possibilities for considering the biosynthesis of various organoselenium compounds. In particular, the biosynthesis of trimethyl selenide and the formation of acid-labile selenide (7) might be expected to proceed through such intermediates, and there is the further possibility that they might also be precursors of other forms of selenium having essential biological functions.

REFERENCES

1. McConnell, K. P., and Portman, O. W. 1952. Excretion of dimethyl selenide by the rat. *J. Biol. Chem.* 195: 277.
2. Byard, J. L. 1969. Trimethyl selenide. A urinary metabolite of selenite. *Archs Biochem. Biophys.* 130: 556.
3. Palmer, I. S., Gunsalus, R. P., Halverson, A. W., and Olson, O. E. 1970. Trimethylselenonium ion as a general excretory product from selenium metabolism in the rat. *Biochim. Biophys. Acta* 208: 260.
4. McConnell, K. P., and Portman, O. W. 1952. Toxicity of dimethyl selenide in the rat and mouse. *Proc. Soc. Exp. Biol. Med.* 79: 230.

5. Obermeyer, B. D., Palmer, I. S., Olson, O. E., and Halverson, A. W. 1971. Toxicity of trimethylselenonium chloride in the rat with and without arsenite. *Toxic. Appl. Pharmac.* 20: 135.
6. Pařízek, J., Ostadalova, I., Kalouskova, J., Babicky, A., and Benes, J. 1971. The detoxifying effect of selenium. Interrelationships between compounds of selenium and certain metals. *In* W. Mertz and W. E. Cornatzer (eds.), *Newer Trace Elements in Nutrition,* Marcel Dekker, New York, p. 85.
7. Diplock, A. T., Caygill, C. P. J., Jeffry, E. H., and Thomas, C. 1973. The nature of the acid-volatile selenium in the liver of the male rat. *Biochem. J.* 134: 283.
8. Ganther, H. E. 1966. Enzymic synthesis of dimethyl selenide from sodium selenite in mouse liver extracts. *Biochemistry* 5: 1089.
9. Ganther, H. E. 1971. Selenium: The biological effects of a highly active trace substance. *In* D. D. Hemphill (ed.), *Trace Substances in Environmental Health,* Vol. 4, University of Missouri, Columbia, Mo., p. 211.
10. Ganther, H. E. 1963. Ph.D. Thesis, University of Wisconsin.
11. Slater, E. C. 1962. Comments on papers, symposium on vitamin E and metabolism. *In* R. S. Harris, I. G. Wool, G. F. Marrian, and K. V. Thimann (eds.), *Vitamins and Hormones,* Vol. 20, Academic Press, New York, p. 521.
12. Ganther, H. E. 1968. Selenotrisulfides. Formation by the reaction of thiols with selenious acid. *Biochemistry* 7: 2898.
13. Ganther, H. E. 1971. Reduction of the selenotrisulfide derivative of glutathione to a persulfide derivative by glutathione reductase. *Biochemistry* 10: 4089.
14. Ganther, H. E. 1970. Selenium metabolism. Mechanisms for the conversion of inorganic selenite to organic forms. *In* C. F. Mills (ed.), *Trace Element Metabolism in Animals,* Livingstone, Edinburgh, p. 212.
15. Massey, V., and Williams, C. H., Jr. 1965. On the reaction mechanism of glutathione reductase. *J. Biol. Chem.* 240: 4470.
16. Tietze, F. 1970. Disulfide reduction in rat liver. I. Evidence for the presence of nonspecific nucleotide-dependent disulfide reductase and GSH-disulfide transhydrogenase activities in the high-speed supernatant fraction. *Archs Biochem. Biophys.* 138: 177.

DISCUSSION

Diplock (London). I wonder about the temperature dependence of your thiol interactions. In our experiments, we keep everything at 4°C and assume that you are doing your experiments at 37°C. Have you looked at a lower temperature to see whether you get the same conversions?

Ganther. The only lower temperature we have used is room temperature, but because of the very great reactivity we observe, I think we would find some effect at lower temperatures.

Petering (Milwaukee). These nonenzymatic conversions are very interesting as they compose a significant fraction of the total conversion from selenite to other products. A few years ago when we were looking at some of the oxidation products of ferridoxin, we found that one form was a labile sulfide, which could liberate sulphur as sulfide in the presence of thiols. Perhaps in your system you are going from selenide to selenodisulfide and then in the presence of excess glutathione converting that to selenide.

Ganther. This is possible. We have shown that dithiols are even more active and will very readily evolve hydrogen selenide out of these systems. However, these systems do respond to TPN and this definitely implies that these are not purely chemical steps but involve reductases of an as yet unidentified nature.

Pařízek (Prague). I'd like to comment on the lag phase you observed in some of your experiments. We have evidence from *in vivo* experiments that dimethylselenide can be converted into trimethylselenonium ions in the body, and I believe that dimethylselenide is an intermediary product on the way to trimethylselenonium ions. Is it possible that you get the production of trimethylselenonium ions in your *in vitro* systems as well, so that during the first few minutes after incubation most of the activity is in this nonvolatile form and then with higher doses of dimethylselenide volatile forms are produced?

Ganther. That is an interesting possibility. We have studied many aspects of this lag phase. We know that part of it is due to the mechanical effect of sweeping the system, but there are several other factors that operate, including the amount of selenium you put into the system.

Shirley (Gainesville). Would a high dietary supplementation of choline have any effect on selenium metabolism?

Ganther. It has been suggested that methyl donors could promote the volatilization of selenium from the body and induce a deficiency. We have studied some nutritional factors that influence volatilization in whole animals, but choline was not effective. Combinations of methionine and high protein will enhance volatilization but methionine by itself is not particularly effective. Once we know the enzymes and cofactors involved in the process, we hope to go back and study the dietary effects.

Schwarz (Long Beach). We reported some time ago that the inclusion of relatively large amounts of methyl donors or methyl precursors in the diet enhances the development of the dietary necrotic liver degeneration by about 30%. This effect is not related to growth. It could be related to the fact that small traces of selenium which were in tissues were removed faster.

NEW ESSENTIAL TRACE ELEMENTS (Sn, V, F, Si): PROGRESS REPORT AND OUTLOOK[1]

KLAUS SCHWARZ

Laboratory of Experimental Metabolic Diseases, Veterans Administration Hospital, Long Beach, California, and Department of Biological Chemistry, School of Medicine, University of California, Los Angeles, California

I had the privilege of presenting a paper entitled "The Role of Environmental Conditions in Trace Element Research–An Experimental Approach to Unrecognized Trace Element Requirements" at TEMA-1 (1). It is my great pleasure now to give a progress report. In 1969 our approach seemed to be full of promises but none of these had yet been fulfilled. Since then tin (2), vanadium (3), fluorine (4), silicon (5), and nickel (6) have been added to the list of elements for which essential roles have been demonstrated (Table 1). Nickel belongs to this list since Nielsen has consolidated his first tentative findings with chicks, as reported below (7).

Our technique, in essence, consists of the application of ultraclean, more or less trace-element-sterile (but not bacteriologically sterile) isolators, highly purified amino acid diets, and a rigid weaning routine which prevents accumulation of trace elements prior to the initiation of the experiment (1, 8, 9). All compounds tested are added to the diet. Fig. 1 shows the three isolators in which four elements, tin, vanadium, fluorine, and silicon, were discovered to be essential for growth.

Rats maintained on amino acid diets under normal, conventional conditions grow well and look normal while those in the trace-element-controlled isolator system are severely deficient (Fig. 2). They are delayed in growth, show alopecia, seborrhea, and other symptoms, such as a lack of pigmentation of the incisors. We first thought that this deficiency was due to only a single element, and at TEMA-1, even though making allowances for the

1. Supported in part by U. S. National Institutes of Health Grant No. AM 08669.

Table 1. Discovery of trace-element requirements

Element	Date	Reference	Element	Date	Reference
Iron	17th Century		Selenium	1957	Schwarz and Foltz
Iodine	1850	Chatin	Chromium	1959	Schwarz and Mertz
Copper	1928	Hart *et al.*	Tin	1970	Schwarz, Milne, and Vinyard
Manganese	1931	Kemmerer and Todd	Vanadium	1971	Schwarz and Milne
Zinc	1934	Todd, Elvehjem, and Hart			Hopkins and Mohr
Cobalt	1935	Underwood and Filmer	Fluorine	1972	Schwarz and Milne
		Marston			Messer, Armstrong, and Singer
		Lines	Silicon	1972	Schwarz and Milne
Molybdenum	1953	deRenzo *et al.*			Carlisle
		Richert and Westerfeld	Nickel	1973	Nielsen

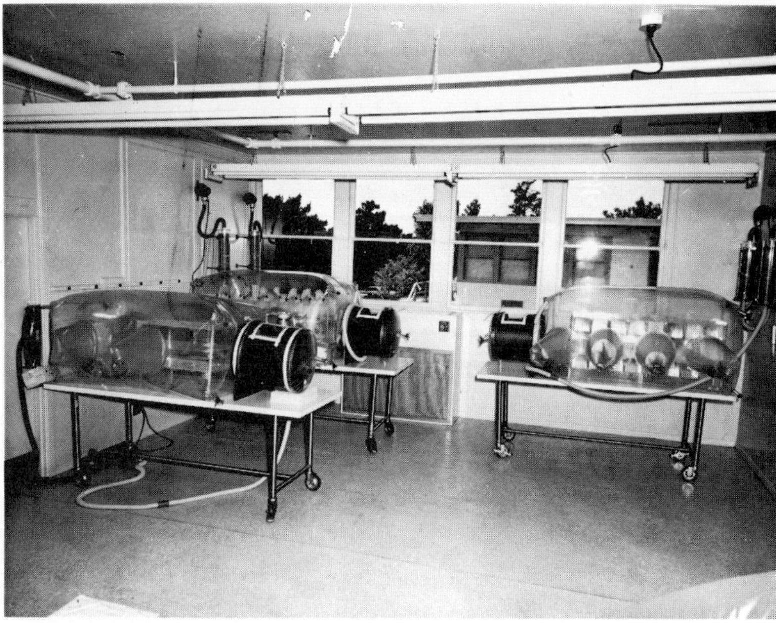

Fig. 1. Trace-element-controlled isolators.

possibility that several elements may be involved, I talked about our attempts to trace down *the* missing element.

Preliminary but significant results were seen in our system as early as 1965 with silicon, and less so with vanadium. J. Cecil Smith and Betty Vinyard contributed substantially to the earlier studies. David Milne has been highly instrumental in the investigation of the last four elements identified by us as essential. I wish to express my appreciation and give due credit to them and others who have assisted in these experiments.

We have set ourselves some rigid standards which we feel should be fulfilled before claims for essentiality can be made.

(1) The experiment should produce highly significant responses.

(2) It should be reproducible at will, and in a series of tests over a lengthy interval of time.

(3) A dose-response curve should be established and the minimum effective dose level of the element should be determined.

(4) Several compounds of the same element should be tested and compared in potency. In some cases, for example silicon, this was not readily possible.

(5) The effect should be physiological, i.e., it must be obtained using amounts which are normally present in foods and tissues.

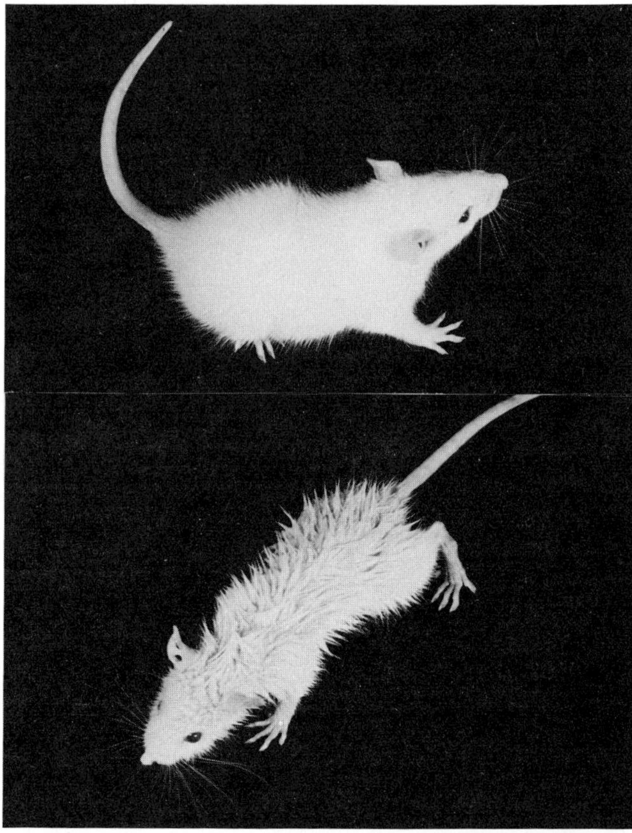

Fig. 2. Trace-element deficiency: upper animal, outside control, kept under conventional conditions on purified diet; lower animal, maintained for 20 days on the same diet in the trace-element-sterile environment system.

TIN

The first clear-cut growth effects with tin were seen in our laboratory in 1968. The data in Table 2, published in 1970, show that tin, added at physiological levels of 1.5–2.0 ppm to the tin-deficient diet, produces a very pronounced enhancement of the growth rate (2). The growth responses exceed 50%. Tin also has an effect on the pigmentation of teeth (10, 11). We were very careful about the effects observed with tin since the element had been discounted in the literature as a potentially essential trace element; it could not be found with rather rough analytical techniques in embryos and the newborn. It also was not detectable in the organs of some adults from Switzerland and those of some tribesmen from Africa (12).

Table 2. Growth effects of various tin compounds in rats in trace-element-controlled environment

Compounds	Dose level (μg Sn/100g diet)	No. of animals	Average daily weight gain (g)	Average increase over controls (%)	p value
Control		5	1.10 ± 0.05[1]		
$Sn(SO_4)_2 \cdot 2H_2O$	50	8	1.37 ± 0.10	24	0.02
$Sn(SO_4)_2 \cdot 2H_2O$	100	8	1.68 ± 0.10	53	< 0.001
$Sn(SO_4)_2 \cdot 2H_2O$	200	8	1.75 ± 0.10	59	< 0.001

[1] Mean ± s.e.

A closer look (Table 3) shows that tin has physicochemical properties which lend themselves in many ways to biological functions. It can undergo two-electron oxidation-reduction reactions from 2+ to 4+, its oxidation-reduction potential is like that of the flavin enzymes, and it can make coordination complexes which are usually hexavalent, but the number of ligands can vary from four to eight. In addition, however, it has a strong tendency to make *covalent* links to carbon since it is in the same group of the periodic system as carbon. Organo-tin compounds are known literally by the thousands. One can, therefore, expect tin to occur in the organism not only in its inorganic form, but also bound through covalent bonds to carbon compounds. Not only proteins, but other body constituents such as fats may contain bound tin.

It is thus understandable that tin has a biological function. Unfortunately, very little has been done to investigate it since we published our results. One possible function of tin is the following, thrown out as a hypothesis for your consideration. After discovering the growth effect of tin, we tried to establish whether or not it is really missing in the embryo. In spite of sensitive analytical techniques, we also found tin levels in embryos and

Table 3. Tin

Chemistry	Atomic number 50, atomic weight 118.7 Group IV element (with C, Si, Ge and Pb) Valence 2+ ⇌ 4+, redox potential −0.13 V Coordination complexes, 6; also 4,5, and 8 ligands Covalent bonds with carbon; organo-tin compounds; even stannic chloride largely covalent Catalyst in polymerization, transesterification, olefin condensation, etc.
Requirement	1.5–2.0 ppm in diet, as stannic sulfate, for growth and normal tooth development Various tin compounds differ in potency
Occurences	Widespread in foods, feeds, and tissues Human kidney 0.23–0.70, liver 0.35–1.00, lung 0.44–1.20 ppm (wet weight)
Human intake	Estimated 3.6–17 mg or more per day, as compared with 5–10 μg for rat
Biological effects	Possible oxidation/reduction catalyst

Table 4. Growth effect of tin compounds in a trace-element-controlled environment[1]

Compound	No. of experiments	No. of animals	Average increase over controls (%)
$Sn(SO_4)_2 \cdot 2H_2O$	3	24	26
$(CH_3)_3SnCl$	1	7	23
$(C_2H_5)_3SnCl$	2	15	17
$(C_3H_7)_3SnCl$	2	16	14
$(C_4H_9)_3SnCl$	2	12	28
$(C_6H_6)_3SnCl$	1	7	29
$(C_6H_6)_3SnCl$	1	7	28

[1]Experimental duration was 26–29 days.

newborn animals hardly measurable. Tin is truly not there to any great extent. It seems to be excluded from the fetus, although shortly after birth tin levels per gram of tissue in the young are similar to those found throughout the lifetime of the animal. They are kept rather steady. The same phenomenon has been reported for the human (12). It appears that the concept that trace elements assumed to be essential must be present in the embryo is not necessarily true. It is possible that tin stimulates the burst of DNA/RNA transcription and translation and of protein synthesis which happens shortly after birth (13). We rarely realize that the organism is lacking in many enzymes at the time of birth, for instance, those which are necessary for glucose metabolism in the liver. A great burst in protein synthesis occurs immediately after birth, and it is possible that tin acts as one of the messengers from the outside which stimulate or induce this particular event.

Different tin compounds have different biological activity. Not only inorganic, but also organic tin compounds promote growth, i.e., the organism is capable of utilizing tin which is covalently bound (Table 4). These compounds are lipid soluble. We found that a rather large proportion of tin in the organism, as a matter of fact, is present in fat. This proves that the general, stereotyped assumption that 'trace metals are bound to protein' does not hold for all cases. Commercial fats are often good sources of tin, and the selection of fats free of tin was part of the reason for success in demonstrating its indispensability. The biochemistry of tin is completely unexplored territory.

VANADIUM

After tin we discovered vanadium to be essential for growth in the rat (3). Vanadium had been implicated before by Hopkins who observed a stimula-

Table 5. Growth response of rats to varying levels of vanadium supplements (sodium orthovanadate, Na_3VO_4)[1]

Dose level (μg V/100 g diet)	Unsupplemented controls: No. of rats[2]	Unsupplemented controls: Average daily weight gain	Supplemented animals: No. of rats	Supplemented animals: Average daily weight gain	Average increase over controls (%)	p value
1	7	1.05 ± 0.08[3]	7	1.27 ± 0.10	21	n.s.
5	16	1.04 ± 0.08	16	1.38 ± 0.08	33	< 0.01
10	6	1.02 ± 0.14	7	1.38 ± 0.07	35	< 0.05
25	14	0.87 ± 0.10	14	1.21 ± 0.09	41	< 0.02
50	6	1.02 ± 0.14	7	1.49 ± 0.12	46	< 0.02

[1] Data are pooled results of five successive experiments.
[2] A total of 37 rats served as controls in five successive experiments.
[3] Mean ± s.e.

tion of the growth of feathers in chicks but did not see general growth effects (14). Further studies on vanadium in chicks have been carried out by Nielsen (15). Since a detailed report on vanadium is presented elsewhere in this symposium (16), we merely show the growth effect in the rat (Table 5) and a summary of characteristic properties of vanadium (Table 6). In our experiment, vanadium is effective at levels between 5 and 10 μg/100 g of diet, i.e., 0.05–0.1 ppm. The growth response is highly significant. As seen with other trace elements, different vanadium compounds have greatly different potencies. Sodium vanadate is probably the best source of vanadium one could supply. The properties of the element support the assumption that it may act in oxidation-reduction reactions in the organism.

Table 6. Vanadium

Chemistry	Atomic number 23, atomic weight 50.9 Group Vb element (with Nb and Ta) Early transition element of first series of metals between Ti and Cr Valence states 5+, 4+, 3+ and 2+ Strong tendency to form coordination complexes
Requirement	0.1 ppm in the diet, as sodium orthovanadate Large differences in potency of various vanadium compounds and different oxidation states Ortho > meta > pyrovanadate 'Physiological' V^{5+}, also V^{4+}, but lower oxidation states possible in organisms
Human intake	2 mg daily on 'well-balanced diet', for 75 kg weight, as compared with 1–2 μg for a 75 g rat
Biological effects	Essential for *Aspergillus niger* and *Scenedesmus obliquus* Found in porphyrin complexes in fossil fuels, and in feathers Catalyses phospholipid oxidation by liver protein *in vitro* Catalyses nonenzymatic oxidation of catecholamines Inhibits cholesterol synthesis Lowers phospholipid and cholesterol in blood Inhibits caries (stimulation of mineralization) Functions most likely as oxidation-reduction catalyst

FLUORINE

Under our experimental conditions fluorine, supplied as fluoride, is essential for growth and general development (4), and not only for anticariogenic effects and bone stability as reported previously. The growth effect (Table 7) is obtained with physiological levels of fluoride which are added to the fluorine-deficient diet, in contrast to Messer and co-workers who applied 50 ppm in the drinking water (17, 18). In our case, 1 ppm gives a partial response; with 2.5 ppm we observe optimal activity. The effect of fluoride was somewhat puzzling because some very qualified investigators, for instance Maurer and Day (19) and Doberenz *et al.* (20), had tried before to determine whether fluorine is essential for growth. They had obtained negative results in spite of elaborate experimental techniques. I have no explanation for the differences between their findings and ours.

Fluorine has many criteria commensurate with a physiological role; it is handled by the organism in a manner which enforces the concept that it is essential (Table 8). Details have been presented elsewhere (4). Messer, Armstrong, and Singer discuss their experiments with fluoride elsewhere in this symposium (18) and Milne reports on some additional work from our laboratory (21). With respect to a possible physiological function I may mention that fluoride may be activating the adenyl-cyclase system, i.e., the enzyme which makes cyclic AMP, as has been observed *in vitro* (22). Even though the levels of fluoride needed *in vitro* are rather unphysiological, it seems possible that at the site of action, the cell membrane, fluoride levels are maintained which are sufficient for this activity.

Table 7. Growth effect of dietary fluoride in rats under trace-element-controlled conditions[1]

Supplemented fluorine (μg F/100 g diet)	No. of rats	Average daily weight gain	Average increase over controls (%)	p value
0	19	1.04 ± 0.07[2]		
100	17	1.22 ± 0.09	17.3	
250	17	1.36 ± 0.09	30.8	< 0.01
750	19	1.33 ± 0.07	27.9	< 0.01

[1]Data are pooled results of three successive 26-day experiments; fluorine was supplied as KF.

[2]Mean ± s.e.

Table 8. Fluorine

Chemistry	Atomic number 9, atomic weight 19 Highly reactive halogen, Group VIIa 0.065% of earth's crust (more abundant than Cl), but only 1–1.5 ppm in sea water May function as ligand to transition element or in relation to phosphate (fluoroapatite)
Requirement	2.5 ppm, as F^-, in diet for optimal growth and normal incisor pigmentation Potency of other F compounds not known
Occurrence	Abundant in food and feeds; also supplied by drinking water Blood plasma 0.1 ppm, homeostatically controlled Liver, heart, kidney, brain, etc., 2–5 ppm of dry weight Bone (storage organ) and teeth, 100–600 ppm, easily reaching 1000–7000 ppm; embyro and newborn (vertebrae and ribs), 50–150 ppm Transfer and accumulation of F^- by placenta
Human intake	'Available' daily intake of 4.4 mg in adult men, as compared with 25 μg in rat Normal animal diets supplying large excess, from 20–100 ppm
Biological effects	Growth and general development, dental caries prevention, maintenance of normal skeleton Stimulating growth in tissue culture Activating citrulline synthesis by liver, and enzymatic decomposition of nitromethane Activating adenyl cyclase (mediator of hormone effects) Required for maintenance of fertility in female mice

SILICON

The last, but not the least, element identified by us as essential is silicon, except for some findings on lead presented below. Our work on silicon was independently and almost immediately confirmed by Carlisle who reports separately on her research elsewhere in this symposium (23). Silicon is second only to oxygen as the most abundant element in the biosphere. It is present not only in sand, rocks, clay, dust, glass, and water, but also in foods and feeds. In fact, it was one of the first elements to which we directed our attention, since the plastic maintenance system excludes opportunities for contact with silicon afforded by conventional methods of animal experimentation. In 1965, during our early phase of work on this element, some silicon was found in the synthetic diet, primarily in salts and crystalline amino acids such as methionine and histidine. Significant growth responses were seen when silicon was added to the ration in relatively large amounts. At first it was suspected that the effects were caused by trace-element contaminants in the silicate supplement. Having identified tin, vanadium, and fluorine as requirements for the rat, we returned in 1971 to silicon. The degree of purity of our diets had been much improved in the meantime.

Very significant growth effects are produced by sodium silicate, but relative to other trace elements the amounts of silicon needed are large. Fifty mg Si/100 g diet, given as metasilicate ($Na_2SiO_3 \cdot 9H_2O$), enhance growth by 25–34% (Table 9). Metasilicate may be poorly retained because it is highly soluble, easily absorbed, and rapidly excreted. Screening of other silicon compounds has led to substances which are five to ten times as effective per atom of silicon (to be published).

Silicon has physical and chemical properties which seem to preclude its active participation in metabolism (Table 10). One is inclined to assume that it may be used as a structural element, even though a catalytic function in intermediary metabolism should not be ruled out at this time. We have indeed discovered a highly important structural function for the element.

The amounts of silicon reported in biological materials over the past 100 years of research vary greatly. A large number of results (24, 25) obtained before the development of suitable methods (26, 27) and the advent of plastic laboratory ware are much higher than those found more recently (28, 29). Many of the older data must be discarded. Silicon levels in blood are very low. They are similar to those in seawater. Values reported differ, the most careful analysts quoting 1–5 ppm. The amounts of silicon in parenchymal

Table 9. Growth effect of dietary silicon in rats on low silicon diet

	No. of animals	Average daily weight gain (g)	Average increase over controls (%)	*p* Value
Control	12[1]	1.51 ± 0.11[2]		
Si supplemented[3]	11	2.02 ± 0.06	33.8	<0.005

[1]Summary of two experiments.

[2]Mean ± s.e.

[3]50 mg % Si, as $Na_2SiO_3 \cdot 9H_2O$

Table 10. Silicon

Chemistry	Atomic number 14, atomic weight 28.1 Group IV element (with C, Ge, Sn, and Pb) Tetravalent; chemistry determined by strong affinity to oxygen Si–O–Si (siloxane) very stable; Si–C bond also stable; Si–Si and Si–H bonds less so; SiO_2 very stable Silicic acid: strong tendency to colloid formation (gel)
Requirement	50 mg % in the diet as $Na_2SiO_3 \cdot 9\ H_2O$ Other forms possibly more potent Necessary for growth and prevention of bone deformation
Occurrence	Second only to O_2 in natural abundance Present in all living beings Blood level approximately 1 ppm (5 ppm?) Liver, muscle, lung, and brain 2–20 ppm Much higher in connective tissue, bone (region of active calcification), and skin
Human intake	Not determined; much of intake possibly not biologically available Clay eaters (geophagy; pregnancy)
Biological effects	Structural, but possibly also 'matrix' for formation of organic compounds Cross linking agent, contributing to stability of acid mucopolysaccharides and other connective-tissue components Forms stable ring structures or networks (Si–O–Si) Catalytic effects on peptides

organs are also comparatively small, with the exception of kidneys, while those in connective tissue are generally high. In parenchymal organs such as liver, lung, muscle, and brain, 2–20 ppm are reported. Results reported before the use of plastic equipment are very much higher. The silicon content of cartilage, ligaments, skin, and other connective tissue is much higher, with values around 100 ppm or more. The amount of silicon in bone may depend on its age. Carlisle has shown by electron probe microanalysis that up to 0.5% of silicon can be found in a very narrow region of active calcification, i.e., in the area of active growth in young bone (30).

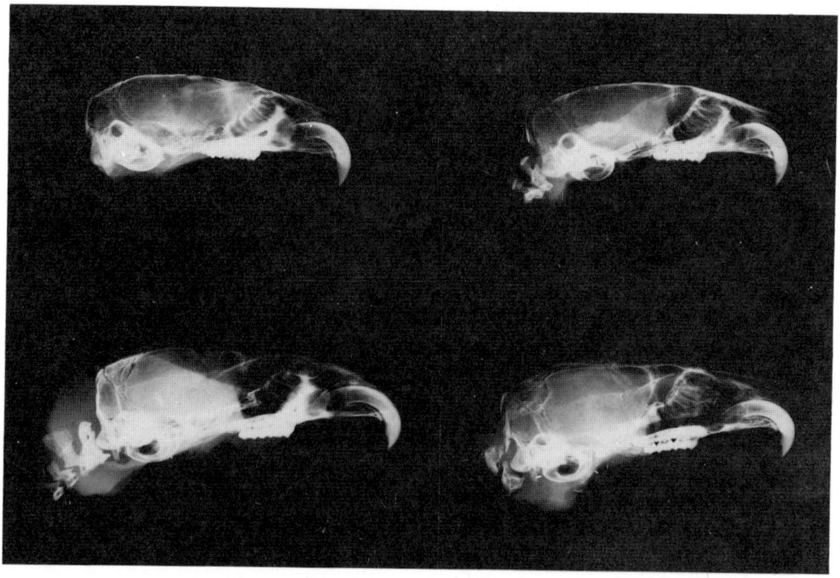

Fig. 3. Skull deformation in silicon-deficient rats. Upper, without silicon; lower, littermates on same diet supplemented with 50 mg % silicon. (Grenz-ray pictures by courtesy of Dr. R. Sognnaes, School of Dentistry, University of California at Los Angeles.)

Silicon deficiency in growing rats leads to a disturbance in the development of bone structures, as shown by grenz-ray pictures of the skull (Fig. 3). Considerable differences in skull size and bone architecture are evident between litter mates raised with and without silicon supplements. Both the distribution of silicon in the organism and the deformities in silicon-deficient rats and chicks suggested that the element is specifically associated with ground-matrix and structural elements of connective tissue, cartilage, skin, and bone.

SILICON AS A CROSS-LINKING AGENT IN CONNECTIVE TISSUE

In connective tissue silicon is not present in free form, as silicate or oligosilicate; it is firmly bound to the organic matrix. During extensive fractionation experiments with cartilage, bovine nasal septum, and other source materials, carried out in early 1972 with the intention of identifying the binding site of silicon, we discovered that the element is primarily connected to the acid mucopolysaccharides (31).

Following the initial discovery, a large variety of glycosaminoglycans have been studied for bound silicon (Table 11). Among the samples analyzed

Table 11. Free, total, and bound Si in glycosaminoglycans, polyuronides, and some glycans

Substance and source	Si (μg/g) Free	Total	Bound (total minus free)
Glycosaminoglycans			
Hyaluronic acid			
(a) Human umbilical cord	25	354	329
(b) Human umbilical cord	1533	1892	359
(c) Bovine vitreous humor	980	949	
Chondroitin 4-sulfate			
(d) Notocord of rock sturgeon	44	598	554
(e) Rat costal cartilage	30	361	331
Chondroitin 6-sulfate			
(f) Human umbilical cord	45	123	78
(g) Human cartilage	36	227	191
(h) Sturgeon cartilage	64	121	57
Dermatan sulfate			
(i) Hog mucosal tissue	46	548	502
Heparan sulfate			
(j) Beef lung	39	466	427
Heparin			
(k) Hog mucosal tissue	33	175	142
Keratan sulfate-1			
(l) Bovine cornea	31	37	
Keratan sulfate-2			
(m) Human costal cartilage	37	105	68
Polyuronides			
Pectin			
(n) Citrus fruit	5	2586	2581
Alginic acid			
(o) Horsetail kelp		43	
(p) Horsetail kelp	5	456	451
Polyglycans			
Glycogen			
(q) Rabbit liver	8	34	26
Starch			
(r) Corn		22	
Dextran			
(s) *Leuconostoc mesenteroides*	19	22	
Inulin			
(t) Dahlia tubers	15	29	14

were a number of Standard References Samples, prepared to serve as standards for research on connective-tissue polysaccharides. Since strong alkali or acid tends to liberate silicic acid, the amount of silicon found in various specimens may be dependent on the method of preparation. High levels of bound silicon, amounting to 330–554 ppm, were detected in purified hyaluronic acid from umbilical cord, chondroitin 4-sulfate, dermatan sulfate, and heparan sulfate. These amounts correspond to 1 atom of silicon per 50,000–85,000 molecular weight or 130–280 disaccharide repeating units. Recently high molecular hyaluronic acids have been found which contain up to 2000 ppm of the element. Lesser amounts, 57–191 ppm, occur in chondroitin 6-sulfate, heparan, and keratan sulfate-2 from cartilage, while hyaluronic acids from vitreous humor and keratan sulfate-1 from cornea were silicon free. High levels of bound silicon are also present in pectin (2580 ppm) and alginic acid (451 ppm). Four polyglycans, glycogen, starch, dextran, and inulin, contained only negligible amounts of silicon, most probably as silicate. All had been manufactured under conditions that avoided exposure to extreme pH conditions. This does not exclude the possibility that starch from other plant species or other storage forms of carbohydrate from elsewhere contain silicon as a structural entity.

Numerous experiments were carried out to characterize the form in which silicon may be present. Bound silicon is not dialyzable, does not react with ammonium molybdate, is not liberated by autoclaving or 8 M urea, and is stable against weak alkali and acid. Strong alkali and acid hydrolyze the silicon-polysaccharide bond. Free, direct-reacting, dialyzable silicate is obtained. Enzymatic hydrolysis of hyaluronic acid or pectin does not liberate silicic acid, but leads to products of low molecular weight still containing silicon in bound form (31).

All data available thus far indicate that the silicon atom is bound over oxygen to the carbon skeleton of the mucopolysaccharides. The bridge between silicon and the mucopolysaccharide would be Si–O–C. Silicon is thus covalently linked to the polysaccharides as a silanolate, an ether- or ester-like derivative of silicic acid. Orthosilicic acid, $Si(OH)_4$, could connect maximally four different binding sites to each other. Two different molecules, or two binding sites within the same macromolecule, would be connected over R_1–O–Si–O–R_2 bridges. Since silicon atoms have a strong tendency to attach over oxygen to each other, two silicon-containing macromolecules could be linked by an oxygen bridge, forming an R_1–O–Si–O–Si–O–R_2 grouping. In this case the two silicon atoms could carry four hydroxyl groups provided they are stereochemically prevented from associating with other silicon atoms. An alternative would be the binding of four additional radicals by way of oxygen bridges.

A large variety of structural, highly organized arrangements between bundles of mucopolysaccharides or proteins are possible in this way. Interlacing structures of this nature, based on the tendency of Si–O to form polymers or ring systems, could hold elements of the ground substance in connective tissue and elements of membranes together in an organized fashion, contributing to the architecture and resilience of connective tissue and membranes. From a physicochemical point of view, this function of silicon makes emminently good sense. The Si–O–C bridge is thermodynamically very stable. The silicon atom is small; it has the same stereochemistry as carbon, but is known to be much more rigid. It neither bends its bond angles nor does it undergo stereochemical conversions as easily as carbon. "Minimization of motion and angle distortion for nonreacting groups is an important factor in organosilicon mechanism" (32).

More recent studies have shown that the proteins in connective tissue, notably collagen, also contain bound silicate (to be published). Silicon, as an organic derivative of silicic acid, therefore, emerges as a cross-linking agent of primary importance for the structure and integrity of connective tissue. The use of silicic acid as a cross-linking agent is not limited to the animal kingdom. Analysis of pectin and other polyuronides shows that plants employ the same principle.

To our knowledge, the occurrence of bound silicate as a structural component in acid mucopolysaccharides has not been mentioned previously even though the literature on these compounds is very extensive. The finding seems to open up a number of fascinating possibilities in biology, biochemistry, pathophysiology, medical research, and therapy.

A POSSIBLE EFFECT OF LEAD ON GROWTH

Past experience has shown that toxicity of an element cannot be used as an argument against the possibility that it may be essential. Typical examples of elements which can be highly toxic but are essential are iron, iodine, selenium, and fluorine. It is noteworthy that the element lead fulfills most of the postulates which can be used to characterize potentially essential trace elements. It is present in the biosphere and is normally found in the organism and in tissues; it is present in egg, milk, and in the newborn, the latter indicating placental transfer; there may be a mechanism to maintain constant levels of lead in blood plasma, but not the erythrocytes; the element is present at 'physiological levels' in normal diets; indeed it is ubiquitous in almost all natural materials of the plant and animal kingdom. The obvious question, whether or not lead is essential, is therefore an important one.

Table 12. Growth effect of lead subacetate in ultraclean environment system[1]

Dose level (ppm)	No. of experiments	No. of animals	Growth (g/day)		Growth effect (% increase)	Significance (p value)
			Control	Supplemented		
0.5	2	12	1.72 ± 0.09	1.72 ± 0.10	0	
1.0	3	23	1.79 ± 0.05	2.08 ± 0.05	16.2	< 0.005
2.5	8	49	1.85 ± 0.04	2.04 ± 0.04	10.3	< 0.005
5.0	1	7	1.53 ± 0.08	1.61 ± 0.10	5.2	n.s.
7.5	1	8	2.11 ± 0.12	2.22 ± 0.09	5.2	n.s.

[1]Preliminary summary of data, June 1973.

In order to work with lead a new filtering system has been developed which removes not only dust particles from the air but also aerosols and substances which are present as vapors. The unit consists of a carbon cartridge, followed by a fine Millipore filter, and a pump. We have not yet succeeded in making the diet as lead deficient as we wish; it still supplies probably 0.2 ppm of the element. If 1.0–2.5 ppm of lead in the form of lead subacetate are added, rather consistent positive growth responses are seen (Table 12). The effects are statistically significant, even though they amount on the average to less than 20% of an increase in growth. The latter fact may be related to the lead supply in the basal diet. Not only lead subacetate, but also lead oxide and lead nitrate produce a response.

Some of our preliminary findings have recently been released by the Environmental Protection Agency, which solicited opinions about the "Proposed Regulations for Lead Reduction Schedule and The Health Rationale for this Action." Their rationale was that lead is a toxic substance for which no beneficial role has been demonstrated. Lead does not contribute to the formation of smog; it enhances combustion and reduces hydrocarbon emission. Since the health argument against lead seems to have been made by implication only, it appeared necessary to inform the agency of our results. Our data thus far show that lead in small amounts may have an essential function, but more clear-cut experimental evidence is needed before a definitive commitment on this point can be made.

OUTLOOK

As to the future of research on new essential trace elements, I want to repeat the statement made at Aberdeen that, for any success, the availability of an accurate and highly sensitive method for the determination of each element in question is of paramount importance. Spark-source mass spectroscopy offers some hope in this respect.

With tin, vanadium, fluorine and silicon, growth rates amount to approximately 80% of those of conventional controls. Without these elements they were only 40% four years ago. Effects of nickel are not seen, most likely since it is present in sufficient amounts. At the present level of refinement of our experimentation, there are still one or two trace elements missing, including the one that is responsible for the loss of hair in rats under our conditions. Attempts have been made to concentrate the missing elements from yeast ash by inorganic fractionation techniques. A concentrate can be prepared which promotes growth strongly at a level of 15 mg per 100 g of diet. Emission spectroscopy of the active sample indicated the presence of lead, cadmium, silver, and barium, among others (Table 12). None of these could account for

Table 13. Comparison of spark-source mass spectrometry to semiquantitative emission spectroscopy (trace element concentrate Y 177)

Element	Emission spectro-scopy (ppm)	Mass spectro-metry (ppm)	Element	Emission spectro-scopy (ppm)	Mass spectro-metry (ppm)
Uranium		0.60	Ruthenium		<0.1
Thorium		1.1	Molybdenum		4.8
Bismuth		0.04	Niobium		<0.1
Lead	160	17	Zirconium		1.3
Thallium		0.60	Yttrium		5.4
Mercury		<0.1	Strontium		31
Gold		0.13	Rubidium		39
Platinum		<0.1	Bromine		3.0
Iridium		<0.1	Selenium		1.2
Osmium		<0.1	Arsenic		<0.1
Rhenium		<0.1	Germanium		<0.1
Tungsten		0.67	Gallium		<0.1
Tantalum		29	Zinc	2900	230
Hafnium		<0.1	Copper	3500	10000
Lutecium		<0.1	Nickel	25	43
Ytterbium		<0.1	Cobalt	43	3.7

Thulium		< 0.1	Iron	900	9800
Erbium		< 0.1	Manganese	110	180
Holmium		< 0.1	Chromium	2.4	21
Dysprosium		< 0.1	Vanadium		1.4
Terbium		< 0.1	Titanium		60
Gadolinium		< 0.1	Scandium		< 0.01
Europium		0.06	Calcium	13000	120000
Samarium		0.34	Potassium	290000	190000
Neodymium		5.5	Chlorine		2500
Praseodymium		1.8	Sulphur		2700
Cerium		7.1	Phosphorus	260000	300000
Lanthanum		2.3	Silicon	500	6500
Barium		53	Aluminum	30	270
Cesium		0.29	Magnesium	2700	18000
Iodine		0.55	Sodium	21000	6000
Tellurium		< 0.1	Fluorine		6000
Antimony		1.8	Oxygen		
Tin		1.5	Nitrogen		
Indium		< 0.1	Carbon		
Cadmium	410	9.0	Boron	110	92
Silver	340	210	Beryllium		
Palladium		< 0.1	Lithium		2.3
Rhodium		< 0.1			

the growth effects observed. Submission of the same sample to spark-source mass spectrometry produced a much more enlightening picture (Table 13, right column). Data were obtained which covered most of the periodic system. The most valuable feature of the method is that it can be made to produce *absolute* values for most elements. Through isotope spiking with a stable mass isotope and comparison of different isotope peaks one can determine how much of the element was present in the analyzed sample.

Several severe obstacles remain and need to be eliminated. One is the so-called sampling error, i.e., the error introduced by our inability to produce an absolutely homogenous, truly representative specimen for the analysis. Another problem to be solved is the ashing process. Here we are badly in need of new ideas and new techniques, the development of which may be just as difficult as that of the analytical methods themselves.

A review of the periodic system reflecting the present state of knowledge (Fig. 4) presents a picture quite different from that seen in 1969 (1).

(a) All essential trace elements known at that time had higher atomic weights than the 11 elements which form the bulk of living matter; the latter belong to the lowest 20 elements of the periodic system. This separation is not valid now since both fluorine and silicon have atomic numbers below 20.

(b) Through the addition of vanadium and nickel, in the first series of transition elements a continuous set of eight elements (nos. 23–30) has arisen. All of them are essential for warm-blooded animals. This series constitutes an area of special emphasis, a point of gravity in the periodic system around which trace element requirements and functions are centered.

(c) The observation, in 1969, that none of the 39 elements beyond iodine (atomic number 53) have been shown to be of physiological significance for animals is still valid, but the emergence of lead as a potentially essential trace element may show that certain elements with higher atomic weights can function in intermediary metabolism similar to those with lower atomic weights.

There are now 14 trace elements for which essential functions in warm-blooded animals have been demonstrated conclusively. According to Fig. 4, 20 additional elements should be taken into serious consideration as potential pretenders for trace-element function. Prospects for the investigation of many of these are good with presently available methods. The demonstration of additional requirements is therefore most likely. However, the number of such elements may be small. Titanium is most attractive at the moment because it is located at the end of the series of eight essential transition metals mentioned above. Titanium, however, is difficult to work with because of its abundance. Our purified amino acid diet still contains up to 2 ppm of the element.

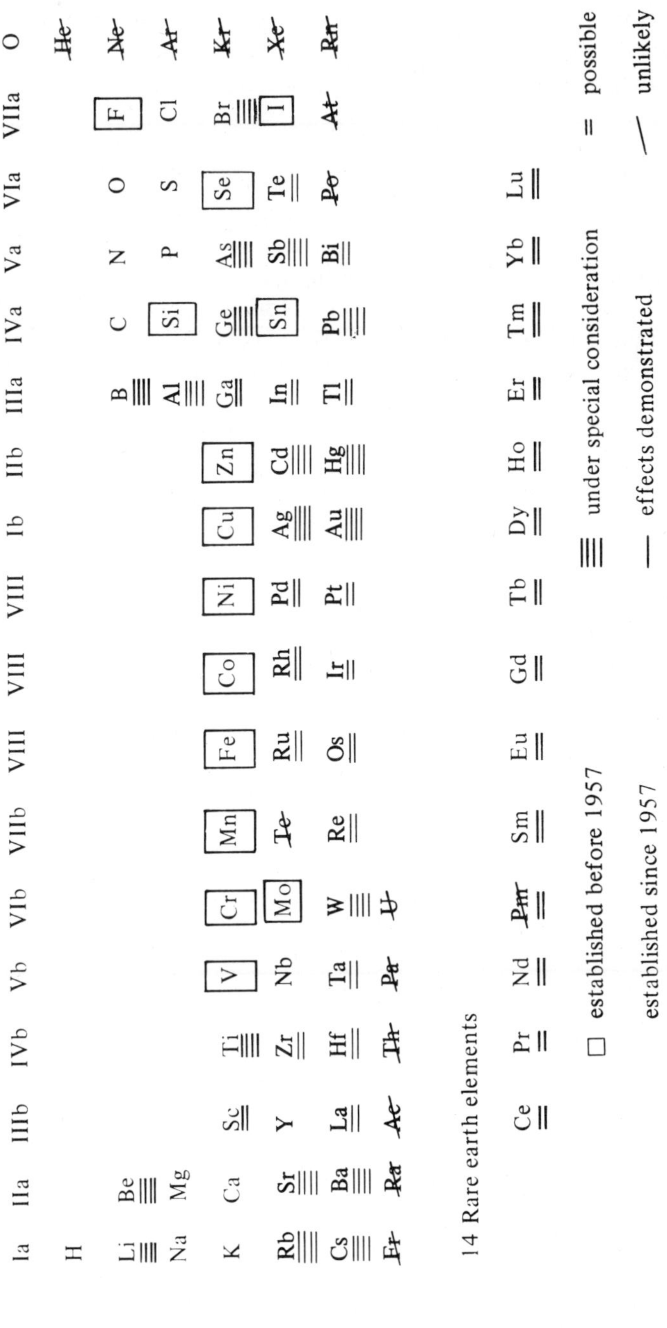

Fig. 4. Periodic system distribution of trace elements of known and potential importance for warm-blooded animals. Status as of 1973.

The development of this field of research has contributed to the emergence of *bioinorganic* chemistry, a new category in biological, chemical, and medical research. Few realize that bioinorganic chemistry is just as complex as organic biochemistry, and in some respects technically more complicated.

One may ask what could be the lowest possible limit of essential trace-element function for warm-blooded animals and other living beings. It can be calculated that one may go down by four to five orders of magnitude before coming to the threshold where only one atom of a specific element would be present for one specific gene in the chromosomes of a cell. I could conceive of this as the ultimate limit for a biological trace-element function.

As we approach lower and lower levels of trace-element concentration, new questions and new problems arise. For instance, homogeneity of the diet becomes an issue, sampling becomes a problem, and so does contamination from the environment. It takes one particle of dust to impart nanogram quantities of lead, cadmium, titanium, etc., to a diet or a sample. Ultraclean techniques may become mandatory not only in the animal room, but also in the analytical laboratory, the diet preparation area, and the purification of dietary ingredients. This should not discourage anyone from working in this field because much can be done with much simpler devices, as our own results show. The potential rewards can be great since each element proven to be essential opens up a new facet in biochemistry and nutrition.

REFERENCES

1. Schwarz, K. 1970. Control of environmental conditions in trace element research: an experimental approach to unrecognized trace element requirements. *In* C. F. Mills (ed.), *Trace Element Metabolism in Animals,* Livingstone, Edinburgh, p. 25.
2. Schwarz, K., Milne, D. B., and Vinyard, E. 1970. Growth effect of tin compounds in rats maintained in a trace element-controlled environment. *Biochem. Biophys. Comm.* 40: 22.
3. Schwarz, K., and Milne, D. B. 1971. Growth effects of vanadium in the rat. *Science* 174: 426.
4. Schwarz, K., and Milne, D. B. 1972. Fluorine requirement for growth in the rat. *Bioinorg. Chem.* 1: 331.
5. Schwarz, K., and Milne, D. B. 1972. Growth promoting effects of silicon in rats. *Nature, Lond.* 239: 333.
6. Nielsen, F. H. 1973. Essentiality and Function of Nickel. *Fed. Proc.*, in press.
7. Nielsen, F. H., and Ollerich, D. A. 1973. Nickel: a new essential trace element. This symposium.
8. Smith, C. J., and Schwarz, K. 1967. A controlled environment system for trace element deficiencies. *J. Nutr.* 93: 182.
9. Schwarz, K. 1972. Elements newly identified as essential for animals. *In Nuclear Activation Techniques in the Life Sciences.* I.A.E.A., Vienna, p. 3.

10. Milne, D. B., Schwarz, K., and Sognnaes, R. 1972. Effect of newer essential trace elements in rat incisor pigmentation. *Fed. Proc.* 31: 700.
11. Schwarz, K. 1973. Recent dietary trace element research, exemplified by tin, fluorine, and silicon. *Fed. Proc.*, in press.
12. Schroeder, H. A., Balassa, J. J., and Tipton, I. H. 1964. Abnormal trace metals in man: tin. *J. Chron. Dis.* 17: 483.
13. Miller, S. A. 1970. Nutrition in the neonatal development of protein metabolism. *Fed. Proc.* 29: 1497.
14. Hopkins, L. L., Jr., and Mohr, H. E. 1971. Essentiality of vanadium. *In* W. Mertz (ed.), *Newer Trace Elements in Nutrition,* Marcel Dekker, New York, p. 195.
15. Nielsen, F. H. 1974. Newer trace elements necessary for nutritional needs. *Food Technol.*, in press.
16. Hopkins, L. L. 1973. Essentiality and function of vanadium. This symposium.
17. Messer, H., Armstrong, W. D., and Singer, L. 1972. Fertility impairment in mice on a low fluoride intake. *Science* 177: 893.
18. Messer, H. 1973. Essentiality and function of fluoride. This symposium.
19. Maurer, R. L., and Day, H. G. 1957. The non-essentiality of fluorine in nutrition. *J. Nutr.* 62: 561.
20. Doberenz, A. R., Kurnick, A. A., Kurtz, E. B., Kemmerer, A. R., and Reid, B. L. 1964. Effect of a minimal fluoride diet on rats. *Proc. Soc. Exp. Biol.* 117: 689.
21. Milne, D. B., and Schwarz, K. 1973. Effect of different fluorine compounds on growth and bone fluoride levels in rats. This symposium.
22. Drummond, G. I., Severson, D. L., and Duncan, L. 1971. Adenyl cyclase–kinetic properties and nature of fluoride and hormone stimulation. *J. Biol. Chem.* 246: 4166.
23. Carlisle, E. M. 1973. Essentiality of silicon. This symposium.
24. King, E., and Belt, T. 1938. The physiological and pathological aspects of silica. *Phys. Rev.* 18: 329.
25. Holt, P., and Yates, D. 1953. Tissue silicon: a study of the ethanol-soluble fraction, using ^{31}Si. *Biochem. J.* 54: 300.
26. King, E., Stacy, B., Holt, P., Yates, D., and Pickles, D. 1955. The colorimetric determination of silicon in the micro-analysis of biological material and mineral dusts. *Analyst* 80: 441.
27. McGavack, T., Leslie, J., and Tang Kao, K. 1962. Silicon in biological material. I. Determinations eliminating silicon as a contaminant. *Proc. Soc. Exp. Biol. Med.* 110: 215.
28. Fregert, S. 1959. Studies on silicon in tissues with special reference to skin. *Acta Dermato-Venereol.* 39 (suppl. 42): 1.
29. Leslie, J., Tang Kao, K., and McGavack, T. 1962. Silicon in biological material. II. Variations in silicon contents in tissues of rat at different ages. *Proc. Soc. Exp. Biol. Med.* 110: 218.
30. Carlisle, E. M. 1970. Silicon: a possible factor in bone calcification. *Science* 167: 279.
31. Schwarz, K. 1973. A bound form of silicon in glycosaminoglycans and polyuronides. *Proc. Natn. Acad. Sci. U.S.A.* 70: 1608.

32. Sommer, L. H. 1965. *Stereochemistry, Mechanism and Silicon.* McGraw-Hill, New York.

DISCUSSION

Reussner (White Plains). Do you know what the silicon content of the semisynthetic diets used by most researchers would be?

Schwarz. It is rather high. We had to take special precautions to eliminate silicon in the preparation of our diets. Silicon is everywhere; you have to exclude glass from your entire laboratory and do everything in plastics that have been carefully screened. For instance, we can pick up silicon from dialysis tubing or ion exchange resin. Methods of analysis are exceedingly sensitive, and we can accurately determine 3 ng of silicon per sample without trying to make a micromethod out of it.

Singer (Minneapolis). Was the diet that you used for the fluorine studies sufficient in vanadium, tin, and silicon?

Schwarz. It was sufficient in tin and vanadium, but not always in silicon. Dr. Milne will discuss this later in the symposium.

Parizek (Prague). In regard to the high content of silicon in the basal membrane of the glomerular wall, have you found any effect on kidney function in animals on the low-silicon diet?

Schwarz. No, we have not, and have done very little on the histology of the deficiency. Dr. Carlisle independently is probably doing quite a bit of histology and she will probably refer to it in her talk.

Mills (Aberdeen). I am somewhat concerned about the possible practical significance of some of the newer trace elements, and wonder if we may not in fact meet situations where these deficiencies are of practical importance. If you go back 20 or 30 years, no one would have believed that most of our copper deficiencies were caused, not by a shortage of the element in the diet, but by the presence of agents that modify copper uptake. We tend to start out by identifying an essential element, and then study the effect of an antagonist. What we often fail to do is to recognize that this essential element may be present in our diet in a highly active biopotent form, and we seldom consider the effect of a similar biopotent form of the antagonist. In this sense I do feel that there may be dietary situations arising where, if we consider the natural biological form of these elements, we may meet a conditioning agent against tin, nickel, etc., that will drastically alter its utilization.

Schwarz. That may well be the explanation for differences in results with respect to fluoride. The low-fluoride diets which others have used are certainly very different in composition from ours. From the results of Dr. Spencer and others, including Dr. Milne and myself, it is clear that there are a variety of dietary ingredients, some of them essential, which interfere with fluoride utilization. Magnesium is one of them, calcium phosphate another, and we found that free silicate would do this. Therefore it may be that other workers didn't see effects because of other factors related to their experimental situation which we were lucky enough not to have.

ESSENTIALITY AND FUNCTION OF NICKEL[1]

F. H. NIELSEN

USDA, ARS, Human Nutrition Laboratory, Grand Forks, North Dakota

Until recently, only indirect evidence has been available to suggest that nickel has an essential physiological role in animals. The data have been obtained from pharmacologic, toxicologic, and *in vitro* biochemical studies. Other evidence has been obtained from analyses which indicate that nickel is consistently present in some biological materials, and which show changes in tissue nickel distribution, or concentrations, in some pathological conditions. Since this evidence has been recently reviewed[2], it will not be discussed here. Instead, evidence which directly shows that nickel has an essential function, or functions, will be presented.

ESSENTIALITY OF NICKEL

Nickel is an essential nutrient for animals. The following criteria for essentiality have been met:

(1) It has a low molecular weight and is a transition element which forms chelates, and thus is chemically suitable for biological functions.

(2) It is ubiquitous on the earth's crust and in sea water; therefore, it has been generally available to plants and animals during their evolution.

(3) It is present in plants and animals.

(4) It is nontoxic to animals orally except in astringent doses.

(5) Homeostatic mechanisms are implied by serum levels, excretion rates, and lack of excessive accumulation.

1. Mention of a proprietary product does not necessarily imply endorsement by the United States Department of Agriculture.

2. F. H. Nielsen and D. A. Ollerich. 1973. Nickel: A new essential trace element. Presented at the 57th Annual Federation of American Societies for Experimental Biology Meeting, Atlantic City, N.J. Submitted for publication in *Federation Proceedings.*

(6) Its deficiency reproducibly results in an impairment of a function, or functions, from optimal to suboptimal.

The first four of these criteria have been adequately reviewed (1–3). The last two criteria have been studied more extensively recently, and therefore will be discussed in more detail.

Nickel is poorly absorbed from ordinary diets and is excreted mostly in the feces. This is apparent from studies with dogs (4) and man (5, 6). Nodiya (5) performed nickel balance studies on 10 Russian males, age 17 years, who were ingesting a mean of 289 ± 23 μg Ni/day (range, 251–309). He found that fecal excretion of nickel averaged 258 ± 23 μg/day (range, 219–278). Horak and Sunderman (6) measured fecal nickel excretion in 10 healthy subjects (age 22–65; four males, six females) who ingested varied diets which were prepared in their own homes. They found that the fecal excretion of nickel averaged 258 ± 126 μg/day (range, 80–540). Most of this nickel was probably that which was not absorbed; however, nickel has been found in the bile of rats (7), so some fecal nickel may come from the bile. Urinary nickel excretion is 10–100 times less than fecal excretion. For example, Perry and Perry (8) found that 24 healthy adults excreted an average of 20 μg/day in the urine, whereas Sunderman (9) noted a mean daily urinary excretion of 19.8 μg in 17 normal adults. In another study, Nomoto and Sunderman (10) found that 26 healthy subjects had a mean daily urinary excretion of 2.4 μg/day. In addition to the feces and urine, there is evidence that sweat may be important in nickel homeostasis. Horak and Sunderman (6) found a mean concentration of 49 μg Ni/liter in sweat collected in plastic bags which encased the arms of five healthy men during sauna bathing. Consolazio *et al.* (11) determined that approximately 8.3 μg Ni is lost in the sweat daily. The concentration of nickel in sweat is approximately 20 times greater than the concentration in serum samples of healthy adults, suggesting that there is active secretion of nickel by the sweat glands. From the previous discussion it is apparent that the absorption of dietary nickel is low, probably in the range of 1–10%, and that fecal excretion is the major route for elimination of ingested nickel from the body. Nickel that is absorbed is apparently lost in the bile, urine, and sweat.

Analyses of tissues indicate that the retained nickel is widely distributed in very low concentrations in the body (2, 12–16). Sunderman, Decsy, and McNeely (17) have found wide variability in the mean concentrations of nickel in sera from several species of animals, but within each species the concentration of serum nickel fell within a relatively narrow range. Examples of mean concentrations and ranges (in μg Ni/liter) include the following: man, 2.6 (1.1–4.6); rats 2.7 (0.9–4.1); chickens, 3.6 (3.3–3.8); rabbits, 9.3 (6.5–14.0). In the serum, nickel exists in three forms: ultrafiltrable, albumin bound, and as a nickel-metalloprotein (17). The amount of nickel in each compartment varies from species to species and this may be due, in part, to species variation in the affinities of albumin for nickel (18, 19). A metalloprotein designated nickeloplasmin has been isolated from the serum of rabbits and man[3] (20). It is a macroglobulin with an estimated molecular weight of 7.0×10^5 and contains approximately 0.8 g atom Ni/mole. Disc gel and immunoelectrophoresis show that purified nickeloplasmin is an α-1 macroglobulin in rabbit serum and an α-2 macroglobulin in man. Soestbergen and Sunderman (21) have found that the ultrafiltrable nickel in serum does not exist primarily as free Ni^{2+}, but as nickel complexes, and that ultrafiltrable nickel receptors play an important physiological role in nickel homeostasis by serving as diffusible vehicles for the extracellular transport and renal excretion of nickel. Further evidence that nickel as an organic complex is important in metabolism is that Tiffin (22) has found that the translocatable form of nickel in plants appears to be a stable anionic amino acid complex. The bioavailable form of nickel is not known, but since Sunderman[4] has found that nickeloplasmin cannot be labeled by feeding $^{63}NiCl_2$, it is possible that nickel is essential as an organic complex, and that inorganic forms of nickel can fulfill the requirement for nickel only when given in large amounts.

Studies which show a consistent impairment in a physiological function when nickel is deficient in the diet have also recently been successful. Nielsen *et al.* (3, 23, 24) found that feeding a diet containing less than 40 ppb nickel resulted in an apparent nickel-deficiency syndrome in chicks. When compared with controls given a supplement of 3–5 ppm nickel, the deficient chicks showed the following: (1) pigmentation changes in the shank skin; (2) thicker legs with slightly swollen hocks; (3) dermatitis of the shank skin; (4) a less friable liver which may have been related to the fat content; (5) an enhanced accumulation of a tracer dose of ^{63}Ni in liver, bone, and aorta. These findings were observed under conditions which produced suboptimal growth. The abnormalities in leg structure and shank-skin dermatitis were inconsistent. Sunderman *et al.* (25) attempted to confirm Nielsen's findings by feeding a diet containing 44 ppb nickel to chicks raised in a slightly different environment. While they found no gross effects, they did observe ultrastructural changes in the liver. These included dilation of the perimitochondrial rough endoplasmic reticulum in 15–20% of the hepatocytes. In a single experiment with chicks, Leach[5] observed a growth response with nickel. Later attempts

3. F. W. Sunderman, Jr., M. I. Decsy, S. Nomoto, and M. W. Nechay. 1971. Isolation of a nickel-α_2 macroglobulin from human and rabbit serum. *Fed. Proc.* 30: 1274 (abstr.).

4. F. W. Sunderman, Jr. 1973. Personal communication.

5. R. M. Leach, 1973. Personal communication.

to obtain the growth response were unsuccessful. Wellenreiter, Ullrey, and Miller (26) fed a diet containing 80 ppb nickel to reproducing quail and saw no gross symptoms except an inconsistent positive effect of nickel on breast feathering in birds which were relatively arginine deficient but on adequate protein intake. In order to clarify and extend the above observations, improvements were made in the experimental environment used to produce nickel deficiency, and a diet was formulated with a nickel content of 3–4 ppb. With this diet and environment, it has been consistently possible to produce a nickel deficiency in animals.

A major difficulty in the production of nickel deficiency in animals is the preparation of a diet low in nickel. Nickel is ubiquitous. Therefore, the conventional methods of diet preparation using purified proteins, or amino acids, carbohydrates, vitamins, and minerals are not suitable because some contain as much as 20,000 ppb nickel. In order to circumvent this problem, the diet must be prepared from natural feedstuffs low in nickel which contain most of the amino acids, vitamins, and minerals essential for the experimental animal. The diet formulations for the production of nickel deficiency in the chick and rat are presented in Table 1. They are based on dried skim milk, ground corn, and corn oil. The added vitamins include A, D, E, K, niacin, folate, and biotin because skim milk and corn were calculated to provide insufficient amounts. The mineral additions are minimal. The major portion is $CaCo_3$ which is relatively low in nickel (approximately 20 ppb). In order to assure adequate concentrations, Mn, Fe, Zn, Cu, I, and Se are added. The corn must be acid washed to obtain consistently a diet containing 3–4 ppb nickel. In these experiments the control chicks were fed the basal diet supplemented with 3 ppm nickel as $NiCl_2 \cdot 6H_2O$.

A second impediment to the production of nickel deficiency, and also for the production of deficiencies of other trace elements required in very minute amounts, is the environment in which the animals are raised. Significant amounts of trace elements may be present in such sources as caging, feed cups, water bottles, dust in the air, and the skin of the investigator's hands. In order to prevent such contamination, it is necessary to employ techniques similar to those developed by Smith and Schwarz (27).

The experimental animals are raised in a controlled environment such as an all-plastic rigid isolator[6], or a laminar-flow animal rack[7] so that the only materials the animals come into contact with are plastic, and the air entering the system is filtered to remove dust. In order to bring out the effects of nickel deprivation in rats, successive generations were raised. Thus, the

6. Germ Free Laboratories, Inc., Miami, Florida.

7. Carworth, Division of Becton, Dickinson, & Co., New City, New York.

Table 1. Composition of the basal diets[1]

Ingredient	Diet (g/kg)	
	Chick	Rat
Skimmed-milk powder[2]	645.00	645.00
Ground corn, acid-washed[2]	142.50	160.00
Non-nutritive fiber[3]	40.00	40.00
Corn oil[2]	100.00	100.00
Glycine[4]	5.00	
Arginine[4]	25.00	25.00
Choline chloride[2]	0.30	
Vitamins, fat-soluble[5]	0.11	0.16
Vitamin mix[6]	4.59	4.84
Mineral mix[7]	37.50	25.00
	1000.00	1000.00

[1]Diet analyzed 3–4 ppb nickel on an air-dried basis.

[2]Nutritional Biochemicals Corp., Cleveland, Ohio.

[3]Solka Floc, SW-40, Brown Co., Boston, Mass.

[4]General Biochemicals Co., Chagrin Falls, Ohio.

[5]The fat-soluble vitamins were α-tocopherol acetate, 0.01g (chick) or 0.06g (rat), and vitamin D_3 (250 IU/drop), two drops or approximately 0.1g (see footnote 4 in text). These were added to corn oil prior to mixing into diet.

[6]The chick vitamin mix contained (in mg) niacin, 20, folic acid, 1.2, menadione, 0.6, biotin, 0.1, B-carotene, 12, vitamin A palmitate (250,000 IU/g), 4, and glucose, 4552.1. The rat vitamin mix contained (in mg) niacin, 20, folic acid, 1.2, menadione, 0.5, biotin, 0.1, B-carotene, 6, vitamin A palmitate (250,000 IU/g), 4, pyridoxine·HCl, 2, thiamine·HCl, 2, and glucose, 4804.2.

[7]The chick mineral mix contained (in the purest form known available) (in g) $CaCO_3$, 12.5, $MnSO_4 \cdot 5H_2O$, 0.2, iron sponge (dissolved in HCl), 0.055, $CuSO_4$, 0.015, ZnO, 0.025, KI, 0.0005, $(NH_4)_6Mo_7O_{24}$, 0.0035, Na_2SeO_3, 0.006, ground corn (acid-washed), 24.7004. The rat mineral mix contained (in g) $CaCO_3$, 7.5, $MnSO_4 \cdot 5H_2O$, 0.2, iron sponge (dissolved in HCl), 0.05 $CuSO_4$, 0.015, ZnO, 0.015, KI, 0.0005, Na_2SeO_3, 0.006, ground corn (acid-washed), 17.2189.

animals were exposed to deficiency throughout fetal, neonatal, and adult life. Day-old Golden Giant Cockerels[8] were used in the chick experiments. All other experimental materials and methods are described elsewhere[2] (3, 23, 24). Statistical analysis was by the t test (28).

After 3½ weeks, the gross appearances of the deficient and control chicks were similar except for the difference in the pigmentation of their shank skin

8. Jack Frost Chicks, Inc., St. Cloud, Minnesota.

Table 2. Liver and heart analysis of nickel-deficient and supplemented chicks[1]

Group	No. of chicks	O_2 uptake[2] (μliter/hr/mg protein)	Total lipid[3] Liver (%)	Lipid P[3] Liver (mg/g)	Total lipid[3] Heart (mg/g)	Lipid P[3] Heart (mg/g)
			Experiment 1			
Ni def (3 ppb)	12	$4.7^{4} \pm 0.2^{5}$	$6.21^{6} \pm 0.10$			
+ 3 ppm Ni	12	5.5 ± 0.2	5.78 ± 0.11			
			Experiment 2			
Ni def (4 ppb)	11	$5.4^{7} \pm 0.2$	$6.27^{6} \pm 0.18$	$1.327^{4} \pm 0.016$	$4.09^{6} \pm 0.06$	$0.942^{7} \pm 0.007$
+ 3 ppm Ni	11	6.0 ± 0.2	587 ± 0.05	1.379 ± 0.016	3.85 ± 0.11	0.898 ± 0.010
			Experiment 3			
Ni def (14 ppb)	12	$5.5^{7} \pm 0.1$	5.71 ± 0.09	1.318 ± 0.016		
+ 3 ppm	12	5.9 ± 0.1	5.55 ± 0.10	1.335 ± 0.015		

[1] In experiments 1 and 3 the corn was not acid washed.

[2] Using liver homogenates and with α-glycerophosphate as the substrate.

[3] Fresh weight basis.

[4] Significantly different ($P < 0.025$) from +3 ppm Ni group.

[5] ± s.e.m.

[6] Significantly different ($P < 0.05$) from +3 ppm Ni group.

[7] Significantly different ($P < 0.10$) from +3 ppm Ni group.

as described previously (3, 23, 24). All chicks weighed 350–400 g. In earlier studies (3, 23, 24), abnormalities in leg structure and a dermatitis were present in the deficient chicks. In the latest experiments, the abnormalities were diminished or inconsistent. It is thought that modification of the diet and improved environmental conditions may have accounted for the decreased incidence or inconsistency of these abnormalities. The other gross sign observed in earlier studies was a decrease in friability of the liver in deficient chicks. This finding was present in chicks raised under the conditions described. Thus, it appears that the change in shank-skin color and the effect on liver consistency are related to nickel status, and that the other gross signs originally described may be less characteristic of nickel deficiency.

In contrast to the gross signs, abnormalities in biochemical indices of metabolism were more consistently found in the nickel-deficient chicks. They included a decreased oxygen uptake by liver homogenates in the presence of α-glycerophosphate, an increase in liver lipids, and a decrease in liver phospholipids (Table 2). On the basis of one experiment, there appears to be an increase in the lipid and phospholipid fraction in the heart.

Ultrastructural abnormalities in the hepatocytes (Figs. 1 and 2) were also a consistent finding. These findings were similar to, though more extensive than, those described by Sunderman *et al.* (25). They included dilation of the cisterns of the rough endoplasmic reticulum and swelling of the mitochondria. The swelling was in the compartment of the matrix and was associated with fragmentation of the cristae. Other ultrastructural changes included a dilation of the perinuclear space and pyknotic nuclei. These ultrastructural abnormalities coupled with the biochemical evidence of deranged metabolism are considered sufficient evidence to indicate that nickel is essential for the chick.

The results of the rat studies are more preliminary, and more experimentation is needed to confirm the results obtained thus far. Reproduction apparently is affected, as seven first-generation nickel-deficient female rats which were mated had a significant number of dead pups (15%) compared with no mortality in the pups of six controls. Nine second-generation nickel-deficient female rats which were mated had a 19% loss of pups. This finding is confounded by the fact that the eight controls had a 10% loss of pups. This was, however, roughly half the loss in the nickel deficient group. The pups of the nickel-deficient dams also weighed slightly less at 4 and 24 (age at weaning) days than those from controls, but the differences were not significant. During the suckling stage, the nickel-deficient pups had generally a less thrifty appearance and were less active. In order to assess this last observation, a Stoelting activity monitor[9] was used to measure the activity of

9. Stoelting Co., Chicago, Illinois.

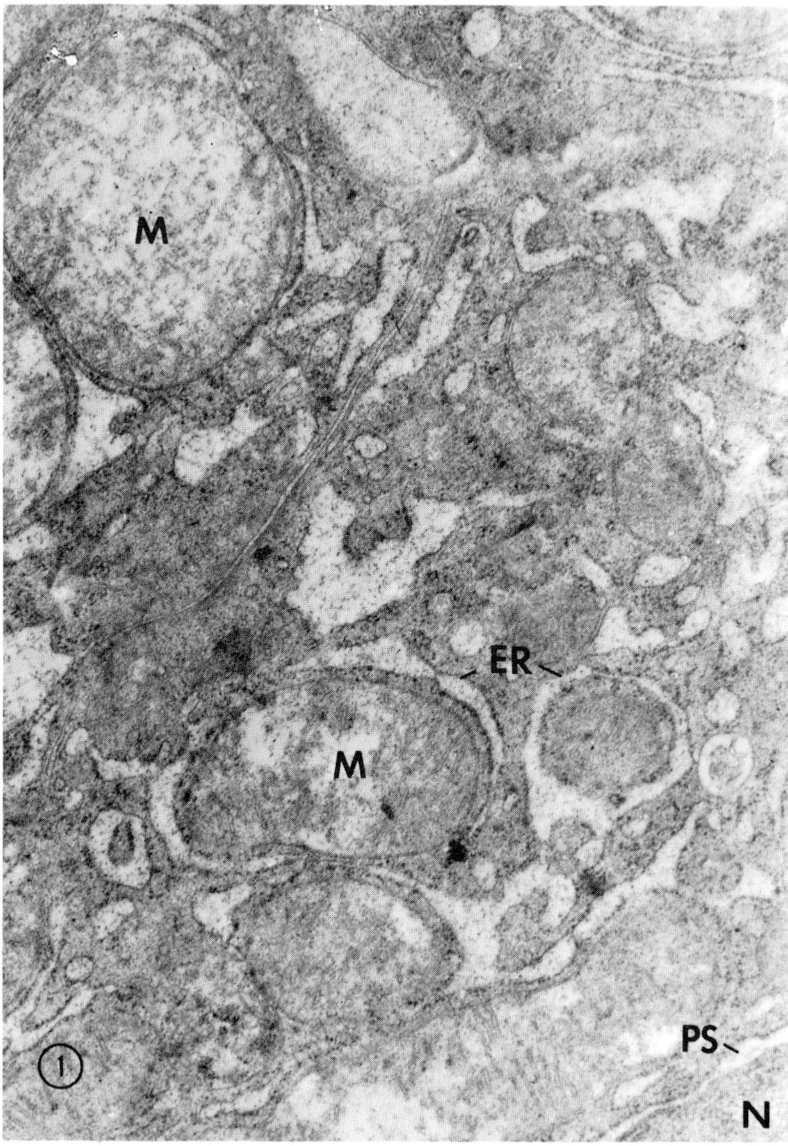

Fig. 1. Hepatic cell from a nickel-deficient chick (4 ppb nickel). Swelling of mitochondria (M) was evident in numerous hepatic cells. The swelling was in the compartment of the matrix and appeared to cause fragmentation of cristae. Note also the dilated cisternae of the rough endoplasmic reticulum (ER) and dilated perinuclear space (PS). Nucleus (N). Uranyl acetate and lead citrate. × 25,500.

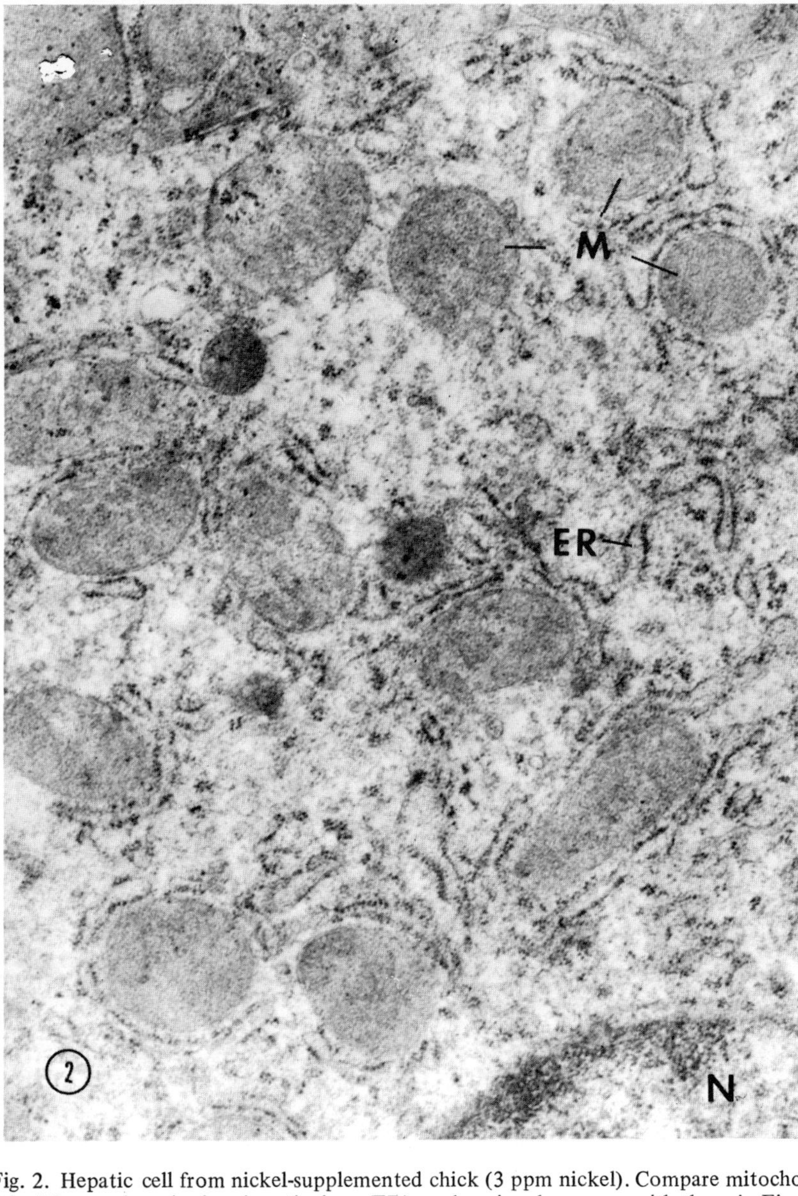

Fig. 2. Hepatic cell from nickel-supplemented chick (3 ppm nickel). Compare mitochondria (M), rough endoplasmic reticulum (ER), and perinuclear space with those in Fig. 1. Nucleus (N). Uranyl acetate and lead citrate. × 25,500.

Table 3. Oxygen uptake of rat-liver homogenates[1]

Group	No. of rats	O_2 uptake (μliter/hr/mg protein)
Ni def (4 ppb)	13	3.20[2] ± 0.08[3]
+ 3 ppm	12	4.17 ± 0.21

[1] α-glycerophosphate as substrate.
[2] Significantly different ($P < 0.001$) from +3 ppm Ni group.
[3] ± s.e.m.

matched litters of the same size from both the deficient and control groups in the second and third generation. The results indicate that the nickel-deficient rats are indeed more lethargic.

Twelve each of deficient and control second-generation male rats were killed for measurement of liver oxidative activity. Grossly, it was observed that the livers of the deficient rats had a muddy brown color (compared with a red-brown color of livers of the controls) and a less distinct substrcuture. As in the chick, the deficient rat-liver homogenates showed a reduced oxidative activity in the presence of α-glycerophosphate (Table 3). In addition, preliminary sucrose density gradients of liver post-mitochondrial supernatants were consistent with a decrease in polysomes and an increase in monosomes in the nickel-deficient rat liver.[2]

FUNCTION OF NICKEL

While the above findings indicate that nickel is essential, they provide only meager insights as to its metabolic function. An attractive hypothesis is that nickel has a role in the metabolism or structure of membranes. The swollen mitochondria suggest a possible abnormality in the mitochondrial membrane. The endoplasmic reticulum is a network of membrane-bound cavities. Dilation of the cisterns of the endoplasmic reticulum suggests a possible abnormality in these membranes. Finally, the dilation of the perinuclear space and the presence of pyknotic nuclei in the nickel-deficient liver suggest an abnormality in the nuclear membrane. The level of phospholipids, an integral part of membranes, is also depressed in nickel deficiency. Further credibility for the membrane hypothesis can be obtained from *in vitro* studies with isolated tissues. Nickel can substitute for calcium in certain steps of the excitation-contraction coupling of isolated skeletal muscle (29, 30). Other investigators have found that nickel can also substitute for calcium in the excitation process of the isolated nerve cell (31, 32). It has been postulated that nickel can substitute for calcium in the binding with a membrane ligand such as the phosphate groups of a phospholipid in the process of nerve transmission and muscle excitation and contraction.

Perhaps the abnormality exists in the relationship between the membrane and DNA and/or RNA. Significant concentrations of nickel are present in DNA (33, 34) and RNA (9, 33, 35) from phylogenetically diverse sources. It has been suggested that nickel and the other metals which are present may contribute to stabilization of the structure of the nucleic acids. Nickel will stabilize RNA (36) and DNA (34) against thermal denaturation and is extraordinarily effective in the preservation of tobacco mosaic virus RNA infectivity (35, 37). Also, it has been reported that nickel may have a role in the preservation of the compact structure of ribosomes against thermal denaturation (38–40) and that nickel will restore the sedimentation characteristics of *Escherichia coli* ribosomes which have been subjected to EDTA denaturation. The preliminary findings on rat-liver polysomes, in addition to this *in vitro* evidence, suggest that nickel does have a structural role in nucleic acids.

In addition to possible roles in membrane and nucleic acid metabolism or structure, nickel has at least one structural role in proteins, that of nickeloplasmin (20). The function of this protein is unclear at present. It also should be noted that nickel can activate numerous enzymes *in vitro,* including arginase (41), tyrosinase (42), desoxyribonuclease (43), acetyl coenzyme A synthetase (44), and phosphoglucomutase (45). However, these studies have not shown nickel to be a specific activator of any enzyme.

SUMMARY

Nickel has met the criteria for essentiality. Recent research has shown that homeostatic mechanisms are present to regulate nickel metabolism. However, more studies are needed to ascertain whether ultrafiltrable organic nickel complexes are important in nickel homeostasis.

Nickel deficiency reproducibly results in an impairment of a function, or functions, as evidenced by the ultrastructural degeneration, reduced oxidative ability, and increased lipids and decreased phospholipid fractions in the liver of chicks. Rats deprived of nickel show a suboptimal reproductive performance, reduced oxidative ability in liver, and abnormalities in the liver polysome profile. These findings are consistent with nickel being an essential nutrient.

It is speculated that nickel has a structural or metabolic role in membrane, DNA, RNA, or protein biochemistry.

REFERENCES

1. Underwood, E. J. 1971. *Nickel. In Trace Elements in Human and Animal Nutrition,* Academic Press, New York, p. 170.
2. Schroeder, H. A., Balassa, J. J., and Tipton, I. H. 1962. Abnormal trace metals in man–Nickel. *J. Chron. Dis.* 15: 51.
3. Nielsen, F. H. 1971. Studies on the essentiality of nickel. *In* W. Mertz and W. E. Cornatzer (eds.), *Newer Trace Elements in Nutrition,* Marcel Dekker, New York, p. 215.
4. Tedeschi, R. E., and Sunderman, F. W. 1957. Nickel poisoning. V. The metabolism of nickel under normal conditions and after exposure to nickel carbonyl. *A.M.A. Archs Ind. Hlth* 16: 486.
5. Nodiya, P. I. 1972. Cobalt and nickel balance in students of an occupational technical school. *Gig. Sanit.* 37: 108.
6. Horak, E., and Sunderman, F. W., Jr. 1973. Fecal nickel excretion by healthy adults. *Clin. Chem.* 19: 429.
7. Smith, J. C., and Hackley, B. 1968. Distribution and excretion of nickel-63 administered intravenously to rats. *J. Nutr.* 95: 541.
8. Perry, H. M., Jr., and Perry, E. F. 1959. Normal concentrations of some trace metals in human urine: changes produced by ethylene-diaminetetraacetate. *J. Clin. Invest.* 38: 1452.
9. Sunderman, F. W., Jr. 1965. Measurements of nickel in biological materials by atomic absorption spectrometry. *Am. J. Clin. Path.* 44: 182.
10. Nomoto, S., and Sunderman, F. W., Jr. 1970. Atomic absorption spectrometry of nickel in serum, urine, and other biological materials. *Clin. Chem.* 16: 477.
11. Consolazio, C. F., Nelson, R. A., Matousch, L. O., Hughes, R. C., and Urone, P. 1964. The trace mineral losses in sweat. *U.S. Army Med. Res. Nutr. Lab. Rep.* 284: 1.
12. Perry, H. M., Jr., Tipton, I. H., Schroeder, H. A., and Cook, M. J. 1962. Variability in the metal content of human organs. *J. Lab. Clin. Med.* 60: 245.
13. Nusbaum, R. C., Butt, E. M., Gilmour, T. C., and DiDio, S. L. 1965. Relation of air pollutants to trace metals in bone. *Archs Envir. Hlth* 10: 227.
14. Schroeder, H. A., and Nason, A. P. 1969. Trace metals in human hair. *J. Invest. Derm.* 53: 71.
15. Schroeder, H. A., and Nason, A. P. 1971. Trace-element analysis in clinical chemistry. *Clin. Chem.* 17: 461.
16. Nechay, M. W., and Sunderman, F. W., Jr. 1973. Measurements of nickel in hair by atomic absorption spectrometry. *Ann. Clin. Lab. Sci.* 3: 30.
17. Sunderman, F. W., Jr., Decsy, M. I., and McNeely, M. D. 1972. Nickel metabolism in health and disease. *Ann. N.Y. Acad. Sci.* 199: 300.
18. Hendel, R. C., and Sunderman, F. W., Jr. 1972. Species variations in the proportions of ultrafiltrable and protein-bound serum nickel. *Res. Comm. Chem. Path. Pharm.* 4: 141.

19. Callan, W. M., and Sunderman, F. W., Jr. 1973. Species variations in binding of ^{63}Ni (II) by serum albumin. *Res. Comm. Chem. Path. Pharm.* 5: 459.
20. Nomoto, S., McNeely, M. D., and Sunderman, F. W., Jr. 1971. Isolation of a nickel α-2 macroglobulin from rabbit serum. *Biochemistry* 10: 1647.
21. Soestbergen, M. V., and Sunderman, F. W., Jr. 1972. ^{63}Ni complexes in rabbit serum and urine after injection of $^{63}NiCl_2$. *Clin. Chem.* 18: 1478.
22. Tiffin, L. O. 1971. Translocation of nickel in xylem exudate of plants. *Plant Physiol.* 48: 273.
23. Nielsen, F. H., and Sauberlich, H. E. 1970. Evidence of a possible requirement for nickel by the chick. *Proc. Soc. Exp. Biol. Med.* 134: 845.
24. Nielsen, F. H., and Higgs, D. J. 1971. Further studies involving a nickel deficiency in chicks. *Proc. Trace Subs. Envirn. Hlth* 4: 241.
25. Sunderman, F. W., Jr., Nomoto, S., Morang, R., Nechay, M. W., Burke, C. N., and Nielsen, S. W. 1972. Nickel deprivation in chicks. *J. Nutr.* 102: 259.
26. Wellenreiter, R. H., Ullrey, D. E., and Miller, E. R. 1970. Nutritional studies with nickel. *In* C. F. Mills (ed.), *Trace Element Metabolism in Animals,* Livingstone, Edinburgh, p. 52.
27. Smith, J. C., and Schwarz, K. 1967. A controlled environment system for new trace element deficiencies. *J. Nutr.* 93: 182.
28. Snedecor, G. W., and Cochran, W. G. 1967. *In Statistical Methods,* Iowa State University Press, Ames, Iowa, pp. 91–94.
29. Frank, G. B. 1962. Utilization of bound calcium in the action of caffeine and certain multivalent cations on skeletal muscle. *J. Physiol.* 163: 254.
30. Fischman, D. A., and Swan, R. C. 1967. Nickel substitution for calcium in excitation-contraction coupling of skeletal muscle. *J. Gen. Physiol.* 50: 1709.
31. Blaustein, M. P., and Goldman, D. E. 1968. The action of certain polyvalent cations on the voltage-clamped lobster axon. *J. Gen. Physiol.* 51: 279.
32. Hafeman, D. R. 1969. Effects of metal ions on action potentials of lobster giant axons. *Comp. Biochem. Physiol.* 29: 1149.
33. Wacker, W. E. C., and Vallee, B. L. 1959. Nucleic acids and metals. I. Chromium, manganese, nickel, iron, and other metals in ribonucleic acid from diverse biological sources. *J. Biol. Chem.* 234: 3257.
34. Eichhorn, G. L. 1962. Metal ions as stabilizers or destabilizers of the deoxyribonucleic acid structure. *Nature, Lond.* 194: 474.
35. Wacker, W. E. C., Gordon, M. P., and Huff, J. W. 1963. Metal content of tobacco mosaic virus and tobacco mosaic virus RNA. *Biochemistry* 2: 716.
36. Fuwa, K., Wacker, W. E. C., Druyan, R., Bartholomay, A. F., and Vallee, B. L. 1960. Nucleic acids and metals. II. Transition metals as determinants of the conformation of ribonucleic acids. *Proc. Natn. Acad. Sci. U.S.A.* 46: 1298.

37. Cheo, P. C., Friesen, B. S., and Sinsheimer, R. L. 1959. Biophysical studies of infectious ribonucleic acid from tobacco mosaic virus. *Proc. Natn. Acad. Sci. U.S.A.* 45: 305.
38. Tal, M. 1968. On the role of Zn^{2+} and Ni^{2+} in ribosome structure. *Biochim. Biophys. Acta* 169: 564.
39. Tal, M. 1969. Thermal denaturation of ribosomes. *Biochemistry* 8: 424.
40. Tal, M. 1969. Metal ions and ribosomal conformation. *Biochim. Biophys. Acta* 195: 76.
41. Hellerman, L., and Perkins, M. E. 1935. Activation of enzymes. III. The role of metal ions in the activation of arginase. The hydrolysis of arginine induced by certain metal ions with urease. *J. Biol. Chem.* 112: 175.
42. Lerner, A. B., Fitzpatrick, T. B., Calkins, E., and Summerson, W. H. 1950. Mammalian tyrosinase: the relationship of copper to enzymatic activity. *J. Biol. Chem.* 187: 793.
43. Miyaji, T., and Greenstein, J. P. 1951. Cation activation of desoxyribonuclease. *Archs Biochem. Biophys.* 32: 414.
44. Webster, L. T., Jr. 1965. Studies of the acetyl coenzyme A synthetase reaction. III. Evidence of a double requirement for divalent cations. *J. Biol. Chem.* 240: 4164.
45. Ray, W. J., Jr. 1969. Role of bivalent cations in the phosphoglucomutase system. I. Characterization of enzyme-metal complexes. *J. Biol. Chem.* 244: 3740.

DISCUSSION

Mills (Aberdeen). Do you have any information on the nickel content of liver subcellular fractions, particularly the mitochondria?

Nielsen. No, but we're working on it now. Sunderman has really developed the analytical technique, and he may have done some of these analyses.

Mills (Aberdeen). The ultrastructural changes that you are getting, including the pycnotic nuclei, are reminiscent of the type of change seen, not as a consequence of changes in the nuclear envelope, but as a change of oxygen tension in the cell. We've seen exactly the same sort of response while working with copper.

Schwarz (Long Beach). On the same issue, I might say that the electron microscopic changes which we see are very reminiscent of those which we described in 1956 and 1954 in the prenecrotic phase of dietary liver necrosis. You see a lack of oxygen consumption; do you also see a respiratory decline?

Nielsen. We have not studied this.

Panić (Zemun). Have you estimated the zinc content of tissues? We have found that the zinc content is changed in a nickel deficiency.

Nielsen. I have not determined the zinc content of the tissues.

Martin (Fort Collins). You mentioned nickel as an enzyme activator, and indicated that it is a nonspecific activator of arginase and tyrosinase. Is nickel more effective than other metal cations?

Nielsen. It is dependent on a number of factors. At certain pH's it is a better activator of arginase than other cations.

Armstrong (Minneapolis). How quickly is the deficiency produced by a nickel-deficient diet reversed on adding nickel to the diet?

Nielsen. I have not tried to reverse it. We find that some of the skin-pigmentation changes you see are evident at about 10 days on the deficient diet. It was also about the 10th to 14th day that we began to see differences in the oxidative ability of liver from deficient chicks.

Smith (Washington). Would you like to speculate on why serum nickel goes up during myocardial infarction, if indeed it does?

Nielsen. Yes, it does. Not only does it go up in certain diseases but it goes down in others, such as chronic urenia or liver cirrhosis. Yesterday, Dr. Burger reported a change in nickel during kwashiorkor.

Mertz (Beltsville). Did you say, Dr. Hennig, that you had produced a nickel deficiency?

Hennig (Jena). We think so, and we think that the pig has a higher nickel requirement than poultry. In the early 1960's we discussed an unidentified growth factor in a material that is a byproduct of vitamin C production. We have now determined the nickel content of this material, and see a close correlation between the growth of the pigs and the nickel content.

Nielsen. In our latest study, the rats from nickel-deficient dams do show about a 10% reduction in growth at 3½ weeks. At the present, I don't feel that this is significant.

ESSENTIALITY AND FUNCTION OF VANADIUM[1]

LEON L. HOPKINS, JR.

Agriculture Research Service, United States Department of Agriculture, Fort Collins, Colorado

No other segment of nutritional science has been as slow in determining the total number of essential nutrients as we have in the trace-element field. The essential amino acids, the essential fatty acids, and the vitamins were discovered within a matter of years after the initial concept was discovered and accepted. Apparently even ancient man knew he needed iron- and iodine-containing foods for good health, yet today most nutritionists would agree that there are essential trace elements yet undiscovered. Of the total number of elements known to man, only about one-fourth have been shown to be essential. This includes carbon, hydrogen, oxygen, nitrogen, and sulfur which make up the organic nutrients. Until it is known what nutrients are needed in the diet, in what amounts and in what chemical form, the food faddists may be right in stating that our diets are nutritionally inadequate.

Since the last TEMA meeting in 1969, reports showing the essentiality of several trace elements have been appearing in the literature. One of these elements, vanadium, is the subject of this presentation. It was theorized early in this century that vanadium was an element of importance to life, and, by the sixties, it had been shown to be essential for lower plants (see Underwood (1) for an excellent review of the earlier literature). Until very recently, only theories involving circumstantial evidence supported the concept that vanadium was essential for animals.

NUTRITIONAL ESSENTIALITY

In 1970 we reported data supporting the view that vanadium is an essential nutrient for animals (2). We had observed for the first time that animals

1. The author is indebted to Mr. Harold Mohr and Miss Beverly Tilton for their valuable assistance in much of the work reported here.

consuming a diet extremely low in the element showed a significant impairment of a normal function that was not seen in animals receiving supplementary vanadium. Since that time, several other investigators have reported similar or additional deficiency symptoms attributable to low levels of vanadium in the diet. In order to achieve these results, various systems have been used to maintain an environment that is consistently and extremely low in vanadium. An entirely closed system such as we have reported (2) has been used where the air is filtered and entry of dust-borne metal is restricted. The use of a laminar-flow environmental system has been reported by Nielsen and Ollerich (3), and conventional animal-room techniques were used by Strasia (4). In all cases, the use of plastic caging systems prevented contact of the animal with metal (5). Distilled water was used by all investigators and diets ranged from standard purified diets (2–4), low in vanadium, to a highly purified amino acid diet that did not permit normal growth (6).

Growth

The initial observation indicative of vanadium deficiency was a significantly reduced wing- and tail-feather growth in chicks consuming a diet containing less than 10 ppb V (2). Reduced body growth has been reported by Strasia (4) and by Schwarz and Milne (6) in rats consuming diets low in vanadium. Nielsen has made a similar observation in chicks consuming a diet containing 30–35 ppb V.[2]

Reproduction

Reproductive performance of Sprague-Dawley rats consuming a diet of less than 10 ppb vanadium was reduced slightly in third-generation females and markedly reduced in the fourth generation (7). Five deficient fourth-generation females produced one litter following the first mating, three following the second, and two were pregnant upon autopsy following the third mating. The five vanadium-supplemented females produced five, four, and five litters respectively. The total number of pups from vanadium-deficient females following the first two matings was 22 compared with 65 for the vanadium-supplemented controls. Thirty-two percent mortality was observed in the vanadium-deficient pups while only 1.5% of the pups from supplemented mothers died. In a second experiment with a different strain of rats (BHE)[3], similar results in mortality were obtained with first-generation females consuming the diet low in vanadium. Although the number of litters from mothers consuming both the deficient and supplemented diets was low, the

2. F. H. Nielsen, personal communication.

3. A special strain of rats maintained at Beltsville, Maryland, for approximately 30 years.

mortality of pups from deficient mothers was again much greater (38%) than those whose mothers were given vanadium supplement (7%).

LIPID METABOLISM

There is considerable evidence that vanadium is involved in lipid metabolism. The Bernheims (8) reported in the thirties that added vanadium markedly increased the oxidation of phospholipids *in vitro* by washed liver suspensions. More recently there have been numerous reports that pharmacological levels of vanadium lowered tissue cholesterol levels (1). Curran and Burch (9) have reviewed this work, much of which originated in their laboratory. They related these altered cholesterol levels to the ability of vanadium to inhibit cholesterol biosynthesis. The site of the inhibition was the microsomal enzyme system referred to as squalene synthetase. Unfortunately, vanadium did not lower blood cholesterol in older animals or in older humans where elevated cholesterol levels are a health problem.

The importance of this earlier work has been emphasized by the reports of altered blood lipid levels in vanadium-deficient chicks. Initially, Hopkins and Mohr reported that chicks consuming a vanadium-deficient diet for four weeks had lowered plasma cholesterol levels when compared with the supplemented animals (2). Similar experiments, continued for a longer period of time, showed that the plasma cholesterol levels of vanadium-deficient chicks again were significantly lower than the supplemented controls at 28 days, but were significantly higher at 49 days (10). The average values were 249 and 224 mg/100 ml for the deficient and control groups respectively at 49 days. Nielsen and Ollerich (3) have reported data which support the view that vanadium-deficient chicks have altered plasma cholesterol levels, but their observations indicate a significantly increased cholesterol level at four weeks of age.

Recent data from our laboratory indicate in addition to altered plasma cholesterol that plasma triglyceride levels are also altered (7). The vanadium-deficient chicks had 48.7 ± 2.4 mg/100 ml plasma triglyceride while the supplemented controls had 25.4 ± 3.0.

Hard Tissues

It has been reported that the highest uptake of radiovanadium injected subcutaneously into young rats was in areas of rapid mineralization in the dentine and bone (11). Radiovanadium injected into adult mice intravenously was also taken up by the teeth, bones, and fetus (12). After 24–48 hr, the bones and teeth were the tissues showing the highest concentration with a maximum in the mineralization zones. Localization of vanadium in the tooth structure supports the hypothesis that this element might have a role in the prevention of dental caries. Underwood (1) has reviewed the evidence that vanadium exerts a beneficial effect on dental-caries incidence and has stated that the data are not conclusive. Nielsen has found that vanadium deficiency retards bone development in chicks.[4] In vanadium-deficient chicks, examination of the tibiae revealed increased epiphyseal plate and decreased primary spongiosa as judged by weight ratios and microscopic examination (3). Histologically, an alteration in the organization of the cells and the amount of matrix in the zones of proliferation and maturation was seen. The deficiency did not significantly change the uptake and distribution of $^{35}SO_4^{2-}$ in the epiphyseal plate or primary spongiosa. Also there was no difference in the hexosamine levels indicating that vanadium does not affect mucopolysaccharide metabolism (Nielsen).[5]

Erythrocytes and Iron Levels

Strasia (4) has reported that rats fed a diet containing less than 100 ppb V had a significantly increased packed cell volume of blood when compared with control groups receiving 0.5, 2.5, and 5.0 ppm supplemental vanadium. Also noted was an increase in blood and bone iron in deficient rats. Nielsen and Ollerich (3) have also reported increased hematocrits in chicks consuming a low vanadium diet (30–35 ppb V).

METABOLISM

To learn more about the metabolism of vanadium, ^{48}V of very high specific activity was injected intravenously (13). By giving approximately 1 ng V per 100 g body weight to rats fed a diet low in available vanadium, it was hypothesized that the radioactivity would distribute itself with time into the areas of specific and physiologically meaningful pools. Following the injection of low levels of ^{48}V, only 30% of the radioactivity observed 10 min after injection remained when the 7-hr sample of blood was taken from rats previously fed a stock diet or a torula-yeast-based diet with supplemental vanadium in the drinking water. Rats consuming the nonsupplemented torula diet containing 210 ppb vanadium retained about twice as much as the 7-hr sampling period, indicating a homeostatic mechanism that clears the vanadium when adequate amounts are in the blood. Of interest were the individual data from each rat, which indicated that there was an actual increase over the 10 min observation in ^{48}V in the plasma of five out of the seven rats under study which had previously consumed the diet low in vanadium. These

4. *Op. cit.*
5. *Op. cit.*

increases were observed in different rats at the ½, 1, 2, and 4 hr intervals. Since most of the radioactivity was found in the noncellular portion of the blood, the plasma proteins were separated using disc gel electrophoresis. In plasma taken 10 min after the injection of nanogram amounts of ^{48}V and separated electrophoretically, 92% of the gel radioactivity was not protein bound. Whereas only a minor amount was initially found in the transferrin region of the gel, within 4 hr the radioactivity increased from 3 to 28% of the total. The amount of unbound ^{48}V decreased proportionately. Attempts to label transferrin by adding ^{48}V *in vitro* and incubating at 37°C for 4 hr were unsuccessful. Apparently, binding of vanadium to the protein required some *in vivo* metabolic change.

Evidence that the ^{48}V was binding to transferrin and not some other protein migrating near the transferrin region was obtained by injecting ^{59}Fe intravenously into rats. This serum was separated identically to serum from rats that had been injected 4 hr previously with ^{48}V. The transferrin region was separated into two pieces, such that a different protein migrating in the fore or aft portion of the transferrin region might be detected. The proportions of isotope in each segment compared favorably between the two isotopes, thus providing evidence that transferrin was binding the metabolically changed vanadium.

In the ^{48}V-distribution studies tissues were sampled at timed intervals following the injection of three different oxidation states of vanadium (13). Although there were marked differences between tissues in their accumulation of the isotope at these very low levels, in general there was very little effect noted on isotope retention within the same tissue owing to the difference in oxidation state. There was a trend noted very early in several tissues to a more rapid accumulation of vanadium in the 5+ oxidation state. Most notable were the lung and sternum, where 10 min after injection V^{3+} was retained the least, V^{4+} moderately, and V^{5+} the most. This trend was not observed 1 hr or more after injection. Four or more hours after injection it was observed that V^{5+} was lost faster, thus indicating a first-in-first-out phenomenon for V^{5+}.

The variation in accumulation of ^{48}V in the tissues of rats fed low levels of available dietary vanadium ranged after 4 days from 5% of the 10-min value in the blood to 121% in the kidney. Those tissues accumulating the highest levels of ^{48}V 4 days after injection include the spleen, liver, and testes as well as the kidney. Of interest from a metabolic viewpoint was the accumulation and retention of vanadium in the liver and testes. Analysis of testes from rats kept in a controlled environment indicated that rats consuming an unsupplemented diet low in vanadium had little vanadium in these organs, 1 and 6 ppb dry weight. However, as demonstrated in the isotope experiments, testes accumulated the element when the diet contained 200 and 250 ppb vanadium.

Liver homogenates from rats intravenously injected were centrifuged in order to observe the distribution of ^{48}V among the subcellular particles (13). Initially, 10 min after injection, most of the isotope was located in the supernatant fraction. With time the ^{48}V migrated into the mitochondrial and nuclear fractions, until after 1 day post injection approximately 40% was in each fraction. Apparently the isotope was incorporated into these organelles since 4 days after injection the same distribution remained and probably accounted for the accumulation and retention previously noted in whole liver samples.

At these low levels vanadium was excreted both through the urine and feces (13). Four days after intravenous injection almost half of the dose was excreted through the urine and 8.6% through the feces.

NUTRITIONAL ADEQUACY

Of concern to nutritionists is the adequacy of vanadium in the feeds and foods consumed by animals and man. In order to determine adequacy, the level required by the organism to maintain health must be known. Owing to limited data, this can only be estimated at this time. In order to demonstrate a vanadium deficiency in chicks and rats, all authors previously cited developed diets that actually were, or were thought to be, less than 100 ppb V. Growth responses were obtained by adding 250–500 (6), 500 (4) and 3000[5] ppb vanadium to these diets.

These data, while helpful, do not specifically determine the requirement level, but do indicate that the minimum requirement that prevents deficiency may be around 100 ppb V when a purified diet is fed. It is possible that the vanadium requirement might be even higher under the conditions of a natural ration or diet. It has been reported that vanadium compounds vary in their ability to influence growth of rats consuming a deficient diet (6). Secondly, the toxicity of vanadium compounds has been reported to be greater when a purified diet was fed than a natural diet (14). This would indicate that not only is the effectiveness of vanadium compounds in feeds and foods important, but also that rations and diets made up of natural ingredients may reduce the absorption and increase the requirement. Therefore, instead of the requirement being approximately 100 ppb, as suggested in studies using purified diets, the amount would be higher when feeds and foods from natural sources are consumed. Requirement levels in this range are not unusually low for trace-element nutrients.

If the lower estimate of 100 ppb as the requirement is assumed to be valid, the question arises whether this requirement is met by the average food and feed intake. Owing to the difficulty of vanadium analysis and the scarcity of reliable data, the answer must be ambiguous. Mitchell (15) has reported the following values for vanadium in feeds using emission spectroscopy: 80 ppb for pasture herbage, 60 ppb for oats, and 120 ppb for oat straw, on a dry matter basis. Söremark (16) reported values for several foods using activation analysis which ranged from less than 0.1 ppb V for pea, beet, carrot, and pear (ash weight) to 52 ppb for radish (wet weight). Milk generally contained less than 0.1 ppb on a wet-weight basis, while liver, fish, and meat contained a few ppb but not over 10 ppb. Mitchell (15) found less than 30 ppb on a dry-matter basis in peas using emission spectroscopy. These limited data indicate that many feed and food ingredients contained levels of vanadium far below the estimated requirement level of around 100 ppb. The contribution of drinking water to the intake is variable, but it may be more available than vanadium from feed or food sources.

Obviously, much more accurate data are needed before any firm conclusions can be drawn. Enough data are available, however, to indicate that the requirement level for vanadium and the level actually present in feeds and foods do not appear to overlap to a degree that will allow us to take adequate vanadium nutrition for granted. Although we go to great pains with isolators, plastic cages, etc., to prevent contamination and ensure consistent results in our deficiency studies, the work of Strasia (4) indicates that vanadium deficiency is not difficult to produce in a normal animal room using plastic cages. Marginal vanadium deficiency could be a practical problem in man.

If inadequate vanadium is a nutritional problem, what type of impairment could be suspected as a result? Most obvious is the effect of vanadium upon the lipid metabolism of animals and man. High levels of vanadium have been reported to lower serum lipids, an effect probably due to the ability of vanadium to inhibit the biosynthesis of cholesterol. The practical importance of this observation increases as a result of the finding that both serum cholesterol and triglyceride levels are altered in vanadium-deficient animals. Marginal human vanadium deficiencies may be responsible for at least part of the altered serum lipid levels observed in our society. Extrapolating animal data to humans or human data resulting from the administration of high levels of vanadium to the deficient state is often misleading. However, if man continues to refine and purify his diet without consideration for replenishing extracted trace elements such as vanadium, problems that are speculation today may be proven to be very real tomorrow.

SUMMARY

Vanadium is an essential nutrient for animals. All of the criteria for essentiality have been met.

(1) It has a low molecular weight, is an excellent catalyst, has the proper atomic structure, is a transition metal, forms chelates, and thus is chemically suitable for biological functions.

(2) It is ubiquitous on the geosphere; in other words, it has been generally available to plants and animals during their evolution.

(3) It is generally present in all plants and animals observed.

(4) It has a low order of toxicity to most living organisms and especially to mammals when taken orally.

(5) Mammalian homeostatic mechanisms are implied by serum levels, lack of excessive accumulation, and excretion rates.

(6) Most important, it is now known that animals consuming a diet low in vanadium demonstrate an impairment of several physiological functions. These include reduced body and feather growth, impaired reproduction and survival of the young, altered red blood cell levels and iron metabolism, impairment of hard-tissue metabolism, and altered blood lipid levels. These data have originated and/or been substantiated in four different laboratories and observed in two different species, the chick and rat.

REFERENCES

1. Underwood, E. J. 1971. Vanadium. *In Trace Elements in Human and Animal Nutrition.* Academic Press, New York, p. 416.
2. Hopkins, L. L., Jr., and Mohr, H. E. 1971. The biological essentiality of vanadium. *In* W. Mertz and W. E. Cornatzer (eds.), *Newer Trace Elements in Nutrition,* Marcel Dekker, New York, p. 195.
3. Nielsen, F. H., and Ollerich, D. A. 1973. Studies on a vanadium deficiency in chicks. *Fed. Proc.* 32: 929.
4. Strasia, C. A. 1971. Vanadium: Essentiality and toxicity in the laboratory rat. *University Microfilms* 1 (Thesis), Ann Arbor, Mich.
5. Mohr, H. E., and Hopkins, L. L., Jr. 1972. An all plastic system for housing small animals in trace element studies. *Lab. Anim. Sci.* 22: 96.
6. Schwarz, K., and Milne, D. B. 1971. Growth effects of vanadium in the rat. *Science* 174: 426.
7. Hopkins, L. L., Jr., and Mohr, H. E. 1974. Vanadium as an essential nutrient. *Fed. Proc.,* in press.
8. Bernheim, F., and Bernheim, M. L. C. 1939. The action of vanadium on the oxidation of phospholipids by certain tissues. *J. Biol. Chem.* 127: 353.
9. Curran, G. L., and Burch, R. E. 1968. Biological and health effects of vanadium. *In* D. D. Hemphill (ed.), *Trace Substances in Environmental Health,* University of Missouri, Columbia, Mo., p. 96.
10. Hopkins, L. L., Jr., and Mohr, H. E. 1971. Effect of vanadium deficiency on plasma cholesterol of chicks. *Fed. Proc.* 30: 462.
11. Söremark, R., Üllberg, S., and Appelgren, L. 1962. Autoradiographic

localization of vanadium pentoxide ($^{48}V_2O_5$) in developing teeth and bones of rats. *Acta Odont. Scand.* 20: 225.
12. Söremark, R., and Üllberg, S. 1962. Distribution and kinetics of $^{48}V_2O_5$ in mice. *In* N. Fried (ed.), *The Use of Radioisotopes in Animal Biology and the Medical Sciences,* Academic Press, New York, p. 103.
13. Hopkins, L. L., Jr., and Tilton, B. E. 1966. Metabolism of trace amounts of ^{48}vanadium in rat organs and liver subcellular particles. *Am. J. Physiol.* 211: 169.
14. Berg, L. R. 1966. Effect of diet composition on vanadium toxicity for the chick. *Poult. Sci.* 45: 1346.
15. Mitchell, R. L. 1957. Emission spectrochemical analysis determination of trace elements in plants and other biological materials. *In* J. H. Yoe and H. J. Koch (eds), *Trace Analysis,* John Wiley, New York, p. 398.
16. Söremark, R. 1967. Vanadium in some biological specimens. *J. Nutr.* 92: 183.

DISCUSSION

Shah (Ottawa). Is there loss of vanadium in producing white flour from whole wheat and is there loss of vanadium when potatoes are peeled?

Hopkins. The loss of vanadium is very similar to that of other trace elements when the various parts of the wheat kernel are extracted away. I am not so sure about potatoes. There's quite a bit of vanadium in soils, and there seems to be some inhibition of vanadium uptake into the plants so a lot of it is found on the skin of roots and tubers. I don't know how available that vanadium is, or whether it does man any good to eat the potato skins.

Henkin (Bethesda). Do you have any comments about the vanocyte of the sea squirt, which contains 4% vanadium?

Hopkins. It really isn't evident what the function of this high level of vanadium is. It has been suggested that the vanadium-protein compound functions as an oxygen carrier, but other evidence would suggest that it does not.

Forth (Homburg). Tunicates accumulate very high levels of vanadium and we worked on this problem a few years ago. This vanadium is present in what we call blood cells, but we have no idea whether these cells have the same function as blood cells in higher organisms. The pH of these cells is very low, and the vanadium is bound to protein and is present in these cells together with sulfate. We could show that vanadium uptake by the cells was impaired by phosphate and arsenate, and one explanation is that the storage of vanadium simply represents vanadium that the cell has been unable to remove from the cell because of a lack of phosphate.

Burch (Omaha). One of the problems we had when working with the interaction between vanadium and cholesterol synthesis was that vanadium seemed to lose its effect as the animals got older. Have you done any studies on older animals?

Hopkins. No.

Horvath (Morgantown). Dr. Hopkins speculated that the low level of vanadium that might be present in foodstuffs could influence the status of the human population. Perhaps the previous speakers would also address themselves to the same point: are there likely to be general population effects that could be conjectured at this time or are there sensitive target population effects that could be conjectured at this time?

Schwarz (Long Beach). Regarding silicon, I think the American population, and populations in all highly civilized countries, are possible target groups. We have measured silicon levels in tortillas which we obtained in Mexico and found about 500 ppm silicon. We also analyzed grain and grain byproducts obtained from the American Baking Institute. While the original grain was quite rich in silicon, the 65% extraction flour contained only 2% of the silicon which was in the original grain. The rest went into the bran, the wheat germ, and other fractions. The same is true if we consider animal sources of silicon, which may be much more important as essential silicon. Silicon would be primarily in the connective tissue elements: cartilage, skin, and blood vessels. In this country at least, these are little utilized for human nutrition, although in primitive countries they may be eaten more extensively. Certain diseases may document this. I am particularly interested in the connection between silicon deficiency and atherosclerosis. There is a very good chance that high incidence of atherosclerosis in civilized countries is really due to silicon deficiency.

Carlisle (Los Angeles). We might also find that as aging goes on, there is decreased absorption of silicon. This has been shown to be true in the rat, and this may also reinforce some of the statements that Dr. Schwarz has made.

Nielsen (Grand Forks). As of yet, there is no evidence of clinical signs of nickel deficiency in humans. However, the human requirement may be similar to that of the chicken and the rat (50–80 ppb or 10–25 μg per 1000 calories of diet), and it's very easy to comprise a diet which contains only 1–4 μg of nickel per 1000 calories of diet. Thus there is the possibility that a nickel deficiency could occur in some populations which exist on restricted diets of meat, milk, refined bread, and nonhydrogenated oils and fats. As yet such a condition has not been recognized.

ESSENTIALITY AND FUNCTION OF SILICON

E. M. CARLISLE

Environmental and Nutritional Sciences Division, School of Public Health, University of California, Los Angeles, California

Although interest in the silica content of animal tissues and the effects of siliceous substances upon animals was evidenced over half a century ago (1, 2), there is relatively little information available even today on the effect of silicon in normal metabolism in higher animals. Emphasis has been placed on the more deleterious aspects of silicon metabolism, including its effects upon forage digestibility, urolithiasis, and especially pneumoconioses (silicosis) caused by dust inhalation. Several important reviews on this work are available (3–6).

Silicon has generally been considered to be nonessential except for some primitive classes of organisms, notably diatoms (unicellular plants) and two groups of animals, the radiolarians (belonging to the Protozoa) and some sponges (Porifera) which utilize silica as a component of body structure. In the diatom, where silica is the major skeletal constituent, an absolute requirement for silicon has been associated with silica-shell formation (7) and more recently with the net synthesis of DNA (8). The structural relationship of silica to the organic constituents of the cell wall has also been established (9, 10).

Silicon has been listed along with 20–30 trace elements which occur more or less invariably, but in widely ranging concentrations, in living tissues as environmental contaminants reflecting contact of the organism with its environment (6). In 1967, in an extensive review of the biological properties of silicon compounds, Fessenden and Fessenden (11) wrote that "although traces of silica are found in all animal tissues, there is no evidence that there is any biological need for silicon in the higher animals." Experiments establishing silicon's essentiality as a trace element for higher animals and demonstrating its mode and site of action were performed between 1964 and 1972. They are discussed here in much the same sequence as they were performed.

PHYSIOLOGICAL ROLE IN BONE CALCIFICATION

In Vitro Studies

Concern for the problem of osteoporotic bone loss and the fact that small amounts of fluoride have been known for some time to change markedly the properties of bone apatite led this investigator to study possible effects of other less-known trace elements in bone metabolism. Fluorine is only one of the elements known to substitute for calcium or for phosphorus in naturally occurring apatites. In mineralogical studies of nonbone (natural) apatite, McConnell (12) has shown, for example, that sodium, potassium, manganese, strontium, magnesium, and carbon can substitute for calcium, and that sulfur, silicon, arsenic, vanadium, and carbon can substitute for phosphorus. In the mineral ellestadite, a crystalline isomorph of apatite, silicon and sulfur have almost completely replaced phosphorus (13).

In order to investigate the possible involvement of trace elements further, techniques were developed in this laboratory for application of the electron microprobe to bone and related tissues during various stages of development and, as a result, silicon, a relatively unknown trace element in nutritional research, was shown to be localized in active growth areas in the bones of young mice and rats (14, 15). The amount of silicon present in specific small sites within the active growth areas appeared to be uniquely related to 'maturity' of the bone mineral. In the earliest stages of calcification both the silicon and calcium content of these sites in the osteoid tissue is very low, but as mineralization progresses the silicon and calcium contents rise congruently. In more advanced stages the amount of silicon in the sites falls markedly and, as calcium approaches the proportions present in bone apatite, silicon is present only at the detection limit; the more 'mature' the bone mineral, the smaller the amount of measurable silicon. Likewise, maximal amounts of silicon are present at Ca/P molar ratios of approximately 0.7, but at Ca/P ratios approaching that of hydroxyapatite (1.67), silicon falls below the detection limit.

In the metaphysis of young bone the distribution of the minute silicon-rich sites corresponds with the margins of trabeculae and bony spicules in the course of formation. Part of a typical traverse across the metaphyseal region of a longitudinal section of a young tibia (Fig. 1a) shows the silicon content of the silicon-rich sites rising from 0.01 to 0.06% in the preosseous border and to 0.12% at the edge of a trabecula, but then declining while the calcium content rises progressively from 0.06 to 27.8%. Calcium contents in the silicon-rich sites, which are abundant in this part of bone, range from 0.15 to 0.70% and Ca/Si molar ratios range from 1/3 to 3. By comparison, on the

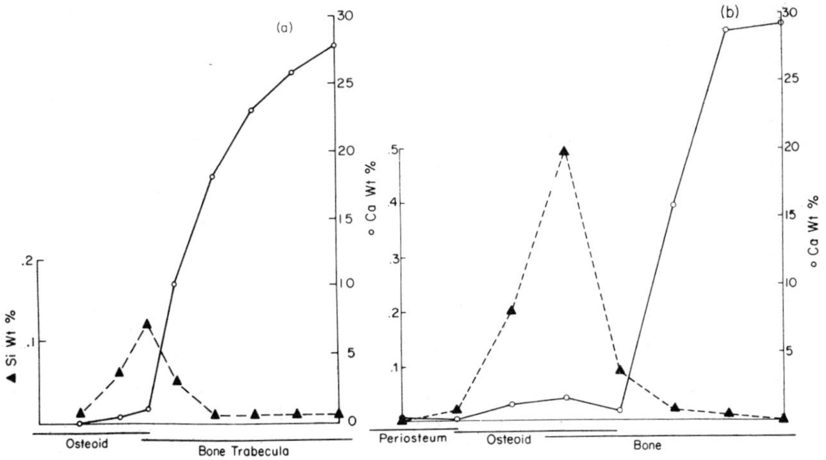

Fig. 1a. Spatial relation between silicon and calcium composition (percentage by weight) in a typical traverse across metaphyseal region of young tibia (longitudinal section) as obtained by electron microprobe techniques.

Fib. 1b. Spatial relation between silicon and calcium composition (percentage by weight) in a typical traverse across the periosteal region of young tibia (cross section) as obtained by electron microprobe techniques.

edge of the trabeculae where the calcium content is about 1%, Ca/Si molar ratios in the silicon-rich sites are typically about 5. The decline in silicon, as the traverse progresses into the trabecula, seem to occur usually at a calcium concentration in the sites slightly above 1% and silicon continues to fall and remains at minimal levels as the calcium concentration progressively increases with distance from the edge of the trabecula.

Silicon has also been found, along with iron, in blood vessels between metaphyseal trabeculae.[1] In this situation, however, silicon is not present with calcium, and the larger-scale and more diffuse localization is quite different from that just described. Iron has not been detected in the silicon-rich sites in preosseus tissue. The fact that silicon occurs both in metaphyseal blood vessels and in the silicon-rich sites, along with the observation by earlier workers that invasion of the metaphysis by blood vessels triggers the sequence of matrix alterations leading to calcification (16), suggests that silicon takes part in the sequence of events leading to calcification.

A relationship also exists between calcium and silicon in the periosteal regions of young bone. In a typical traverse across this region, for example, it was found that in the fibrous layer of the periosteum both the calcium and

1. E. M. Carlisle, 1973 (manuscript in preparation).

silicon values are invariably low, whereas in the adjacent osteoid layer silicon-rich sites appear containing up to 25 times as much silicon and nine times as much calcium as the fibrous layer (Fig. 1b). The silicon content in the sites falls again to the original extremely low value at calcium levels in excess of 15%. Several other traverses across the periosteal region of this same sample yielded similar results.

Evidence that silicon is involved in calcification at an early stage is provided by Ca/P ratios. In hydroxyapatite, and in mature bone generally, the Ca/P ratio is approximately 1.67. In the silicon-rich sites, even though calcium contents range appreciably, Ca/P molar ratios are typically in the range 0.6–0.8 and invariably are less than 1.0. From this it follows that in these sites phosphorus is present as organic phosphate and is not yet combined with calcium in an inorganic calcium phosphate precursor to bone apatite or as bone apatite itself. This conclusion is based on the fact that the lowest possible Ca/P ratio in any known calcium phosphate salt is 1.0, the ratio found in secondary calcium phosphate ($CaHPO_4$). Thus, silicon appears to be involved in a series of events leading to calcification at a time when calcium and phosphorus are not yet combined to form the mineral phase of bone.

Neither the initiating nor limiting factor in the mineralization of bone in the living animal is known. It has been thought that crystallization in matrices must occur on sites which form specific nucleation centers and several recent investigators (17–21) have regarded calcium binding as a most important and first event in calcification. The data presented above suggest that silicon may be associated with calcium in this process.

In Vivo Studies

Experiments were next undertaken to determine whether an effect of silicon could be demonstrated on *in vivo* calcification (22). Weanling rats obtained from mothers placed on a low-silicon diet prior to mating were maintained on diets containing three levels of calcium and three levels of silicon using conventional trace-element procedure to keep silicon contamination to a minimum. The experiments were designed to ascertain whether a relationship exists between the level of dietary silicon and the mineralization of bone.

The effect of different dietary levels of silicon (10, 25, and 250 ppm) and calcium (0.08, 0.40, and 1.20%) intake on the bone ash content of the tibia is shown in Figure 2. It is apparent that increases of silicon in the low-calcium (0.08%) diet results in a highly significant increase in the percentage ash content, and, therefore, mineralization of the tibia during the first three weeks of the experiment. Calcium contents reveal the same relationship between mineralization and dietary silicon intake.

Although, on the low-calcium diet the percentage of ash, at all dietary silicon levels, falls markedly after the third to fourth week, the absolute ash

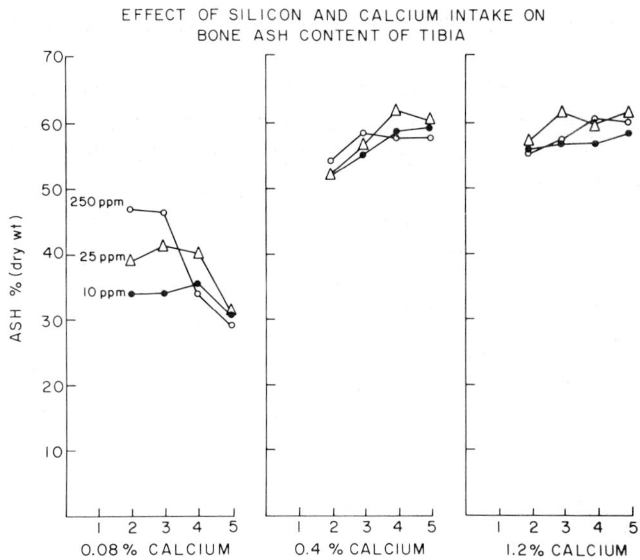

Fig. 2. Effect of silicon and calcium intake on bone-ash content of tibia. Dietary levels of silicon, 10, 25, and 250 ppm, and calcium, 0.08, 0.4, and 1.2%. Values represent mean ash percentages of 12 rats sacrificed from each treatment group per week.

content (Table 1) either remains approximately constant (high-silicon diet) or increases (medium- and low-silicon diets), and after five weeks the ash content is about the same on all diets. Thus, at the end of two weeks on the high-silicon level, the mean total ash content of the tibia is 88 mg compared with 89 mg at the end of five weeks. On the lowest silicon level (10 ppm) the

Table 1. Effect of silicon intake on absolute ash content of tibia of rats on a low-calcium (0.08%) diet

	Ash content[1]	
Diet silicon (ppm)	2 weeks (mg)	5 weeks (mg)
250	88[a]	89
25	69[a]	90
10	66[a]	85

[1]Mean ash values of 12 animals per treatment group.

[a]Indicates a difference at the 5% confidence level.

total ash content of the tibia is 66 mg at week two compared with 85 mg at the end of week five. On the middle silicon level (25 ppm), it is 66 mg at week two increasing to 90 mg at the end of week five. What has been found for the tibia applies also to four other groups of bones, namely, humerus, radius, ulna, and femur. It appears, therefore, that silicon hastens the rate of bone mineralization, the tibia on the 250-ppm silicon level reaching a higher degree of mineralization in a shorter period of time than tibia on the medium- and low-silicon diets. By the same token, the Ca/P ratio of the tibia of rats on the low-calcium diet with 250 ppm silicon is 1.41 at the end of three weeks, in contrast with 1.32 and 1.31 on the low- and medium-silicon diets. This effect on bone maturity is also indicative of the tendency of silicon to accelerate mineralization. The concept of an agent affecting the speed of chemical maturity is not new. Muller and co-workers (23) found that the chemical maturity of D-deficient bone (measured in this case by the amount of heat-produced pyrophosphate), although it is inferior to control bone during the period of maximum growth, approaches the control level at the end of the experiment.

ESSENTIALITY OF SILICON

The work demonstrating the requirement of silicon for normal growth and development in the chick when a low-silicon diet is fed in a trace-element-controlled environment, and thus establishing silicon as an essential trace element, was made possible through certain improvements in the experimental conditions (24, 25). In animal studies involving trace elements special precautions must be observed to avoid contamination, and silicon offers an unusual challenge in this respect because of its abundance in the environment and in the laboratory. Silicon contamination has been kept to a minimum in the present work by the use of silicon-low plastics and by using a specially constructed environmental chamber in which the trace-element content of the air is greatly reduced. A diet considerably lower in silicon content, based on an optimal mixture of L-amino acids for the chick, replaced the earlier diets. The chick was chosen as the experimental animal because of its more rapid skeletal growth, because of the likelihood of obtaining an earlier depletion of silicon since the experimental diet could be fed at an earlier age, and because the chick can be deutectomized to reduce the initial silicon supplied by the yolk reserves.

It should be pointed out that in an attempt to formulate a diet as low as possible in silicon and thus to produce a silicon deficiency, it was necessary, in the early experiments, to make it borderline with respect to certain nutritional requirements, and, as a consequence, the growth rate of the

Table 2. Growth response of chicks to silicon supplementation

Study no.	No. of chicks	Average daily weight gain in 23 days (g)[1] Unsupplemented group	Supplemented group	Difference (%)	P
1	36	2.37 ± 0.11	3.10 ± 0.10	30.0	< 0.01
2	30	3.25 ± 0.09	4.20 ± 0.09	30.0	< 0.02
3	48	2.57 ± 0.09	3.85 ± 0.11	49.8	< 0.01

[1]Mean ± s.e.m.

control animals (silicon-supplemented group) was less than usual. Deutectomy also contributed to the reduced rate of the controls. The attempt to produce a silicon deficiency was successful, and in more recent experiments it has been found possible to improve the diet without affecting the deficiency.

Differences between the chicks on the basal and silicon-supplemented diets were noted after approximately 1–2 weeks. As shown in Table 2, silicon was found to exert a significant effect on growth, the unsupplemented silicon group showing a markedly decreased growth rate. The percentage difference in the average daily weight gain in study 3 (Table 2) over the 23-day period is 49.8%. The chicks on the basal diet are much smaller but in proportion (Fig. 3). They appear stunted. On subsequent examination all organs appeared relatively atrophied. The legs and comb of the deficient chick are particularly pale. Macropathologic examination showed that the skin and mucous membranes are somewhat anemic. The subcutaneous tissue seems muddy to yellowish compared with a white-pinkish color of the supplemented bird. The deficient chick has no wattles and the comb is severely attenuated. Significantly retarded skeletal development in the unsupplemented silicon group was evidenced by reduced circumference, thinner cortex, and relatively less flexible leg bones. In addition, the skulls were smaller and abnormally shaped with the cranial bones appearing somewhat flatter. This effect of silicon on skeletal development strengthens the earlier postulate that silicon is involved in an early stage of bone formation. The essentiality of silicon as a trace element for growth is not limited to its role in bone development. Other tissues and organs are affected.

The essentiality of silicon for growth has been confirmed in rat studies by Schwarz (26). Inbred litter-mate, male weanling Fisher 344 rats were used in a trace-element isolator preventing access to silicon. Chemically defined diets based on amino acids in place of protein were also used here. The addition of 50 mg silicon per 100 g diet produced a 33.8% increase in growth

Fig. 3. Picture of 4-week-old chicks on silicon-supplemented diet (left) and low-silicon basal diet (right).

on Diet A and a 25.2% increase on Diet B. Skeletal changes were also evident in these studies. The skulls were shorter and the bone structure surrounding the eye appeared distorted. Pigmentation of the incisors was also affected, indicating a disturbance of enamel development or pigment deposition in teeth. Significant improvements of tooth pigmentation were produced by silicon supplements. Silicon was, however, only partly effective in preventing the impairment of pigment deposition. Significant effects on incisor pigmentation were also produced by physiological levels of tin, vanadium, or fluorine.

MODE OF ACTION

Dietary deficiency studies substantially identical with those through which the essentiality of silicon has been established have provided evidence of its metabolic role (27, 28). As noted above, skeletal and other abnormalities involving the formation of cartilage and connective tissue are associated with silicon deficiency in the chick. Experiments were designed, therefore, specifically to examine these tissue abnormalities in greater detail. The diet and experimental conditions were the same as those used in the previous studies. As in earlier experiments, differences between the low-silicon and silicon-supplemented groups were noted after approximately 1–2 weeks. Chicks in the silicon-supplemented group had thicker legs and larger combs in proportion to their size. The incremental weight gain, 18% over a 23-day period, was slightly less than in the previous studies but significant. On macropathologic examination the tibial-metatarsal and tibial-femoral joints appeared markedly smaller in the chicks on the low-silicon diet. The epiphyseal (proximal) ends of the tibia (Fig. 4) and metatarsus had less articular cartilage and were narrower and less shaped. Joints at the distal extremities of the bones also were smaller and less well formed. However, although, as shown in Table 3, the tibia and femur are heavier in the supplemented group, analyses of the bones showed no difference in dry weight, amount of organic matter, or absolute amount of ash between the groups. The sole analytical difference is that tibia and femur from the deficient group have 34–35% less water.

The larger proportions of cartilage and of water in bones of the silicon-supplemented group, the fact that cartilage is known to be rich in mucopolysaccharides, and the observation that mucopolysaccharides exist in solution as highly viscous materials which avidly bind water suggested that studies should be made of mucopolysaccharide content. Moreover, analyses of mucopolysaccharide-rich tissues in this laboratory had demonstrated appreciable amounts of silicon and, by extraction and purification, this silicon was shown to be

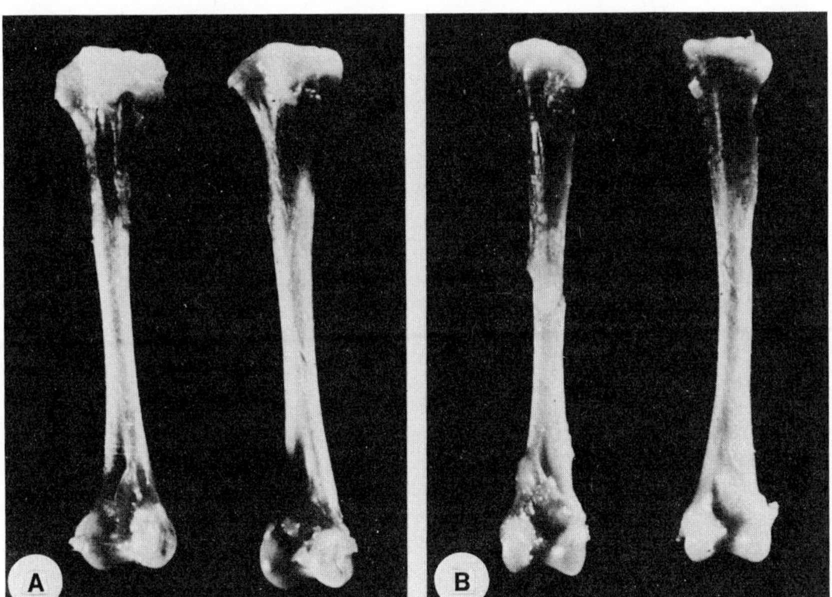

Fig. 4. Tibial bones from 4-week-old chicks on (a) a silicon-supplemented diet and (b) a low-silicon basal diet.

Table 3. Effect of silicon intake on tibia and femur composition[1]

Diet	Tissue (wet wt) (g)	Water content (g)	Tissue (dry wt) (g)	Organic matter (g)	Ash (g)
			Tibia		
Low silicon	0.85 ± 0.09^a	0.37 ± 0.03^a	0.48 ± 0.02	0.34 ± 0.03	0.15 ± 0.01
Supplemented	1.03 ± 0.02	0.56 ± 0.04	0.48 ± 0.02	0.34 ± 0.04	0.14 ± 0.02
			Femur		
Low silicon	0.63 ± 0.06^a	0.28 ± 0.02^a	0.36 ± 0.01	0.25 ± 0.02	0.11 ± 0.01
Supplemented	0.78 ± 0.02	0.43 ± 0.03	0.36 ± 0.01	0.25 ± 0.02	0.11 ± 0.01

[1] There were 12 chicks per group.
All values reported as mean ± s.d.
[a] Significantly different from the supplemented animals at $P < 0.01$.

Table 4. Effect of silicon intake on articular cartilage composition[1]

Diet	Tissue (wet wt) (mg)	Total hexosamine (wet wt) (mg)	Percentage hexosamine (wet) (%)
Low silicon	63.32 ± 8.04[a]	0.187 ± 0.023[a]	0.296 ± 0.009[a]
Supplemented	86.41 ± 4.82	0.310 ± 0.031	0.359 ± 0.011

[1]There were 12 chicks per group.
All values reported as mean ± s.d.
[a]Significantly different from the supplemented animals at $P < 0.001$.

localized in the mucopolysaccharide fraction. Articular cartilage was removed from the tibia of the experimental animals, therefore, and by analysis was shown (Table 4) not only to be present in larger quantities in the silicon-supplemented group but also to contain a greater proportion of hexosamine. Thus, the greater water content in bones of the silicon-supplemented chicks coincides with a larger content of mucopolysaccharide in the articular cartilage.

The comb of the cockerels was then examined because a difference had been noted in comb size and because cockerel comb is a so-called 'target connective tissue'. Here, too, the amount of connective tissue, the total amount of hexosamines, and also the proportion of hexosamine in the comb was found to be larger in the supplemented group. Additional analysis also revealed a significantly higher silicon content (Table 5).

These observations on the amounts of mucopolysaccharides in connective tissues in the bone and in the comb of silicon-deficient and silicon-supplemented chicks, and the relationships between silicon and mucopolysac-

Table 5. Effect of silicon intake on comb composition[1]

Diet	Tissue (wet wt) (mg)	Total hexosamine (wet wt) (mg)	Percentage hexosamine (wet) (%)	Silicon (dry wt) (ppm)
Low silicon	90.30 ± 4.99[a]	0.085 ± 0.012[a]	0.094 ± 0.003[a]	11.4 ± 0.36[a]
Supplemented	134.80 ± 10.20	0.175 ± 0.020	0.130 ± 0.009	21.2 ± 3.02

[1]There were 12 chicks per group.
All values reported as mean ± s.d.
[a]Significantly different from the supplemented animals at $P < 0.001$.

charides in these and other connective tissues provided the first indication that silicon is involved in mucopolysaccharide synthesis.

SITE OF ACTION

As mentioned above, epithelial and connective tissues are rich in silicon. These so-called collagenous tissues are a complex of cells, fibrous structures, and an amorphous ground substance or matrix composed mainly of mucopolysaccharides. By extraction of several of these connective tissues and further purification, silicon has been shown to be chemically combined in the mucopolysaccharide fraction, most likely in ester linkage (27, 28). The silicon content of the mucopolysaccharide-protein complex extracted in this laboratory from bovine nasal septum hyaline cartilage, for example, is 87 ppm compared with 13 ppm in the original dried cartilaginous tissue. Chondroitin sulfate A, the major component of mucopolysaccharides in bovine septum, was found to contain significantly more silicon than chondroitin sulfate C, a minor component. Other extracted and purified mucopolysaccharides have also been found to contain amounts of silicon in the same range as those of chondroitin sulfate A.[2] These include hyaluronic acid and dermatan sulfate extracted from skin (pig and rat), and hyaluronic acid extracted from rooster comb and umbilical cord. Chondroitin sulfate C and keratan sulfate from cartilage contain lesser amounts. This silicon is not removed from the mucopolysaccharide except by strong acid or alkaline extraction. Similar results on isolated mucopolysaccharides, which included some research reference standards, have been reported (29).

The above data provide substantial proof that silicon, in addition to being involved in mucopolysaccharide synthesis, is an essential component of animal mucopolysaccharides and that its site of action in connective-tissue metabolism is in the mucopolysaccharide-protein complexes of the ground substance. It should be understood that the term 'mucopolysaccharide' is used for macromolecules isolated from proteolytic digests of connective tissues or after extraction with alkali. The mucopolysaccharides obtained are thus artifacts, the breakdown products of protein complexes, in part free of peptides, in part still covalently linked to peptide chains of differing lengths (30).

This finding has important implications because, apart from bone formation, silicon must also participate in other processes in which mucopolysaccharides are involved. Foremost among these are growth and maintenance of connective tissues, as in embryonic development and wound healing. Degener-

2. E. M. Carlisle, "Silicon and Polymeric Carbohydrates" (manuscript in preparation).

ative conditions such as atherosclerosis and osteoarthritis, and the overall aging process are also associated with significant changes in mucopolysaccharides.

SILICON AND AGING

The silicon content of the aorta, skin, and thymus is found to decline significantly with age of the animal, in contrast with other analyzed tissues which showed little or no change.[3] This relationship occurs in several species, including rabbit (Fig. 5), rat, chicken, and pig. For example, in the rabbit between 12 weeks and 18–24 months of age (Fig. 5), silicon diminishes in the aorta by 84%, in the thymus by 96%, and in the skin by 83%. Silicon content of rat skin has been shown to decrease by 60% between five weeks and 30 months of age in contrast with other tissues, such as brain, liver, spleen, lung, and femur, which showed an increase (31).

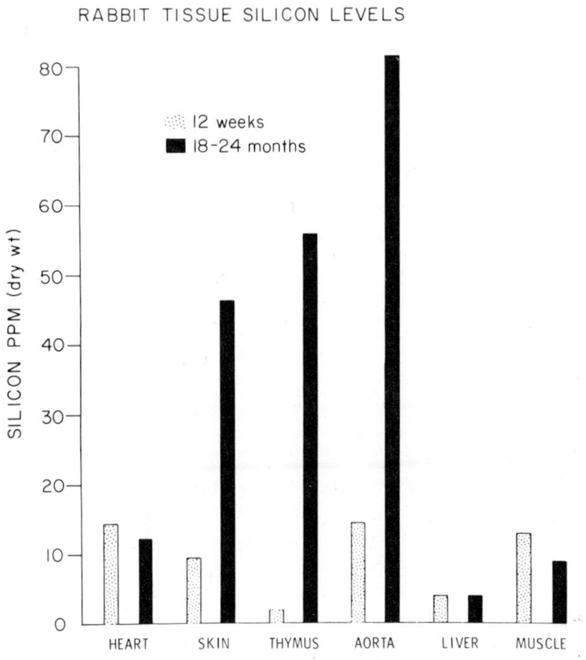

Fig. 5. Effect of age on rabbit-tissue silicon levels. Values represent mean silicon levels of 18 rabbits expressed as ppm dry weight of tissue.

3. E. M. Carlisle, 1973 (unpublished data).

In humans, also, the silicon content of the dermis has been reported to diminish with age (32, 33). In contrast with an earlier finding (34), French investigators (35) have reported that the silicon content of the normal human aorta decreases considerably with age and, furthermore, that the level of silicon in the arterial wall decreases with the development of atherosclerosis. Of possible significance here is the report of another French worker (36) on changes in absorption and resulting levels of silicon in the blood and intestinal tissues of rats in relation to age, sex, and various endocrine glands. It is suggested that the decline of hormonal activity in senescence may well account for the modifications in silicon observed in aged animals.

CONCLUDING REMARKS

Silicon has been shown to be an essential trace element for higher animals, and a mechanism and site of action for this element has been identified. It is suggested that silicon may contribute to the structural framework of connective tissue by forming links or bridges within and between polysaccharide chains and perhaps linking the polysaccharide chains to proteins. In this way, silicon may aid in the development of the architecture of fibrous elements and contribute to structural integrity by providing strength and resilience.

REFERENCES

1. Gonnerman, M. 1919. Beiträge zur kenntnes der Biochemie der Kieselsaure und tonerde. *Biochem. Z.* 88: 401.
2. Gonnerman, M. 1919. Der quantitative ausscheidung der kieselsäure durch den menschlichen ham. *Biochem. Z.* 94: 163.
3. King, E. J., and Belt, T. H. 1938. The physiological and pathological aspects of silica. *Physiol. Rev.* 18: 329.
4. Jones, L. H. P., and Handreck, K. A. 1967. Silica in soils, plants and animals. *Adv. Agron.* 19: 107.
5. Van Soest, P. J. 1970. The role of silicon in the nutrition of plants and animals. *Proceedings of Cornell University Conference for Feed Manufacturers,* Cornell University, Ithaca, N.Y., p. 103.
6. Underwood, E. J. 1971. *Trace Elements in Human and Animal Nutrition.* Academic Press, New York.
7. Reimann, B. E. F. 1964. Deposition of silica inside a diatom cell. *Exp. Cell. Res.* 34: 605.
8. Darley, W. M., and Volcani, B. E. 1969. Role of silicon in diatom metabolism. *Exp. Cell. Res.* 58: 334.
9. Reimann, B. E. F., Lewin, J. C., and Volcani, B. 1965. Studies on the biochemistry and fine structure of silica shell formation in diatoms. I. The structure of the cell wall of *Cylindrotheca fusiformis. J. Cell. Biol.* 24: 39.
10. Reimann, B. E. F., Lewin, J. E., and Volcani, B. 1966. Studies on the

biochemistry and fine structure of silica shell formation in diatoms. II. The structure of the cell wall of *Navicula pelliculosa* (Bréb.) *Hilse, J. Phycol.* 2: 74.
11. Fessenden, R. J., and Fessenden, J. S. 1967. The biological properties of silicon compounds. *Adv. Drug Res.* 4: 95.
12. McConnell, D. 1938. A structural investigation of the isomorphism of the apatite group. *Am. Mineral.* 23: 1.
13. McConnell, D. 1937. The substitution of SiO_4 and SO_4 groups for PO_4 groups in the apatite structure: ellestadite, the end member. *Am. Mineral.* 22: 977.
14. Carlisle, E. M. 1969. Silicon localization and calcification in developing bone. *Fed. Proc.* 28: 374.
15. Carlisle, E. M. 1970. Silicon a possible factor in bone calcification. *Science* 167: 279.
16. Sledge, C. G. 1966. Some morphologic and experimental aspects of limb development. *Clin. Orthop.* 44: 241.
17. Gersh, I. H., and Catchpole, H. R. 1960. The nature of ground substance of connective tissue. *Prospects Biol. Med.* 3: 282.
18. Kashiwa, H. K. 1966. Calcium in cells of fresh bone stained with glyoxal bis (2-hydroxy yanil) *Stain Tech.* 41: 49.
19. Mathews, J. L., Martin, J. H., and Collins, E. J. 1968. Metabolism of radioactive calcium by cartilage. *Clin. Orthop.* 58: 213.
20. Sobel, A. E., Laurence, P. A., and Burger, M. 1960. Nuclei formation and crystal growth in mineralizing tissues. *Trans. N.Y. Acad. Sci.* 22: 233.
21. Urist, M. R. 1966. Origins of current ideas about calcification. *Clin. Orthop.* 44: 13.
22. Carlisle, E. M. 1970. A relationship between silicon and calcium in bone formation. *Fed. Proc.* 29: 565.
23. Muller, S. A., Posner, A. S., and Firschein, H. E. 1966. Effect of vitamin D-deficiency on the crystal chemistry of bone mineral. *Proc. Soc. Exp. Biol. Med.* 121: 844.
24. Carlisle, E. M. 1972. Silicon an essential element for the chick. *Fed. Proc.* 31: 700.
25. Carlisle, E. M. 1972. Silicon an essential element for the chick. *Science* 178: 619.
26. Schwarz, K., and Milne, D. B. 1972. Growth-promoting effects of silicon in rats. *Nature, Lond.* 239: 333.
27. Carlisle, E. M. 1973. A skeletal alteration associated with silicon deficiency. *Fed. Proc.* 32: 930.
28. Carlisle, E. M. 1974. Silicon as an essential element. *Fed. Proc.*, in press.
29. Schwarz, K. 1973. A bound form of silicon in glycosaminoglycans and polyuronides. *Proc. Natn. Acad. Sci.* 70: 241.
30. Meyer, K. 1966. Mucopolysaccharides. *Fed. Proc.* 25: 1032.
31. Leslie, J. G., Kung-Ying, T. K., and McGavack, T. H. 1962. Silicon in biological material. II. Variations in silicon contents in tissues of rats at different ages. *Proc. Soc. Exp. Biol. Med.* 110: 218.
32. Brown, H. 1927. The mineral content of human skin. *J. Biol. Chem.* 75: 789.
33. MacCardle, R. C., Engman, M. F., Jr., and Engman, M. F., Sr. 1943.

XCIV, Mineral changes in neurodermatitis revealed by microincineration. *Archs Dermat. Syph.* 47: 335.
34. Kvorning, S. A. 1950. The silica content of the aortic wall in various age groups. *J. Gerontol.* 5: 23.
35. Loeper, J., Loeper, J., and Lemaire, A. 1966. Étude du silicum en biologie animals et au cours de l'atherome. *Presse Med.* 74: 865.
36. Charnot, Y., and Peres, G. 1971. Contribution a l'étude de la regulation endocrinienne du métabolisme silicique. *Anal. Endocr.* 32: 397.

DISCUSSION

Izatt (Provo). Many silicon compounds are very insoluble. Do you have any ideas about how silicon is transported to the mineralization site, and what form it is in?

Carlisle. We have found that the availability, and presumably solubility, of different silicon compounds varies widely in the chick, and Schwarz has seen the same in rats. There has been some work done with bacteria which shows that silicon esters of sugars are found which might be a type of transport form.

Aughey (Glasgow). Some workers in our veterinary school have been studying changes in dog articular cartilage leading to arthritis, and they have shown a difference in metachromasia of the chondroitin mucoprotein of the articular cartilage. Could you comment on a possible relationship between this and silicon status?

Carlisle. We are just starting work on the possible changes in polymerization of hyaluronic acid in this condition. We have isolated a fine zone of developing cartilage from calf fetus and it seems to be very high in silicon.

Gedalia (Jerusalem). What in your opinion would be the effect of silicon administration to pregnant experimental animals? Would it increase the calcium:phosphorus ratio in the newborn?

Carlisle. It would probably not increase the calcium:phosphorus ratio because if you form bone, the ratio is usually the same. It did seem, however, when we were doing the rat reproduction studies, that the animals on silicon-supplemented diets were more mineralized; that is, they had more bone ash.

Armstrong (Minneapolis). I was particularly interested that you found a calcium:phosphorus ratio of 0.8. I can't imagine a calcium-phosphorus compound with a calcium:phosphorus ratio of less than 1 in the biological tissues.

Carlisle. What I meant by a calcium:phosphorus ratio of less than 1 might be a little confusing. I was measuring this with microprobe analysis, so what we were really measuring was organic phosphorus and this was at a very early stage before mineralization starts.

Erway (Cincinnati). We've been looking for some reason why mucopolysaccharides are involved in the formation of calcite crystals in the inner ear.

There is an integral relationship between the mucopolysaccharide matrix and the formation of these crystals. Perhaps it is acting as a nucleation site, and whenever the matrix is abnormal because of manganese deficiency, or in some of the mutants we study, crystals don't form. I wonder how the silicon might be involved?

Carlisle. It's very interesting that the polysaccharide-type matrix of all calcareous tissues does contain silicon. From the work that I've done, it seems that silicon may be involved as a calcium concentrating agent, and then phosphate is released to combine with calcium or take the place of the silicon at the calcium-silicon site.

ESSENTIALITY AND FUNCTION OF FLUORIDE[1]

H. H. MESSER,[2] W. D. ARMSTRONG, and L. SINGER

Department of Biochemistry (Health Sciences), University of Minnesota, Minneapolis, Minnesota

Although fluorine (F) constitutes only a small percentage (0.065%) of the elements of the earth's crust (1), its remarkably wide distribution throughout nature suggests some biological role for the element. Whether this role is that of an essential trace element, or merely a beneficial effect under specific conditions, has been a contentious point for many years. It is now 40 years since one of the earliest laboratory investigations (that of Sharpless and McCollum in 1933) failed to demonstrate any characteristic defects in rats raised on a diet "very low in F but not quite free" (2). Since that time a majority of studies have yielded negative or at best equivocal results, and only very recently has more conclusive evidence of an essential role for F been produced.

A fluorine-deficiency state has not been observed under natural conditions, and experimental studies of animals on an artificially restricted F intake have been necessary. This approach has by no means been confined to the study of F. For trace elements in this category it is important to define very precisely the criteria used in assessing essentiality, since both the experimental design and the conclusions drawn depend on the criteria used. For our own studies we have used two criteria as the minimum requirements that must be met for essentiality.

(1) A specific deficiency state should be produced by a diet lacking the element in question, but which is otherwise adequate and satisfactory.

(2) The deficiency should be prevented or cured by addition to the diet of that element alone.

1. Supported by Grant No. DE-01850, National Institute for Dental Research, Bethesda, Maryland.

2. Present address: Physiology Department, University of British Columbia, Vancouver, B.C., Canada.

Three additional criteria may provide more conclusive evidence of essentiality.

(1) In general, the deficiency should be correlated with subnormal tissue levels of the element.

(2) The deficiency should be accompanied by pertinent biochemical or physiological changes, which will be prevented or cured when the deficiency is prevented or cured.

(3) A homeostatic control mechanism regulating tissue levels of the element should be demonstrable.

Very few 'essential' trace elements have been shown to satisfy all of these criteria, and F is no exception. However, we feel that at least the first two of these have been met, allowing us to include F tentatively in the list of essential trace elements.

LOW FLUORIDE DIETS

Before considering the different lines of evidence justifying this conclusion, it is pertinent to review the characteristics of low-F diets, since a number of studies appear to have been prejudiced by nutritionally poor diets or by inadequate control diets. The wide natural distribution of F makes preparation of diets low in F but otherwise adequate nutritionally quite difficult. Grains and milk products generally have a low F content (3), and have been used without purification in the preparation of diets with a basal F level down to 0.1 ppm (4, 5). Extensive extraction of individual components to remove F, and the use of hydroponically grown crops, have yielded diets estimated to contain as little as 0.005 ppm F (6–8). In some cases, the resultant diet has failed to support normal growth or reproduction.

The diet of Taylor *et al.* (5), which we have used in studies of reproduction and hemopoiesis, consists largely of milk and grain products used without purification, and has an F level of 0.1–0.5 ppm. Thus it ranks in the upper range of F concentrations for a low-F diet, but has the advantage of bulk preparation. It supports growth of experimental animals at a rate at least equal to that obtained with standard laboratory chows, and when supplemented with F in the drinking water supports normal reproduction. Hence it satisfies the criterion of a diet nutritionally adequate in all respects other than the trace element in question (9).

GROWTH RATES AND GENERAL HEALTH

Numerous studies during the past 40 years have failed to show any impairment of growth or health in experimental animals raised on low-F intakes (2,

8, 10–12). McClendon and Gershon-Cohen (13) observed a retardation in the growth of rats fed a low-F diet based on hydroponically grown grains. However, the control (F-supplemented) group received field-grown crops rather than hydroponically grown crops, plus 20 ppm F in the drinking water. Thus the possibility cannot be excluded that dietary factors other than F were lacking in the hydroponically grown diet. More recently, Schroeder *et al.* (14) described an increased body weight in female mice and increased longevity in male mice supplemented with F, compared with mice on the same basal diet (stated to have "no detectable F") plus distilled water. Schwarz and Milne (15), using a controlled 'isolator' system and highly purified amino acid diets, showed an increase of up to 30% in the daily weight gain of rats fed F-supplemented diets (basal F level: 0.04–0.46 ppm; supplemented with 1.0, 5.0, or 7.5 ppm F as KF). As reported elsewhere in the symposium (16), this isolator technique has proven very successful in demonstrating growth responses to a number of trace elements. However, addition of F alone to the basal diet did not prevent the development of other signs of a deficiency, such as loss of hair and seborrhea.

MINERALIZED TISSUES

The predilection of F for mineralized tissues in the body has raised the question of a requirement for F in calcified tissue metabolism. In his opening address at the first Trace Element Symposium in Aberdeen in 1969, Dr. Underwood (17) suggested that the very high incidence of osteoporosis in women, which is significantly reduced by a high-F intake, can be regarded as a sign of F deficiency. However, he also pointed out the paradox that the desirable F intake to prevent this disease in adults is undoubtedly toxic in children. Fluorine is known to increase crystal size and crystal perfection in biological apatites, and to reduce mineral solubility (18). As a consequence of these effects F inhibits bone resorption and promotes the stabilization of newly synthesized bone matrix, at least under organ-culture conditions (19, 20). The beneficial effects of F in reducing dental caries (21) and osteoporosis (22) have come to be widely regarded as pharmacological effects (23).

Based on *in vitro* studies in aqueous systems, several workers have postulated that F may be required in biological calcification involving apatitic structures (24, 25). Fluorine has been shown to promote precipitation of crystalline apatites from metastable calcium phosphate solutions, and at concentrations of F found *in vivo* in extracellular fluid (26). In fact, under biological conditions of low temperature and neutral pH, the precipitation of hydroxyapatite crystals from solutions containing calcium and phosphate ions appears to require the presence of F (24). It has been suggested that fluorapatite crystals are first formed, which then act as nucleation centers for the growth of hydroxyapatite crystals (25).

If F is required for biological calcification, then defective mineralization should occur when F intake is restricted to a minimum. To our knowledge, there has been no such demonstration of impaired calcification. A reason for failures to demonstrate a requirement for F in calcification may be the extremely low F concentrations capable of promoting crystallization. Brown (27) has calculated that a F concentration of 10^{-8}–10^{-7}M (0.0002–0.002 ppm) should be sufficient to promote apatite formation *in vitro.* The widespread occurrence of F may prevent its complete elimination from animal systems, and thus preclude a convincing demonstration of its requirement in calcification.

REPRODUCTION

Several studies of reproduction have yielded inconclusive evidence concerning an effect of fluoride. Impaired reproduction or poor viability of offspring in both low- and high-fluoride groups has been common (4, 6, 7), suggesting other nutritional inadequacies of the highly refined diets. Thus, even though Maurer and Day (6) showed that reproduction could be maintained in rats over four generations with a diet estimated to contain only 0.007 ppm F, the viability of pups in both low-F and F-supplemented groups was only approximately 50%. Weber (28) found no differences attributable to fluoride intake in a breeding trial of mice over three generations, although his report did not include details of litter production. All previous studies have used very small experimental groups, because of the difficulties involved in preparing diets. Even when the studies were conducted over several generations, the length of time during which each generation was observed appears to have been small.

In a study involving much larger numbers of animals (40–50 per treatment group) and four pregnancies per generation, we found a progressive infertility in mice on a low-F intake (29). Weanling female albino mice were fed the low-F diet (5) described earlier, plus either deionized water or water containing 50 ppm F as NaF. Breeding performance was assessed over 25 weeks of continuous breeding from eight weeks of age, up to a maximum of four litters. A second generation study used pups from the third or fourth litter of the first generation, with an identical breeding regimen. In the group receiving 50 ppm F, 96.0% of mice in the first generation, and 90.0% of mice in the second generation produced four litters in the 25-week period (Fig. 1). These values indicate that reproduction was essentially complete in this group. In the low-F group a highly significant ($p < 0.001$) impairment in breeding capacity was evident by the second or third litter in both generations, and less than 50% produced four litters (Fig. 1). Based on the number of mice 'at risk' for each litter, an average of almost 20% of animals in the low-F group became infertile after each litter.

The low-F intake did not exert a marked effect on the time intervals between litters, frequency of *post partum* conceptions, litter size, body weight of pups at birth, or the growth rates of either pups or mothers. The only parameter other than infertility which was significantly affected by the restricted F intake was the age at delivery of the first litter. In the second generation, mice of the low F-group exhibited a significant ($p < 0.005$) delay in the time taken to produce their first litter (average age, 16.0 ± 0.9 weeks, compared with 13.0 ± 0.5 weeks for the high-F group). We consider this evidence for a delayed onset of sexual maturity.

Recovery from infertility was also investigated by transferring low-F mice with defective reproduction to a high-F intake. Mice which were transferred from a low- to a high-F intake after demonstrating impaired fertility showed a restoration of breeding capacity (Fig. 2). Mice retained on the low F intake continued to show impaired reproduction.

The basis of the infertility remains unknown, but the infertility constitutes a deficiency state produced by a diet which is otherwise adequate nutritionally. The deficiency was both prevented and cured by the addition of F alone (as NaF) to the diet. Thus the two basic criteria which we defined earlier have been met, and support the conclusion that F is an essential trace element. Unfortunately, infertility is probably a nonspecific manifestation of the deficiency, and provides little indication of the basic defect in these animals.

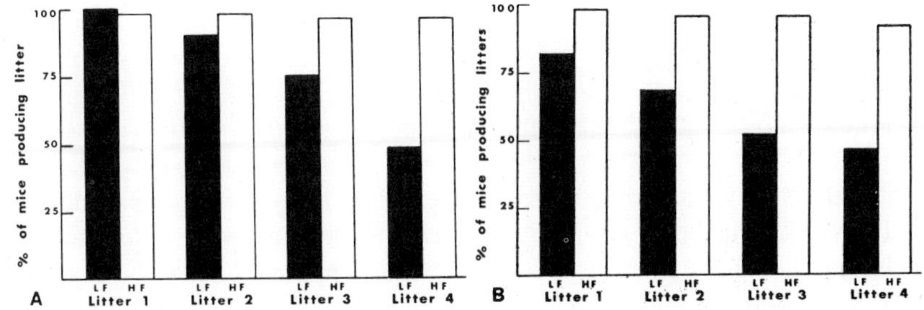

Fig. 1. Influence of F intake on litter production by two generations of mice. The percentage of mice producing a given litter in the high-F group was not significantly less than 100% for either generation. In the low-F group a significant impairment in litter production occurred for (A) litters 3 and 4 of generation 1, and (B) for all litters of generation 2.

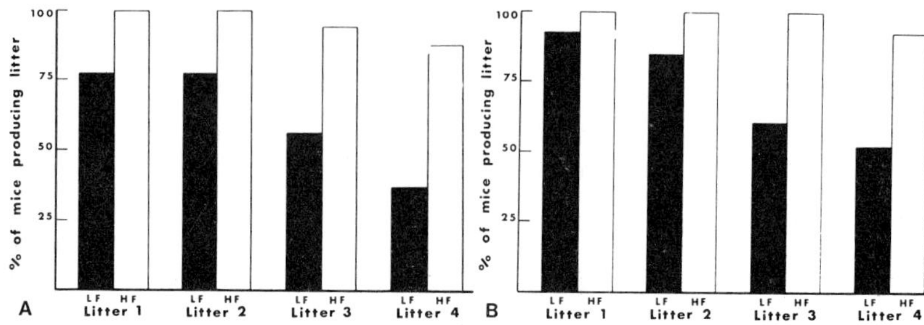

Fig. 2. Restoration of fertility by a high-F intake. All mice had first demonstrated impaired reproduction on a low-F intake. The figures depict litter production in a 20-week period following transfer to a high-F intake or retention on the low-F intake. Mice were placed on the recovery study after failing to produce (A) any litter in 5 weeks, or (B) only 1 or 2 litters in 15 weeks.

HEMOPOIESIS

In an attempt to find a more specific physiological or biochemical defect in F-deficient animals, we investigated hemopoiesis in these mice. In particular, two forms of anemia of a largely 'physiological' nature (30) were studied: the anemias of pregnancy and of infancy. We have recently reported (31) that a high-F intake affords protection against these two forms of anemia.

Female albino mice, which had received a low-F diet (5) plus deionized water or water containing 50 ppm F (as NaF) since weaning were time-mated for 12 hr at 10 weeks of age. Packed cell volumes were determined on tail blood samples taken two days before mating and 5, 10, 15, and 19 days after mating. The initial hematocrit values were not significantly influenced by F intake (low F, 49.9 ± 0.9%; high F, 48.7 ± 1.2%). Both groups showed a progressive anemia of pregnancy (Fig. 3). By the nineteenth day, the severity of the anemia was significantly ($p < 0.05$) greater in the low-F group than in the high-F group. The F intake did not influence body weight before mating or 19 days after mating, the number of pups per litter, or the duration of pregnancy.

Hematocrit values for tail blood taken from pups born to these mothers (and to additional mice on identical dietary regimens) were determined at birth and at 5, 10, 15, 20, and 60 days of age. Newborn mice of both groups possessed comparable hematocrit values (low F, 38.5 ± 1.4%; high F, 38.1 ± 1.0%). Pups born to mothers with a high-F intake showed only a slight decline prior to weaning, followed by an increase to 51.5 ± 0.9% at 60 days of age (Fig. 4). In the low-F group, the hematocrit values had decreased to approximately 25% by 10 days of age, and remained low until after weaning.

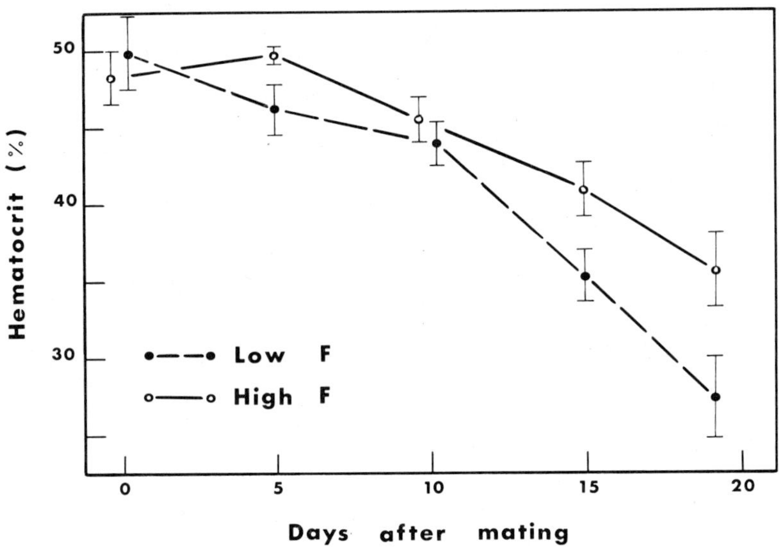

Fig. 3. Hematocrit values of tail blood of pregnant mice fed a low-F diet plus deionized water (low F) or water containing 50 ppm F (high F).

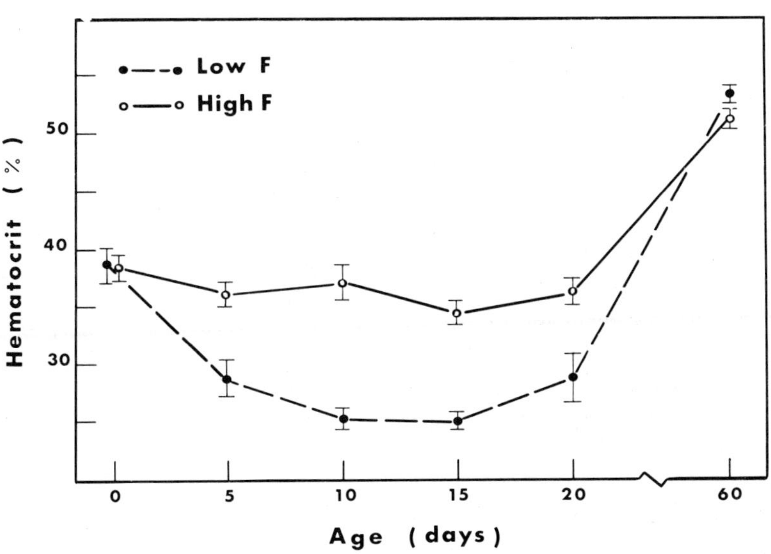

Fig. 4. Effect of maternal F intake on hematocrit values of tail blood of mouse pups. For the 60-day values, pups were weaned and given the same F intake as their dams.

The values for both groups at 60 days of age were similar to those of the mothers before mating.

Both of these forms of anemia are considered to be associated, at least in part, with an inadequate iron intake in situations of an increased physiological demand (30). Preliminary evidence[3] suggests that the anemia of infancy in these mouse pups is of the microcytic, hypochromic type, which supports the conclusion of a secondary deficiency of iron in the low-F pups. Some degree of anemia also occurred in the high-F animals, but the high-F intake provided a highly significant ($p < 0.001$) protection. The low-F diet appears adequate to meet normal adult requirements, but the enhanced requirement of pregnancy and the rapid growth of infancy cannot be met, resulting in the anemia. Since a high-F intake is known to enhance intestinal absorption of iron (32), the protection against anemia conferred by the high-F intake can be explained on this basis. Milk is a poor source of dietary iron (33), and F may also act to increase the secretion of iron or other trace elements in milk, or the absorption of trace elements from milk.

ESSENTIALITY AND FUNCTION

In view of the more recent (and more positive) evidence, we must reconsider the question: Is F an essential trace element?

Undoubtedly, a deficiency state has been demonstrated, based on the retarded growth rates in rats reported by Schwarz and Milne (15) and the infertility and anemia which we described (29, 31). Thus the first criterion defined earlier has been met, although decreased growth rates and infertility are both relatively nonspecific in that they are common to many trace-element and vitamin deficiencies. The second criterion, concerning prevention or cure of the deficiency by F alone, has also been satisfied in the case of prevention of infertility and restoration of reproductive capacity. In the case of anemia and in the growth studies of Schwarz and Milne, inclusion of F in the diet was not sufficient to overcome completely all signs of deficiency.

Identification of a specific biochemical function for F remains elusive. Decreased growth rates and infertility are sufficiently nonspecific that they do not suggest an experimental approach to identifying the underlying defect. The anemia studies provide a more promising approach, particularly since they implicate F in some aspect of iron metabolism. However, unless it can be shown that F is essential to the absorption or utilization of iron (or other trace metals), then we may have observed only a secondary deficiency state.

3. M. E. Wegner, personal communication.

At the very least, the study has reaffirmed the importance of interactions between elements in trace-element metabolism.

Decreased enzyme activities in a number of organ systems of low-F animals have been reported, including liver isocitric dehydrogenase (8), and bone acid and alkaline phosphatases (34, 35). None of these changes in enzyme activity have been correlated with specific signs of deficiency, and their significance is unknown. Similarly, the inhibition of a wide variety of enzymes by F *in vitro* (36) and the activation of adenyl cyclase (37) both require such high concentrations of F that any physiological significance is unlikely. The only known chemical reaction of physiological significance involving F is the displacement of hydroxyl ions from hydroxyapatite to form fluorapatite. However, even though the incorporation of large quantities of F into biological apatites reduces both dental caries and bone resorption, skeletal and dental abnormalities are not seen in F-deficient animals. Thus an essential role for F in calcified tissue metabolism has not been established.

CONCLUSIONS

A deficiency state has been demonstrated in animals with a restricted F intake, characterized by retarded growth rates, infertility, and anemia. These signs of deficiency are relatively nonspecific (with the possible exception of the anemia), and a specific biochemical lesion accompanying these signs has not been identified. The demonstration of a deficiency state, and its prevention and cure by F alone, justify the tentative inclusion of F in the list of essential trace elements.

REFERENCES

1. Fleischer, M. 1953. Recent estimates of the abundance of elements in the earth's crust. *U.S. Geol. Survey Circ.* No. 285, Washington, D.C.
2. Sharpless, G. R., and McCollum, E. V. 1933. Is fluorine an indispensable element in the diet? *J. Nutr.* 6: 163.
3. Muhler, J. C. 1970. Supply of fluorine to man: ingestion from foods. *In Fluorides and Human Health,* W.H.O. Monograph Series, No. 59, Geneva, p. 32.
4. Muhler, J. C. 1954. Retention of fluorine in the skeleton of the rat receiving different levels of fluorine in the diet. *J. Nutr.* 54: 481.
5. Taylor, J. M., Gardner, D. E., Scott, J. K., Maynard, E. A., Downs, W. L., Smith, F. A., and Hodge, H. D. 1961. Toxic effects of fluoride on the rat kidney. *Toxic. Appl. Pharmac.* 3: 290.
6. Maurer, R. L., and Day, H. G. 1957. The non-essentiality of fluorine in nutrition. *J. Nutr.* 62: 561.

7. Venkateswarlu, P. 1962. Studies on fluoride metabolism and transport. Ph.D. Thesis, University of Minnesota.
8. Doberenz, A. R., Kurnick, A. A., Kurtz, E. B., Kemmerer, A. R., and Reid, B. L. 1964. Effect of a minimal fluoride diet on rats. *Proc. Soc. Exp. Biol. Med.* 117: 689.
9. Underwood, E. J. 1971. *Trace Elements in Human and Animal Nutrition,* 3rd edn, Academic Press, New York, p. 1.
10. Phillips, P. H., Hart, E. B., and Bohstedt, G. 1934. The influence of fluorine ingestion upon the nutritional qualities of milk. *J. Biol. Chem.* 105: 123.
11. Evans, R. J., and Phillips, P. H. 1939. A new low fluorine diet and its effect upon the rat. *J. Nutr.* 18: 353.
12. Lawrenz, M. 1945. *Cited by* H. H. Mitchell and M. Edman, *Soil Sci.* 60: 81.
13. McClendon, J. F., and Gershon-Cohen, J. 1953. Trace element deficiencies. Water-culture crops designed to study deficiencies in animals. *J. Agric. Food Chem.* 1: 464.
14. Schroeder, H. A., Mitchener, M., Balassa, J. J., Kanisawa, M., and Nason, A. P. 1968. Zirconium, niobium, antimony and fluorine in mice: effects on growth, survival and tissue levels. *J. Nutr.* 95: 95.
15. Schwarz, K., and Milne, D. B. 1972. Fluorine requirement for growth in the rat. *Bioinorg. Chem.* 1: 331.
16. Milne, D. B., and Schwarz, K. 1973. Effect of different fluorine compounds on growth and bone fluoride levels in rats. This symposium.
17. Underwood, E. J. 1970. Progress and perspectives in the study of trace element metabolism in man and animals. *In* C. F. Mills (ed.), *Trace Element Metabolism in Animals,* Livingstone, Edinburgh, p. 5.
18. Zipkin, I. 1970. Physiological effects of small doses of fluoride. 3. Effects on the skeleton of man. *In Fluorides and Human Health,* W.H.O. Monograph Series, No. 59, Geneva, p. 185.
19. Messer, H. H., Armstrong, W. D., and Singer, L. 1973. Fluoride, PTH and calcitonin: inter-relationships in bone calcium metabolism in organ culture. *Calc. Tiss. Res.* 13: 217.
20. Messer, H. H., Armstrong, W. D., and Singer, L. 1973. Influence of *in vivo*-incorporated fluoride on collagen metabolism by mouse calvaria in organ culture. *Archs Oral Biol.,* 18: 1393.
21. Adler, P. 1970. Fluorides and dental health. *In Fluorides and Human Health,* W.H.O. Monograph Series, No. 59, Geneva, p. 323.
22. Bernstein, D. S., Sadowski, N., Hegsted, D. M., Guri, C. D., and Stare, F. J. 1966. Prevalence of osteoporosis in high- and low-fluoride areas in North Dakota. *J. Am. Med. Ass.* 198: 499.
23. National Academy of Sciences. 1971. Is fluorine an essential element? *In Fluorides,* National Academy of Sciences, Washington, D.C., p. 66.
24. Newesely, H. 1961. Changes in crystal types of low solubility calcium phosphates in the presence of accompanying ions. *Archs Oral Biol.* 6 (suppl.): 174.
25. Perdok, W. G., 1963. Crystallographic aspects of calcification. *Archs Oral Biol.* 8 (suppl.): 85.

26. Brudevold, F., and Messer, A. C. 1961. Seeding effect of hydroxyapatite in calcifying solutions containing different levels of fluoride. *J. Dent. Res.* 40: 728 (abstr.).
27. Brown, W. E. 1966. Crystal growth of bone mineral. *Clin. Orth.* 44: 205.
28. Weber, C. W. 1966. Fluoride in the nutrition and metabolism of experimental animals. Ph.D. Thesis, University of Arizona.
29. Messer, H. H., Armstrong, W. D., and Singer, L. 1972. Fertility impairment in mice on a low fluoride intake. *Science* 177: 893.
30. Wintrobe, M. M. 1961. *Clinical Hematology,* 5th edn, Lea and Febiger, Philadelphia, p. 731.
31. Messer, H. H., Wong, K., Wegner, M., Singer, L., and Armstrong, W. D. 1972. Effect of reduced fluoride intake by mice on haematocrit values. *Nature New Biol.* 240: 218.
32. Ruliffson, W. S., Burns, L. V., and Hughes, J. S. 1963. The effect of fluoride ion on ^{59}Fe levels in blood of rats. *Trans. Kansas Acad. Sci.* 66: 52.
33. Jacobs, M. B. 1951. *The Chemistry and Technology of Food and Food Products,* 2nd edn, Vol. 2, Interscience, New York, p. 835.
34. Lai, C. C. 1970. Phosphatase activities of bone and calcification mechanism in the rat. Ph.D. Thesis, University of Minnesota.
35. Messer, H. H., Armstrong, W. D., and Singer, L. 1973. Fluoride, PTH and calcitonin: effects on metabolic processes involved in bone resorption. *Calc. Tiss. Res.,* 13: 227.
36. Cimasoni, G. 1970. Fluoride and enzymes. *In* T. L. Vischer (ed.), *Fluoride in Medicine,* Hans Huber, Bern, p. 14.
37. Marcus, R., and Aurbach, G. D. 1969. Bioassay of parathyroid hormone *in vitro* with a stable preparation of adenyl cyclase from rat kidney. *Endocrinology* 85: 801.

DISCUSSION

Unidentified. It is very difficult to separate iron absorption from iron utilization because you automatically get an increase in the absorption under any circumstances where utilization is increased. Have you done any studies to determine whether you are getting an increase in absorption or the utilization of iron?

Messer. No, but there is a literature report that the administration of fluoride either previously or at the same time as a dose of ^{59}Fe did increase the intestinal absorption of iron in the rat.

Suttie (Madison). Have you shown that your effect on reproduction is specific for fluoride, or can you also reverse it completely by simply adding iron to the diet?

Messer. We were able to reverse the impaired reproduction completely by simply adding fluoride to the diet. We have not investigated if it can be reversed by iron.

Suttie (Madison). Have you analyzed your diet for iron or copper?

Messer. Yes, the diet contains about 25 ppm of iron which is somewhat marginal. I don't recall the exact concentration of copper but it is also marginal.

Venkateswarlu (Minneapolis). I would like to comment on the fluoride content of bones from the low-fluoride animals. We have different ways of making low-fluoride calcium phosphate, and if we diffuse or extract the fluoride from calcium phosphate without ashing, we get a lower value than if we ash the sample first. My feeling is that all of us are reporting higher fluoride values than we should be in bone ash.

Messer. Our values are in the same range reported by others, and Dr. Weber will be reporting fluoride levels approximately half of what we have found. I believe, however, the diet he uses is much lower in fluoride than ours.

Klevay (Grand Forks). I should like to make a general comment regarding criteria of essentiality. If you consider the usual intakes of water and the need for vitamin B_{12}, the ratio between the two is something like 10^8. However, if you increase the intake of water due to high temperatures and consider only the cobalt portion of the B_{12} molecule, then the ratio becomes something like 10^9. Consequently, I don't believe you can tell anything about essentiality of a material by considering how much of that material is present in the diet.

Schwarz (Long Beach). I must say I have some misgivings, to put it mildly, with the first criterion of essentiality mentioned by Dr. Messer. If we were to stick to that, neither riboflavin, pantothenic acid, vitamin B_6, or vitamin A would be recognized, because the discovery of these certainly was made with diets which did not contain all of the essential ingredients. This criteria of essentiality is, in my book, absurd.

Messer. I should just like to say that I have merely restated Underwood's criteria, and I defer to him as my authority.

Suttie (Madison). Let us return to the point of what the word 'essentiality' means. I would agree that when you first demonstrate that something may have an essential function, it might be necessary to do it in the presence of a diet that's not adequate, but I would think that, for something to be called 'essential,' it should essentially be shown that the element produces a response in nutritionally adequate diets. With regard to the fluoride problem, workers in the past have produced diets as low in fluoride as those used by Dr. Schwarz and Dr. Messer, and I think it is significant that in these recent reports the addition of 1 ppm F is not sufficient for maximum response. There certainly has been a number of people who have produced diets which are probably somewhere between 0.1 and 0.3 ppm fluoride and have not obtained any response from additional fluoride. Because of this, I think that it is premature to call fluoride essential at this time.

Klevay (Grand Forks). The important point seems to be whether or not a material x which we are considering as possibly essential is not either substituting for y, or can be replaced by y. In the initial experiments you can't always do this, but sooner or later you must carefully look at it from this viewpoint.

Schwarz (Long Beach). I should comment on that. We are testing not just these elements which we have shown to be essential, but are continuously testing out a whole spectrum of trace elements at different stages of our experimentation. I don't think any of the other agents which we have tested will substitute for fluoride. Therefore from that point of view, I think the criterion of essentiality is fulfilled. However, I can probably repeat what I said in Atlantic City: at the moment, essentiality, just like beauty, is pretty much in the eyes of the beholder.

RAPID MULTI-ELEMENT ANALYSIS OF BIOLOGICAL SAMPLES USING AN ENERGY-DISPERSIVE X-RAY FLUORESCENCE METHOD

N. F. MANGELSON, G. E. ALLISON, J. J. CHRISTENSEN,
D. J. EATOUGH, M. W. HILL, R. M. IZATT, and K. K. NIELSON

Center for Thermochemical Studies and Departments of Chemistry, Chemical Engineering, and Physics, Brigham Young University, Provo, Utah

Elemental analysis by the method of X-ray fluorescence has been used as a quantitative, analytical tool for many years. However, with increasing awareness of the role of trace elements in plant and animal health and as a result of recent technological advances, analysis by X-ray fluorescence is receiving increased attention.

The method of X-ray fluorescence requires excitation of the inner electronic structure of atoms and energy analysis of the resultant X-ray fluorescence. Excitation can be accomplished by several means: (a) photons from an isotopic source or from an X-ray tube; (b) heavy ions, especially protons from a particle accelerator; (c) electrons such as are used in a microprobe. A fraction of excited atoms upon de-excitation emit X-ray photons with the energy or wavelength of each emitted photon being characteristic of the element from which it was emitted. Measurement of emitted X-ray energies, therefore, allows qualitative identification of elements. The method may be made quantitative by measuring the intensity or number of X-rays identifiable with an element. Calibration curves are obtained by use of standard samples.

Determination of X-ray energy and intensity is accomplished by one of two competitive techniques. The traditional technique disperses (or separates) X-rays of different energies by diffracting them through a crystal [Birkes, X-Ray Spectrochemical Analysis, 2nd edn, Interscience, New York (1969)]. This method has the advantage of high energy resolution.

The present paper is concerned with the newer technique of energy dispersion. In recent years small, high-resolution, solid-state si(Li) and Ge detectors and low-noise amplifiers have been developed [Giauque *et al.*, Lawrence Berkeley Lab. Rep. No. LBL-647, (July 1972)]. When an X ray enters the detector system, a small electronic pulse is created, whose magnitude is proportional to the X-ray energy. Each pulse is analyzed for magnitude and a record of the event is stored in a multichannel analyzer. A typical analysis spectrum is shown in Fig. 1. Several laboratories have checked their X-ray fluorescence equipment and methods by determining trace-metal concentrations in N.B.S. SRM No. 1571, Standard Orchard Leaves.

Cooper [Nucl. Instrum. Methods 106: 525 (1973)] has compared X-ray fluorescence data generated by several means of excitation and analyzed by the energy-dispersive method. Although data taken by monochromatic photon or by low-energy proton excitation are generally comparable, those seeking new fluorescence installations are, probably, best advised to use monochromatic X rays from an X-ray tube.

Present X-ray fluorescence results indicate that the limit of detectability in the sample presented to the system ranges from 1–20 ppm for many biologically significant elements. Elements lighter than potassium are not well determined by this method for most sample materials. Technological improvements and imaginative methods will continue to lower the limits of detectability and increase the number of elements for which quantitative data

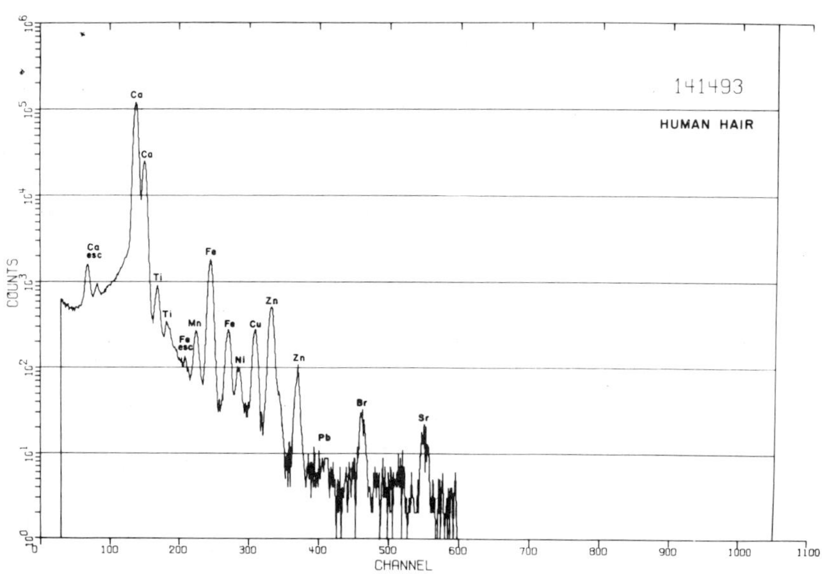

Fig. 1. X-ray fluorescence spectrum of human hair. The sample was bombarded with 2-MeV protons. Increasing channel number corresponds to increasing X-ray energy.

may be obtained. Advantages of the X-ray fluorescence method of trace element analysis are (a) limits of detectability in a range useful for many biological samples, (b) rapid, multielement analysis, (c) adaptability to automation, and (d) simplified sample preparation. The sample of human hair whose spectrum is shown in Fig. 1 was prepared by oxygen plasma ashing, dissolution of the ash in nitric acid, and evaporation of this solution on a Nucleopore filter. Elemental concentrations in ppm of the washed hair are as follows: Ti (12.4 ± 0.2), Mn (5.0 ± 0.1), Fe (44.7 ± 0.4), Ni (3.6 ± 0.1), Cu (11.0 ± 0.3), Zn (10.9 ± 0.2), Pb (3.0 ± 0.4), Br (1.9 ± 0.2), and Sr (14.7 ± 1.0).

X-ray excited X-ray fluorescence has been used in the M. D. Anderson Hospital and Tumor Institute, Houston, Texas, in studies where it was shown that serum copper levels may be used as an early indicator of remission of Hodgkin's disease during radiotherapy [Ong *et al., Adv. X-ray Anal.* 16: 124 (1973); Tessmer *et al., Radiology* 106: 635 (1973)].

Rudolph *et al.* [*Trans. Am. Soc. Artif. Intern. Organs,* 19: 456 (1973)] have used the X-ray fluorescence method for comparing trace-element content of tissue from uremic and control patients. Abnormalities of trace-element concentration in tissues of uremic patients were observed.

DISCUSSION

Kubota (Ithaca). How do you calculate concentrations from the spectrum?

Mangelson. We run a series of standard solutions in the X-ray beam and determine the number of counts per microgram of the element by integrating the peak above the background count. In an unknown sample we simply determine the number of counts under the peak per unit beam and compare that with our standard curve to determine the number of micrograms of that element and from that calculate the ppm in the sample.

Smith (Washington). How did you prepare the samples for analysis? We have had some problems in this regard.

Mangelson. The ash of hair samples and many others are received in solution form, and we simply deposit 10–100 μl on a filter paper, dry it, and expose it to the beam. Another method we have used for lung ash is to weigh about 100 μg of the ash on a foil and bind the ash onto the foil with some polystyrene in toluene solution, or make a little sandwich by putting foil above and below the ash.

REVERSE-EXTRACTION TECHNIQUE FOR THE DETERMINATION OF FLUORIDE IN BIOLOGICAL MATERIALS

P. VENKATESWARLU

Department of Biochemistry, Medical School, University of Minnesota, Minneapolis, Minnesota

Fluoride is extracted from an acidified sample into an organic solvent containing a silanol such as diphenyl silanediol, back extracted into an aqueous phase (NaOH), and then determined spectrophotometrically employing cerium-alizarin complexan reagent. Recovery of 0.2 μg fluoride is 99.8 ± 1.38 (s.e.) %. The fluoride content of selected samples obtained by the present procedure (and the conventional method involving diffusion) are as follows: low-F bone ash, 0.0032% (0.0031%); high-F bone ash, 0.400% (0.401%); two urine samples, 0.84 and 1.1 ppm (0.88 and 0.95 ppm).

However, fluoride in blood serum is not amenable to determination by the extraction-spectrophotometric procedure because of the low levels of fluoride involved and also of the carry over, from the serum to the back extract, of traces of some substances which interfere with the spectrophotometric procedure. Instead of the colorimetric reagent, a fluoride electrode was employed advantageously in this situation with a two- to three-fold gain in the concentration of fluoride compared with that in the original sample. The recoveries, by the reverse-extraction technique, of radioactive fluoride (^{18}F) added to bovine serum, human urine, and an extract of pea plant were 97.8 ± 0.55 (4)%, 99.7 ± 0.91 (4)% and 100.1 ± 0.81 (7)%, respectively. Two samples of bovine serum and one of human serum, analyzed by the reverse-extraction fluoride-electrode procedure, were found to contain 0.010, 0.011, and 0.025 ppm F. Recoveries of 0.01 μg fluoride added per milliliter of these sera were 99.3 ± 1.37 (8), 100.7 ± 1.71 (4), and 98.7 ± 3.95 (6)%, respectively. The present evidence suggests that the silanol-extractable fluoride in normal sera represents ionic (plus ionizable) fluoride which is also determined by the calcium phosphate adsorption technique [Venkateswarlu, Singer, and Armstrong, Analyt. Biochem. 42: 350, (1971)]. However, since the extraction is carried out at a very low pH (about 20% $HClO_4$) and should the sample being analyzed contain any acid-labile organofluoro compounds, the fluoride determined by the extraction procedure would be expected to be a measure of the sum total of ionic and acid-labile fluoride fractions, as in the cease of procedures involving diffusion for isolation of fluoride.

Fluoride in serum adsorbed onto calcium phosphate (the serum supernatant containing the bulk of the interfering substances was discarded after adsorption), and then isolated by the reverse-extraction technique, was determined with thorium-morin fluorimetric reagent and also with the fluoride electrode. The results of the fluoride content of two samples of bovine sera, 0.015 ± 0.0005 (21) ppm and 0.021 ± 0.0005 (12) ppm were rather higher than those obtained with the fluoride electrode, which were 0.013 ± 0.0005 (11) ppm and 0.017 ± 0.0003 (4) ppm, respectively.

If ammonium hydroxide is used for back extraction instead of sodium hydroxide, the volume of the back extract could be reduced to as low as 30 μliters. In order to take advantage of this further gain in concentration of fluoride, which is 30–50 times that in the original sample, a simple hanging-drop fluoride electrode assembly has been devised. This permits measurement of fluoride in samples as small as 5 μliters, containing an absolute amount of fluoride no greater than 2.5 pM.

The reverse-extraction and hanging-drop electrode techniques were employed to determine fluoride in pooled microtome sections of rat epiphyseal plates (tibia). In three normal weanling rats the fluoride contents were 47, 54, and 69 ppm, and in three rachitic rats 5, 6, and 29 ppm of dry weight of cartilage. However, the lower fluoride levels in rachitic cartilage observed in this study are certainly not to be interpreted as resulting from the rachitic condition of the rats, because the fluoride content of the synthetic rachitogenic diet was found to be indeed very low, 0.8 ppm F, compared with 50–80 ppm F generally found in commercial rat feeds. Employing these techniques, in another study the fluoride contents of tail tendon of rats raised on low-fluoride diet (0.5 ppm F) for two months following weaning and concomitantly receiving 0, 10, and 50 ppm F^- in the drinking water were found to be 0.03, 0.09, and 0.26 ppm dry weight, respectively. [Supported by Grant DE-01850 from the National Institute of Dental Research, Bethesda, Maryland.]

DISCUSSION

Mangelson (Provo). What have you done to check the adequacy of the procedure?

Venkateswarlu. As long as our recoveries are good, I am satisfied.

SEASONAL VARIATIONS IN HAIR MINERAL LEVELS OF THE ALASKAN MOOSE

ARTHUR FLYNN and ALBERT W. FRANZMANN

Trace Element Center, Case Western Reserve University School of Medicine, Cleveland Metropolitan General Hospital, Cleveland, Ohio, and Kenai Moose Research Center, Alaska Department of Fish and Game, Soldatna, Alaska

Variations in hair mineral levels in the Alaska moose, *alces gigas,* indicate a great fluctuation in dietary uptake of elements with seasonal nutrition change. Hair mineral values have previously been used in tracing the elemental status of restricted-fed ruminants [Arch. Tier. 15: 469 (1965)]. Definite seasonal patterns were observed in several species of cattle which indicated hair as a monitor of elemental change. Therefore, an initial study was undertaken to evaluate seasonal hair levels of the moose in 10 mineral elements as a monitor of nutritional deficiencies and geochemical variations. The selection of the Alaskan moose afforded the opportunity to study hair changes with almost exclusive natural mineral exposure in a nonrestricted 'wild' animal.

Moose hair was collected from 275 animals, calves to 18-year-old adults, from the Kenai Moose Research Center on the Kenai peninsula of southern Alaska. Samples were taken over a 12-month period under field conditions and were comprised of plucked hair from the mane of each animal. Hair was placed in labeled plastic bags with no further preparation and sent for analysis to the Trace Element Center, Case Western Reserve University School of Medicine at Cleveland Metropolitan General Hospital, Cleveland, Ohio.

From each sample a small bundle of hair was selected, the first 2.0 cm including the follicle were saved and washed twice with anhydrous diethyl ether. After drying, the bundle was weighed so that approximately 200 mg of hair was included in each subsample. The hair was then digested in 10.0 ml of a quartenary ammonium compound, 24% methanolic tetramethylammonium hydroxide, with the digestion being carried out at 55°C for a 2-hr period. Ten mineral elements were determined: four macroelements (calcium, magnesium,

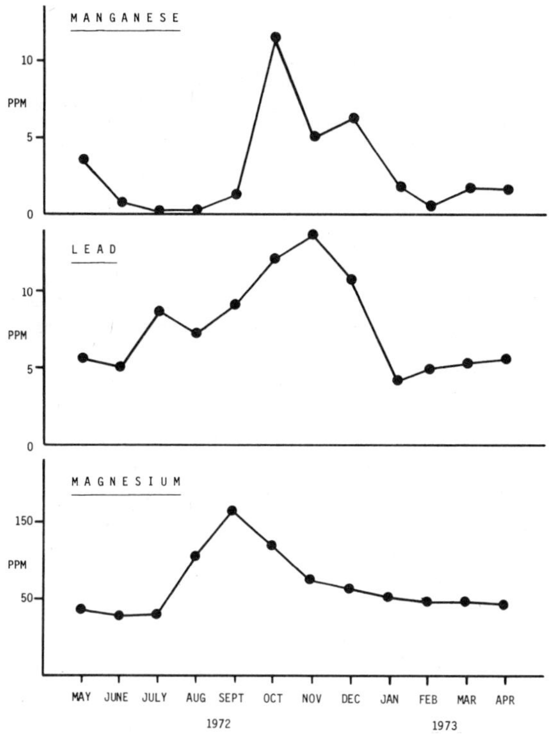

Fig. 1. Seasonal hair levels of manganese, lead, and magnesium in Alaskan moose from May 1972 through April 1973.

potassium, and sodium) and six trace elements (cadmium, copper, iron, lead, manganese, and zinc). All determinations were done by flame atomic absorption spectroscopy on a semi-automated Perkin-Elmer Model 403 spectrometer using plastic utensils throughout all sample handling.

Three outstanding mineral element changes were noted with the progression of the seasons, as is illustrated in Fig. 1. Severe manganese deficiency, hair levels of < 1.0 ppm, was determined in the moose population with the late-winter/early-spring samples. A definite lag between diet improvement and sufficient hair manganese was noted, with normal hair levels, values > 8.0 ppm, appearing in the autumn samples. Subsequent to this normalization, a steady decrease in hair manganese was measured over the next four months again attaining levels of < 1.0 ppm manganese.

Magnesium demonstrated sharp increases much earlier with diet improvement, rising from 30 ppm in the winter samples to 105 ppm in early summer. A gradual decrease in hair magnesium levels occurred over the next five months with levels again dipping below 50 ppm. The changes in hair magnesium closely relate to the enrichment of diet and the gradual decrease of the enriching food with the snows of late autumn.

In similar fashion to these two essential elements, the toxic element lead increased in the moose hair from the winter levels of < 5.0 ppm, more than doubling its levels in the summer values of > 12.0 ppm hair lead. A large number of hair lead samples were in excess of 15.0 ppm, demonstrating an increased natural exposure to this toxic element.

The very significant seasonal variations in the moose hair values of the three elements, manganese, magnesium, and lead, are indicative of the unstable mineral nutrition of this nonrestricted 'wild' animal. Differences among elements in the lag time between betterment of diet and increases in hair mineral levels is also of interest with regard to the establishment or maintenance of body pools. The great seasonal differences measured, especially the deficiencies of late winter and early spring, may play roles in maintaining population growth. Therefore, in monitoring animals with such varied diets, a series of samples would give better indication of mineral status.

The uptake and retention of minerals in the hair make it an effective monitor of an elemental pool. The hair shaft itself can be thought of as a recording filament of present and past mineral nutriture. The portion of the shaft that includes the follicle, where the important biochemical transformations of the hair occur, is the segment that most closely relates to the current nutritional status [Trace Substance in Environmental Health 5: 383 (1972)]. In order to monitor seasonal or monthly changes in mineral levels we have used this initial 2.0 cm section. Further examination will be carried out to relate the remainder of the shaft mineral levels to both past elemental nutrition and environmental exposure.

The seasonal changes in hair manganese, magnesium, and lead reflect great fluctuation that can occur with natural exposure. The dramatic increases reflect the positive influence of the dietary chain in biogeochemical transfer and indicate the caution to be taken in hair sampling.

DISCUSSION

Wolf (Beltsville). Is growth rate of the hair correlated with the mineral concentrations?

Flynn. Hair growth was about 0.8 cm per month except in late summer and autumn when this increased to about 1.2 cm per month. We're looking further into the association of hair growth rate with mineral uptake, but it seems that the peaks in mineral concentrations occur when we have the greatest, not the lowest, hair growth.

Eatough (Provo). Did you see any effect of sex on the elements measured?

Flynn. Most of our samples were from females. We separated the data from the males, and so far have not found a significant difference due to sex.

Hidiroglou (Ottawa). Would it be better to express the results on a pigment or melanin base, because the mineral composition of hair changes according to the color.

Flynn. We sampled nonpigmented areas of hair. There is some pigmentation in calves but they were not included in the results.

Hidiroglou (Ottawa). Would sampling of the front of the head not be better than sampling of the sides?

Flynn. The selection was made by Dr. Franzmann because of the ease of sampling this area and the extensive growth of hair in this particular area.

Mertz (Beltsville). How long a period of hair growth do your hair samples represent?

Flynn. The first 2 cm of hair would represent about 2 months.

Mertz (Beltsville). Did you find a similar pattern of periodicity for zinc?

Flynn. No, zinc fluctuates but we did not find the great seasonal variation that we found with the specified elements.

Shah (Ottawa). What variation did you find for cadmium compared with lead?

Flynn. Cadmium was initially at 2 ppm in hair and increased to about 8 ppm in the same type of pattern as for lead. There was a continual increase and then a decrease with the coming of winter.

Bell (Oak Ridge). How did calves compare with adults, with copper for example?

Flynn. The adults showed a very low level of copper in the spring and then an increase in the summertime followed by a decrease in winter. The calves were never low and maintained their hair copper level during this period. We also saw some difference in manganese but the rest of the elements were similar in calves and adults.

Cary (Ithaca). Have you analyzed plants in conjunction with hair?

Flynn. We're currently analyzing forage and blood samples for comparison with hair.

Hennig (Jena). We have found that the manganese content in pig hair is 12 ppm in the summer and 2 ppm in winter. Have you seen such influences?

Flynn. We see exactly that; levels in the moose were less than 1 ppm in the wintertime and approximately 10 ppm in summer.

INGESTED SOIL AS A SOURCE OF ELEMENTS TO GRAZING ANIMALS

W. B. HEALY

Soil Bureau, Department of Scientific and Industrial Research, Lower Hutt, New Zealand

Grazing animals ingest soil along with herbage and this ingested soil can be a source of elements, over and above that contributed by herbage. Since Ti is relatively low in herbage and relatively high in soils, its presence in feces can be used to measure soil content [N.Z. J. Agric. Res. 11: 487 (1968)].

Soil type, stocking rate, earthworm population, and management will all affect the amount of soil ingested. Under New Zealand conditions, with animals wintered outdoors, annual ingestion of soil can reach 75 kg for sheep and 600 kg for dairy animals [Chemistry and Biochemistry of Herbage, Vol. 1, Academic Press, London, pp. 567–588 (1973).] Probably about half the annual intake occurs other than in the winter period. Ingestion of soil by sheep has been reported in Scotland [Proc. Nutr. Soc. 23: 24 (1964)], in Australia [Aust. J. Exp. Agric. Anim. Husb. 6: 753 (1966)], and in Ireland [Ir. J. Agric. Res. 9: 187 (1970)].

Ingested soil appears to be taken in accidentally along with herbage and does not appear to be the result of depraved appetite. Some soil appears to be taken in at all times of the year, although there is a strong winter peak [N.Z. J. Agric. Res. 8: 737 (1965)]. Although annual intakes can be large they are still probably less than 2% of fresh herbage consumed. Individual animals in a flock [N.Z. J. Agric. Res. 13: 940 (1970)] or herd [Chemistry and Biochemistry of Herbage, Vol. 1, Academic Press, London, pp. 567–588 (1973)] may differ by a factor of two or more in soil intake.

While soil ingestion may have some harmful effects on animal nutrition, e.g., excessive wear of incisor teeth in sheep [N.Z. J. Agric. Res. 8: 737 (1965); Aust. J. Exp. Agric. Anim. Husb. 6: 753 (1966); Ir. J. Agric. Res. 9: 187 (1970)], it may also be a source of elements. Element intakes assessed from analysis of clean herbage from cages can be only minimal. Herbage from grazed areas can contain up to 25% soil (D.M. basis), and soil content of feces indicates this is what is being eaten. A soil content in feces of almost 70% indicates that animals are grazing herbage contaminated at this level. Comparison of element 'content' of grazed and caged herbage shows large differences in element 'content.' Comparison of element intakes by sheep from herbage and from ingested soil, making assumptions as to composition and intakes, shows that ingested soil may supply more of various elements than does herbage [Chemistry and Biochemistry of Herbage, Vol. 1, Academic Press, London, pp. 567–588 (1973)].

In vitro studies using digestive liquors from experimental animals show that soil can substantially alter their element composition and that the effect varies with soil type [N.Z. J. Agric. Res. 15: 289 (1972)]. Short-term animal studies using radioisotopes show that ingested soil can be a source of Co, Mn, Se, and Zn [N.Z. J. Agric. Res. 13: 503 (1970)]. Longer-term feeding trials with soil show that ingested soil can increase blood Se by approximately 50% over a six-week period [Chemistry and Biochemistry of Herbage, Vol. 1, Academic Press, London, pp. 567–588 (1973)]. Early work on Co-deficient animals [Empire J. Exp. Agric. 2: 1 (1934)] showed that Co deficiency could be prevented by regular small drenches of soil and that the effect depended on soil type. Recent work [N.Z. J. Agric. Res. 15: 778 (1972)] has correlated presence and absence of goiter in lambs with low and high levels of soil ingestion by their respective mothers. Iodine levels were higher in feces where soil content was high. Recent balance studies on sheep by Grace and Healy (in press) show that, at low dry-matter intakes, ingested soil significantly increased retention of Ca and Mg.

It is usually considered that soils influence animal nutrition by the quantity and quality of the herbage they produce, so that the usual sequence is soil-plant-animal. It is suggested a direct soil-animal effect also needs to be considered because of ingestion of soil.

DISCUSSION

Hidiroglou (Ottawa). What is the form of selenium in soil and have you measured the apparent absorption of this selenium?

Healy. We have no detailed information on the actual forms in soil, but we can say that the animal was capable of extracting the Se at least to some extent.

Bell (Oak Ridge). Was there stratification of the soil particles in the rumen liquor and in the abomasal contents? In our simulated fallout research using relatively large particles (about 44 μm), we got 100 times as much in the stratified pockets as we got in the rumen liquor.

Healy. The rumen papillae probably tend to hold on to these particles and particle size probably plays a part. In the studies we used a uniform particle size and most of the particles undoubtedly passed through, although there could be considerable buildup in the gastrointestinal tract.

BIOGEOCHEMICAL AND SOIL INGESTION STUDIES IN RELATION TO THE TRACE-ELEMENT NUTRITION OF LIVESTOCK

IAIN THORNTON

Applied Geochemistry Research Group, Imperial College of Science and Technology, London, U.K.

Geochemical reconnaissance surveys based on stream-sediment sampling have recently been completed for England, Wales, and Northern Ireland. Maps showing the distribution of some 20 trace elements indicate broad-scale regional patterns, relating both to the composition of bedrock and soil and to contamination from past and present-day industry. These maps are of primary use in the delineation of possible suspect areas of trace-element imbalance, wherein to concentrate more detailed soil, plant, and animal investigation.

Widespread patterns of high molybdenum have been detected in many areas [J. Sci. Food Agric. 23: 879 (1972)], and both clinical and subclinical bovine copper deficiency have been confirmed within some of the geochemically defined areas. The possible extent of the problem is large as shown by the provisional molybdenum map for England and Wales (Fig. 1), and, in the light of appreciable responses in live-weight gain to copper supplementation in one of the areas [J. Agric. Sci. Camb. 78: 165 (1972)], may well be of considerable economic significance. Alloway [J. Agric. Sci. Camb. 80: 521 (1973)] has suggested that molybdenum-induced hypocuprosis in sheep may be a contributory factor in the occurrence of swayback under farm conditions, and geochemical maps for molybdenum and copper are currently being evaluated as a possible means of focusing attention on those areas where the relative abundance of the two elements may be the predominant cause of the disease.

These national geochemical maps also provide a useful catalog of baseline information on the natural variation of elements which are contaminants or pollutants, and at the same time indicate areas where contamination may be a potential hazard to agricultural production. For example, stream-sediment data [Inst. Geol. Sci. Rep. 70/2 (1970)] in the Southern Pennines reflect lead contamination due to mining and smelting dating back to Roman times, and lead poisoning in livestock is occasionally recorded in this area. In Southwest England past mining and smelting is again reflected in areas where agricultural

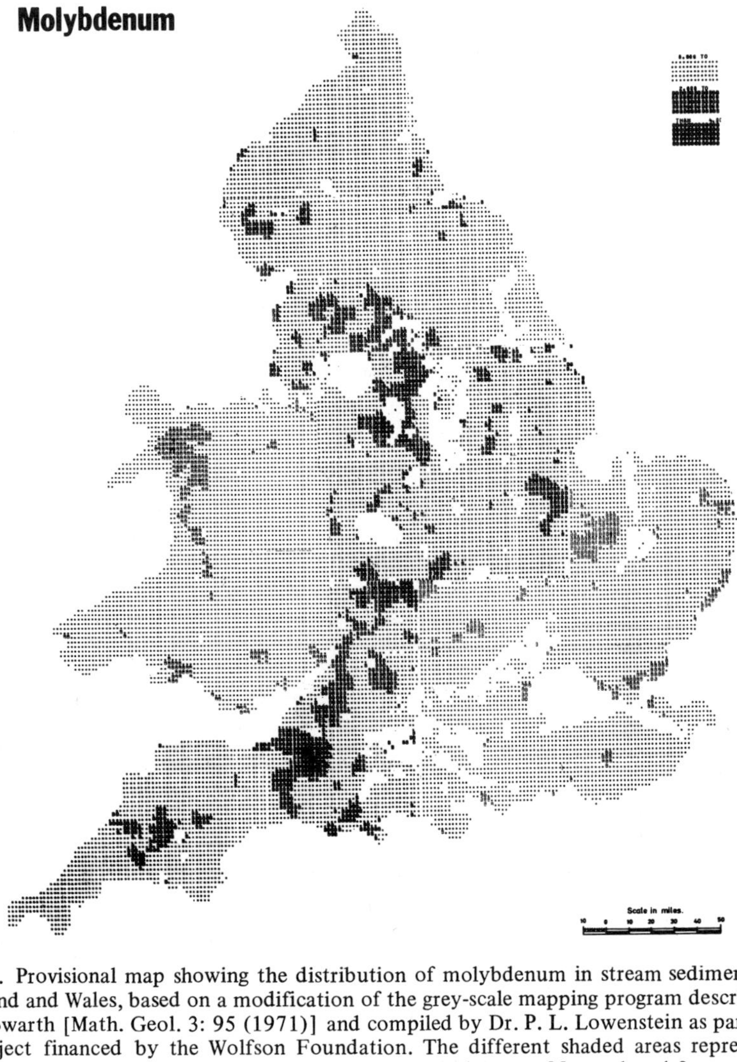

Fig. 1. Provisional map showing the distribution of molybdenum in stream sediment in England and Wales, based on a modification of the grey-scale mapping program described by Howarth [Math. Geol. 3: 95 (1971)] and compiled by Dr. P. L. Lowenstein as part of a project financed by the Wolfson Foundation. The different shaded areas represent sediments containing < 2.0 ppm Mo (light area), 2.0–4.0 ppm Mo, and > 4.0 ppm Mo (darkest area).

Table 1. Soil ingested by grazing cattle on selected farms in Southwest England[1]

Month	Soil ingested (g/day)	Soil as percentage dry-matter intake
November (10)	620 (140–1480)	4.5 (1.1–10.9)
January (6)	460 (210–820)	3.3 (1.6–6.0)
March (4)	780 (660–880)	5.7 (4.9–6.4)

[1]The amount of soil ingested was calculated from the ratio of titanium in fresh feces to that in surface soil (0–5 cm), assuming a digestibility of 70% and an intake of 30 lb dry matter per day. The data are expressed as the mean and range of values found for the number of farms indicated in parenthesis.

soils contain up to 2500 ppm arsenic, 1000 ppm tin, 1500 ppm copper, 1700 ppm lead, and 1200 ppm zinc.

Recent measurements in Southwest England indicate that soil ingestion by grazing cattle over the winter months varied from 140 to 1400 g/day (Table 1). Cattle may thus ingest up to ten times the amount of copper, lead, and arsenic in the form of soil to that in herbage. These data confirm previous reports from the United Kingdom and New Zealand that appreciable amounts of soil may be involuntarily ingested. They suggest that soil may be an important source of trace elements in the diet, and perhaps the main source of elements such as cobalt, which are present in relatively small amounts in herbage compared with soil. As indicated by recent *in vitro* studies [N.Z. J. Agric. Res. 15: 289 (1972)], the forms of the elements and their extractability and absorption in the alimentary tract must also be considered. It is unlikely, however, that any one chemical extractant may be used to reflect 'availability', and measurement of the total leve in the soil may well prove useful.

If, as seems likely, the source of dietary trace elements is a function of a soil-animal as well as a soil-plant-animal relationship, this would help to explain the good degree of correlation recorded between geochemical reconnaissance data and the distribution of livestock disorders relating to trace-element imbalance in the United Kingdom. [Research supported by the Agricultural Research Council.]

DISCUSSION

Suttle (Edinburgh). We've taken three soils, two of them high in Mo, and added them at a rate of 10% to the diet of initially hypocupremic sheep which were being repleted with a copper supplement, and to our surprise we found that each of the soils, regardless of Mo content, was effective in reducing the utilization of dietary copper by about 50%. Molybdenum was being absorbed from the high Mo soils as evidenced by an elevated Mo excretion in the urine, but it's obvious that the problem of soil ingestion by the grazing animal is going to be as complicated as many of our other trace-element problems.

Smith (Washington). We found that 20% of an Iranian clay in the diet would prevent rats from becoming zinc deficient when they were fed an otherwise zinc-deficient diet. Also, in the poultry industry some benefits have been reported from adding clay to the diet. However, other reports show detrimental effects such as the report that clay when fed to humans in a study in Turkey decreased iron absorption. Do people sometimes eat clay because they need it?

Thornton. The soil ingestion that Dr. Healy and I have been referring to has been an involuntary soil ingestion rather than an intentional one. I have watched cattle grazing first-year lay pasture and they were pulling up the plants by the roots and taking a lot of soil involuntarily. Although trying to spit it out, they no doubt swallowed a lot of it in the process.

Kubota (Ithaca). While animals may ingest soil, there are many plants that accumulate trace elements at a much more rapid rate than the soil might indicate. For example, a plant may accumulate 10- to 50-fold or more Mo from a wet or poorly drained soil despite no difference in soil Mo content, so I think that the plant factor is still important.

Lee (Adelaide). We have found in Australia that we have soils which contain as much as 50 ppm of total cobalt which produced both cobalt-deficient pastures and cobalt-deficient stock which grazed them, and we have dosed the animals with this soil until it 'runs out of their ears' and it doesn't cure a cobalt deficiency. Have you observed any close relationship between soil cobalt and cobalt-responsive conditions in livestock in Britain?

Thornton. We have in fact observed a relationship between the distribution of cobalt in stream sediments and the distribution of known cobalt deficiency in sheep in the United Kingdom in specific granitic areas. In some areas we do hear of a failure of sheep to respond to cobalt therapy where they apparently are cobalt deficient. Not only is cobalt deficient in the areas concerned, but one may have a whole range of other elements deficient or at a marginal level, and one could put forward the hypothesis that this perhaps is why the animals don't respond to cobalt alone.

MINERAL COMPOSITION OF TROPICAL GRASSES

HENRY KAYONGO-MALE, J. W. THOMAS, and D. E. ULLREY

Michigan State University, East Lansing, Michigan

Indiscriminate application of knowledge about mineral nutrition of livestock in temperate regions to livestock fed in the tropics may lead to failure to achieve anticipated livestock productivity. Limited reports show a wide spectrum of mineral levels in tropical forages ranging from adequate [E. Afr. Agric. J. 20: 168 (1955)] to deficient [J. Agric. Food Chem. 10: 171 (1962)]. Environmental factors may affect both plants and animals differently in tropical than in temperate areas.

Grass plots from a grass collection at the University of Puerto Rico, Mayaguez, were treated with ammonium nitrate (350 kg/ha) and after 15 days cut at 10 cm above soil level. After 30 days of regrowth, 101 grass samples were harvested, dried, ground, and transported to the U.S.A. Samples were ashed and minerals estimated with a direct-reading Jarrell-Ash 'Atom Counter' spectrograph.

Most of the genera contained over 0.2% Ca (Table 1) which should be adequate to maintain mature ruminants, but 70% were deficient in Ca for milking cows. Phosphorus levels found in this study agree with values found by some [Rev. Ceres 13: 344 (1967)] but not by others [E. Afr. Agric. For. J. 38: 191 (1963)]. Of the grass species studied, 62% were deficient in P for milking dairy cows, but 98% were adequate for beef cattle, except for young calves. The Ca/P ratios ranged from 1.3:1 to 2.4:1 which is satisfactory for ruminants. Magnesium values tended to be high, as has been noted by others in tropical forages [E. Afr. Agric. J. 20: 168 (1955)]. However, *Chloris gayana,* a common pasture grass, was low in Mg (0.11%).

Many Puerto Rican grasses were deficient in Na but were more than adequate in K. Animal requirements for these elements in the tropics are high [J. Aust. Inst. Agric. Sci. 33: 254 (1967)]. Low Na and high K values have

Table 1. Mineral content of 101 30-day-old tropical grasses (dry basis)

	Macrominerals (%)			Microminerals (ppm)	
Element	Mean	Range[1]	Element	Mean	Range[1]
P	0.29	0.20–0.42	Fe	416	113–2717[a]
Ca	0.37	0.13–0.99[b]	Cu	30	13–97[b]
Mg	0.32	0.11–0.78[b]	Mn	160	53–402[b]
K	3.2	1.4–5.0[b]	Zn	34	24–60[b]
Na	0.08	0.01–0.27[b]	Mo	1.7	0.8–6.1

[1]Significant ($P < 0.05$) differences between genera are denoted by [a]; differences with $P < 0.01$ are denoted by [b].

been reported by others [J. Agric. Sci. 59: 251 (1962)], and thus supplementation with NaCl becomes essential [J. Agric. Sci. 63: 373 (1964)].

Iron, copper, manganese, and molybdenum deficiency in Puerto Rican grasses was not apparent. Of the grass species studied, 82% did not meet established Zn requirements for dairy cattle. Zinc deficiency has been demonstrated in other tropical areas [E. Afr. Agric. J. 25: 121 (1959); Rev. Ceres 13: 344 (1967)] and supplementation with Zn has been found beneficial [Nature, Lond. 186: 1061 (1960)].

All minerals in herbage are not necessarily available, and biological availability depends on many factors which must be considered in interpreting adequacy of herbage mineral levels for livestock. These data indicate that, under practical livestock feeding, the elements most likely to be first limiting under Puerto Rican grazing conditions would be P, Ca, Na, and Zn.

Paspalum species were low in Cu/Mo ratio, and high levels of Mo in relation to Cu have been shown to depress Cu utilization in ruminants. The $(Na^+ + K^+)/(Ca^{2+} + Mg^{2+})$ ratio was high in all the genera studied except *Paspalum* (means of 3.1 for *Paspalum versus* 4.1–4.7 for *Bracharia, Panicum,* and *Pennisetum*).

Analysis of variance (Table 1) showed that most of the elements important in livestock nutrition were significantly different between genera ($P < 0.01$), indicating that these grasses could be genetically selected, or selected genera used, to obtain desired mineral content for livestock feeding. Examination of variation within genera showed that only *Pennisetum* had reasonable potential for changes in mineral content through genetic selection of individual species.

Simple correlations were calculated for all determinations on 101 samples. Correlations were signifcant ($P < 0.05$) if $r \geqslant 0.20$. Phosphorus content was positively related to K, Cu, Zn, and Mo content (r = 0.18–0.22).

Potassium and sodium contents were inversely related to Ca, Mg, Mn, and Mo contents ($r = -0.16$ to -0.45). Calcium and magnesium contents were highly correlated ($r = 0.72$), and both were positively related to Mn and Mo ($r = 0.30$–0.65). The Cu/Mo ratio was negatively related to Ca, Mg, Si, Mn, Bo, and Fe ($r = -0.18$ to -0.42).

Phosphorus and potassium were negatively related to fibrous constituents and field dry matter (−0.12 to −0.76) but positively related to protein content and digestibility estimates (0–0.50). Both Ca and Mg were positively related to lignin and lignin/hemicellulose ratios (0.18–0.37), and these two elements had similar but variable relationships to various estimates of digestibility. Concentration of Mn was positively related to fibrous constituents and field dry matter, but negatively related to crude protein and digestibility estimates, while Zn had the reverse relationship.

DISCUSSION

Unidentified. I do not agree that calcium deficiency is likely in tropical conditions, because legumes will be high in calcium.

Kayongo-Male. I am saying that for grasses alone at 30 days of regrowth we would expect calcium deficiency when compared with National Research Council requirement figures.

Thornton (London). How do you explain the very wide range of Mo values? Do the high values relate to specific soils?

Kayongo-Male. I can't answer that.

Hartmans (Wageningen). We found that Mo varies considerably during the season with the autumn values being 3- or 4-fold the values in spring.

Ullrey (East Lansing). Mr. Kayongo-Male's samples were cut at approximately the same time so there shouldn't have been a seasonal effect, and they were taken from the same soil so that neither of those explanations for the variation of Mo content is likely.

ACCUMULATION OF TOXIC HEAVY METALS BY PLANTS IN MISSOURI'S LEAD BELT

DELBERT D. HEMPHILL

Environmental Trace Substances Center, University of Missouri, Columbia, Missouri

Missouri has had a long history of lead and zinc mining and smelting; however, the discovery of a new lead-ore deposit, the 'Viburnum Trend', in Southeastern Missouri has resulted in a marked increase in lead production. Presently, Missouri supplies approximately 75% of the lead mined and smelted in the U.S.A. Appreciable amounts of copper and zinc ores are associated with the lead ore, but no mercury ore has been detected. The new lead belt (approximately 12 miles wide and 35 miles long) has five mines and mills, but only one smelter; consequently, much of the ore (lead sulfide) is transported by rail or truck to smelters outside the area. Seven counties have been involved in these activities in the past or present.

In the spring of 1970, horses began to die in the vicinity of one of Missouri's three lead smelters, and the cause of deaths was diagnosed as lead poisoning. Local residents along the ore-truck routes, and in the vicinity of the smelters, became alarmed about the possible hazards of consuming their fruits and vegetables, and about the grazing of the forage grasses by their animals. Miscellaneous samples of fruits, vegetables, and forage grasses collected by the Agricultural Extension agents and analyzed by atomic absorption spectrophotometry by our laboratory personnel revealed elevated levels of lead, and indicated the need for a more systemic sampling and analysis of the toxic-metal content of crops used for human and animal consumption.

Toxic heavy-metal content of forage plants and soils along ore-truck routes was compared with that of plants and soils along routes with comparable traffic but no ore haulage. Several studies have shown that soil and vegetation near highways contain elevated levels of lead and other toxic metals which decrease with increasing distance from the highway.

Concentrations of lead in or on unwashed vegetation along the ore-truck routes averaged 280 μg/g dry wt on the road right of way, 34 μg/g at 100 yd, and 11.6 μg/g at 200 yd from the highway. Concentrations along control routes were 18 μg/g on the road right of way, 8.1 μg/g at 100 yd and 7.2 μg/g at 200 yd. The cadmium content of unwashed forage grasses along ore-truck routes ranged from 0.5 to 4.32 with a mean of 1.51 μg/g dry wt for samples collected from the road right of way, 0.5 to 0.78 μg/g with a mean of 0.58 at 100 yd, and less than 0.5 μg/g for distances of 200 yd or greater. Maximum values along control routes were less than 1.0 μg/g.

In order to compare the accumulation of toxic heavy metals by vegetable crops grown in the old and new lead-belt counties with those grown in North Central Missouri in areas with little automobile traffic and no industrial activity, cooperators were selected by Agricultural Extension agents in three counties in North Missouri and seven counties in the old and new lead-mining areas. Lettuce, radishes, and greenbeans were grown by each cooperator. Seed and cultural directions were furnished to all cooperators, and samples were collected by personnel from our laboratory. All samples were washed in distilled water soon after harvest. Analyses were made by atomic absorption spectrophotometry.

For the control counties in North Central Missouri, the lead content of lettuce leaves ranged from 6.9 to 33.9 with a mean of 20.6 μg/g dry wt. The edible root of the radish contained considerably lower levels with a mean of 7.7 μg/g. The pod of the greenbean was relatively low in lead with less than 5 μg/g. In the 'Old Lead Belt' counties lead content of lettuce leaves ranged from 10.3 to 742.0 with a mean of 83.8 μg/g dry wt; values for the radish root ranged from 5.0 to 518 with a mean of 33.4 μg/g and greenbean pods 5.0 to 10.1 with a mean of 5.4 μg/g. Lead content of vegetables in the 'New Lead Belt' counties varied widely depending upon their proximity to the mines and mills, smelter, and ore-truck routes. Mean values were lettuce 114, radishes 22.3, and greenbeans 8.8 μg/g dry wt. Lead levels in vegetables grown in home gardens in a small town within 2 miles of a smelter that had operated more or less continuously since the 1890's were lettuce 47 to 1324 with a mean of 284 μg/g dry wt, radishes 7.4 to 50.0 with a mean of 22.3, and greenbeans 5.0 to 136.0 with a mean of 25.1 μg/g dry wt.

Cadmium levels in the vegetable crops were normally less than 2 μg/g dry wt; however, in the small town with the smelter that had operated since the 1890's, cadmium content of lettuce ranged from 3.3 to 34.5 with a mean of 13.02 μg/g dry wt. Values for radishes ranged from 1.18 to 13.7 with a mean of 4.72 and for greenbeans from 0.5 to 8.5 with a mean of 2.36 μg/g dry wt.

The accumulation of toxic levels of lead in forage plants is evidenced by the deaths of several horses beginning in 1970. Mining, milling, transporting, and smelting the ore are the major sources of the contamination. The consumption of vegetable crops with elevated levels of lead and cadmium is of concern to some of the local residents. Although we are attempting to measure health parameters (birth defects, morbidity, and mortality) we have, to date, been unable to establish any positive correlations.

DISCUSSION

Thornton (London). We've been surveying lead in garden soils in two towns, one that we feel is contaminated from old mining and smelting activities, and one control nearby. We find quite large concentrations of lead in the garden soils in the control town, and we wonder if this may depend on the cultural practices of our gardeners and the various materials they may add to their garden soils.

Hemphill. This is certainly a factor. Lead arsenate was used as a primary insecticide for many years. Unfortunately, I cannot distinguish between what may have resulted from the use of lead arsenate and what has resulted from contamination through mining operations. In general we tend to get good correlation between the level in plants and that in soil, but we cannot distinguish between that from root uptake and that resulting from fallout from the air.

Unidentified. Tender young tea leaf contains about 20–40 ppm fluoride whereas the mature leaf may contain 2000–3000 ppm fluoride. Did you consider the effect of maturity?

Hemphill. The lettuce was harvested at the edible stage and was of the same variety planted at the same date and harvested as near the same date as possible. However, we did not distinguish between the oldest leaf and the youngest leaf.

THE STRONTIUM CONCENTRATION IN DEVELOPING MANDIBLES AND TEETH AND IN ENAMEL OF NON-ERUPTED AND ERUPTED TEETH OF HUMAN BEINGS

I. GEDALIA, S. YARIV, H. NAYOT, and E. EIDELMAN

Department of Preventive Dentistry and Department of Geology, The Hebrew University-Hadassah School of Dental Medicine, Jerusalem, Israel

Determinations of strontium (Sr) were carried out on ashed mandibles and teeth of 119 stillborn fetuses at various stages of development. The fetuses had been born to women who, during pregnancy, lived in three regions in which the drinking-water sources contained average Sr contents of 1.12 ppm (range 0.98–1.22) and about 0.1 ppm fluoride (F), 1.07 ppm Sr (range 1.04–1.10) and about 0.55 ppm F, and 1.53 ppm Sr (range 1.43–1.60) and 0.6–1.0 ppm F, respectively [Wolf *et al.*, Archs Oral Biol. 18: 233 (1973)]. In addition Sr estimations were carried out in the enamel of matched pairs of nonerupted and erupted third molars of the same persons. The teeth were extracted from 15 individuals, aged 21–29 years, born in different countries.

Sr was determined by atomic absorption spectroscopy using a Zeiss model M 20 spectrophotometer. The accuracy of individual analyses was estimated to be ±5%.

The differences between the Sr concentrations in the fetal mandibles and teeth in the areas studied (Table 1) are presumably due to the difference in the Sr content of the drinking water ingested by the mother during pregnancy. Lower mean Sr values were found by us in the mandibles of the nine-month-old fetuses as compared with the levels reported for the bones of adult subjects in different geographic areas. The lack of increase in the mean Sr concentration of the mandible and tooth ash between the seventh and ninth months *in utero* in the increased-Sr drinking-water area is probably due to the high growth rate of the fetal skeleton at this period so that mineral deposition in the fetal skeleton occurs at a faster rate than acquisition of Sr.

Table 1. Strontium concentration of ashed human fetal skeletal tissues[1]

Months *in utero*	Average Sr content of drinking water					
	1.12 ppm		1.07 ppm		1.53 ppm	
	Mandible	Teeth	Mandible	Teeth	Mandible	Teeth
5	71 ± 7(3)				68 ± 15(4)	
6	84 ± 13(6)				84 ± 18(6)	
7	83 ± 16(7)	95 (1)	75 ± 17(10)		118 ± 7(6)	116 ± 13(6)
8	72 ± 13(8)	73 ± 15(8)	76 ± 17(6)		115 ± 17(4)	117 ± 12(4)
9	71 ± 20(18)	76 ± 17(18)	71 ± 19(24)	71 ± 16(20)	118 ± 60(17)	118 ± 60(17)

[1]Values are ppm Sr in tissue ash for the number of cases in parentheses ± s.d.

It is also possible that, with crystal perfection due to phase development, a discrimination against Sr incorporation takes place [Likins *et al.*, Archs Biochem. Biophys. 101: 215 (1963)]. The total Sr quantity in the fetal skeleton increased with development (Fig. 1).

The deposition of fluoride in the hard tissues of the growing fetus [Gedalia *et al.*, Archs Oral Biol. 9: 331 (1964)] follows a similar pattern of deposition to that of strontium. It has to be pointed out, however, that more fluoride was incorporated in bones than in teeth, whereas the Sr incorporation was similar.

There was no significant difference in the Sr concentrations in the enamel of nonerupted and erupted third molars from the same individual: mean values 145.8 ± 41.0 ppm (range 93.7–240) and 138.8 ± 38.0 ppm (range 94.1–214.0) respectively. The correlation between the strontium concentrations of matched pairs of teeth was highly significant. The variation in the Sr concentration in the enamel of the teeth from different geographical areas is probably due to variations in the Sr content of the different drinking-water supplies.

Our findings and those of the literature [Steadman *et al.*, J. Am. Dent. Ass. 57: 340 (1958)] indicate conclusively that Sr is mainly laid down in the

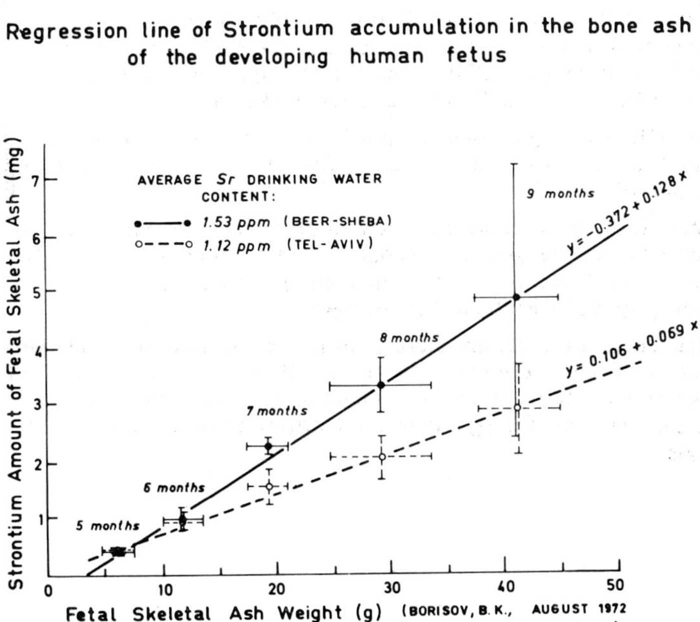

Fig. 1. Relationship between ash weight of fetal skeleton and total strontium content.

enamel before eruption, probably during formation of the enamel by substitution of calcium. Fluoride is incorporated continuously, throughout the lifetime, in the surface enamel.

Kubota (Ithaca). Have you noted any difference in caries incidence?

Gedalia. These teeth were caries free. Because the effect of strontium on caries remains controversial, we should like to analyze teeth which are carious and noncarious, and we are planning to do animal experiments.

Shah (Ottawa). Have you any data on the amount of fluoride in the teeth or bone of women and fetuses in areas where the drinking water has 5–6 ppm fluoride.

Gedalia. In our studies to date the drinking water contained no more than 1 ppm fluoride.

Shah (Ottawa). Some claim that natural fluoride, even at high levels, is not harmful but 1 ppm F^- from sodium fluoride in the water is harmful. Is there a difference in fluoride availability?

Gedalia. Perhaps someone else can answer that.

Shirley (Gainesville). A natural fluoride such as that in rock phosphate is available. At the University of Florida they found that cattle grazing pastures fertilized with raw rock phosphate over a 15-year period accumulated twice as much fluoride (i.e., 2700 *versus* 1200 ppm) in the bones as cattle on pasture fertilized with defluorinated rock phosphate.

Spencer (Hines). I agree that natural fluoride is as available as the fluoride in sodium fluoride. Also, phosphate in the diet of humans does not inhibit absorption of fluoride.

Schwarz (Long Beach). We find that fluorophosphate and hexafluorosilicate do not give us the growth response in rats which we find with sodium fluoride. The form of fluoride in drinking water may be very different depending upon the geological conditions.

Gedalia. There are certainly dietary influences on fluoride availability. If the diet is high in calcium, more of the dietary fluoride is excreted in the feces. In another study people who were malnourished had fluorosis, while people consuming the same fluoride source but who were on an adequate diet had no fluorosis.

MULTI-MEDIA INDICES OF ENVIRONMENTAL TRACE-METAL EXPOSURE IN HUMANS

CECIL PINKERTON, J. P. CREASON, D. I. HAMMER, and A. V. COLUCCI

Bioenvironmental Laboratory Branch, Human Studies Laboratory, National Environmental Research Center, Research Triangle Park, North Carolina

Sets of soils, house dust, water, and hair were collected from families residing in each of three neighborhoods with different environmental exposures, but within the same metropolitan area. Low, intermediate, and high metal-exposure rankings were based on available dustfall measurements. Samples were acid extracted and analyzed for Cd, Cu, Mn, Ni, Pb, and Zn by atomic absorption spectrophotometry. Hair was washed in nonionic detergent. These data are summarized in Tables 1 and 2. Clearly these tables show the feasibility of collection and analysis of a variety of environmental media. In addition, these studies provide insight into the possible importance of each medium in contributing to human-body burdens of pollutants. Metals in dustfall, soil, and house dust follow a similar gradient, indicating a close environmental relationship. Water-metal concentrations do not follow this gradient.

In this study no corresponding gradient was shown between metal levels in hair and those in the environmental media tested (Table 2). Several reasons for these observations are probable. The most obvious explanation is that hair does not reflect exposure to metals; however, a number of studies have established this relationship exists, providing differences are great enough, such as in smelter communities. There is the possibility of hair metalloprotein breakdown over time and subsequent loss of metals by detergent washing prior to analysis.

Based on these results, we feel that accurate estimation of environmental metal exposure resulting in human trace-metal burden should consider the possible contribution from media such as ambient air, soil, dustfall, water, and, if possible, food and house dust. This approach allows for some limited estimation of the relative contribution from all sources to human-body

Table 1. Metal content of dustfall, soil, house dust, and water in three New York cities of differing environmental exposure[1]

Element	Low Riverhead, Long Island	Intermediate Queens, New York	High Bronx, New York	Tests of significance		Ratio Bronx/ Riverhead
				Linear	Quadratic	
			Dustfall			
Ni	0.015	0.254	0.517	< 0.05	n.s.	34.5
Cd	0.026	0.077	0.079	< 0.001	< 0.05	3.0
Pb	1.64	8.42	14.5	< 0.001	< 0.01	8.8
Zn	2.89	7.31	13.0	< 0.001	n.s.	4.5
Cu	1.64	2.28	13.2	< 0.005	n.s.	8.1
Mn	0.246	0.994	2.40	< 0.001	n.s.	9.7
			Soil			
Ni	1.28	5.01	9.24	< 0.001	< 0.01	7.2
Cd	0.32	1.02	1.41	< 0.001	< 0.05	4.4
Pb	39.1	181.8	348.3	< 0.001	< 0.05	8.9
Zn	25.8	123.9	224.1	< 0.001	< 0.05	8.7
Cu	6.93	34.7	44.0	<0.001	< 0.005	6.3
Mn	38.2	118.1	167.1	< 0.001	< 0.005	4.4

			House dust			
Ni	13.2	25.4	40.9	<0.001	n.s.	3.1
Cd	4.16	12.7	14.0	n.s.	n.s.	3.4
Pb	269.8	608.6	771.5	<0.001	<0.05	2.7
Zn	625.2	672.7	1484.2	<0.001	<0.01	2.4
Cu	106.7	192.2	228.2	<0.001	n.s.	2.1
Mn	78.3	102.6	150.4	<0.001	n.s.	1.9
			Water			
Ni	0.0014	0.0009	0.0008	n.s.	n.s.	0.57
Cd	0.0002	0.0001	0.0001	n.s.	n.s.	0.50
Pb	0.0006	0.0020	0.0120	<0.01	n.s.	20.0
Zn	0.038	0.028	0.026	n.s.	n.s.	0.68
Cu	0.125	0.125	0.124	n.s.	n.s.	0.99
Mn	0.006	0.007	0.013	<0.005	n.s.	2.2

[1]Values are the geometric means of the concentration of Ni, Cd, Pb, Zn, Cu, and Mn found in dustfall (mg/m^2/month), soil, house dust ($\mu g/gm$), and water (ppm) in three New York communities; n.s., not significant.

Table 2. Metal content of hair obtained from individuals living in three New York cities of differing environmental exposure[1]

Element	Low Riverhead, Long Island	Intermediate Queens, New York	High Bronx, New York	Tests of significance		Ratio Bronx/ Riverhead
				Linear	Quadratic	
Ni	0.596	0.849	0.726	n.s.	n.s.	1.3
Cd	0.715	1.26	0.599	n.s.	< 0.01	0.84
Pb	9.90	14.8	12.0	n.s.	n.s.	1.2
Zn	33.9	30.2	24.0	< 0.05	n.s.	0.71
Cu	13.9	18.0	11.3	n.s.	n.s.	0.81
Mn	1.97	1.25	0.827	< 0.05	n.s.	0.42

[1]Values are geometric means of the concentrations of Ni, Cd, Pb, Zn, Cu and Mn (μg/g) found in hair obtained from individuals living in three New York communities.

burdens and thus is an important input into the planning of control measures.

It was concluded from these studies that metal levels in house dust, soil, and dustfall followed a similar gradient in three New York communities indicating difference between these areas in the environmental metal exposure. Water levels in these communities did not, however, follow this gradient, nor did hair metal levels of residents in each community reflect an exposure gradient.

DISCUSSION

Thornton (London). Collection of the hair samples from people having haircuts does not seem to be a well-standardized procedure and may explain your failure to find a correlation of metals in hair with the exposure indices.

Pinkerton. In this study we simply wanted to know if typical haircuts would reflect the various exposure indices. I agree that the portion of the hair is an important consideration, and we did not control this.

EXCRETION OF CADMIUM AND LEAD INTO MILK

G. P. LYNCH, D. G. CORNELL, and D. F. SMITH

ARS, Nutrition Institute, Ruminant Nutrition Laboratory, U.S. Department of Agriculture, Beltsville, Maryland, and ARS, Diary Products Laboratory, U.S. Department of Agriculture, Washington, D.C.

The excretion of cadmium (Cd) and lead (Pb) into milk was measured from three pairs of Holstein cows in early lactation. One pair was designated controls, the second pair received 11.0 mg $Pb(CO_3)_2$/kg body wt, and the third pair received 9.0 mg $CdCl_2$/kg body wt as daily oral doses in gelatin capsules. These particular dose rates were selected to ensure some measurable quantity of Cd and Pb in milk and not produce any serious physiological changes in the cows. Doses were adjusted weekly to account for body-weight changes. All data were collected for a 14-day predose, 14-day dose, and 14-day postdose period. Total dry-matter intake (hay, grain, and silage), milk production, and milk samples were obtained daily. Milk samples were pooled for weekly determination of butterfat (BF), nonfat solids (SNF), and Cd and Pb contents. Analysis for Cd and Pb was by differential pulse anodic stripping, using a polarographic analyzer and a hanging-drop mercury electrode [J. Dairy Sci., 56: 1579 (1973)]. Three blood samples were obtained weekly from each cow for analysis of packed cell volume (PCV), red-cell count (RBC), and differential counting of leukocytes and hemoglobin (Hb). The data were analyzed by unequal subclass analysis of variance [Harvey, USDA Publ. 20-8 (1966)].

Significant differences among animals by period were shown for BF, RBC, differentials, and PCV. These differences began during the predose period and therefore could not be attributed to cadmium and lead treatment (Table 1).

Dosing with Cd or Pb reduced body weight, dry-matter intake, and milk production. Cd treatment gave the greatest depression in dry-matter intake (−10% for lead, −26% for cadmium) and milk production (−8% for lead, −25% for cadmium). Neither BF or SNF were changed by Cd and Pb treatment (Table 2).

Table 1. Significant predose period changes in variables tested

Variables	P	Control	Pb	Cd	S.E.
Butterfat (%)	< 0.03	4.5	4.0	3.8	0.08
Erythrocytes (10^6/ mm^3	< 0.01	6.09	4.73	5.31	0.06
Leukocytes (10^3/ mm^3)	< 0.01	8.46	7.23	9.39	0.23
Neutrophils (%)	< 0.01	36.7	40.5	29.6	0.98
Lymphocytes (%)	< 0.01	44.7	39.7	60.1	0.87
Monocytes (%)	< 0.05	6.1	5.7	5.3	0.47
Eosinophyls (%)	< 0.01	12.0	13.7	4.8	0.54
PCV (%)	< 0.05	32.1	33.1	31.2	0.26

Milk solids from control cows contained an average of 1.6 ppb Cd and 43.6 ppb Pb for all periods. The value for Cd converted to a whole-milk basis (0.21 ppb) is in approximate agreement with values of 1.5–3.6 ppb Cd in whole milk reported by Schroeder *et al.* [J. Chron. Dis. 24: 236 (1961)]. The lead content on a whole-milk basis (5.9 ppb) is lower than the national average of 49 ppb Pb reported by Murthy *et al.* [J. Dairy Sci. 50: 651 (1967)]. Dosing with either element increased levels of both elements in milk solids (Table 2). At these dose levels 0.0005% of the Cd during dose week 1

Table 2. Dose-period treatment effects of Cd and Pb on lactating dairy cows[1]

Variables	P	Control	Pb	Cd	S.E.
Body weight (kg)	< 0.01	639.5[a]	613.2[b]	621.5[b]	4.56
Daily dry-matter intake (kg)	< 0.01	17.90[a]	16.14[b]	13.19[c]	0.20
Daily milk production (kg)	< 0.01	19.37[a]	17.79[b]	14.54[c]	0.22
Butterfat (%)	n.s.	4.5	4.2	4.2	0.15
Solids, nonfat (%)	n.s.	9.09	8.76	8.86	0.06
Hemoglobin (gm %)	n.s.	11.0	11.1	10.6	0.11
Erythrocytes (10^6/ mm^3)	n.s.	6.13	5.87	5.46	0.06
Leukocytes (10^3/ mm^3)	n.s.	8.56	7.70	9.29	0.28
Neutrophils (%)	n.s.	36.8	40.6	33.3	1.11
Lymphocytes (%)	n.s.	46.9	41.3	56.8	0.93
Monocytes (%)	n.s.	4.9	6.0	5.2	0.45
Eosinophyls (%)	n.s.	10.8	11.3	4.2	0.65
PCV (%)	n.s.	31.8	32.5	30.4	0.32
Cd in milk solids (ppb)	< 0.01	1.6	3.1	11.2	1.30
Cd dose recovered week 1 (%)				0.0005	
Cd dose recovered week 2 (%)				0.0006	
Pb in milk solids (ppb)	< 0.01	43.6	1520	70.5	154.2
Pb dose recovered week 1 (%)			0.05		
Pb dose recovered week 2 (%)			0.16		

[1]Treatment means with different superscripts ([a], [b], [c],) are statistically ($P < 0.01$) different.

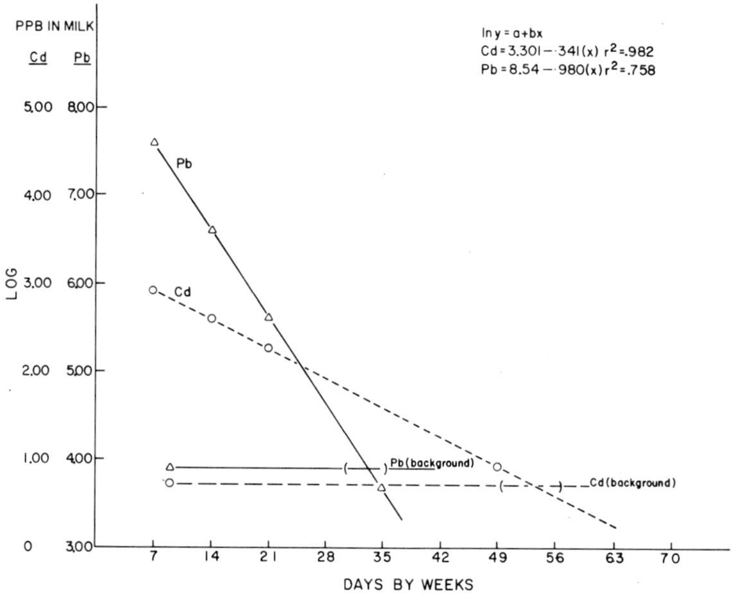

Fig. 1. Regressions of cadmium and lead levels in milk solids with time during the postdose period. Background values were averages of analytical values from control cows, cadmium values of lead-treated cows, and lead values from cadmium-treated cows.

and 0.0006% of the Cd during dose week 2 was recovered in milk solids. Lead recovery in milk solids was 0.05% during dose week 1 and 0.16% during dose week 2. Increased retention of lead with length of dosing has also been reported in sheep (Blaxter, J. Comp. Path. 60: 140 (1950)]. Estimates of the time required to reduce the milk levels of Cd and Pb to background levels were calculated from the regression ($1n\ Y = a + bx$) of elemental concentrations with time during the postdose period (Fig. 1). Cd levels fell to background in 54 days, whereas Pb fell to background in 34 days. These results indicate a greater potential for Pb accumulation in cow's milk but a shorter washout period than for Cd. [The authors wish to acknowledge the assistance of Dr. B. T. Weinland, Biometrics Staff, NER, ARS, Beltsville, Maryland, in statistical analysis.]

NORMAL AND PATHOLOGICAL EXCRETION OF TRACE ELEMENTS BY URINE IN MAN

J. MA. CULEBRAS-POZA, M. DEAN-GUELBENZU,
M. SANTIAGO-CORCHADO, and A. SANTOS-RUIZ

Department of Biochemistry, Faculty Pharmacy, University Complutense of Madrid, Spain

The relation which can exist between the biochemical metabolism of trace elements and pathological processes in human medicine has been the object of special attention during the last few years, and zinc has received particular attention. In the case of some pathologic processes, such as diabetes mellitus and various liver diseases, it has been possible to observe variations in the level of blood and urine trace elements, and to compare them with normal individuals [Pidduck *et al.*, Diabetes 19: 240 (1970); Meltzer *et al.*, Am. J. Med. Sci. 244: 283 (1962)]. It was of interest to us to examine the urinary excretion of seven trace elements in cases of pathological processes and compare the results with those of normal people.

Urine samples were collected for 24 hr, treated to obtain oxinates, calcined, and mixed with spectrochemically pure graphite. The sample was then excited with a voltaic arc and analyzed by a method of automatic emission spectrometry. Urine samples from 20 healthy individuals and 100 individuals with different types of diabetes and liver cirrhosis were studied. Control studies indicated that none of the sources (diet or medicaments) supplying these individuals with trace elements had a significant influence on the excretion values which were observed in pathological cases.

The mean values of urinary Co, Cu, Fe, Mn, Ni, Pb, and Zn excretion in the various groups studied are presented in Table 1. Little or no excretion of trace elements was observed in normal individuals, nor was there any appreciable influence of age on the urinary excretion of most elements. In the case of zinc, however, the highest excretion always corresponded to people of 50 years or more. Individuals with diabetes showed an increased excretion of all the trace elements examined. Juvenile and maturity diabetes had different

Table 1. Mean values of urinary excretion of trace elements in diabetes mellitus, diabetic coma, and hepatic cirrhosis[1]

	Normal people (20)	Diabetes mellitus (38)	Juvenile diabetes (12)	Maturity diabetes (26)	Diabetic coma (8)	Hepatic cirrhosis (17)
Mean age (years)	31	43	24	59	45	59
Mean diuresis/24hr	1160	1700	2129	1400	2244	1276
Co	0	0.570 ± 0.140	0.670 ± 0.110	0.530 ± 0.130	1.820 ± 0.330	0.520 ± 0.490
Cu	0.140 ± 0.050	3.230 ± 3.970	2.890 ± 0.810	3.390 ± 0.280	6.900 ± 0.630	2.180 ± 1.550
Fe	0.990 ± 0.320	3.870 ± 1.540	2.180 ± 0.540	4.660 ± 1.170	14.950 ± 0.950	12.650 ± 8.910
Mn	1.570 ± 0.880	8.520 ± 2.670	4.990 ± 0.710	10.150 ± 1.260	20.260 ± 1.200	16.200 ± 3.260
Ni	0.570 ± 0.320	1.270 ± 0.550	2.000 ± 0.340	0.930 ± 0.150	4.030 ± 0.380	1.370 ± 0.510
Pb	0.028 ± 0.034	0.890 ± 0.660	1.800 ± 0.350	0.470 ± 0.150	2.920 ± 0.360	0.970 ± 0.650
Zn	0.042 ± 0.033	0.310 ± 0.260	0.026 ± 0.038	0.440 ± 0.210	1.760 ± 0.710	1.290 ± 1.300

[1] Values are mean excretions of the element in mg/24 hr ± s.d. The number in brackets denotes the number of samples per group.

effects on trace-element excretion. In the juvenile diabetes, Co, Ni, and Pb increased significatively, whereas Fe, Mn and Zn remained normal. On the contrary, in maturity diabetes there was a significant increase of Fe, Mn, and Zn excretion, while the eliminated quantities of Co, Ni, and Pb remained lower. Individuals in a diabetic coma were observed to have the highest level of excretion for all the trace elements studied. The level of diuresis had little or no influence on the amount of trace-element excretion. There was an increase in urinary excretion of trace elements in individuals with liver cirrhosis, but the excretion observed was less than that seen during diabetic coma.

Our results can be summarized in the following manner. Normal urine contains very low levels of the trace elements studied and all of the diseases studied significantly increased this excretion. Variations in trace-element composition of the diet did not seem significantly to alter excretion. The greatest excretion of trace elements was seen in cases of diabetic coma, followed by liver cirrhosis, and diabetes mellitus. Increased excretion of Zn was seen only in adult cases of diabetes mellitus.

DISCUSSION

Unidentified. Some of your values are very surprising, which leads me to question the analysis. For example, some of your lead values are very high and indicate lead poisoning. On the other hand, your normal value for zinc of 42 μg/day is only one-tenth of the normal value that's been reported in the literature.

ACTIVATION OF SEROTONIN AND 5-HYDROXYINDOLACETIC ACID IN BRAIN STRUCTURES AFTER APPLICATION OF CADMIUM

B. RIBAS-OZONAS, M. C. OCHOA ESTOMBA, and A. SANTOS-RUIZ

Department of Biochemistry, Coordinated Center of C.S.I.C., Faculty of Pharmacy, University Complutense of Madrid, Spain

It is known that cadmium is a toxic agent which produces arterial hypertension and that industrial development leads to an increase of cadmium in the biosphere. Cadmium, as well as some other trace elements, can also be a factor in the development of atherosclerosis [Bull. World Hlth Org. 40: 305 (1969)]. Certain synthetic products used in households can increase the toxic fetal and teratogenic effects of cadmium by altering the properties of the 'defense barrier' of the placenta [Nature, Lond. 239: 231 (1972)]. The permeability of the 'blood-brain barrier' may also be altered as a consequence of the action of atmospheric pollution, medication, or various pathological states. We therefore undertook a study of the influence of cadmium upon brain 5-hydroxytryptamine (5-HT), and 5-hydroxyindolacetic acid (5-HIAA).

Groups of eight female 'Wistar' rats weighing 180–220 g were supplied a single dose of 33, 250, 500, 600, and 100 μg of Cd^{2+} as cadmium chloride in saline serum through a cannula implanted into the side ventricle 1–3 days before the experiments. The 5-HIAA levels in the hypothalamus, thalamus, cortex, medulla oblongata, mesencephalon, and cerebellum were measured by the extraction method of Curzon and Green [Br. J. Pharmac. 39: 653 (1970)] and determined with a Perkin-Elmer 204 spectrofluorometer.

The rats were killed and serotonin levels determined 60 min after the administration of the single dose of cadmium. The data in Fig. 1 indicate that the concentration of 5-HT (ng/g fresh tissue) diminished with the lowest dose, but increased with the other doses of cadmium in all the cerebral structures. The increase in the hypothalamus was most pronounced.

The concentration of 5-HIAA increased in all tissues as a function of the supplied dose, whereas 5-HT concentrations diminished from the 600 μg dose

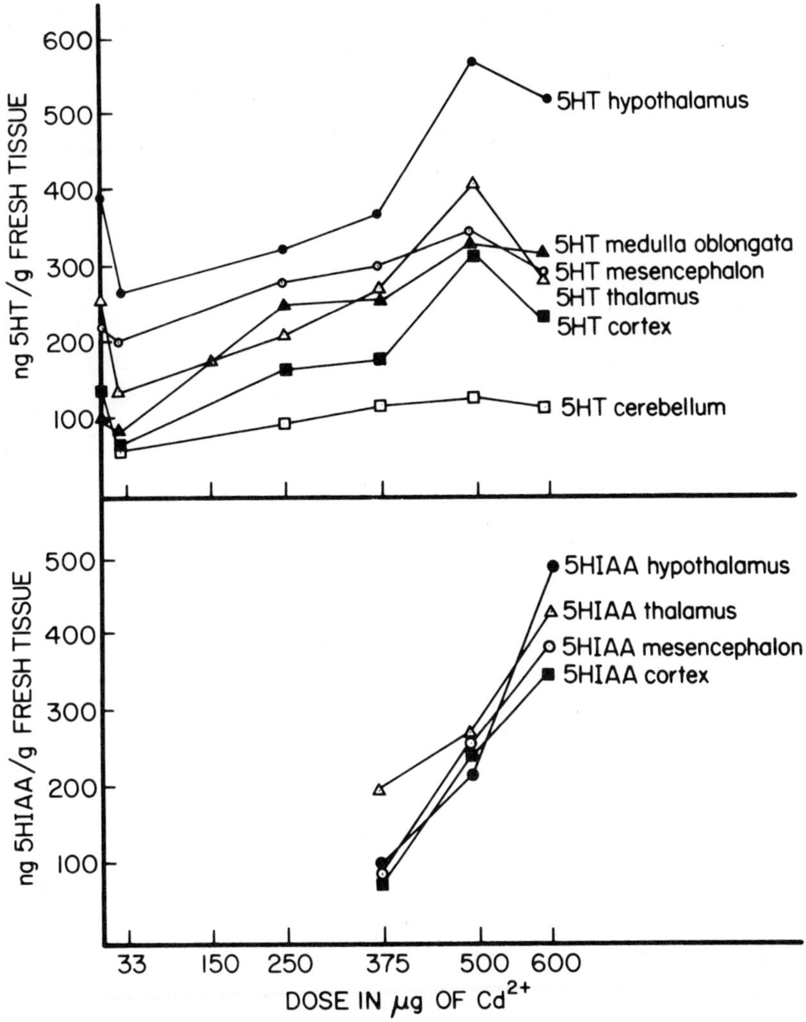

Fig. 1. Effect of cadmium administration on the concentration of 5-hydroxytryptamine and 5-hydroxyindolacetic acid in various parts of the brain.

of Cd^{2+} and coincided with the quietness of the animal and the lowering of body temperature. The initial decrease in 5-HT concentration is due to a nonspecific union between small doses of cadmium and the enzymatic system which synthesizes the 5-HT, whereas the increase following higher doses certainly arises from its interaction with the active group of the aromatic amino acid decarboxylase. Doses of up to 250 μg of cadmium induce an

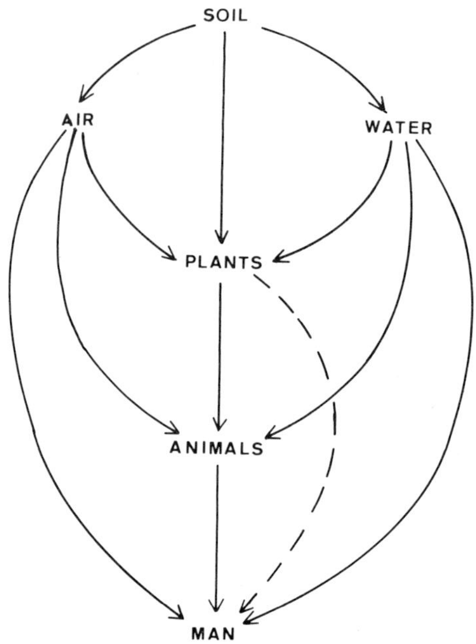

Fig. 2. The environmental cycle of trace elements.

increase of the motor activity of the animal and there is an accentuated activity until the 1000 μg level at which point a coma develops. We suggest that cadmium at high doses inhibits the aromatic amino acid decarboxylase and not the monoaminooxidase, which provokes a great accumulation of 5-HIAA.

Some of the effects of intravenous Cd^{2+} can possibly be compared with those which result from intravenous injections of rubidium in monkeys [Nature, Lond. 233: 321 (1969)]. This element has been shown to have effects similar to those of antidepressant drugs [Science, 172: 1355 (1971)]. In cases of cadmium contamination, the geochemical analysis of the soils would be of great interest as a conditioning factor (Fig. 2) in the development of the pathological response.

DISCUSSION

Pařízek (Prague). Are these effects really specific for cadmium, or are they mainly the result of infusion or injection of a solution of a metal which could be a quite nonspecific effect?

Theuer (Evansville). A recent report shows that when cobalt is infused into the brain, it results in an epileptiform seizure which is amenable to taurine. In this case a different metal gives a different effect.

HEPATIC CUPROPROTEINS IN COPPER-LOADED RATS

L. C. BLOOMER and T. L. SOURKES

Laboratory of Chemical Neurobiology, Department of Psychiatry, McGill University, Montreal, Quebec, Canada

In spite of the massive amounts of information which have been accumulated about copper and its involvement in Wilson's disease and other metabolic disorders, many questions are still left unanswered. The precise nature of the genetic defect in Wilson's disease is not known, and only a small amount of the copper in the liver has been adequately defined. The purpose of this work has been to examine both quantitatively and qualitatively the nature of the copper in the cytosol of rat liver.

Sephadex G-75 was used as a tool to study the copper distribution in the cytosol of normal, newborn, and copper-loaded rats. Male Sprague-Dawley rats were copper loaded by giving daily or twice-daily injections of copper sulfate in saline (2.5 mg Cu^{2+}/kg body wt, i.p.). Rats were sacrificed by decapitation and their livers perfused with sucrose. The livers were then homogenized with thee volumes of ice-cold, isotonic sucrose in a Potter-Elvehjem homogenizer with a Teflon pestle. The homogenate was centrifuged at 27,000 $\times$ *g* for 30 min and at 105,000 $\times$ *g* for 45 min. The resulting supernatant (diluted cytosol) was then chromatographed using Sephadex G-75 (5 mM phosphate, pH 7.0).

Analysis of the results showed that copper is unevenly distributed among the proteins in the cytosol. In normal adult rats most of the copper in the cytosol cochromatographs with proteins of approximately 40,000 molecular weight. No significant amount of copper could be detected in other fractions from the column.

The distribution of copper within the cytosol is altered by copper loading; the amount of copper which cochromatographs with proteins of 40,000 molecular weight does not change appreciably, but a new peak of copper concentration appears with proteins of approximately 12,000 molecular weight. In neonatal rats the distribution of copper within the cytosol is similar to adult copper-loaded rats, i.e., the copper cochromatographs with proteins of both 40,000 and 12,000 molecular weight.

In rats whose total liver copper has been elevated by intraperitoneal injection of copper, 60% of the copper in the cytosol cochromatographs with proteins of 12,000 molecular weight. This relationship is valid whether the total liver copper is elevated (1) by short-term loading, (2) by multiple injections of copper, or (3) by increasing the number of injections from 1 to 2 per day. It also holds for increases in hepatic copper brought about by physiological mechanisms operating in the neonatal rat. Similar results cannot be obtained by addition of copper to the cytosol *in vitro.* The ratio of copper to protein in the 12,000 molecular weight region reaches a maximum when the total liver copper approaches 80–100 μg/g wet weight. This ratio at saturation is 0.65 mg copper/100 mg total protein.

After the low molecular weight protein becomes saturated with copper the metal begins to appear with proteins which are excluded by Sephadex G-75. At very high levels of total liver copper ($>$ 200 μg copper/g wet weight), all sizes of soluble proteins contain appreciable amounts of copper.

If the copper injections are terminated, the amount of copper which cochromatographs with proteins of 12,000 molecular weight (C-peak copper) decreases exponentially with a half-life of 3–4 days.

Various experiments were carried out to define further the nature of the copper binding in the C-peak. This copper did not react directly with the copper chelating agent, zinc dibenzyldithiocarbamate. It was released from the protein by strong acid and by 4.4 mM ammonium persulfate, pH 6.0. The copper in the C-peak was analyzed by ESR spectroscopy. The copper in the C-peak gave no signal, but after oxidation proved to be divalent; it thus may be cuprous in the native protein. Free sulfhydryl groups could not be detected in significant quantity in the adult rat liver preparation.

The protein content of the C-peak region was determined by the Lowry method and by polyacrylamide gel electrophoresis. The results demonstrate that the copper-binding protein is constitutive rather than inducible. Injection of copper intraperitoneally causes an increase in the copper in the C-peak without an accompanying qualitative or quantitative change in the protein [Bloomer and Sourkes, Biochem. Med. 8: 78 (1973)].

The analysis of the polyacrylamide gel electropherograms suggested that one of the proteins, band-4 protein, binds the major portion of the copper in the C-peak. This protein has been isolated by Sephadex G-75 chromatography followed by separation on hydroxyapatite. The amino acid content of the band-4 protein is the same whether it was prepared from normal or copper-loaded rat liver.

Thus, copper is taken up into the liver and bound to a discrete low molecular weight protein. It has been suggested that the protein functions (1) as an acceptor of copper from serum albumin, and (2) in the removal of copper from the liver by excretion into the bile [Bloomer, Ph.D. Thesis, McGill University (1973)]. Proteins with similar structure and function may be present in human [J. Clin. Invest. 40: 1081 (1961)] and bovine liver [Bloomer, Ph.D. Thesis, McGill University (1973)]. [Supported by Medical Research Council, Canada.]

DISCUSSION

Erway (Cincinnati). Presumably there was no net synthesis of the protein in 1 hr.

Bloomer. Loading varied from a single injection to multiple injections over six weeks. Following a single injection, copper in the C-peak reached a maximum about 12 hr after the injection but was measurable at 1 hr, the earliest period studied.

Erway (Cincinnati). What was the schedule of copper objections?

Bloomer. Our data indicate that the protein was not synthesized, but the copper was picked up by the liver and bound to the pre-existing protein.

THE HALF-CYSTINE-RICH COPPER PROTEIN OF NEWBORN LIVER. PROBABLE RELATIONSHIP TO METALLOTHIONEIN AND SUBCELLULAR LOCALIZATION IN NON-MITOCHONDRIAL PARTICLES POSSIBLY REPRESENTING HEAVY LYSOSOMES

HUNTINGTON PORTER and JOHN R. HILLS

Department of Neurology, New England Medical Center Hospitals and Department of Medicine (Neurology), Tufts University School of Medicine, Boston, Massachusetts

An insoluble crude copper protein containing more than 4% copper and more than 25% cystine has been isolated from newborn bovine and newborn human liver under copper-free conditions [Biochim. Biophys. Acta 299: 143 (1971)]. Its extraordinarily high copper content suggests that this copper protein has a storage or detoxifying function for copper in the newborn animal analogous to that of ferritin for iron storage.

The amino acid composition (in residues per cent) of the half-cystine-rich polypeptide isolated from the bovine crude copper protein after sulfitolysis was as follows: Cys, 39.3; Lys, 10.9; Arg, 1.9; Asp, 1.5; Thr, 2.6; Ser, 11.6; Glu, 2.0; Pro, 8.2; Gly, 9.3; Ala, 8.8; Val, 4.0. This composition has some striking similarities to that of more purified preparations of metallothionein but differs from the latter in some respects. The bovine neonatal polypeptide contains no methionine and only one-third as much aspartic acid as found in adult equine metallothionein. The neonatal copper protein, as isolated from the mitochondrial fraction, also differs in form from adult equine metallothionein in its high copper content, its insolubility, and in containing predominantly cystine rather than cysteine. It seems probable that the insoluble neonatal copper protein represents a polymerized form of a metallothionein which sequesters copper in the newborn animal.

The present study confirms selective localization of the bovine neonatal copper protein in the heavy mitochondrial fraction prepared by the method of De Duve *et al.* [Biochem. J. 60: 604 (1955)]. The heavy mitochondrial fraction contained more than 50% of the particulate tissue copper with almost a four-fold increase in copper concentration and a three-fold increase in succinic dehydrogenase specific activity. β-glucuronidase activity, used as a marker for lysosomes, was decreased in this fraction.

The heavy mitochondrial fraction was further fractionated on sucrose or glycogen-sucrose density gradients. The sucrose gradient was from 35% to 60% (w/w) sucrose. The glycogen-sucrose gradient was from 0 to 33.7 g glycogen in 43.5 g sucrose per 100 g water. Centrifugation was at 50,000 × g for 2 hr. Centrifugation of the heavy mitochondrial fraction through either gradient yielded a sediment which contained 70% of the heavy mitochondrial fraction copper with a 15-fold increase in copper concentration and a two- to three-fold increase in β-glucuronidase specific activity (e.g. Table 1). Succinic dehydrogenase activity in the sediment was reduced to less than 2% of the total activity and less than 25% of the specific activity present in the heavy mitochondrial fraction. On electron microscopy the sediment contained some

Table 1. Distribution of copper and β-glucuronidase activity among subfractions obtained by centrifugation of the heavy mitochondrial fraction from newborn bovine liver through a sucrose density gradient

Particulate subfraction	Protein[1] (mg)	Copper Total[1] (μg)	Copper Ratio[2] (μg/mg)	β-glucuronidase Total units[1]	β-glucuronidase Specific activity[3]
Total heavy mitochondrial fraction	16.0	43.2	2.7	1445	90.4
Top subfraction	1.8	1.0	0.6	43	24.1
Main yellow band, upper 2/5	6.2	2.3	0.4	125	20.2
Main yellow band, lower 3/5	3.7	8.2	2.2	433	117
Sediment	0.76	32.9	43.2	198	260
(Soluble material)	(2.0)	(2.0)	(1.0)	(560)	(280)

[1]Protein, total copper, and total β-glucuronidase activity expressed as units/g fresh tissue.

[2]μg Cu/mg protein.

[3]β-glucuronidase expressed as nmoles of phenolphthalein released from phenolphthalein glucosiduronic acid per hr per mg protein.

small irregular dense structures consistent with a lysosomal origin but could not be definitely identified morphologically. In contrast, the mitochondria and succinic dehydrogenase activity were found in an upper band of the gradient containing only a small proportion of copper.

Amino acid analyses showed that the sucrose gradient sediments contained the cystine-rich protein moiety of the copper protein as well as its copper. The amino acid composition (in residues per cent) of the gradient sediment without treatment with detergents was as follows: Cys, 22.2; Lys, 7.5; His, 1.3; Arg, 3.5; Asp, 6.0; Thr, 4.0; Ser, 8.2; Glu, 7.1; Pro, 5.1; Gly, 8.9; Ala, 7.6; Val, 5.2; Met, 1.2; Ile, 3.2; Leu, 5.4; Tyr, 1.6; Phe, 2.5.

In conclusion, the insoluble half-cystine-rich copper protein isolated from newborn liver probably represents a polymer of a metallothionein. Although concentrated in the heavy mitochondrial fraction, it is not a true mitochondrial constituent. The increased β-glucuronidase specific activity suggests that some of this copper protein may be localized in lysosomes. If so, the selective concentration in the heavy mitochondrial fraction in the preceding subcellular fractionation step suggests that the particles containing the copper protein in newborn liver may represent a distinct population of heavy lysosomes. If applied to newborn liver, the original method for preparation of metallothionein, involving freezing of the tissue, slow thawing overnight, and extraction with hypotonic buffer, might be expected to rupture lysosomal membranes. [Supported by Research Grants NS 01733 and RR 05598 from the U.S. Public Health Service and by the Iannessa Wilson's Disease Fund.]

DISCUSSION

Saltman (La Jolla). Does this polypeptide bind copper?

Porter. Can we rebind the copper to it? I don't know because I can't dissolve the protein without either reducing the disulfide bridges with mercaptoethanol or treating it with sulphite and tetrathionate. As I have it prior to sulfitolysis, it contains about 5% copper and accounts for the largest part of the copper in the newborn liver.

Mills (Aberdeen). Have you degraded with trypsin and obtained peptides still bearing copper?

Porter. If you degrade the detergent-insoluble copper protein with trypsin, most of the copper becomes bound to new negatively migrating peptides which contain large amounts of lysine. I suspect that this is an artifact and that the copper is rearranged after you split the protein.

Theuer (Evansville). Some premature infants have a requirement for cystine. By changing the amount of dietary cystine, can you modify the amount of this protein?

Porter. I haven't done that.

Weser (Tubingen). Are you able to obtain this cuprein also from adults?

Porter. Liver copper concentration in the adult rat is less than 1/30 that in the newborn. In the human, liver copper in the adult is less than 1/80 that in the newborn. This is total copper. The protein may be present in the adult but not binding copper.

Weser (Tubingen). What is the molecular weight of the polymer?

Porter. The polymer is insoluble so I can't assess its molecular weight, but the solubilized material has a molecular weight of about 6000.

SUPEROXIDE DISMUTASES: PROPERTIES, DISTRIBUTION, AND FUNCTIONS

EUGENE M. GREGORY and IRWIN FRIDOVICH

Department of Biochemistry, Duke University Medical Center, Durham, North Carolina

Superoxide dismutase, which catalyzes the disproportionation of O_2^- to H_2O_2 + O_2, is ubiquitous among oxygen metabolizing organisms but absent from obligate anaerobes. The O_2^- ion appears to be a common intermediate in the reduction of oxygen, and this enzyme provides a defense against the deleterious reactivities of this free radical. Three distinct types of superoxide dismutase have been discovered. One of these is a cupro-zinc enzyme found in the cytosols of eukaryotes. The second is a mangano-enzyme found in prokaryotes and in the matrix of mitochondria. The third is a ferri-enzyme found in some bacteria and blue-green algae. The characteristics of these enzymes and their biological functions will be discussed below.

The cupro-zinc superoxide dismutase has a molecular weight of 32,600, and is composed of two equal subunits which are known to be identical in the case of the bovine enzyme. There is one Cu^{2+}, one Zn^{2+}, one SH, and one S–S per subunit. The metals can be removed reversibly. The apoenzyme is more labile than the holoenzyme and is devoid of catalytic activity. The Cu^{2+} ion restores full activity to the apoenzyme and considerably enhances its thermal stability, but both Cu^{2+} and Zn^{2+} are required to effect a full restoration of native stability. No metal has been found capable of binding in place of Cu^{2+} and imparting activity to the apoenzyme, but both Co^{2+} or Hg^{2+} can bind in place of Zn^{2+} and can enhance the stability of this protein. The cupro-zinc superoxide dismutases are inhibited by cyanide and are unaffected by the Tsuchihashi treatment with chloroform plus ethanol. The latter property has facilitated its isolation from such diverse sources as bovine blood, wheat germ, *Saccharomyces cerevisiae,* and *Neurospora crassa.* Multiple electrophoretically distinct forms of the enzyme have been seen in numerous sources. In the case of wheat germ these forms have been separated on a preparative scale and found to differ in numerous properties and even in amino acid composition. Because these forms of the enzyme were found in the seeds but not in the leaves of a single nonhybrid strain of wheat, there is no doubt that they are true isozymes.

Analysis of the superhyperfine splitting of the EPR spectrum of the bovine cupro-zinc superoxide dismutase has suggested that there are three nitrogenous ligands on the Cu^{2+}. Since the apoenzyme was susceptible to photosensitized oxidation and to coupling with diazotized sulfanilic acid, whereas the holoenzyme was totally resistant, and since histidine was the only residue diminished during these treatments of the apoenzyme, it appears that histidine residues may constitute the ligands of the Cu^{2+} in this enzyme. The validity of this proposal should soon be subjected to the ultimate test, since both the amino acid sequence and the X-ray crystallographic analyses are well on their way to completion.

The rate constant for this enzyme has been determined with the aid of pulse radiolysis and has been found to be 2×10^9 M^{-1} sec^{-1} and to be independent of pH in the range 5.0–10.0. This rate constant was not susceptible to saturation with the substrate (O_2^-) in the range tested and appears to be diffusion limited. The mechanism of the cupro-zinc enzyme appears to involve alternate reduction and reoxidation of the Cu^{2+} during successive interactions with O_2^-. The rate constants for both half reactions of the catalytic cycle are similar.

The mangano-enzyme isolated from *Escherichia coli* has a molecular weight of 40,000, is composed of two subunits of equal size, and contains one Mn^{2+} per molecule. It is not inhibited by cyanide and is inactivated by the Tsuchihashi treatment, and so can easily be differentiated from the cupro-zinc enzyme even in complex mixtures. The mangano-enzyme from chicken liver mitochondria was found to be very similar to the corresponding bacterial enzyme, but was twice as large and contained four subunits instead of two. The ferrisuperoxide dismutase isolated from *E. coli* was similar to the mangano-enzyme in several respects and is related to it as shown by sequence homologies. The mangano-enzyme was found in the matrix of *E. coli* and was induced by oxygen. The ferri-enzyme was found in the periplasmic space of *E. coli* and was not induced by oxygen. These enzymes appear to play different roles in this organism. Thus the level of mangano-enzyme was manipulated by changing the pO_2 during growth and was thus shown to impart resistance towards hyperbaric oxygen, whereas the level of the ferri-enzyme was manipulated by varying the level of iron in the growth medium and was thus shown to impart resistance towards O_2^- generated outside the cells. High levels of the mangano-enzyme also correlated with enhanced resistance towards the oxygen enhancement of the lethality of streptonigrin.

Escherichia coli were compared with *Bacillus subtilis* and the two were found to differ in instructive ways. Thus oxygen induced superoxide dismutase in *E. coli* but not in *B. subtilis* and induced catalase in *B. subtilis* but not in *E. coli. Bacillus subtilis,* unlike *E. coli,* did not gain resistance towards hyperbaric oxygen from being cultured in oxygen. It thus appears that superoxide dismutase is important in imparting resistance towards oxygen whereas catalase is not.

[Owing to a shortage of space this abstract has not provided references to the many talented investigators, both in our own laboratory and elsewhere, whose work we have summarized above. We would, however, like to mention a few of our colleagues whose contributions have been invaluable: C. O. Beauchamp, S. A. Goscin, B. B. Keele, Jr., J. M. McCord, H. P. Misra, F. J. Yost, and R. A. Weisiger.]

DISCUSSION

Weser (Tubingen). There are about four or five different pathways with different rate constants concerning superoxide dismutation, molecular oxygen, and peroxides. Which of these do you believe has physiological relevance? Also, what about singlet oxygen?

Gregory. We're proposing that the dismutation reaction, $O_2^- + O_2^-$ going to peroxide and oxygen, is the protective mechanism in the cell system. We're quite aware that the back reaction occurs, but the peroxide and molecular oxygen concentrations in cells are not high enough to favor the back reaction. We have seen no firm data that singlet oxygen is produced and, if it is produced, it is either before or after the catalytic events of superoxide dismutase. Thus, we are focusing on O_2^-.

Rotruck (Cincinnati). What is the mechanism by which superoxide kills cells?

Gregory. That is a very fertile area of investigation. If supercoiled DNA is exposed to an *in vitro* system containing streptonigrin and the ferridoxin-TPNH reductase system which produces a flux of O_2^-, you can get nicks in the DNA strand. The O_2^- also participates in the autoxidation of sulfhydryls which can be damaging to sulfhydryl-requiring enzymes and other proteins. Under an extreme oxygen tension of 46 atm, *Streptococcus faceales* produced aberrant colonies with much slower growth, and we were able to plate such colonies through several generations indicating a genetic type of damage. Therefore, there is probably a wide spectrum of things happening in the cell with O_2^-.

COPPER AND ZINC PROTEINS IN RUMINANT LIVER

IAN BREMNER

Rowett Research Institute, Bucksburn, Aberdeen, U.K.

An investigation of the distribution of Cu and Zn amongst hepatic proteins in animals of varying Cu and Zn status was considered of interest for two reasons: (a) characteristic changes might occur during the onset of deficiency or toxic states; (b) insight might be obtained into the Cu-Zn interaction. Livers were obtained from sheep and calves whose Cu and Zn status ranged from the deficient to the near toxic. These were homogenized in 10 mM Tris-acetate, pH 8.2, and the 105,000 × *g* supernatants fractionated on Sephadex G-75. Three main fractions (I–III) containing both Cu and Zn and with approximate molecular weights of > 75,000, 35,000, and 12,000 were isolated. A further fraction containing Zn but no Cu (fraction Ia) with a molecular weight of approximately 65,000 was also present (Fig. 1). Fraction II was identified as hepatocuprein on the basis of its molecular weight, Cu and Zn content, and superoxide dismutase properties [McCord and Fridovich, J. Biol. Chem. 244: 6049 (1969)]. Fraction III was similar to metallothionein [Kägi and Vallee, J. Biol. Chem. 235: 3460 (1960)]. Further purification of one sample on Bio-Gel P-10 gave a main Zn protein which was homogeneous on gel electrophoresis, had a molar ratio of –SH:Zn of 2.9, contained > 2% Zn, and had an amino acid composition similar to that of liver metallothioneins [Kägi, Abstracts, 8th International Congress on Biochemistry, p. 130 (1970); Nordberg *et al.*, Biochem. J. 126: 491 (1972)].

There were large variations in the distribution of metals amongst fractions I–III. In Zn-deficient animals Cu and Zn were virtually absent from fraction III, with an abnormally high concentration of Cu in fraction I. Zinc was also absent from fraction III in most high-Cu livers, where the ratio of Cu in fractions I and III ranged from 0.2–5.0. Usually about 4% and 10% of the hepatic Cu and Zn respectively occurred in fraction II, although in Cu-deficient livers as much as 20% of the Cu could be in this form.

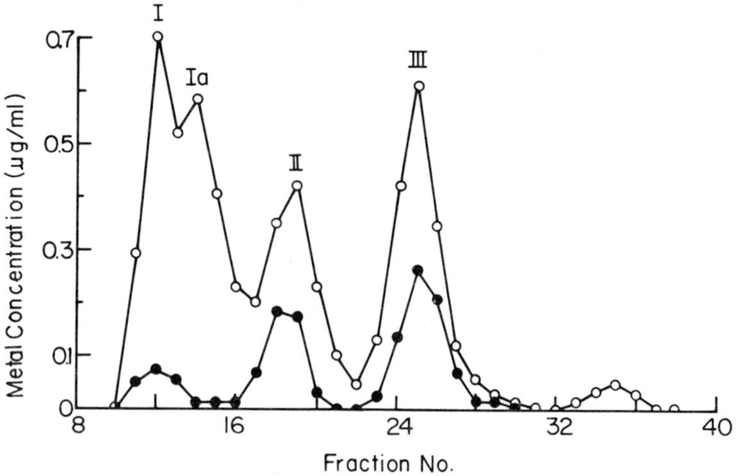

Fig. 1. Typical separation on Sephadex G-75 of Cu (•) and Zn (○) hepatic proteins from calf liver with Cu and Zn concentrations of 13 and 35 μg/g. The positions of fractions I–III are shown.

There was no direct relationship between the hepatic concentration of Cu or Zn and the amount of that metal in fraction III. However, the total amount of metal in this fraction was closely related to the hepatic concentration of Zn, but not of Cu (Fig. 2). Any increase in liver Zn concentration over 20 μg/g wet weight caused an 0.75 X equivalent increase in the amount of metal in fraction III, regardless of whether it was Cu or Zn. The proportion of binding sites in fraction III occupied by Cu was in turn related to the Cu/Zn ratio in the liver (Fig. 2).

The variations in Cu and Zn distribution amongst hepatic proteins can be explained by these observations. In Zn-deficient animals the virtual absence of fraction III and the resultant occurrence of much of the soluble Cu in fraction I is associated with the low liver Zn concentration. In high-Cu livers, there is displacement of most, if not all, of the Zn from fraction III by Cu. Given a sufficiently high liver Zn concentration, enough binding sites may be available on fraction III for most of the soluble Cu. If the Zn concentration is low and little 'thionein'-protein is present, however, Cu occurs mainly in fraction I.

Liver Zn concentrations are therefore of considerable importance in controlling the distribution of both Cu and Zn amongst hepatic proteins. These results do not unequivocally explain the Cu-Zn interaction but do clearly illustrate the interdependence of the metals in liver. Furthermore, they suggest that Zn may play an important role in the synthesis of metallothionein.

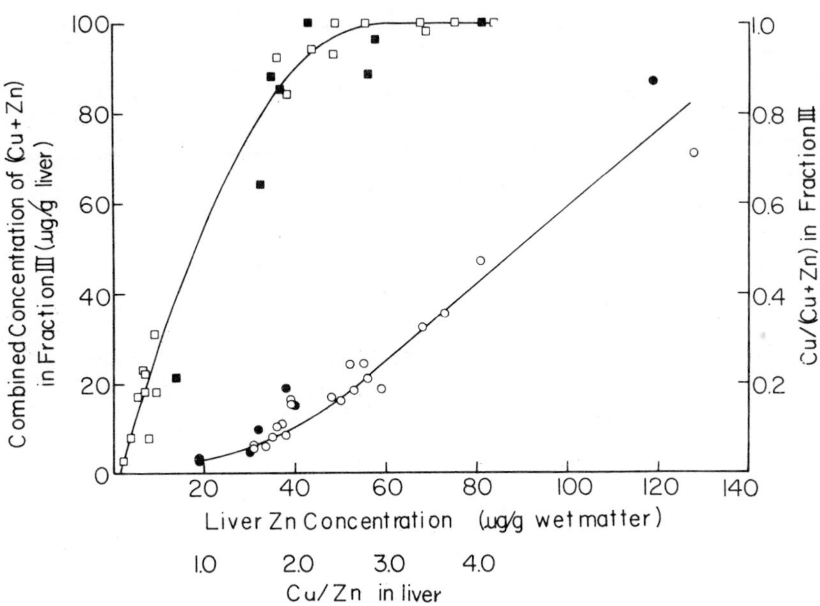

Fig. 2. Relationship between (a) total amount of Cu + Zn in fraction III and liver Zn concentration (○, ●), and (b) proportion of binding sites in fraction III occupied by Cu and Cu/Zn ratio in liver (□, ■). Results are shown for both calves (open symbols) and sheep (closed symbols).

DISCUSSION

Saltman (La Jolla). Have you one protein that binds either of two metals or two proteins, one for each metal, or a combination of both?

Bremner. The amount of each metal occurring in fraction III is a function of liver copper:zinc ratio at that particular time, so you have a definite interaction between copper and zinc for the binding sites in this particular protein.

Mills (Aberdeen). Dr. Saltman raised a question that does bother us. Are these interrelationships occurring on the one protein? Do we have a cadmium protein, do we have a copper protein, do we have a zinc protein, or do we have a mixed protein? We really don't know.

Vallee (Boston). Are you suggesting that there is only one metallothionein or are there multiple metallothioneins?

Bremner. By further purification, I have obtained the main fraction III which is 95% pure on the basis of acrylamide gel electrophoresis.

Vallee (Boston). We are reasonably sure that there are at least two hepatic metallothioneins which do not differ necessarily in their zinc content but are separable on the basis of amino acid composition. They might be binding different metals.

Petering (Cincinnati). How much of the liver copper and zinc are you accounting for in these three protein peaks?

Bremner. Usually about 30% of the copper and 50–70% of the zinc is in this soluble fraction.

ZINC PROTEINS IN RAT LIVER

IAN BREMNER and NEILL T. DAVIES

Rowett Research Institute, Bucksburn, Aberdeen, U.K.

It has been suggested that Zn plays an important part in the synthesis of a metallothionein-like protein in ruminant liver [Bremner, this symposium]. A study has now been made of the changes in Zn binding in rat liver after Zn injection and during the onset of Zn deficiency.

Livers from male rats (150 g) were fractionated by gel filtration on Sephadex G-75 as described earlier [Bremner, *loc. cit.*]. Three heterogeneous Zn fractions (I–III) with approximate molecular weights of > 65,000, 35,000 and 12,000 were separated (Fig. 1). Fraction II had both superoxide dismutase and carbonic anhydrase activities. Fraction III showed only small absorbance at 280 nm and may correspond to the metallothionein-like protein.

Only 0.4 μg Zn/g liver occurred in fraction III in *ad libitum* fed rats. On intraperitoneal injection of 300 μg Zn (as $ZnSO_4$ in saline) liver Zn concentrations increased for 18 hr and then returned to near normal over the next 2 days. The concentrations of Zn in fractions I and II were unaffected by Zn injection and remained at around 10.5 and 4.5 μg/g respectively. There was a large increase in the concentration of Zn in fraction III, however, with 70% of the increased Zn concentration in this form at all times (Fig. 2). The appearance of Zn in this fraction was apparently synchronous with the deposition of Zn in the liver, unlike the equivalent Cd situation where the injected metal occurred initially in other proteins of higher molecular weight [Webb, Biochem. Pharmac. 21: 2751 (1972)].

The failure to produce more Zn in fraction III by addition of Zn to a homogenate suggested that the associated apoprotein was normally absent from liver, and that Zn-induced *de novo* synthesis of the metal-binding protein might occur. This was confirmed by simultaneous injection of Zn (300 μg) and sufficient cycloheximide (2 mg/kg) to cause 90% inhibition of protein synthesis. None of the injected Zn was found in fraction III 4 hr after

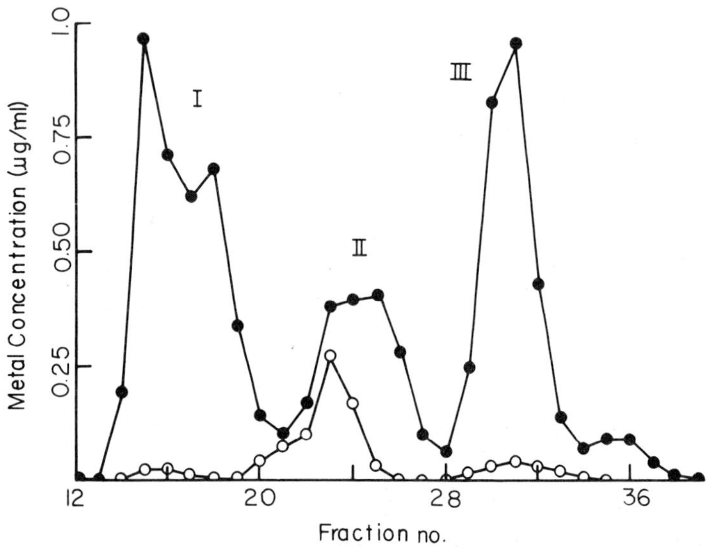

Fig. 1. Fractionation on Sephadex G-75 of liver supernatant from rats 7 hr after injection of 300 μg Zn. Concentrations of Zn (•) and Cu (○) are shown.

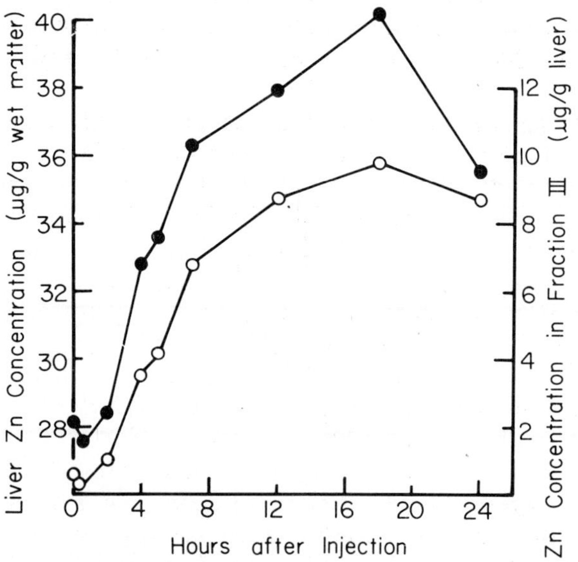

Fig. 2. Changes in liver Zn concentration (•) and in Zn concentration in fraction III (○) after injection of rats with 300 μg Zn.

injection. The distribution of Zn in fractions I and II was unaffected, and it is significant that uptake of Zn by the liver was greatly reduced by the cycloheximide. It was also found that, whereas Zn injection stimulated [^{14}C] lysine incorporation into hepatic proteins by 25–30%, there was > 50% stimulation in fraction III.

Earlier studies [Bremner, *loc. cit.*] showed that the equivalent fraction III was absent in the liver of Zn-deficient sheep. Separation patterns were similar in Zn-deficient and *ad libitum* fed rats, with only traces of Zn in fraction III. It was found, however, that hepatic Zn concentrations increased in pair-fed control rats, with about 35% of the additional Zn in fraction III. On transfer of rats from restricted food intake of Zn-supplemented to Zn-deficient diets, the concentration of Zn in fraction III returned to normal within a few days.

These findings suggest that this Zn-binding protein may serve as a temporary store for Zn when hepatic concentrations are greater than approximately 30 μg/g or may be involved in some detoxication mechanism.

DISCUSSION

Vallee (Boston). I'd like to clarify certain points. Metallothionein was initially identified in kidney and preponderantly contained cadmium, almost as much zinc, some copper, and some iron. A metallothionein was next identified in horse and human liver. It contained almost entirely zinc with minor amounts of cadmium and copper. There is much confusion in the literature concerning resolution of metallothionein into one or several proteins, into those which are high in cysteine or low in cysteine, and into those with different amino acid compositions, and now we are superimposing on that metal deficiencies and excesses which influence the metal content of whichever moiety we don't know. Before we make assignments as to the native state of this material and its purpose, which we still have to do without the benefit of clear evidence, we ought to purify these proteins more adequately so that one can really assess what's going on.

Bremner. I agree.

Vallee (Boston). Concerning the function of metallothionein, of which we really know very little, I think it's very important that we don't lead each other a merry chase on the basis of presumptions that aren't necessarily the only ones.

Bremner. Webb has also shown that zinc can induce its synthesis, and he has shown that this particular fraction has a high sulfhydryl content.

Vallee (Boston). I want to caution against the word induction because it is notoriously difficult in the laboratory to restore a metal to thionein once it has lost it. The idea could easily be gained that there is new thionein synthesis, but thionein itself would not be recognized as a protein by the normal means, since it does not absorb at 280 nm. There is a real problem of identifying the protein in the absence of metal.

Saltman (La Jolla). I'm concerned, too, with the use of the word induction; it's equally valid to say that you're stabilizing a protein.

Bremner. I would accept that.

Mills (Aberdeen). The points that Dr. Vallee has made are well taken. The problem at the moment is that we have a new appreciation of the possible functions of metallothionein. We certainly accept all the points that he has made regarding the need to characterize properly, but, as nutritionists, how much effort do we put into it? We're interested in this as a physiological problem; is this protein stabilized by changes within the physiological limits of zinc, etc? At the moment it's an option of what we do. We have many points that we should still like to look at in relation to deficiency states, toxicity states, etc.

Vallee (Boston). The absorption characteristics of the various metal mercaptide chromophores in metallothionein vary greatly between zinc, copper, and other metals, and one can fortunately make progress by measuring absorption at each of these maxima simultaneously.

Bremner. We do find that if we replace other metals by cadmium, we have an increase in absorbance at 250–260 nm indicative of a cadmium mercaptide.

Vallee (Boston). But lack of absorption doesn't mean that thionein isn't there.

ALBUMIN AS A POSSIBLE SITE FOR COPPER-ZINC INTERACTION

G. W. EVANS and CAROLE HAHN

ARS, USDA, Human Nutrition Laboratory, University Station, Grand Forks, North Dakota

The interaction between copper and zinc in mammalian systems is well documented, but the site of interaction between these elements remains obscure. Van Campen [J. Nutr. 97: 104 (1969)] demonstrated that copper interferes with ^{65}Zn absorption in rats by impeding the movement of ^{65}Zn from the epithelial cells to the blood. Since both copper and zinc are transported in the blood bound to albumin, we examined the possibility that high levels of copper may compete with zinc for binding sites on plasma albumin.

In zinc-deficient rats copper had no significant effect ($P > 0.05$) on either carcass absorption or intestinal uptake of ^{65}Zn (Table 1). In the zinc-supplemented rats, copper significantly decreased ($P < 0.01$) carcass absorption but the intestinal uptake of ^{65}Zn was not significantly altered ($P > 0.05$). These results suggest that zinc uptake by the intestinal mucosa is not the rate-limiting step in zinc absorption. Moreover, the serum zinc concentration of the zinc-deficient rats was significantly less ($P < 0.01$) than that of the zinc-supplemented rats, which suggested that the number of metal-binding sites available on serum albumin may be a rate-limiting step in zinc absorption and thereby influence the effect of copper on zinc absorption.

As shown in Table 1, when copper was injected intraperitoneally into control rats the carcass absorption of ^{65}Zn was significantly decreased ($P > 0.01$), whereas the intestinal uptake of ^{65}Zn was again unaffected. In addition, injection of copper produced a significant elevation ($P < 0.01$) in the serum copper level. Since the copper was injected at half-hour intervals, the major portion of the metal in the serum was probably associated with albumin rather than ceruloplasmin. Thus, since both zinc and copper are

Table 1. The effect of copper on ^{65}Zn absorption

Group	Carcass absorption[1] (% dose/hr)	Intestinal uptake[2] (CPM/g)	Serum Zn[3] (μg/100 ml)	Serum Cu (μg/100 ml)
Zn deficient[4]	49.2 ± 5.1	8244 ± 2090	62 ± 5	124 ± 6
Zn deficient + oral Cu[5]	51.2 ± 2.9	10360 ± 300	63 ± 4	130 ± 6
Zn supplemented[6]	21.7 ± 1.5	3773 ± 120	112 ± 6	126 ± 5
Zn supplemented + oral Cu[7]	8.7 ± 2.6	3791 ± 150	114 ± 5	132 ± 4
Stock diet	16.1 ± 1.0	3450 ± 350	124 ± 4	126 ± 3
Stock diet + i.p. Cu[8]	4.0 ± 2.5	3284 ± 1000	123 ± 5	192 ± 12

[1]Carcass, excluding the entire intestinal tract, 1 hr after administration of 0.065 μg[^{65}Zn]Zn. All values in the table are the mean ± s.d. of four observations.

[2]15-cm segment of the small intestine weighed fresh.

[3]Serum zinc and copper were determined by atomic absorption spectrophotometry.

[4]Rats fed a zinc-deficient diet for 13 days.

[5]65 μg Cu administered orally with the [^{65}Zn]Zn.

[6]Zinc-deficient rats injected i.p. with 200 μg Zn 24 hr prior to administration of [^{65}Zn]Zn.

[7]65 μg Cu administered orally with the [^{65}Zn]Zn.

[8]Rats were injected i.p. with 100 μg Cu immediately after the oral administration of the [^{65}Zn]Zn solution and again 30 min later.

Table 2. The effect of copper on [^{65}Zn]Zn binding by albumin[1]

^{65}Zn-Zn added (μM)	Cu added (μM)	[^{65}Zn]-Zn recovered (μM)
0.01	0	0.01
0.01	2.0	0.01
0.01	3.0	0.0016
0.01	5.0	0
1.0	0	1.0
1.0	2.0	0.4

[1]Metal-free bovine serum albumin was prepared by dialysis against EDTA followed by dialysis against distilled water. Solutions of [^{65}Zn]Zn or [^{65}Zn]Zn and Cu were added to 1.0 μM albumin, incubated for 10 min and chromatographed on Bio-Gel P-2. Recovery of [^{65}Zn]Zn was calculated from the radioactivity in the fractions eluted with albumin.

transported by albumin, these results suggested that copper may inhibit zinc absorption by competing with zinc for binding sites on albumin.

The results shown in Table 2 demonstrate that copper inhibits the *in vitro* binding of zinc to albumin and suggest that albumin contains a limited number of copper-zinc binding sites. When copper and 10^{-2} μM [^{65}Zn]Zn were added *in vitro* to EDTA-treated albumin, the copper had no effect on [^{65}Zn]Zn binding to albumin until the ratio of copper to albumin was increased to 3:1. In addition, when 2.0 μM of copper was added to albumin simultaneously with 1.0 μM [^{65}Zn]Zn, only 40% of the [^{65}Zn]Zn was recovered with albumin. These results suggest that albumin may possess three sites which are involved in copper-zinc binding to albumin. Obviously, the data in Table 2 are not sufficient to determine the exact number of copper-zinc binding sites or the order of binding, but the results certainly demonstrate that albumin is a potential site for antagonism between copper and zinc.

CADMIUM ACCUMULATION IN RAT LIVER: CORRELATION BETWEEN BOUND METAL AND PATHOLOGY

D. WINGE, J. KRASNO, and A. V. COLUCCI

Department of Biochemistry, Duke University, Durham, North Carolina, and Environmental Protection Agency, Human Studies Laboratory, Research Triangle Park, North Carolina

Rat-liver metallothionein is a low molecular weight protein containing a high percentage of cysteine residues. This protein has been previously reported to bind zinc, cadmium, copper, and mercury. In addition, metallothionein appears to be induced by the administration of cadmium and to be localized predominantly in the cytoplasmic portion of the cell.

Table 1. Relationship between cadmium-injection dose and pathologic changes in rats

	Injection dose group (mg Cd/kg body wt)					
	3.0	2.5	2.0	1.0	0.5	Control
Average change in body weight (g)	−50	−44	−34	+40	+50	+60
Roughening of coat	+	+	+	−	−	−
Lethargy	+	+	+	−	−	−
Diarrhea	+	+	+	−	−	−
Impairment of coordination	+	+	+	−	−	−
Accumulation of abdominal and pleural fluid	+	+	+	−	−	−
Cyanosis	+	+	+	−	−	−
Nose bleeding	+	+	+	−	−	−
Final liver wet weight (g)	10.7	10.3	11.7	15.9	16.5	18.0
Presence of degenerating hepatocytes	+	+	+	−	−	−

In rats administered various doses of $CdCl_2$ the accumulation of cadmium in the liver was shown to be roughly proportional to dosage level. Control rats had 0.5 μg Cd/g of liver (wet weight) while rats injected with 0.5, 1.0, 2.0, 2.5, and 3.0 mg Cd/kg body wt had respectively 8.5, 27.0, 60.0, 60.0, and 73.0 μg Cd/g of liver. Similarly, the appearance of pathologic changes in both the whole animal and in the liver was shown to be dependent on the liver cadmium level (Table 1).

Sephadex G-75 chromatography of the cytoplasmic fraction of liver homogenates revealed an elution pattern that reflected exposure to cadmium (Fig. 1). At low liver levels the metal was found to be bound solely to metallothionein (fractions 90–110). However, as accumulation in the liver increased, spillage of cadmium to other protein fractions was noted (fractions 30–40 and 73–79). Inasmuch as this spillage of cadmium was shown to coincide with the appearance of pathologic changes in both the whole animal and liver as indicated in Table 1, an overriding of cellular protective mechanisms is suggested. It is thus clear that the synthesis of liver metallothionein does increase coincident with cadmium accumulation and thus can be protective in function.

However, these studies demonstrate that on acute exposure it is possible to override this protective mechanism and produce pathological changes. Similarly it is concluded that these effects are clearly correlated with specific amounts of this metal in the liver cell.

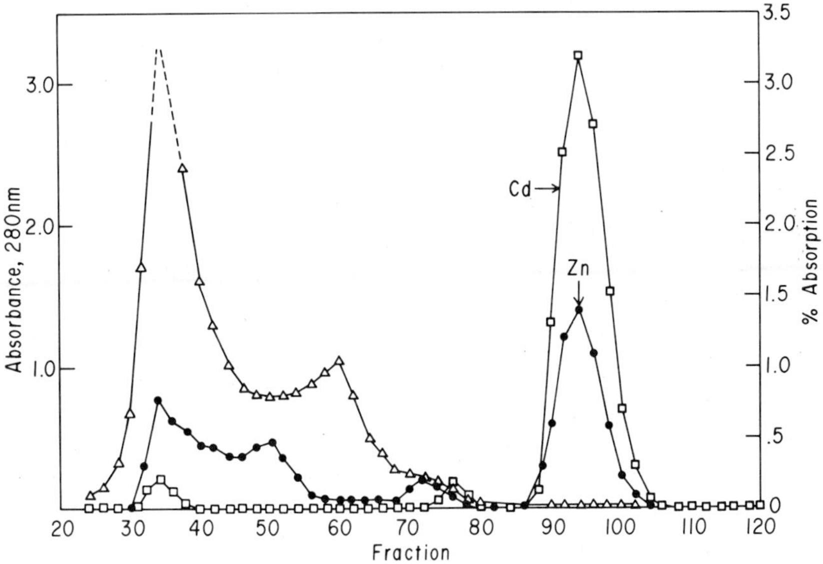

Fig. 1. Chromatography of rat liver cytoplasmic proteins on Sephadex G-75: A_{280} △———△, Zn ●———●, Cd □———□.

DISCUSSION

Vallee (Boston). Part of the reason that I took occasion earlier to caution against some deductions is the common presumption that a particular metallothionein species is the 'active' species. It doesn't seem to occur that it may be thionein, the metal-free fraction, which might be the active biological principle. Its interaction with metals in this instance might but be a diversionary maneuver.

Colucci. I agree. We're merely concerned with spillage and possible protecting mechanisms.

Vallee (Boston). But there is no evidence at all that it has any protective properties.

Colucci. There is some indication that you have to exceed a certain level of cadmium in the liver before you begin to see toxicity.

Vallee (Boston). I don't think so; I am unaware of that literature.

Colucci. We were unable to see any real frank toxicity in these animals until the cadmium concentration reached around 30 μg/g in the liver.

Vallee (Boston). I want to point out again that at present there is nothing known about the function of this protein either in regard to biological action or toxicity, and I am speaking again largely to avoid any propagation of that idea in the press.

INFLUENCE OF CADMIUM ON THE SYNTHESIS OF LIVER AND KIDNEY CADMIUM-BINDING PROTEIN

ROBERT J. COUSINS

Department of Animal Sciences, Cook College–College of Agriculture and Environmental Science, Rutgers University, New Brunswick, New Jersey

A cadmium-binding protein (CdBP) has been reported to be synthesized in response to Cd^{2+} injections [Nordberg *et al.*, Acta Pharmac. Toxic. 29: 456 (1971); Shaikh and Lucis, Fed. Proc. 29: 301 (1970)]. We fed young male pigs 0, 50, 150, 450, and 1350 ppm Cd^{2+} for six weeks to investigate the influence of dietary Cd on renal CdBP synthesis. Signs of a toxicity included decreased hematocrit and decreased renal leucine-aminopeptidase activity [R. J. Cousins *et al.*, J. Nutr. 103: 964 (1973)]. CdBP was prepared by chromatography of 105,000 g supernatant fractions of renal cortex homogenates on Sephadex G-75, DEAE-Sephadex, and QAE-Sephadex. The protein exhibited strong UV absorption at 250 nm and was measurable by the Lowry method. Renal CdBP was not detected in the control animals; however, in the Cd-fed animals the amount of CdBP was directly related to the total kidney Cd level (Fig. 1a), which by regression analysis was found to plateau at a dietary Cd level of greater than 342 ppm. At this level the total renal Cd content was 263 μg/g.

Purification of the porcine CdBP on DEAE-Sephadex separated the protein into three components. Two of the components, designated B and C, were homogeneous on disc gel electrophoresis. These components contained either 24 or 45 mg Cd/g protein and varied in molecular weight from 13,200 to 14,700. The maximal cadmium binding capacity was 5.9 g atom Cd per mole of protein. Component B predominated in the cytoplasmic fractions from the animals from the lowest Cd intake groups, and conversely component C predominated in those fractions from the highest Cd intake groups. The third component (A) could be further fractionated on QAE-Sephadex into four zones ranging in molecular weight from 10,100 to 13,800.

Since CdBP was not detected in those animals that were not fed Cd it appeared that this protein was synthesized *de novo* in response to Cd ingestion. In order to investigate this, an *in vivo,* double-label isotope technique using ^{115}Cd and [^{3}H]cystine was developed so the location of Cd binding could be followed simultaneously with the biosynthesis of the binding protein in rat liver. Forty-two hours after i.p. administration of 10 μCi of [^{3}H]cystine and s.c. administration of 4.5 μM Cd^{2+} to 150 g rats, substantial ^{3}H was detected in the CdBP isolated by chromatography of liver supernatants on Sephadex G-75 (Fig. 1b, middle profile). CdBP labeled with ^{3}H was not detected in the soluble fractions from the control rats that did not receive the Cd^{2+} injection. Simultaneous administration of [^{3}H]cystine, ^{115}Cd, and 4.5 μM of Cd^{2+} yielded CdBP labeled with both ^{3}H and ^{115}Cd (Fig. 1b, bottom profile), while without the Cd^{2+} injection only trace amounts of ^{115}Cd and ^{3}H were detected in the CdBP (Fig. 1b, top profile). These data suggest that CdBP is normally present in rat liver at low levels; however, Cd^{2+} exposure enhances CdBP synthesis resulting in a 16,000-fold increase in Cd-binding ability. The level of protein synthesis at which Cd^{2+} exerts its marked stimulation of CdBP stimulation of CdBP synthesis was investigated with cycloheximide (1 μg/g body wt) and actinomycin D (0.8 μg/g body wt). Both drugs inhibited the ability of 4.5 μM Cd^{2+} to stimulate

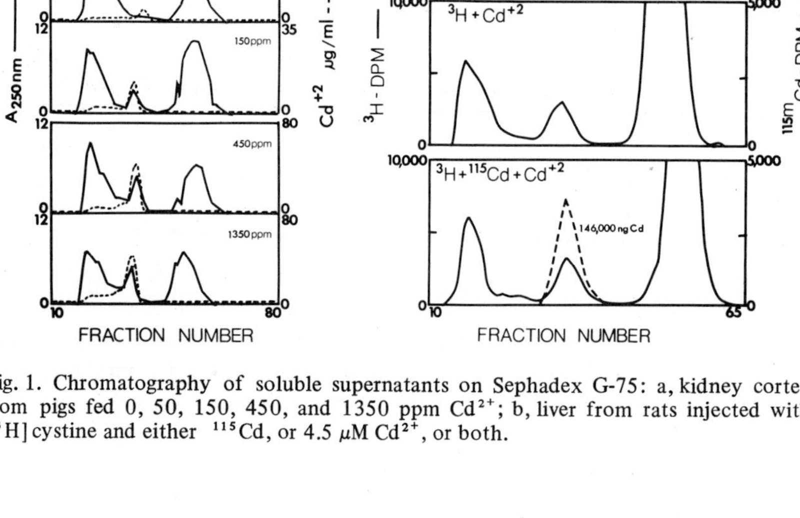

Fig. 1. Chromatography of soluble supernatants on Sephadex G-75: a, kidney cortex from pigs fed 0, 50, 150, 450, and 1350 ppm Cd^{2+}; b, liver from rats injected with [^{3}H]cystine and either ^{115}Cd, or 4.5 μM Cd^{2+}, or both.

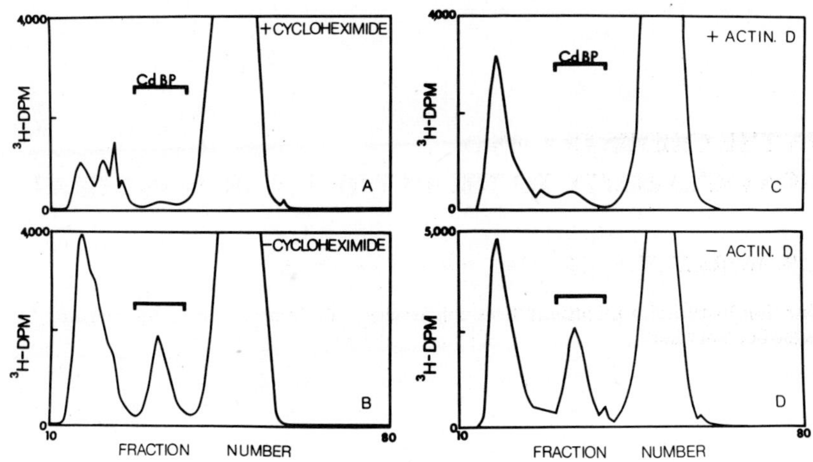

Fig. 2. Effects of cycloheximide and actinomycin D on rat liver CdBP synthesis. Profiles a and b represent G-75 Sephadex column separations of soluble proteins from (a) cycloheximide-treated and (b) control rats. Profiles c and d represent separations obtained from actinomycin-D-treated and control rats, respectively.

the incorporation of [^{3}H]cystine into CdBP (Fig. 2). These results demonstrate that new protein synthesis is required for the production of liver CdBP and that Cd^{2+} enhances the synthesis of its binding protein by controlling the transcription of DNA into messenger RNA. [Supported in part by USPHS, NIH Grant ES-00777 and The Nutrition Foundation Inc., Future Leader Grant 458. Paper of the Journal Series, New Jersey Agricultural Experiment Station, Rutgers University–The State University of New Jersey, New Brunswick, New Jersey.]

ON THE CHEMISTRY AND BIOAVAILABILITY TO THE RAT OF THE IRON IN WHEAT

E. R. MORRIS, F. E. GREENE, and A. C. MARSH

Nutrition Institute, Agricultural Research Service, U.S. Department of Agriculture, Beltsville, Maryland

Anemia continues to be a health problem in the United States population; as much as 25% of some population groups may have low hemoglobin values. Low dietary iron intake and the consumption of poorly assimilable forms contribute to the incidence of anemia. Cereals, of which wheat is a major portion, may supply as much as one-third the daily intake of iron by the U. S. population and an even larger fraction in other nations. Investigation of the bioavailability of wheat iron and its chemical nature was undertaken because of the significant contribution by wheat to the iron nutriture of humans.

Bioassay of iron availability was by the prophylactic maintenance of hemoglobin in the growing rat (J. Nutr. 102: 901 (1972)]. Relative biological values (RBV) were calculated by slope ratio [J. Agric. Food Chem. 20: 246 (1972)] with the response to a reference salt, ferrous ammonium sulfate, equal to 100. Commercial milling blends of wheat and milling fractions were used.

The RBV ranged from 90 to 115 for the various milling fractions of the first sample of hard wheat tested. A second sample of hard-wheat fractions was obtained, and the results with that of some soft-wheat fractions shown in Table 1. The RBV for this second sample were 80–100, slightly lower than previously obtained, but still highly available. Why the whole soft wheat had a lower RBV than the germ and bran fractions is not known. One percent ascorbic acid did not cause a statistically significant increase in hemoglobin response to 8 ppm wheat iron.

Cytochromes are present in the germ of wheat, but the chemical nature of the bulk of the iron is not known. Extraction studies showed that only a trace of the wheat iron was extracted by butanol or water, but approximately two-thirds was extracted in aqueous 1 M NaCl. The different milling fractions

Table 1. Relative biological value of iron in wheat

	Relative biological value(%)[1]	
Fraction	Hard wheat	Soft wheat
Whole grain	81	65
Bran	85	80
Germ	97	103
Shorts	81	

[1]Reference ferrous ammonium sulfate, 100%.

were similar in extractability of iron. The amount of iron extracted by the salt solution was reduced 50% if the molarity was reduced to 0.5. The pH of the extracting solution had little influence on the amount of iron extracted over the range 1–7, but increase to 8.5 greatly reduced the extractability. The iron in the NaCl extract eluted following the void volume when chromatographed on Sephadex G-25 gel permeation medium (Fig. 1). The iron compound or complex would thus be less than 5000 molecular weight. The phosphorus peak overlapped by the iron has not been characterized. Since the maximum fractions for the iron and phosphorus do not coincide it is possible they represent two different substances. It seems unlikely that a ferric phytate complex exists in the wheat, since the bioassay of a synthetic ferric phytate preparation gave an RBV of less than 40, considerably lower than obtained for the wheat fractions. Ferrous-ion color reaction with ferrozine could be obtained only when reducing agent was added to the extract (atmosphere equilibrated or N_2 purged).

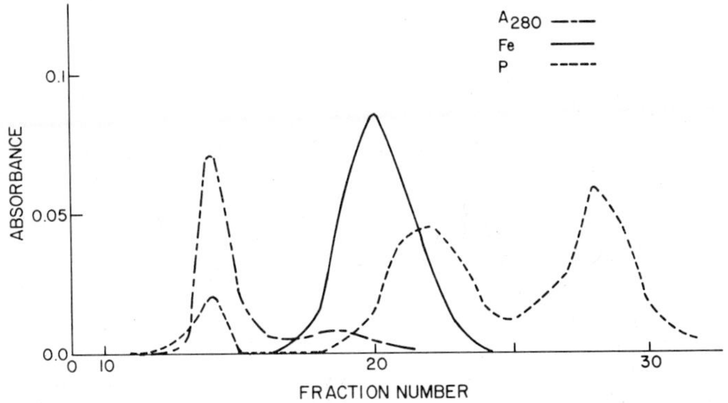

Fig. 1. Gel permeation chromatography of 1 M NaCl extract of wheat bran. 2 × 50 cm column of Sephadex G-25 with 1 M NaCl eluent.

It was concluded that (1) wheat and its milling fractions are sources of readily available iron for hemoglobin formation by the rat, and (2) a major portion of the iron in wheat is 1 M NaCl extractable as a low molecular weight ferric complex.

ZINC-METALLOENZYME ACTIVITIES IN RESPONSE TO DEPLETION AND REPLETION OF ZINC

P. ROTH and M. KIRCHGESSNER

Institute of Animal Nutrition of the Technical University of Munich, Freising-Weihenstephan, Federal Republic of Germany

In order to develop a procedure for measuring the availability of zinc in intermediary metabolism, 1700 enzyme analyses in various rat tissues were evaluated. The response curve of the activity of the following enzymes was studied in response to depletion and repletion of zinc at different levels: lactate, malate, alcohol, and glutamate dehydrogenase; alkaline phosphatase; pancreas carboxypeptidases A and B.

Our results on the activities of zinc metalloenzymes in response to depletion and repletion of zinc are partly in agreement with the studies of Hsu *et al.* [Proceedings of the Seventh International Congress of Nutrition, Vol. 5, p. 735 (1967)], Oberleas *et al.* [Am. J. Clin. Nutr. 22: 1304 (1969)], Prasad *et al.* [J. Appl. Physiol. 31: 842 (1971)] and Mills *et al.* [Biochem. J. 102: 712 (1967)], but, on the other hand, are also contrary to observations made by Luecke *et al.* [J. Nutr. 94: 344 (1968)] and Hsu *et al.* [Science 153: 882 (1966)].

Young weaned Sprague-Dawley rats were fed a semisynthetic casein diet containing the following zinc levels: 2.2 ppm and 19 ppm of zinc in experiment I; 1.2 ppm, 4.5 ppm, 12 ppm, and 96 ppm zinc in experiment II. In serum the activity of the alkaline phosphatase of depleted rats (2.2 ppm Zn) was lowered by 27% after 2 days compared with *ad libitum* fed control animals (19 ppm Zn), by 48% after 4 days, and by 75% at the end of the 30-day experiment. On the other hand, the serum lactate and malate dehydrogenase did not show any changes in activity throughout the experiment. In experiment II (see Table 1) the activity of the alkaline phosphatase fell to 25% of control values after 14 days on the 1.2 ppm Zn diet and increased only slightly after feeding a diet with 4.5 ppm Zn, but reached the value of pair-fed control animals (96 ppm) within 3 days after feeding a diet with 12

Table 1. Activities of alkaline phosphatase in serum of depleted, repleted, and control rats

Days[1]	Activity of alkaline phosphatase (mU/ml serum)[2]				
	Control rats (96 ppm Zn)	Pair-fed rats (96 ppm Zn)	Deficient rats (1.2 ppm Zn)	Repleted rats (4.5 ppm Zn)	Repleted rats (12 ppm Zn)
0	←	88 ± 11	→		
14	217 ± 23	246 ± 32	55 ± 3		
17/3	222 ± 23	255 ± 42	51 ± 7	77 ± 14	209 ± 46
22/8	182 ± 37	238 ± 29	55 ± 17	83 ± 21	218 ± 45
29/15	201 ± 27	215 ± 16	50 ± 7	96 ± 17	221 ± 38

[1]The days are (total days/days of repletion).
[2]The values represent the mean of 6 animals per group ± s.e.

Table 2. Activities of pancreatic carboxypeptidase A of depleted, repleted, and control rats

	Activity of pancreatic carboxypeptidase A (OD/min, mg protein)[2]				
Days[1]	Control rats (96 ppm Zn)	Pair-fed rats (96 ppm Zn)	Deficient rats (1.2 ppm Zn)	Repleted rats (4.5 ppm Zn)	Repleted rats (12 ppm Zn)
0	←	0.079 ± 0.015	→		
14	0.058 ±0.005	0.053 ± 0.008	0.030 ± 0.005		
17/3	0.051 ± 0.010	0.045 ± 0.006	0.030 ± 0.006	0.027 ± 0.004	0.042 ± 0.008
22/8	0.061 ± 0.014	0.062 ± 0.014	0.031 ± 0.003	0.028 ± 0.005	0.053 ± 0.006
29/15	0.051 ± 0.004	0.066 ± 0.009	0.023 ± 0.002	0.028 ± 0.004	0.052 ± 0.005

[1]The days are (total days/days of repletion).
[2]The values represent the mean of 6 animals per group ± s.e.

ppm Zn. The activity of the alkaline phosphatase in rat femurs showed a similar response to Zn depletion and repletion similar to that of the serum enzyme, except that the activity increased and decreased somewhat slower.

The activities of malate dehydrogenase in liver and of lactate and alcohol dehydrogenase in muscle remained unaltered during zinc depletion. The liver alcohol dehydrogenase, however, was depressed by 26% and the muscle malate dehydrogenase by 24% in the state of severe zinc deficiency. A significantly lower activity of the liver lactate dehydrogenase, amounting to 34% of the activity of *ad libitum* fed controls (96 ppm Zn) after 10 days of the experiment and to 58% after 30 days, was also observed in both pair-fed control animals (96 ppm Zn) and in the Zn-depleted animals. Therefore, this was not a consequence of zinc deficiency per se.

The pancreas lost 24% of its carboxypeptidase A activity after 2 days of Zn depletion, 37% after 10 days, and 47% at the end of the experiment (29 days). Dietary repletion at a level of 4.5 ppm Zn did not increase the activity of the pancreas carboxypeptidase (see Table 2), whereas after 3 days of feeding a diet with 12 ppm Zn, an activity was reached which was comparable to that of pair-fed animals (96 ppm Zn). Furthermore, after 29 days of depletion in this experiment, it was possible to demonstrate for the first time a reduction in the activity of the pancreas carboxypeptidase B compared with *ad libitum* (by 52%) and pair-fed (48%) control animals.

Based on our observations, only the alkaline phosphatase in serum and femur as well as the carboxypeptidase A and presumably also B in pancreas appear commendable for measuring Zn availability in future studies.

TRACE-ELEMENT CONTENT AND ENZYMATIC ACTIVITIES IN TISSUE OF ZINC-DEFICIENT PIGS

R. E. BURCH, R. V. WILLIAMS, H. HAHN, R. V. NAYAK, and J. F. SULLIVAN

Veterans Administration Hospital and Creighton University School of Medicine, Omaha, Nebraska

We have been confused and perplexed by the data appearing in the literature on zinc deficiency in various animal species, including man. Zinc deficiency is certainly associated with poor growth and development. An attractive hypothesis to explain these findings has been that there is a deficiency of zinc metalloenzymes or zinc-activated enzymes in these animals. However, deficiency of known zinc-related enzymes has usually been demonstrated only with extreme deprivation of zinc. Less extreme deprivation of zinc has not been associated with decreased activity of important enzymes but still resulted in poor growth and development of the animals being studied.

We do not believe that the zinc-deficiency syndrome is adequately explained by decreased activity of zinc-related enzymes. The enzyme data presented in this study support this viewpoint. Further, we show that tissue trace elements, in addition to zinc, are dramatically affected by zinc deficiency. We propose the theory that the altered tissue-trace-metal content occurring in association with zinc deficiency may result in alteration of metalloenzymes or metal-activated enzymes which are not related to zinc, per se.

Ten pigs were fed a diet containing 7.3 ppm zinc for six weeks. Five of these animals received 120 ppm zinc supplementation of their diet and served as controls. Each animal received 600 g of diet daily. Experimental animals failed to gain weight. At the end of 6 weeks experimental pigs had serum zinc values of 0.62 ± 0.07 μg/ml while control pigs had 1.0 ± 0.25 μg/ml. Serum copper was decreased with zinc deficiency but this decrease was not statistically significant. Serum calcium and magnesium were not affected by zinc deficiency. After 6 weeks there was slight elevation of serum glutamic pyruvic transaminase (22 ± 10 *versus* 51 ± 17 Karmen units, $p < 0.025$), and a decrease in serum alkaline phosphatase (88 ± 11 *versus* 63 ± 17 I.U., $p < 0.025$) and cholesterol (118 ± 17 mg/100 ml *versus* 75 ± 14 mg/100 ml, $p < 0.005$) associated with zinc deficiency. Tissue selenium and manganese were determined by neutron activation, and tissue zinc, magnesium, copper, and calcium were determined by atomic absorption. Zinc deficiency was associated with diminished tissue levels of zinc only in liver, kidney, and pancreas; selenium was decreased in liver, heart, spleen, pancreas, and muscle. Manganese was diminished in liver and heart; copper was increased in liver but decreased in kidney and heart; magnesium was diminished in liver, spleen, and muscle. Histochemical staining for the activity of malic dehydrogenase, lactic dehydrogenase, succinic dehydrogenase, NADH diaphorase, alkaline phosphatase, and alcohol dehydrogenase in heart, kidney, liver, spleen, aorta, muscle, pancreas, pituitary, prostate, adrenal, testes, and thyroid revealed only decreased alcohol dehydrogenase in liver from zinc-deficient pigs and decreased succinic dehydrogenase in zinc-deficient spleen and pituitary. Otherwise, only minor differences were noted. In order to compare the histochemical stain with a direct enzyme assay we compared isocitric dehydrogenase in three different tissues with these two methods. These data indicated that histochemical stains were relatively unreliable in determining small differences in enzyme activity.

We next evaluated several enzymes in hepatic tissue by more conventional biochemical assay procedures. Isocitric dehydrogenase, glutamic dehydrogenase, and arginase activities in liver were not affected by zinc deficiency. However, both leucine aminopeptidase ($2.75 \times 10^6 \pm 0.28$ *versus* $2.18 \times 10^6 \pm 0.18$ units/g/hr) and ornithine carbamoyltransferase (5154 ± 1413 *versus* 2178 ± 668 μM/g/hr) activities were decreased when normal liver was compared with zinc-deficient liver ($p < 0.005$). Ornithine carbamoyltransferase activity was not affected by added zinc, manganese, or a combination of the two. Leucine aminopeptidase from normal liver was not affected by addition of these cations but manganese did activate this enzyme slightly when liver from a zinc-deficient pig was used. This slight activation was insufficient to explain the difference between control and zinc-deficient leucine aminopeptidase activities. A comparison of leucine aminopeptidase values and trace-metal content in liver, heart, and kidney indicated that decreased tissue zinc content seemed to be most closely related to decreased leucine aminopeptidase activities in zinc-deficient pig tissues.

Decreased serum cholesterol and decreased hepatic ornithine carbamoyltransferase activity in zinc-deficient pigs are extremely interesting findings. The mechanism of lowered serum cholesterol remains to be elucidated. Increased excretion, decreased synthesis, or both could be involved. Circumstantial evidence in humans with cirrhosis indicates that the activity of hepatic ornithine carbamoyltransferase may be decreased. These patients have low serum zinc, elevated blood ammonia, and decreased blood urea. Also, the administration of arginine in hepatic coma has produced disappointing results. Obviously, circumstantial evidence and extrapolation of zinc-deficient pig data to humans are not acceptable. Therefore, a study to determine urea cycle enzymes and trace-metal content in human cirrhotic liver is presently in progress.

Basically, we find it difficult to accept the hypothesis that the zinc-deficiency syndrome is solely related to a deficiency of zinc-related enzymes. A marked deficiency of these enzymes is difficult to demonstrate in animals that do not grow well and exhibit other manifestations of zinc deficiency. However, changes of other tissue trace metals do occur in the present study. Thus, the zinc-deficiency syndrome may well be related to the tissue-trace-metal changes, other than zinc, which are associated with zinc deficiency.

DISCUSSION

Sandstead (Grand Forks). Were the animals pair fed and what did they weigh at the end of the experiment?

Burch. Yes, the zinc-deficient and the control animals were each given 600 g of diet per day. All the animals consumed their diet. At the end of the experiment the zinc-deficient animals weighed slightly less than they did at the start of the experiment which was 6.4 kg; the control animals gained weight.

Sandstead (Grand Forks). That's not exactly pair feeding. In my laboratory, a zinc-deficient rat loses its appetite and stops eating. Do swine have a decrease in appetite?

Burch. They don't eat normal quantities of diet, but they do eat.

Sandstead (Grand Forks). Well, the problem is that you don't really have controls for starvation because your so-called pair-fed controls gained weight.

Burch. I don't think that in order to make the control animals comparable with the zinc-deficient animals we should have given them less food and actually starved them. That's not a good control either.

Sandstead (Grand Forks). I suggest you do just that because one of the problems is that some of these enzymes are affected by starvation and you can't conclude that these effects are related to zinc deficiency per se.

Hurley (Davis). It's always possible to use two controls, one pair fed and the other weight paired, but I am sure that's harder to do with pigs than with rats.

Burch. We intend to include that in the next series.

REPRODUCTION AND PRENATAL DEVELOPMENT IN RELATION TO DIETARY ZINC LEVEL

LUCILLE S. HURLEY and GLADYS COSENS

Department of Nutrition, University of California, Davis, California

Previous reports from this laboratory have shown that a diet severely deficient in zinc even for relatively short periods of time causes a high incidence of fetal death and congenital malformation in rats. These abnormalities affected all organ systems, and were accompanied by a low level of maternal plasma zinc at term, and a low concentration of zinc in the fetus [Hurley, Am. J. Clin. Nutr. 22: 1332 (1969)]. The finding of congenital malformations resulting from zinc deficiency has since been confirmed by several other groups. The present investigation concerned the effect of milder deficiencies of zinc during pregnancy, conditions more nearly simulating those pertinent to humans. The influence of various levels of dietary zinc during pregnancy was therefore examined on several parameters of reproductive capacity.

Virgin adult female Sprague-Dawley rats, 210 ± 10 g, were purchased from a commercial source and were mated with normal stock-fed males. The day of finding sperm in the vaginal smear was considered day zero of gestation. At the beginning of pregnancy and continuing to term, the rats were given purified diets containing EDTA-washed isolated soybean protein, and varying only in their zinc content. Eight diets were used, with levels of zinc at 0.4, 1.25, 2.5, 6.5, 9, 14, 18, and 100 ppm, as measured by atomic absorption spectroscopy. Individual stainless-steel cages were used and rigorous precautions were taken to prevent contamination with zinc in the diet or the environment. At term, day 21 of gestation, maternal blood samples were taken for zinc analysis and fetuses were removed for examination.

The proportion of rats with living young at term was reduced (86 and 77%, respectively) in the groups receiving 0.4 or 1.25 ppm zinc (as compared with 100% in all other groups), while the proportion of dead or resorbed fetuses was greatly increased in both these groups and also in the group receiving 2.5 ppm zinc. The incidence of malformed fetuses was very high in all three of these groups (from 88 to 67%), and in the group receiving 6.5 ppm of zinc as well (21%). Between 1.25 and 9 ppm zinc there was a straight-line relationship between incidence of malformed fetuses and dietary zinc. The proportion of total implantation sites affected was also high in all groups getting 6.5 ppm dietary zinc or less.

Other parameters of reproduction that also correlated with dietary zinc were change in maternal net body weight (mother's body weight exclusive of litter weight), fetal weight at term, and weight of the total litter at term. Change in maternal net body weight and mean fetal weight correlated with dietary zinc between 2.5 and 14 ppm, while total litter weight correlated with dietary zinc between 1.25 and 14 ppm zinc.

Measurement of zinc in fetal tissue showed that zinc concentration of ash from full-term fetuses was significantly lower in all experimental groups than in controls. However, there were no differences between any of the experimental groups and therefore no linear correlation between fetal content and incidence of malformation. Likewise, the maternal plasma zinc level at term did not correlate with the incidence of malformation.

The effect of maternal dietary zinc level on maternal plasma zinc was also measured during the course of pregnancy. Under the same conditions as in the previous study, groups of females were killed at days 0, 1, 2, 7, 14, and 21 of gestation for analysis of plasma zinc. In the groups that had the highest incidence of malformations, namely, those receiving 1.25 and 0.4 ppm of dietary zinc, the maternal plasma zinc was very low during the first two trimesters of pregnancy (below 50 μg/100 ml during the second trimester), while in the groups in which few or no malformations were seen the maternal plasma zinc level was relatively high during the second trimester, even though it was low at term. The level of maternal plasma zinc during the second trimester thus appears to be the critical factor in determining whether or not the fetuses will be malformed. On the other hand, maternal plasma zinc at term seems to be the result of a balance between the removal of zinc from maternal plasma in order to provide for fetal tissue, and the addition to maternal plasma of zinc from fetal tissue undergoing resorption.

The total maternal body weight gain during pregnancy (including litter) was also related to zinc intake, and there was an excellent correlation between total body weight and incidence of malformations and other indices of reproduction. The two groups in which there was the highest incidence of malformations, 1.25 and 0.4 ppm, showed markedly different curves of body weight from the other groups. The level of 9 ppm, which appeared to be borderline by several criteria, showed a lack of weight gain in the third trimester.

In seeking measures for the evaluation of zinc nutriture in pregnant women, it would appear that determination of plasma zinc level during the second trimester would provide a better index than such measurements at term. [Supported in part by NIH research grant HD-01743.]

DISCUSSION

Petering (Cincinnati). The absolute levels of zinc you're speaking about may not hold with all types of diet. Using a 2.5 ppm zinc diet containing 20% egg white and cornstarch we got no abnormalities but only smaller fetuses, but we have shown that a three-day absolute zinc deficiency, in which there was essentially no zinc in the diet, produced a statistical increase in abnormalities in the rat.

Hurley. Obviously, what is important is the amount of zinc that is getting to the fetus and our values are for our particular conditions with our diets. I think a similar relationship will hold under your conditions, although the absolute zinc levels may be different.

Mills (Aberdeen). We have been trying two approaches. One is the use of an egg-white protein diet with zinc concentrations down to about 0.5 ppm, and we've had problems producing fetal abnormalities. The other is use of a dietary protein source rich in phytic acid with which we get good confirmation of your effects. This may suggest that phytate causes a very severe zinc deficiency, perhaps because it prevents endogenous zinc in the intestine from being reabsorbed.

Hurley. Again, what goes into the mouth isn't really the important thing; it's what gets into the bloodstream that's important. When we first started working with zinc we were using a diet that contained 1 ppm zinc. That was as low as we could get at that time and we did not get malformations. Now we get abundant abnormalities with 1.25 and with 6.5 ppm zinc. The difference is that our other conditions have become so much better and the environment less contaminated that the animals are actually getting less zinc now than they were then.

Mills (Aberdeen). I still wonder if there is an effect of soybean protein on this problem other than its effect on zinc availability.

Hurley. In Dr. Petering's work and in work by Dr. Dreosti in South Africa, soybean protein was not used and malformations were observed.

INTESTINAL ABSORPTION OF COPPER, ZINC, AND IRON AFTER DIETARY DEPLETION

F. J. SCHWARZ and M. KIRCHGESSNER

Institute of Animal Nutrition, Technical University of Munich, Freising-Weihenstephan, Federal Republic of Germany

The level of trace elements in the body can greatly influence their absorption. For trace-element homeostasis the intestinal wall appears to be of particular importance. Studies were therefore conducted to investigate the absorption of copper, zinc, and iron using isolated intestines from Cu-, Zn- and Fe-depleted rats.

Weaned male Sprague-Dawley rats were fed *ad libitum* a casein diet which contained 0.5 ppm of copper for Cu depletion, 2 ppm of zinc for Zn depletion, and 8 and 4 ppm of iron for Fe depletion. All other trace elements were added to these diets at adequate levels, and the control groups contained 4 ppm of copper, 20 ppm of zinc, and 20 and 50 ppm of iron. After the dietary pretreatment, the jejunum was isolated from the trace-element-depleted and control rats. The absorption of copper, zinc, and iron was measured in two different experimental series by the everted-sac method [Wilson and Wiseman, J. Physiol. 123: 116 (1954); Schwarz and Kirchgessner, Int. J. Vit. Nutr. Res. 42: 592 (1972)].

In the first experimental series [Kirchgessner *et al.*, Bioinorg. Chem. 2: 255 (1973); Schwarz and Kirchgessner, Z. Tierphysiol. Tierernahrung Futtermittelk. 31: 91 (1973)], the intestinal transfer and uptake of copper from a mucosal solution containing 10^{-4} M of $CuSO_4$ was measured during a 60-min incubation of intestinal sacs from rats depleted in Cu, Zn, or Fe and from control animals. The results are summarized in Table 1.

The Cu transfer and Cu uptake by the intestinal sacs of Cu- and Zn-depleted rats were significantly higher than by those of control groups ($P < 0.05$). During iron deficiency, however, there were no such differences. Furthermore, Table 1 shows that zinc depletion also caused a several-fold increase in the Zn transfer and a significantly higher Zn uptake ($P < 0.01$) from a mucosal solution containing 10^{-4} M $ZnSO_4$.

Table 1. Transfer into the serosal solution and uptake of Zn and Cu by the intestinal wall of Zn- and Cu-depleted rats

	Dietary pretreatment[1]		
	Zn depletion	*Ad libitum* fed control	Pair-fed control
Zn transfer (μg/g of dry weight)	12.09 ± 5.62	2.97 ± 0.46	6.25 ± 1.63
Zn uptake (μg/g dry weight)	399 ± 43	262 ± 24	328 ± 34
	Cu depletion	*Ad libitum* fed control	
Cu transfer (μg total)	0.71 ± 0.25	0.41 ± 0.14	
Cu uptake (μg/g dry weight)	236 ± 27	190 ± 14	

[1]Values are mean ± s.e.

In the second experimental series 10^{-5} M $CuSO_4$, $ZnSO_4$, or $FeCl_3$ was added to the mucosal solution of intestinal sacs from rats depleted of Cu, Zn, or Fe. The solutions were labeled with ^{64}Cu, ^{65}Zn, or ^{59}Fe. The results are presented in Table 2 together with the results of the first experimental series. When Zn, Cu, or Fe was deficient, this element was significantly better

Table 2. Intestinal transfer and uptake of Cu, Zn, or Fe after Cu, Zn, or Fe depletion

	Dietary pretreatment[1]					
	Cu depletion		Zn depletion		Fe depletion	
Experimental series	I	II	I	II	I	II
Intestinal transfer						
Cu	↑	↑	↑	↑	±0	±0
Zn		±0	↑	↑		±0
Fe		↓		±0		↑
Uptake by intestinal wall						
Cu	↑	±0	↑	↑	±0	±0
Zn		±0	↑	↑		±0
Fe		↓		±0		↑

[1]↑increase, $P < 0.01$; ↓ decrease, $P < 0.01$; ±0 no significant difference.

absorbed. Thus, it is not only the *in vitro* absorption of iron that adapts to a change in supply [Forth, *In* Mills (ed.), Trace Element Metabolism in Animals, Livingstone, Edinburgh, p. 298 (1970)], but also the *in vitro* absorption of zinc and copper. Evans and Grace [Fed. Proc. 32: 895 (1973)] postulated that the mucosal zinc content may regulate zinc absorption. The results also show that during Zn deficiency the absorption of copper is increased. Under the conditions of the present study, an increased zinc absorption was not obtained during Cu deficiency. An analysis of hepatic Cu revealed a lower liver Cu content for Zn-depleted rats than for controls. This can be explained on the basis of a severely reduced feed intake and therefore low Cu intake during Zn deficiency.

As to the effect of zinc status upon Fe absorption and, in turn, the effect of iron status upon Zn absorption, there was only a tendency toward an increase in absorption during deficiency as compared with controls. The markedly increased zinc transfer during Fe deficiency as observed by Pollack *et al.* [J. Clin. Invest. 44: 1470 (1965)] and by Forth [*loc. cit.*] may be due to a difference in the nutritional and depletion status of their animals and those in the present investigation. The absorption of Cu was not affected by Fe status, but Fe absorption was significantly lowered by Cu deficiency. During an extended Cu depletion, hemoglobin and serum ceruloplasmin were reduced [Grassmann and Kirchgessner, Z. Tierphysiol. Tierernährung Futtermittelk. 31: 113 (1973)]. However, iron storage was higher than normal, and Lee *et al.* [Proc. Soc. Exp. Biol. Med. 127: 977 (1968)] found that the mucosa of the small intestine are filled with iron. This could be the cause for the reduced iron absorption.

DISCUSSION

Saltman (La Jolla). What theory unifies your data?

Schwarz. When any of the three elements was deficient, the intestine showed higher *in vitro* transfer of that element into the serosal solution and the *in vitro* uptake by intestinal tissue was higher, indicating increased absorption. We also looked for interactions between the elements and found only that in copper deficiency iron absorption is reduced, and in zinc deficiency copper absorption is increased. We cannot give a unified theory for these interactions but must look specifically at each one to assess the mechanism.

Forth (Homburg). I believe that there is a transfer system for iron which may be shared by some heavy metals, and a second transfer system for copper which may be shared at least by zinc. Although only a tendency here, we showed some years ago that in iron deficiency there is an increase in zinc absorption, but not in copper absorption. In the case of zinc deficiency, there was an increase in copper absorption but not in iron absorption. These two systems may be important in metal absorption and help explain the interactions.

Saltman (La Jolla). I question the use of ferric chloride in such experiments, because Fe(III) has a ridiculously low solubility. In assessing transport mechanisms, it must be clear whether we're studying hydrolyzed species or complexes that can donate their metals.

Schwarz. We were looking for relative differences in absorption between the deficiency state and the normally supplied state, and we did show absorption of each salt used, whether small or not. Our question was not whether the complexes would show higher absorption, but whether the animals showed differences in relative absorption. If you did the same experiments with different complexes you would probably get the same differences but at a higher absorption level.

Forth (Homburg). We would perhaps prefer these experiments to be done in the presence of a 10-fold excess of citrate, but you may not find any difference between the results provided that you added the iron in an acid solution, and then it depends on the uptake of iron in the first 5 min of the experiment. If you add your iron as ferrous sulphate, for example, there is no way to prevent its hydroxylation.

Evans (Grand Forks). We found no difference in copper uptake and iron uptake in zinc-deficient animals by using an *in vivo* system, so there is a difference here. I suggest that an *in vivo* system provides a more realistic picture of what's going on because there are homeostatic mechanisms regulating the absorption of these elements, and they are manifest within the animal as well as in the intestinal mucosa.

ON THE METABOLIC AVAILABILITY OF ABSORBED COPPER AND IRON

E. GRASSMANN and M. KIRCHGESSNER

Institute of Animal Nutrition, Technical University of Munich, Freising-Weihenstephan, Federal Republic of Germany

An adequate trace-element supply for the body depends not only upon the dietary level but also upon the utilization of the trace element offered. In TEMA-1 we reported on a relationship between the absorption rate of copper bound in amino acid complexes and the absorption rate of the amino acids themselves. This *in vivo* observation [Kirchgessner and Grassmann, Z. Tierphysiol. Tierernährung Futtermittelk. 26: 3 (1970)] could be confirmed *in vitro* for several Cu-amino acid complexes [Grassmann *et al.*, Z. Tierphysiol. Tierernährung Futtermittelk. 28: 28 (1971); Schwarz *et al.*, *ibid.* 31: 98 (1973)].

After absorption, the availability of a trace element in metabolism also affects its utilization. Criteria which require a synthesis for the utilization of the trace element as an additional step after absorption are used as a means to measure availability. For copper, ceruloplasmin evidently meets this requirement. In order to determine the influence of complex formation on availability, studies were conducted to measure copper absorption from Cu complexes, using the liver storage test [Kirchgessner and Weser, Z. Tierphysiol. Tierernährung Futtermittelk. 20: 261 (1965)]. Concurrently, the influence of these complexes on the ceruloplasmin activity (Table 1) was examined [Kirchgessner and Grassmann, Z. Tierphysiol. Tierernährung Futtermittelk. 26: 340 (1970)].

The copper storage in the liver was increased by all compounds compared with $CuSO_4$, particularly by Cu-EDTA. The response of the ceruloplasmin activity shows that the availability also depends upon the type of ligand. Thus, the absorption rates and ceruloplasmin activities follow about the same trend whether copper is added as sulfate, oxalate, or EDTA. Therefore, it may be assumed that the ceruloplasmin synthesis, in these instances, only

Table 1. Influence of complex formation on the utilization of copper and iron[1]

Copper			Iron		
Compound added	Liver Cu content (μg total)	Ceruloplasmin activity (mV/min)	Compound added	Hemoglobin (g/100ml)	Catalase activity (U/μl)
	14.4^{a} ±4.6	0.040^{a} ±0.010		3.5^{a} ±0.4	0.32^{a} ±0.01
Cu-sulfate	34.9^{b} ±5.5	0.189^{b} ±0.057	Hemin	3.8^{a} ±0.5	0.37^{a} ±0.05
Cu-fumarate	37.8^{c} ±1.8	0.143^{b} ±0.044	Hemoglobin	4.9^{b} ±1.0	0.49^{b} ±0.06
Cu-L-leucinate	38.0^{c} ±2.7	0.310^{d} ±0.050	Fe(III)-EDTA	5.8^{c} ±0.4	0.56^{bc} ±0.08
Cu-oxalate	39.1^{c} ±4.2	0.257^{c} ±0.050	Fe(II)-fumarate	6.0^{c} ±1.0	0.59^{cd} ±0.08
Cu-EDTA	42.2^{c} ±6.1	0.313^{d} ±0.065	Fe(III)-L-leucinate	6.8^{cd} ±0.9	0.61^{cd} ±0.09
			Fe(III)-citrate	7.2^{d} ±1.1	0.68^{d} ±0.08

[1]Values are mean ± s.e.
Means with the same superscript within a column are not different from each other ($P > 0.05$).

depended upon the amount of copper absorbed; i.e., availability of copper for synthesis was similar. Copper from fumarate and L-leucinate, however, produced significantly different ceruloplasmin activities in spite of comparable absorption rates. For fumarate the values are greatly reduced in comparison to all the other complexes, and for L-leucinate they are particularly high. The inference, therefore, is that there is a higher metabolic availability for copper from L-leucinate and a very low one from fumarate.

For estimating the availability of iron in metabolism, the hemoglobin content as well as the blood catalase activity evidently are suitable criteria, because here too a synthesis is necessary as an additional step after absorption. For iron, however, no method as simple as the liver-storage test for copper is at hand to determine relative absorption rates. In this instance, therefore, the influence of the ligand can at present be evaluated only by the overall utilization. In Table 1 the utilization of various Fe compounds by anemic rats is shown. As judged by the synthesis of hemoglobin, iron was utilized best from citrate and L-leucinate and not so well from fumarate and EDTA. Hemin iron was practically not utilized, while iron from hemoglobin produced values significantly higher than in depleted animals. Essentially the same results are obtained using the catalase activity of the blood as criterion for utilization.

With regard to availability there is an interaction between iron and copper. The hemoglobin level and also the catalase activity can be used to detect a lowered availability of iron during Cu deficiency. Ceruloplasmin activity is suited, at least temporarily, to demonstrating a reduced availability of copper during Fe deficiency. It is known that in the liver of the rat iron accumulates very rapidly in response to a deficient supply of copper, and this is true even during a 'mild' Cu deficiency [Grassmann and Kirchgessner, Arch. Tierernährung 23: 261 (1973); Z. Tierphysiol. Tierernährung Futtermittelk. 31: 113 (1973)].

This disturbance of the Fe utilization, which in recent studies is attributed to impeded mobilization of iron due to a ceruloplasmin deficiency [Frieden, Nutr. Rev. 28: 87 (1970)], slowly results in a small, but indeed significant, decline in the hemoglobin level and catalase activity of the blood. The response of the catalase activity was slower than that of the hemoglobin [Grassmann and Kirchgessner, Arch. Tierernährung 23: 261 (1973)]. On the other hand, the Cu content of the liver greatly increases as a result of Fe depletion. This increase in the liver content temporarily occurs at the expense of a reduced ceruloplasmin activity [Grassmann and Kirchgessner, Tierphysiol. Tierernährung Futtermittelk. 31: 38 (1973)].

This interaction between these two trace elements certainly must be considered in any study on the influence which ligands have on their metabolic availability in order to avoid misleading interpretations.

DISCUSSION

Henkin (Bethesda). I should like to emphasize the important human model for defective copper absorption, which is Menke's kinky hair syndrome and which is a much more common childhood abnormality than we previously thought. This fatal disease represents an abnormality of transport of copper across the gut, in which copper bound to EDTA and other attempted treatments have been totally worthless in promoting any copper transport across the gut wall. Studies such as you have described with the rat may ultimately help us understand and treat this disease, but I think we should also channel our thinking directly on the human problem. Curiously, there is no abnormality in iron or zinc metabolism in these children.

Gubler (Provo). We have to be careful in using such criteria as you have described for iron uptake, because one gets complete normalization of hemoglobin long before you normalize iron storage. This has been one of the big problems in therapy in the past.

Grassmann. Yes, absorption of iron in anemic rats would be quite different from the normal.

ZINC, IRON, AND CHELATE INTERACTION IN POULTS AND CHICKS

F. H. KRATZER

Department of Avian Sciences, University of California, Davis, California

High levels of zinc have been shown to interfere with iron utilization in the rat by Cox and Harris [J. Nutr. 70: 514 (1960)], Kinnamon [J. Nutr. 90: 315 (1966)], and Settlemire and Matrone [J. Nutr. 92: 153 (1967)]. Chelating agents such as EDTA improve the availability of zinc [Kratzer *et al.*, J. Nutr. 68: 313 (1959); Nielsen *et al.*, J. Nutr. 89: 35 (1966)], but interfere with iron utilization [Davis *et al.*, J. Nutr. 77: 217 (1962); Fritz *et al.*, Poult. Sci. 50: 1444 (1971)]. The purpose of the present work was to determine whether other chelating agents could improve iron availability in the presence of EDTA, and whether a zinc-iron antagonism exists at low dietary levels of these elements.

In three experiments various chelates were fed to poults in a zinc-deficient soy protein diet. EDTA corrected the zinc deficiency by improving growth (Table 1) and reducing the perosis score, but the packed cell volume (PCV) was markedly decreased. NTA, citric acid, and ascorbic acid were ineffective in counteracting this effect. Ascorbic acid was also ineffective in improving the growth of poults fed a zinc-deficient diet.

In two experiments in which the zinc and iron were varied in a purified diet, body-weight gain was decreased by the addition of iron to a zinc-deficient diet (Table 2). The PCV was improved as the zinc was increased in a low-iron diet. The perosis score was not influenced until adequate zinc was added.

Two experiments were conducted in which chickens were fed the low-zinc and low-iron purified diet with various levels of added zinc and iron. There was a slight reduction or no response in growth when iron was added to the low-zinc diet (Table 3). The addition of zinc at suboptimal iron levels caused reduced PCV values.

Table 1. Effect of EDTA, NTA, Fe, and Zn on growth, perosis and packed cell volume in poults (experiment 1)

Supplement[1]	18-day body-weight gain (g)[2]	Perosis score	Packed cell volume (%)[2]
None	222.1 ± 7.2	2.84	32.6 ± 1.0
EDTA	291.9 ± 11.6	0.50	21.9 ± 1.9
EDTA + NTA	301.1 ± 9.5	0.23	24.4 ± 1.8
EDTA + Fe	311.3 ± 8.4	0.18	29.5 ± 1.3
EDTA + NTA + Fe	302.5 ± 10.6	0.47	27.7 ± 1.0
Zn	269.3 ± 11.1	0.30	34.4 ± 1.3
EDTA + NTA + Fe + Zn	293.7 ± 13.1	0.30	25.7 ± 1.2

[1]Supplements were added to the zinc-deficient diet at the following concentrations: Na_2EDTA, 0.668%; NTA, 0.47%; Fe citrate, 0.1075%; ZnO, 0.0125%.
[2]Values are mean ± s.e.

NTA, citric acid, and ascorbic acid were all ineffective in counteracting the adverse effect which EDTA has on iron availability. In poults the addition of iron to a low-zinc diet reduced growth, while the addition of zinc to a low-iron diet caused improved PCV values. Chicks showed a marked reduction in PCV with the addition of zinc to low-iron diets.

Table 2. Effect of Zn and Fe on growth, perosis and packed cell volume in poults (experiment 5)

Supplement added (ppm) Zn	Supplement added (ppm) Fe	24-day body-weight gain (g)[1]	No. of survivors	Perosis score	Packed cell volume (%)[1]
0	0	106.5 ± 5.9	6	2.0	30.4 ± 1.1
10	0	273.2 ± 11.1	10	2.8	33.7 ± 0.6
20	0	342.7 ± 13.3	10	2.2	32.9 ± 1.5
0	10	46.8 ± 8.4	4	2.0	28.9 ± 2.1
10	10	239.8 ± 11.5	10	2.1	31.9 ± 1.4
20	10	364.5 ± 13.3	10	1.7	35.6 ± 0.6
60	10	413.0 ± 14.0	10	0.2	33.7 ± 0.7
10	30	287.2 ± 10.6	10	2.8	37.1 ± 0.7
20	30	424.9 ± 16.6	10	2.0	36.8 ± 0.8
60	100	450.3 ± 18.9	9	0.2	36.2 ± 0.7

[1]Values are mean ± s.e.

Table 3. Effect of Zn and Fe on growth and packed cell volume in chicks (experiment 7)

Supplement added (ppm)			
Zn	Fe	28-day body-weight gain (g)[1]	Packed cell volume (%)[1]
0	0	192.9 ± 17.0	29.8 ± 0.7
0	20	173.3 ± 10.3	32.9 ± 0.7
10	0	345.7 ± 9.2	24.5 ± 0.7
10	20	429.4 ± 10.5	27.4 ± 0.4
25	20	450.8 ± 15.1	25.9 ± 0.5
25	80	502.8 ± 22.8	28.0 ± 0.8

[1]Values are mean ± s.e.

DISCUSSION

Smith (Washington). We have also found no alleviation of zinc deficiency in rats by ascorbate, but plan to extend these studies to the zinc-deficient guinea pig.

Sandstead (Grand Forks). When you gave iron to zinc-deficient chicks, growth decreased, yet giving iron to zinc-deficient human beings allows them to grow better. Why would giving iron to a zinc-deficient animal impair growth?

Kratzer. The degree of deficiency is probably the key to it, but it's difficult to compare degrees of deficiency in different animals. If there is a severe deficiency of both of these, we may find adverse effects if supplementation with only one is made.

Forth (Homburg). How did you decide on the concentrations of chelating agents?

Kratzer. I am not certain why we chose those levels. The EDTA level was set at the level used in our previous work and some of the others were on an equal molar basis to that. I believe that NTA was at a somewhat higher level than the equivalent EDTA, because its activity in improving zinc availability was less than that for EDTA. Actually HEDTA was more effective than EDTA.

EFFECT OF PARENTERALLY ADMINISTERED EDTA ON ENDOGENOUS SECRETION AND METABOLISM OF ZINC IN THE RAT

H.-J. LANTZSCH and K. H. MENKE

Fachgruppe 9, Abteilung Tierernährung, Universität Hohenheim, Stuttgart, Federal Republic of Germany

In two experiments with a total of 36 growing male Wistar rats, maintained on 15 g per day of a 25 ppm Zn semisynthetic glucose-casein diet, the influence of daily i.p. injections of 25, 50, or 100 mg $Ca\text{-}Na_2\text{-}EDTA \cdot 6\,H_2O$ for one, two, or six days on urinary and intestinal excretion, intestinal absorption, body retention, and organ and tissue content of Zn was investigated. Rats were housed in single plexiglass cages which permitted a separate and quantitative collection of urine and feces. Three weeks before EDTA treatment the rats in experiment 2 were labeled with a single dose of $[^{65}Zn]$glycine. Fecal Zn of endogenous origin was calculated from the specific activities of feces and urine during the 12-day collection period.

EDTA increased urinary Zn excretion (Table 1) in proportion to the injected dose in the range of 2.4–3.0 μg Zn/mg EDTA injected for up to two injections of 50 mg EDTA on two consecutive days. Only 2% of the theoretical binding capacity of EDTA for Zn (135.5 μg/mg) was used. At higher doses (100 mg), or with more frequent injections (Table 2), urinary excretion increased absolutely but decreased relative to the amount of EDTA injected, indicating reduced zinc concentration in the extracellular space. This decreased concentration is replenished more slowly, through subsequent delivery of Zn from the intracellular space, than EDTA is excreted through the kidneys.

Zn concentration in blood serum 24–48 hr after the last EDTA injection was reduced (1.4–1.7 μg Zn/ml, experiment 1; 0.8–1.2 μg Zn/ml, experiment 2) under control levels (2.2 μg/ml, experiment 1). An inexplicably low serum Zn level (1.0 μg/ml) was found in the control group of experiment 2.

Table 1. Effect of EDTA on urinary zinc excretion[1]

	Group I	Group II	Group III	Group IV	Group V	Group VI
EDTA injection						
1 day (mg/day)		25	50	25	50	100
2 days (mg/day)				25	50	100
Food intake (g/day)	13.5 ± 0.5	13.9 ± 0.0	13.4 ± 0.9	13.5 ± 0.7	13.7 ± 0.4	13.3 ± 0.8
Zn intake (μg Zn/day	320.7 ± 10.6	329.0 ± 0.3	317.7 ± 22.6	320.9 ± 17.5	324.6 ± 9.3	314.4 ± 18.7
Zn in urine (μg Zn/day	26.5 ± 4.7	88.4 ± 2.0	162.9 ± 34.0	86.4 ± 14.2	151.4 ± 23.0	193.6 ± 11.0
Zn/EDTA (μg/mg)		2.5	2.7	2.4	2.5	1.7
Significant differences of urinary Zn excretion	Group I : II,III,IV,V,VI	$P < 0.001$				
	Group II : III,V	$P < 0.01$				
	Group II : VI	$P < 0.001$				
	Group III : IV	$P < 0.01$				
	Group IV : V	$P < 0.01$				
	Group IV : VI	$P < 0.001$				

[1]Growing male rats were given i.p. injections of $CaNa_2$-EDTA for one or two days. There were four rats per group and urine was collected for two days. Values are means ± s.d.

Table 2. Effect of EDTA on zinc metabolism

	Group I	Group II	Group III	Group IV
EDTA injection (mg/day)		25	50	100
Weight, begin (g)	93.5	89.5	90.0	91.0
Weight, end (g)	254.0	277.0	256.5	257.5
Food intake (g/day)	13.9	14.0	14.0	13.8
Zn intake (μg/day)	351.3	351.5	351.5	348.3
Urine-Zn (μg/day)	12.1	86.3	107.7	123.5
Urine-^{65}Zn (nCi/day)	3.8	43.2	54.0	64.9
Urine-Zn sp. act. (nCi/mg)	318.0	500.5	501.4	525.5
Urine-Zn (Δ μg/day)		+ 74.2	+ 95.6	+111.4
Urine-^{65}Zn (Δ nCi/day)		+ 39.4	+ 50.2	+ 61.1
Urine-Zn sp. act. (nCi/mg)		531.0	525.1	548.5
Urine-Zn/EDTA (μg/mg)		3.0	1.9	1.1
Feces-Zn, total (μg/day)	242.3	160.8	163.8	199.8
Feces-^{65}Zn (nCi/day)	59.1	36.3	36.3	35.4
Feces-Zn sp. act., total (nCi/mg)	243.9	225.7	221.6	177.2
Feces-Zn, endogenous (%)	76.7	45.1	44.2	33.7
Feces-Zn, endogenous (μg/day)	185.8	72.5	72.4	67.3
Feces-Zn, endogenous (Δ μg/day)		−113.3	−113.4	−118.5
Feces-^{65}Zn (ΔnCi/day)		− 22.8	− 22.8	− 23.7
Feces-Zn, exogenous (μg/day)	56.5	88.3	91.4	132.5
Zn absorption (%)	83.9	74.9	74.0	62.0
Zn absorption (μg/day)	294.8	263.2	260.1	215.8
Zn retention (μg/day)	+ 96.9	+104.4	+ 80.0	+ 25.0

Growing male rats were given i.p. injections of $CaNa_2$-EDTA for six days. Urine and feces were collected for 12 days for all groups. In EDTA groups values are calculated for days 7–12 of the collection period when EDTA was administered.

After six EDTA injections, urinary output of Zn and ^{65}Zn (Table 2) was increased to the same extent. Calculated specific activities of Zn excreted above the control level were equal to the mean specific activities of organ and tissue Zn. Therefore we can assume that increased urinary Zn originates from body pools and not from enhanced intestinal absorption.

Intestinal secretion of Zn is drastically reduced after six injections of EDTA (Table 2). When EDTA is present in the extracellular space, Zn normally secreted in the gut is partially chelated, and in this form it can leave the body only through the kidneys. Because the reduction in intestinal Zn secretion was independent of injected dose, we assume that only a limited part of the Zn normally secreted in the gut is able to be chelated by EDTA. This may be complexed by 25 mg EDTA. The increased urinary Zn output seen when increased amounts of EDTA were injected indicates an elimination of Zn from body stores. Surprisingly there was a greater reduction of the intestinal Zn secretion than there was an increase in urinary output (Table 2). Since intestinal absorption of Zn is also reduced after EDTA injections, it is possible that reduced absorption was responsible for this difference.

Of the organs and tissues examined, there were significant changes in Zn concentration of skin and blood cells after EDTA injections. After six injections of 100 mg EDTA (experiment 2) Zn concentration in skin declined from 4.6 (control) to 3.0 μg Zn/g skin. A fall in Zn concentration of blood cells from 12.4 (control) to 10.4 and 10.2 μg Zn/g was seen after six injections of 25 and 50 mg EDTA. After injection of 100 mg EDTA, the Zn concentration rose again to 11.1 μg/g. Similar results were obtained in experiment 1. In kidneys Zn was accumulated after two or six injections of 50 or 100 mg EDTA.

DISCUSSION

Spencer (Hines). We have given EDTA and I agree with you that the zinc does come from the body stores. We also saw a decrease in the effect after one gives multiple injections of EDTA, but urinary zinc excretion continued to be greater than normal.

ZINC STATUS IN DEPLETION AND REPLETION AND ITS RELATION TO VITAMINS AND TRACE ELEMENTS

J. PALLAUF and M. KIRCHGESSNER

Institute of Animal Nutrition, Technical University of Munich, Freising-Weihenstephan, Federal Republic of Germany

Weaned male rats were fed a semisynthetic casein diet (1.8–2.0 ppm zinc) and used in a series of three experiments. First, the extent and rate of zinc depletion in individual organs and in the whole body were studied. Secondly, depleted rats were repleted by supplying graded levels of zinc to determine the levels at which plateaus in the zinc status were reached. Finally, the influence of different levels of B vitamins and trace elements on the zinc status and deficiency symptoms was investigated in zinc-deficient animals. The objectives of this study were to determine metabolic interactions and to detect if these vitamins and trace elements could replace zinc in some of its functions.

Groups of rats were zinc depleted for 35 days. Every five days six animals were killed for zinc analysis of liver, bones, hair, and whole body [Kirchgessner and Pallauf, Z. Tierphysiol. 29: 65 (1972)]. Within the experimental period the absolute zinc content of the total rat decreased to an average of 86%, that of the total liver to 64%, and that of bone samples (femora and humeri) to 68% of the initial value. The depletion of the liver was particularly rapid, and the decrease in zinc normally followed a logarithmic function. In all organs the effect of the depletion was most pronounced with reference to the ash content and least obvious with reference to the fresh matter. When compared with the zinc content of animals that died during the depletion, those rats which were decapitated at the end of the experiment were already in a sublethal state of depletion.

Weaned rats were also depleted for 10 days and then repleted for 21 days with diets containing graded levels of zinc (2–500 ppm) [Kirchgessner and Pallauf, Z. Tierphysiol. 29: 77 (1972)]. Optimum zinc contents were obtained in the serum at 12 ppm, and in the liver, bones, and whole body at and above about 15 ppm dietary zinc. Within the range of about 5–12 ppm dietary zinc there was a nearly linear relationship between the dietary zinc supply and the zinc concentration of the organs studied. For optimum growth, however, a dietary level of 8 ppm zinc was adequate. Homeostatic regulation prevented higher dietary levels of zinc from altering the zinc concentrations. Accumulation was apparent only at the 500 ppm level. Blood serum, as already suggested by Luecke *et al.* [*In* Mills (ed.), Trace Element Metabolism in Animals, Livingstone, Edinburgh (1970)], and more particularly liver and bones are evidently good indicators of a suboptimal zinc supply [Pallauf and Kirchgessner, Int. J. Vit. Nutr. Res. 41: 543 (1971); Z. Tierphysiol. 30: 193 (1972); Zbl. Vet. Med. A, 19: 594 (1972)]. Hair, however, cannot be considered a general indicator of zinc supply as there was a positive correlation only within the range of 10–15 ppm dietary zinc [Pallauf and Kirchgessner, Zbl. Vet. Med. A, 20: 100 (1973)].

In view of similarities between the deficiency symptoms of zinc and vitamins, additions of biotin (2, 5, 10, 20, and 50 mg/kg of diet) and folic acid (1, 2.5, and 5 mg/kg of diet), and additions of the vitamins B_1, B_2, B_6, B_{12}, pantothenic acid, and niacin (2, 5, and 10 times the requirement) were studied in zinc-deficient rats. The vitamin additions did not affect the symptoms of zinc deficiency nor the zinc depletion in serum and liver. There was no evidence for any interactions between the vitamins tested and zinc [Pallauf and Kirchgessner, Int. J. Vit. Nutr. Res. 42: 555 (1972); *ibid.* 43: 339 (1973)]. Contrary to findings by Chu *et al.* [Nutr. Rep. Int. 1: 11 (1970)], even biotin had no effect.

In vitro unspecific activator functions of zinc in zinc-enzyme complexes can also be assumed by other metals. Hoekstra [*In* Mills (ed.), Trace Element Metabolism in Animals, Livingstone, Edinburgh (1970)] reported an alleviation of porcine parakeratosis when 50 ppm cobalt replaced zinc in a practical type of diet high in calcium. Therefore, the effect of dietary additions of graded levels of Mn, Cu, and Fe (3, 10, and 50 times the requirement) and of Co and Ni (3, 10, and 50 ppm) on zinc deficiency was studied. Zinc-deficiency symptoms and live weights were not significantly affected. The influences on the zinc status are shown in Table 1. The additions of Mn predominantly increased the zinc concentration of the liver. The results with copper were inconsistent [Pallauf and Kirchgessner, Zbl. Vet. Med. A (1974), in press]. Iron apparently increased the zinc content of the liver without influencing the serum zinc. A high dosage of cobalt reduced the serum zinc level. Nickel at all levels tested reduced serum zinc [Kirchgessner and Pallauf, Z. Tierphysiol. 31: 268 (1973)]. Thus there are evidently interactions between zinc and the trace elements studied in intermediary metabolism. There were no indications, however, that addition of these trace elements to the zinc-deficient diet could replace the function of zinc.

Table 1. Influence of various trace elements added at different levels to a low-zinc diet on the zinc status of liver and serum[1]

		Level of trace element added (mg/kg diet)																
		Mn				Cu				Fe			Co			Ni		
Criteria	Experiment	120	300	1500	3000	12	20	40	200	120	400	2000	3	10	50	3	10	50
Liver																		
μg total Zn	1	0	++	++		+++		0	++	++	++	++	0	-	0	0	0	0
	2	++	+++	0	+++	0	0											
μg Zn/g	1	++	++	++		0		0	0	0	++	+++	0	0	0	++	0	+++
	2	+++	0	0	+++	0	0											
Serum																		
μg Zn/ml	1	0	0	+++		+++		---	0	0	0	0	0	0	--	--	--	--
	2	0	0	0	+++	0	0											

[1] 0 Not significantly different from the control.
+ Significantly different ($P < 0.05$), increase < 15%.
++ Significantly different ($P < 0.05$), increase 15–30%.
+++ Significantly different ($P < 0.05$), increase > 30%.
– Significantly different ($P < 0.05$), decrease < 15%.
–– Significantly different ($P < 0.005$), decrease 15–30%.
––– Significantly different ($P < 0.05$), decrease > 30%.

DISCUSSION

Hoekstra (Madison). With reference to your last comment, we did believe for a time that there may have been some substitution of cobalt for zinc in the pig. In three swine experiments, we have been able consistently to reproduce the marked alleviation of zinc deficiency by cobalt supplementation. The practical-type diet used was high in calcium and contained natural phytate. In the two experiments subsequent to the first, which was mentioned at TEMA 1, we did find that cobalt increased the zinc content of some of the tissues, so that our speculation now is that the cobalt, instead of replacing zinc in critical functions, may be sparing the availability of zinc under these high-calcium conditions. We do not have an experiment to prove that, but I think it's the most likely explanation. Similarly to your results with the rat, we were unable to obtain any sparing effect of cobalt on zinc in chicks fed semipurified diets normal in calcium.

ZINC IN THE RAT PANETH CELL

MARGARET E. ELMES

Department of Physiology, The Queen's University of Belfast, Northern Ireland

Zinc was first detected histochemically in the granules of Paneth cells by Okamoto [Trans. Soc. Path. Japan 32: 99 (1942)] using diphenylthiocarbazone (dithizone) which produced a purple-red coloration of the granules. This technique was modified by Midorikawa and Eder [Histochemie 2(6): 444 (1962)], who used alcohol-fixed tissues stained at pH 8.0–8.5 to demonstrate zinc in the Paneth cells of the rat. Autoradiographic studies by Millar, Vincent and Mawson [J. Histochem. Cytochem. 9: 111 (1961)] showed that injected ^{65}Zn could be demonstrated in Paneth cells in the rat ileum. They suggested that Paneth cells may act as an excretory pathway for zinc, and also stated that in zinc-deficient rats zinc cannot be demonstrated histochemically in the Paneth cell.

It was decided to investigate the following:

(1) The effect of fasting and zinc deficiency on (a) the total Paneth cell count as performed on rat tissues fixed either in formol saline or Bouin's solution without acetic acid and stained with hematoxylin and eosin, and (b) the histochemically detectable zinc in the rat Paneth cell using the technique of Midorikawa and Eder.

(2) The effect of fasting on the total zinc content of scraped-off intestinal mucosa as estimated by neutron activation analysis of dried samples.

(3) Any relationship between the appearance of Paneth cell granules under the electron microscope and their zinc content as determined by electron microprobe analysis. Rat ileum fixed in glutaraldehyde, postfixed in osmium tetroxide, and embedded in Araldite was used. After sectioning the tissues were either put on copper grids for routine electron microscopy, or on titanium and aluminum grids for microprobe examination.

The number of Paneth cells per crypt in 100 well-orientated crypts counted in samples from the duodenum, jejunum, and ileum increased from approximately 1 in the duodenum to 4 or more in the ileum. Not all cells contained dithizone-reactive zinc and the distribution of those that did was

Table 1. Effect of fasting on dithizone-positive Paneth cells

	Intestinal Segment[1]		
Treatment	Duodenum	Jejunum	Ileum
Control	24.586	20.960	63.775
24-hr fast	16.186	11.929	43.997
48-hr fast	15.365	11.478	41.433
72-hr fast	12.638	15.643	49.316
All groups	17.194	15.002	49.633

[1]Values are mean percentage of Paneth cells which were dithizone positive. Between sites $F = 33.46$, $df = 2.48$, $P < 0.001$.

patchy. The highest proportion of dithizone-reactive cells was found in the ileum and neither this nor the total count were affected by fasting (Table 1). Feeding a diet containing less than 1 ppm of zinc for 4–6 weeks caused a complete absence of dithizone-reactive cells in all areas (Table 2), but had no significant effect on the total Paneth cell count.

Correlation of the secretory status of the Paneth cell as seen in Bouin's fixed tissue with its dithizone reactivity was only successful in the duodenum where all animals with a high proportion of empty and vacuolated cells had no dithizone-reactive cells. The results in the jejunum and ileum were inconclusive.

Neutron activation analysis of the total zinc in scraped-off mucosa of fasted rats showed a fall in the ileum at 72 hr, but the overall pattern of results indicates that the significance of this finding is questionable (Table 3).

Table 2. Effect of zinc deficiency on dithizone-positive Paneth cells

	Intestinal Segment[1]		
Treatment	Duodenum	Jejunum	Ileum
Zn deficient	0.000	0.000	0.000
Pair-fed controls	0.000	0.569	24.723
Ad lib. controls	6.712	4.680	19.935
All groups	2.237	1.750	14.553

[1]Values are mean percentage of Paneth cells which were dithizone positive. Between groups $F = 4.22$, $df = 2.12$, $P < 0.05$. Between sites $F = 7.85$, $df = 2.24$, $P < 0.01$.

Table 3. Zinc concentration of intestinal mucosa

	Intestinal Segment[1]		
	Duodenum	Jejunum	Ileum
24-hr fast	18.5 ± 5.14	14.8 ± 2.77	26.6 ± 1.34
48-hr fast	19.3 ± 2.66	12.9 ± 4.33	27.7 ± 5.14
72-hr fast	20.9 ± 5.08	22.9 ± 4.27	13.1 ± 1.50
96-hr fast	12.4 ± 1.77	21.1 ± 4.67	27.9 ± 3.56

[1] Values are μg Zn/g wet weight (mean ± s.e.). For 96–72-hr fast (ileum) $F = 3.80$, $df = 3$, $P < 0.05$.

Electron microscopic examination of the granules of rat Paneth cells showed three types in which preliminary electron microprobe estimations indicated the presence of zinc in dense homogenous granules, and in lesser quantities in moderately dense and pale granules. [Neutron activation analysis was performed by Dr. B. W. East, Scottish Universities Research and Reactor Centre, East Kilbride, Scotland, and electron microprobe estimations by Dr. P. R. Lewis, Physiological Laboratory, Cambridge, England.]

DISCUSSION

Spencer (Hines). During fasting, the dithizone reactive cells remained about the same, but neutron activation indicated that you had less zinc in the intestinal mucosa. What is your interpretation of this?

Elmes. I don't put much reliance on the neutron activation analysis because our sample size was so small and we could use only a small number of animals. I did get the one significantly lower result in the ileum after 72 hr, but I doubt if this is biologically significant.

Kasarskis (Madison). Have you shown by ^{65}Zn autoradiography that Paneth cells which are not dithizone positive are in fact devoid of zinc?

Elmes. Not as yet.

Aughey (Glasgow). I think you would have great difficulty in interpreting in any statistical fashion ^{65}Zn autoradiography on Paneth cells unless you did a lot of background counting on pure samples of ^{65}Zn, because ^{65}Zn is both a gamma and a beta emitter.

Elmes. Yes, I have heard of considerable difficulties with ^{65}Zn.

Beisel (Frederick). You're postulating that the Paneth cell functions in excreting zinc, but could the zinc have some metabolic role within the cell in producing secretions that are perhaps unique?

Elmes. Yes, that is one of my suggestions, but I think it may be concerned with granule synthesis. On the other hand, very little is known about granule synthesis at present and a point against this idea is that the granules appear normal in an animal which has dithizone-negative Paneth cells.

INTESTINAL ABSORPTION AND SECRETION OF ^{65}Zn IN THE RAT

ALFRED H. METHFESSEL and HERTA SPENCER

Metabolic Section, Veterans Administration Hospital, Hines, Illinois

As part of an investigation on the effect of age on mineral metabolism experiments were conducted on intestinal absorption and secretion of ^{65}Zn with time in rats of different ages. A single dose of $^{65}ZnCl_2$ was given by intubation to the intact rat or was instilled into the lumen of ligated intestinal sacs to determine the absorption, or injected intravenously via the tail vein to determine intestinal secretion of ^{65}Zn. The absorption of ^{65}Zn was very rapid in the intact rat as shown by the maximal absorption of 25% ^{65}Zn within 30 min. The maximal ^{65}Zn activity in plasma also occurred at 30 min when most of the ^{65}Zn was still located in the proximal portion of the small intestine. Studies in which the ^{65}Zn was instilled into ligated intestinal sacs demonstrated that maximal absorption of ^{65}Zn occurred from the duodenum, moderate absorption from the midjejunum and ileum, and slight absorp-

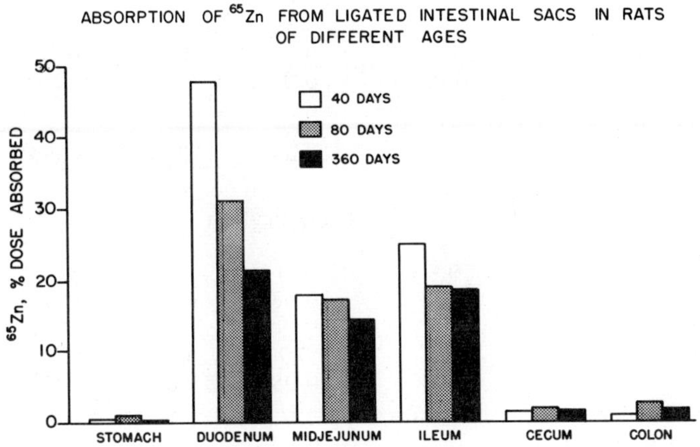

Fig. 1. Absorption of ^{65}Zn from the gastrointestinal tract.

Table 1. Secretion of ^{65}Zn into the gastrointestinal tract of the rat

Time after injection	^{65}Zn (% per segment)			
	Stomach	Small intestine	Large intestine	Total
		Intact rat		
10 min	0.6	5.9	0.9	7.4
30 min	0.8	8.3	1.1	10.3
1 hr	0.8	9.4	1.5	11.7
3 hr	0.8	12.5	2.1	15.4
24 hr	0.6	6.6	4.8	12.0
		In vivo ligation		
1 hr	1.0	10.4	1.5	12.8
3 hr	0.8	10.1	1.3	12.2

Table 2. Secretion of injected ^{65}Zn into the gastrointestinal tract of the rat

Time after injection (hr)	^{65}Zn (% dose per g wet tissue)[1]					
	Stomach	Duodenum	Jejunum	Ileum	Cecum	Colon
			Intact rat			
1	0.2 ±0.03	1.1 ±0.03	0.8 ±0.05	0.8 ±0.05	0.1 ±0.01	0.2 ±0.03
3	0.2 ±0.02	1.1 ±0.06	1.0 ±0.04	1.1 ±0.12	0.2 ±0.01	0.3 ±0.01
			In vivo ligation			
1	0.2 ±0.02	0.9 ±0.07	1.0 ±0.08	0.7 ±0.11	0.1 ±0.01	0.3 ±0.02
3	0.1 ±0.02	0.6 ±0.04	0.9 ±0.08	0.9 ±0.06	0.1 ±0.01	0.3 ±0.03

[1]Values are means ± s.d.

tion from other portions of the gastrointestinal tract (Fig. 1). The absorption of ^{65}Zn from the duodenum was significantly decreased with increasing age, whereas there was little age effect on the absorption of ^{65}Zn from other portions of the intestine. Intestinal secretion of ^{65}Zn in the intact rat was also rapid, with 6% or more of the injected ^{65}Zn found in the small intestine from 10 min to 24 hr after the injection of ^{65}Zn (Table 1). Injected ^{65}Zn was readily detected in all segments of the gastrointestinal tract, and was secreted uniformly and rapidly throughout the entire small intestine of the intact rat and in rats with the ligated gastrointestinal tract (Table 2). The ^{65}Zn secreted into the small intestine represented 80% or more of the total ^{65}Zn secreted into the gastrointestinal tract within 2 hr after injection. This study indicates that the small intestine plays a major role in the absorption and secretion of zinc, and that aging decreases the absorption of zinc for the duodenum of the rat. [Supported by Contract AT(11-1)-1231-91 from the U.S. Atomic Energy Commission.]

DISCUSSION

Miller (Athens). Your data agree with the generally accepted conclusion that the duodenum is the area from which zinc absorption is most rapid, but does this really represent the situation in the intact animal? For example does it not specifically ignore the rates at which zinc normally passes through the different sections?

Methfessel. This is very possible and I don't know the various passage rates.

Strain (Cleveland). A number of years ago we found that the female absorbs much more of many common radioisotopes than the male, particularly when she is pregnant. The old male absorbed very little. Your work reinforces some of our observations.

Elmes (Belfast). How do you think the zinc gets into the lumen? It comes in far too fast to be from desquamated cells from the tips of the villi. You've excluded the pancreatic secretions.

Methfessel. The intestinal zinc appears almost immediately and I presume that the zinc is first going into the intestinal wall, but I cannot say how much is secreted because the samples we assayed included the intestinal wall.

Elmes (Belfast). You think that in some way the zinc ion diffuses through the cells?

Methfessel. Not necessarily; a considerable amount of zinc is retained in the cells.

Forth (Homburg). I do not believe that we have to assume that all secretions occurring from the mucosal epithelium must go across the cells. We have some evidence that even heavy metals go in the spaces between the cells.

TURNOVER OF ^{65}Zn IN DOMESTIC FOWL

C. GARCIA DEL AMO, R.MA ZUNZUNEGUI, and A. SANTOS-RUIZ

Department of Biochemistry, Faculty of Pharmacy, University Complutense of Madrid, Spain

The turnover of ^{65}Zn in domestic fowl has been studied. The use of an oviparous animal made it easy to follow the whole cycle of evolution hen-egg-embryo in a short period. Two series of hens of the Honeger breed were used. They were given 20 μCi/kg of ^{65}Zn orally or intramuscularly, and were killed between 1 and 60 days after administration. The radioactivity in the organs selected was measured in a liquid scintillation counter and the results obtained are compared in Table 1.

We found that, 10 days after oral isotope administration, the organ which retained the most radioactivity was the liver, whereas the pancreas has the highest ^{65}Zn content following intramuscular injection. In the other selected organs, except the gizzard, there was always higher radioactivity when ^{65}Zn was given intramuscularly rather than orally.

Another group of 12 hens was injected intramuscularly with the same dose of isotope as used before. The total radioactivity of each egg was determined each day in a 60-day experiment. The radioactivity of the yolk, albumin, shell, and shell membrane was also measured with a liquid scintillation counter. We found that the ^{65}Zn inoculated to the hen passes quickly to the eggs; 4.5% of the dose was found in the eggs after 10 days and 9% at the end of the experiment.

The data in Fig. 1 show that the highest content of isotope in the egg was found at the 6th day. It diminished quickly until the 15th day and then slowly until the end of the experiment. The radioactivity in the yolk was 92% of that in the whole egg [Khovanskikh, Sel'skokhoz. Biol. 6(6): 894 (1971)]. The nonmineralized shell did not contain any isotope and the albumin and the shell membrane only 8% together.

Fourteen hens were placed with cocks of the same breed, and, when it was evident that the eggs were fertile, the hens were given ^{65}Zn in the same

Table 1. Content of ^{65}Zn in hen organs

Organs	d.p.m./g dry organ							
	Oral				Intramuscular			
	10 days	20 days	30 days	50 days	10 days	20 days	30 days	50 days
Brain	21,000	17,250	19,476	16,666	163,272	153,462	160,000	74,000
Eye	12,061	9487	6516	11,016	105,071	76,224	93,000	52,000
Crest	18,064	6610	2761	11,163	130,936	57,995	64,000	53,000
Nails	15,257	2742	3464	3919	71,132	110,735	152,500	85,000
Suc. ventricle	60,021	15,948	12,194	9567	141,900	123,063	130,000	80,000
Gizzard	52,607	40,285	36,181	36,383	1959		7500	19,000
Long bone	10,172	33,187	29,914	30,152	426,950	234,401	198,000	110,000
Beak	20,383	23,190	24,033	21,828	162,691	174,384	255,000	130,000
Skin	14,484	6602	5972	4733	57,936	53,424	58,000	36,000
Small feather	15,052	41,283	6163	4910	20,489	13,318	30,000	4000
Liver	259,265	38,693	18,035	20,583	50,310	46,406	28,000	12,500
Pancreas	109,030	33,090	18,485	15,850	474,707	259,142	190,000	110,000
Kidney	84,238	25,018	16,651	20,623	160,560	146,118	51,000	62,000
Spleen	70,210	26,218	16,474	15,311	405,188	181,256	85,000	40,000

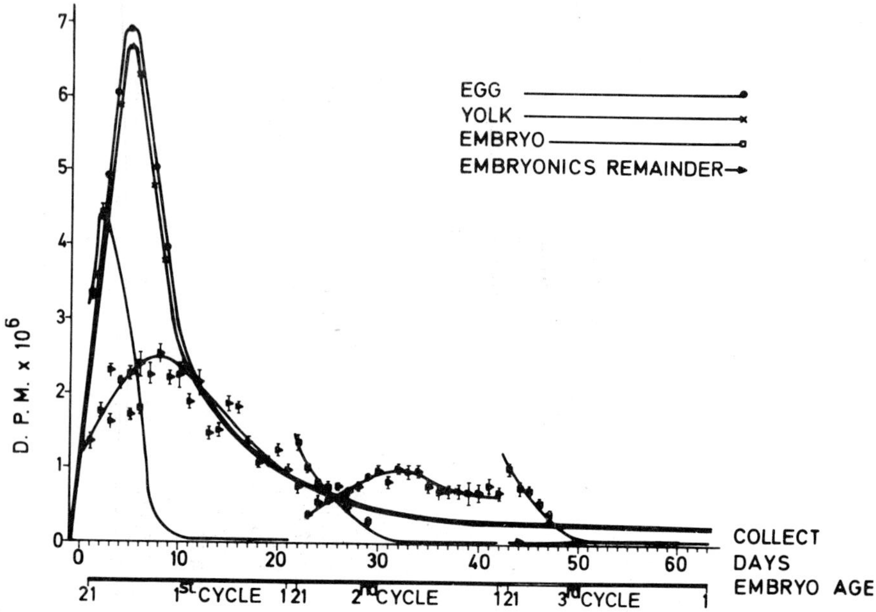

Fig. 1. Content of ^{65}Zn in various parts of the egg.

manner as before. The eggs were marked with the corresponding day and were put into an incubator day by day. At the end of the period (21 days) they were taken out together, so that the eggs laid on day 1 were totally developed and the last eggs had only one day of development. The radioactivity was determined as a function of the ages of the embryos in three consecutive laying cycles. It can be seen in Fig. 1 that during the first 10 days of embryo development hardly any ^{65}Zn is detected in the embryo. The highest embryonic concentration of ^{65}Zn took place on the 21st day, except in the first cycle where it was found on the 19th day [Zunzunegui *et al.*, Ann. Real Acad. Farm. 39(2): 237 (1973)]. The development of the embryo in all three periods followed similar patterns but with diminishing intensity. The embryo's weight gain was directly proportional to the content of ^{65}Zn. The remainder of the fertilized egg gave similar curves, but with diminishing intensity and a maximum of concentration of ^{65}Zn towards the middle of the cycle.

THE EFFECT OF MICRO-ORGANISMS UPON ZINC-65 RETENTION USING GERMFREE RATS

J. CECIL SMITH, JR., E. G. MCDANIEL, and LOIS D. MCBEAN

Veterans Administration Hospital, Washington, D.C., and National Institutes of Health, Bethesda, Maryland

We have previously reported a lower zinc requirement for germfree rats [Smith *et al.*, J. Nutr. 102: 711 (1972)]. The studies presented here investigate the possibility that this lower requirement might be the result of an increased zinc retention in the germfree rats.

Our investigations were conducted in two experiments. In both experiments zinc as $^{65}ZnCl_2$ was mixed in the diet and fed over a period of 5 days. The total cpm/g of diet was 25,391 ± 1016 (mean ± s.d.) for experiment 1 and 17,744 ± 3825 for experiment 2. The corrected specific activity of the isotope was 3.97 μCi/μg. The total dietary intake for the 5-day period when radioactive diet was fed averaged 46 ± 6.9 g for the germfree rats and 33 ± 3.6 g for the exgermfree rats fed the zinc-deficient diet in experiment 1 and 100 ± 21 g for the germfree and 100 ± 16 g for the exgermfree rats fed the zinc-sufficient diet.

After sacrifice, the radioactivity of the intact whole body was counted using a whole-body counter. The gastrointestinal tract was removed and the ingesta flushed from the tract. The ingesta was counted separately. The data are presented as percentage of dose retained in the whole body excluding the gastrointestinal ingesta.

The purpose of the first experiment was to compare radioactive zinc retention in germfree and exgermfree rats fed either a zinc-deficient (1.6 ppm zinc) or a zinc-sufficient (14 ppm zinc) diet. The experiment included one group of 13 germfree animals subdivided so that seven animals were fed the zinc-deficient diet and six were fed the sufficient diet. In addition, another group consisted of 11 exgermfree animals, with five rats fed the zinc-deficient diet and six fed the sufficient ration. The animals were allowed to eat *ad libitum.* The final body weight (mean ± s.d.) for each of the four groups was as follows: germfree, zinc-deficient, 106 ± 18 g; exgermfree, zinc-deficient, 67 ± 16 g; germfree, zinc-sufficient, 266 ± 53 g; exgermfree, zinc-sufficient, 233 ± 45 g.

A second experiment was designed to allow the comparison of zinc retention in eight germfree and seven exgermfree rats of similar body weight. Both groups were fed the zinc-deficient diet containing 1.6 ppm zinc. The germfree group was restricted in dietary intake to prevent them from becoming larger in body weight than the exgermfree rats. The final body weights (mean ± s.d.) were as follows: germfree, 73 ± 4.6 g; exgermfree, 72 ± 4.3 g.

The results of these experiments are shown in Tables 1 and 2. As shown, the whole-body ^{65}Zn retention was not significantly different between the germfree or exgermfree rats when either the zinc-deficient or zinc-sufficient diet was fed. However, there was a significantly greater percentage of the dose retained by the animals fed the zinc-deficient diet, irrespective of whether they were germfree or exgermfree.

In the second experiment, involving germfree or exgermfree animals fed the zinc-deficient diet, but which were pair weighted so that there would be little difference between the body weight, there again was no significant difference between the percentage of dose retained in the whole body.

Table 1. Zinc-65 retention in germfree and exgermfree rats fed a zinc-deficient or zinc-sufficient diet

	Body weight[1] (g)	Dose retained in whole body[1,2] (%)
Zinc deficient (1.6 ppm)		
Germfree	87 ± 21	39 ± 4.1(7)[a]
Exgermfree	63 ± 14	36 ± 8.6(5)[b]
Zinc sufficient (14 ppm)		
Germfree	241 ± 54	19 ± 2.9(6)
Exgermfree	233 ± 43	22 ± 4.9(6)

[1]Minus gastrointestinal ingesta.

[2]Values are mean ± s.d.; number of animals in parentheses.

[a]Significantly different from germfree animals fed the zinc sufficient diet, $p < 0.001$.

[b]Significantly different from exgermfree animals fed the zinc sufficient diet, $p < 0.01$. There was no significant difference between germfree and exgermfree rats fed the same diet.

Table 2. Zinc-65 retention in germfree and exgermfree rats fed a zinc-deficient diet (pair weighted)

	Total body weight (g)	Dose retained in whole body[1] (%)
Zinc deficient (1.6 ppm)		
Germfree	73 ± 4.6	37 ± 3.1(8)
Exgermfree	72 ± 4.3	43 ± 9.9(7)
Level of significance		n.s.

[1]Minus gastrointestinal ingesta. Values are mean ± s.d.; number of animals in parentheses

It was reported earlier [Reddy *et al.*, J. Nutr. 102: 101 (1972)] that the germfree state had no significant effect on the absorption and retention of zinc. Those investigators used animals fed a zinc-sufficient diet. Our results also demonstrate that the absence of a microflora does not affect the whole-body zinc retention, thus indicating that net absorption in the germfree rat is not different from animals possessing a microflora (exgermfree). This is true irrespective of whether a zinc-sufficient or zinc-deficient diet is fed. These data would support the conclusion that the decreased requirement for zinc in germfree rats cannot be attributed to an increased retention or net absorption. [Supported in part by Grant No. NIH 5 R01 AM 15703-02 and VA Research Funds.]

DISCUSSION

Weinberg (Bloomington). What is your theory as to why germfree animals need less zinc?

Smith. I think that at the metabolic, cellular, or subcellular level the germfree animal is able to utilize zinc more efficiently.

Elmes (Belfast). You don't consider that the bacteria themselves could be using up some dietary zinc?

Smith. In a former publication I put forth that hypothesis, but calculations showed that the microflora could be solid zinc pellets and it would still not account for the difference. In other words, the microflora used so little zinc that competition between microflora and host for zinc could not account for this effect.

ADAPTATIONS IN ZINC METABOLISM BY LACTATING COWS FED A LOW-ZINC PRACTICAL-TYPE DIET

W. J. MILLER, M. W. NEATHERY, R. P. GENTRY, D. M. BLACKMON, and P. E. STAKE

Departments of Dairy Science and Veterinary Medicine, University of Georgia, Athens, Georgia

In young ruminants a low-zinc diet causes several changes in zinc metabolism, indicating considerable homeostatic control [Am. J. Clin. Nutr. 22: 1323 (1969); J. Dairy Sci. 53: 1123 (1970)]. These metabolic alterations include a rapid increase in percentage of dietary zinc absorbed and decreased endogenous fecal excretion. However, for most tissues, total zinc content declines slightly or not at all, whereas labile zinc in most soft tissues decreases sharply.

Since effects of a low-zinc diet on zinc metabolism in lactating ruminants had been studied relatively little, an experiment was conducted with 10 lactating Holstein cows fed a low-zinc practical-type diet (16.6 ppm Zn) with and without supplemental ZnO (39.5 ppm total Zn) for six weeks. The low-zinc diet did not adversely affect milk production, fat-corrected milk, solids-corrected milk, milk fat, milk nonfat solids, milk protein, feed consumption, or body-weight changes, indicating adequate zinc intake for these functions [J. Dairy Sci. 56: 212 (1973)].

The low-zinc diet had little effect on stable zinc content of body tissues. Average tissue zinc based on samples representing 84% of the total body (fresh) was 23.1 and 22.1 ppm for cows fed control (39.5 ppm Zn) and low-zinc diets (Table 1). Even though gastrointestinal (GI) tract digesta of low-zinc fed cows contained considerably less zinc than controls, average GI tract tissue zinc was not affected. Thus homeostasis was effective in preventing marked depletion of total body zinc, even in tissues most closely associated with the dietary change.

In contrast to the lack of effect on performance and very minor influence on tissue stable zinc, the low-zinc diet caused major changes in zinc

Table 1. Effects of low-zinc diet on zinc metabolism in lactating cows

	Control (39.5 ppm)	Low zinc[1] (16.6 ppm)
Stable Zn		
Whole body (ppm fresh)	23.1	22.1
GI tissues (ppm dry)	71	71
Plasma (ppm)	0.92	0.78
Milk (ppm)	4.22	3.26[a]
Milk (% of intake)	8.7	17.8[a]
Net absorption (%)	30	45[a]
^{65}Zn		
Net absorption (% of dose)[2]	34.8	53.4[a]
Milk (% of dose)[2]	6.3	14.3[a]
Liver (% of dose/kg fresh)	0.059	0.159[a]
Heart (% of dose/kg fresh)	0.029	0.088[a]
Kidney (% of dose/kg fresh)	0.025	0.072[a]
Small Int. (% of dose/kg fresh)	0.026	0.083[a]

[1]Values marked [a] are significantly different from control at 1% level of probability.
[2]Percentage of dose after 14 days.

metabolism. A single oral tracer dose of ^{65}Zn was given four weeks after initiation of dietary treatments (Table 1). Net stable zinc and ^{65}Zn absorption each increased 50% with the low-zinc diet. The 53.4% net ^{65}Zn absorption of low-zinc cows indicates that zinc absorption can be comparatively high in cattle approaching maturity. Whether lactating cows could absorb as high a percentage of dietary zinc as young ruminants has not been established.

Milk zinc of cows fed the low-zinc diet was reduced 23%, with most of the effect occurring during the first week. Low-zinc-fed cows secreted twice as high a percentage (17.8 *versus* 8.7%) of the dietary zinc into milk. Milk ^{65}Zn, during two weeks after dosing, increased even more (6.3 *versus* 14.4% of the dose) (Table 1). Analyses of the relationship between stable zinc and ^{65}Zn indicated that zinc secreted into milk of low-zinc-fed cows had been in the body pool a shorter average time than with controls. In low-zinc-fed cows only 19.7% of the milk zinc had been in the body pool more than two weeks compared with 28.5% for controls.

Zinc-65 retained by the biologically more active tissues (such as GI tract, liver, lung, heart and kidney) increased more than ^{65}Zn absorption, indicating depletion of labile zinc (Table 1). This agrees with data from young ruminants.

The higher percentage of ^{65}Zn going into milk and the shorter average time after dosing suggest that milk secretion functions in much the same way as biologically active tissue in having a strong affinity for labile zinc. Thus, when a low-zinc diet is fed, zinc secretion into milk, to a considerable degree, obeys principles that are opposite to those of fecal excretion. Reducing dietary zinc to a low but nondeficient level causes rapid and major adjustments in zinc metabolism of lactating cows. These changes indicate considerable homeostatic control, with milk zinc being a major factor. However, milk zinc for individual cows, within treatments, was highly repeatable ($r = 0.92$) at different periods. [Supported in part by NIH Grant AM-07367.]

DISCUSSION

Kubota (Ithaca). Why did you use beet pulp in your diet? If you want to find zinc deficiency in corn, you always plant corn after sugar beets. This is an almost sure way to get zinc deficiency in plants.

Miller. In most summary reports on feed composition beet pulp is erroneously reported to contain about 0.7 ppm zinc, which is very deficient. This must have been a decimal point error. That is why we originally selected it and fortunately it was a suitable, rather low-zinc feed, although the latest beet pulp is considerably higher in zinc than that which we used initially. It still produces a rather low-zinc, but otherwise adequate, diet for dairy cows.

Kubota (Ithaca). I was thinking that maybe the beet pulp would cause low zinc availability in the same way as sugar beets apparently affect the zinc in the soil.

Miller. In our earlier studies young bulls got along very nicely on a total dietary zinc content of about 9 ppm and beet pulp made up about 90% of the diet. Certainly, the zinc was not unavailable.

Spais (Thessaloniki). You had lower production with your supplemented animals. We have had a lot of experience under practical conditions with zinc-deficient cows, and we always increase milk production by supplementing zinc.

Miller. Obviously our cows were not zinc deficient, and the decreased milk production with zinc supplementation must be an odd-ball chance occurrence.

INTERACTIONS AMONG DIETARY COPPER, ZINC, AND THE METABOLISM OF CHOLESTEROL AND PHOSPHOLIPIDS

LESLIE M. KLEVAY

USDA, Human Nutrition Laboratory, Grand Forks, North Dakota

Male rats weighing about 50g were fed *ad libitum* a purified diet, deficient in copper and zinc, based upon sucrose, corn oil, and egg-white protein, and containing no cholesterol or bile acid. Control animals drank water containing 10 μg Zn/ml and 2 μg Cu/ml. Experimental animals drank water containing either 10 μg Zn/ml and 0.25 μg Cu/ml or 20 μg Zn/ml and 0.5 μg Cu/ml. All animals were raised either under clean conditions [Klevay *et al.*, Envir. Sci. Technol. 5: 1196 (1971)] in a room with filtered air or using identical equipment in a conventional animal room.

The concentrations of cholesterol in plasma of animals raised under clean conditions are shown in Figure 1; those of conventional animals are shown in Figure 2. The data shown are those first found to demonstrate statistically significant differences. Differences between experimental and control groups increased when experiments were continued for 270 days. In one experiment the concentrations of phospholipids in experimental and control groups were 385 (± 35.5 s.e.) and 274 (± 20.6 s.e.) mg/dl ($p < 0.01$).

Decreased consumption of sucrose [Yudkin, Angiology 17: 127 (1966)], increased consumption of vegetable fiber [Trowell, Am. J. Clin. Nutr. 25: 926 (1972)], consumption of hard water [Nutr. Rev. 25: 164 (1967); *ibid.*, 26: 295 (1968)], and exercise [Nutr. Rev. 21: 178 (1963)] are associated with decreased risk of coronary heart disease. Foods with low concentrations of sucrose and high concentrations of vegetable fiber also have high concentrations of phytic acid [Averill and King, J. Am. Chem. Soc. 48: 724 (1926); McCance and Widdowson, Biochem. J. 29: 2694 (1935)], a material that decreases the availability of zinc, but apparently not of copper [Vohra *et al.*, Proc. Soc. Exp. Biol. Med. 120: 447 (1965)], from the intestinal tract.

There is a high correlation ($r = 0.95$) between the hardness of drinking water and the concentration of calcium in the water [Crawford *et al.*, Lancet

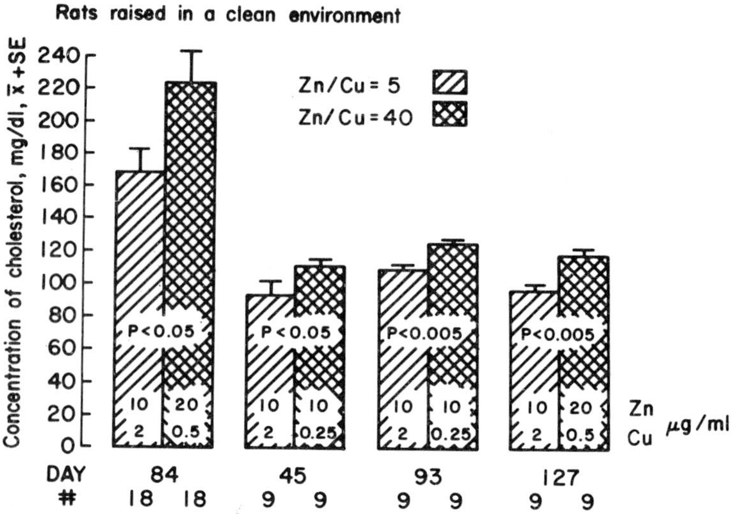

Fig. 1. Effect of dietary zinc and copper on the plasma cholesterol of rats raised in a clean environment. The concentrations of zinc and copper in the drinking water of the animals are shown at the base of each bar. The number of animals per group and the experimental day on which the plasma samples were obtained are shown beneath the bars.

1: 827 (1968)]. Calcium added to the diet of rats [Heth *et al.,* J. Nutr. 88: 331 (1966)] caused a shift of zinc from liver to bone. Calcium added to the diets of men [Yacowitz *et al.,* Br. Med. J. 1: 1352 (1965)] and rats [Yacowitz *et al.,* J. Nutr. 92: 389 (1967)] resulted in decreased concentrations of cholesterol in serum.

The concentrations of zinc and copper in sweat have been shown to be 0.93 µg/dl [Prasad *et al.,* J. Lab. Clin. Med. 62: 84 (1963)] and 0.058 µg/dl [Mitchell and Hamilton, J. Biol. Chem. 178: 345 (1949)], respectively, the ratio of zinc to copper being approximately 16.

It is hypothesized that the phytic acid in the diets associated with low risk of coronary heart disease has the protective effect of reducing the amount of zinc relative to that of copper available for absorption from the intestinal tract. Also the calcium in hard water has the protective effect of causing reduction in the amount of zinc in the liver, the major site of cholesterol synthesis [Dietschy and Wilson, New Engl. J. Med. 282: 1128 (1970)]. Further, the sweating associated with exercise causes a relatively greater loss of zinc than of copper from the body. These factors cause a decrease in the ratio of zinc to copper, resulting in a lower concentration of cholesterol in plasma similar to that found in these rats. This reduction in the ratio of zinc to copper presumably results in a lower risk of coronary heart

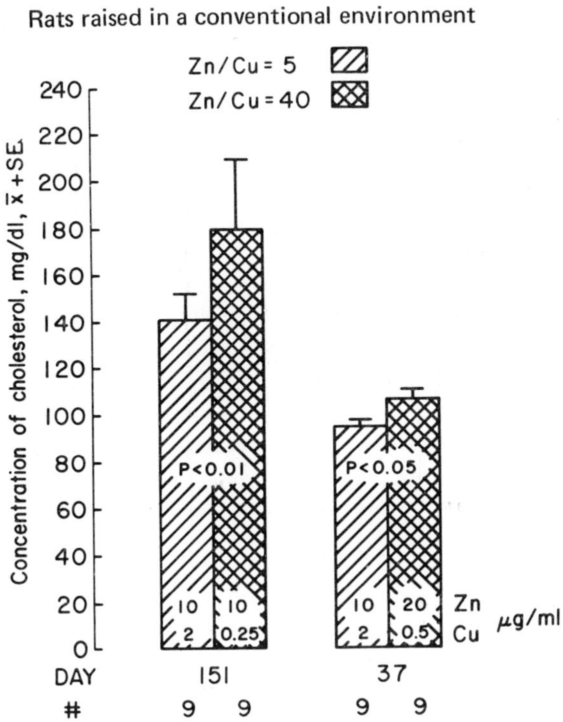

Fig. 2. Effect of dietary zinc and copper on the plasma cholesterol of rats raised in a conventional environment. The concentrations of zinc and copper in the drinking water of the animals are shown at the base of each bar. The number of animals per group and the experimental day on which the plasma samples were obtained are shown beneath the bars.

disease. [Partial support was received from PHS Grant No. 5 P10 ES00159 and NIOSH Grant No. 1 R01 OH 00349-01.]

DISCUSSION

Quarterman (Aberdeen). I've published data showing that in zinc-deficient rats plasma cholesterol is identical with pair-fed controls. Your animals were not pair fed, but did you record the differences in food intake between the groups?

Klevay. In some experiments; I could find no difference.

Hurley (Davis). Hard water usually contains large amounts of magnesium and this element may be the important one. There is also some experimental evidence which appears to link magnesium to cholesterol levels.

Peters (Madison). Intravenous EDTA has been used in the treatment of atherosclerotic heart disease. EDTA is quite selective in that it removes far more zinc than copper, and this would fit in with your hypothesis.

Klevay. I believe they found that zinc was increased 11-fold, while copper was increased about 5% in urine, at a time when cholesterol levels dropped about 36%.

Horvath (Morgantown). In this country where much copper plumbing is used, the water actually consumed may have a different copper:zinc ratio than the source and also be different from areas where other kinds of plumbing are used.

Klevay. Perhaps that is true; however, as far as I know all epidemiology has been done on the water supplies. Presumably atherosclerosis is a disease that begins some time after birth and progresses throughout the lifetime, so 15 or 20 years of increased copper in the water for the person who's dying at age 50 may not make that much difference. I should like to hope that it would, but I don't think we know the answer to that.

Sandstead (Grand Forks). Would you comment on how the presentation by the previous speaker on milk and zinc might relate to your present work?

Klevay. I was interested to note that he could reduce the amount of zinc in cow's milk by making the animal marginally deficient in zinc because the zinc:copper ratio in normal cow's milk is about 38. The zinc:copper ratio of human milk is about 6.

MANGANESE METABOLISM AND HOMEOSTASIS IN CALVES AND RATS

J. W. LASSITER, W. J. MILLER, M. W. NEATHERY, R. P. GENTRY, E. ABRAMS, J. C. CARTER, JR., and P. E. STAKE

Departments of Animal Science and Dairy Science, University of Georgia, Athens, Georgia

Previous studies using injected Mn (manganese) have led to the widely accepted conclusion that variable excretion is the regulator of Mn homeostasis [Am. J. Physiol. 211: 203 (1966)]. This was based, however, on more rapid excretion following injected ^{54}Mn when high dietary Mn levels were fed. The injected ^{54}Mn in their research was not exposed to conditions within the gastrointestinal tract, and thus the effect of absorption on regulation of Mn homeostasis was not studied. In our rat studies dietary Mn levels varying from 4 to 2000 ppm greatly affected ^{54}Mn tissue concentrations from a single oral dose. For example, duodenal tissues of rats fed 4 ppm Mn contained 25 times as much ^{54}Mn 4 hr after dosing as did tissues of rats fed 1000 ppm Mn in experiment 1 or 2000 ppm Mn in experiment 2 (Table 1). Liver ^{54}Mn in rats fed 4 ppm Mn was 15 times that of rats getting 1000 or 2000 ppm Mn. Such great differences in uptake only 4 hr after dosing evidently show a major effect on absorption. For the great differential effects to have resulted primarily from effects on excretion would mean little or no effect on absorption occurred, and effects on excretion (of the ^{54}Mn after absorption) would have had to occur within the 4 hr period. The observation of such great effects on Mn excretion within 4 hr is not supported by the studies of Britton and Cotzias [Am. J. Physiol. 211: 203 (1966)], whose highest level of dietary Mn (3.2×10^{-2}M) only diminished to two days the biological half-life of the fastest component of injected ^{54}Mn in mice. Bile is the principal excretory route for Mn [Am. J. Physiol. 211: 211 (1966)], excretion within seven days by routes other than bile being less than 25% of total injected ^{54}Mn without, and 60% with, injected stable Mn. Bertinchamps *et al.* [Am. J. Physiol. 211: 217 (1966)], however, reported that only 0.3%

Table 1. Effect of dietary manganese on ^{54}Mn distribution

	Experiment 1[1]		Experiment 2[1]	
	Diet Mn		Diet Mn	
	4 ppm	1000 ppm	4 ppm	2000 ppm
Duodenum[2]	44.80[a]	1.62[b]	14.13[a]	0.57[b]
Stomach[2]	10.05[a]	1.40[b]	7.97[b]	1.02[b]
Jejunum[2]	10.30[a]	1.82[b]	7.33[a]	0.94[b]
Liver	1.89[a]	0.12[b]	0.75[a]	0.05[b]
Spleen	0.055[a]	0.008[b]	0.048	0.003
Kidneys	0.038[a]	0.007[b]	0.166	0.004

[1]Values are percentage of ^{54}Mn dose per gram of tissue dry matter 4 hr after dosing. In each experiment, values for each tissue not followed by same superscript letter are significantly different ($P < 0.05$).
[2]Contents washed out.

of a 1 mg dose of Mn injected into the duodenal lumen was re-excreted into bile within 2 hr. Greenberg *et al.* [J. Biol. Chem. 147: 749 (1943)] had earlier indicated only 1% of an oral dose accumulated in bile within 48 hr. Thus, in the light of these biological half-lives and excretion rates, the present study of orally administered ^{54}Mn, showing 25-fold differences in duodenal-tissue concentrations and 15-fold differences in liver tissues, could not be expected to have resulted mainly from effects on excretion within the 4-hr period.

Studies with baby calves fed whole milk indicate far higher ^{54}Mn absorption than the generally regarded typical 3–4%. For example, three days after a single oral dose, total-body ^{54}Mn retention of low-Mn control calves was 9 times greater (18.2% *versus* 2.2%) than in manganese-supplemented animals. Intravenously dosed controls retained four times more ^{54}Mn (60.1% *versus* 16.3%) than those given 15 ppm supplemental Mn in the milk. These data indicate that the dietary Mn level has a greater effect on absorption than on endogenous excretion, and that both variable excretion and absorption play important roles in homeostasis of Mn. In orally dosed calves supplemented with Mn, 65% of the retained dose of ^{54}Mn remained in the GI tissues and contents, and less than 10% in the liver. In controls over half the total-body ^{54}Mn was in the liver. Such a high percentage of total-body ^{54}Mn in GI tissues and contents of supplemented calves compared with controls also indicates a great effect of dietary Mn level on absorption.

It is well established that animals retain a far smaller proportion of dietary Mn than of zinc. Following duodenal dosing of calves with ^{54}Mn, uptake by intestinal mucosa was observed to be similar to that for ^{65}Zn in an earlier study. However, ^{54}Mn concentrations were far lower in other tissues than observed earlier for ^{65}Zn. Turnover rate of ^{54}Mn in liver was much faster than that of ^{65}Zn. Thus, important ways in which animals discriminate against Mn relative to Zn appear to include transfer from intestinal mucosa to blood rather than mucosal uptake and to rate of endogenous excretion of Mn. [Supported in part by Public Health Service Research Grant No. AM-07367-NTN from the National Institute for Arthritis and Metabolic Diseases.]

COMPARTMENTAL ANALYSIS OF $^{63}Ni(II)$ METABOLISM IN RODENTS

CLAUDE ONKELINX, JOEL BECKER, and F. WILLIAM SUNDERMAN, Jr.

Departments of Oral Biology and Laboratory Medicine, University of Connecticut Health Center, Farmington, Connecticut

A single i.v. injection of $^{63}NiCl_2$ was administered to 37 female Wistar rats (age ≅ 85 days), in dosages of 17 μg Ni/rat (≅ 82 μg Ni/kg), and to 18 New Zealand albino rabbits (12 male, 6 female, weight ≅ 3.4 kg), in dosages of 816 μg Ni/rabbit (≅ 240 μg Ni/kg). At intervals from 1 hr to 8 days after the injection, serial blood samples were collected, and concentrations of $^{63}Ni(II)$ were determined in plasma or serum by liquid scintillation counting. The elimination of ^{63}Ni in excreta was also measured. The experimental methods were described previously by van Soestbergen and Sunderman [Clin. Chem. 18: 1478, (1972)]. The data indicate that, in both species, $^{63}Ni(II)$ is rapidly cleared from plasma or serum during the first two days after the injection, and that it disappears at a much slower rate from three to eight days. In rats, the percentage of administered $^{63}Ni(II)$ which was excreted in urine averaged 68% during the first day after injection and 78% during three days after injection. Fecal excretion of $^{63}Ni(II)$ in rats averaged 15% of the administered dose during three days after injection. In rabbits the percentage of administered $^{63}Ni(II)$ which was excreted in urine averaged 78% during the first day after injection. Biliary excretion of $^{63}Ni(II)$ in two rabbits averaged 9% of the administered dose during 5 hr after injection. Fecal excretion of $^{63}Ni(II)$ in rabbits was not measured. In both rats and rabbits the time course of ^{63}Ni concentrations in plasma or serum was closely approximated by a sum of two exponential terms. The following equations were derived by a computerized curve-fitting procedure, using the data for all of the animals:

$$\text{Rabbits } S = 1165 \exp(-0.092\, t) + 4.94 \exp(-0.0084\, t) \qquad (1A)$$

$$\text{Rats } S = 226 \exp(-0.11\, t) + 0.57 \exp(-0.014\, t). \qquad (1B)$$

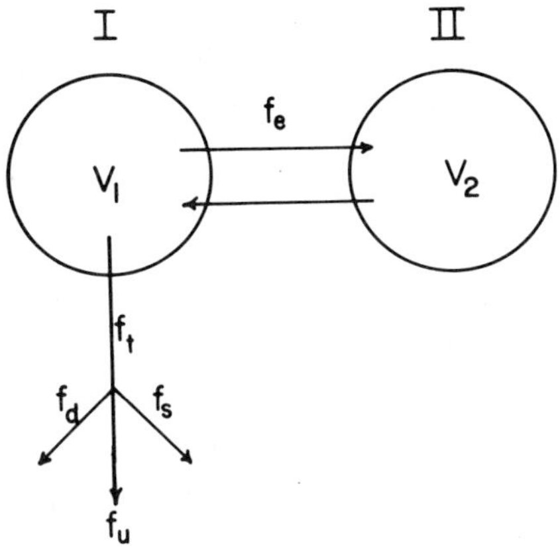

Fig. 1. Schematic diagram of the two-compartment model of $^{63}Ni(II)$ metabolism in rodents (see text for explanation of symbols).

In equations (1A) and (1B), S stands for $^{63}Ni(II)$ concentrations in plasma or serum (μg/liter), t stands for time after injection (hr), and exp is the base of natural logarithms. This kinetic pattern strongly suggests that $^{63}Ni(II)$ becomes diluted within a distribution volume comprising two compartments, and that $^{63}Ni(II)$ is eliminated according to first-order kinetics. We therefore propose the two-compartment model which is represented in Fig. 1. Compartment I, which includes the serum or plasma, is the central compartment into which $^{63}Ni(II)$ is injected and from which elimination takes place. Compartment II represents a hypothetical volume related to the first compartment by an exchange process. V_1 and V_2 stand respectively for the volumes of compartments I and II (ml); f_e is the clearance of $^{63}Ni(II)$ by exchange between I and II (ml/hr) and f_t is the total clearance of $^{63}Ni(II)$ by excretion out of I (ml/hr). The total excretory clearance f_t can be subdivided into at least three components such that

$$f_t = f_u + f_d + f_s \tag{2}$$

in which f_u stands for the urinary excretory clearance, f_d stands for the fecal excretory clearance, and f_s stands for that portion of the total excretory clearance of $^{63}Ni(II)$ which is not accounted for by the first two components. When $f_s > 0$, this component would correspond to deposition of $^{63}Ni(II)$ into a nonexchangeable or very slowly exchangeable body reservoir, acting as a 'sink' (e.g. hair). If one assumes (a) uniform mixing within each compartment and (b) transfer of $^{63}Ni(II)$ by first-order kinetics, the two-compartment model is described mathematically by a set of two first-order differential equations which can easily be solved. The urinary, fecal, and biliary clearances of $^{63}Ni(II)$ from compartment I can be calculated by use, in each case, of the following equation:

$$f = \frac{R\Big]_{t_0}^{t_1}}{\int_{t_0}^{t_1} S\,dt} \qquad (3)$$

in which R is the amount of $^{63}Ni(II)$ measured in urine, feces, or bile during the interval between t_0 and t_1, and f_s is calculated by difference using equation (2). Based upon equations (1A), (1B), (2), and (3), the calculated parameters for the two-compartment model of $^{63}Ni(II)$ metabolism in rats and rabbits are given in Table 1. The two-compartment model has been tested and verified by its ability correctly to predict the concentrations of $^{63}Ni(II)$ in serum or plasma of rodents which were given continuous or daily injections of $^{63}Ni(II)$, as summarized in Table 2. [Supported by NSF GH-34444, AEC AT (11-1)-3140, and HSM 99-72-24.]

Table 1. Parameters of the two-compartment model of $^{63}Ni(II)$ metabolism

Parameter	Symbol	Units	Rat	Rabbit
Compartment I	V_1	ml	75.1	697
Compartment II	V_2	ml	8.3	265
Exchange I↔II	f_e	ml/hr	0.12	2.31
Total excretory clearance	f_e	ml/hr	8.14	61.2
Urinary excretory clearance	f_t	ml/hr	6.39	54.1
Fecal excretory clearance	f_u	ml/hr	1.28	–[1]
Clearance by excretion	f_d			
into 'sink'	f_s	ml/hr	0.47	–[1]
Biliary excretory clearance	f_{bile}	ml/hr	–[2]	6.6
Excretion constant	f_t/V_1	hr^{-1}	0.108	0.088

[1]Measurements of fecal $^{63}Ni(II)$ were not pertormed in rabbits; hence f_d and f_s could not be calculated.

[2]Measurements of biliary $^{63}Ni(II)$ were not performed in rats; hence, f_{bile} could not be calculated.

Table 2. Testing of the two-compartment model of ^{63}Ni(II) metabolism

Species	Dosage of ^{63}Ni(II)	Duration	^{63}Ni concentration in serum/plasma (μg/liter)	
			Observed	Predicted
Rat	0.21 μg/hr[1]	6.6 hr	12.8	12.7
Rat	0.21 μg/hr[1]	23.5 hr	21.2	23.2
Rat	0.21 μg/hr[1]	30 hr	25.1	24.2
Rabbit	15 μg/day[2]	14–16 days	2.8 (2.6–3.3)[3]	3.1

[1]Continuous intravenous infusion.
[2]Daily intravenous injection.
[3]Serum obtained immediately prior to daily injection of ^{63}Ni(II); (mean and range of values in three rabbits).

THE BINDING OF SELENIUM TO SERUM ALBUMIN

JOHN L. HERRMAN and KENNETH P. McCONNELL

Veterans Administration Hospital and University of Louisville Medical Center, Louisville, Kentucky

Dogs which have been injected subcutaneously with $^{75}SeCl_4$ exhibit a large amount of radioactivity associated with the albumin fraction [McConnell *et al.*, Tex. Rep. Biol. Med. 18: 438 (1960)]. Within 1 hr after injection about 40% of the total serum radioactivity was found in the albumin fraction, with a steady drop during the next 48 hr. This suggested to us that albumin may be serving a role in the transport of selenium by accepting the label in areas where the concentration is high and then transferring it to other proteins at later times. In order to investigate protein-selenium interactions and to assess further the potential of albumin to serve as a carrier of Se, we have initiated a series of *in vitro* experiments with bovine serum albumin (BSA).

When either labeled selenite or selenomethionine (SeMet) was incubated with BSA at neutral pH, the protein precipitated with 5% trichloroacetic acid (TCA), washed with TCA, and counted, radioactivity was retained with the precipitate. When measured as a function of ligand concentration, the retention of selenite saturated at less than 0.20 M ^{75}Se/M BSA; with SeMet retention was linear with increasing concentration up to molar ratios of SeMet to BSA of 2:1. Binding of both ligands was reduced by preincubation of the protein with *p*-mercuribenzoate (PMB); maximal inhibition occurred at an approximately 1:1 molar ratio of PMB to BSA.

It was found that after selenite was allowed to react with 2-mercaptoethanol at pH 4.0 to form the selenotrisulfide, selenodimercaptoethanol, as described by Ganther [Biochemistry 7: 2898 (1968)], incorporation of ^{75}Se was enhanced up to six-fold. Saturation occurred at 0.55 M ^{75}Se/M albumin. This binding was inhibited by PMB.

Scatchard plots drawn from data obtained in equilibrium dialysis experiments studying the interaction of selenodimercaptoethanol and BSA were biphasic. A high-affinity site, with maximal binding of 0.53 (mole/mole) and association constant of $9.4 \times 10^3\ M^{-1}$, plus a class of six low-affinity sites per protein molecule with an association constant of $1.9 \times 10^2\ M^{-1}$, were found. In a peripheral experiment 0.50 –SH group per albumin molecule was obtained by titration of the protein with 5,5′-dithiobis-(2-nitrobenzoic acid). This value agrees well with the maximum amount of high-affinity binding site in the equilibrium dialysis experiment and in the preceding acid precipitation experiment, indicating that the binding of the selenium compound may occur via the sulfhydryl group of the protein. Further evidence for –SH group involvement was that the high-affinity binding site was eliminated by the prior addition of PMB to the protein.

Neutral ligands such as propane, butane, octanol, and decanol are known to bind to albumin [for a review see Steinhardt and Reynolds, Multiple Equilibrium in Proteins, Academic Press, New York (1969), pp. 85–124], so it seems reasonable to suppose that selenodimercaptoethanol may also have that capability. We suggest that the ligand binds in a position adjacent to the free sulfhydryl group and then participates in a 'selenotrisulfide exchange' reaction to form a mixed selenotrisulfide with the protein. The overall process could be designated thus (A, albumin):

$$HOCH_2CH_2{-}S{-}Se{-}S{-}CH_2CH_2OH + A{-}SH$$
$$\downarrow$$
$$A{-}S{-}Se{-}S{-}CH_2CH_2OH + HS{-}CH_2CH_2OH$$

Biologically significant sulfhydryl reducing agents such as CoASH, cysteine, and glutathione can also apparently form selenotrisulfides upon reaction with selenious acid [Ganther, Biochemistry 7: 2898 (1968); Sandholm and Sipponen, Archs Biochem. Biophys. 155: 120 (1973)]. Presumably these products could also participate in selenotrisulfide exchange reactions with albumin in a manner analogous to that proposed with selenodimercaptoethanol, thus presenting a mechanism for the reduction of ingested selenite and subsequent covalent binding of its reduction product to albumin.

DISCUSSION

Levander (Beltsville). Could you rationalize the interaction of selenomethionine with serum albumin in terms of its sensitivity to sulfhydryl agents?

Herrman. I have no good explanation. It would not actually have to bind to the sulfhydryl, but could bind in the region of the sulfhydryl.

Levander (Beltsville). Did you try other sulfhydryl blocking agents?

Herrman. No.

Ganther (Madison). Your work was done at pH 4 and I think that this points up the high reactivity of the selenotrisulfide group in comparison with its sulfur analog. It would be difficult to demonstrate sulfhydryl-disulfide interchange at a pH below 6, unless it was at a pH very much below 4.

EFFECT OF SELENICALS ON THE PRIMARY IMMUNE RESPONSE IN MICE

JULIAN E. SPALLHOLZ, JOHN L. MARTIN, MARLENE L. GERLACH, and ROLLIN H. HEINZERLING

Departments of Biochemistry and Microbiology, Colorado State University, Fort Collins, Colorado

The primary immune response (PIR) of mice is characterized by the appearance of circulating antisheep red blood cell (SRBC) hemagglutinating antibody (Ab) two days post-SRBC sensitization [J. Immunol. 95: 26 (1965)]. Anti-SRBC Ab initially consists of predominately mercaptoethanol (ME) sensitive IgM (19S) Ab followed by the appearance of ME-resistant IgG (7S) Ab [J. Immunol. 95: 39 (1965)].

For mice, dietary Se as selenite enhances the PIR. The enhanced PIR in these mice is indicated by increased anti-SRBC hemagglutinating titers of IgM Ab [Proc. Soc. Exp. Biol. Med. 143, 685 (1973); Fed. Proc. 32: 886 (1973) (abstr.)] and IgG Ab [Inf. Immunity 8: 841 (1973)]. The serum Ab titers of mice fed diets containing Se generally show increased ME-labile IgM Ab four days and seven days post-SRBC sensitization followed by increased ME-resistant IgG Ab on day 14. Maximum Ab titers were obtained from mice fed dietary Se as selenite at 1–3 ppm Se. Ab titers of mice fed diets containing Se at < 1 ppm or > 3 ppm were greater than those of controls, but were 2–8 times less than the Ab titers of mice fed 1–3 ppm dietary Se.

It has been briefly reported [Fed. Proc. 32, 886 (1973) (abstr.)] that injectable selenite also enhances the PIR in mice sensitized with the SRBC antigen. Similar results have ostensibly been reported in which rabbits were administered typhoid vaccine [Zdrav. Belosuss. 18: 34 (1972), *In* Se Te Abstr. 14: 1 (1973)]. In the former report it was shown that Se administered i.p. to mice as selenite (5 μg Se/mouse) simultaneously with the SRBC antigen ($\sim 5 \times 10^8$ cells i.p.) enhanced IgM Ab titers on day 5 post-SRBC sensitization. An organic selenical, diphenyl selenide, was biologically much less active when administered at a 5 μg Se equivalent.

Subsequently, more recent data indicate that Se as selenite must be administered prior, simultaneously, or within 24 hr of the SRBC antigen to obtain the enhanced PIR. The groups having comparable Ab titers were those which were administered selenite 24 or 48 hr pre-SRBC sensitization, simultaneously, or 24 hr post-SRBC sensitization. Mean Ab titers of these groups were 4–14 times greater than those of controls. However, mice administered selenite 48 or 72 hr post-SRBC sensitization had titers equal to control animals. Sera for analysis were collected on days 4–7 post-SRBC sensitization. Of the selenicals tested by injection the two most effective in enhancing Ab titers were selenite and selenide. Selenate and diphenyl selenide exhibited much less biological activity than either selenite or selenide. Hemagglutinating Ab of normal and ME-treated sera of mice injected with a particular selenical or sulfite are shown in Figs. 1 and 2.

Results of dietary and injectable Se studies of mice indicate that some selenicals function pharmacologically to potentiate the amount of circulating

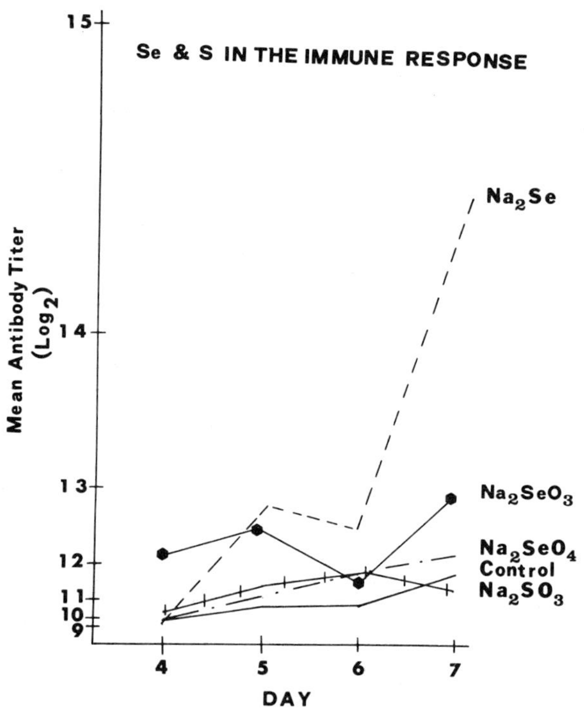

Fig. 1. Anti-SRBC hemagglutinating titers of normal mouse sera. Mice were injected i.p. with 5 μg of Se or S simultaneously with SRBC. Mean titers are values of four samples assayed in duplicate on days 4–7 post-SRBC sensitization.

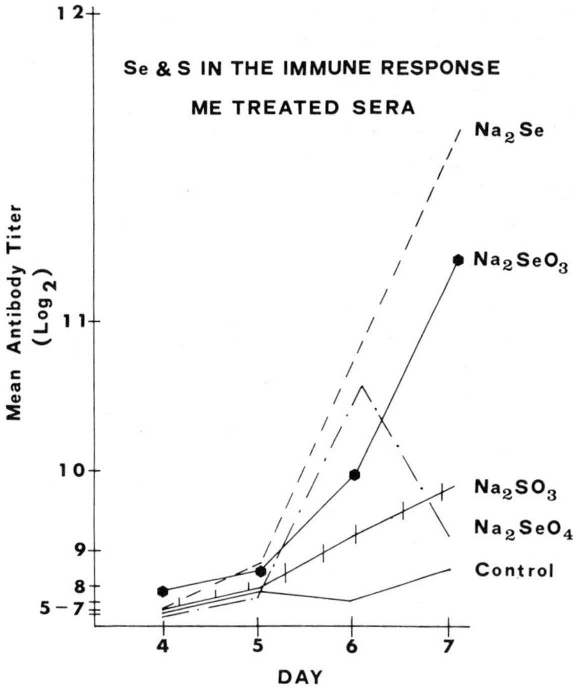

Fig. 2. Anti-SRBC hemagglutinating titers of ME-treated mouse sera of Fig. 1. Mean titers are values of four samples assayed in duplicate on days 4–7 post-SRBC sensitization.

Ab to an antigen. These experiments relating to the effect of Se compounds upon immunologic responsiveness have been predicated, in part, by many investigations which have indicated that selenite, and perhaps selenide, possess anticarcinogenic properties. [Supported by NIH Grant ES 00254-12.]

DISCUSSION

Frost (Schenectady). Ubiquinone is involved in the host defense system and in immune response. Am I right on that? Could this be related to your effect?

Spallholz. You're correct, but I don't know if these effects are related. Could I make another point? Because of toxicity, most investigators using the rat seem to stop at 5 ppm Se, but we have been able to raise mice with very little evidence of toxicity up to 40 ppm Se. In our early experiments with commercial chows, we used only two levels of Se, 2.8 and 0.7 ppm, but the present experiments clearly confirm that selenium, at least as selenide or selenite, has a marked effect on the synthesis of circulating antibodies.

Huckabee (Oak Ridge). How does the selenium exist in the organism and how does it stimulate antibody synthesis?

Spallholz. According to Dr. Diplock, Se exists, at least in part, in the 2–oxidation state as selenide. In our experiments the Se had to be present prior to or during immunization; thus it probably acts, not in stimulating protein synthesis, but apparently in the processing of antigen.

Ganther (Madison). Your studies imply that feeding a lot of selenium may help animals resist infection. Must you go above 0.5 or 1 ppm Se to get the effect?

Spallholz. That is the implication. A level of 1 ppm Se supplemented to commercial chow was not effective in increasing IgG. We had to supplement above 1 ppm but no more than 3 ppm Se to get the effect.

Ganther (Madison). Because certain leucocytes are high in glutathione peroxidase and humans with a hereditary deficiency of glutathione peroxidase are more susceptible to infection, we have been comparing survival of animals infected with *Salmonella* and fed either a selenium-deficient diet or the diet supplemented with 0.5 ppm Se. We have preliminary evidence that selenium supplementation improves survival, but whether this relates to glutathione peroxidase or your antibody effect is unresolved.

Spallholz. We have done experiments with *Dipplococcus pneumoniae* which causes a peritonitis. We find increased survival by feeding or injecting selenicals. May I suggest that in your studies you go up to 1–3 ppm Se for a comparison.

Diplock (London). Sodium selenide is very rapidly oxidized in solution. Do you have some trick for getting around this?

Spallholz. I am uncertain about the form at the time of injection. All I know is that selenide as it comes out of the bottle is effective when we inject it.

SELENIUM-BINDING PROTEINS OF OVINE TISSUES

P. D. WHANGER, N. D. PEDERSEN, and P. H. WESWIG

Department of Agricultural Chemistry, Oregon State University, Corvallis, Oregon

White muscle disease (WMD), a selenium-responsive myopathy [Muth, J. Am. Vet. Med. Ass. 142: 1379 (1963)], is characterized pathologically by muscular degeneration and calcium accumulation [Muth *et al.,* Am. J. Vet. Res. 20: 231 (1959)]. Biochemically, a significant increase in activity of some plasma enzymes, such as lactic dehydrogenase, glutamic-oxalacetic transaminase, malic dehydrogenase [Whanger *et al.,* J. Nutr. 97: 553 (1969); Am. J. Vet. Res. 31: 965 (1970); Nutr. Rep. Int. 6: 21 (1972)], creatine phosphokinase, and fructose diphosphate aldolase [Whanger *et al.,* Nutr. Rep. Int. 6: 21 (1972)], has been observed in WMD lambs. Conversely, a decrease in the activity of some cytoplasmic enzymes, such as lactic dehydrogenase, glutamic-oxalacetic transaminase, and peroxidase (Whanger *et al.,* J. Nutr. 99: 331 (1969)], and an increase in the free activity of the lysosomal enzymes [Whanger *et al.,* J. Nutr. 100: 773 (1970)] in affected muscles of WMD lambs have been observed. In addition, alterations in the tissue sulfhydryl levels with a significant increase of glutathione levels in muscle of WMD lambs was found [Broderius *et al.,* J. Nutr. 103: 336 (1973)].

Since this information had provided only limited data on the metabolic role of selenium, the distribution of this element between the different molecular weight tissue proteins in normal and WMD lambs was studied [Pedersen *et al.,* Bioinorg. Chem. 2: 33 (1972)]. In both normal and WMD lambs selenium was bound to two different molecular weight proteins in the kidney (> 500,000 and 200,000), and three different molecular weight proteins in the liver (> 500,000, 120,000 and 20,000), pancreas (>500,000, 140,000 and 70,000), and plasma (> 500,000, 120,000 and 50,000). However, four different molecular weight selenium-binding proteins were found in heart and muscle of normal lambs (> 500,000, 120,000, 40,000, and 10,000), but only three in these tissues (10,000 absent) of WMD lambs. Therefore, to obtain information on the functions of selenium the 10,000 molecular weight selenium-binding proteins were isolated and characterized from muscle extract of normal lambs by ammonium sulfate fractionation, gel filtration chromatography, and hydroxylapatite chromatography [Whanger *et al.,* Fed. Proc. 31: 691 (1972)]. In the final step of purification, four selenium-binding proteins were eluted, two of which contained the majority of the selenium.

Since one of the major selenium-containing proteins has been studied extensively, the discussion will be limited only to its properties. Glutamate, aspartate, glycine, and lysine are the predominant amino acids in this protein, with only traces of cysteine and tryptophan and a low methionine content. Evidence has been obtained that selenium is incorporated into as well as required for its synthesis [Whanger *et al.,* Fed. Proc. 31: 691 (1972)]. The absorption, circular dichroic, and magnetic circular dichroic spectra of this protein with and without dithionite markedly resemble those reported for the reduced and oxidized spectra of cytochrome C [Whanger *et al.,* Biochem. Biophys. Res. Comm. 53: 1031 (1973)]. Thus, it contains a heme group identical to cytochrome C, and may be a selenium-containing cytochrome. However, this protein is not cytochrome C, since the amino acid composition and molecular weights are more like that of cytochrome B_5 [Structure and Function of Cytochromes, University Park Press, Baltimore, Md. (1967), pp. 581–593]. The relationships of selenium and hemoproteins (cytochromes) are presently under investigation in our laboratory. [Supported in part by Public Health Service Research Grant Number NS 07413 from the National Institute of Neurological Disease and Stroke.]

DISCUSSION

Hoekstra (Madison). What is the stoichiometry of Se in your purified protein?

Whanger. It's greater than 1 g atom Se/mole. We don't have exact figures on this.

Levander (Beltsville). Several years ago McConnell's group isolated cytochrome C from liver and found a very low content of Se. You find your protein only in muscle; is that right?

Whanger. Muscle and heart. It's not present in the liver or kidney.

Ganther (Madison). Do you have any idea on the form of Se in the protein?

Whanger. My guess is that it's there as a seleno-amino acid, but we don't know. The Se is tightly bound and not dialyzable at alkaline pH or acid pH.

Martin (Fort Collins). At what pH was your alkaline dialysis?

Whanger. 12.

Levander (Beltsville). Do you think that you don't find it in liver because the primary selenium-deficiency lesion in the lamb is not in the liver but rather the muscle?

Whanger. We'd like to speculate that at this stage.

Frost (Schenectady). I'd like to call attention to the fact that selenium-vitamin E has been cleared in Mexico for relief of angina. I think you are a little closer to the answer when you talk about cytochrome which is an important enzyme in the heart having to do with oxygen transfer metabolism.

Whanger. We hope something fruitful comes out of this. It makes a lot of sense that the oxygen generation or energy utilization of the tissues is inpaired in Se deficiency. It may also be very fruitful for someone to take a look at cytochromes in genetic muscular dystrophy.

FATE OF [^{75}Se-] SELENOMETHIONINE IN THE GASTROINTESTINAL TRACT OF SHEEP

M. HIDIROGLOU and K. J. JENKINS

Animal Research Institute, Agriculture Canada, Ottawa, Canada

A total of 40 sheep were used in four experiments to investigate the metabolism of [^{75}Se] selenomethionine in the rumen and its absorption in various sections of the gastrointestinal tract. In the first experiment the fate of selenomethionine, incubated *in vivo* in the rumen of sheep for 96 hr, was studied. The results show (Table 1) that up to 6 hr after dosing with the selenoamino acid 50% of the ^{75}Se activity in the rumen liquor was associated with the bacterial fractions, with the remainder in the cell-free and protozoa-plus-ingesta fractions. In the second experiment a higher radioactive dose was given, the sheep sacrificed after 2 hr, and their ruminal bacteria isolated for paper chromatographic identification of selenocompounds occurring in the bacterial protein. The results of chromatographic treatments of enzymatic protein hydrolysates (pronase) are shown in Table 2. Selenium-75 labeled selenomethionine and selenoystine were identified in the samples by spot elution, cochromatography with authentic standards, and autoradiography. In a third experiment identification of selenocompounds in rumen bacterial protein was also conducted using an *in vitro* procedure and in exchange column chromatography (Table 3). As found in the previous experiment, [^{75}Se] selenomethionine was metabolized by the bacteria to form [^{75}Se] selenocystine with both compounds incorporated into the bacterial protein. In experiment 4, [^{75}Se] selenomethionine absorption in sheep was studied. After administration, some [^{75}Se] selenomethionine secretion occurred in the forestomach, but the highest secretion occurred in the mid-jejunum with subsequent absorption in the lower part of the small intestine. Most of the labeled selenomethionine administered was secreted during the first 5 hr after dosing. When [^{75}Se] selenomethionine was given orally, the midjejenum also was the site of maximum ^{75}Se absorption but at 24 hr after

Table 1. Fate of L-[^{75}Se]selenomethionine in the rumen of sheep[1]

Hours after dosing	Counts/min/ml rumen liquor (RL)	Radioactivity in RL fractions (%)				
		Bacteria			Protozoa + plant particle	
		TCA insoluble	TCA soluble	Cell free	TCA insoluble	TCA soluble
1	9597 ± 7262	31	13.6	2.4	42	11
2	18020 ± 1562	29	13.1	2.3	43	12
3	14715 ± 5055	25	17.5	2.5	43	12
4	11337 ± 2958	29	17.3	2.7	42	9
5	10918 ± 2261	31	18.9	2.1	39	9
6	11009 ± 2451	30	15.6	3.4	43	8
7	11865 ± 3645	26	16.8	2.2	44	8
24	5777 ± 1533	30	12.7	2.3	46	8
36	3147 ± 779	25	11.2	2.8	52	8
48	2266 ± 783	24	9.3	2.7	53	11
60	1451 ± 570	22	8.8	3.2	52	13
72	1382 ± 948	20	9.5	2.5	55	13
84	421 ± 45	24	9.4	2.6	51	13
96	369 ± 99	20	8.9	2.1	54	13

[1]Five crossbred sheep (30–35 kg) were administered one single oral dose of 250 μCi of ^{75}Se/35 kg body weight. Values are mean ± s.d.

Table 3. Distribution of ^{75}Se activity after *in vitro* incubation of rumen fluid[1]

Compound	Bacterial protein hydrolyzate		Bacteria TCA-soluble material	
	Percentage of total digest ^{75}Se	Percentage of incubation sample ^{75}Se	Percentage of total TCA-sol. ^{75}Se	Percentage of incubation sample ^{75}Se
Selenocystine	10.1	3.0	nil	nil
Selenomethionine	42.2	12.4	18.0	12.7
Unidentified (left on column)	47.7	14.1	82.0	57.8
Total	100.0	29.5	100.0	70.5

[1]Rumen fluid was incubated with [^{75}Se]selenomethionine and distribution into bacterial protein; material soluble in 20% TCA was determined.

Table 2. Distribution of radioactivity in rumen bacteria of sheep[1]

Compound	R_f[2]	Proteolysate	TCA-soluble material
Immobile		10–18	17–22
Selenoglutathione	0.15	2–5	
Selenocystine	0.22	8–12	
Selenomethionine	0.65	15–25	16–21
Unidentified	0.80–0.87	40–50	60–71

1Five sheep (20–25 kg body weight) were administered a single oral dose of [^{75}Se]selenomethionine (50 μCi/kg) and killed after 2 hr. Values are the percentage of the total radioactivity in proteolysate or TCA-soluble material in each compound.

2Descending chromatography on Whatman 3; solvent, n-butanol:pyridine: water (1:1:1 v/v).

dosing. [The authors are indebted to Dr. J. E. Knipfel for valuable assistance in the ion exchange, chromatographic analyses.]

DISCUSSION

Martin (Fort Collins). We have evidence in *Escherichia coli* that the Se in selenomethionine is oxidized to metabolites such as selenocysteic acid. Did you find any higher oxidation selenium compound?

Hidiroglou. Yes, we found selenomethionine selenoxide which is the main oxidation product.

SOME EFFECTS OF SMALL CHANGES OF SELENIUM INTAKE BY SHEEP

H. J. LEE

Division of Nutritional Biochemistry, Commonwealth Scientific and Research Organization, Adelaide, Australia

Unexpectedly large and prompt increases in the selenium concentrations in the tissues of grazing sheep following an increase from 0.03 to 0.06 ppm Se (dry matter) in pasture [Lee and Kuchel, unpublished] led to a more precise evaluation of the influences of small changes in selenium intake on blood selenium concentrations in sheep.

Twenty-five Merino wethers of adequate selenium status were fed wheaten hay chaff containing 0.03 ppm Se and 8% crude protein. After 6 weeks the animals were matched according to blood Se concentrations in groups of five which were dosed daily for 16 weeks with (1) 0, (2) 0.01, (3) 0.02, (4) 0.04, and (5) 0.08 mg Se as Na_2SeO_3. Observations were continued for 12 weeks after treatments ceased. Mean daily food consumption fell below 700 g in only two cases, one each in groups 1 and 2, which died of pyloric stenosis. Group mean body weights ranged from 38 to 43 kg at the outset and from 36 to 41 kg at the end.

In jugular blood selenium concentration was determined [Watkinson, Analyt. Chem. 38: 92 (1966)] at weekly intervals and in feces after 0, 5, and 16 weeks of treatment and at the end (Fig. 1). Blood selenium levels in the untreated sheep fell progressively and at no time differed significantly from those in group 2. Treatment induced significant increases ($P < 0.05$) after four weeks in group 5, 6 weeks in group 4, and 16 weeks in group 3, while the differences between groups 2 and 3, 3 and 4, 4 and 5 were significant ($P < 0.05$) after 14, 6, and 5 weeks respectively. Only the differences between groups 1, 4, and 5 remained significant 12 weeks after treatments ceased.

After 5 weeks of treatment the group mean fecal selenium concentrations had risen in the order of the size of the supplements and were similar after 16 weeks. The increases due to treatment were significant ($P < 0.05$ or P

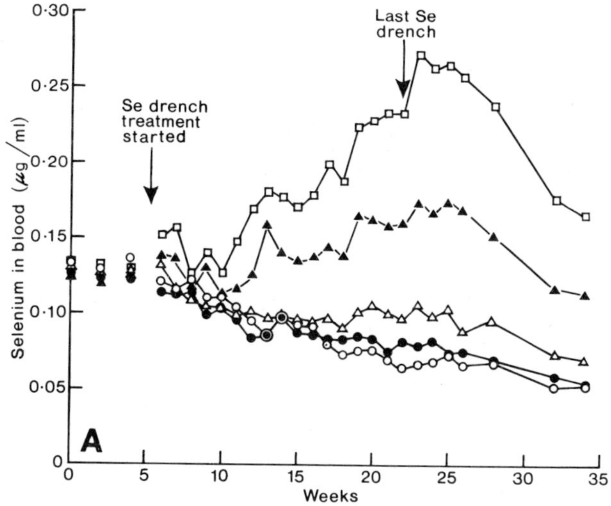

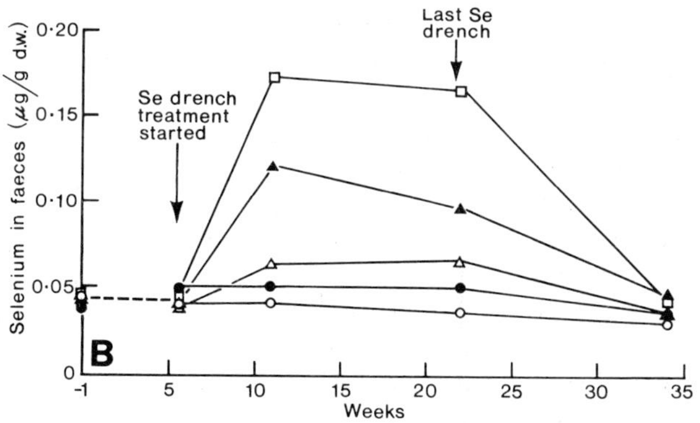

Fig. 1. Selenium in (a) blood and (b) feces of sheep following selenium treatment: ○——○ group 1, 0.0 mg Se/day; ●——● group 2, 0.01 mg Se/day; △——△ group 3, 0.02 mg Se/day; ▲——▲ group 4, 0.04 mg Se/day; □——□ group 5, 0.08 mg Se/day.

< 0.01) on both of these occasions. The mean values fell markedly after treatments ceased but, in groups 4 and 5, the residual effects remained significant ($P < 0.05$) at the end. The figures for groups 1 and 2 did not differ significantly at any sampling.

While the measurement of selenium concentrations in blood and pasture or feces will at any one time indicate the current state of the selenium nutrition of grazing sheep, the results presented suggest that the measurement of two or more of these concentrations on at least two occasions would permit the assessment of the nutritional history and future prospects with respect to the selenium status of the animals.

DISCUSSION

Spallholz (Fort Collins). Are these small increases of Se in blood and feces associated with any effect on the concentration of Se in muscle?

Lee. We did not examine the tissues of these animals, but we have previous data on that sort of thing.

Spallholz (Fort Collins). The reason I ask the question is that mice fed up to 40 ppm Se had little retention of selenium in muscle after selenium was withdrawn from the diet.

Lee. That's a lot of Se. When we deliberately induced fatal selenosis in sheep by feeding a high level of selenite for months, the liver contained over 100 ppm Se. I've forgotten what muscle contained but it was greatly increased.

Andrews (Upper Hutt). In New Zealand a critical Se level in pasture would be about 0.02 ppm and a critical level for blood in the grazing animal would be about 0.01–0.015 ppm, so we differ on these points.

Lee. Yes, this is one point on which we've agreed to differ.

Bell (Oak Ridge). It is well to keep in mind in interpreting whole-blood Se values that the Se in red blood cells will stay there for the lifespan of the cells, which is 143 days for sheep, while the plasma Se will respond very quickly.

Lee. We considered using plasma instead of whole blood, but as most of our previous results were with whole blood we stayed with it.

SELENIUM METABOLISM IN PIGS

J. LEIBETSEDER, A. KMENT, and M. SKALICKY

Institute of Physiology, School of Veterinary Medicine, Vienna, Austria

In several areas of Austria a selenium deficiency seems to exist in different kinds of domestic animals. Therefore, studies of the dynamics of selenium as well as clinical investigations were carried out in pigs.

For these studies 28 boars with an average body weight of 21.5 kg (s.d., 0.88 kg) were used. The animals were fed a commercial fattening ration initially *ad libitum* and 0.5 kg twice daily per animal for one week prior to beginning the trial. At the beginning of the trial the boars were fasted for twelve hours. After a single intravenous injection of 1 mCi ^{75}Se as sodium selenite (specific activity 1–10 mCi/mg Se) shaken with 20 ml porcine plasma for 20 min, blood samples were taken through a jugular vein catheter in intervals starting 4 min and finishing maximally 14 days after the injection. In addition to that, the animals were killed by bleeding under narcosis at successive post-injection intervals. After removing and separating the internal organs, one-half of the pigs' bodies were carefully dissected into their individual tissue parts which were weighed and homogenized. Aliquots of the homogenates were used for all determinations. The activity was measured with a well-type scintillation detector; the quantitative determinations of selenium were performed by neutron activation analysis described earlier [Leibetseder, *In* Trace Mineral Studies with Isotopes in Domestic Animals, I.A.E.A., Vienna (1969)].

The equation

$$\gamma_M = 0.39850 \exp(-44.543t) + 0.13538 \exp -4.725t + 0.02144 \exp -1.650t + 0.09456 \exp 0.042t$$

where γ_M is percentage of injected ^{75}Se per μg Se in plasma and t is time (measured in hours) gave a sufficiently accurate fitting of the curve to the experimental data of the specific selenium activity in the plasma. The equation indicates one central (M) and three peripheral (M_1, M_2, M_3) compartments of a mammillary compartment model with transfer rates ν_1, ν_2, and ν_3. The rate of excretion from M (urine, expiration, endogenous Se in feces, deposition in the body) is called ν_T. The size of the compartments and the transfer rates can be calculated from the equation above [Aubert and Milhaud, Biochim. Biophys. Acta 39: 122 (1960)]. The results of this calculation are given in Table 1, which shows a relatively small plasma compartment and three peripheral compartments of different sizes with a particularly high transfer rate for compartment M_3. The sum of the compartments M, M_1, M_2, and M_3 (the Se pool) represents the amount of easily exchangeable selenium, which amounts to an average of 1.14 mg in the animals tested.

In order to check the accuracy of the calculated transfer rates, the total transfer from the central compartment (ν_M) can also be determined:

$$(\nu_M)_0 = \frac{M}{-(\gamma_M)_0} \frac{(d\gamma_M)}{(dt)_0}.$$

The total transfer must be equal to the sum of single transfers. The results are $\nu_M = 4915.2$ µg/hr, and $\nu_1 + \nu_2 + \nu_3 + \nu_T = 4918.8$ µg/hr. This total transfer rate is remarkable with regard to the relatively small central compartment. The total Se content of the plasma must be exchanged in slightly more than 2 min. Within 24 hr about 1.17 mg Se leaves the central compartment for excretion or deposition in the body. Therefore, the same amount of absorbable Se should be available daily for pigs of this body weight.

The whole-body compartment analysis, which indicates the number of compartments, their sizes, and the transfer rates, allows no anatomical coordination of the compartments. Therefore we also studied the dynamics of exchange of various organs and tissues. By determining the Se content and the ^{75}Se activity in intervals after the ^{75}Se injection, we were able to analyze the time course of the specific activity in several organs and to calculate the transfer rate. We investigated 34 tissues in this manner. Table 2 shows the results of some selected organs.

Table 1. Selenium compartment sizes and transfer rates

Compartment	µg Se	Transfer rate	µg Se/hr
M	173.3	ν_T	48.3
M_1	453.3	ν_1	510.0
M_2	325.9	ν_2	785.7
M_3	187.7	ν_3	3574.3

Table 2. Content, transfer rate, and biological half-life of selenium in various organs

Organ	Content μg/g wet tissue	Transfer Rate μg/g/hr	Biological half-life (days)
Kidney	1.27	1.02×10^{-3}	52.19
Liver	0.72	2.01×10^{-4}	149.14
Heart	0.21	1.77×10^{-4}	48.47
Bones	0.07	6.32×10^{-5}	47.50
Skeletal Muscle	0.03	2.11×10^{-5}	67.05
Cerebrum	0.10	8.07×10^{-6}	506.30
Skin	0.04	2.84×10^{-6}	557.71

After establishing these physiological parameters of the selenium metabolism, we shall start to investigate animals with clinical symptoms of selenium deficiency in order to find out which metabolic parameters are mainly affected by selenium deficiency. [Supported in part by Fonds zur Förderung der wissenschaftlichen Forschung in Österreich. J. Leibetseder was an I.A.E.A. Fellow (1972-73) in the Department of Physiological Sciences, School of Veterinary Medicine, University of California, Davis, Cal.]

DISCUSSION

Levander (Beltsville). What do the various compartments represent in physiological terms?

Leibetseder. I pointed out that it is impossible to correlate the compartments with anatomical structures. Therefore, we tried to establish transfer rates between plasma and several organs in addition to that.

Binnerts (Wageningen). In calculating specific activities, you analyzed the samples for radioactivity and divided by the quantity of selenium present. Is that right? You did not make a distinction between different fractions of selenium?

Leibetseder. By neutron activation analysis we could only determine total selenium, not the different chemical forms.

Binnerts (Wageningen). For instance, if half the selenium was in another form and not as easily available, it would have a big effect on the calculation of transfer rates. That's the point I am worrying about.

THE RELATIONSHIP BETWEEN THE SELENIUM-CATALYZED SWELLING OF RAT-LIVER MITOCHONDRIA INDUCED BY GLUTATHIONE (GSH) AND THE SELENIUM-CATALYZED REDUCTION OF CYTOCHROME *c* BY GSH

O. A. LEVANDER, V. C. MORRIS, and D. J. HIGGS

Nutrition Institute, Agricultural Research Service, U.S. Department of Agriculture, Beltsville, Maryland

Weanling male rats fed a torula-yeast diet deficient in vitamin E and Se die of liver necrosis in about four weeks. Approximately two weeks before the onset of necrosis, liver slices or homogenates prepared from such animals are not able to support normal respiration rates during prolonged incubations. The purpose of the studies reported here was to attempt to explain this 'respiratory decline' of vitamin-E- and Se-deficient rat liver preparations and the events leading up to the massive acute necrotic episode.

The approach taken in this work was to examine the effect of various chemical agents on the *in vitro* swelling of liver mitochondria taken from rats fed diets adequate or deficient with regard to vitamin E or Se. Previous experiments showed that dietary Se, unlike dietary vitamin E, had no protective effect against those reagents (ascorbate, iron, or GSH plus GSSG) which caused mitochondrial swelling associated with lipoperoxidation [Higgs and Morris, Fed. Proc. 32: 885 (1973)]. Rather, dietary Se accelerated the mitochondrial swelling induced by certain thiols such as cysteine or GSH. Moreover, the accelerating affect of dietary Se on thiol-induced swelling could be mimicked by the direct addition of selenite *in vitro* to suspensions of mitochondria from rats fed diets adequate in vitamin E but deficient in Se ('Se-deficient mitochondria') [Morris *et al.,* Fed. Proc. 32: 886 (1973)]. This enhanced swelling of Se-deficient mitochondria due to the *in vitro* combination of GSH plus selenite was totally blocked by cyanide but only partially blocked by antimycin A or amytal. This finding suggested that the acceleration of GSH-induced mitochondrial swelling by *in vitro* Se could be accounted for by a Se-catalyzed reduction of cytochrome *c* by GSH, and indeed Se was found to be an excellent catalyst for such a reaction [Levander *et al.*, *loc. cit.*].

It has now been found that 10^{-5} M Cd^{2+}, Hg^{2+}, or arsenite are potent inhibitors of the mitochondrial swelling caused by GSH plus selenite. This result is of interest since respiratory decline can be induced in vitamin-E- and Se-deficient mitochondria by the addition of either Cd^{2+} or arsenite *in vitro* [Corwin and Schwarz, Arch. Biochem. Biophys. 100: 385 (1963)]. The GSH plus selenite induced swelling was much less sensitive to inhibition by arsenate, organic mercurials, tellurium oxyanions, or monothiol blocking agents such as N-ethyl maleimide or iodoacetate.

The Se-catalyzed reduction of cytochrome *c* by GSH in a chemically defined system was rather insensitive to inhibition by Cd^{2+} or Hg^{2+}, since 10^{-3} M was needed for > 90% inhibition. Also, arsenite was a very ineffective inhibitor for this process since 10^{-2} M inhibited this reaction by only 11%. The fact that these dithiol inhibitors were relatively poor inhibitors for cytochrome *c* reduction but were quite good inhibitors of GSH plus selenite induced mitochondrial swelling indicates that a vicinal dithiol is involved in the latter process. Although blockade of vicinal dithiols is the most thoroughly understood mechanism of inhibition by arsenite, Massey and Edmondson [J. Biol. Chem. 245: 6595 (1970)] have suggested that arsenite can form complexes with persulfides and neighboring sulfhydryl groups:

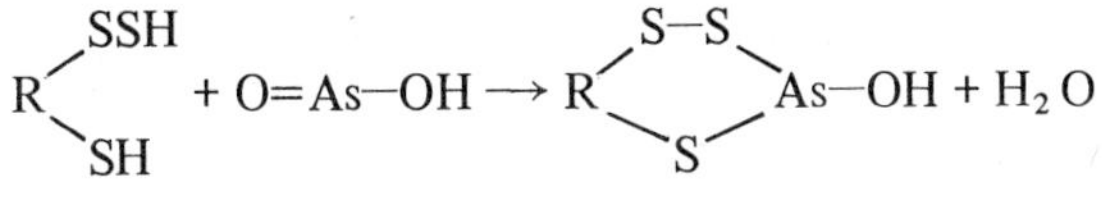

Similarly, one could postulate a complex of arsenite with a selenopersulfide and a nearby sulfhydryl group. Therefore, the catalysis of GSH-induced mitochondrial swelling by Se might involve such a grouping necessary for the reduction of cytochrome *c in situ.* A dithiol-sensitive factor needed for the transfer of electrons from cytochrome *b* to cytochrome *c* has been described [Slater, Biochem. J. 45: 14 (1949)].

A role for Se in catalyzing the transfer of electrons from GSH to cytochrome *c in vivo* could provide a rationale for the nutritional interrelationship among Se, vitamin E, and sulfur amino acids. Mitochondrial substrates entering the respiratory chain before the antimycin-A-sensitive site have been shown to generate H_2O_2 [Boveris *et al.*, Biochem. J. 128: 617 (1972)]. In an animal lacking vitamin E, such endogenous H_2O_2 production could result in widespread peroxidative damage to polyunsaturated lipids in membranes due to lack of antioxidant protection. In a rat lacking both vitamin E and Se, this destructive process would eventually terminate in dietary liver necrosis. In an animal receiving Se, however, the Se could 'short-circuit' the respiratory chain and permit the transfer of electrons directly from GSH to cytochrome *c*. The potentially damaging H_2O_2-producing step could thus be bypassed and the animal would not succumb to liver necrosis.

DISCUSSION

Shirley (Gainesville). Have you tried higher concentrations of Se than you reported here? Your observations may relate more to Se toxicity than to its biological function.

Levander. No, although I expect you might get a lot of elemental selenium formed in the cuvettes if you try to go much higher.

Ganther (Madison). Isn't it xanthine oxidase that has persulfide grouping according to Massey?

Levander. Yes.

Ganther (Madison). Are you suggesting that arsenite complexes with that and prevents xanthine oxidase from bringing about swelling?

Levander. No. We suggest the need for some other selenopersulfide acting in concert with a sulfhydryl to cause this reduction of cytochrome *c*. This is an extrapolation from Massey's results; we do not invoke xanthine oxidase. We consider the glutathione-induced swelling which is catalyzed by Se to be nonenzymatic. We're working with a model system.

Frost (Schenectady). Isn't it possible that some of these things you're observing are due to the oxidant effect of selenite? Some of these experiments might be better done with sodium selenide.

Levander. Selenocystine is much more active than selenite in accelerating this reaction, and we feel that it is because the Se is at the selenide valence level. I was surprised that Dr. Spallholz was able to inject selenide into his animals, because when we try to dissolve selenide we end up with a mess, and have therefore avoided it in our *in vitro* system.

GROWTH AND REPRODUCTION WITH RATS FED SELENITE-Se

A. W. HALVERSON

Station Biochemistry Department, South Dakota State University, Brookings, South Dakota

Male and female Sprague-Dawley albino rats were fed a vitamin-E-containing torula-yeast or casein diet with and without different levels of added selenite-Se. The basal yeast and casein diets contained 0.014 and 0.080 ppm of selenium, respectively. The additions of selenium were at 0.1, 1.0, 2.5, and 5.0 ppm in the yeast diet, and at 1.25, 2.50, and 3.75 ppm in the casein diet. The animals fed the yeast diet were put on experiment at 70 days of age and allowed to reproduce at 90 days of age. The growth and survival of the young were observed until 77 days of age. When a study was for two life cycles, as in the feeding of the casein diet, the procedure was similar except that the animals were allowed to reproduce three times in life cycle 2.

The use of a selenite-Se supplement in the diet was beneficial to reproduction under some conditions. As shown in Table 1, life-cycle-2 animals which were fed the basal casein diet bore fewer litters in a third reproduction than in a first or second reproduction. Of the litters which were produced, those of the third reproduction also showed a poor survival during the first two days of life. A selenium addition in the diet at a level of 1.25 ppm, and especially at a level of 2.50 ppm, caused improvement in the litter production but not in the survival. An addition at a level of 3.75 ppm was effective against both measurements, however.

The use of added selenium in the diet had a noticeable effect on the growth of postweanling male animals on the yeast diet. An addition of 0.1 or 1.0 ppm in the diet caused a considerable increase in growth with these animals; albeit, an addition at 2.5 ppm was less effective and one at 5.0 ppm gave no increase. The effect with the female animal on the same diet was mainly limited to a slight growth depression because of selenium at the highest level, i.e., 5.0 ppm. With the casein diet during two life cycles, the

Table 1. Effect of added selenite-Se in the casein diet on successive reproductions with life-cycle-2 rats[1]

Added Se in diet (ppm)	No. of females observed	Original litters in each reproduction			Data on surviving litters at 2 days after birth					
					Percentage of original litters			Average number of live members		
		1	2	3	1	2	3	1	2	3
0.00	10	10	9	2	80	78	50	8.1	9.6	4.0
1.25	13	13	11	7	77	82	43	8.3	6.1	3.3
2.50	14	13	13	11	92	92	27	8.6	9.1	6.0
3.75	14	13	13	10	92	100	100	8.0	6.2	5.9

[1]Combined results of two experiments. The results do not include animals which became sick.

result from a selenium addition at a level of 3.75 ppm was a slight growth depression with either male or female animals. An addition at a lower level in the casein diet, i.e., 1.25 or 2.50 ppm, also resulted in some degree of growth depression but this was mainly with life-cycle-2 males.

The results of other measurements showed little or no impairment in the postweanling survival when added selenium was fed in either diet, with the exception of when the addition was at 5.0 ppm in the yeast diet. A further observation was that selenium deposition in muscle or liver tissue was somewhat higher with female than with male animals when selenium was fed as in the yeast diet.

The present findings were in general agreement with those of McCoy and Weswig [J. Nutr. 98: 383 (1969)], who reported that rats require selenium in reproduction as well as in growth, and that the requirement for growth appears to be somewhat higher for males than for females. The additional finding with the rats in the current study of growth-inhibitory effects, however slight, because of added selenite-Se in the diet at a level somewhat below 3.75 ppm was in agreement with work by Chavez and Jaffe [Archs Latinoam. Nutr. 17: 69 (1967)] who fed selenium as seleniferous sesame cake.

DISCUSSION

Spallholz (Fort Collins). Did you see a difference in the percentage mortality between 1 and 5 ppm Se?

Halverson. Yes. There was a slight indication that the 5 ppm Se level was associated with an improved survival in both males and females between 2 and 21 days of age; mortality was about 30% when Se was less than 5 ppm, and perhaps 18% at 5 ppm Se. However, we fed 5 ppm Se in only the one experiment with the yeast diet.

Spallholz (Fort Collins). In a single experiment with mice fed a torula yeast diet which I have not repeated, I found a greater mortality at 5 ppm Se than I did at 10 ppm Se. Survival and growth in this experiment increased between 1 and 2.5 ppm Se, declined at 5 ppm Se and then proceeded to increase at 7.5 and 10 ppm Se. Why didn't you feed levels of Se above 5 ppm?

Halverson. My experience with rats is that 5 ppm Se in the diet over a life cycle is deleterious. With rats on the yeast diet containing 5 ppm Se, two of eight males and one of eight females died at 6–7 weeks from anemia characteristic of selenite toxicity. We also observed mild liver damage with 3.75 ppm Se in the casein diet, and I would anticipate substantial mortality if we fed 5 ppm Se in a casein diet.

Baumann (Madison). I am perplexed by mice surviving levels as high as 40 ppm Se for a long period of time, and I wonder if there is an explanation in terms of dietary protein, etc., or is the mouse really that much more resistant to Se toxicity than the rat.

Martin (Fort Collins). The mice seem to adapt. We start out at maybe 20 ppm Se and see an initial toxic response, but once the mice adjust we can continue to increase Se up to 40 ppm with different selenicals with no apparent deleterious effects. The diet is commercial chow with the typical protein content. Dr. Burk told me that he has observed rats growing well at 10 ppm Se, but, upon withdrawing selenium, the rats died.

CHANGES IN EATING AND DRINKING BEHAVIOR OF RATS CAUSED BY SELENIUM

T. R. SHEARER

Department of Preventive Dentistry, University of Oregon Dental School, Portland, Oregon

Changes in the feeding behavior of experimental animals have been shown to be an important variable in studies on the biological effects of excesses or deficiencies of trace minerals. The present experiment was designed to provide more detailed data on the changes occurring in the eating and drinking habits of rats exposed continuously to drinking water containing increased amounts of sodium selenite.

Sixteen weanling, male Holtzman rats were equally divided into a control group receiving distilled drinking water and into a selenium group receiving distilled drinking water containing 3 ppm Se, as Na_2SeO_3. Ground Purina Chow was available *ad libitum.* Over an experimental period of 19 days, 13 parameters concerning the amount and pattern of food and water consumption were measured for each rat by the use of easily constructed electronic measuring cages modified after a device by Spengler [Helv. Physiol. Acta 18: 50 (1960)]. Our device utilizes electronic load cells (model FTD-G-100, Schaevitz Engineering, Camden, N.J.) to measure the change in weight of food and water cups, which the rats reach by means of external, right-angle plastic access tubes mounted on the front of standard galvanized, screen-bottom rat cages. The rats were maintained in a temperature-controlled (22°C) and light-controlled (12 hr dark, 12 hr light) animal room.

The data in Table 1 show how selenium acts to decrease food and water intake. Selenium decreased total food consumption by 27% due to a 19% decrease in the number of meals, and caused a 19% decrease in body weight. Furthermore, the average time needed to consume each meal and the rate of eating (min/g) were increased by 30% and 39%, respectively. Total water intake was reduced by 28% due to a 32% decrease in the number of drinks per day and a 39% decrease in the total number of minutes spent drinking per

Table 1. Effect of dietary selenium on the pattern of food and water intake in the rat

Measurement	Control group	Selenium group[1]	Percentage change
Food intake			
Total amount (g/day)	17.2 ± 0.3[2]	12.6 ± 0.4[b]	−27
No. of meals per day	20.0 ± 1.0	16.2 ± 1.4[a]	−19
Min per meal	5.6 ± 0.2	7.3 ± 0.6[a]	+30
Min per gram	6.7 ± 0.2	9.3 ± 0.6[b]	+39
Water intake			
Total amount (ml/day)	22.5 ± 1.0	16.3 ± 0.9[b]	−28
No. of drinks per day	27.3 ± 1.5	18.7 ± 1.2[b]	−32
Min drinking	56.1 ± 3.5	34.4 ± 2.5[b]	−39
Body weight (g)	128.0 ± 1.0	104.3 ± 3.8[b]	−19

[1]Values with superscript [a] are statistically significant from the control, $P < 0.05$; values with superscript [b] are statistically significant from the control, $P<0.01$.
[2]Mean ± s.e. for 8 animals per group; values are averages per day for each rat observed on days 7, 13, and 19 of the experiment.

day. Selenium did not change the average size of each meal, total eating time per day, the volume of each drink, time per drink, or the rate of drinking.

Previous studies have also shown that selenium can alter the physiological ratio of food to water consumption [Hadjimarkos, Caries Res. 3: 14 (1969)] and also change some other parameters of eating and drinking behavior in rats [Konig *et al.*, Helv. Odont. Acta 8: 82 (1964)], all of which may influence the oral clearance of food particles from the teeth and subsequent caries activity. In this regard it should be noted that the length of time food is available to rats and the rate of food consumption have been directly related to caries activity in the rat [Rosen *et al.*, J. Dent. Res. 38: 177 (1959); Larson *et al.*, Archs Oral Biol. 7: 463 (1962)].

Epidemiological studies among children, reviewed elsewhere, and experiments with rats [Buttner, J. Dent. Res. 42: 453 (1963)] and monkeys [Bowen, J. Irish Dent. Ass. 18: 83 (1972)] have demonstrated that ingestion of increased amounts of selenium during the period of tooth development enhances significantly the susceptibility to dental caries. We recently demonstrated that most of the radioactive ^{75}Se incorporated into enamel and dentine during tooth development is present in the protein fraction of these two dental tissues [Shearer and Hadjimarkos, J. Nutr. 103: 553 (1973)]. It has been suggested that incorporation of selenium into the protein components of enamel during its development inhibits mineralization which apparently accounts for the increased susceptibility to caries [Eisenmann and

Yaeger, Archs Oral Biol. 14: 1045 (1969); Newesely, *In* Staple (ed.), Advances in Oral Biology, Vol. 4, Academic Press, New York (1970), p. 11].

Studies on the post-developmental effects of selenium on dental caries in animals have yielded conflicting results [Hadjimarkos, *loc. cit.;* Buttner, Caries Res. 3: 1 (1969)]. A basic factor responsible for this is the lack of control animals whose eating and drinking patterns are the same as animals receiving selenium. These data indicate that definitive studies of the post-developmental effects of selenium on dental caries would not be possible. There appears to be no available method for simultaneously imposing both food- and water-intake patterns characteristic of selenium-fed animals on a suitable group of control animals to compare directly caries activity between the two groups. Even in animal experiments where total food intake and weight gain between selenium and control animals is similar, it has been shown that selenium in the drinking water causes significant decreases in the water intake [Hadjimarkos, Nutr. Rep. Int. 1: 175 (1970)]. These previous results and the results in the present experiment are apparently due to the offensive taste and/or odor of selenium.

DISCUSSION

Levander (Beltsville). Do your data show that the amount of time the +Se group spent eating was greater than for the −Se group, yet the +Se group spent much less time drinking water?

Shearer. The total amount of time spent eating each day was actually the same as the control group, but the number of meals was less. Thus, the +Se rats slowed their eating process, and this is important from the dental caries standpoint as is the water intake.

Ullrey (East Lansing). The concentration of Se in either food or water would be important in this apparent interrelationship of Se with the development of dental caries. Has anyone established that a level such as 0.1 ppm Se in water would induce dental caries?

Shearer. The main route of intake of selenium by man is food rather than water. A level of 0.1 ppm Se in the urine appears to be a maximum allowable level in terms of dental caries.

Ullrey (East Lansing). But what is the quantitative relationship between urinary Se concentration and water or diet Se concentration?

Shearer. I don't know. Urine is probably the most useful measure of relative total Se intake.

Mertz (Beltsville). I should like to emphasize again that one can have a perfect correlation between any two factors, but this does not necessarily imply any cause-effect relationship.

Shearer. There is no question about that.

ANTIOXIDANTS AND CANCER. III. SELENIUM AND OTHER ANTIOXIDANTS DECREASE CARCINOGEN-INDUCED CHROMOSOME BREAKAGE

RAYMOND J. SHAMBERGER

Cleveland Clinic and Cleveland Clinic Educational Foundation, Cleveland, Ohio

The antioxidants, selenium and vitamin E, are known to prevent carcinogenesis in animals [Shamberger, J. Natn. Cancer Inst. 44: 931 (1970)], possibly through a decrease of peroxidative damage of DNA [Shamberger, J. Natn. Cancer Inst. 48: 1491 (1972)]. The current belief about the mechanism of carcinogenesis is that a carcinogen, virus, or other factors alter a macromolecule such as DNA, and change the inherent information that can be transmitted to daughter cells or change the encoded information needed for metabolic function and control during the life of the cell. If this theory is correct, antioxidants might prevent carcinogen-induced chromosome break-

Leukocytes from a 38-year-old male volunteer were incubated in chromosome media 1 A (Grand Island Biological Co., Grand Island, N.Y.) at 37°C for three days. About 15 hr before chromosome harvest, the antioxidants and carcinogens were added separately in 0.2 ml of 2% bovine serum albumin. The antioxidants tested were ascorbic acid, BHT (butylated hydroxytoluene), Na_2SeO_3, and dl-α-tocopherol. The carcinogens used were sodium cyclamate and 7,12-dimethylbenz(γ)anthracene. The concentrations of the antioxidants and the carcinogens are listed in Table 1. Two hours before harvest, 0.2 ml Velban (0.5 μg/ml) was added to each tube and mixed thoroughly. The techniques were those used routinely for chromosomal analyses at the Cleveland Clinic. After slides were made, the metaphase figures were evaluated. Most of the metaphase figures evaluated contained 46 chromosomes, but in a few cases either 45 or 47 chromosomes were observed. Cells that were grossly defective or contained only a partial number of chromosomes were not evaluated. All chromosomal abnormalities were scored. Most of the aberrations were gaps and breaks. Exchange figures and unusual chromosomes were

Table 1. Effect of antioxidants on carcinogen-induced chromosomal breakage[1]

Carcinogen	Antioxidant	No. cells	No. breaks	Percentage breaks	Percentage breaks minus control	Percentage reduction
None	None	211	23	10.9		
DMBA[2]	None	290	82	28.3	17.4	
DMBA	10 μM ascorbic acid	127	30	23.6	11.9	31.7
DMBA	0.21 μM BHT[3]	157	29	18.4	6.3	63.8
DMBA	0.20 μM Na_2SeO_3	171	37	21.6	10.1	42.0
DMBA	10 μM dl-α-tocopherol	156	28	17.9	6.4	63.2
Cyclamate	None	222	26	11.6		
Cyclamate	0.2 μM Na_2SeO_3	80	9	11.2		

[1]Chromosome breakage was induced by 1.6 μM DMBA or by 100 μM sodium cyclamate.
[2]DMBA, 7,12-dimethylbenz(γ)anthracene.
[3]BHT, butylated hydroxytoluene.

observed infrequently (less than 1%). In all experimental groups chromosomes from at least 100 metaphase figures were observed. Controls containing dimethylbenzanthracene and controls without any addition were evaluated throughout the experiment.

The group of cells treated only with dimethylbenzanthracene had 17.4% more chromosomal breaks than the untreated controls. Chromosomal breakage was reduced by all of the antioxidants tested (Table 1). The reductions were as follows: ascorbic acid, 31.7%; B-hydroxytoluene, 63.8%; Na_2SeO_3, 42.0%; dl-α-tocopherol, 63.2%. The number of cells with multiple chromsomal breaks and with either one extra or one deleted chromosome were distributed equally throughout the experimental groups. The groups of cells treated with 100 μM sodium cyclamate had only a slightly higher percentage of broken chromosomes than the controls. The cyclamate group had only a slightly higher percentage of breaks (11.6% than the cyclamate group treated with selenite (11.2%).

The protection against chromosomal breakage by antioxidants also seems to be consistent with epidemiological evidence that antioxidants prevent cancer. Shamberger *et al.* [C.R.C. Crit. Rev. Clin. Lab. Sci. 2: 211 (1971)] have observed an inverse relationship between selenium exposure and mortality from cancer in the United States. In some unpublished data we have observed a similar trend in Canada and New Zealand. Mortality from carcinoma of the gastrointestinal tract is much lower in the high-selenium cities and shows the greatest difference in mortality from carcinomas of various types. Shamberger *et al.* [Cleveland Clin. Quart. 39: 119 (1972)] have postulated a relationship between the declining American death rate from gastric carcinoma and the public acceptance in 1930 of wheat cereals rich in selenium and vitamin E and the introduction of food preservative antioxidants in 1947. Dungal *et al.* [Br. J. Cancer 21: 270 (1967)] have postulated that the high incidence of gastric carcinoma in Iceland may be related to a large intake of smoked food, perhaps in association with a low intake of vitamin C. Schlegel *et al.* [Trans. Am. Ass. Genito-Urinary Surg. 61: 85 (1969)] have observed that vitamin C reduces uroepithelial carcinoma in mice and have postulated a similar mechanism in humans. Recently, Wattenberg [Proc. Am. Ass. Cancer Res. 14: 7 (1973)] has observed that the antioxidants butylated hydroxyanisole and thiuram disulfide in the diet have greatly reduced the number of tumors per animal in several carcinogenic test systems.

Although our finding that antioxidants prevent chromosomal breakage in tissue culture is consistent with numerous carcinogenic experiments in animals, biochemical studies, and epidemiological studies, the anticarcinogenic effect in tissue culture cannot be differentiated from mutagenic, teratogenic, or cellular toxic effects.

DISCUSSION

Spallholz (Fort Collins). Your prior publications have correlated Se status with cancer and pointed to an anticarcinogenic property of Se. From your present data, vitamin E and BHT were apparently more effective than selenite in preventing chromsome breakage. Perhaps such compounds more effectively prevent chromosome breakage in people if the diet contains a higher level of Se.

Shamberger. I consider this effect to be a universal antioxidant effect. Selenite was present at an extremely low concentration compared with the other antioxidants and I consider it equally or more effective than other antioxidants.

Levander (Beltsville). One tendency of tumors is to switch from an aerobic to an anaerobic type of metabolism. Do you think the Se-cancer relationship relates to a role for selenium in respiratory change?

Shamberger. I believe that these antioxidants are reducing the binding to DNA because there apparently is some kind of peroxidation involved in this process. I have not seen any difference in glutathione peroxidase between a small number of cancer patients and normals but more work is needed on this.

Levander (Beltsville). How do you relate dimethylbenzanthracene, lipid peroxidation, and tumorigenesis? Is it just the destruction of membrane structure and opening it up to peroxidation?

Shamberger. Yes, I suspect that there is also formation of malonaldehyde which may react with DNA.

Frost (Schenectady). This matter of Se in relation to cancer is frustrating. Se was first impugned as a cause of cancer but that has been pretty well disproven now and the FDA accepts it. Yet, letters at the Hearing Examiners are about 3 to 1 against the addition of Se to feeds on the unproved basis that Se is a cause of cancer. Now there is great difficulty in getting experiments done to test the other side of the question, that selenium may have value against cancer. Dr. Shamberger and I have been trying to shed some light on this anticancer value. The biochemical rationale for a possible anticancer value of Se seems valid to me; the problem is in getting the experiments done.

Schwarz (Long Beach). Immunosuppressant agents enhance the development of cancer up to 10,000-fold and it has been established that subtoxic doses of Se have strong immunosuppressant properties. It may be that the so-called carcinogenicity of selenium is simply due to its immunosuppressant effect.

Spallholz (Fort Collins). Some of our very preliminary experiments indicate that delayed hypersensitivity in animals, which is the effective immunological response against cancer, is interfered with at 3 ppm Se despite the increase in circulating antibody. Slight enhancement of delayed hypersensitivity seems to occur at 1–2 ppm Se.

Andrews (Upper Hutt). What level of Se is necessary to show the anticarcinogenic effect?

Shamberger. A dietary supplement of 0.1 ppm Se from sodium selenite had virtually no effect, but 1 ppm had a marked effect against one of the most potent carcinogens known.

FACTORS INFLUENCING IRON UTILIZATION BY THE BABY PIG

J. P. HITCHCOCK, P. K. KU, and E. R. MILLER

Animal Husbandry Department, Michigan State University, East Lansing, Michigan

Iron supplementation is essential for the nursing pig reared in confinement. This study was undertaken to determine the influences of intestinal microflora, type of dietary protein, and form of diet on oral-iron utilization by the baby pig.

Twenty germfree pigs farrowed and reared by techniques described by Waxler and Whitehair [Swine in Biomedical Research (1966), p. 611] and 16 conventionally farrowed pigs reared by techniques described by Miller *et al.* [J. Nutr. 95: 278 (1968)] received a sterilized diet of condensed cow's milk (2 ppm Fe) with 0, 50, or 100 mg Fe from ferrous sulfate added per kilogram of milk solids. A positive control group in both major plots received 100 mg Fe from iron-dextran intramuscularly at the start of the study (Table 1).

Table 1. Influence of environment on 4-week weight gain, hemoglobin, and serum iron.

Iron (ppm)	28-day gain (kg)	Hemoglobin (g/100 ml)	Serum Fe (ppm)
Germ free environment			
0	2.1	5.9	1.1
50	2.8	10.1	1.8
100	2.8	12.1	2.0
Injected[1]	2.7	11.5	1.6
Conventional environment			
0	–[2]	–[2]	–[2]
50	4.9	8.6	0.9
100	4.8	9.8	1.8
Injected[1]	4.9	7.7	0.9

[1]A single intramuscular injection of 100 mg of iron from iron dextran.
[2]None alive.

Table 2. Influence of type of protein and iron level upon hematology and iron balance

Casein	Casein			Soy		
	101 ppm Fe[1]	148 ppm Fe[1]	189 ppm Fe[1]	95 ppm Fe[1]	147 ppm Fe[1]	189 ppm Fe[1]
Final hematology						
Hemoglobin (g/100 ml)	5.2	4.9	7.1	5.7	6.5	9.9
Serum iron (ppm)	0.94	0.89	1.84	0.97	0.98	2.12
TIBC (ppm)	1.74	1.98	2.86	1.94	1.75	3.19
Daily iron balance[2]						
Iron intake (mg)	19.7	32.5	44.6	20.8	34.5	44.2
Fecal iron (mg)	10.0	16.7	21.2	5.5	10.5	10.2
Urine iron (mg)	0.1	0.1	0.1	0.1	0.1	0.1
Percentage retention	49	48	52	73	69	77

[1]By analysis.

[2]Conducted during final two weeks of trial.

The results presented in Table 1 are representative of the trends observed for all variables measured in this study. All unsupplemented conventional pigs died by 4 weeks of age while all unsupplemented germfree pigs survived. Fifty ppm of iron supported maximum gain while 100 ppm iron resulted in significantly higher hemoglobin and serum iron levels in both germfree and conventional pigs.

On the basis of comparison with positive control groups, it is concluded that the iron requirement of both germfree and conventional pigs reared on condensed cow's milk lies between 50 and 100 ppm.

The influence of protein type was studied in a 5-week, 2 × 3 factorial-designed experiment utilizing 24 pigs on dry purified diets containing casein or isolated soy as protein sources and containing 100, 150, or 190 ppm iron from ferrous sulfate. Conditions of rearing, allotment, and diet compositions were as described by Miller *et al.* [J. Nutr. 95: 278 (1968)]. Results of this study are presented in Table 2. The data indicate a better utilization of iron on the soy protein diet as measured by growth and hematology. Iron-balance data clearly indicate greater fecal loss of iron, thus poorer iron absorption, by pigs receiving casein diets, but no difference in urinary losses of iron. Thus, the baby pig utilizes iron contained in soy diets better than that in casein diets of similar levels of iron.

The effect of liquid *versus* dry diet was compared utilizing 32 baby pigs fed a casein synthetic purified diet (similar to that described above), either in the dry form or mixed with water as 20% of solids and homogenized. Ferrous sulfate was added to provide iron levels of 50, 100, or 150 ppm of the solids.

At each common dietary iron level pigs receiving the liquid diet had significantly higher weekly hemoglobin values than pigs receiving the dry diet. The three-week hemoglobin (g/100 ml) values for liquid *versus* dry diets for each increment in dietary iron were as follows: 8.36, 5.80; 9.42, 7.73; 11.70, 8.06. Thus, iron from liquid diets was better utilized by the baby pig than iron from a similar dry diet.

DISCUSSION

Saltman (San Diego). There is substantial nutritional evidence that casein is deleterious to iron absorption. Our studies *in vitro* have shown that casein will bind up to about 10–12% by weight of iron, if the casein is in intact micellar form. Upon dissociation of casein into subunits with chelating agents, it will not bind iron. Reconstituted casein micelles will again bind iron. From electron microscopy it appears that the iron is held in micellar units in a manner analogous to that in ferritin. This unique property of casein as a metal-binding protein deserves further attention.

THE EFFECTS OF IRON AND COPPER SUPPLEMENTATION OF THE DIET OF SOWS DURING PREGNANCY AND LACTATION ON THE IRON AND COPPER STATUS OF THEIR PIGLETS

R. G. HEMINGWAY, NORA A. BROWN, and J. LUSCOMBE

Glasgow University Veterinary School, Bearsden, Glasgow, U.K., and Harper Adams Agricultural College, Newport, Shropshire, U.K.

It is well established that the inclusion of 125–250 μg Cu/g in fattening diets significantly improves the growth rate and the efficiency of food conversion for pigs. It is also appreciated that such rates of Cu inclusion reduce the concentration of Fe and increase the concentration of Cu in the liver of the pig [Cassidy and Eva, Proc. Nutr. Soc. 70, 31 (1958)]. High-level Cu inclusion is widely practiced even in creep feeds offered to very young pigs. The widespread production of concentrates described as being nutritionally suitable for both sows and weaned pigs suggests that high-Cu diets may often be fed to sows. It is important to establish the effects of such a diet fed to sows on the Fe and Cu status of their piglets.

During pregnancy and lactation four groups, each of six sows, were fed a basal diet consisting of 42.5% barley, 40% wheat, 7.5% soybean meal, and 7.5% fish meal plus 2.5% mineral/vitamins mixture, and containing 17% crude protein, 120 μg/g Fe, 20 μg/g Cu, and 85 μg/g Zn. Three groups housed on concrete were additionally fed either (A) no supplement, (B) 600 μg/g Fe as ferrous sulphate or (C) 600 μg/g Fe as ferrous sulphate plus 250 μg/g Cu as copper sulphate. Group D also received no supplement but the sows were kept at grass and farrowed in huts in the field. The housed pens were cleaned each day as in a normal system of husbandry, but otherwise no attempt was made to prevent piglets having access to the feces of the sows.

The two middle-weight piglets in each litter were removed at birth for the determination of Fe and Cu. Each alternate piglet in each litter received

Table 1. Effect of dietary Fe and Cu and Fe injections on body weight, hemoglobin, total body Fe and Cu, and liver Fe and Cu[1]

Group	A	B	C	D	
Environment	Housed	Housed	Housed	At grass	Standard error of mean
Fe supplement (μg/g)	0	600	600	0	
Cu supplement (μg/g)	0	0	250	0	
Birth weight (kg)	1.34	1.34	1.23	1.13	0.088
Total body Fe (mg)	36.0	39.2	39.1	34.5	2.80
Total body Cu (mg)	3.20	3.24	3.07	3.13	0.21
Liver Fe (μg/g)	816	765	1090[a]	1053[a]	79.7
Liver Cu (μg/g)	203	186	218	235	13.5
Hemoglobin at 21 day (%)					
With iron injection[b]	9.8	9.0	10.3	n.d.[2]	0.41
No iron injection	5.0	5.9[a]	6.0[a]	n.d.	0.28
Live weight at 42 day (kg)					
With iron injection[b]	14.3	14.5	13.4	n.d.	0.40
No iron injection	10.3	10.3	11.0	n.d.	0.30

[1]Values for Fe and Cu content of body and liver dry matter at birth are means for 12 piglets from 6 sows per group; values for hemoglobin and live weight after injection are means for 24 piglets from 6 sows per group. Values marked with superscript [a] are significantly greater ($P = 0.05$) than the other 2 groups. All injected values marked with superscript [b] are significantly greater ($P = 0.01$) than for noninjected pigs.

[2]n.d., not determined.

200 μg Fe by injection (Ferrofax, Crookes Laboratories Ltd.) at three days. Hemoglobin concentrations were examined on days 7, 14, and 21, and the piglets were weighed on day 42.

Inclusion of Fe (diet B) and Fe + Cu (diet C) marginally but nonsignificantly increased the total Fe content of the whole piglet at birth. The rather lower apparent Fe content of the piglets born to unsupplemented sows at grass (group D) reflects their slightly reduced mean birth weight.

The mean concentration of Fe in the liver at birth was significantly increased by the inclusion of Fe + Cu (diet C) but not by inclusion of Fe alone (diet B). Keeping the sows at grass also increased the concentration of Fe in the liver even though no supplement was given.

Piglets which suckled sows supplemented with either Fe alone (diet B) or with Fe + Cu (diet C) had improved hemoglobin concentrations at day 21 when no Fe injection was given. Injection of Fe, however, had a much greater incremental effect on hemoglobin concentrations and also had a markedly significant effect on live-weight gain irrespective of dietary supplementation.

Inclusion of a high level of Cu in the diet did not increase either the total Cu content of the piglet at birth or the concentration of Cu in the liver. Piglets born to sows supplemented with Fe + Cu responded normally to a Fe injection in respect of both hemoglobin concentration and growth rate.

These results are associated with high-level Fe + Cu supplementation for one period of pregnancy and lactation. Dammers and van der Grift [Tijdschr. Diergeneesk. 88: 346 (1963)] have demonstrated that a comparable high-level Cu inclusion in the diet of sows had no adverse effect over six consecutive pregnancies, but some increase was recorded in the Cu concentration in the liver of the piglets.

DISCUSSION

MacPherson (Auchincruive). We have found that, in contrast to piglets, the liver copper in sows builds up to 1000–2000 ppm over 2 years on 250 μg Cu/g diet and believe that there could be a risk because there were more fatalities in the piglets born to these sows.

Hemingway. This could happen, but with the many growing pigs in Britain which have received high copper diets there hasn't been any widespread copper poisoning reported.

Schwarz (Munich). Do you have any explanation for the higher iron contents in the liver of the piglets from the copper-supplemented sows?

Hemingway. If you increase the liver copper content of the sow by feeding copper and thereby displace some iron, some of the iron may end up in the piglet, but this is only speculation.

Andrews (Upper Hutt). Does your data suggest that the site of reaction is in the gut rather than in the body?

Hemingway. I don't think we have any large effect of copper on iron. The only pronounced beneficial effect on the piglet was that seen by giving an iron injection.

AN ASSOCIATION BETWEEN COPPER CONCENTRATION IN BLOOD AND HEMOGLOBIN TYPE IN SHEEP

GERALD WIENER, J. G. HALL, and SUSAN HAYTER

ARC Animal Breeding Research Organisation, Edinburgh, U.K.

Differences in concentrations of copper in the blood, liver, and brain of different breeds of sheep have been demonstrated in recent years by Wiener, Field, and others [J. Agric. Sci. Camb. 72: 93–101 (1969); *In* Mills (ed.), Trace Element Metabolism in Animals (1970), p. 92; J. Agric. Sci. Camb. 76: 513–520 (1971); Anim. Prod. 16: 261–270 (1973)]. Two of these investigations concerned flocks containing purebreds and crosses between them. The first flock (BCW) was composed of the Scottish Blackface, Cheviot, and Welsh Mountain breeds together with all possible crosses among these. Close inbreeding had also been practised in this flock both with the pure breeds and the crossbreds mated *inter se.* The second flock was composed of the Finnish Landrace and (Tasmanian) Merino breeds together with the first-generation reciprocal crossbreds. In both flocks large breed differences were found in blood copper levels. In the BCW flock determinations were on whole blood while in the Finn-Merino flock they were on plasma. However, the copper levels of the crossbreds equalled those of the higher parental breed in the first flock, but were midway between the two parental breeds in the second. It was hoped that the research described below would throw some light on this anomaly.

Hemoglobin (Hb), which depends on copper for its formation, exists in two electrophoretically distinct forms in normal British sheep. Animals were classified as homozygous A or B or heterozygous AB. In the BCW flock sheep of type B had significantly higher whole-blood copper concentrations (+16 μg/100 ml) than those of type A, with AB sheep lying midway between the two. Although Hb-type frequencies differed among breeds, this accounted for

Table 1. The effect of hemoglobin type on copper levels in the plasma, whole blood and red blood cells of sheep (BCW flock)

	Total plasma Cu		Whole blood Cu		Red blood cell Cu[1]	
Hb type	Fitted value[2]	S.E.[3]	Fitted value	S.E.	Fitted value	S.E.
AA	49.3		52.6		61.4	
AB	54.8	6.1	55.0	3.7	57.8	4.3
BB	70.7	5.0	69.9	4.2	70.7	4.9

[1]Indirect estimate.

[2]Fitted by least squares analysis to Blackface breed.

[3]Standard error of difference from Hb type AA.

only a small part of the previously recorded breed differences in whole-blood copper. In the Finn-Merino flock no association was found between Hb type and plasma copper level [Wiener *et al.,* Anim. Prod. 17: 1–7 (1973)].

These findings suggested that the association of Hb type with whole-blood copper levels may be attributable to concentration differences in red-cell copper, and also that plasma and red-cell copper levels may show genetic independence.

In order to test this hypothesis copper levels of plasma and whole blood were estimated by Dr. N. F. Suttle (Moredun Research Institute) in some 200 10-month-old female sheep of the BCW flock, but of a later generation than previously. Red-cell copper levels were estimated indirectly from these determinations. The breed effect was significant and so was the effect of Hb type as is shown in Table 1. The table also shows that, while plasma copper levels were affected by Hb-type differences, corpuscular copper levels were not especially implicated. This suggests that the association of Hb type with whole-blood and plasma copper is more likely to be explained by a genetic relationship than a physiological one.

Possibly aided by the inbreeding in the BCW flock to which these results apply, it may be that linkage has occurred between the genes controlling Hb type with those controlling blood copper. This would further suggest that there is a single gene (or a very closely linked group of genes) which exerts a large effect on blood copper levels.

If further work confirms this view it might be possible to change the copper level within a breed fairly readily to make it better able to withstand conditions of copper deficiency or excess, which create problems in sheep production.

DISCUSSION

Scoggins (Melbourne). Have you looked at the relationship between copper and the red-cell potassium, which is known to be genetically determined in sheep?

Wiener. We found no association between the copper level and the high *versus* low potassium types.

Lewis (Weybridge). Was there any difference in the proportions between the A and B hemoglobin type in your different breeds?

Wiener. The different frequencies of the A and B alleles in our different breeds accounts for a small proportion of the breed difference which we previously reported, but most of the breed difference is additional to the hemoglobin-type association.

Thornton (London). Do you have any information on differences in susceptibility to swayback between these hemoglobin types?

Wiener. No.

Hartmans (Wageningen). Is there a relation between the blood copper level in the different breeds and susceptibility to copper toxicity?

Wiener. In our past work, the breeds differed significantly from each other in the incidence of death resulting from therapeutic copper injections. Recent evidence suggests a significant difference in copper poisoning resulting from voluntary intake of concentrate feeds, but as yet this information on hemoglobin type is only a month or two old and we haven't looked at the various permutations of the data.

Spais (Thessaloniki). Did you investigate the copper levels in different breeds of cattle?

Wiener. Our cattle work is at a very early stage; however, we have found very marked differences both in the copper levels of the different breeds and in the changes which take place in copper levels during pregnancy and early lactation. From preliminary calculations, about 50% of the total variation in copper level or the changes in copper level over a period of 3 weeks before to 9 weeks after calving is attributable to breed differences.

Thornton (London). Have you information on the frequencies of the different hemoglobin types within the individual breeds of sheep, and are these the same between the breeds?

Wiener. Yes, there are big differences. For example, the Blackface breed has a frequency of the A allele of about 0.8, whereas the Welsh breed has a frequency of the A allele of about 0.4.

COMPARATIVE RESPONSES OF HYPOCUPREMIC EWES TO ORAL AND INTRAVENOUS Cu AND THEIR USE IN ASSESSING THE AVAILABILITY OF DIETARY Cu

N. F. SUTTLE

Moredun Research Institute, Edinburgh, U.K.

Knowledge of the biological variation in availability of dietary Cu to ruminants is limited by technical difficulties in measuring deviations from the generally low level of Cu availability using conventional balance techniques. A technique is required which measures directly the amount of Cu absorbed, and, theoretically, the relative responses in plasma Cu of hypocupremic ewes to oral and intravenous supplies of Cu could provide such an assessment. This method for measuring Cu availability has, therefore, been investigated.

Hypocupremia was induced in a group of Scottish Blackface ewes, aged 7–9 years, by feeding 0.8 kg/day of the Cu-deficient diet described by Suttle and Field [J. Comp. Path. 78: 351 (1968)]. In experiment 1 the uniformity and repeatability of responses in plasma Cu to a standard dietary supplement, 4 mg Cu/kg, was examined. Thirty-six ewes with initial plasma Cu values of 0.09–0.46 mg/liter were repleted for 33 days on two occasions.

Individual responses in plasma Cu after the first repletion ranged from 0.08 to 0.76 mg/liter about a mean of 0.36 mg/liter, the coefficient of variation being 50%. The mean and range of responses for the second repletion were similar, and the responses of individual ewes on the two occasions were highly correlated ($R = 0.68; P < 0.001$; 34 d.f.). Responses in plasma Cu were linear with time but unrelated to live-weight and initial plasma Cu concentration.

In experiment 2 the sensitivity of the response in plasma Cu was investigated by repleting 5 groups of 7 hypocupremic ewes with the basal diet supplemented with Cu to provide 2.7, 4.2, 5.7, 7.2, or 8.7 mg total Cu/kg.

Ewes were allocated to treatments at random within strata formed by ranking them in order of the responses shown in plasma Cu in experiment 1. Responses in plasma Cu after 33 days were -0.058 ± 0.031, -0.009 ± 0.024, 0.140 ± 0.045, 0.271 ± 0.031, and 0.341 ± 0.077 mg/liter, respectively. Each increment in dietary Cu above 4.2 mg/kg significantly increased plasma Cu at some stage during repletion and the responses were linearly related to Cu intake. The relationship between plasma Cu response (y, mg/liter) and Cu intake (x, mg/day) after 21 days was given by the equation

$$y = 0.0871\,x - 0.250 \quad (R = 0.99;\ 3\ \text{d.f.}). \tag{1}$$

In experiment 3 the response to intravenous Cu was investigated by repleting 6 hypocupremic ewes with 0.05–0.35 mg Cu/day on 3 or 4 occasions. Copper as $CuSO_4$ in saline was infused continuously into the jugular vein, using a multichannel peristaltic pump. The relationship between plasma Cu response (y, mg/liter) and intravenous Cu (x, mg/day) was linear, the equation being

$$y = 2.315\,x - 0.156 \quad (R = 0.86;\ 22\ \text{d.f.}) \tag{2}$$

after 17 days. The equivalent equation for ceruloplasmin Cu was

$$y = 1.450\,x - 0.220 \tag{3}$$

indicating that ceruloplasmin accounted for approximately 68% of the response in plasma Cu.

Assuming that the jugular infusion of Cu adequately simulated the postabsorptive supply of Cu, then the ratio of the coefficients in equations (1) and (2) measures the availability of Cu. For experiment 2 this was 4.1% (0.0871/2.135); this compares favorably with values obtained for a similar diet of 2.6 ± 1.8% s.e. ($n = 7$) by stable Cu balance and 3.7 ± 2.1 s.e. ($n = 3$) by a radioactive marker-ratio technique [Suttle, Proc. Nutr. Soc. 32, 10 A (1973)].

Increments of 0.045 mg/day in available Cu in experiment 2 significantly increased responses in plasma Cu; stable-Cu balance techniques would not have detected such small differences between ingested and fecal Cu. It is also unlikely that these differences would have been detected by monitoring liver Cu stores using a liver biopsy technique. If the retention of apparently absorbed Cu by the liver was 100%, liver Cu would increase by only 0.2 mg/kg/day in response to 0.045 mg Cu/day. The repletion technique, therefore, appears to be a relatively sensitive method for determining Cu availability.

The availability of Cu in an unknown diet can be estimated from the response elicited in hypocupremic ewes by first predicting from the intravenous Cu relationship the available Cu input required for the observed increment in plasma Cu. Several precautions must be taken, however. Firstly, the response in plasma Cu to increases in dietary Cu is sigmoidal and the response must be measured in the linear part of the response curve. Secondly, it is essential to standardize the animal component in any determination of dietary Cu availability. Individuals similar in age, breed, and dietary treatment showed an estimated range of availability of 4–12% in experiment 1. Individual differences should, therefore, be randomized between treatments and corrections made for changes in the breed or population on experiment. Finally, systemic effects of the diet on Cu metabolism should be checked.

DISCUSSION

Reynolds (Beltsville). Did those animals that accumulated more copper during repletion deplete faster or slower?

Suttle. These animals depleted at a uniform rate which reflects the fact that the differences are in absorption, and during depletion all we are measuring is the constancy of the endogenous loss and urinary excretion of copper.

Matrone (Raleigh). The rat absorbs about 28% of the copper, but only about 5% is retained, the rest of it being excreted via the bile. Have you given thought to these considerations in your work with sheep which indicates about 4% 'availability' of dietary copper?

Suttle. I regard retention to be much the same as availability. Dr. Hemingway has shown, using a liver copper repletion technique to measure copper utilization, that the sheep retains about 5% of its dietary copper intake in the liver. Urinary excretion is very small.

Spais (Thessaloniki). To what do you attribute the individual variation?

Suttle. This appears to be due to differences in the efficiency with which individual animals absorb copper.

Spais (Thessaloniki). We have observed that individual animals which can't form much hydrogen sulfide, even in the presence of high levels of sulfate, have a high availability of copper, while animals with large production of hydrogen sulfide have a low availability. This probably depends on the microflora in the rumen.

Suttle. We have observed that the variability in responses on high-sulfur diets is much greater than that on low-sulfur diets, but we got no correlation between rumen sulfide concentration determined at a given time and the individual differences in response. A correlation may have been missed because of the dynamic nature of sulfide in the rumen.

Hemingway (Glasgow). During repletion was the copper sulfate absorbed on the food or was it drenched into the animal?

Suttle. Copper sulfate was added to the diet as a crystalline mixture.

Schwarz (Munich). You measured copper availability with the copper plasma values; do you have any results on ceruloplasmin copper and are the results similar?

Suttle. We measured ceruloplasmin in all the experiments except the initial uniformity trial. There has been no difference in the interpretation of the effect of dietary treatment using ceruloplasmin values and using total plasma copper.

Lee (Adelaide). We have found that the retention of copper in the liver of Merino sheep dosed daily with 10 mg of Cu as copper sulfate has been about 4% of the total dose irrespective of the interval at which the dose was given. This may provide some validation for your procedure.

Hartmans (Wageningen). I would expect that the efficiency of copper utilization may decrease when the dietary content is higher or when the animals are less depleted, and that your values might be lower had you used animals of normal copper status. Would you agree with that? Your method may be applicable only if you start with the same type of depletion on every occasion.

Suttle. Yes, we have already seen in the work from Germany that copper status can affect the efficiency of copper absorption. In my studies the animals were repleted as soon as they become hypocupremic (0.02–0.04 mg Cu/l); they were not in a state of clinical copper deficiency. If there was enhancement of availability due to the low copper status, one would have expected a fairly early change in the response as they became normocupremic. This did not appear to happen.

EFFECTS OF MOLYBDENUM AND SULPHUR AT CONCENTRATIONS COMMONLY FOUND IN RUMINANT DIETS ON THE AVAILABILITY OF COPPER TO SHEEP

N. F. SUTTLE

Moredun Research Institute, Edinburgh, U.K.

The Cu × Mo × S interrelationship in ruminant nutrition is believed to involve only the smaller inorganic component of dietary S [Dick, Aust. Vet. J. 29: 18; 29: 233 (1954)]. The interrelationship is not held to be widely involved in the etiology of hypocuprosis, because variations in SO_4 and Mo contents of pasture within the respective normal ranges are not thought to affect Cu metabolism in ruminants [Allcroft and Lewis, J. Sci. Food Agric. 8: 97 (1957); Hartmans *In* Mills (ed.), Trace Element Metabolism in Animals, Livingstone, Edinburgh (1970), p. 441)]. These suppositions have not been tested experimentally, however, and they have therefore been investigated using a new technique for measuring Cu availability.

Groups of 5–7 hypocupremic ewes were repleted for 35 days by feeding a semipurified diet [Suttle and Field, J. Comp. Path. 78: 351 (1968)] supplemented with 6 mg Cu/kg and containing various amounts of Mo and organic or inorganic S. Ewes were allocated to treatments on the basis of their ranking in a uniformity trial. The approximate levels of Mo and S used, the final responses in plasma Cu, and the estimated availabilities of dietary Cu are given in Table 1.

Within each experiment the response with both Mo and S added to the diet was significantly less than that for any other treatment; Mo thus interacted with organic as well as inorganic S. Mo per se had no effect, but S per se significantly reduced the response in plasma Cu in each experiment, the effect being most marked for cysteine S. Cu availability was reduced by 16–44% by S and by 42–70% by Mo + S.

Table 1. Effect of dietary sulphur and molybdenum on the availability of dietary copper[1]

	Dietary Mo (mg/kg) / Dietary S (g/kg)	0.5 / 1.0	0.5 / 4.0	4.5 / 1.0	4.5 / 4.0
Experiment	Source of S	Change in plasma Cu after 35 days repletion[2] (μg/liter)			
1	Na_2SO_4	562 ± 53 (4.7)	387 ± 90 (3.7)	497 ± 67 (4.7)	−35 ± 38 (1.6)
2	Methionine	405 ± 77 (3.8)	367 ± 117 (3.2)	522 ± 114 (4.6)	50 ± 33 (2.2)
3	Cysteine	864 ± 11 (7.0)	284 ± 84 (3.9)	872 ± 216 (7.8)	104 ± 15 (2.1)

[1]Hypocupremic ewes were given 8 mg Cu/kg, varying amounts of Mo and S, and the sulphur source indicated.
[2]The figures in brackets are the calculated availability of Cu estimated by the method of Suttle (previous paper).

When the ranges of S and Mo were narrowed by comparing 2.0 or 3.0 g S/kg with 2.5 or 4.5 mg Mo/kg in a 2 × 2 factorial design, the response in plasma Cu was highest for the group given 2.0 g S and 2.5 mg Mo/kg, 342 ± 72 μg/liter, and equally low for the remaining three groups, the means ranging from 38 ± 42 to 53 ± 71 μg/liter. A marked Mo × S interaction was again apparent, Cu availability being reduced from 4.3% to an average of 2.6%. Distribution of Cu in the plasma was largely unaffected by the treatments.

The site of the Mo × S interaction was investigated by adding 4 mg Mo/kg to the high-S (3 g/kg) diet of three hypocupremic ewes repleted by a continuous intravenous infusion of 0.3 mg Cu/day; after 16 days, the mean response in plasma Cu was similar to that of three unsupplemented ewes similarly infused (554 ± 87 *versus* 621 ± 101 μg/liter). There was, however, some evidence of a partial inhibition of ceruloplasmin synthesis, the response being 149 ± 61 and 428 ± 108 μg/liter for Mo-supplemented and unsupplemented ewes, respectively. Bypassing the gastrointestinal tract, therefore, considerably lessened the effect of Mo and S on the utilization of Cu.

Provided that the Cu, Mo, and S inherent in natural foodstuffs interact in a manner similar to that shown in the above studies, the following statements can be made. Mo interacts with organic S as well as inorganic S in limiting the utilization of dietary Cu by sheep. The interaction is so marked that variations within the normal ranges for Mo and S herbage will have significant effects on the Cu status of the grazing animal. It may prove possible to correlate the incidence of clinical Cu deficiency with the ratio of Cu:Mo × total S in the diet. The observed interaction between Cu, Mo, and S takes place predominantly in the gut; it may involve the formation of an unavailable compound containing Cu and Mo, similar to that described by Dowdy and Matrone [J. Nutr. 95: 191, 197 (1968)], since plasma Mo concentrations at a given Mo intake were inversely related to dietary Cu intake.

DISCUSSION

Thornton (London). Have you found evidence of molybdenum-induced hypocuprosis in sheep under natural farm conditions with these levels of sulfur and has it been related to swayback?

Suttle. I have done no field surveys. This will be part of the follow-up program to these studies. I'm hoping that if you use the product of molybdenum concentration multiplied by sulfur concentration, you might find a better relationship with the incidence of clinical copper deficiency.

Hill (London). Can't you argue rather in the opposite direction; that is, that other factors are more likely to be involved in the field since we would expect more problems if the level of Mo and S were more important than we find?

Suttle. Undoubtedly molybdenum and sulfur are not the only factors involved. Soil ingestion is one other complicating factor. In field surveys, we just take a superficial look at the total mean molybdenum and total mean sulfate concentrations in particular areas. We are going to have to take a far more detailed look, and, in the light of the present results, it is worth re-examining the possibilities of an interrelationship.

Hartmans (Wageningen). We must look for factors interfering with copper as well as factors interfering with molybdenum availability.

Binnerts (Wageningen). I found that grass feeding gives a much lower uptake by the liver of intravenously injected copper than hay feeding, but I found no effect or only a small effect of sulfate and molybdenum. Other factors seem to be involved.

Matrone (Raleigh). In working with molybdenum, you should not only measure serum copper but also ceruloplasmin activity, because we have some evidence for a copper-molybdenum complex in the blood which is not utilizable.

Suttle. In these studies, the interpretation of treatment effects were no different using ceruloplasmin and using total plasma copper. At 4 ppm Mo we are not getting systemic effects or any increase in direct reacting copper. Beyond 8 ppm Mo in the diet, you do get disturbances in the distribution of copper in the plasma including a fraction which appears to include both copper and molybdenum which we include in 'residual copper'.

ACTION OF CYANIDES AND THIOCYANATES ON COPPER METABOLISM IN SHEEP

A. G. SPAIS, A. C. AGIANNIDIS, N. G. YANTZIS, A. A. PAPASTERIADIS, and T. C. LAZARIDIS

Department of Medicine, Veterinary Faculty, Aristotelian University of Thessaloniki, Greece

In salty areas in Greece, where conditioned copper deficiency occurs [Spais, Rec. Med. Vet. 135: 161 (1959)], *Cynodon dactylon,* a cyanogenetic plant, is very common and consumed along with halophytes rich in sulfates by sheep. Furthermore, cyanides, normally detectable in the rumen liquor and blood of sheep, can be found in the latter in considerable quantities under certain conditions and especially after administration of thiocyanates [Spais *et al.,* unpublished data]. On the other hand, cyanides constitute enzymic inhibitors and they can form with copper, at least *in vitro,* either soluble or insoluble compounds in proportion to the ratio existing between them [see, e.g., Treadwell *et al.,* Analytical Chemistry, John Wiley, New York (1951)].

In view of these facts three experiments were conducted to study the effect of cyanides alone or in combination with sulfates and also of thiocyanates on copper status in sheep. The experimental conditions and the results are summarised in Table 1. Blood copper values have been omitted since monthly determinations showed that there were no significant differences between groups, except in those of sulfates.

On the basis of the results obtained, the following conclusions were made.

(1) Cyanides (40 mg CN^- as KCN *per os* for a year) reduce liver copper reserves in lambs and sheep.

(2) Cyanides administered with sulfates do not influence the anticipated action of the latter on copper status in sheep, but it seems that they favor transfer of copper from dam to embryo.

Table 1. Effect of cyanides, cyanide-sulfate combinations, and thiocyanates on copper status in sheep.

Experiment no.	Duration (months)	Experimental animals	No. of animals	Substances administered daily per animal[1]	Administration mode	Copper concentration in liver (mean ± s.e. of each group in ppm D.M.): Sheep, Start biopsy	Sheep, End biopsy[2]	Lambs born and sacrificed on birth[2]
1	12	Lambs of native breed 4 months old	6	Controls	Drenching every 12 hr	293.7 ± 44.53	260.8 ± 46.41	
			6	40 mg CN^- as KCN		285.7 ± 34.25	180.3 ± 21.02[a]	
			6	600 mg SCN^- as KSCN		282.2 ± 8.33	331.8 ± 13.94[b]	
			6	6 g SO_4^{2-} as Na_2SO_4		283.8 ± 32.40	81.7 ± 15.46[c]	
			6	6 g SO_4^{2-} + 40 mg CN^-		277.5 ± 30.42	93.0 ± 14.24	
2	12	Ewes of native breed 14 months old	6	Controls	Capsules every 12 hr	122.0 ± 22.07	149.2 ± 23.53	251.3 ± 22.96
			6	40 mg CN^- as KCN		115.8 ± 23.26	96.2 ± 17.07[d]	252.8 ± 4.83
			6	6 g SO_4^{2-} as Na_2SO_4		115.0 ± 15.18	43.0 ± 3.25[e]	56.1 ± 14.85[f]
			6	6 g SO_4^{2-} + 40 mg CN^-		120.0 ± 7.91	45.5 ± 4.97	148.5 ± 39.97[g]
3	6	Ewes of native breed 2 yr old	7	Controls	Incorporated in food as pellets	167.4 ± 33.17	136.4 ± 30.17	
			7	60 mg SCN^- as KSCN		167.7 ± 23.50	183.0 ± 18.08	
			7	600 mg SCN^- as KSCN		167.4 ± 22.70	266.1 ± 38.24[h]	
			7	1800 mg SCN^- as KSCN		168.3 ± 35.03	226.0 ± 22.47[i]	

[1]Daily ration: 70% alfalfa meal (Cu, 10.5–13.0 ppm D.M.; Mo, 3.7–4.5 ppm D.M.; SO_4^{2-}, 0.40–0.60% D.M.); 30% barley (Cu, 5.0–6.1 ppm D.M.; Mo, 0.5–1.0 ppm D.M.; SO_4^{2-}, 0.15–0;20% D.M.).

[2]Significant difference (decrease or increase, $P<0.05$) from control denoted by [a b d i]; significant difference (increase, $P<0.05$) from SO_4^{2-} group denoted by [g]; significant difference (decrease or increase, $P<0.01$) from control group denoted by [c e f h].

(3) Thiocyanates increase liver copper reserves in lambs and sheep. The most considerable increase is achieved with daily doses of 600 mg SCN^-.

The results show that cyanides and thiocyanates can be included in the list of substances that may influence copper metabolism in sheep. Of particular interest are cyanides, which reduce copper reserves. This fact is of a more general interest, since, apart from *C. dactylon*, many species of *Trifolium* (e.g., *T. repens*), frequently found in abundance in pastures, are also cyanogenetic. In addition, fresh and juicy plants contain, in various quantities depending on the species, cyanides that perish during ripening and drying [Spais *et al.*, unpublished data].

On the basis of the above limiting action of cyanides on the copper reserves of sheep, one would expect that the action of thiocyanates would be similar, because it was found that thiocyanates, given *per os*, produce large amounts of cyanides in the rumen liquor and blood [Spais *et al.*, unpublished data]. Despite the above, it was found that thiocyanates facilitate the concentration of copper in the liver. This can be explained on the basis of the ability of cyanides to produce either soluble or insoluble compounds with copper, in proportion to the quantitative ratio between them.

DISCUSSION

Lewis (Weybridge). When you gave thiocyanate and cyanide, did you find any effect on the thyroids? Secondly, Mrs. Allcroft and I fed two groups of sheep, one on a cock's foot straw and another on kale, and the difference between the two in copper metabolism was from sulfate.

Spais. We measured thyroid weight but there was little effect. We know of your work, the results of which can be explained, in part, by our findings.

Hidiroglou (Ottawa). Could you elaborate on the mechanism of copper and cyanide antagonism?

Spais. Cyanide may bind to the copper and even take copper from the insoluble copper sulfide formed in the rumen through formation of cupric cyanide and then cuprous cyanide, which is very insoluble. With an excess of cyanide, a complex is formed which is very soluble. Therefore, with thiocyanates, we have extensive production of cyanide in the rumen and some quantity of copper is made soluble which may be responsible for the increased liver Cu we found with thiocyanate.

INDUCED COPPER DEFICIENCY IN BEEF CATTLE; THE EFFECTS OF EARLY AND LATE SUPPLEMENTATION ON COPPER STATUS AND ANIMAL PRODUCTION THROUGHOUT LIFE

D. B. R. POOLE, P. A. M. ROGERS, and D. D. MACCARTHY

Agricultural Institute, Dunsinea, Castleknock, Co. Dublin, Ireland

It is known that induced copper deficiency may cause reduced growth rate in suckled beef calves in Ireland under conditions of elevated pasture molybdenum. Copper-deficient cattle of over 1 year of age seldom show a response to supplementary copper, indicating a strong effect of age on the severity of the clinical syndrome. The present experiment was undertaken to study the effect on cattle of copper deficiency and copper therapy at different stages from age 5 months to slaughter at 2 years old.

Suckler calves, born in November 1969, were randomized into treatment and control groups at turnout to pasture in April 1970. The treatment group contained 26 calves and these received 100 mg copper edetate (Glaxo, Coprin); 28 calves constituted the control group. The pasture contained 10–13 ppm Cu, 6–8 ppm Mo, and 0.06% SO_4. At weaning in September, early and late supplementation groups were established. Half of the calves in the treatment group were treated in September and again on three occasions in 1971 (+ + group), treatment being withheld from the other half (+ − group); half of the control group received similar copper treatment (− + group), the other half remained as control (− − group). After weaning, the calves remained on the same pasture. From December to April they were wintered on silage (Mo, 1.2 ppm) and barley meal. From May to slaughter in October 1971 the cattle were grazed on high-Mo pastures, pasture means being 9.2 ppm Cu, 6.3 ppm Mo, and 0.08% SO_4. The + − group was removed in July 1971 to adjust stocking rate.

Blood and liver biopsy samples were collected periodically. Blood values were low in the dams of the calves (mean 0.03 mg %). In April 1970 blood levels in the calves were subnormal (mean 0.06 mg %). In September the control calves had low blood copper (0.02 mg %), low liver copper (5 ppm DM), and low liver cytochrome oxidase activity, 5 μm/min/g wet tissue. Blood levels in the treated calves were only marginally higher at 0.03 mg %, indicating rapid depletion of the injected copper over the period April to September.

During the winter some recovery in copper status occurred; however, after turnout to pasture in May 1971, blood and liver copper and liver cytochrome oxidase activity in the untreated groups fell quickly, while copper therapy maintained normal levels in the treated groups.

In Table 1 animal performance data are shown. The severe copper depletion from 5 to 10 months of age affected live-weight gains, the treated group gaining an extra 30 kg. From age 10 to 12 months the further copper treatment maintained the higher growth rates in previously supplemented calves, and induced a certain degree of compensatory growth in previously deficient calves. Therefore at 1 year of age, the difference between the continuously treated and the control groups was 54 kg, with the other two groups almost equal in an intermediate position. During the winter-feeding period no further effect of copper was seen, nor was there any response

Table 1. Effect of copper supplementation of cattle on live-weight increases, final live weights, and carcass components[1]

	++ Cu	+− Cu	−+ Cu	−− Cu	Significance[2]
Live-weight increase (kg)					
From 4/25/70					
to 9/9/70	130	128	102	102	**
to 12/10/70	163	139	134	109	***
to 5/4/71	223	204	204	177	***
to 7/20/71	299	278	274	247	**
to 10/20/71	328		321	286	*
Final live weight (kg)	489		465	434	*
Hot-carcass weight (kg)	280		264	237	**
Killing out (%)	57.2		56.7	54.7	*
Separable lean (kg)	167.8		161.6	148.6	n.s.
Separable fat (kg)	68.6		61.0	49.6	**
Bone (kg)	35.8		34.0	33.4	n.s.

[1]See text for details of Cu supplementation.
[2]Level of significance refers to an f test.

subsequently on high-Mo pastures. As a result the initial differences between the groups were maintained through to slaughter at 2 years of age.

Hot carcass weights were significantly affected by the copper treatment, both treatment groups being significantly heavier than the controls. Killing-out percentages were higher for the treatment groups. The right side of the carcass was dissected into closely fat trimmed, commercial boneless cuts. On this basis the difference in carcass weight is shown to be due in part to an increase in fat cover and in part to the better yield of separable lean meat.

It appears that a phase of clinical copper deficiency, from 5 months to 1 year of age, affected the growth pattern of the animals, resulting in lower yield of fat and muscle, but not of bone, at slaughter. It would appear that copper depletion after 1 year of age had no effect on performance. The results of this experiment go some way to explain aspects of bovine copper deficiency. They also indicate clearly the severe economic effects of early clinical copper deficiency in calves.

DISCUSSION

Mills (Aberdeen). Have you looked at connective tissue in the muscle, and what is the basis of the lesion? Perhaps connective tissue in muscle is not growing at an early stage while an animal is copper deficient?

Poole. We have not yet looked at connective tissue in muscle. However, we did measure strength of the metacarpal bone by a downward pressure on the center of the bone. The copper-deficient animals had significantly weaker bones in spite of the fact that the bone was not detectably different by weight.

Andrews (Upper Hutt). A feature of molybdenum-conditioned copper deficiency in New Zealand has been very excessive scouring in the springtime. Did you observe any scouring in your unsupplemented calves and, if so, for how long did it persist?

Poole. There was intermittent scouring at 8–10 months of age, during the late summer of the first year. In our experience, scouring is an intermittent and not a severe symptom unless molybdenum levels in the pasture are above 10 ppm.

Suttle (Edinburgh). Do you think the poor appearance of the supplemented calves was due to symptoms of copper deficiency that weren't responding to copper, or was it something else that looked like copper deficiency?

Poole. The symptoms I was referring to were those of the unsupplemented animals. I think it is quite possible that had we given copper at a year and a half of age, we might have improved the appearance of the animals. But, from the evidence on growth rates, I don't think you could expect to have improved their growth rate.

EFFECTS OF HIGH MOLYBDENUM INTAKE ON PROTEIN COMPONENTS AND HISTOCHEMICAL PROPERTIES OF WOOL

R. KAWASHIMA and N. ISHIDA

Department of Animal Science, Kyoto University, Kyoto, Japan

It has been known that a high dietary intake of molybdenum induces some abnormalities in wool, such as depressed crimp frequency, decrease in wool strength, abnormal elasticity, and loss of pigment. These wool abnormalities are quite similar to those found in copper-deficient sheep. While the role of molybdenum and copper in wool formation is unknown, Marston [Proceedings of Symposium on Fibrous Proteins, Society of Dyers and Colourists, Leeds (1946), p. 207] suggested that copper seems to play a role as a coenzyme in the oxidation of thiol groups on the process of keratinization of wool fiber.

The purpose of this experiment was to examine the effects of high molybdenum intake on the cortical segmentation and distributions of some enzymes at various levels of wool follicles and on the components of proteins in wool.

Each of four Japanese Coridail wethers was fed daily 200 g of hay and 700 g of mixed ration which consisted of alfalfa meal, wheat bran, and ground barley. Two sheep received 200 mg of molybdenum ammonate per day per head for eight months. Wool and skin samples were taken from the mid-shoulder of the sheep. Wool samples were embedded in gelatine and sectioned. Skin samples were fixed in ethanol-formalin solution, wax embedded, and sectioned. These sections of wool and skin were stained with 0.1% methylene blue in Tris buffer at pH 7.4 and studied under the microscope. To examine alkaline phosphatase and leucine aminopeptidase, fresh-frozen skin was sectioned, fixed in cold acetone, and then stained by Gomori's method [Microscopic Histochemistry, University of Chicago Press,

Chicago, Ill. (1952), p. 175] and by the method of Burstone *et al.* [J. Histochem. 4: 217 (1956)].

In order to examine the components of wool protein, proteins were extracted from wool samples by the method of O'Donnell *et al.* [Aust. J. Biol. Sci. 17: 973 (1964)]. Clean dried wool was reduced with mercaptoethanol in 8M urea-Tris buffer at pH 8.5. Thiol groups were carboxymethylated with iodoacetic acid at pH 10.5. The protein extracts were separated chromatographically on a DEAE-cellulose column in 8M urea-Tris buffer at pH 7.4. The pooled fractions from the chromatography were dialyzed against water, freeze dried, and subjected to disc electrophoresis, which was carried out at pH 8.0 on 7.5% polyacrylamide gel in Tris-diethylbarbituric acid buffer containing 8M urea.

In the cross sections of control wool the cortices showed that the bilateral structure consisted of ortho and para segments. Uptake of basic dye was found in the ortho cortex which was generally situated at the convex side of the crimp waves. The differential staining of the cortex started at the upper part of the keratogenous zone of the wool follicle.

In molybdenum-treated sheep wool crimp became less distinct. The cortical segmentations of this wool varied from the bilateral arrangement of the two segments to more irregular ones. These abnormal staining patterns were also observed in the cortex as well as in the Huxley layer of the wool follicle at the keratogenous zone.

In wool follicles from treated sheep alkaline phosphatase positive portions were found in the dermal papillae and blood capillaries around the follicle. Leucine aminopeptidase activities were observed in the outer root sheath, germinal matrix, and dermal papillae. Any differences in activity and localization of these enzymes were not observed between control and treated sheep.

In general, wool from molybdenum-treated sheep was more difficult to dissolve and to extract than that from the controls. Seven chromatographic fractions were separated on a DEAE-cellulose column. The number and position of the peaks were similar for control and treated wool, but the relative amounts of individual fractions were different. In the treated wool, the relative amount of fractions eluted at lower salt concentration decreased and those eluted at higher salt concentration increased.

The gel patterns obtained by disc electrophoresis of control and molybdenum-treated sheep showed some discrepancies, both in the number of bands and in their relative amounts.

Histological and chemical studies of wool from molybdenum-treated sheep showed that molybdenum administration induced irregularity in the cortical segmentation and alteration of the protein components in the wool extracts. Gillespie [Aust. J. Biol. Sci. 17: 282 (1964)] reported that copper-deficient wool shows a decrease in relative proportions of sulphur-rich components in wool extracts, although there is little difference in the starch-gel electrophoresis of protein fractions between normal and copper-deficient wool. The results of our experiment seem to be in partial agreement with Gillespie's results with the exception of gel-electrophoresis observations. This disagreement may be due to the different method used to induce copper-deficient status in sheep or to the different chemical procedures of fractionation.

It may be concluded that excess molybdenum feeding in sheep may induce some changes in protein synthesis and consequently may change the protein conformation in wool.

DISCUSSION

Unidentified. Did you study the copper status of these animals?

Kawashima. I used a normal diet with a copper content of about 10 ppm. In treated sheep the copper content of liver was 60 ppm compared with the normal control of 440 ppm.

EFFECTS OF LONG-TERM MAINTENANCE OF SHEEP ON A LOW-COBALT DIET AS ASSESSED BY CLINICAL CONDITION AND BIOCHEMICAL PARAMETERS

A. MacPHERSON and F. E. MOON

West of Scotland Agricultural College, Auchincruive, Ayr, U.K.

Down Cross wethers housed and fed on a low-cobalt ration exhibited clinical symptoms of deficiency within 4–6 months. Death occurred in three sheep, two of which had cerebrocortical necrosis and the third a fatty-liver degenerative condition. Abnormally low hemoglobin (4.9 g/100 ml), plasma glucose (15 mg/100 ml), plasma alkaline phosphatase (9.6 U/liter), and abnormally high GOT concentrations (2300 S.F./ml) were recorded in the deficient state. Vitamin-B_{12} treatment produced rapid responses while cobalt treatment was more slowly effective.

For further investigation of these effects 12 sheep were randomized into three groups and fed on a low-cobalt ration of hay, flaked maize, and maize gluten meal, providing 0.044 mg Co/day. The animals in each group were supplemented weekly as follows: nil; 100 mg vitamin B_1 + 100 mg vitamin C (subcutaneously); 0.7 mg Co (orally). Vitamin-B_1 supplementation was examined as rumen-bacterial synthesis of the B-complex vitamins would be adversely affected by B_{12} deficiency and as B_1 deficiency has been implicated in the occurrence of cerebrocortical necrosis. Vitamin C was included firstly to try and prevent the susceptibility of Co-deficient sheep to skin bruising, secondly to see if its reported effect of elevating plasma alkaline phosphatase in scorbutic animals would operate in Co-deficient animals, and thirdly to counteract the susceptibility of the deficient animals to infectious conditions.

The results are summarized in Table 1 and show the most abnormal level recorded for each of the parameters measured for the two Co-deficient groups during 14 weeks on the deficient ration and also the percentage recovery

Table 1. Effect of cobalt deficiency and its treatment in sheep[1]

Parameter measured	No treatment group: Max. or min. pretreatment level	No treatment group: Percentage recovery Percentage recovery within 4 weeks of B_{12} treatment	Treated with vitamins B_1 + C: Max. or min. Pretreatment level	Treated with vitamins B_1 + C: Percentage recovery Percentage recovery within 4 weeks of B_{12} treatment	Treated with Co: Normal level
Live weight (kg)	33.8 ± 3.8	34.0	34.6 ± 4.5	39.0	44.9 ± 1.10
Feed residues (g/sheep/day)	572 ± 29.7	77.0	530 ± 38.3	76.0	nil
Hb (g/100 ml)	8.9 ± 1.05	26.0	7.7 ± 0.50	24.0	11.2 ± 0.13
PCV (%)	26.6 ± 2.60	7.0	23.9 ± 2.99	19.0	34.2 ± 0.52
Glucose (mg %)	38 ± 6.19	100.0	26 ± 3.54	100.0	67 ± 0.93
Alk. Phos. (U/liter)	43.7 ± 6.64	100.0	18.1 ± 4.55	52.0	74.9 ± 2.0
Pyruvic acid (mg %)	1.69 ± 0.25	86.0	1.94 ± 0.26	94.0	0.83 ± 0.04
GOT (S.F./ml)	581 ± 195	57.0	395 ± 142	98.0	67 ± 2.10
Vitamin B_{12} (pg/ml)	108 ± 9.0	19.0	108 ± 6.0	33.0	556 ± 54.1
Ascorbic acid (mg/liter)	2.64 ± 0.30	17.0	0.91 ± 0.29	74.0	5.79 ± 0.60

[1]Sheep were fed a Co-deficient diet and supplemented with vitamin B_1 and C (see text) or with cobalt. After 14 weeks the sheep which were not receiving cobalt were given 500 μg of vitamin B_{12}. All measured values are mean ± s.d., and the percentage recovery as the result of vitamin B_{12} treatment is also indicated. The values for the Co-treated sheep constitute normal levels for the various parameters.

following subcutaneous injection of 500 μg vitamin B_{12}. The mean normal values for the Co-supplemented group are also recorded. The vitamin B_1 + C treatment delayed the rise in GOT concentration by some seven weeks compared with the nil treatment, but had no ameliorating effect on any other parameters; in fact most of these, but alkaline phosphatase in particular, were consistently more abnormal in this group. This effect was found to be due to the B_1, which tests showed to lower the alkaline phosphatase concentrations of two sheep from 90 to 50 U/liter while the levels for two untreated sheep remained between 85 and 95 U/liter.

Perhaps the most significant findings were the low plasma ascorbic acid concentrations, indicating a vitamin-C-deficient condition in both the Co-deficient groups. Injection of 100 mg ascorbic acid boosted plasma concentrations from 0.91 to 3.82 mg/liter within 4 hr in the B_1 + C group. Vitamin-B_{12} treatment appeared to be effective in slowly increasing plasma ascorbic acid concentrations towards normal.

Two sheep in the untreated group developed severe infections which, however, responded well to oxytetracycline. No such conditions arose in the other two groups. One sheep in the B_1 + C group died as a result of severe fatty infiltration of the liver.

In summary, the Co-deficiency state appears to involve a complexity of reactions in which the identification of primary and secondary factors is difficult. Whilst the hitherto unsuspected vitamin-C deficiency in the cobalt-deficient sheep may be the result of impaired synthesis following liver damage, further work is needed to clarify the rapidity with which the vitamin-C-deficient state develops and its significance to the animal.

DISCUSSION

Lewis (Weybridge). We have found thiaminase in the rumen and feces of animals affected with cerebrocortical necrosis. Did you measure thiaminase in any of these groups?

MacPherson. No, but I don't think that our two theories are necessarily in opposition to each other. Cobalt deficiency may produce conditions in the rumen in which the symbiotic bacteria do not produce sufficient thiamine or perhaps thiaminase could be produced.

Wiener (Edinburgh). Did you assess copper status in these animals, because recently we found that sheep which died from cerebrocortical necrosis had only half the brain copper concentrations of contemporary sheep which died for other reasons.

MacPherson. We measured blood copper, which was normal, but not brain copper. We also examined liver and kidney copper concentrations in sheep which died, and found in liver from 28 ppm copper to about 160 ppm which is within the normal range.

Wiener (Edinburgh). In our case, the liver copper concentrations were also normal; it was only brain copper that differed.

Kienholz (Fort Collins). A few years ago I reported that ascorbic acid would help prevent cobalt toxicity; you report here that by adding cobalt or B_{12} you increase ascorbic acid levels. Can you comment further on this relationship.

MacPherson. I suspect that ascorbic acid deficiency is arising from the fact that we are getting severe liver damage, and this is the site of synthesis of ascorbic acid. In only one case was there a low level of ascorbic acid but no indication of liver damage. It was reported in 1961 that in dogs with the cerebral cortex removed ascorbic acid metabolism was affected and the dogs could not be saturated with ascorbic acid. Possibly there is some effect on the cerebral cortex from cobalt deficiency that is affecting ascorbic acid levels.

Shirley (Gainesville). Have you given any consideration to propionic and methylmalonic acids?

MacPherson. Presumably the effect of cobalt deficiency on plasma glucose, as an example, is mediated via propionic acid because this is the main precursor of glucose in the ruminant. We haven't examined propionate levels in blood, but in two isolated cases we didn't find any significant difference in rumen fluid.

Andrews (Upper Hutt). A few years ago we reported seven cases of polyencephalomalacia, as we called it, among 19 of our cobalt-deficient sheep. The cobalt-supplemented sheep were unaffected. This was associated with a stress condition when we drove them.

MacPherson. The cerebrocortical necrosis did not appear to be associated with abnormal stress in our animals. We normally bled them weekly or fortnightly. This was the only stress they had apart from somebody bringing feed to them.

ZINC DEFICIENCY IN CATTLE UNDER GREEK CONDITIONS

A. G. SPAIS and A. A. PAPASTERIADIS

Department of Medicine, Veterinary Faculty, Aristotelian University of Thessaloniki, Greece

Zinc deficiency in cattle, and especially in calves, under experimental conditions and on semisynthetic or purified diets has been produced by many workers [Miller *et al.*, J. Dairy Sci. 43: 1854 (1960); Ott *et al.*, J. Anim. Sci. 24: 735 (1965); Mills *et al.*, Br. J. Nutr. 21: 751 (1967)]. However, as shown from the available literature, only a few cases under natural conditions have been reported so far [Leag *et al.*, Nature, Lond. 186: 1061 (1960); Haaranen *et al.*, Feedstuffs 33: 28 (1961); Dynna *et al.*, Acta Vet. Scand. 4: 197 (1963)].

In Greece zinc deficiency under natural conditions was first diagnosed by clinical and laboratory examinations and by response to therapy in spring 1967 [Papasteriadis, Sci. Yearbk. Vet. Fac. 14: 169 (1973)]. The research undertaken was soon extended practically throughout the country. For this purpose (a) 150 herds were clinically examined and the zinc content in 1000 representative blood plasma samples was determined, (b) 1500 feed samples coming from the above and various other herds were also examined for zinc content, and (c) the response to zinc administration in 24 feeder bullocks was studied.

The clinical examination showed that (a) approximately 4% of the animals had evident and severe symptoms of the deficiency consisting mainly of unthrifty appearance, rough coat, loss of hair, and eczematous lesions with thickening and folding in various parts of the skin, (b) 56% showed similar signs but of milder intensity, and (c) only 40% were apparently healthy. The deficiency was observed in cattle of all ages. Cases seem to be more frequent late in winter and early in spring.

The results of zinc determination in bovine plasma are shown in Fig. 1: (a) 1.5% had a concentration of Zn in their plasma between 40 and 60 μg/100

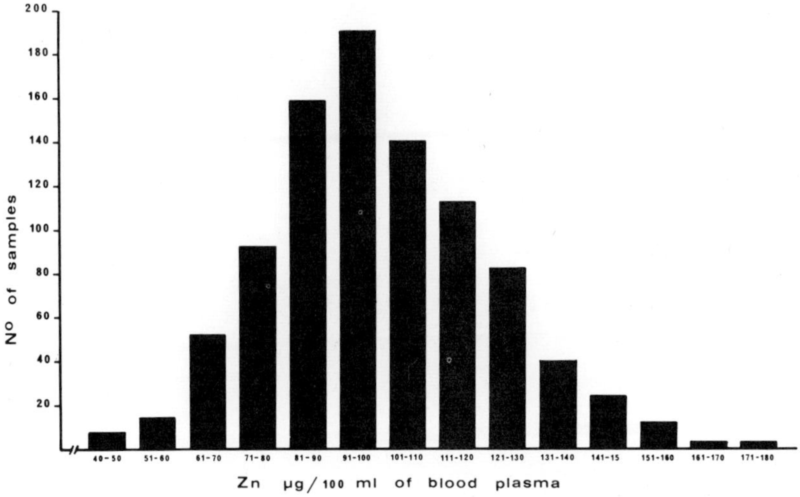

Fig. 1. Distribution of plasma zinc levels in the cattle studied.

ml; these levels were observed in the seriously affected animals with characteristic lesions; (b) 52.3% had values of 60–100 μg/100 ml, corresponding usually to animals with slight signs; (c) 46.2% had values of more than 100 μg/100 ml, which can, in accordance with the data of most workers, be considered as normal. In the last group most animals showed no symptoms at all; however, the presence of after effects from a previous zinc deficiency could not be excluded.

The zinc content in the 1500 samples of the cattle feeds ranged between 20 and 40 ppm. Some grass and alfalfa samples from plants growing in sandy soils of Northwestern Macedonia had an even lower zinc content (5 ppm). In herds where symptoms of zinc deficiency were observed together with low concentration of plasma zinc, the animals were fed on rations of average zinc content ranging usually between 20 and 30 ppm.

It is not possible to confirm whether the zinc deficiencies observed were simple, because the rations of the animals usually contain more than 20 ppm Zn. According to Miller *et al.* [J. Dairy Sci. 46: 715 (1963)] and Mills *et al.* [Br. J. Nutr. 21: 751 (1967)], these concentrations of zinc meet the requirements of the animals under experimental conditions. However, our data and those of other workers show that zinc deficiencies, under natural conditions, can be observed even with zinc concentrations of up to 40 ppm. The influencing factors which cause conditioned deficiencies in cattle seem to be still unknown.

Supplementation of feeds given to bullocks, which had mean zinc plasma levels below 100 μg/100 ml and mean body weight 200 ± 3.2 kg, with

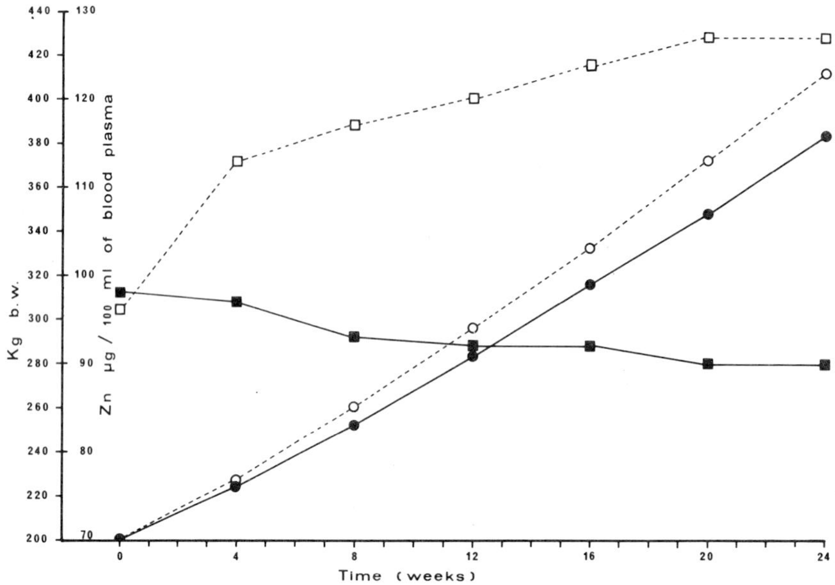

Fig. 2. Changes in group mean plasma zinc concentration and live weight of feeder calves. Zinc levels: ■——■ controls; □——□ supplemented. Live weight: ●——● controls; ○——○ supplemented.

50 ppm Zn resulted in an increase in body weight amounting to 7%, as compared with controls fed with a basal ration containing 32 ppm Zn. Animals that showed an increase in their production had final mean levels of zinc in their plasma of 126 ± 3.8 μg/100 ml and mean body weight 413 ± 8.2 kg, while that of the controls had 90 ± 3.1 μg/100 ml of zinc in their plasma and 384 ± 5.5 kg body weight (significant difference from experimental group, $P < 0.001$). Further details on the results are given in Fig. 2. From these results one can conclude that values below 100 μg/100 ml in the blood plasma can be considered as subnormal.

According to the results of our survey and experimental work we can consider that zinc deficiency in cattle, despite the limited number of cases described so far in other countries, is important under Greek conditions and more extensive that one could have anticipated.

DISCUSSION

Lee (Adelaide). What would you regard as a normal zinc concentration in Greek pastures? If cattle will show signs of zinc deficiency when consuming pastures containing 40 ppm, I am concerned because in Australia we have 250 million sheep and 20 million cattle which never receive that amount of zinc; it's usually half that.

Spais. Our animals are grazing dry areas with rather sparse grass, so we supplement with concentrates which may lower the overall dietary zinc concentration. But I emphasize that there were no zinc-deficiency signs in animals with 40 ppm zinc in the grass; signs appeared between 20 and 30 ppm zinc in the grass.

Mills (Aberdeen). In the first reported cases of zinc deficiency in ruminants in Guinea, the zinc content of the available forage was between about 18 and 32 ppm. However, we still do not understand the factors influencing zinc availability in a ruminant animal other than saying it is not phytate. Recently we have been able to improve wound healing in bulls, but not steers or heifers, by increasing dietary zinc from 50 ppm up to 100 ppm. Therefore sex may be influencing zinc availability in the ruminant; however, plasma zinc concentration was not particularly low.

Hartmans (Wageningen). In the Netherlands we did not get any response to zinc at any level in plasma higher than 60 μg/100 ml, and I was surprised by seeing that in these cases there was a response.

Mills (Aberdeen). We regard plasma zinc levels of less than 40 μg/100 ml as a risk in ruminants fed semisynthetic diets, but in the light of the recent work with practical diets, we may have to revise that upward. In addition, plasma zinc may be a poor index of overall zinc status.

Hennig (Jena). Did you measure zinc content of the hair? In our work the mean zinc content of cow's hair was 115 ppm, and in cows with less than 100 ppm the skin was crusty. We think that extensive feeding of cabbage and rape seed meal decreased the zinc status of the animal. Do you feed rape and similar crops?

Spais. Yes, we feed cabbage to some extent. We began to estimate the hair zinc content but it responded very slowly compared with the plasma zinc level and was a poor index so we abandoned it.

CHANGES IN CIRCULATING COPPER, MANGANESE, AND ZINC WITH THE ONSET OF LAY IN THE PULLET

R. HILL

The Royal Veterinary College (University of London), Boltons Park, Potters Bar, Hertfordshire, England

Manganese is present in very low concentration in blood and is difficult to determine, even now, in small volumes of blood or plasma. On the other hand, the uptake of ^{54}Mn by red cells *in vitro* is very readily determined and may, it was thought, provide data related to total Mn in plasma or red cells. Experiments with pullets given a very low- or a high-Mn diet showed that the uptake of ^{54}Mn by red cells was significantly greater for birds given the low-Mn diet than for those given the high-Mn diet. However, the data were complicated by other differences associated with sexual maturation. These were investigated more closely, and at the same time determinations of total plasma Mn, Cu, and Zn were made, as well as of ^{65}Zn uptake by red cells. Earlier work had shown there were changes in the Mn content of blood [Bolton, Br. J. Nutr. 9: 170 (1955)] and the Fe content of plasma [Panić, *In* Mills (ed.), Trace Element Metabolism in Animals, Livingstone, Edinburgh (1970), p. 324] with sexual maturation.

Twelve light hybrid pullets were reared on a commercial-type grower diet, then changed at 18 weeks of age to a layer diet. Both diets contained supplements of Mn and Zn as well as an appropriate group of vitamins. Until the birds were 19 weeks of age, the first blood-sampling day, light was restricted to 10 hr in every 24 hr, and at this age the pullets were unmistakably immature in appearance. Day length was increased to 18 hr in two steps during the following week, and all of the birds except one came into lay when between 25 and 28 weeks of age. This other bird laid its first egg at 29 weeks of age.

Blood samples were taken from these twelve pullets and from four pullets of the same hybrid that had been in lay several months; the samples were taken when the young pullets were 19, 22, 25, 28, and 32 weeks of age. Plasma from four birds was pooled to provide a sufficiently large sample to determine total plasma Cu, Mn, and Zn, and red-cell uptake of ^{54}Mn and ^{65}Zn *in vitro*. The uptake determinations were made with red cells in plasma and in saline.

Plasma Cu values were 11–15 μg/100 ml in 19-week-old immature pullets and increased during the following six weeks to 25–31 μg/100 ml. No further increase occurred when egg laying began just after 25 weeks of age. Values for the mature layers were 22–34 μg/100 ml, similar to those of the young pullets from 25 weeks of age.

The Mn content of plasma increased quite markedly from 3.0 to 4.8 μg/100 ml when the birds were 19 weeks of age to 8.5–9.1 μg/100 ml at 25 weeks of age. There was a tendency for the content to fall slightly during the following weeks when eggs were being laid, but the data were too few to show if this was a real trend. Values for mature laying pullets were at all times greater than those for young birds, even when both groups were in lay.

The *in vitro* uptake of ^{54}Mn by red cells changed inversely with plasma Mn; values of 13–16% at 19 weeks fell to 7–9% at 25 weeks and, unlike plasma Mn, the percentage uptake by red cells continued to fall to 3–4% at 32 weeks of age. In the mature layers uptake was 3–4% throughout the 12 weeks in which observations were made. The inverse relationship between plasma Mn and red-cell uptake of ^{54}Mn may, it was considered, result from the production during sexual maturation of a plasma protein that binds Mn, in which case the uptake of ^{54}Mn by red cells when plasma was replaced by saline would not change during this period of observation. This proposition was not sustained by the results; uptake of ^{54}Mn by red cells in saline was greater in plasma but in both media marked decreases occurred near the point of lay and they were almost parallel. The decrease in uptake of ^{54}Mn was therefore clearly a property of the red cells and not simply an indirect effect from plasma proteins.

Plasma Zn concentration was more variable than that of either Cu or Mn and no clear pattern of change was detected. The *in vitro* uptake of ^{65}Zn by red cells from whole blood was considerably lower, about one-third, than the corresponding uptake of ^{54}Mn, but, as with ^{54}Mn, a decrease in uptake occurred just before and during the start of egg laying. The decrease for ^{65}Zn was smaller than for ^{54}Mn but was well defined and statistically significant. When plasma was replaced by saline, the uptake of ^{65}Zn by red cells again decreased with advancing maturity.

Numerous and very large changes in the composition of plasma of pullets as they come to the point of lay have been demonstrated over a long period of study [Bell and Freeman, Physiology and Biochemistry of the Domestic

Fowl, Vol. 3, Academic Press, New York (1971)], and they reflect enormous changes in the metabolism of birds associated with the production of well-shelled eggs. These clearly include certain aspects of trace-element transport and metabolism. Panić, at the first of these symposia, described the appearance at the start of egg-laying of a very strong iron-binding protein in plasma, and of a corresponding large increase in plasma Fe concentration. The changes in plasma proteins that occur at this time are numerous and complex, and no doubt some changes in quality or quantity of plasma protein are associated with the increases in Cu and Mn described above. Much less attention has been given to the metabolism and composition of red cells than of plasma at the time of sexual maturation, but the results given here, for uptake *in vitro* of ^{54}Mn and ^{65}Zn, suggest that quite marked changes occur in red cells that could well be studied more closely.

DISCUSSION

Hidiroglou (Ottawa). How did you determine manganese?

Hill. By atomic absorption after extraction with a chelating agent into acetylbutylketone.

Weber (Tucson). Did you study blood calcium levels in relationship to the trace minerals?

Hill. No.

Panić (Zemun). We got very similar results, but your explanation that the increased manganese is connected with shell formation seems unlikely because manganese goes mainly to the egg yolk, not the shell. We suppose that this increased manganese is bound to some phosphoproteins. Also, we got no effect on zinc.

Hill. If the manganese is associated with a phosphoprotein, it isn't bound as tightly as iron, because our chelation method removed manganese and zinc but not iron. We checked this with radioactive manganese and zinc. Relative to the low manganese values in the shell of the egg, manganese is important for the formation of shell and the quantities in the shell gland might be quite comparable or larger perhaps than those of the egg.

Lassiter (Athens). Do you think the increased blood manganese comes from increased absorption, or do you think it is shifting from some other tissue?

Hill. It must be an increased absorption, although we haven't carried out experiments to demonstrate that this is so.

SOME CHARACTERISTICS OF TRACE-ELEMENT METABOLISM IN POULTRY

B. PANIĆ, LJ. BEZBRADICA, N. NEDELJKOV, and A. G. ISTWANI

Institute for the Application of Nuclear Energy in Agriculture, Veterinary Medicine and Forestry, Zemun, Baranjska 15, Yugoslavia, and Department of Animal Production, Agricultural Faculty, University of Damascus, Syria

Plasma iron concentrations are several times higher in laying hens than in immature pullets and mammals [Panić, *In* Mills (ed.), Trace Element Metabolism in Animals, Livingstone, Edinburgh (1969), p. 324]. Much of the plasma iron of laying hens is firmly bound to phosvitin. This investigation is concerned with a study of manganese, zinc, and copper in poultry with special emphasis on factors which influence their metabolism in laying hens.

The radioisotopes ^{54}Mn, ^{65}Zn, and ^{64}Cu were used in experiments on chicks, immature and mature hens, and cockerels. Tissue zinc and copper concentrations were determined by atomic absorption spectroscopy (Unicam SP-90). The results show that the metabolism of these elements in poultry is affected by the estrogenic hormones secreted by the ovary.

It was found that the level of radioactive manganese in the blood plasma of poultry did not depend on the amount of its retention. The retention of ^{54}Mn 4 hr after intramuscular injection was similar in laying hens and immature pullets (Table 1). However, plasma radioactivity at the same time was about 15 and 70 times higher in laying hens and estrogen-treated immature pullets, respectively, than in control immature pullets. Twenty-four hours after i.m. administration of $^{54}MnCl_2$ the amount of ^{54}Mn retained (measured by a whole-body counter) was considerably higher in the control immature pullets (55.96%) than in the laying hens (45.43%) and the estrogen-treated immature pullets (45.30%). The level of the total radioactive manganese in the liver was also found to be highest in the control pullets (14.47% dose), then in estrogen-treated ones (13.71%), and lowest in laying hens (10.55%). However, the level of radioactive manganese in the blood plasma of control immature pullets was, over the same period, much lower

Table 1. Retention and level of ^{54}Mn in plasma and liver of laying hens and immature pullets after i.m. injection of $^{54}MnCl_2$ [1]

	Hr after dosing	Laying hens	Immature pullets: Control	Immature pullets: Estradiol treated
^{54}Mn retained (% dose)	4(2)	92.19	94.09	93.50
	24(5)	45.43	55.96	45.30
	120(5)	24.05	38.67	26.37
^{54}Mn in plasma (% dose/ml X 10^3)	4	9.65	0.63	46.80
	24	6.60	0.32	27.76
	120	1.69	0.034	11.42
Phosphoprotein P in plasma (μg/g)	4	52.30	0.48	278.00
	24	61.19	0.14	260.80
	120	98.15	0.35	154.56
Plasma proteins (g%)	24	4.19	3.98	6.63
	5	4.83	4.53	6.25
^{54}Mn in liver (% dose per whole organ)	4	28.54	21.83	26.40
	24	10.55	14.46	13.71
	120	3.05	8.66	6.47

[1]Values are means for the number of birds shown in parentheses.

than in estrogen-treated pullets (about 90 times), and laying hens (about 20 times). The pattern was similar five days after injection of $^{54}MnCl_2$ (Table 1). The lowest level of ^{54}Mn in blood plasma was found in control pullets.

The quantity of phosphoprotein phosphorus in the plasma of control pullets was very low, whereas in laying hens and especially in the estrogen-treated pullets high concentrations were found (Table 1).

Investigations of zinc content in the blood plasma have shown that plasma zinc concentrations in laying hens are about two to three times higher than in immature chicks of both sexes. Moreover, it was found that in laying hens with normal ovulation and oviposition the plasma zinc content was two to three times higher than in nonlaying hens of the same age which had ovaries with undeveloped follicles. However, in nonlaying hens with ovaries containing many follicles at different stages of development (but no oviposition) the plasma zinc concentration was as high as in normal laying hens.

The injection of estradiol to immature pullets increased the level of radioactive and stable zinc in blood plasma significantly [Panić *et al.*, J. Sci. Agric. Res. Beograd. 24: 70 (1971)]. These results show that the plasma zinc content in sexually mature hens depends on the functional (and estrogenic) activity of the ovary.

The concentration of stable and radioactive zinc in liver was found to be much higher in nonlaying hens (with undeveloped follicles in the ovary) than in normal laying hens. This can be explained by the mobilization of zinc from the liver and its deposition in ovocytes and egg yolks in laying hens. After oral administration of ^{65}Zn, it was shown that this element, like iron and manganese, is deposited almost exclusively in the egg yolk, while egg white and shell contained insignificant amounts. The level of ^{65}Zn in egg yolks and ovocytes was positively correlated with the level of ^{65}Zn in the blood plasma.

The concentration of copper in the blood plasma of poultry is several times lower than in mammals. In two experiments with three groups of five birds, it was found that the mean plasma Cu concentration in immature pullets was 16.75 and 18.44 μg/100 ml, in laying hens 23.35 and 27.43 μg/100 ml, and in immature pullets treated with estradiol 36.30 and 40.81 μg/100 ml. After injecting radioactive copper intramuscularly, the highest plasma activity was found in estradiol-treated immature pullets (4.43% dose), then in laying hens (2.05%), and the lowest activity in control immature pullets (1.95%). The level of radioactive copper in the liver was lowest in laying hens and highest in control immature pullets, as was found for zinc.

The copper concentration in the plasma of five 3-month-old birds varied from 11.00 to 15.62 μg/100 ml (mean 13.19 μg/100 ml); at the onset of laying of the same birds it considerably increased and amounted to 21.87–29.63 μg/100 ml (mean 25.53 μg/100 ml). After i.m. injection of ^{64}Cu into laying hens the level of ^{64}Cu in egg yolks was positively correlated with the level of ^{64}Cu in the blood plasma. Contrary to the other trace elements investigated, copper was also deposited in a considerable quantity in the egg white. Copper concentration in the egg yolk in this experiment was about two-thirds and in egg white about one-third of the total copper concentration found in the whole egg.

The results obtained suggest that the transfer of all elements from the blood into the egg yolk is facilitated by plasma phosphoproteins.

THE ROLE OF THE ABOMASUM IN RECYCLING OF IODINE IN THE BOVINE

J. K. MILLER, E. W. SWANSON, G. E. SPALDING, W. A. LYKE, and R. F. HALL

University of Tennessee Agricultural Research Laboratory, Oak Ridge, Tennessee, and Department of Animal Science, Knoxville, Tennessee

The abomasum is a major site for re-entry of circulating iodine into the bovine digestive tract [J. Dairy Sci. 47: 539 (1964)], with gastric concentration of iodide from plasma exceeding that of chloride in calves [Proc. Soc. Exp. Biol. Med. 121: 291 (1966)]. This investigation was designed to quantitate abomasal recycling of iodine.

Ten cows were surgically fitted with 2 cm inside diameter polyvinyl cannulae in the abomasum and duodenum which were separated by ligatures. The cannulae were connected with a metal tube until 22 hr later when they were stoppered and each cow was dosed intravenously with Na ^{131}I. Abomasal contents were allowed to flow into containers which were changed every 10 min during the first hour and at 30-min intervals until 6 hr after dosing. Jugular-vein blood was sampled when the container was changed, and excreta were collected when voided. Cows were slaughtered immediately after the 6-hr sampling. Thyroid glands were removed and digestive-tract contents, bile, and urine were weighed and sampled for ^{131}I measurement.

Abomasal flow decreased from an initial 5.1 kg/hr to 1.6 kg/hr during the last hour. Percentages of administered ^{131}I per kilogram of abomasal flow averaged 1.0 at 10 min, peaked at 6.7 at 2 hr, and decreased to 3.2 by 6 hr. The ratio of abomasal to plasma radioiodine concentrations increased from 1.0 at 10 min to 35.3 at 6 hr. Total percentages of the ^{131}I dose recovered during 6 hr averaged the following: thyroid, 1.2%; rumen, 2.9%; omasum, 0.5%; abomasum (drainage plus contents at slaughter), 65.2%; small intestine, 0.1%; large intestine, 0.3%; bile, 0.09%; feces, 0.05%; urine, 5.4%.

Cumulative recovery of the ^{131}I dose in abomasal drainage is shown in Fig. 1. These data were fitted by a linear compartmental model (Fig. 2) in

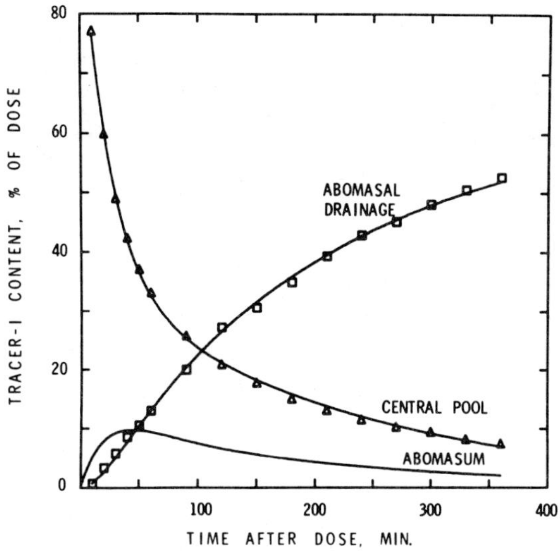

Fig. 1. Disappearance of ^{131}I from blood plasma of cattle fitted with abbomasal cannulae and its recovery in abomasal drainage.

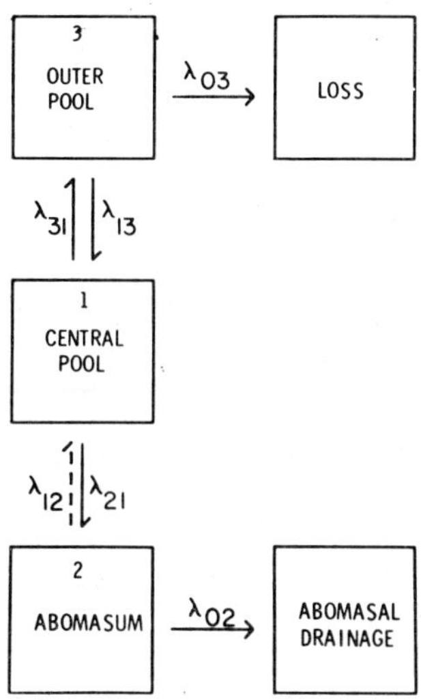

Fig. 2. Model describing ^{131}I movement between compartments in cattle fitted with abomasal cannulae.

which the central pool represents blood plasma plus other body fluids which quickly equilibrate with plasma. Turnover rates describing movement of ^{131}I between compartments with their standard errors (min^{-1}) were as follows: λ_{21}, 0.00688 ± 0.00007; λ_{02}, 0.028 ± 0.001; λ_{31}, 0.022 ± 0.001; λ_{13}, 0.0146 ± 0.0009; λ_{03}, 0.0031 ± 0.0003. Data obtained in this experiment were inadequate to resolve λ_{12}, but results of other studies [J. Dairy Sci. 54: 397 (1971)] indicate it is very small relative to λ_{21}. Curves were fitted assuming that $\lambda_{12} = 0$.

Flow from the ligated abomasum did not enter the small intestine where reabsorption would normally have occurred [J. Dairy Sci. 47: 539 (1964)]. Thus, blood-plasma ^{131}I concentrations during the last 3 hr dropped 19%/hr. Previously it had been observed that ^{131}I concentrations in blood of three intact cows dropped at the rate of 5%/hr. The loss of iodine in urine is largely dependent on the level of plasma iodide [J. Dairy Sci. 56: 378 (1973)]. The iodine-concentrating action of the abomasum may promote conservation of iodine by creating an extravascular iodine pool, thus preventing its excessive loss in urine. [Published with the permission of the Dean of the University of Tennessee Agricultural Experiment Station, Knoxville; the Oak Ridge laboratory is operated by the Experiment Station for the U.S.A.E.C. under contract No. AT-40-1-Gen-242.]

DISCUSSION

Hoekstra (Madison). Dr. Moir from Perth, Australia, recently described to me his recent work on ruminants in which he shows a very profound cycling of chloride, being secreted into the omasum, then passing to the abomasum, and finally being reabsorbed. He thinks chloride has a role in the absorption of the volatile fatty acids produced in the rumen. Perhaps your observations on iodide might simply reflect this huge flux of chloride.

Miller. Work in other species has suggested that iodide enters the abomasum or the gastric stomach by a different mechanism than chloride, but it could also include a reflection of that.

Binnerts (Wageningen). Can you remark on monogastric animals?

Miller. I have seen publications on secretion of iodide in the gastric stomach of frogs, dogs, humans, and, I believe, other species. We wondered if HI had anything to do with digestion, but the amount of iodine in the body is so small that this is doubtful.

THE INFLUENCE OF IRON, COPPER, ZINC AND EDTA ON FOWL TYPHOID IN CHICKS

R. HILL and I. M. SMITH

The Royal Veterinary College (University of London), Boltons Park, Potters Bar, Hertfordshire, England

A number of studies have shown the close involvement of Fe in certain diseases of bacterial origin [Glynn, *In* Smith and Pearce (eds), Microbial Pathogenicity in Man and Animals, Symposium 22, Society of General Microbiology (1972), pp. 75–112]; increasing the amount of available Fe has increased the severity of these diseases, and a single study suggests that it may be possible to decrease their severity by decreasing the amount of available Fe using strong Fe chelators. From this work it appears that a major factor restricting bacterial growth in animal tissues is the availability of Fe. As part of a project on the influence of the diet on fowl typhoid in chicks, observations on the effects of Fe, as well as Cu, Zn, and ethylenediaminetetraacetic acid (EDTA), were made.

The experimental details were similar to those described earlier [Hill and Smith, J. Comp. Path., 79: 469 (1969)]; the diets described here were all given from 12 days of age, which was three days before oral inoculation with *Salmonella gallinarum.* The diets to which Fe, Cu, Zn, or EDTA were added were designated 26 CG, a diet of semipurified nutrients containing casein and gelatin, and MPD, a commercial-type diet with several protein sources. Birds that were gaining weight at the end of the experiment, 18 days after inoculation, were recorded as survivors.

The addition of 113 mg Fe/kg to diet 26 CG (57 mg Fe/kg) had no effect on survival, but omitting Fe to provide a diet containing about 7 mg Fe/kg (26 CG-Fe) decreased survival significantly from 20 to 5%. This low-Fe diet given from 12 days of age did not cause anemia in control birds during the course of the experiment. The addition of 200 mg Fe/kg to diet MPD increased survival slightly, from 21% with MPD to 29% with MPD + 200 mg Fe/kg. This increase was not significant statistically.

These results implied that Fe may, in certain circumstances, not have the effect on the outcome of bacterial disease suggested in the review of Glynn [*loc cit.*], and this was confirmed using injections of Fe; the survival of birds inoculated orally with *S. gallinarum* was significantly increased by Fe injected intraperitoneally or intramuscularly. This result in birds, contrary to most of the evidence for rats and mice discussed by Glynn, would occur if, in the chick, anemia were an earlier and more critical aspect of fowl typhoid than of the bacterial diseases studied in rats and mice, and the additional Fe allowed the bird to maintain an adequate blood hemoglobin concentration, thereby increasing its chances of survival.

Ethylenediaminetetraacetic acid, though not a specific chelator of Fe, decreases the absorption and retention of Fe [Suso and Edwards, Poult. Sci. 47, 1417 (1968)] and was added to four diets to observe its effect on survival of chicks inoculated orally with *S. gallinarum.* With each diet EDTA (the disodium salt added at 0.1% of the diet) increased survival, and the overall effect, while modest in magnitude (–EDTA 15.6%; + EDTA 28.0%), was highly significant statistically. As increasing the amount of available Fe increased survival, it seems unlikely that the beneficial effect of EDTA can be explained by its effect on Fe.

The distribution of Cu and Zn in the body is modified in a number of diseases [McCall *et al.*, Fed. Proc. 30, 1011 (1971)], and in birds with fowl typhoid plasma Cu increased markedly and Zn decreased slightly; plasma Fe also decreased. The effects of these elements when added individually or with EDTA were determined on the survival of chicks inoculated orally with *S. gallinarum*; the elements were added at 200 mg/kg and EDTA at 0.1%, as the disodium salt (Table 1). Copper alone markedly and significantly increased survival, Zn had hardly any effect, and Fe increased survival slightly. The inclusion of EDTA invariably increased survival and the overall effect was

Table 1. The influence on the survival of chicks inoculated orally with *S. gallinarum* of adding Cu, Fe, or Zn either alone or with EDTA to a commercial-type chick diet

	− EDTA		+ EDTA	
Diet	No. survived / No. inoculated	Survival (%)	No. survived / No. inoculated	Survival (%)
MPD	14/101	14	29/102	28
MPD + Cu	37/101	37	46/102	45
MPD + Fe	27/104	26	52/104	50
MPD + Zn	18/103	17	44/103	43

highly significant, but there were interactions with the elements, particularly Cu and Zn. The Cu effect was only marginally increased by EDTA, while the very small effect of Zn alone was increased markedly by EDTA. The survivors of infected birds given Cu, with or without EDTA, were appreciably heavier at the end of the experiment than the survivors of other treatments. This may mean that the effect of Cu occurred in the digestive tract at the time of dosing, and that in a fairly high proportion of the birds very few or no organisms passed across the wall of the digestive tract.

In the experiments described here EDTA consistently increased the survival of chicks inoculated orally with *S. gallinarum,* as did Cu, while Fe and Zn increased survival when EDTA was present. [Supported by a grant from the Wellcome Foundation.]

DISCUSSION

Peters (Madison). I think that one should take into account the action of the endotoxin of Salmonella. If you inject the toxin of gas bacillus into rats, it will kill them, but if you inject the endotoxin along with EDTA, it doesn't hurt them at all. If you then add an excess of zinc or copper or some other trace mineral, the endotoxin will again cause death. I suggest that EDTA may inactivate the endotoxin of Salmonella.

Hill. The endotoxin of this Salmonella is not very toxic to the chick. In addition to what I described, we have two experiments involving selenium, but we didn't get any improvement in survival, probably because we used only 0.1 ppm instead of 3 or 4 ppm as suggested by Dr. Spallholz. However, this disease is an acute condition and the antibodies are not developed until after most of the birds have died, so we may not be able to show a protection by Se.

Beisel (Frederick). I agree that armchair predictions about what a bacterium is going to do or what its toxin is going to do can never be made from studies in the test tube alone, but have to be done in the infectious situation in which the whole animal is involved. Is this an intracellular infection when it does produce death in the birds, for most of the typhoid infections of man are considered to be intracellular rather than extracellular? This must be considered in evaluating whether the metal in plasma is influencing the organism or whether it's an intracellular metal-organism interaction.

Hill. I can't answer your question.

EXCRETION OF TRACE ELEMENTS IN BILE

WILLIAM H. STRAIN, WILLIAM L. MACON, WALTER J. PORIES, CARLOS PERIM, FRANCES D. ADAMS, and ORVILLE A. HILL, JR.

Department of Surgery, Cleveland Metropolitan General Hospital, Case Western Reserve University School of Medicine, Cleveland, Ohio

The recent upsurge in interest in gallbladder disease has prompted us to investigate the excretion of trace elements in bile. Gallbladder disease has a prevalence rate of about 10% and gallbladder surgery is probably the most common surgical problem throughout the world. Gallstone formation is associated with biliary infection and stasis, great variations in chemical composition, geographic variations, and racial and sex differences. The variations in chemical composition of 481 gallstones from 11 countries [Suter and Wooley, Gut 14: 215 (1973)] indicate that 59.9% are cholesterol stones, 13.1% are calcium salts, and 27.0% are mixtures of cholesterol and calcium salts. Compared with Caucasians, stone formation seems to occur with increased frequency in the American Indian and with decreased incidence in Asiatics. In most societies cholesterol stones are formed more frequently in females than in males. Our present report is exploratory in nature and consists of data obtained by analysis of gallbladder bile removed during 50 consecutive autopsies of individuals free of gallstones.

Gallbladder-bile specimens were collected from 50 consecutive autopsies free of gallstones and analyzed by atomic absorption spectroscopy for both major and minor elements. Disposable plastic syringes equipped with stainless-steel needles were used to collect the bile prior to opening the gallbladder. The bile specimens were transferred to plastic polyvials, the plastic cover closed, and the specimen stored at 0°C. Prior to analysis, each vial was warmed to room temperature and appropriate dilutions made with distilled water free of trace elements. The analyses were carried out with the aid of an automated 403 Perkin-Elmer atomic absorption spectrometer. Determinations were made of calcium, magnesium, sodium, and potassium among the major elements, and of cadmium, copper, iron, lead, manganese, and zinc

Table 1. Mean levels of major and minor elements of gallbladder bile from 50 consecutive autopsies

Element	Mean (ppm)	S.D. (ppm)	S.E. (ppm)
Major elements			
Calcium	25,190	± 15,862	1896
Magnesium	7859	± 2778	332
Potassium	97,930	± 23,159	2788
Sodium	395,536	± 113,473	13,661
Minor elements			
Cadmium	2.2	± 0.5	0.1
Copper	147.7	± 141.6	16.9
Iron	499.0	± 566.4	67.7
Lead	21.3	± 8.5	1.1
Manganese	7.3	± 3.3	0.4
Zinc	170.8	± 105.9	12.4

among the minor elements. The data obtained are presented in Table 1 as mean values, standard deviations, and standard errors, all expressed in parts per million (ppm).

The analytical results of Table 1 are consistent with previously reported data on the excretion of major and minor elements in gallbladder bile. The mean values, standard deviations, and standard errors show that the composition of gallbladder bile is extremely variable. Similar variations are evident in the most recent data on the determinations of the major elements by Crawford [J. Med. Lab. Tech. 13: 304 (1955)] who has included a good review of the published values in the literature. Underwood [Trace Elements in Human and Animal Nutrition, 3rd edn, Academic Press, New York (1971)] has summarized the information on trace elements in gallbladder bile. Publications since the appearance of his book seem to focus largely on organic moieties which favor the excretion of the various trace elements.

The great variability of the values suggests that numerous factors govern the excretion of the mineral elements in the bile. Accordingly, it seems probable that study of the associations between gallbladder-bile composition and the causes of death may be profitable. In addition, comparisons of bile composition with and without gallstones may reveal factors that favor stone formation. Finally, study should be made of the excretion of toxic elements, such as cadmium, lead, and manganese, which are present in the blood serum in such low concentration and are excreted in the gallbladder bile at relatively high levels of 2.2, 21.3, and 7.3 ppm, respectively.

DISCUSSION

Klevay (Grand Forks). Have you data on the prevalence of stones in these subjects?

Strain. They are all stone-free. We plan to look at that next.

Armstrong (Minneapolis). The concentration of these components would be expected to vary with degree of concentration of the bile.

Strain. They certainly did.

Lassiter (Athens). Since bile is the principal excretory route for manganese and since the manganese leaves the blood rapidly, wouldn't you expect it to be much higher in bile than in blood plasma?

Strain. Because we couldn't find Mn clinically in the serum, we were surprised that the level in bile was so high. You have a good point.

Petering (Cincinnati). Your lead:cadmium ratio in the bile is about the same ratio that we are getting in hair.

Stake (Athens). Have you looked into the form in which these elements occur in the bile; with what they are complexed?

Strain. This is of interest, but we haven't looked.

ON THE ROLE OF ADRENOCORTICOSTEROIDS IN THE CONTROL OF ZINC AND COPPER METABOLISM

ROBERT I. HENKIN

Section on Neuroendocrinology, National Heart and Lung Institute, Bethesda, Maryland

Changes in zinc and copper metabolism in some patients with abnormalities of adrenocorticosteroid metabolism have been previously described [J. Clin. Invest. 48: 38, (1969)]. Changes in zinc and copper metabolism following physiological changes in adrenocorticosteroid metabolism which occur during a period of over 24 hr have also been described previously [J. Appl. Physiol. 31: 88 (1971)]. In the present studies serum zinc and copper concentration and urinary zinc and copper excretion were measured [Clin. Chem. 17: 369 (1971)] in eight patients with adrenal cortical insufficiency (ACI) and with panhypopituitarism on and off therapeutic hormonal replacement, and in eight patients with Cushing's syndrome without treatment and after treatment with drugs which suppressed endogenous secretion of adrenal corticosteroids or after surgical adrenalectomy. These same parameters were also measured in normal volunteers before and after administration of adrenocorticotropin (ACTH), 40 units given intravenously over 8 hr each day for 1–4 days, or after administration of the carbohydrate-active steroid prednisone, 50 mg orally, daily for 5 days. Serum, urine, and biliary zinc and copper concentrations were also measured in cats prior to and after surgical adrenalectomy and/or hypophysectomy and after replacement with either carbohydrate-active steroids, Na-K active steroids, or both hormones. Diffusible and nondiffusible zinc and copper were also measured in each patient group by ultrafiltration through a membrane which allowed the passage of proteins of molecular weight greater than 20,000 but retarded proteins of heavier molecular weight [Henkin, *In* Newer Trace Elements in Nutrition, Marcel Dekker, New York (1971)].

Patients with untreated ACI exhibited serum concentrations of zinc and copper significantly higher than controls, whereas urinary excretion of zinc and copper were significantly lower (Fig. 1). After hormonal replacement

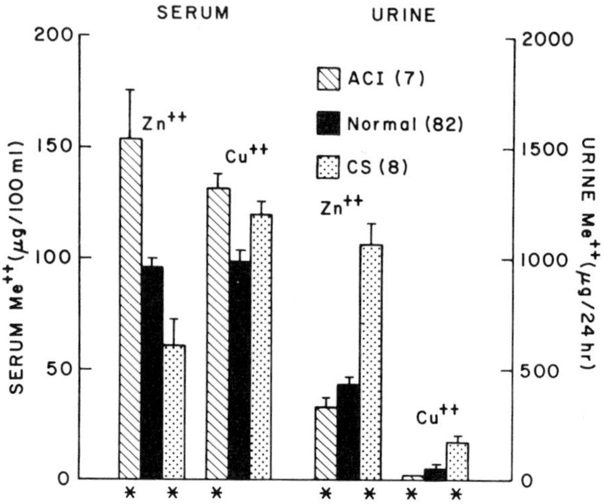

Fig. 1. Serum levels and urinary excretion of zinc and copper in patients with untreated adrenal cortical insufficiency (Addison's disease and panhypopituitarism, ACI) and untreated Cushing's syndrome (CS) compared with normal volunteers (Normal). Numbers in parentheses indicate the number of subjects in each group. * indicates $p < 0.05$ with respect to normal.

both serum zinc and copper concentrations decreased significantly toward or to normal while urinary zinc and copper excretion increased. Diffusible serum zinc and copper excretion decreased significantly in patients with untreated ACI and returned to normal following reinstitution of hormonal replacement therapy.

Significant increases in serum and biliary zinc and copper also occurred after adrenalectomy or hypophysectomy in cats, and urinary zinc and copper excretion decreased significantly (Figs. 2 and 3). After replacement with carbohydrate-active steroids and Na-K active steroids serum and biliary zinc and copper concentrations decreased toward normal levels and urinary zinc and copper excretion increased.

In normal volunteers ACTH administration decreased serum zinc and copper concentrations during the first day of administration and for each subsequent day, and increased urinary zinc and copper excretion. Oral administration of prednisone decreased serum zinc and copper concentrations and increased urinary zinc and copper excretion (Fig. 4). Diffusible serum zinc and copper increased significantly following administration of prednisone and returned to normal following the discontinuation of this hormone. Patients with Cushing's syndrome exhibited serum zinc concentrations which were

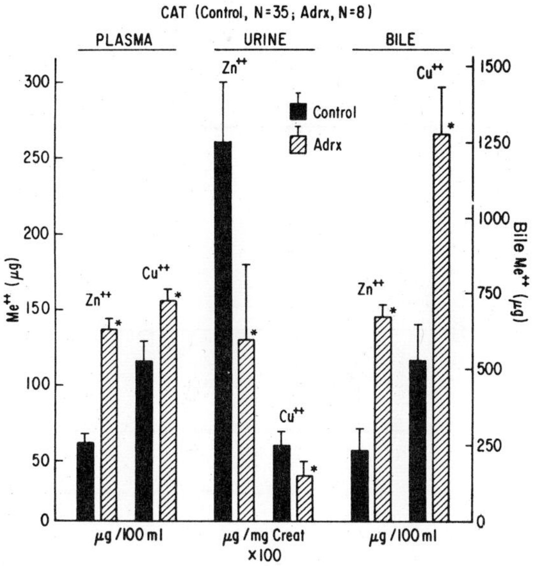

Fig. 2. Serum and biliary levels of zinc and copper and urinary zinc and copper excretion in cats before (Control) and after surgical adrenalectomy (Adrx). Urine values are expressed per milligram creatinine. Serum and urine levels are referred to the scale on the left ordinate, biliary levels to the scale on the right ordinate. * indicates $p < 0.01$ with respect to control.

significantly lower than nowmal, and urinary zinc and copper excretion which were significantly greater than normal (Fig. 1). Treatment resulted in a return of serum zinc concentrations toward normal and lowering of urinary zinc and copper excretion toward or to normal. Diffusible serum zinc and copper were significantly increased in patients with untreated Cushing's syndrome and returned toward or to normal following treatment.

These studies indicate an inverse relationship between plasma cortisol and serum zinc and copper concentration, and a direct relationship between plasma cortisol and urinary excretion of zinc and copper. These changes can be related to direct effects of cortisol on the production of increases in diffusible serum zinc and copper, subsequent decreases in both serum zinc and copper concentrations, and an increase in urinary zinc and copper excretion. Changes in glomerular filtration produced by the presence of excessive amounts or absence of adrenocorticosteroids also influence the renal excretion and subsequently serum and urine levels of both zinc and copper.

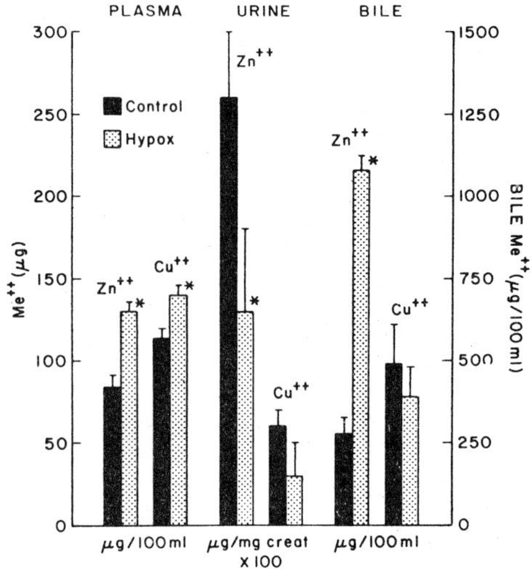

Fig. 3. Serum and biliary levels of zinc and copper and urinary zinc and copper excretion in cats before (Control) and after surgical hypophysectomy (Hypox). Urine values are expressed per milligram creatinine. Serum and urine levels are referred to the scale on the left ordinate, biliary levels to the scale on the right ordinate. * indicates $p < 0.05$ with respect to control.

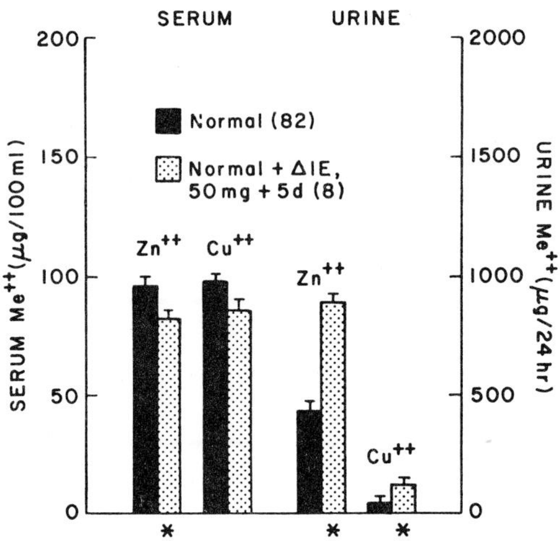

Fig. 4. Serum levels and urinary excretion of zinc and copper in normal volunteers before (Normal) and after administration of prednisone (Δ1E), 50 mg daily, orally for 5 days (Normal + Δ1E). Numbers in parentheses indicate number of subjects in each group. * indicates $p < 0.01$ with respect to normal.

DISCUSSION

Unidentified. A report by Pories *et al.* indicated that ACTH administration causes plasma zinc values to increase. Do you have any comment on that discrepancy from your results?

Henkin. Our studies show consistently through a variety of approaches a decrease in serum zinc. It is difficult to explain how a large dose of ACTH could result in an increase in serum zinc unless there was a transient change that would not be seen with repeated doses over a longer span of time. Possibly, something else about the condition or treatment of their subjects could have produced this anomalous result.

Strain (Cleveland). Relative to our work, I don't see how we can agree because we are not using standardized material.

Sandstead (Grand Forks). Have you looked at zinc-deficient animals regarding a steroid response? We got very confusing data on our human subjects in Egypt.

Henkin. We studied adrenals of zinc-deficient and pair-fed rats to see if we could demonstrate any specific block to the production of corticosterone, but there was no difference between the pair-fed animals and the animals with zinc deficiency.

GROWTH-HORMONE-DEPENDENT CHANGES IN ZINC AND COPPER METABOLISM IN MAN

ROBERT I. HENKIN

Section on Neuroendocrinology, National Heart and Lung Institute, Bethesda, Maryland

Since zinc and probably copper play important roles in various aspects of protein synthesis, changes in the concentrations of these metals in the serum and urine of patients with known abnormalities of growth hormones were studied.

Serum zinc, copper, and plasma growth-hormone concentrations and urinary zinc and copper excretion were measured [Clin. Chem. 17: 369 (1971)] in 18 patients with well-documented acromegaly before and after treatment with X irradiation and/or surgery. Mean concentration of growth hormone in the untreated patients was significantly elevated above normal and above the values obtained following therapy (Table 1). Following therapy, mean values fell toward normal levels. In the untreated patients mean serum zinc concentration was significantly lower than in the controls and rose significantly following therapy to levels which were within normal limits (Table 1). In the untreated patients mean urinary zinc excretion was significantly greater than in the controls or in the patients following therapy. Following therapy, urinary zinc excretion fell to levels which were within normal limits. In the untreated patients mean serum copper concentration was significantly higher than in the controls and levels decreased somewhat following therapy.

Serum and urine levels of zinc and copper and plasma growth hormone were also measured in four patients with well-documented isolated growth-hormone deficiency (without any other endocrinopathy) before and after treatment with human growth hormone, 2.5 I.U. given three times weekly. In the untreated patients levels of plasma growth hormone could not be measured and there was no growth-hormone response to intravenous insulin administration (0.1 U/kg) which produced significant hypoglycemia, sweating, and clinical symptoms in each patient.

Table 1. Zinc, copper, and growth hormone in controls and in clinical states metabolism.

Condition	No. of patients	Serum Zn^{2+} (μg/dl)	Serum Cu^{2+} (μg/dl)	Urine Zn^{2+} (μg/24 hr)	Urine Cu^{2+} (μ/24 hr)	Plasma growth hormone (ng/ml)
Acromegaly						
Untreated	18	62 ± 3[a]	128 ± 7[e]	1318 ± 154[b]	41 ± 9	47 ± 15[b]
Treated	8	82 ± 5	121 ± 6	580 ± 222	34 ± 5	11 ± 4
Isolated growth hormone deficiency						
Untreated	4	110 ± 11[c]	146 ± 13[de]	136 ± 24[f]	16 ± 0.5[f]	< 2[f]
Treated	4	69 ± 4	115 ± 16	284 ± 52	42 ± 8	HGH 2.5 I.U. 3 times/week
Controls	96	92 ± 2	106 ± 2	456 ± 23	36 ± 5	

[a]Significantly lower than control or treated values, $p < 0.001$, t test.
[b]Significantly higher than control or treated values, $p < 0.01$, t test.
[c]Significantly higher than treated values, $p < 0.001$, t test.
[d]Significantly higher than treated values, $p < 0.01$, paired t test.
[e]Significantly higher than control values, $p < 0.001$, t test.
[f]Significantly lower than control or treated values, $p < 0.01$, t test.

In the untreated patients mean serum zinc concentration was significantly higher compared with values observed following treatment with growth hormone and was higher than values in controls, but not statistically so. Following treatment with growth hormone, serum zinc levels fell to values significantly below normal ($p < 0.001$, t test) which were in the same range as those observed in patients with untreated acromegaly (Table 1). Mean urinary zinc excretion in the untreated patients was significantly lower than in the controls or in the patients following treatment with growth hormone. Following treatment, urinary zinc excretion rose significantly but levels were still below those of normal controls.

Mean serum copper concentration in the untreated patients was significantly higher than in the controls or in the treated patients. Following treatment, levels fell to within normal limits. Mean urinary copper excretion in the untreated patients was significantly lower than in the controls or in the patients following therapy; after therapy, values rose to control levels.

These data demonstrate that growth-hormone levels in plasma are inversely related to serum levels of zinc and directly related to urinary zinc excretion. These changes are probably related to the manner by which growth hormone either directly and/or indirectly affects the binding of zinc to macromolecular and micromolecular ligands in blood, thereby affecting the urinary excretion of zinc bound to micromolecular ligands [Biochim. Biophys. Acta 273: 64 (1972)]. Circulating zinc ion is primarily bound to the histidine moieties of albumin (the major macromolecular ligand). If there is a significant increase in the concentration of circulating histidine (the major micromolecular ligand), there is a shift in the zinc from albumin to the amino acid histidine [Bioinorg. Chem. 2: 125 (1972)]. Since albumin does not normally cross the renal glomerular membrane owing to its large size whereas histidine does, zincuria is produced with a subsequent hypozincemia. Clinical states in which there are increased levels of circulating growth hormone are associated with increased cellular turnover and increased levels of serum and urinary amino acids, including histidine. These changes lead to the zincuria and hypozincemia observed in acromegaly; correction of the growth-hormone excess leads to a return to normal levels of zinc in serum and urine. Conversely, clinical states in which decreased levels of circulating growth hormone occur are associated with decreased cellular turnover and concomitant decreases in urinary zinc excretion, and subsequent increases in serum zinc levels.

DISCUSSION

Mills (Aberdeen). Has anyone actually isolated zinc-histidine, copper-histidine, or copper-cysteine from biological systems? Secondly, where there is a discrepancy between different workers on effects on plasma zinc and plasma copper, might these be related to such things as protein adequacy, ambulatory *versus* resting patients, and differences in catabolic state?

Henkin. Some work has indeed appeared which strongly suggests the existence of several zinc and copper amino acid complexes in biological systems. In our studies we have determined the association constant for histidine-zinc to be 10^7 and for histidine-copper to be 10^9. *In vitro* we can demonstrate stripping of Zn and Cu from albumin by histidine and mathematically prove that all the metal came from albumin, not from ceruloplasmin or α-2-macroglobulin. The second question is more difficult to answer. There are many factors which influence zinc and copper metabolism in man. However, in the studies which I have just reported the patients with Addison's disease or Cushing's syndrome and the normal volunteers were studied under strictly controlled conditions on a metabolic ward. Thus, even though there are many problems, I attempted, as much as possible, to change one variable at a time, mainly the concentration of adrenal cortical hormone, either directly or indirectly. However, it is obvious that changing that one variable does alter several other factors. When one has patients leaving the hospital, coming back, no control of diet, antibiotics being administered, and other procedures that obviously are going to create very marked changes, particularly in the pituitary-adrenal axis, one has a difficult problem in interpreting one's results.

Beisel (Frederick). Have you made any measurements of ceruloplasmin and any direct analyses of the specific amino acids which you are considering as microliganders either in serum or in urine or in both?

Henkin. When we measure serum Cu, we also measure ceruloplasmin. We have also measured and calculated how much Cu is bound to ceruloplasmin and how much is bound to other ligands. While we have not systematically measured amino acids in blood or urine of our patients, we and others have observed elevations of blood and urinary histidine in patients with Cushing's syndrome and during the acute phase of acute viral hepatitis. However, we must also consider the renal picture and demonstrate to our satisfaction that there is no change in renal clearance of these metals. The best model which we have used in this respect occurs with the administration of certain drugs under controlled conditions. For example, administration of 6-azauridine causes no change in renal function but does cause these same changes in zinc metabolism which correlate with increases in urinary histidine and blood histidine.

INHIBITION OF BIOLOGICAL ACTION OF ZINC IN MAN BY OTHER TRACE ELEMENTS

WALTER J. PORIES, EDWARD G. MANSOUR, FRED R. PLECHA, ARTHUR FLYNN, and WILLIAM H. STRAIN

Department of Surgery, Cleveland Metropolitan General Hospital, Case Western Reserve University School of Medicine, Cleveland, Ohio

Excessive cadmium, copper, and manganese appear to inhibit the action of zinc in the promotion of healing and other biological roles. Elevated serum or hair levels of cadmium and manganese usually indicate toxic exposure to these elements, whereas high copper levels appear to be frequently associated with inborn or acquired metabolic errors. Wilson's disease is a well-known example of an inborn metabolic copper disease.

Several, and perhaps most, degenerative diseases represent acquired metabolic errors of copper metabolism, and these are usually characterized by high copper and low zinc levels in blood and hair. Diseases which exhibit this high copper/zinc ratio include most forms of cancer, cardiovascular disease, and some forms of schizophrenia. Zinc therapy, sometimes combined with manganese medication, furnishes a satisfactory way of treating patients with high copper/zinc ratios. The therapy usually has to be continued for the rest of the life of the patient since the medication does not correct the basic metabolic fault. The application of zinc therapy to the treatment of schizophrenia has been summarized by Pfeiffer and Iliev [International Review of Neurobiology, Academic Press, New York (1972), Suppl. 1, p. 141].

Possible application of zinc therapy to other degenerative diseases with high copper levels is conveniently illustrated by data from 46 patients with carcinoma. These comprise 13 females with breast cancer, 6 gastrointestinal tumors, 3 genitourinary-tract carcinomas, 1 cervical tumor, 3 patients with Hodgkin's disease, 13 lung cancers, 4 melanomas, 2 sarcomas, and 1 xanthoma. Since no sex effect has been noted as yet, the data are presented in Table 1 as a composite of mean blood serum levels for calcium, magnesium, copper, and zinc. Standard deviations and standard errors are included to indicate the

Table 1. Serum calcium, magnesium, copper, and zinc of 46 patients with carcinoma[1]

Element	Mean ± s.d.	S.E.
Calcium	9213 ± 1469	219
Magnesium	2235 ± 363	54
Copper	199 ± 60	8.9
Zinc	75.5 ± 20.5	3.0
Cu/Zn Ratio	2.8 ± 1.0	0.14

[1]Values are means expressed as μg of element per 100 ml serum. Normal Cu/Zn ratios are from 0.8 to 1.0.

variations in the mean levels. From these data the mean copper/zinc ratio ± s.d. is 2.8 ± 1.0. This ratio is very high when contrasted with the normal range for this ratio of 0.8–1.0 based on the blood serum values for normals reported in the literature, or of 0.82 when based on our own normal values for serum copper and zinc [Strain *et al.*, Lancet 1: 1021, (6 May 1972)].

Zinc therapy has been shown to normalize the copper/zinc ratio in one patient with bowel cancer, as shown by the data given in Table 2. The therapy consisted of zinc sulfate USP 80 mg plus 300 mg of ascorbic acid plus other minerals and vitamins t.i.d. At the start of the medication the copper zinc ratio was 1.9, and this ratio declined over a period of six weeks to 0.89. Zinc therapy was then discontinued, whereupon the serum copper/zinc ratio returned to 1.8 after three weeks.

More information on the toxic effects of high levels of copper is sorely needed. Several of the papers presented at TEMA-1 pointed out the need for specific answers to copper toxicity, but such problems are always difficult to resolve. The interrelationship of copper and zinc appears to be one of the more important aspects of copper metabolism because the result may be

Table 2. Serum copper and zinc levels and Cu/Zn ratios in a patient with bowel cancer treated with zinc therapy[1]

	Weeks on Zn therapy				No Zn Rx
Element	0	1	3	6	9
Copper	140	130	100	80	110
Zinc	75	75	85	90	60
Cu/Zn Ratio	1.9	1.7	1.2	0.89	1.8

[1]Values for serum copper and zinc are in μg per 100 ml of serum. Zn therapy was 80 mg of $ZnSO_4 \cdot 7H_2O$ USP given three times a day.

critical for both man and animals. Possibly the two elements compete for positions in similar enzymatic systems and excess copper displaces zinc.

DISCUSSION

Spencer (Hines). You published a paper a few years ago which showed that tumor growth was inhibited in rats with zinc deficiency. Now you gave zinc to patients with cancer; do you think it affected tumor growth?

Strain. The zinc-deficient rats were on the verge of death; they did not grow well nor did the tumor grow. We need more work on this.

Shah (Ottawa). We were told earlier that a high Zn:Cu ratio is not good for us because it tends to increase the levels of cholesterol, and today you are telling us that a high Zn:Cu ratio is good for us. I wonder what I should do?

Strain. I think there is an ideal ratio. When it's very much different from this ideal ratio then you can expect abnormal things to happen.

Smith (Washington). Although two papers have reported low plasma Zn levels in bronchogenic cancer, we are going to report soon that in 105 untreated patients we could find no significant difference from normal in plasma zinc levels.

Strain. It doesn't matter where the cancer is; they all have high copper and low zinc. Sometimes the ratio is as high as 5.

Henkin (Bethesda). I must stress that when one administers any drug to a patient, there has to be a rationale. Both the medical profession and the Food and Drug Administration (FDA) require this in great detail. At present, the FDA does not recognize any therapeutic indication to administer zinc beyond the 15 mg daily in your diet. The only indication to administer zinc is as an emetic. There are some investigations of drug applications which allow use of zinc under very special conditions, but there is no official recommendation to give zinc for any condition.

FERRIC FRUCTOSE AND TREATMENT OF ANEMIA IN PREGNANT WOMEN

D. J. EATOUGH, W. A. MINEER, J. J. CHRISTENSEN, R. M. IZATT, and N. F. MANGELSON

Center for Thermochemical Studies and Departments of Chemical Engineering and Chemistry, Brigham Young University, Provo, Utah, and Wasatch Medical Clinic, Orem, Utah

The relationship of iron to blood formation has been recognized since the seventeenth century; however, iron deficiency still remains a nutritional problem of world-wide medical concern. Assimilability of iron supplements is a major problem in the treatment of anemia. It has been assumed that a mucosal block exists, preventing the transfer of Fe(III), with Fe(II) being transported by an active mechanism which regulates the intestinal uptake of iron. Studies by Saltman and co-workers have shown that Fe(III) given to animals in the form of certain chelate complexes is readily absorbed and a mechanism has been proposed for uptake of these chelate complexes [Spiro and Saltman, Struct. Bonding 6: 116 (1969)]. The most efficient Fe(III) complex studied to date is ferric fructose. It has been previously demonstrated in female albino guinea pigs that retention of iron administered as ferric fructose is three times that of iron administered as ferrous sulfate [Gates *et al.*, Am. J. Clin. Nutr. 25: 983 (1972)]. This study was initiated to see if ferric fructose would be effective in the treatment of anemia in pregnant women.

We have conducted a preliminary clinical study of the treatment of anemia in 15 pregnant women using ferric fructose supplements. All patients were in the third trimester of pregnancy and had hematocrit levels of 35 or lower which did not improve upon supplementation for 4 to 8 weeks with normal dosages of either ferrous sulfate or ferrous gluconate (300 mg iron/day) and folic acid. Treatment with the ferrous salt supplement was then replaced with supplementation using only a ferric fructose solution (40 mg iron/day). The solution was prepared by adding $FeCl_3$ to an invert sugar

Table 1. Summary of results for each patient[1]

Patient	Age	No. pregnancy	Month sample taken[2]	Hair iron (ppm)	Serum iron (μg %)	Serum iron (% saturation)	Hematocrit	Change in hematocrit with supplementation
1	21	1	8	24	72	27	35	5
2	23	3	8	16	51	8	34	7
3	21	1	7	13	77	21	36	6
4	19	1	8	15	123	27	35	4
			9	27				
5[3]	20	2	4	12	55	13	33	5
			5		129	31		
			9	19				
6	30	3	8	18	134	30	33	3
7	22	1	7	17	185	22	32	6
8	33	6	8	18	92	24	34	5
9	20	1	7	20	73	12	33	7
			9	25				
10	27	2	6	32	87	21	35	0
11	24	5	7	19	112	25	32	5
12	28	4	7	16	92	18	35	3
			8	14	107	23		
13	24	5	7	13	102	23	34	0
			8		121	25		
14	23	1	4	9	94	24	32	8
			8	16				
15	20	2	8		57	11	25	10

[1] Iron supplementation was 40 mg Fe as ferric fructose per day. The change in hematocrit was calculated 4–8 weeks after supplementation with ferric fructose. Normal values are as follows: hair iron, 29 ± 10 ppm; serum iron, 65–175 μg %; serum iron, 20–55% saturation; hematocrit, 38–45.

[2] First month listed for each patient is at the time ferric fructose supplementation was initiated.

[3] A total of 400 mg of iron was given (Imferron iron-dextran) by two separate injections during the ferric fructose supplementation period.

solution (10 mg Fe(III)/ml) and neutralizing to pH 7 with NaOH. At the initiation of the ferric fructose supplementation, hematocrit levels, serum iron, iron-binding capacity, and hair iron levels were measured. Hematocrit levels were determined using a conventional centrifugal spin technique. Hematocrit levels were monitored throughout the experiments. The serum iron levels and binding capacities (using ferric ammonium citrate to saturate the serum) were measured by Utah Valley Hospital, Provo, Utah, using conventional spectroscopic techniques. Hair iron levels were determined using a Perkin-Elmer Model 305A flameless atomic absorption spectrophotometer. Where possible these tests were repeated 6–8 weeks after initiating supplementation if delivery had not occurred.

Results of the study are summarized in Table 1. In the 15 patients studied a positive improvement in the anemic condition was seen in 13 individuals with an average increase in the hematocrit levels of 5 ± 1 unit. Improvement in the hematocrit levels was seen as early as two weeks after supplementation with ferric fructose. Of those patients who responded to ferric fructose supplementation and on whom a second test series was run 6–8 weeks after initiation of supplementation, improvement in the serum percentage saturation level was seen in 3 out of 3 cases and an increase in the hair iron level was seen in 4 of 5 cases. Changes in the hematocrit levels with time for two typical patients are given in Fig. 1.

A preliminary investigation of treatment of anemia in pregnant women has therefore demonstrated that ferric fructose may be successfully used as an iron supplement in many cases where poor response to ferrous salts is seen. The results suggest that ferric fructose should be more thoroughly investigated as an iron supplement. Changes in the hematocrit levels after supplementation correlated equally well with both percentage saturation levels and hair iron levels, suggesting that hair iron levels may be a good measure of body iron stores.

DISCUSSION

Kasarskis (Madison). Can you comment on the hematological and especially the iron status of the infants at term from the mothers receiving ferric fructose?

Eatough. We made no measurements on the infants.

Saltman (San Diego). I can't comment about people, but baby pigs had significantly higher iron if their mothers were fed ferric fructose than when they were given other iron preparations.

Mertz (Beltsville). Did you have controls who continued to receive ferrous gluconate and were not switched over to the fructose complex?

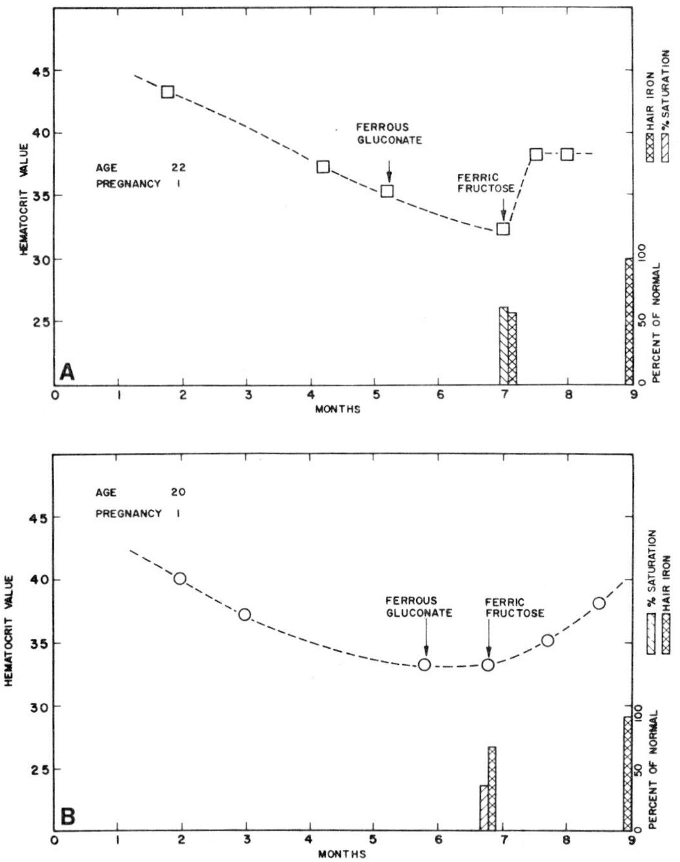

Fig. 1. Changes in hematocrit level with time for two typical patients. Initiation of supplementation with ferrous gluconate or ferric fructose is indicated. Hair iron levels and percentage saturation in the serum are given as bar graphs.

Eatough. No. The criterion for inclusion in the study was that they did not respond to ferrous gluconate or sulfate during a period of about 6 weeks. Some were continued as long as four months with no response, but to date this pretreatment period has been our only control.

Forth (Homburg). What is the thermodynamic stability constant of the ferric fructose complex, and is it effective by keeping iron available for absorption or by absorption of the intact complex?

Saltman (San Diego). The issue here is one of stabilizing polynuclear iron with sugar, and with a considerable excess of fructose you obtain a complex which has a molecular weight of about 2000 and is quite stable for long periods. If you isolate a complex of iron with fructose by precipitating with various organic solvents, you get a unit which has a molecular weight of about 60,000 and has different properties. I think that what's in the gut is the lower molecular weight form. However, the issue is not one of stability constants, but one of kinetic availability of the iron. The rate at which you can depolymerize the iron in ferric fructose, molecular weight 2000, is about 5–10 times faster than for ferric fructose, molecular weight 60,000. The complex itself is not absorbed, but it solubilizes the iron in a form which can be bound by endogenous chelates or dietary chelates and thus make that iron available.

Forth (Homburg). I agree. We showed that the complex formed with an 800-fold excess of fructose over iron was a very good donor of iron to the iron-transporting system.

Eatough. In our studies there was a great excess of fructose.

Scott (Glasgow). Do you think that pregnant women become relatively deficient in folate and that is the reason the two patients failed to respond?

Eatough. I don't know; that is possible.

ABSORPTION AND DISTRIBUTION OF THE ^{59}Fe FROM Fe-TARTRATE BY BABY PIGS

B. GROPPEL, M. ANKE, G. DITTRICH, M. GRÜN, and A. HENNIG

Karl-Marx-University Leipzig, Jena, German Democratic Republic

Anemia in baby pigs can be prevented by intramuscular injection of iron-dextran. The iron must, however, be injected by a veterinarian, and the costs are relatively high. Preliminary tests have shown that iron-tartrate is effective in preventing anemia in baby pigs. The results of these experiments are presented in this report.

Twenty-six baby pigs of normally fed sows were used. They were allowed to suck every hour, but did not have any starter feed. Sows and baby pigs were kept in polystyrene pens with wood-wool litter which was changed every day. At three days old the baby pigs were given 750 mg of ^{59}Fe-tartrate by stomach tube. To be successful, this single oral application of 150 mg of iron in the three-day-old pigs would have to keep the hemoglobin level of the animals above 8 g/100 ml up to the age of three weeks. Under practical conditions, this oral dose of iron-tartrate proved to be more effective in preventing anemia than the application of iron-dextran. Iron-tartrate is relatively insoluble, but good absorption was expected because of the favorable effect of iron-tartrate mixed in the feed of baby pigs on the hemoglobin levels.

Two days after the administration of ^{59}Fe (Table 1), 87% of the applied ^{59}Fe dose remained in the baby pigs. Of this, 50% was in the alimentary duct (wall and chyme). The percentage of ^{59}Fe in the gut gradually decreased until the eighth day after administration, and the alimentary duct was nearly free of ^{59}Fe by the tenth day. The high activity in the alimentary duct of the baby pig several days after the administration of ^{59}Fe was surprising. It was not determined if the iron was in the chyme or in the wall of the alimentary duct. The Fe-tartrate was well tolerated and there were no cases of death

Table 1. Distribution of ^{59}Fe in the baby pig after oral administration of iron-tartrate[1]

Day after administration	^{59}Fe in the whole carcass	^{59}Fe in the alimentary duct	^{59}Fe incorporated
2	87	50	37
4	60	26	34
6	56	19	37
8	57	18	39
10	26	1	25

[1]Values are the percentage of the administered dose in each portion measured.

(250 tested baby pigs), contrary to the investigations of Tollerz (Stockholm 1965) on oral doses of 80 mg Fe as Fe-fumarate.

From the second to the eighth day after administration of ^{59}Fe-tartrate between 34% and 39% of the applied ^{59}Fe, which would represent more than 50 mg of ^{59}Fe, was measured outside the alimentary duct. Only after the concentration of ^{59}Fe in the alimentary duct was less than 1% of the dose, was there less than 50 mg of iron per baby pig. The absorbed ^{59}Fe was incorporated in the different organs at different rates (Fig. 1). The liver contained the highest percentage of body ^{59}Fe for the first six days of the experiment; then this reserve of iron decreased to 5% of the body ^{59}Fe by 10 days. The ^{59}Fe content of the skeleton fluctuated between 10% and 19% during the whole duration of the experiment.

By the tenth day of the test blood and muscle contained 37% and 28% of the body's ^{59}Fe, respectively, indicating that the main quantity of the ^{59}Fe in the baby pigs' bodies was incorporated into hemoglobin and myoglobin. Therefore, by the tenth day of the experiment, skeleton, muscles, and blood contained about 80% of the incorporated ^{59}Fe and 11% was contained in the liver and the gastrointestinal tract, making a total of 90% of the ^{59}Fe activity in these five parts of the body. The accumulation in these five organs was even greater at earlier sampling times. At all times, there were only negligible quantities of ^{59}Fe in the lungs, heart, brain, kidneys, and spleen (Table 2), though during the period of the experiment a significant increase of the storage of ^{59}Fe took place in lungs, kidneys, and spleen. The importance of various organs in iron metabolism is expressed more clearly if the data are expressed on a dry-matter basis. During the period of the experiment, the Fe level of blood, kidneys, heart, lungs, and brain did not change, demonstrating that the incorporation of ^{59}Fe in these tissues is of no specific importance. On the other hand, the ^{59}Fe level of the liver decreased significantly during

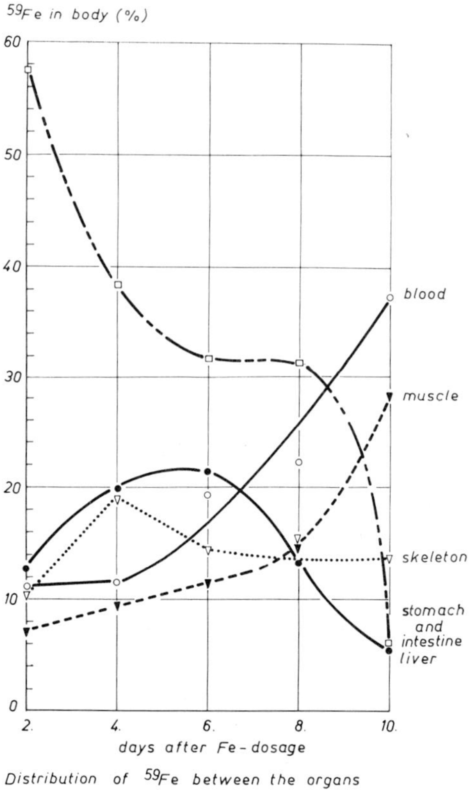

Fig. 1. Distribution of ^{59}Fe in the body following oral administration of iron-tartrate.

Table 2. Accumulation of ^{59}Fe in various organs after oral administration of iron-tartrate[1]

	Day after application of ^{59}Fe				
Organ	2	4	6	8	10
Lungs	0.6	1.4	1.3	1.1	5.4
Heart	0.2	0.2	0.6	0.8	1.2
Brain	0.3	0.5	0.6	0.4	0.6
Kidneys	0.2	0.4	0.6	0.3	0.6
Spleen	0.04	0.1	0.3	0.4	2.8

[1]Values are the percentage of the administration dose in each organ.

the experiment and dropped to its lowest value on the last day of the experiment. Thus the iron stored in the liver was used by the tenth day after the application of ^{59}Fe. The specific activity of the skeleton also decreased between the fourth and tenth day after the beginning of the experiment.

The present results of experiments with ^{59}Fe-tartrate show that a single oral dose of 150 mg of iron in the form of tartrate provides a reserve of 50 mg of iron in the carcass of the baby pig. This reserve prevents the occurrence of anemia in baby pigs during the first three weeks of their lives. [In the German Democratic Republic the Fe compound was admitted by the Central Commission of Experts for Veterinary Medicine. The results have been presented in detail by Anke *et al.* [Arch. Tierernähr. 22: 357 (1972)].

DISCUSSION

Forbes (Urbana). Did you try to dose the pigs orally with Fe-tartrate before they were three days old?

Hennig. No.

EFFECTS OF DEFICIENCY IN MANGANESE UPON PLASMA LEVELS OF CLOTTING PROTEINS AND CHOLESTEROL IN MAN

E. A. DOISY, JR.

Departments of Biochemistry and Internal Medicine, St. Louis University School of Medicine, St. Louis, Missouri

Initially, effort was expended to define the adult human's requirement for vitamin K. Design of a ration suitable for humans low in this vitamin was facilitated by appropriate modification of the 'chemically defined' diet (AAF) of Olson *et al.* [Am. J. Clin. Nutr. 23: 1614 (1970)], yielding a 2400 kcal diet furnishing 28 μg of vitamin K_1 per day.

Animals and human infants deficient in vitamin K show lowered Quick prothrombin and clotting factors II, VII, IX, and X; these clotting factors respond promptly after administration of the vitamin, increasing to normal in a few hours, as noted in our patient 2. He was maintained on formula diets for 4 months. When Quick prothrombin reached 60% of normal, he received 1 μg/kg body weight of phylloquinone (vitamin K_1) i.v. (45 μg in homologous serum). Twelve hours later his Quick prothrombin had risen from a low of 52% to a high of 103% of normal.

In contrast to patient 2 the clotting response of the first volunteer fed this regimen limited in vitamin K (K_1) was markedly delayed; his lowered clotting proteins would not rise in response to increasingly large amounts of K_1 i.v. and i.m. (45, 500, and 10,000 μg). Ten days after restoration to the normal hospital diet, Quick prothrombin had slowly risen to 80% of normal and in another 3 weeks to 100%. Other changes noted during the 6.5 months of AAF diet were decreases in serum cholesterol from 206 to 80 mg%, paralleling the nadir of his blood-clotting proteins, reddening of his black hair and beard, slowed growth of hair and nails, and scaly dermatitis. This atypical, delayed response to K_1 suggested the possibility of coincident deficiency in another essential nutrient. Retrospective analyses of all records and minutes finally revealed that $MnSO_4$ had been accidentally omitted from the trace-element supplement after the first 3 months, the deficiency being continued until normal diet was fed 3.5 months later.

Authorities generally agree on the following points.

(1) Mn is an essential element for all higher animals, abetting survival, growth, and reproduction.

(2) Mn deficiency has not been recognized in man.

(3) Irreversible sterility in the species examined results from deficiency of this element; therefore, planned repetition of this experiment in man at this time is unethical.

(4) The biological availability of Mn is determined by many factors [Cotzias, Physiol. Rev. 38: 503 (1958)] : pH, valence of Mn, type of chelating agents present, and the pattern and amounts of other elements. Each factor may alter the availability of Mn^{2+}.

Owing to the acidity of the AAF diet (hydrochloride salts), 2–3 teaspoons of antacid were given before each meal, at bedtime, and on request, and $MgCO_3$ as a laxative every other night; these would reduce the absorption of Mn by alkalinizing the chyme and by mineral imbalance. The AAF diet contributed 800 mg Ca, 600 mg P, and 20 mg Fe^{2+} per day; full doses of Fe^{2+} were given intramuscularly three times weekly, owing to the frequency of blood sampling. Thus, these four factors should limit even further the biologic availability of the already low levels of Mn in the ration, calculated to supply 0.345 mg/day. Normal intakes range from 1 mg to 20 mg/day (high-meat *versus* high-nut-cereal diet, respectively).

Analyses by neutron activation analysis (NAA) of samples from patient 1 during his deficiency compared with his values on normal diets revealed the following: (1) a 55% decrease in serum Mn during deficiency; (2) a decrease of 85% in stool Mn; (3) after 5 days of refeeding normal diet, a 46% increase in urinary Mn excretion to 5 μg/day, about 4% of normal urinary output.

In order to explain these findings it is proposed that Mn acts to complete biosynthesis of the glycoprotein clotting factors, activating the glycosyl transferases which add the carbohydrate moieties to the apoprotein 'prepro-thrombin' as shown by Leach for other glycoproteins [Fed. Proc. 30: 991 (1971)].

The sterility seen in Mn deficiency in animals and the hypocholesterolemia noted in patient 1 are probably related; lack of Mn would reduce the biosynthesis of sterols and sex hormones at several steps [Benedict *et al.*, Archs Biochem. Biophys. 110: 611 (1965)].

Presumptive evidence for the first case of Mn deficiency in man is presented. [The author wishes to thank the USPHS for support, HL-11595, and his assistants, Mss. Carter, Hutcheson, and Johnson.]

DISCUSSION

Hennig (Jena). In our experiments with manganese-deficient kids, we found a lower fat content of the body. Have you estimated the fat content of chickens, or indirectly of man? Manganese has been reported to be necessary for the synthesis of mevalonic acid.

Doisy. The chickens will be analyzed for cholesterol and fat in both serum and liver, but we have no data yet. We did not measure fat levels in man, but serum phospholipids and total triglycerides in the serum of the first volunteer paralleled his serum cholesterol and fell to quite low levels. Manganese is reported to be necessary to biosynthesize both mevalonate and isoprenoid units.

Spencer (Hines). To what extent do antacids inhibit manganese absorption, because antacids are widely used and manganese deficiency may be more common than we think?

Doisy. The first volunteer consumed a lot of antacids containing magnesium and aluminum silicates which I am confident would lower manganese absorption, but I don't know how much.

Lassiter (Athens). Early studies with manganese in poultry implicated manganese as an antilipotropic agent; does this relate to your work?

Doisy. We haven't seen fatty livers in our chickens which were seriously manganese deficient and we did not biopsy the liver in humans.

CHANGES IN THE TRACE-ELEMENT CONCENTRATION IN THE SERA AND HAIR OF KWASHIORKOR PATIENTS

F. J. BURGER

National Institute for Nutritional Diseases of the Medical Research Council, Pretoria, South Africa

Kwashiorkor is a disease of young children, associated with a deficiency of protein and a relative excess of carbohydrates in the diet. The clinical features are well known and the main ones are as follows: edema, hyperpigmentation and scaling of the skin, mucous-membrane lesions, apathy and irritability, hair changes (sparse, thin, lusterless hair which may take on a red or occasionally even a grey color), and retardation of growth. Other features of the disease are anemia (various types), fatty liver, gastrointestinal changes, changes of the pancreas, electrolyte imbalance, and defective protein, lipid and carbohydrate metabolism.

Taking into consideration the features of the disease and the role of trace elements in physiological processes, it seems probable that a trace-element imbalance is implicated. Trace-element studies in kwashiorkor are limited mainly to iron, zinc, and copper determinations in serum, and copper determinations in hair.

The purpose of the present study was to use multielemental techniques to determine several trace elements of physiological importance in the sera and hair of controls and of kwashiorkor patients on admittance to hospital and three weeks later after recovery. Although the emphasis was on trace-element analysis, measurements on electrolytes such as sodium, potassium, calcium, and magnesium were performed. The changes that take place in these electrolytes during kwashiorkor are fairly well known, but recent analyses of nails of kwashiorkor patients have shed new light on the picture.

Sera and hair specimens of 20 control Bantu children (ages 1–4 years), and of 20 kwashiorkor patients on admission and on recovery, were analyzed

Table 1. Effect of kwashiorkor on macro- and microelements in serum[1]

	Control	Kwashiorkor Adm[2]	Kwashiorkor Disc		Control	Kwashiorkor Adm[2]	Kwashiorkor Disc
Na	313.0	312.6	300.0	Mn	0.027	0.011[a]	0.014
K	18.5	17.2	18.2	Mo	0.017	0.020[a]	0.026
Ca	9.7	8.1[a]	9.9	V	0.083	0.050[a]	0.064
Mg	2.1	1.7[a]	2.2	Ni	0.081	0.036[a]	0.052
Fe	2.54	0.58[a]	1.50	Co	0.062	0.065	0.064
Zn	2.35	0.82[a]	1.49	Sr	0.066	0.061	0.058
Cu	2.49	0.95[a]	0.80	Cr	0.044	0.049	0.054

[1]Elemental composition was determined by emission spectrography at admission for treatment (Adm) and at discharge (Disc). Concentration of Na, K, Ca, and Mg in mg/100 ml, and of microelements in ppm.

[2]Statistical significant difference from both the control value and the value obtainable at the time of discharge is indicated by superscript[a].

applying atomic absorption spectrophotometry, emission spectrography, and neutron activation analysis. These patients were the same as those studied by Roode and Prinsloo [S. Afr. Med. J., 46: 1334 (1972)] in their investigations of acidified milk. The results of these analyses are summarized in Tables 1, 2, and 3. [Emission spectrography analyses were obtained with the help of Dr. L. R. P. Butler, National Physics Research Laboratory, Council for Scientific

Table 2. Effect of kwashiorkor on serum proteins

	Controls	Kwashiorkor Adm[1]	Kwashiorkor Disc
Total protein (g/100 ml)	7.1	4.3[a]	6.7
Albumin (%)	49.8	46.4[a]	55.2
Globulin α_1 (%)	4.4	5.2[a]	3.0
Globulin α_2 (%)	15.3	13.0	11.4
Globulin β (%)	12.0	11.5	12.5
Globulin γ (%)	18.7	23.9[a]	17.9
Transferrin (mg/100 ml)	321.7	85.0[a]	318.4
Ceruloplasmin (mg/100 ml)	48.8	20.6[a]	14.6

[1]Statistical significant difference from both the control values and the values obtained at the time of discharge is indicated by superscript [a].

Table 3. Effect of kwashiorkor on elemental composition of hair[1]

	Control	Kwashiorkor				Control	Kwashiorkor		
			Disc					Disc	
		Adm	Old	New			Adm	Old	New
Na	1304.9	1592.3	1189.6	1020.8	Mo		2.83	3.56	4.48
Cl	4805.3	6725.3	5133.1	4932.8	Cr		3.20	6.40	7.40
Br	53.3	38.3	51.4	49.9	Se		2.53	2.31	1.77
V	0.53	0.77	0.74	0.58	Cs		0.37	0.46	0.40
Mn	3.80	7.61	4.40	3.73	Cd		16.29	6.89	11.74
Cu	39.00	30.00	30.01	45.00	Fe		30.09	28.67	30.21
I	1.00	0.92	0.72	0.87	Zn		125.50	88.07	92.93
Ba		4.20	3.60	3.60	Co		0.46	0.82	0.67

[1]Values were obtained by neutron activation analysis and are expressed as ppm. Samples were taken at admission for treatment, and of both old and new hair at the time of discharge.

and Industrial Research, South Africa, and neutron activation analyses with the help of Dr. J. Turkstra, Department of Chemistry, Atomic Energy Board, South Africa. Professor Prinsloo, Department of Pediatrics, University of Pretoria, South Africa, obtained the blood and hair specimens used.]

DISCUSSION

Sandstead (Grand Forks). The serum zinc levels on your controls and of the infants with kwashiorkor seemed high in relation to other reports in the literature.

Burger. Yes, I agree with you. I repeated some analyses by atomic absorption spectrometry and again got these high values. This is probably due to a racial difference. The Bantus have very high values of these elements.

Hambidge (Denver). Do you have information on the rate of hair growth at the different stages?

Burger. Hair growth is very very slow, so after about three weeks you get very little hair. From the whole head you will get about 20 mg.

Mertz (Beltsville). What was your sample preparation on hair samples before analysis?

Burger. The hair samples were washed with a 50:50 mixture of acetone and alcohol.

Elmes (Belfast). In view of the low transferrin values you obtained, what thoughts do you have on the relative importance of the timing at which trace elements *versus* protein are introduced in therapy? Presumably protein would have to be given first.

Burger. The very low values of transferrin are due to protein deficiency and there is no sense in introducing iron before protein. As a matter of fact, I think you can harm the patient.

ABNORMAL MINERAL METABOLISM IN PATIENTS WITH A 'RARE' JOINT DISEASE IN ZULULAND, SOUTH AFRICA

F. J. BURGER and S. A. FELLINGHAM

South African Medical Research Council, Pretoria, South Africa

Wittmann and Fellingham [Lancet 1: 842 (1970)] first drew attention to an unusual hip disease which has been found in the Mseleni area, Northern Zululand. During 1970 a multidisciplinary program was initiated to define the disease and attempt to determine its etiology. Results of analyses mainly on blood and urine obtained from these patients and controls are presented.

In November, 1970 blood specimens were obtained from 152 sufferers from the Mseleni area and from five controls from a nearby area where no case of the disease has been observed. On the basis of X rays taken each patient was classified into one of five groups [Lockitch *et al.*, in press], i.e., normal (normal X-ray pattern and no clinical signs), JSI only (joint surface irregularity), JSI and OA (joint surface irregularity plus osteoarthrosis), OA only (osteoarthrosis only), and protrusio (protrusio acetabuli), the numbers in each group being 5, 23, 18, 86, and 25, respectively. The percentage age distribution of cases for the four groups of sufferers peaked at 15.5, 25.5, 55.5, and 15.5 yr for JSI, OA and JSI, OA only, and protrusio, respectively.

The following blood parameters were measured: hematocrit, erythrocyte sedimentation rate, rheumatoid factor, alkaline phosphatase, inorganic phosphate, total calcium, total magnesium, sodium, potassium, chloride, bicarbonate, total protein, and protein electrophoresis.

Statistical analysis of the data included a stepwise discriminant analysis in which the effect of age on the other variables was first removed. It was concluded that alkaline phosphatase, inorganic phosphate and the product of inorganic phosphate, and total calcium concentration, taken individually, could each bring about a significant discrimination between the groups.

Alkaline phosphatase discriminated best. Alkaline phosphatase fell progressively from the normal groups through to the osteoarthrosis only group; a fall-off in alkaline phosphatase may, therefore, coincide with the disease progression.

It appeared from the radiological survey that the disease manifests itself initially by joint surface irregularity (JSI). Since this picture is predominant in children, it was decided to repeat some of the blood analyses on small groups of child controls and patients, and also to measure certain new parameters that could not be measured previously owing to the laborious nature of the techniques.

The following parameters were measured in the serum (fasting blood samples): alkaline phosphatase, acid phosphatase, ionic calcium, and several trace elements, viz. fluoride, iron, copper, zinc, manganese, molybdenum, vanadium, nickel, cobalt, strontium, chromium, barium, and lead. It was demonstrated that alkaline phosphatase, cobalt, chromium, and barium were lower in the affected children than in the controls. The serum fluoride concentration was low in both groups; hence it would probably be worthwhile to study a group of children outside the affected area.

The concentration of the plasma free amino acids (alanine, aspartic acid, glutamic acid, glycine, hydroxyproline, isoleucine, leucine, methionine, proline, phenylalanine, serine, threonine, tyrosine, and valine) was determined. No consistent abnormality was found.

At the end of 1972 four patients were admitted to the hospital for remedial orthopedic surgery. Three of them were diagnosed radiologically as suffering from osteoarthrosis and one suffering from a form of bilateral Perthe's disease. The blood was analyzed for all the parameters already mentioned, but nothing could be added to the observations made. Twenty-four-hour urine specimens were analyzed for excretion of calcium, phosphorus, hydroxyproline, and 11-hydroxycorticosteroids. Of these parameters only the phosphorus excretion in all four patients was below normal. In single specimens of urine several qualitative tests for mucopolysaccharides were performed but the results were negative.

The density and composition of trabeculae bone from the heads of the femora were determined. There was an excess (varying in degree from one sample to another) in the organic material and water content of the bone. This observation is supported by the chemical analysis. The macroelements such as calcium, phosphate, and especially magnesium appeared to be low (expressed on a dry-weight basis). The concentration of trace elements (iron, copper, zinc, strontium, manganese, and fluoride) also appeared to be low compared with values found in the literature.

DISCUSSION

Lassiter (Athens). It appears that this condition might be due to manganese deficiency which would not be apparent from urine and blood analyses, although alkaline phosphatase sometimes decreases in manganese deficiency. Analyses of hair would have been interesting.

Burger. We can't get hair, because these people are very superstitious. I was especially interested in manganese and we analyzed this in bone. An analysis of cartilage which is being done may be more informative.

Spencer (Hines). I assume that the bone showed osteomalacia and am therefore interested in the vitamin-D status, serum calcium, serum phosphorus, and alkaline phosphatase.

Burger. The histological slides did show osteomalacia. Alkaline phosphatase was low. Vitamin-D status is very difficult to measure in a survey but serum alkaline phosphatase did not indicate vitamin-D deficiency. The calcium level of serum, total as well as the ionic, was normal. Only a slight decrease in serum inorganic phosphate was found.

STUDIES IN THE GENETICALLY DIABETIC MOUSE: EFFECT OF GLUCOSE TOLERANCE FACTOR (GTF) AND CLOFIBRATE (CPIB) ON THE DIABETIC SYNDROME

ROBERT W. TUMAN and RICHARD J. DOISY

Department of Biochemistry, State University of New York Upstate Medical Center, Syracuse, New York

Chromium is thought to be required by animals and man for maintenance of normal carbohydrate metabolism. Chromium deficiency leads to impaired glucose tolerance which can be normalized by oral chromium supplementation. The biologically active form of chromium has not been identified as yet, but was called glucose tolerance factor (GTF) by Schwarz and Mertz.

Diabetic mice from the C57 BL/KsJ-db/db strain were obtained from Jackson Laboratories. For chronic studies 18 male diabetic mice were separated into three groups. One group received Clofibrate (CPIB) orally (0.3% of diet), one group received daily injections of GTF (5 mg i.p.) and oral CPIB, and the third group served as untreated controls.

Blood and plasma glucose concentrations were measured by the glucose oxidase method and the Beckman glucose analyzer, respectively, on blood obtained from the orbital sinus. Plasma chromium levels were determined by atomic absorption spectroscopy. Mice used in chronic studies were sacrificed at 29 weeks of age and liver glycogen was determined by the phenol-sulfuric acid procedure.

A single injection of GTF lowers plasma glucose in normal fed mice by 68 mg/dl, and raises the circulating chromium (Table 1). In diabetic mice GTF lowers glucose levels by 243 mg/dl, insulin lowers them by 157 mg/dl, while combined injection lowers them by 468 mg/dl. Endogenous plasma chromium of diabetic animals is approximately 50% higher than in normal animals. Although not shown, a single injection of GTF lowers plasma triglycerides from 323 to 170 mg/dl and cholesterol from 206 to 133 mg/dl.

Table 1. Acute effect of a single injection of GTF and insulin on nonfasting plasma glucose and plasma chromium in normal and genetic diabetic mice[1]

Group	Treatment	Plasma glucose (mg/dl)		Plasma Cr (ng/ml)	
Normal	Saline	179 ± 6 (6)		4.1	
C57BL/KsJ – +/db	GTF	111 ± 3(6)† (p < .001)		7.5	
Diabetic		Pre-Rx	Post-Rx[2]	Pre-Rx	Post-Rx
C57BL/KsJ – db/db	Saline	785 ± 100 (6)	812 ± 79 (6)[a]	6.1	6.6
	GTF	830 ± 74 (5)	587 ± 111 (5)[b]	5.9	7.9
	Insulin	867 ± 110 (6)	710 ± 68 (6)[d]		
	GTF + insulin	928 ± 37 (12)	460 ± 44 (12)[c]		

[1]Glucose values shown represent the mean ± s.e.m. for *n* animals in each group, as indicated by the number in parentheses. GTF (5.0 mg) was administered i.p. 4 hr prior to collection of blood and exogenous insulin (1 U) was given s.c. ½ hr prior to bleeding.
[2]Statistical significance between pre- and post-treatment glucose levels is indicated as follows: [a]N.S., $P < 0.05$; [b]$P < 0.05$; [c]$P < 0.001$; [d]$P < 0.05$.

Both treatments effectively lower the elevated glucose levels in diabetic mice. The mean glucose levels during 23 weeks of treatment were CPIB 249 ± 13 mg/dl and CPIB + GTF 255 ± 12 mg/dl *versus* control 361 ± 8 mg/dl. Treatment was more effective early in the course of the disease. During the first 11 weeks the average values were CPIB 211 mg/dl, CPIB + GTF 221 mg/dl, and controls 355 mg/dl, compared with CPIB 302 mg/dl, CPIB + GTF 301 mg/dl, and controls 368 mg/dl during the final 12 weeks.

Both CPIB and CPIB + GTF significantly lower food and water consumption throughout the entire period of treatment compared with the control group: food (g/day/mouse), CPIB 5.1, CPIB + GTF 5.3, control 6.0; water (cm^3/day/mouse), CPIB 6.4, CPIB + GTF 6.1, control 9.0.

CPIB has no effect on blood glucose and liver glycogen, whereas GTF + CPIB significantly lowers both levels compared with controls (Table 2).

Both treatments effectively lower the elevated blood glucose levels. Treatment is more effective early in the disease when insulin levels are high and less effective later when the pancreas becomes exhausted and insulin levels fall to normal or below normal levels. Treated groups are significantly less hyperphagic and polydipsic than the untreated controls, and both treatments normalized food and water consumption. GTF has an ameliorating effect on the diabetic state over and above the effect of CPIB. The possibility exists that the diabetic animal is unable to convert chromium into biologically active GTF. [Supported by U.S.P.H.S.–N.I.A.M.D.D.15,100-03. Clofibrate was supplied by W. Applin, Ayerst Laboratories, and glucose tolerance factor (5.5 μg cr/mg dry wt.) by W. Mertz.]

Table 2. Chronic effect of CTF and CPIB on the diabetic syndrome in genetic diabetic mice (data at sacrifice)[1]

Treatment	Blood glucose[2] (mg/dl)	Liver glycogen[2] (mg/g wet wt liver)
CPIB	326 ± 19 (6)[a]	38.3 ± 3.4 (6)[a]
CPIB + GTF	225 ± 14 (6)[b]	26.6 ± 2.8 (6)[c]
Control	321 ± 37 (4)	46.8 ± 2.0 (4)

[1]Glucose values shown represent the mean ± s.e.m. for *n* animals in each group as indicated by the number in parentheses. Liver glycogen was determined in duplicate on *n* animals in each group.

[2]Statistical significance is indicated as follows: [a] $P > 0.05$ (n.s.) for CPIB *versus* control db/db; [b] $P < 0.05$ for CPIB + GTF *versus* control db/db; [c] $P < 0.01$ for CPIB *versus* control db/db.

DISCUSSION

Unidentified. Could you describe the source of GTF?

Tuman. In the acute and later chronic studies, the GTF was prepared from brewer's yeast and in the four-week chronic study with GTF alone, the source of the GTF was pork kidney powder.

Hambidge (Denver). I assume this was not a pure form of GTF, and am interested to know if there were any undesirable side effects and whether it can be administered to human subjects.

Tuman. In my studies at the dose level used, there were no apparent side effects.

Mertz (Beltsville). This preparation is not yet as pure as we would want it for human studies and we are not satisfied of an absence of toxicity with very high doses.

Doisy (Syracuse). At 30 min after injection, did you not see a rise in blood sugar.

Tuman. Yes, with some preparations. A rise in blood sugar was even observable up to 2 hr; then at 4 hr we saw the hypoglycemic effect.

Sandstead (Grand Forks). Did you give any of these animals the GTF orally to see if you get the same effect? Your lesion could be a problem in gastrointestinal absorption analogous to the vitamin B_{12} problem in pernicious anemia.

Tuman. None of these animals received oral GTF.

TRACE-MINERAL METABOLISM IN PORPHYRIA AND OTHER NEUROPSYCHIATRIC CONDITIONS– ROLE OF CHELATION THERAPY

HENRY A. PETERS

Neurology Department, University of Wisconsin Medical School, Madison, Wisconsin

Since 1954 we have reported the efficacy of chelation therapy in the treatment of acute intermittent porphyria (AIP) as well as in chronic porphyria (CTP). Systematic screening of urine and feces for abnormal porphyrin excretion has enabled us to identify over 100 cases of AIP since 1953. Symptoms in AIP include psychosis in the form of toxic, delerioid states, reduced levels of consciousness, and even catatonic and paranoid schizophrenic patterns. Other patients frequently appear hysterical, and exacerbations of symptoms are frequently precipitated by exposure to sulfa and barbiturate drugs and their derivatives, exposure to paints and solvents, and even fasting. Neurological symptoms include grand mal and psychomotor seizures, homonomous hemianopsia, bulbar signs, and peripheral neuropathy with tetra-paresis often leading to respiratory paralysis. Mortality rates as high as 90% have been reported in the past when neurological and psychiatric symptoms were present.

We have reported [Neurology 4: 477 (1954); Fed. Proc. 20: 3 (1961); Neurology 8: 621 (1958); Ann. Int. Med. 47: 889 (1957)] on the efficacy of chelation therapy with ethylenediaminetetraacetic acid (EDTA) administered intravenously during acute exacerbations of AIP, and dimercaptopropanol (BAL) has also been used with success when lead stores were not elevated. Other authors have confirmed these results [Painter *et al.,* Tex. J. Med. 55: 811 (1959); Luby *et al.,* Psycho-som. Med. 21: 34 (1959); Gulbrandsen *et al.,* Tidskrift. Norske Lagef. 9: 420 (1958); Galambos and Peacock, Ann. Int. Med. 50: 1056 (1959); Roman, Am. J. Clin. Nutr. 22: 1290 (1969); Roman, Enzymol. Acta Biocatalyt. 32: 31 (1967)].

Woods *et al.* [A.M.A. Archs Derm. 77: 559 (1958); Archs Derm. 84: 920 (1961)] also demonstrated that chelation therapy with oral EDTA is effective in the treatment of CTP, and Peters *et al.* [Am. J. Med. Sci. 251: 104 (1966)] reported chelation therapy with EDTA as being effective in the treatment of Cutanea tarda prophyria in an epidemic of hexachlorobenzine-induced porphyria in Turkey. Other therapeutic measures that have been reported of benefit in the treatment of AIP include administration of histidine [Nielson *et al.,* Trans. Pacific Coast Obst. Gyn. Soc. 25: 71 (1957); Porteous, J. Obst. Gyn. 70(2): 311 (1963)] and adenosine monophosphate [Gajdos *et al.,* Lancet, p. 175 (1961)]. All of these compounds possess the ability to complex zinc. In addition to oral EDTA for the treatment of CTP, desferrioximine was reported by Holzmann [Dt. Med. Wschr. 86: 127 (1961)] to be of value as well as phlebotomy as reported by Ippen [Dt. Med. Wschr. 86: 127 (1961)]. Cholestyramine, as reported by Stathers [Lancet, p. 780 (1966)], and chloroquine have been employed.

Peters has reported on increased urinary excretion of zinc, and at times copper, during the active stage of AIP when the patient was developing neuropathy, seizures, psychosis, etc. Administration of EDTA or BAL grossly enhanced the excretion of zinc, copper, and other heavy metals, and this was accompanied by clinical improvement. Abnormal urinary heavy-metal excretion returns to normal in remission and the abnormal tryptophan metabolic pattern during exacerbations as reported by Price *et al.* [Neurology 9: 456 (1959)] also tends to normalize following chelation therapy and recovery.

Porphyrins make up an important part of the white matter in the central nervous system and peripheral nerves. We have postulated that the disease of AIP is a reflection of a porphyrin deficit in the nervous system due to withdrawal of porphyrins complexed with zinc plus a block in the enzymatic pathway of porphyrin synthesis which would normally replenish these essential molecules. Meyer *et al.* [New Engl. J. Med. 286: 1277 (1972)] have identified such a block in uroporphyrinogen synthetase activity which is reduced in activity by 50% in erythrocytes of AIP patients. Other conditions in which there is a marked diuresis of zinc and/or copper include polyarteritis nodosa, delirium tremens, and toxicity due to lead, mercury, or arsenic. Metal diuresis in porphyria and these other conditions may represent an exhaustion of the body's natural chelating ability with a depletion of SH levels (glutathione) and withdrawal of porphyrins in a last-ditch effort to make up this deficit. The abnormal urine tryptophan metabolite excretion may also act in a chelating capacity. Uncomplexed metals may be responsible for an enzymatic block in porphyria synthesis as well as disturbance of other metabolic pathways. Chelation therapy may act by removal of these enzymatic blocks.

THE EFFECT OF A TWO-DAY EXPOSURE TO DIETARY CADMIUM ON THE CONCENTRATION OF ELEMENTS IN DUODENAL TISSUE OF JAPANESE QUAIL

R. M. JACOBS, M. R. S. FOX, B. E. FRY, JR., and B. F. HARLAND

Division of Nutrition, Food and Drug Administration, Department of Health, Education, and Welfare, Washington, D.C.

Cadmium, an increasingly ubiquitous pollutant of the biosphere, poses a potential environmental hazard of major importance. Presently man's daily dietary intake of cadmium approximates 50 μg. This level may occasionally be punctuated by brief intakes of higher magnitudes. Cadmium is known to affect the metabolism of several essential elements, namely iron, zinc, copper, and calcium. The effects of brief intakes of elevated levels of cadmium on the metabolism of these elements has not been studied extensively. The duodenum, the site of absorption of many elements, was the tissue of primary interest in these studies.

Newly hatched Japanese quail of both sexes were fed an adequate 35% soybean protein diet. Groups of birds were fed graded levels of cadmium as $CdCl_2$ (1.25, 2.5, 5, 10, 20, 40 mg/kg diet) from day 0 to 14 or from day 12 to 14 and then were killed. In other experiments birds received a 10 mg Cd/kg diet from day 12 to 14 only. Groups were killed on days 14, 15, 16, 18, and 21. The duodenum was excised, washed with cold 0.9% saline and blotted dry. The tissues were wet ashed and assayed for iron and cadmium by atomic absorption spectrophotometry. Diet consumption was measured throughout the experimental periods.

In birds receiving the graded levels of cadmium continuously to day 14 hemoglobin was depressed in those receiving 5 mg Cd/kg or more. A decrease in iron concentration of the duodenum was observed in birds receiving 2.5 mg Cd/kg diet. Birds fed the various levels of cadmium for two days had concentrations of cadmium in their duodenums approaching those of birds fed cadmium continuously. The iron concentration of the duodenums was decreased significantly in those birds receiving a level of cadmium equal to and in excess of 10 mg Cd/kg diet. The retention of the ingested cadmium by the duodenum during this two-day period was high for each of the intake levels, 40, 47, 38, 28, 15, and 8% respectively.

In birds fed 10 mg Cd/kg diet from day 12 to 14 and killed on days 14, 15, 16, 18, and 21, the concentrations of cadmium in the duodenum were 57, 45, 35, 12, and 6.9 μg/g, respectively. The corresponding iron concentrations were 14, 14, 15, 20, and 19 μg/g duodenum. Control values on days 14 and 21 were 20 and 22 μg/g tissue, respectively. Retentions of the ingested cadmium by the duodenum on the same days were 26, 21, 20, 8.5, and 4%, respectively.

The duodenum is an active accumulator of dietary cadmium. It is this affinity for dietary cadmium which apparently affects iron metabolism, causing decreases in iron concentration (absorption) of the duodenum. As suggested by the continued high retention of cadmium after a two-day exposure, duodenal cadmium is relatively stable and the excretion of duodenal cadmium is primarily due to epithelial exfoliation (estimated life span 2–4 days). The duodenum acts as an effective barrier to cadmium at intermediate levels of exposure and perhaps prevents internal damage from cadmium per se.

It is difficult to say at this time whether or not the decrease in iron concentration of the duodenum after the two-day exposure to cadmium reflects a residual deficiency in iron absorption or a more rapid iron transport after a transient period of deficiency. These questions are currently under investigation. From past experience we know that inhibition of iron absorption by cadmium can be observed at much lower levels of dietary cadmium than used here when the dietary iron intake is suboptimal.

Certainly it would be difficult to extrapolate these results to humans in terms of what the effect of low or periodic exposure to dietary cadmium would be in a person with suboptimal iron nutrition. These questions and others await further investigation.

DISCUSSION

Armstrong (Minneapolis). Could iron fortification overcome the effects of cadmium which you observed?

Jacobs. We haven't tried iron fortification at these low cadmium levels, but in previous work with 75 ppm cadmium in the diet, 300 mg of ferric citrate in addition to the 100 mg already there did not eliminate all of the anemia. If we supplied ascorbate at the 1% level, or lower, it was effective in eliminating most of the anemia. We do seem to have some residual toxicity from the cadmium; maybe toxicity due to the zinc-cadmium interaction was not alleviated completely. Ferrous iron was more effective than ferric iron in eliminating the anemia.

Singer (Minneapolis). What was the basal level of cadmium in the diet?

Jacobs. It was very low, and in the range you can't determine accurately, but probably below 0.02 ppm.

Piscator (Stockholm). Have you looked for duodenal mucosa pathology?

Jacobs. Not in short-term experiments; we have looked at the 75 ppm Cd level over a period of 4–6 weeks and at 6 weeks the pathology is severe and characterized by a shortening of the villi. Dr. Richardson in the audience might comment on that more accurately than I can.

Richardson (Washington). The majority of our studies were at 6 weeks, but by 4 weeks we had a striking change in the villi of the duodenum. There was a dramatic shortening and blunting of villi with marked cellular infiltration, and a marked increase in goblet cells.

EFFECTS OF LONG-TERM CADMIUM EXPOSURE IN CALCIUM-DEFICIENT RATS

M. PISCATOR and S. E. LARSSON

Department of Environmental Hygiene, Karolinska Institute, Stockholm, Sweden, and Department of Orthopedic Surgery, Umea University, Umea, Sweden

Abundant human and animal evidence indicates that long-term exposure to cadmium will cause renal tubular dysfunction. Eventually, disturbances in mineral metabolism due to the reabsorption defects may lead to osteomalacia. Several studies have been performed during the past few years to investigate the influence of cadmium on calcium metabolism. Larsson and Piscator [Israel J. Med. Sci. 7: 495 (1971)] found that after two months exposure (25 ppm Cd in drinking water) in a group of female rats on low calcium intake there was a significant decrease in the mineral content of bone compared with rats without cadmium exposure but on the low-calcium diet. It was concluded that the osteoporosis was caused by increased bone resorption to maintain plasma calcium. There was no evidence for cadmium having a direct action on bone metabolism and an indirect influence from endocrine organs seemed to be conceivable. In order to study these questions further long-term exposure experiments were undertaken.

Groups of female rats were put on low (0.04%) and normal (1%) calcium diets. Cadmium was given in deionized water at concentrations of 0, 2.5, 5, and 10 ppm for seven months. For another six months half of these concentrations were given. At the end of the exposure period urine was collected for analysis of total protein, ribonuclease, creatinine, and electrophoretic distribution of urine proteins. Blood samples were taken for determination of plasma calcium. Calcium-45 was injected and its disappearance from plasma followed for 72 hr. The animals were then killed. Cadmium was determined in liver and kidneys by atomic absorption spectrophotometry. Calcium-45 activity, water, and ash content were determined in whole tibia and metaphyses and diaphyses.

At all exposure levels calcium-deficient animals had retained about twice as much cadmium in liver and kidneys as rats on normal calcium intake. In the calcium-deficient group exposed to 10–5 ppm cadmium the mean cadmium concentration in the renal cortex was 371 ppm dry weight (90 ppm wet weight). In this group the excretion of ribonuclease was significantly increased ($P < 0.01$), whereas it was not possible to show an increase in total urine protein. In Table 1 results from some bone studies in the highest exposure groups are given. A significant decrease in the mineral content of the metaphyseal bone was found in cadmium-exposed calcium-deficient rats. Both in rats on low and normal calcium intake cadmium was found to cause a significant increase in the calcium accretion rates.

The observed increase in cadmium retention in calcium-deficient animals confirms that calcium deficiency will cause an increased absorption of cadmium. The highest average cadmium concentration in the renal cortex was about half of that calculated to cause renal tubular dysfunction (200 ppm wet weight). Since an increase in ribonuclease excretion is one of the signs of renal tubular dysfunction it cannot be excluded that in the calcium-deficient rats a critical level was reached at a renal cortex concentration of 90 ppm wet weight. The decrease in the mineral content of bone confirms that cadmium will accelerate the osteoporosis caused by calcium deficiency. The increased accretion rates are a sign of bone resorption for maintaining plasma levels of calcium. In rats with increased ribonuclease excretion it cannot be excluded that slight renal dysfunction may have contributed to the changes in mineral metabolism. Earlier conclusions about cadmium not having a direct influence on bone still seem to hold true, and the indirect action from endocrine organs and the kidney is probably the most important. More details, especially on the parathyroid, will be given in future reports. [Supported by grants from the Swedish Medical Research Council (Project No. 13-7750).]

Table 1. Ash content and calcium accretion rate in tibial bone of cadmium-exposed rats on normal and low calcium intake[1]

Ca intake	Cd in water (ppm)	Tibia metaphyses ash percentage[2]	Ca accretion rate in tibia (mg/hr $\times 10^{-7}$)[2]
Normal	0	42.2 ± 1.8 (6)	1.70 ± 0.20 (6)
Normal	10–5	41.0 ± 0.8 (6)	2.48 ± 0.40 (6)[b]
Low	0	38.9 ± 2.0 (4)	2.60 ± 0.07 (3)
Low	10–5	36.6 ± 1.3 (4)[a]	2.98 ± 0.06 (4)[b]

[1]Values are mean ± s.d. for the number of animals indicated in parentheses.

[2]Values which differ significantly from the animals receiving the same calcium intake, but different cadmium intake are indicated as follows: [a] 0.05 probability level; [b] 0.01 probability level.

DISCUSSION

Spencer (Hines). Do you have any thoughts on the interaction of cadmium and calcium metabolism at the intestinal level?

Piscator. It is easier to understand for lead which can be transported at a higher rate if you have calcium deficiency because lead and calcium are so related. In calcium deficiency we have a large increase in calcium absorption, but only a doubling in cadmium. I don't know exactly what happens but certainly cadmium in some way gets a free ride through the membranes. It is also interesting that later, when we have some accumulation of cadmium and when the body burden of cadmium has increased, we get the same effect on the intestinal mucosa as observed by Jacobs and Fox.

Cousins (New Brunswick). Although there hasn't been much clinical work on the osteomalacia induced by cadmium, what would you think about the observation that a patient with cadmium-induced osteomalacia showed a decreased calcium absorption that was responsive to pharmacological doses of ergocalciferol? Would you think this would be indicative of a possible vitamin-D defect, since the enzyme that appears to activate vitamin D is a kidney enzyme which is also very sensitive to dietary calcium?

Piscator. We certainly have an effect on vitamin D when we have advanced renal tubular damage. We may have a reabsorption defect as some of the vitamin-D binding proteins are low molecular weight and may pass out and fail to be reabsorbed. Furthermore, the transformation to the more active vitamin D in the kidney may also be affected by the kidney disease. With regard to the situation you mentioned, there was a very high exposure to cadmium and there might have been a local effect on the intestinal wall which caused an impairment in absorption. These workers inhale a lot of cadmium and they also get high concentrations of cadmium in the intestines because they swallow a lot of cadmium cleared from the respiratory tract. They can get high local concentration in the mucosa both from the ingested cadmium and from cadmium which has been inhaled. It is also known that if you have severe tubular disease, such as that seen in congenital tubular diseases, membranes in other parts of the body might be affected. It is possible that this happens in cadmium poisoning.

Singer (Minneapolis). What is the effect of the low dietary calcium:phosphorus ratio on the bone ash of the low-calcium rats?

Piscator. We get a low bone ash because, first of all, we have the same phosphorus content in both groups and only 0.04% calcium in that group, which would result in an osteoporosis, and then we also get an acceleration of the osteoporotic process by giving cadmium.

Singer (Minneapolis). Was the bone ash on the high- and low-calcium diets significantly different?

Piscator. Yes, but this was a very marginal effect and it was not possible to show it in whole tibia, just the metaphysis.

HISTOLOGICAL AND ULTRASTRUCTURAL OBSERVATIONS ON THE EFFECTS OF CADMIUM ON THE VENTRAL AND DORSOLATERAL LOBES OF THE RAT PROSTATE

ROY SCOTT, ELIZABETH AUGHEY, AND IAN MCLAUGHLIN

Department of Urology, Royal Infirmary, Glasgow, and Department of Histology and Embryology, University of Glasgow Veterinary School, Glasgow, U.K.

Subcutaneous injection of cadmium in the rat induces local sarcoma [Haddow *et al.*, Br. J. Cancer 18: 667 (1964)] and in the testis destruction of the seminiferous epithelium and subsequent regeneration of interstitial tissue [Pařízek *et al.*, Nature, Lond. 177: 1036 (1956)]. These effects on rat testis can be prevented by giving zinc, selenium, or certain thiol compounds [Gunn *et al.*, Am. J. Path. 42: 685 (1963)]. It is known that cadmium substitutes for zinc at the cellular level [Cotzias *et al.*, Am. J. Physiol. 201: 631 (1961)]; this may be the basis of its toxic effects [Hill *et al.*, J. Nutr. 80: 227 (1963)]. Sequestration of zinc is characteristic of the dorsolateral lobe of the rat prostate [Gunn *et al.*, Proc. Soc. Exp. Biol. Med. 88: 556 (1955)]. An experiment was designed in which cadmium in one of two doses 0.17 (group 3) or 0.34 mg $CdCl_2$ (group 4) was injected into the left dorsolateral lobe of the rat prostate, the noninjected right lobe being used as a control. Gunn *et al.* [Archs Path. 83: 493 (1967)] injected cadmium into the ventral lobe of the rat prostate with no obvious effects. This experiment was repeated here, using the same dosage as the dorsolateral lobe and injecting only the left ventral lobe (group 1, 0.17 mg and group 2, 0.34 mg). In addition a group of 50 rats were injected subcutaneously with cadmium chloride and the prostate lobes and testis examined after subcutaneous tumor development (10–14 months). A further group of 54 rats were injected with the carcinogen 20-methylcholanthrene to observe standard cancerous change in prostate. Tissue from the ventral and dorsolateral lobes of the prostate and testis was prepared for histological and ultrastructural examination.

In some acini of the ventral lobes of all rats in groups 1 and 2 killed between 3 and 22 months after injection histological changes in addition to normal acini were observed as follows: (a) a reduction to cuboidal or squamous epithelium; (b) the appearance of clear cells causing a localized increase in height; (c) a stratified epithelium with clear cells and cuboidal or squamous superficial cells; (d) in animals killed at over 17 months acini lined by a stratified squamous or cuboidal epithelium with increased periacinar connective tissue and a generalized leucocytic invasion. Eighteen percent of group-1 and 22% of group-2 animals showed castrate regression.

The stratified epithelium with infiltration of leucocytes and increased periacinar connective tissue was observed after 17 months in five left ventral prostates (L.V.P.), one right ventral prostate (R.V.P.), and four dorsolateral prostates (D.L.P.) in group 1. In group 2 these changes were observed in seven L.V.P., four R.V.P., and twelve D.L.P. Normal testes were found in four animals in group 1 and three in group 2; otherwise testicular changes were those of postcadmium necrosis.

In group 1 four animals developed prostatic tumors: two were squamous cell carcinomas (one at 13 months, one at 20 months); two were sarcomata (19 months) with one nonkeratinized squamous cell carcinoma. In group 2 one anaplastic sarcoma was identified and four nonkeratinized squamous cell carcinoma (electron microscopy of the epithelium showed tonofibrils in the cells) suggesting early progressive changes towards keratinization were observed only after 17 months exposure. Similar histological changes were observed in the dorsolateral lobes, but to a lesser degree.

The right and left dorsolateral lobes of the group 3 and 4 animals between 3 and 22 months showed basically the same histological changes as in groups 1 and 2. Thirty-three percent of group 3 and 27% of group 4 showed castrate regression. Stratified epithelium with leucocytic invasion was observed in seven LDLP, five RDLP, and eight VP in group 3, and in thirteen LDLP, five RDLP, and nine VP in group 4. Normal testes were found in three group-3 and three group-4 rats; otherwise the testicular changes were as already described.

There were no tumors in group 3, but in group 4, one fibrosarcoma, three adenocarcinoma, and three nonkeratinized squamous cell carcinoma were observed.

A study of 50 animals injected subcutaneously with cadmium chloride showed the cytological changes in the direct injection groups only in the dorsolateral prostate of the high-dosage animals.

In the electron microscopy study the cuboidal and low columnar cells showed a reduction in cell organelles and an increase in dense bodies of the lysosome/phagosome type; in the stratified epithelium after 17 months the cells were often attenuated and tonofibrils could be identified. The squamous epithelium showed marked degenerative change, and basal attenuated cells were regularly present. The 'clear' cells of light microscopy had few organelles in a dielectronic cytosome and appeared as undifferentiated cells. Tonofibrils were only identified in the squamous cell carcinoma and foci classified as 'nonkeratinized' squamous cell carcinoma.

The gradual replacement of the simple parenchymatous epithelium by a stratified squamous epithelium appears to indicate a series of progressive precancerous changes particularly evident about 17 months (cf. 20-methylcholanthrene). It is possible that some of the changes are mediated indirectly either by testicular dysfunction or as a result of diminished prostatic blood flow and with reduced androgen supply, but a recent study by Madlafousek *et al.* [J. Reprod. Fert. 26: 189 (1971)] showed normal copulatory activity in rats following cadmium-induced testicular necrosis.

The thesis that cadmium substitution of zinc is a carcinogenic factor is difficult to substantiate in the direct-injection experiments which show that the ventral lobe (low zinc content) is more vulnerable than the dorsolateral lobe (high zinc content), though the zinc may have a protective effect here; the cytological changes induced in the dorsolateral prostate of the repeated subcutaneous-injection animals suggest that this may be a quantitative result and the zinc protection of the dorsolateral prostate could be overcome. [Supported by the British Cancer Campaign.]

DISCUSSION

Pařízek (Prague). What was the concentration of cadmium, the ionic strength, and pH of your solution?

Aughey. That I don't know. It was prepared by the biochemists in Glasgow so that we could deliver 0.17 mg of cadmium chloride in 0.01 ml of solution.

Pařízek (Prague). This is a very concentrated solution of a metallic compound, and are you sure that these results are specific for cadmium? I am sorry to ask this question, but there is interest in the relationship of cadmium and cancer of the prostate and I feel that your results might be misleading. I must say that I disagree with a number of your conclusions and think we must be more careful, as results obtained by injecting highly concentrated solutions to an organ are often compared with and extrapolated to those seen following parenteral administration of the compound.

Elmes (Belfast). You observed clear cells going into the connective tissue. Did you observe breakdown of the basement membrane and, if you did, at what stage?

Aughey. We found breakdown of the basement membrane at a stage where we observed fibrils with the electron microscope.

Scott (Glasgow). Mrs. Aughey and I are very aware of the high dosage of cadmium which we used in this experiment. We simply copied the dosage which has previously been used, and we have planned to repeat these experiments with a very much reduced dosage of cadmium. The basic points were inspired by some observations in the British literature that cadmium workers have a higher incidence of carcinoma than the normal population, and we thought it was reasonable initially to try this high dosage to see what happened. I agree with you that it is a high dosage, and we will repeat this with a smaller dosage of cadmium.

Mason (Bethesda). This dosage is almost that which, when injected as a single subcutaneous dose, will destroy the two testes, which I presume would weigh 50 times what this dorsal and ventral prostatic lobe would weigh.

Pařízek (Prague). If I might just briefly review what I've said, I don't think that reducing the dose would help. You have to show that this effect is specific for cadmium and I am quite sure that it isn't. You may get the same effects with a number of other compounds.

Piscator (Stockholm). I should like to mention that long-term experiments at the Chester Beatty Institute in London, where relatively low levels of cadmium were injected, did not result in a higher incidence of cancer in the cadmium-exposed animals.

Aughey. We did find changes after about 12–14 months of subcutaneous injections and we should like to do some feeding experiments in the future.

EFFECT OF TOXIC LEVELS OF CADMIUM AND ITS INTERRELATIONSHIP TO IRON, ZINC, AND ASCORBIC ACID IN GROWING CHICKS

C. W. WEBER and B. L. REID

Department of Poultry Science, College of Agriculture, University of Arizona, Tucson, Arizona

Toxic levels of cadmium were fed in three separate experiments to young growing chicks. The duration of the experimental period was for four weeks with feed and water supplied *ad libitum.* The toxic level of cadmium was determined and the metabolic effects of the cadmium toxicities were evaluated. The first experiment employed a practical type of diet using cadmium levels of 100–700 ppm. A 25% reduction in growth was obtained at the 100 ppm level, while a 50% mortality was obtained at the 400-ppm level. A significant reduction in tibiae, expressed as a percentage of body weight, were observed for the 100-ppm level of cadmium fed.

Table 1. Effects of cadmium and ascorbic acid on tibiae[1]

Dietary treatments	Tibiae[1] (% body wt)	Tibiae length[1] (mm)	Tibiae citric acid[1] (μg/mg)
Controls	0.64[a]	72.1[a]	4.75[a]
50 ppm Cd	0.61[ab]	69.4[b]	7.27[b]
50 ppm Cd + 0.5% vit. C	0.61[ab]	70.7[ab]	8.46[cd]
50 ppm Cd + 1.0% vit. C	0.63[ab]	70.8[ab]	7.89[bc]
100 ppm Cd	0.57[b]	64.9[c]	8.05[bc]
100 ppm Cd + 0.5% vit. C	0.58[ab]	66.7[c]	10.41[ef]
100 ppm Cd + 1.0% vit. C	0.60[ab]	66.8[c]	9.75[f]

[1]Values in the same vertical column with different superscripts are significantly different.

Table 2. The effect of dietary cadmium and ascorbic acid on trace elements of the tibia

	Tibia ash[1]			
Dietary treatments	Fe (ppm)	Zn (ppm)	Mn (ppm)	Cd (ppm)
Control	459^{c}	207^{d}	2.5^{e}	0^{a}
50 ppm Cd	392^{bc}	160^{c}	1.7^{a}	6^{b}
50 ppm Cd + 0.5% vit. C	302^{a}	154^{bc}	2.0^{bc}	3^{ab}
50 ppm Cd + 1.0% vit. C	410^{bc}	142^{b}	2.3^{de}	7^{b}
100 ppm Cd	362^{ab}	103^{a}	1.9^{bc}	7^{b}
100 ppm Cd + 0.5% vit. C	420^{bc}	112^{a}	2.0^{bc}	5^{b}
100 ppm Cd + 1.0% vit. C	413^{bc}	108^{a}	2.0^{bc}	5^{b}

[1]Values in the same vertical column with different superscripts are significantly different.

The second and third experiments involved using purified diets and evaluating the effects of dietary ascorbic acid, zinc, and iron upon the toxicity of cadmium of 50 and 100 ppm in combination with two levels of ascorbic acid of 0.5% and 1.0% of the diet. The ascorbic acid succeeded in partially overcoming the toxicity of cadmium by the increase of testes alkaline phosphatase activity to that of the controls when fed 100 ppm Cd and 1% ascorbic acid. However, the ascorbic acid failed to restore body weight or tibiae citric acid levels to those of the control chicks (Tables 1 and 2). The hematocrits were not significantly different for the various treatments and indicated no interferences with iron absorption by dietary cadmium. The third experiment involved feeding 100 ppm Cd in combinations with ascorbic acid, zinc, and iron. The various combinations of ascorbic acid, Zn, and Fe could not restore body size, tibiae length, or tibiae citric acid levels to that of the controls. However, plasma citric acid and liver alkaline phosphatase were returned to levels of the controls. The hematocrit values for the various treatments were not significantly different.

DISCUSSION

Piscator (Stockholm). You noted a decrease in zinc content of the tibia, and zinc is certainly important for bone formation. Our studies also included a determination of bone zinc, but we had a lower exposure to cadmium over a longer time than you had. I should just like to ask what method you used for the bone analysis?

Weber. We used atomic absorption analysis of the zinc and used standard procedures to eliminate interference from the high concentrations of calcium and phosphorus present.

EFFECT OF DIFFERENT IONS ON FLUORIDE METABOLISM IN MAN

HERTA SPENCER

Metabolic Section, Veterans Administration Hospital, Hines, Illinois

The effect of different ions, notably of calcium and phosphate, on the intestinal absorption of fluoride has been studied in animals [J. Dent. Res. 39: 49 (1960); Proc. Soc. Exp. Biol. Med. 115: 295 (1964); J. Nutr. 96: 60 (1968)], but little information is available on these interactions in man. In the present study the effect of different amounts of calcium and of phosphorus, used alone and combined, on fluoride excretions was studied in man.

The studies were carried out under strictly controlled dietary conditions during a low and high fluoride intake in adult males who received a constant, analyzed, basal diet which contained an average of 220 mg calcium, 750 mg phosphorus, and 1.9 mg fluoride per day. The intake of fluoridated drinking water increased the fluoride intake to about 4.5 mg/day. Fluoride supplements, given as NaF, increased the fluoride intake to an average of 13.8 mg/day. Three calcium intakes were used: 220, 1200, and 2200 mg/day. The 200-mg calcium intake was due to the intake of the diet and the higher intakes were due to the intake of calcium gluconate tablets. Two intake levels of phosphate were used (800 and 1400 mg/day); the diet contained 800 mg phosphorus and the higher phosphorus intake was attained by adding glycerophosphate. Complete collections of urine and stool were obtained. Fluoride balances were determined by analyzing aliquots of 6-day urine and stool pools and aliquots of the diet in each 6-day metabolic period. Fluoride in the diet, excreta, and drinking water was analyzed by the method of Singer and Armstrong [Analyt. Biochem. 10: 495 (1965)].

Increasing the calcium intake from 200 to 1200 or 2200 mg/day increased the fecal fluoride to a similar extent. Adding 600 mg phosphate to these calcium intakes increased the fecal fluoride excretion more than calcium alone. Despite the increase in fecal fluoride excretion, the fluoride

Table 1. Effect of calcium and phosphorus intake on fluoride balances

Intake (mg/day)		Fluoride (mg/day)			
Ca	P	Intake	Urine	Stool	Balance
220	930	14.6	9.6	0.24	+4.7
220	1450	14.6	9.7	0.37	+4.5
2300	930	15.0	9.0	0.72	+5.3
2300	1450	14.9	8.9	1.22	+4.9

balances changed little as the fecal fluoride excretion was very low in relation to the fluoride intake (Table 1) and as compared with the considerably higher urinary fluoride excretion [Am. J. Med. 49: 807 (1970)]. The effects of calcium and phosphorus on the fecal fluoride excretion were similar during a fluoride intake of 4.5 or 14 mg/day. The urinary fluoride excretion did not change significantly during the intake of calcium or phosphorus.

The increase in fecal fluoride was not proportional to the increase in fecal calcium. For instance, the fecal fluoride did not increase further although the fecal calcium increased by several hundred milligrams. Despite the increase in fecal fluoride the intestinal absorption of fluoride was only slightly decreased. A further increase in stool fluoride and a further decrease of the net absorption of fluoride was attained when both calcium and phosphate were given together. The increase of the fecal phosphorus excretion was more often associated with an increase of the fecal fluoride excretion than with an increase of the fecal calcium excretion. Despite the increase in fecal fluoride excretion during the various studies the plasma fluoride levels did not change.

In conclusion, both calcium and phosphorus increased the fecal fluoride excretion; however, the increase was greater when calcium and phosphorus were given together. The increase in fecal fluoride excretion did not alter the fluoride balance. Neither calcium nor phosphate increased the urinary fluoride excretion significantly. [Supported by U. S. Public Health Grant DE-02486.]

DISCUSSION

Shah (Ottawa). Have you tried to determine in any of your studies what the error associated with your positive or negative balance calculation is?

Spencer. I can only say that the coefficient of variation in our analysis of stool fluoride is about 15% on a low-calcium and low-fluoride intake. It is only 3% on a high-fluoride intake. Inasmuch as the urinary fluoride is very high, our error in analysis is about 3%.

THE DISTRIBUTION OF IONIC AND BOUND FLUORINE IN PLASMA AND SERUM

LEON SINGER and W. D. ARMSTRONG

Department of Biochemistry, University of Minnesota, Minneapolis, Minnesota

Taves was first to point out that the fluorine in plasma or serum exists in both ionic and bound forms, and reported evidence of association of fluoride to albumin in plasma [Nature (Lond.) 220: 582 (1968)]. We have indicated the possibility that plasma lipids bind fluorine in plasma and, from ultrafiltration studies, have found that a considerable amount of fluorine of bovine plasma is bound to an unidentified ultrafilterable substance with a molecular weight less than 25,000 [abstr. no. 385, 49th General Session, IADR, p. 149, (1971)]. In order to investigate the distribution of fluoride in the ionic and bound forms in the plasma of various species, we have measured the total fluorine and the ionic fluoride in plasma (or serum) obtained from man, bovine, rat, and rabbit, and have calculated the bound fluorine in the specimen.

The results of the analyses of human plasma from donors who have lived in a low-fluoride community (0.03 ppm F in the water) for at least five years demonstrate that the ionic fluoride is maintained among these individuals at a relatively low and uniform concentration (0.013–0.020 ppm). Increasing total plasma fluorine content up to 0.18 ppm, when it occurs in the persons using nonfluoridated water, is due mainly to increases of the bound fluorine component. The human specimens were analyzed for total fluorine after ashing with a calcium phosphate fixative.

The fluorine contents of plasma of four persons living in a fluoridated area (~1.0 ppm) had a plasma ionic fluoride about twice that observed in plasma of persons from the community with low fluoride concentration in the drinking water. The ionic-fluoride concentration of persons from a fluoridated community remains reasonably constant at 0.03–0.05 ppm, while there is a trend for an increase in bound fluorine as the total fluorine of the plasma becomes greater. Several specimens were drawn on different dates from three of these persons, and the results of the analyses indicate that variations in bound fluorine can occur under physiological circumstances in a single individual.

The bovine apparently has a somewhat similar mechanism for handling increases in total serum fluorine as the human. The ionic-fluoride concentration is low and constant (0.01–0.02 ppm) as the total fluorine level approaches 0.40 ppm. At values of total fluorine higher than this concentration there is a sharp increase in the ionic fluoride contents of the specimen. As the total fluorine content approaches 1 ppm there is nearly equal distribution of fluorine between the ionic and bound forms.

To investigate the distribution of fluorine in plasma of other species, heparinized blood samples were obtained from rats who had considerable differences in fluoride intake. These animals were, after weaning, fed a commercial laboratory diet (Purina Laboratory Feed) containing 20–80 ppm fluoride and given an aqueous fluoride intake of 0, 50, 100, or 150 ppm F as sodium fluoride, added to their drinking water for 34 days before bleeding. The results clearly indicate that, as plasma total fluorine increases, it is ionic fluoride rather than bound fluorine that reflects increased total plasma fluorine values. There is a remarkable constancy in the mean concentration of bound fluorine for each group (0.10–0.12 ppm). This is contrary to what was observed in the human and bovine specimens reported above. The rabbit was used as an investigational animal after being fed deionized water or deionized water containing 50 ppm F as sodium fluoride and given a low-fluoride (8 ppm) commercial diet (Nutrena Rabbit Pellets) for 109 days from weanling age before being bled. The results of the analyses of the plasma indicate, as was observed with the rat, that ionic fluoride increases reflect elevations in plasma fluorine concentration and that the bound fluorine remains low and fairly constant (0.11 ± 0.010 and 0.11 ± 0.023 ppm for 0 and 50 ppm F groups, respectively) and at the same level as that reported for the rat.

There are apparent differences in the distribution of fluorine between the ionic and bound forms in the plasma of the rat and rabbit as compared with human or bovine plasma or sera. The explanations for these differences cannot be definitely stated, although there may be speculation that unknown variations in composition of the sera may be the result of dietary intake of some substance(s) or compound(s), or due to some basic metabolic differences between species. [Supported by a grant from NIDR.]

DISCUSSION

Gedalia (Jerusalem). At which pH do you carry out the determinations with the fluoride ion electrode?

Singer. We use an acetate buffer, 0.1 M in saline at pH 4.8.

Suttie (Madison). Do you get these same values for total fluoride if you assay the diffusate colorimetrically or with the electrode?

Singer. If you do not ash the sample, then do a diffusion, you get a value with the electrode which is about half of what you'll get colorimetrically. But if you ash the diffusate, rediffuse it, collect the diffusate, and again assay it with the electrode, you'll get a higher number than you did the first time with the electrode. There is something that is present in the diffusate of an unashed sample that prevents the electrode from responding the first time.

Suttie (Madison). Which number do you believe, the colorimetric or electrode value of the diffusate?

Singer. I want to believe a number which is between the two.

Shah (Ottawa). I think one lesson seems to be clear—that the rat may be a bad model if we want to study some aspects of fluoride metabolism in humans.

BEHAVIORAL, NEUROCHEMICAL, AND METABOLIC CORRELATES OF METHYLMERCURY IN MICE

PAUL SALVATERRA, EDWARD J. MASSARO, BRADLEY LOWN, and JOHN MORGANTI

Department of Biochemistry, State University of New York at Buffalo, Buffalo, New York, and Department of Psychology, State University College at Buffalo, Buffalo, New York

Open-field behavioral parameters (rearings and ambulations) were investigated in 6–8-weeks-old male Swiss-Webster mice after single intraperitoneal (i.p.) injections of methylmercury (MeHg) (1, 5, or 10 mg Hg/kg) dissolved in 0.14 M NaCl. Data on mean rearings and ambulations as a function of dose and time after dose are illustrated in Fig. 1. Analysis of variance of these data using a 4 × 3 factorial design (4 doses: 0, 1, 5, or 10 mg Hg/kg; 3 times: 1, 3, or 72 hr after dose) yielded significant main effects both of dose ($F = 26.58$, df 3/108, $P < 0.001$) and time ($F = 24.88$, df 2/108, $P < 0.001$). Individual comparisons between groups (Duncan multiple range test) showed that the 1 mg Hg/kg groups were not significantly different from control groups, while the 5 and 10 mg groups were significantly different at 1 and 3 hr. By 72 hr none of the experimental groups differed significantly from one another or the controls, indicating complete recovery. Ambulation data were similar to those of rearings, but yielded lower levels of statistical significance.

The same behavioral parameters were monitored 72 hr after each dose of a series in which single doses (2.5 mg Hg/kg, i.p.) were administered every 72 hr. Analysis of variance of the rearings data yielded significant main effects both of dose (MeHg *versus* saline) ($F = 4.549$, df 1/180, $P < 0.05$) and number of injections (1–10) ($F = 2.915$, df 9/180, $P < 0.01$). Individual comparisons showed that the source of the dose effect lay in a highly significant drop in rearings following the sixth MeHg injection. Analysis of variance of ambulations revealed no significant effects of experimental treatment.

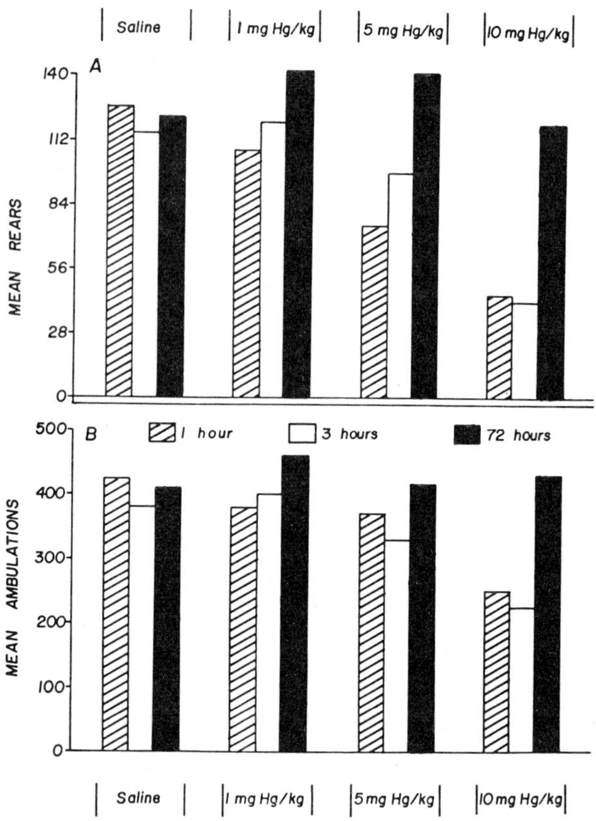

Fig. 1. Mean number of (a) rears and (b) ambulations for each dose and time after dose. Each bar represents the mean score for 10 animals for a 10-min observation period. The 'field' was a white 3-ft square divided into 36 6-inch areas with black lines. It was enclosed by a 2-ft high wall and was illuminated from above uniformly with a 30 W fluorescent lamp.

In order to investigate neurochemical correlates of the behavioral alterations in the single-dose study, levels of selected brain metabolic intermediates (glucose-1-phosphate (G1P), glucose-6-phosphate (G6P), fructose-1, 6-diphosphate (FDP), dihydroxyacetone phosphate (DHAP), α-glycerol-phosphate (α-GOP), pyruvate (pyr), phosphocreatine (PC), ATP, ADP, and AMP) were investigated. Figure 2 illustrates the dose-response pattern in cerebral cortex plus cerebellum both at 1 and 3 hr post MeHg administration. Inhibition of glycolysis, as well as ATP accumulation and significant reduction in AMP levels, was observed. The pattern of change for the glycolytic intermediates could be explained by allosteric inhibition of phosphofructokinase elicited by

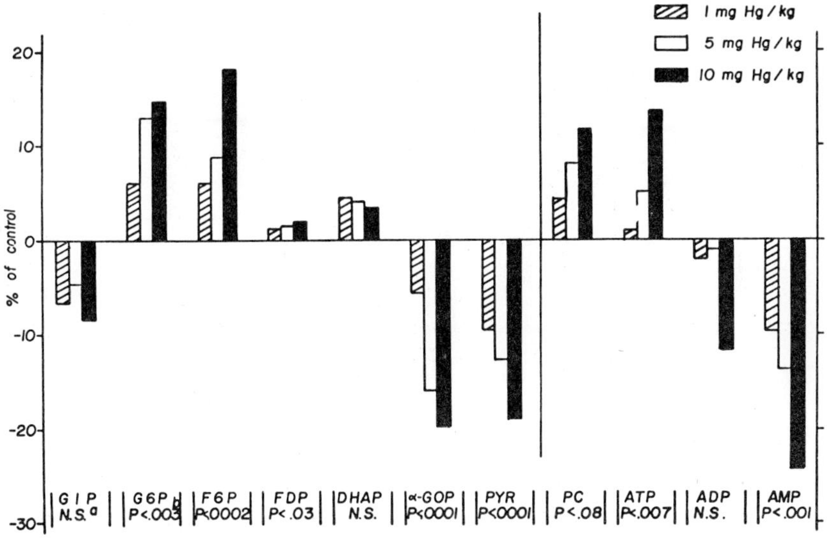

Fig. 2. Changes in mouse brain (cerebral cortex plus cerebellum) levels of glycolytic intermediates, α-GOP, PC, and adenine nucleotides induced by varying doses of MeHg (both 1 and 3 hr post i.p. injection): [a]not significantly different from control; [b]level of significance from analysis of variance.

a large ATP/ADP-AMP ratio. Inhibition of phosphorylase *a* by such a ratio also would result in (the observed) decreased G1P levels. Figure 3 illustrates the time-dependent effects of a 10 mg Hg/kg dose of MeHg. The pattern is one of larger changes at 1 or 3 hr followed by a recovery (return to control values) at 72 hr for all intermediates that changed significantly with dose (Fig. 2) except ATP and PC.

The recovery of the neurochemical parameters in the single-dose study correlated well with the recovery of normal open-field behavior, but not with the amount of MeHg actually present in the brain. Using ^{203}Hg-labeled MeHg, maximum brain uptake was observed three days post administration; approximately a quarter maximum uptake occurred within 60 min, while half maximum uptake required 3 hr.

In the repeated dose study, the rearings measure correlated significantly both with the cerebellar ($r = -0.707$, $P < 0.025$) and the cortical ($r = -0.791$, $P < 0.01$) ^{203}Hg concentrations, but not with the whole body, hair, or blood levels. [Supported by USPH grants 5R01FD0466-02, RR05400, GM-01459, and by the sponsored Programs Grant Awards of the State University at Buffalo.]

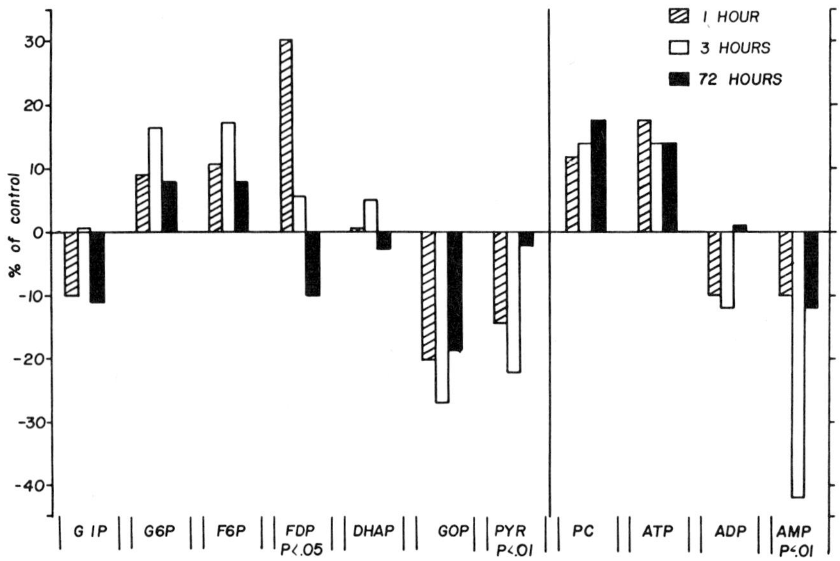

Fig. 3. Changes in mouse brain (cerebral cortex plus cerebellum) levels of glycolytic intermediates, α-GOP, PC, and adenine nucleotides induced by 10 mg Hg/kg MeHg at various times after i.p. injection: level of significance from analysis of variance.

NEUROLOGICAL ASSESSMENT OF METHYLMERCURY TOXICITY IN CATS

R. F. WILLES, S. M. CHARBONNEAU, E. A. NERA, and I. C. MUNRO

Toxicology Division, Food Research Laboratories, Health Protection Branch, Ottawa, Ontario, Canada

Studies were undertaken to assess the chronic neurological effects of methylmercury in adult cats (1) to determine the minimum dietary level of methylmercury which produces a toxic response in cats, (2) to determine the nature of the toxicological effects following chronic exposure, and (3) to compare the biological availability and toxicity of methylmercury given in two different forms, as pure methylmercury chloride and as methylmercury-contaminated fish.

Methylmercury was administered *per os* to five groups of eight cats as pure methylmercury chloride and to an additional five groups of eight cats as methylmercury-contaminated fish. Comparable groups of each regime received daily doses of 8, 20, 46, 74, or 176 μg Hg/kg. A control group of 10 cats received a diet containing uncontaminated fish. Blood mercury levels, hematology, urinalysis, and BUN were conducted every four weeks. Neurological examinations were conducted monthly and then daily when signs of neurotoxicity became apparent. Food consumption was monitored daily and body weight was determined each week. The study has now been underway for one year.

Signs of methylmercury toxicity were observed at 176 μg/kg/day after 14 weeks, and at 74 μg/kg/day after 36 weeks. Loss of balance was the first sign of toxicity observed in both dose groups and was evident when the animals walked a narrow wooden bar. The animals gradually developed an abnormal gait consisting of a widened stance and crouching of the rear legs.

The concentration of Hg in the brain was in the range of 10 ppm at both dose levels. Pathological changes were observed in the cerebellum, and in the sensory and occipital cortices. The change in the cerebellum included a decrease in the number of Purkinje and granular cells which was most evident in the deep cerebellar vermis. There was degeneration of granular cells, some proliferation of microglial cells, and formation of glial nodules in the cortices.

During the course of this study clinical signs of toxicity were restricted to the nervous system. The pattern of clinical signs was similar to that reported for subhuman primates and humans [Norberg *et al.,* Proceedings of 16th International Congress on Occupational Health, Karolinska Institute, Sweden, p. 234; Hunter *et al.,* Quart. J. Med. 33: 193 (1940); Tokuomi, *In* Minimata Disease, Kumamoto University, Japan (1968), p. 37]. The extensive loss of cellular elements in the cerebellum was similar to that previously reported by Grant [*In* Mercury, Mercurials and Mercaptans, Charles C Thomas, Springfield, Ill. (1972)]. Similar effects have been noted in humans who died from methylmercury poisoning [Takeuchi, *In* Minimata Disease, Kumamoto University, Japan (1968), p. 141].

The results of this study to date are in agreement with studies conducted in Sweden, and suggest that there is no marked difference in the toxicity or biological availability of methylmercury when administered either as pure methylmercury chloride or as methylmercury-contaminated fish.

DISCUSSION

Ganther (Madison). Have you analyzed the fish you are feeding for total selenium content, and, secondly, have you found any effects of feeding less than 1 ppm of mercury in the form of fish in the diet?

Willes. The diet contained 0.12 ppm selenium, and the lowest level of methylmercury at which we found an effect in these studies was 46 μgHg/kg. This would be equivalent to about 0.76 ppm in the diet. These signs were not clearcut at all, and I am not sure whether or not they are real.

EFFECT OF LOW-FLUORIDE DIETS FED TO MICE FOR SIX GENERATIONS

C. W. WEBER and B. L. REID

Department of Poultry Science, College of Agriculture, University of Arizona, Tucson, Arizona

Soybean and sorghum were hydroponically grown and incorporated into a low-fluoride diet for mice. Three treatments were used consisting of a low-fluoride diet, the same diet plus 6 ppm fluoride, and a basal diet using field-grown ingredients. Groups of five females per treatment were fed for a period of 120 days. The females were bred during this period and offsprings saved for continuation of the study. The average body weights (Table 1) through six generations were 17.7, 17.9, and 18.7 g for the low-fluoride, plus 6 ppm fluoride added, and the control group, respectively. The feed intakes were 2.5, 2.5, and 2.6 g per treatment. The average bone fluoride values (Table 2) for six generations were 22, 267, and 96 μg/g, while the bone citric acid values were 8.0, 8.3, and 8.0 μg/mg for the three dietary treatments of low-fluoride, plus 6 ppm fluoride, and controls. The average values (Table 3) for number born and weaned per treatment were seven born and five weaned for all three treatments. Further parameters were investigated consisting of

Table 1. The effect of fluoride diets fed through six generations upon growth and feed intake

Dietary treatments	Third-week body weights[1] (g)	Feed intake per mouse per day[1] (g)
Low fluoride	17.7	2.5
Low fluoride + 6 ppm fluoride	17.9	2.5
Control	18.7	2.6

[1]None of the treatments caused significantly different results.

Table 2. The effect of fluoride diets upon femur concentrations of citric acid and fluoride[1]

Dietary treatments	Femur citric acid levels[2] (μg/mg)	Femur fluoride levels[2] (ppm)	Small intestinal lipase[2]
Low fluoride	8.00[a]	22[a]	7.34[b]
Low fluoride + 6 ppm Fl.	8.30[a]	267[c]	7.28[b]
Control	7.95[a]	96[b]	6.68[a]

[1]The data represent four generations of mice.

[2]Values in the same vertical column with different superscripts are significantly different.

enzyme determinations of liver, heart, and kidney homogenates which showed no significant differences for the following enzymes: GOT, GPT, alkaline phosphatase, and isocitric dehydrogenase. The intestinal lipase activity for the low-fluoride diet was not significantly different from that for the 6 ppm diet (7.34 and 7.28, respectively), while the control diet was 6.68 delta ml. No significant differences were observed between the low-fluoride and 6-ppm fluoride-treated mice through six generations. [Supported in part by the grant No. 2274-07 from Public Health Service.]

Table 3. The effect of fluoride upon number of mice born and weaned[1]

Dietary treatments	No. of mice born	No. of mice weaned
Low fluoride	7	5
Low fluoride + 6 ppm Fl.	7	5
Control	8	5

[1]The data represent four generations of mice.

DISCUSSION

Gedalia (Jerusalem). I note that you increased bone fluoride levels from 20 ppm to 260 ppm by adding only 6 ppm fluoride to the low-fluoride diet. Does that seem to be too large an increase?

Shah (Ottawa). Dr. Weber has indicated that the low-fluoride diet contained 0.2 ppm F, and if you take the ratio between 0.2 and 6 you have a factor of

30, and if you take the ratio of the bone fluoride content it is over 10, so it is not completely out of order.

Weber. We also did a fluoride retention study which would indicate that fluoride retention from the grain was only about 50%, and that fluoride retention from the sodium fluoride was high.

Messer (Vancouver). How many litters per generation did you observe?

Weber. We usually bred them only twice. We introduced the males for a period of 3–5 days and after the female conceived we removed the male. We usually had mice weaned for the next generation while we were finishing the 120 days of the previous generation.

Venkateswarlu (Minneapolis). Did you ash the bone samples?

Weber. No.

EFFECT OF DIFFERENT FLUORINE COMPOUNDS ON GROWTH AND BONE FLUORIDE LEVELS IN RATS

DAVID B. MILNE and KLAUS SCHWARZ

Laboratory of Experimental Metabolic Diseases, Medical Research Programs, Veterans Administration Hospital, Long Beach, California, and Department of Biological Chemistry, School of Medicine, University of California, Los Angeles, California

Fluorine, supplied as fluoride, has been found to be essential for optimal growth and general development of rats raised on purified amino acid diets in trace-element-sterile isolators [Schwarz and Milne, Bioinorg. Chem. 1: 331 (1972)]. It also may be essential for reproduction and prevention of anemia in mice [Messer *et al.*, Science 177: 892 (1972); Nature, Lond. 240: 218 (1972)].

Various biological activities of monofluorophosphate, hexafluorosilicate, and monofluoropyruvate, all supplied at 2.5 ppm F in the diets, were compared with that of fluoride (Table 1). Weanling Fisher 344 rats were maintained in trace-element-sterile isolators and fed purified amino acid diets as described. On the basal diet, containing < 0.04 ppm F, the animals grew poorly and exhibited bleached incisors. Of the four compounds tested, only fluoride produced a pronounced significant growth effect. The response to monofluoropyruvate was only about half that seen with fluoride, while monofluorophosphate and hexafluorosilicate resulted in slight, but not statistically significant, increases. Thus, different F compounds showed distinctly different potencies. Fluoropyruvate may be effective since it is easily decomposed in the presence of sulfhydryl groups or during transamination reactions, releasing fluoride ions [Carbon-Fluorine Compounds, Elsevier, Amsterdam (1972), p. 73].

Discrepancies in the effects of various F compounds were also seen with respect to incisor pigmentation, but these variations did not run parallel to

Table 1. Effect of different fluorine compounds on growth, incisor pigmentation, and bone fluoride levels in rats maintained in a trace-element-sterile isolator[1]

Compound	Growth (% increase over control)	Incisor pigmentation (% change over control)	Fluoride concentration in femur (μg/g fresh bone)
Control			3.1 ± 0.6 (8)
KF	+23.1	+17.0 (24)	29.5 ± 3.1 (10)
K_2PO_3F	+ 3.7	+41.4 (17)	21.6 ± 0.9 (8)
K_2SiF_6	+ 8.0	−48.3 (10)	23.3 ± 2.8 (5)
$FCH_2COCOONa$	+13.0	−39.4 (12)	36.5 ± 1.2 (8)

[1] All fluorine compounds were supplemented at levels supplying 2.5 ppm F for 26 days. The values are mean ± s.e. for the number of rats indicated in parentheses.

the differences seen in growth. Monofluorophosphate consistently improved incisor pigmentation, as did fluoride [Milne *et al.,* Fed. Proc. 31: 700 (1972)]. Both fluoropyruvate and hexafluorosilicate, on the other hand, appeared to have a detrimental effect on the deposition of the iron-containing pigment in the incisors.

After 26 days on the deficient diet, bone fluoride levels (femur) were only 3.1 ppm, on a fresh weight basis, as compared with 83.5 ppm for controls fed laboratory chow (22 ppm F). Regardless of the compound used, and despite large differences in the biological activities, all four compounds under study led to the deposition of similar amounts of F (between 20 and 30 ppm) in the femurs of these rats. However, the levels of F were slightly lower in rats supplemented with monofluorophosphate or hexafluoropyruvate than in those fed fluoride or fluoropyruvate.

In the course of these experiments, after silicon was found to be necessary for growth [Schwarz and Milne, Nature, Lond. 239: 333 (1972)], sodium silicate, $Na_2SiO_3 \cdot 9H_2O$, was added to the diets to supply 500 ppm Si. In comparing the data obtained prior to Si addition to those obtained on diets with the Si supplement, it was apparent that high levels of sodium metasilicate interfere with the utilization of fluoride for growth and incisor pigmentation. Silicate, however, had no effect on the incorporation of fluoride in the femurs (to be published elsewhere). Under acidic conditions, as in the stomach, fluoride will react rapidly with silicate to form the hexafluorosilicate ion. Our data indicate that this compound may not be utilized for growth or promotion of incisor pigmentation, even though it is deposited in the bone.

Zipkin and McClure reported long ago [Pub. Hlth Rep. 66: 15–23 (1951)] that bone uptakes of fluorine were similar when potassium fluoride, sodium hexafluorosilicate, or sodium monofluorophosphate were added to the drinking water. They also found that fluoride and hexafluorosilicate were slightly better than monofluorophosphate in preventing dental caries when added to the water. These compounds were ineffective, however, when they were injected, even though all of them led to the deposition of F in bone, in analogy to our experience. Thus, caries prevention appears to be the result of a topical effect on the tooth surface.

One may conclude that considerable differences exist between various F-containing compounds with respect to their biological effects. Some compounds, like fluoride or fluoropyruvate, might be utilized in processes relating to growth, or possibly iron metabolism. Others, such as hexafluorosilicate and monofluorophosphate, may not be readily available to the rat as sources of metabolically active fluoride even though they lead to F deposition in bone. [This work was supported in part by USPHS Grant AM 08669.]

DISCUSSION

Venkateswarlu (Minneapolis). You have suggested direct deposition of fluorosilicate ion in calcified tissue. Did you determine the silicon contents of the bone in the two groups?

Milne. No, we did not, but it is something we plan to look at.

Suttie (Madison). You indicated that in the presence of silicon, the fluoride response is somehow suppressed. Isn't it just as logical to say that in the presence of fluoride, small amounts of silicon are utilized more adequately, and therefore you got a response or a much better response without silicon than with it.

Milne. It is possible but I think we are getting an independent effect from both compounds.

Suttie (Madison). Is the growth effect of fluoride that you now get with silicon in the diet statistically significant?

Milne. The growth effect of fluoride in the presence of silicon is still statistically significant. One thing that I do want to emphasize is that this is silicon as silicate, and other forms of silicon may not have this effect.

Schwarz (Long Beach). Silicate was added at what we think were unphysiological levels, because we have silicon compounds which are about 10–20 times as effective. Silicate is absorbed very fast; it has a half-time in the blood stream of only a few minutes and is rapidly excreted. It was the only soluble silicon compound which we had available at that time. If you take some of the organosilicon compounds which are biologically more active, I am sure you will get the fluorine response, but our data are not yet satisfactory enough to really prove it.

Suttie (Madison). Do you have any idea of what's responsible for the rather significant differences in growth rate of rats fed your basal diet from one experiment to the other? These differences are in fact sometimes as large as the difference within one experiment between plus and minus fluoride?

Milne. We do change to different lots of various dietary components, and either changes in dietary fluoride or some unidentified factors in the diet could have an influence on this.

Schwarz (Long Beach). May I comment on that. We are continually refining our diet, and we are now at animal experiment number 251 or 252. Not all the data which we are reporting here were obtained with the same basal diet. We are particularly careful with the calcium phosphates, and we keep on refining those. Different baseline growth levels are much more homogeneous, if you look at one time span in our experimentation. We have also tried to standardize rigidly the weaning routine of the animals we use.

Singer (Minneapolis). When you fed commercial lab chow containing 22 ppm F for four weeks and compared that with animals that received 2.5 ppm F, you found a difference of only about 2- or 3-fold in femur fluoride.

Milne. No, it was about four times.

Singer (Minneapolis). It was a difference of between 20–30 and 80 ppm. I find it hard to believe one of those values, because I would believe that if you fed the low-fluoride diet and a high-fluoride diet you should have a very extensive difference in femur retention.

Schwarz (Long Beach). Dr. Singer, this is not a matter of belief, but a fact. We have observed these facts and we present them here. You must realize that not all the fluorine which you measure in laboratory chow is truly available to the animal. That is a very important point which I think is grossly neglected in fluoride research.

Shah (Ottawa). Would you want to speculate on what level of fluoride is required to influence deposition of the iron-containing pigment in the incisor?

Milne. We discussed this at Atlantic City last year. When determining a dose response with fluoride, we seem to get optimal response between 1 and 2.5 ppm fluoride. When we went up to 7.5 ppm fluoride, the level of incisor pigmentation started to decrease. It was still higher than the control but lower than at the 2.5 ppm level. I would say that the optimal level would be in the order of 2.5 ppm. The decrease at higher levels is probably due to the bleaching and changing of the structure of the incisors, which is a well-recognized effect of excessive fluoride ingestion.

LOW NICKEL RATIONS FOR GROWTH AND REPRODUCTION IN PIGS

M. ANKE, M. GRÜN, G. DITTRICH, B. GROPPEL, and A. HENNIG

Karl-Marx-University Leipzig, Jena, German Democratic Republic

In the last ten years several attempts to prove the essentiality of the element nickel have been made. In experiments with rats and poultry Schroeder *et al.* [J. Chron. Dis. 15:51 (1961)] and Wellenreiter *et al.* [*In* Mills (ed.), Trace Element Metabolism in Animals, Livingstone, Edinburgh (1970)] obtained no difference in growth with rations containing between 0.1 and 1.0 ppm Ni. We have used a semipurified ration containing 100 ppb or 10 ppm nickel fed to pigs, goats, and their offspring.

Male and female pigs and goats fed the low-Ni diet had a significantly decreased rate of gain (Fig. 1), and both sexes developed slower than the control animals fed 10 ppm Ni. Because of the faster gain in body weight the pigs and goats sufficient in nickel were in estrus sooner and were bred earlier than the nickel-deficient animals (Table 1).

Because of the delayed sexual maturity and return of estrus after mating, the sows fed low Ni farrowed an average of 44 days later, and had one baby pig per litter less than the controlled animals. On the 28th day post partum, 65% of the baby pigs from the sows fed 10 ppm Ni, but only 37% of the pigs from the sows fed low nickel were still alive. The birth weight of these baby pigs was not influenced by the dietary Ni level. On the 28th day after birth, pigs from deficient sows weighed 15% less than controls. About 20–30% of the baby pigs and young goats from the deficient group developed a scaly and crusty skin during the lactation period. These epithelial changes were similar to parakeratosis, and also were seen in growing animals and nursing sows.

The ash content of skeletal tissue from the Ni-deficient pigs and goats was slightly, but not significantly, decreased from controls. The decreased ash content of the bones analyzed was based on the low Ca content of the samples (Table 2). The nickel-deficient pigs seem to incorporate less calcium

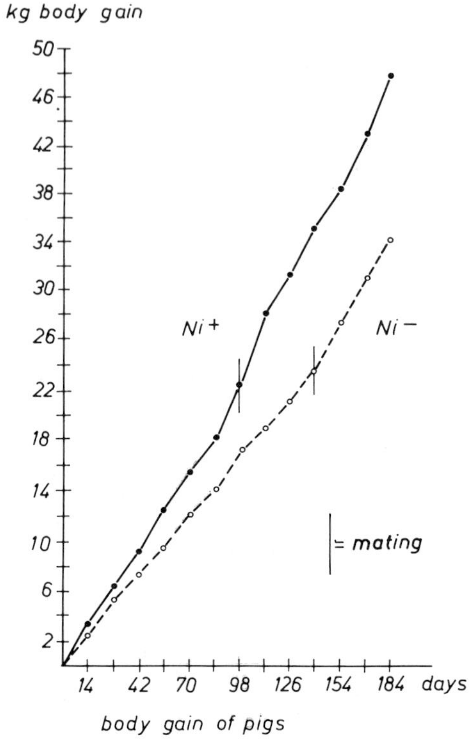

Fig. 1. Effect of diets rich and poor in nickel on body-weight gain of pigs.

in their skeleton. Therefore the Ca:P ratio in the bones of the deficient-fed animals was decreased. Further experiments will be needed to determine which cations, instead of calcium, will be incorporated in the skeleton of the nickel-deficient animals.

Alteration of the Ni status of animals was found to have no effect on the content of most elements analyzed in various tissues. In addition to the

Table 1. Effect of diets rich or poor in nickel on reproduction in pigs

	No. of animals	Birth date delay in days	Baby pigs per sow		Birth weight (g)	
			Birth	28th day	Birth	28th day
Rich Ni	7		11.3	7.3	375	2279
Poor Ni[1]	7	44[b]	10.1	3.7[a]	358	1986[a]

[1]Values significantly different than Ni rich are shown by superscripts [a] ($P < 0.05$) and [b] ($P < 0.001$).

Table 2. Effect of diets rich and poor in nickel on the calcium content of skeletal tissue of growing pigs[1]

	Poor Ni (g Ca/kg dry matter)	Rich Ni (g Ca/kg dry matter)
Rib	172(20)	202(18)
Carpal bone	163(6)	174(5)
Skeletal	152(7)	163(8)

[1]Values are means for the number of animals indicated in parentheses. All bones from the poor Ni group were significantly decreased ($P < 0.05$) from the rich Ni group.

change in skeletal Ca content, the zinc content of several organs changed in agreement with the epithelial changes noted (Table 3). The organs of both male and female nickel-deficient animals usually contained a little less zinc than those of control animals fed sufficient nickel. The differences in liver, bristle, ribs, and brain Zn content were statistically different from controls.

These results, in connection with the parakeratotic changes observed, indicate that a nickel level of 100 ppb obviously affects zinc metabolism. On the other hand, it is possible that the nickel deficiency influences only calcium metabolism, and that this disorder induces the zinc dislocation and the zinc deficiency.

DISCUSSION

Hoekstra (Madison). This reminds me of work we have done in recent years on the cobalt-zinc relationship. If we produce a zinc deficiency with practical diets containing elevated levels of calcium, we can completely suppress the zinc deficiency with 10–50 ppm cobalt. Contrary to what we thought at first, there are some increases in the zinc content of the tissues. We have one experiment where nickel rather than cobalt was tried, and it worked the same as cobalt. What was the zinc, calcium, and phosphorus content of your diet?

Table 3. Effects of diets rich and poor in nickel on the zinc content of tissues from pigs and goats[1]

	Liver		Hair		Rib		Brain	
	Sows	Goats	Sows	Goats	Sows	Goats	Sows	Goats
Poor Ni	235	62	161	105	107	71	48	48
Rich Ni[2]	315[b]	110[a]	176[a]	116[a]	116[a]	80[a]	55[a]	57[a]

[1]Values are mean ppm Zn.
[2]Values significantly different than Ni poor are shown by superscripts [a] ($P < 0.05$) and [b] ($P < 0.01$).

Hennig. The calcium content was 0.6%, the phosphorus content was 0.5%, and the ration contained 100 ppm zinc. The cobalt content was not estimated.

Hoekstra (Madison). One-hundred ppm zinc would appear to be sufficient to prevent a deficiency. Do you think you had a deficiency because the diet was low in nickel, or was there some other reason for a zinc deficiency at that high a level of dietary zinc?

Hennig. As I indicated the changes in the skin are similar to parakeratosis, and we think that the distribution of zinc in the body is indicative of zinc deficiency. When we have supplemented pig rations with high calcium, we have noted a similar distribution of body zinc. The zinc content of skin and hair is lowered, and the zinc content of the inner organs is increased.

Nielsen (Grand Forks). Your results on reproduction are very interesting and we have had similar results with rats. Our baby rats were unthrifty, and, besides not being active, they had hair that was much finer and the coat appeared much rougher.

Hartmans (Wageningen). Have you tried to correct the parakeratotic appearance in these animals by giving zinc?

Hennig. No, we haven't.

CONNECTIVE-TISSUE DISORDERS AND ADRENAL ASCORBIC ACID DURING COPPER DEPLETION AND SUBSEQUENT REPLETION IN THE GROWING RAT

P. A. ABRAHAM and J. L. EVANS

Departments of Animal Sciences and Nutrition, Rutgers University, New Brunswick, New Jersey

The metabolism of copper (Cu) and ascorbic acid are closely related. Ascorbic acid can increase the severity of a Cu deficiency [Hill and Starcher, J. Nutr. 85: 271 (1965)], and discussions of the role of Cu in the cross linking of collagen and elastin are presented by Carnes *et al.* and Rucker *et al.* [*In* Hemphill (ed.), Trace Substances in Environmental Health, Vols 2 and 4, University of Missouri, Columbia, Mo. (1968, 1970), pp. 29, 255]. In addition, reviews on Cu (connective-tissue formation and iron metabolism) and ascorbic acid (tyrosine hydroxylase formation) were published [Nutr. Rev. 31: 28, 41, 93 (1973)]. The objectives of this report are to determine the time required to replete adrenal ascorbic acid, abdominal skin elastin, and solubility of collagen following Cu depletion subsequently with Cu repletion via the diet in the growing rat.

Weanling rats were divided into two groups and fed either a skim-milk basal diet which contained no ascorbic acid supplement, less than 1 ppm Cu, and no lathyrogen, or the basal diet plus 35 ppm Cu as $CuCO_3$. Repletion was started on depletion day 42 with part of the rats (Fig. 1 and Table 1). The diet, rats, and repletion procedure were described by Evans and Abraham [J. Nutr. 103: 196 (1973)] and Abraham and Evans [*In* Hemphill (ed.), Trace Substances in Environmental Health, Vol. 5, University of Missouri, Columbia, Mo. (1972), p. 335]. In addition, the procedures cited were used for ascorbic acid [Roe, *In* Glick (ed.), Methods of Biochemical Analysis, Vol. 1, Interscience, New York (1954), p. 115], elastin [McGavack and Kao, Exp.

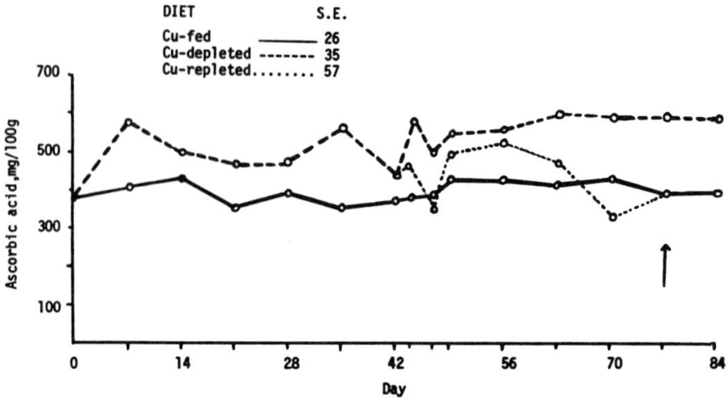

Fig. 1. Change in total adrenal ascorbic acid associated with dietary copper. Recovery of copper-repleted rats indicated by arrow. The least significant difference was 51 ($P < 0.05$). The pooled standard errors (S.E.) from analyses of variance are given. Each point represents the mean of the six rats per diet assayed.

Med. Surg. 18: 104 (1960)], and hydroxyproline times 6.94 to represent collagen [Bergman and Huxley. Analyt. Chem. 35: 1961 (1963)].

Evidence for the development of Cu deficiency during the 42-day depletion was provided by reduced plasma ceruloplasmin, liver Cu, and hemoglobin, and elevated liver iron, heart weight, and spleen weight.

Cu deficiency resulted in an increased ascorbic acid content in the adrenal gland (Fig. 1). However, the relative amounts of ascorbic, dehydroascorbic, and diketogulonic acids did not show significant variations in Cu deficiency. The increase in adrenal ascorbic acid developed during Cu deficiency was corrected to the control level after 35 days Cu repletion. The elastin content in the skin of the Cu-fed control rats increased during the experiment (Table 1). In the Cu-deficient rats the skin elastin tended to increase less than in the Cu-fed controls. However, following repletion for 42 days (day 84), the skin elastin content from the Cu-repleted rat was closer to the Cu-fed control than to the Cu-depleted rat. Skin collagen showed a higher solubility in three solvents by day 84 in Cu-depleted rats (Table 1). Upon Cu repletion of the 42-day Cu-depleted rats, the soluble collagen content of the Cu-repleted skin at day 84 tended to be lower than the 42-day Cu-depleted rat, was lower than the 84-day Cu-depleted rat, but tended to be higher than the 84-day Cu-fed control. From these data it appears that recovery of the defect in collagen cross linking developed during Cu depletion was effected during Cu repletion.

Table 1. Elastin and solubility of collagen in rat skin

Time (days)	0	42		84		
Cu treatment	Fed	Fed	Depleted	Fed	Depleted	Repleted
			Elastin (mg/g dry skin)			
	6.7	8.5[1]	7.6[1]	10.7[2]	9.7[2]	10.4[2]
			Hydroxyproline[3] (mg/g dry skin)			
Solvent						
0.28 M NaCl	4.2	3.0[4]	3.5[4]	2.8[5]	4.3[5]	3.3[5]
1.00 M NaCl	6.5	5.8[6]	6.0[6]	5.6[7]	6.9[7]	5.9[7]
0.5 M Acetic acid	12.3	13.5[8]	13.8[8]	12.7[9]	14.9[9]	13.4[9]
Total	23.0	22.3	23.3	21.1	26.1	22.6

1,2 Least significant difference was 1.3 ($P < 0.05$). Standard errors of means were 0.4. Each mean represents the mean of the six rats per diet assayed.

3 Hydroxyproline times 6.94 equals collagen.

4–7 Standard errors of means were 0.2.

4–9 Least significant difference was 0.6 ($P < 0.05$). Each mean represents the mean of the six rats per diet assayed.

8,9 Standard errors ranged from 0.1 to 0.5; mean was 0.3.

DISCUSSION

Henkin (Bethesda). Have you noted any priapism in your copper-depleted rats? One of the common observations we made when we were first looking at copper deficiency was a hemorrhagic erection among the males, and I've never seen this described by others. Have you ever seen anything like that?

Evans. No, we didn't.

Davis (Gainesville). Did you notice any hemorrhage on the surface of the hearts when you were working with your copper-deficient animals?

Evans. Normally about 15% of the copper-deficient rats died, and most of the deaths occurred due to a hemorrhage during the first four weeks of the experiment. I don't think we can say just where the hemorrhage was localized.

Erway (Cincinnati). I should like to comment upon the change in pigmentation seen in copper deficiency. It is quite reasonable that alteration in activity of the copper-containing enzyme, tyrosinase, might be one of the first signs. I have asked several people here, and most of them confirmed that for sheep loss of pigmentation in a pigmented breed or of black hair on the muzzle of other breeds is one of the first signs of the deficiency. We've seen the same thing in copper-deficient mice. Strain C57 blacks maintained on a copper-deficient diet started to turn brown. When we repleted the diet, plucked the hair and allowed the hair to grow back, it came in as black, as is characteristic of these mice.

COPPER AND CARDIOVASCULAR CHANGES

GEORGE K. DAVIS, H. R. CAMBEROS, and CHARLES E. ROESSLER

Ministerio de Agricultura y Ganaderia, Buenos Aires, Argentina, and Division of Sponsored Research and Department of Environmental Engineering, University of Florida, Gainesville, Florida

In cattle and sheep in southeastern Buenos Aires province in Argentina the widespread occurrence of massive cardiovascular calcification [Carrillo and Worker, Rev. Inv. Agric. I.N.T.A., Argentina, Ser. 4, Pat. An. IV: 9 (1967)] has led to controlled studies in these species and in laboratory animals (rabbits and guinea pigs), which have demonstrated that comparatively small amounts of the leaves of *Solanum malacoxylon* (SM), either fresh or dried, will result in a deranged calcium and phosphorus metabolism [Camberos and Davis, Gac. Vet., Argentina 32: 466 (1970)]. The principal clinical change in cattle and sheep has been a rapid elevation of calcium and phosphorus in the blood followed by metastatic ossification of the elastic fibers of the cardiovascular system. In rabbits a somewhat similar pattern of calcium-phosphate deposition occurs, but in guinea pigs [Camberos *et al.*, Am. J. Vet. Res. 31: 685 (1970)] the kidneys have been the primary target unless the animals have been rendered copper deficient, in which case the elastic fibers have shown degeneration and provide foci for metastatic calcification [Camberos *et al.*, *In* Mills (ed.), Trace Element Metabolism in Animals, Livingstone, Edinburgh (1970), p. 369].

In attempts to follow the sequence resulting in cardiovascular calcification in sheep and cattle, electron microscopic examination was made of tissues from the aortas of cattle with liver copper values of 9–13 ppm in the dry matter. These examinations have shown degeneration of the elastic fibers with beginning deposits of calcium phosphate at the points of fiber breakdown. Similar changes occur in copper-deficient guinea pigs with a following or concurrent deposition of calcium phosphate with daily inclusion of 100 mg of dried leaves of *S. malacoxylon* or a water-extract equivalent in the diet of 350-g animals.

Hill *et al.* [Fed. Proc. 26: 129 (1968)] have postulated that with copper deficiency the amine oxidase, necessary for deaminating the ε-amino group of lysine, is reduced, thus preventing conversion of lysine to desmosine. We are suggesting that, with reduction of desmosine and the cross-linkage group of elastin, lesions occur which, in the presence of the elevated calcium and phosphate, associated with feeding *S. malacoxylon* or a water extract of this plant, provide foci for metastatic deposition of calcium phosphate.

Our evidence suggests that the lesions which occur as a result of copper deficiency in cattle and sheep are independent of the calcium phosphate deposition. On the other hand, the substance present in the *S. malacoxylon,* causing elevated blood calcium and phosphorus, may result in ossification of tissues with primary location in the elastic fibers of the cardiovascular system owing to the presence of foci which may or may not be associated with copper deficiency.

In experiments with guinea pigs ^{47}Ca was administered orally and intraperitoneally in an effort to delineate the action of a water extract of *S. malacoxylon,* which contained 90% of the activity of the dried leaves. Significantly, greater amounts of the ^{47}Ca was absorbed from the oral dose given to SM-treated animals than was absorbed from the dose given the controls. Urinary excretion was also increased in the SM-treated animals, but the net retention indicated that calcium storage was greater as a result of SM administration (300 mg equivalent per day per 300 g animal), although this increased retention was slight. Intraperitoneal injection of ^{47}Ca resulted in a lower ^{47}Ca excretion in the feces and a greater urinary loss of ^{47}Ca in the SM-treated animals than in the controls.

The augmented absorption and excretion of calcium as a result of the administration of SM suggests that the action of the active substance in SM is similar to that observed where high levels of vitamin D are administered. One outstanding difference has been the rapidity of action. The effects of SM have been observed to occur within hours in guinea pigs and in less than 48 hr in ruminants.

DISCUSSION

Sandstead (Grand Forks). What was the zinc content of your diet, and the ratio of zinc to copper?

Davis. I really can't tell you. We were feeding guinea pig chow and we did not actually analyze it.

Hill (Potters Bar). In areas where this disease occurs, is there always a sufficiently low copper status of the animals, or copper content of the herbage, to be sure that this is the explanation for all field cases?

Davis. I think not. Where we were looking, we found low liver copper, but the disease occurs, I feel, in broader areas where there is not necessarily a severe copper deficiency. The severity of the situation appears, however, to be much less.

Poole (Dublin). The use of massive doses of vitamin D in the prevention of milk fever has in some cases been associated with calcification of blood vessels. Do you feel that your results have anything to offer with regard to that particular problem?

Davis. When we first saw this condition, we thought of excessive vitamin D, and we fed 20,000,000 units of vitamin D per day to sheep for a period of 6 weeks. It took almost 6 weeks, but we did produce a calcification. I don't feel it's the same thing because the elevation of calcium and phosphorus in the blood of our animals occurs within 48 hr of the time that we give really quite small levels of plant material: in sheep, somewhere around 100 gm of the leaf; in cattle, about a quarter of a pound of the dry matter.

SECONDARY COPPER DEFICIENCY IN RUMINANTS

A. HENNIG, M. ANKE, B. GROPPEL, and H. LÜDKE

Karl-Marx-University Leipzig, Jena, German Democratic Republic

Cattle and especially sheep down wind from industrial sources have been shown to suffer from resorption sterility and noninfectious abortions. We have noted skeletal damage, particularly in the front legs, and the body gain of kids and calves is decreased. In spite of an adequate Cu supply [Anke *et al.*, Tierzucht 26: 55 (1972); and Anke, Monat. Vet. Med. 28: 294 (1973)], the symptoms described resemble those caused by Cu deficiency. Cu-deficient goats had only 45% of the Cu concentration of the control animals in their brains. In addition to the brain, the Cu content of the liver, blood serum, hair, testes, and heart of the Cu-deficient ruminants (Table 1) significantly differed from that of the control groups.

Under our circumstances sulphur and cadmium presumably are the antagonists of Cu. In a series of tests with young male goats fed 10–15 g S and 10 ppm Cu/kg ration (dry weight) the growth-depressive influence of an S abundance (Fig. 1) could be significantly demonstrated in comparison with

Table 1. Effect of Cd feeding or Cu deficiency on Cu in organs of young goats[1]

Groups	Liver	Hair	Heart	Carcass
Kids of control goats (n = 24)	86	8.3	11.6	4.8
Kids of Cd goats (n = 5)	5.1	5.1	6.0	3.3
Kids of Cu-deficient goats (n = 6)	3.3	6.1	7.0	3.9

[1] Values are ppm Cu in each organ.

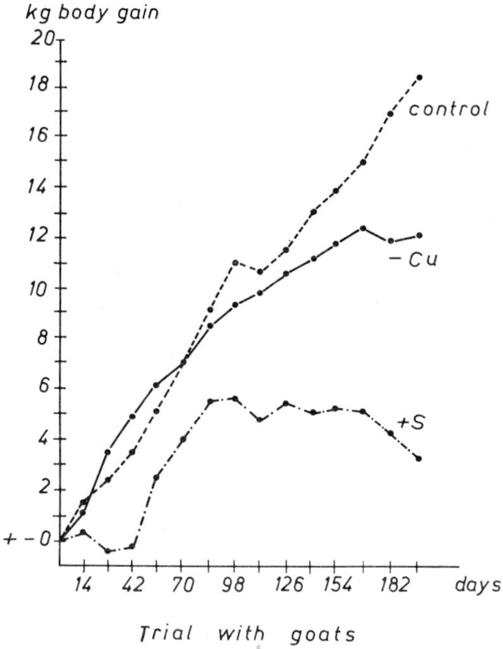

Fig. 1. Effect of a low Cu or high S ration on growth of male goats.

control groups. The amount of S was approximately the same as we usually find in the feed of emission areas.

High Cd levels [Hennig and Anke, Arch. Tierern. 14: 55 (1964); Anke *et al.*, Arch. Exp. Vet. Med. 25: 799 (1971)] block the protein which is necessary for the Cu absorption and decrease the amount of Cu in maternal and fetal tissues. Feeding of Cd caused low liver Cu and death of infant goats.

The Cu status of 181 cattle and 58 sheep in various places was analyzed and compared with their S and Cd status. In addition to the brain, the animals' livers, kidneys, ribs, serum, and pigmented hair were analyzed. We found that a Cu concentration of 9–18 ppm in the brain was normal. Cu-deficiency symptoms were not found in these cows; however, such symptoms were found in cows whose brain showed a Cu concentration of less than 9 ppm. Cows raised in 'fen' areas suffered from a primary Cu deficiency, which had been aggravated by the relatively high amount of Mo in their rations. In some villages the proportion of S in the fodder amounted to 6–15 g/kg dry matter. These were situated in the main wind direction from SO_2-emitting industrial centers. In two places the Cd emission of a zinc works was responsible for the low concentration of Cu in the brain and for the secondary Cu deficiency.

The Cd status of animals can be best demonstrated by the Cd content of the kidneys (Cd toxicosis). In Cd-emission areas an analysis of the plants indicated that the Cu supply was sufficient. The high content of Cd in kidneys and liver of cows from these areas can be explained by the contamination of air and feed (1–2.5 ppm Cd). These concentrations of Cd are very easily absorbed and retained in the body, as a ^{115}Cd test with goats proved. Four days after a ruminal application of ^{115}Cd, 40% of the Cd was in the body. This is much higher than the absorption found by Miller *et al.* [J. Dairy Sci. 52: 2029 (1969)] for nonphysiological Cd administration. In Cd-emission areas, which also have an increased emission of other elements such as Cu and Zn, the damage caused by Cd toxicosis (pulmonary emphyseme, exitus during wet weather, changes in the skeleton) will be masked and become indistinct. The content of Cd kidney and liver of these Cd-poisoned ruminants is significantly increased over the Cd content of the same organs of cattle from SO_2-emission areas (< 0.001).

Similar secondary Cu-deficiency symptoms and low brain Cu were seen in sheep. In SO_2-emission areas about 45% of the lambs suffered from endemic ataxia. Ewes had less than 5 ppm in their brains and lambs less than 3.0 ppm. The relatively high Cu concentration in the livers of these animals probably had its origin in the nonusable Cu, which, owing to a decreased sulfuroxidase activity, does not take part in metabolic changes [Kowalski and Ris, Dokl. Akad. Nauk SSSR, Moscow]. Cows and sheep with less than 8.9 ppm Cu in the brain had a decreased Cu content of brain, liver, kidneys, ribs, serum, and hair (Table 2). Kidneys, ribs, serum, and hair of Cu-deficient cows had about 20–30% less Cu than those with a sufficient Cu supply. The liver

Table 2. Effect of high and low Cu status of cattle on Cu, Cd, Zn, and Mn content of various tissues[1]

	Cu content (ppm)		Cd content (ppm)		Zn content (ppm)		Mn content (ppm)	
Tissue	High	Low	High	Low	High	Low	High	Low
Brain	11.4	5.5	0.43	0.42	50	53	2.9	2.5
Liver	157	13	0.75	1.32	128	119	10.2	10.4
Kidneys	17.1	13.4	4.19	10.50	90	92	5.7	5.3
Ribs	5.9	5.0	3.03	3.06	58	56	7.2	6.8
Serum	0.83	0.68			1.39	1.31		
Hair	6.8	5.7			113	118	7.1	11.4

[1]Analyses were performed on 188 cattle which were divided into two groups on the basis of the Cu content of the brain. The 'high' group had over 8.9 ppm brain Cu, and the 'low' group less than 8.9 ppm brain Cu.

showed a Cu deficiency only if the Cu content was < 10 ppm, and a higher liver Cu is not always an indicator of Cu supply. Pollution-control laws in the German Democratic Republic have already resulted in some decrease in the emission of these toxic materials.

DISCUSSION

Spais (Thessaloniki). Did you measure the total sulfur or sulfate?

Hennig. Total sulfur.

Lewis (Weybridge). Are your cadmium figures on the kidney based on kidney cortex or whole kidney? The concentration is usually much higher in the cortex than in the whole kidney, and unless you have an even proportion in all your samples you can get widely differing kidney cadmium levels.

Hennig. The cadmium content was determined by atomic absorption on the whole kidney.

Suttle (Edinburgh). Could you make any measurements of the molybdenum concentrations in your foodstuffs, and could you indicate total sulfur concentration that you found?

Hennig. We did not measure molybdenum concentrations, but we have determined molybdenum in 2000 samples of fodder in the German Democratic Republic, and in the fen areas we have seen a high content of molybdenum, 1–3 ppm. The sulfur concentration in the fodder was 8–15 g/kg dry matter.

EFFECT OF ZINC ON THYMIDINE KINASE ACTIVITY AND DNA METABOLISM

D. OBERLEAS and A. S. PRASAD

Medical Research Service, Veterans Administration Hospital, Allen Park, Michigan, and Department of Medicine, Wayne State University, School of Medicine, Detroit, Michigan

Many enzymes are known to require zinc for their physiological functions [Parisi and Vallee, Am. J. Clin. Nutr. 22: 1222 (1969)]. No enzyme, however, has been shown to be sufficiently sensitive or of sufficient metabolic significance to account for the very early and marked effect zinc depletion has on growth and anorexia. It would appear that such an enzyme would have to be involved in cell division or early nucleic acid synthesis preceding cell division. This report briefly describes a suitable animal model which allows the induction of thymidine kinase activity in early zinc deficiency. Thymidine kinase is essential for DNA synthesis and subsequent cell division. Therefore thymidine kinase may be the enzyme which accounts for the symptoms of early zinc depletion.

Male albino rats (Holtzman strain) weighing 60–80 g for 13- and 17-day experiments and 100–120 gm for 6-day experiments were allotted by weight and fed appropriate diets as previously described [Prasad and Oberleas, J. Appl. Physiol. 31: 842 (1971)]. At zero or seven days on dietary treatment a polyvinyl chloride sponge thoroughly washed in deionized water, dried, cut, and autoclaved, and approximately 3 cm X 6 cm X 3 nm was implanted between the skin and muscle on the backs of these animals. Six or ten days after implantation the animals were sacrificed, the sponges isolated by blunt dissection, and the rapidly developing capsule surrounding the sponge was collected for subsequent study.

Carbon-labeled thymidine (2.5 μCi/sample) in Hank's balanced salt solution (pH 7.3) and with 50 μg penicillin and streptomycin per ml were used to measure DNA incorporation. After 24-hr incubation, the tissue was washed three times with Hank's solution, homogenized in 4 ml Tris buffer (pH 7.4), and dialyzed against deionized water to remove all soluble radioactivity. The

Table 1. [^{14}C] Thymidine incorporation into DNA

	Deficient	Paired-Fed	t
Experiment 1	$18.0 \pm 5.7 \times 10^3$ (4)	$136.2 \pm 15.2 \times 10^3$ (4)	7.24[1]
Experiment 2	$14.4 \pm 2.9 \times 10^3$ (6)	$54.7 \pm 7.5 \times 10^3$ (5)	4.99[1]

Values are mean DPM/mg DNA ± s.e. for number of rats in parentheses.
[1]$p < 0.001$.

homogenate was precipitated with 0.6 M perchloric acid (2:1 vol. ratio) for 15 min, centrifuged, and the precipitate washed twice with 0.2 M perchloric acid. The precipitate was suspended in 1 ml deionized H_2O and 10 ml of Scintisol-Complete (Isolab Inc., Akron, Ohio) was added for counting. Tissue for thymidine kinase activity was prepared according to the method of Bollum and Potter [J. Biol. Chem. 233: 478 (1958)]. Assay for thymidine kinase activity was a modification of that used by Stirpe and LaPlaca [Biochem. J. 122: 347 (1971)].

The zinc content of the rapidly regenerating tissue from zinc-deficient rats was significantly lower than that from zinc-supplemented rats after only six days. Table 1 shows the incorporation of labeled thymidine into DNA of rapidly regenerating collagen tissue after only six days of dietary treatment. In both experiments the incorporation of thymidine was significantly reduced in the zinc-deficient animals when compared with paired-fed animals. Though the incorporation was fairly consistent in the deficient animals between the two experiments there was considerable difference in the response of the paired-fed animals. The reason for this difference is not readily available.

The activity of thymidine kinase, shown in Table 2, measured after as little as six days of dietary treatment was decreased significantly below that of paired-fed or *ad libitum* fed rats. The activity was reduced even further by

Table 2. Thymidine kinase activity in regenerating tissue[1]

	Deficient (A)	Paired-Fed (B)	*Ad libitum* (C)
6 day	1.04 ± 0.14 (12)	3.57 ± 0.36 (13)	3.37 ± 0.36 (12)
13 day	0.58 ± 0.02 (6)	2.40 ± 0.59 (5)	1.65 ± 0.12 (5)
17 day	None (5)	2.70 ± 0.6 (5)	2.68 ± 0.7 (5)

[1]Values are mean nmoles TMP formed per hr per mg protein ± s.e. for number rats in parentheses.
At 6 days the level of significance for deficient *versus* either control was $p < 0.001$.
At 13 days the level of significance for deficient *versus* either control was $p < 0.025$.

extended deficiency of zinc and was reduced to an immeasurable level by 17 days of dietary treatment and 10 days after implantation.

These results indicate that either thymidine kinase or the processes necessary for the induction of thymidine kinase activity are extremely sensitive to the depletion of zinc. The activity of this enzyme, in metabolic sequence, precedes the polymerization of DNA and cell division. The decreased activity or induction of the activity of thymidine kinase could be the earliest and most sensitive metabolic process which could account for the rapid decrease in growth rate, anorexia, delayed wound healing, etc., in zinc-deficient subjects. [Supported in part by USPHS, DGH Research Corporation, Michigan Heart Association, and NHLI, NIH.]

DISCUSSION

Sandstead (Grand Forks). I think you'll have to measure thymidine kinase earlier than you did, because when we studied nuclear DNA-dependent RNA polymerase in the liver of suckling rats, there was an effect within a couple of days.

Pallauf (Munich). Do you have any explanation why your pair-fed rats don't grow as well within the first week as the deficient animals?

Oberleas. I have the feeling that in the process of pair feeding, there is a little bit of diet wastage that we haven't accounted for.

Chesters (Edinburgh). I would suggest there might be a difference in thymidine kinase in your two pair-feeding experiments associated with the random chance that your animals have eaten a lot or little on the day before you actually used them for experiment. If you were to go back and check your data, you might find such an effect.

INCORPORATION OF [^{14}C] GLYCINE-2 INTO SKIN COLLAGEN IN ZINC-DEFICIENT RAT

JENG M. HSU and WILLIAM L. ANTHONY

Veterans Administration Hospital, Baltimore, Maryland

Previous work in this laboratory has demonstrated that zinc is essential in the regulation of several amino acids [Hsu *et al.*, J. Nutr. 101: 445 (1971)]. This study was undertaken to investigate further the effect of zinc deficiency on glycine incorporation into skin collagen.

Zinc-deficient rats and zinc-supplemented controls were prepared according to the procedure as described previously [Hsu *et al.*, *loc. cit.*]. Following 18 hr fasting, each rat was weighed, injected intramuscularly (5 μCi/100 g body weight) with [^{14}C] glycine-2 (Sp.A.10 μCi/mM), and sacrificed 4 and 24 hr thereafter.

One gram of dorsal skin was shaved, dissected clean, and dried to a constant weight. The weight difference before and after drying was calculated as water content of the specimen. Total lipid was determined by extracting the skin with chloroform-methanol (1:1) in a Soxhlet apparatus for 60 hr. The hydrolysates were prepared from the dried skin by sealing them in glass ampules with 6N HCl and heating them in an autoclave at 18 psi and 110°C for 18 hr. The hydrolysates were diluted and analyzed for nitrogen by the Kjeldahl technique with Nessler's reagent and for hydroxyproline by the method of Kivirikko *et al.* [Analyt. Biochem. 19: 249 (1967)].

Tissues were shaken in the cold for three days with daily changes of 0.15 M NaCl, then for three days with 0.50 M NaCl, and finally for three days with 0.50 M Na-citrate (pH 3.6). Extracts from a given solvent were combined and centrifuged. The supernatants were dialyzed for five days in the cold against distilled water, evaporated, and hydrolyzed with 6N HCl for 16 hr at 115°C. Hydrolysates were analyzed for hydroxyproline by the method mentioned previously.

Shaved dorsal skin was homogenized for 1 min with 0.2 M NaCl to make a final 5% homogenate. The tissue was shaken for 16 hr at 4°C in a Dubnoff shaker, and the mixture centrifuged at 25,000 × *g* for 1 hr in a Spinco ultracentrifuge. Soluble and insoluble collagens were extracted from the supernate and the precipitate fractions according to the method of Kao *et al.* [Proc. Soc. Exp. Biol. Med. 110: 538 (1962)]. Radioactivity in all specimens was determined by mixing with diotol and counted in a liquid scintillation spectrometer [Hsu *et al.* J. Nutr. 101: 445 (1971)].

Water content in the skin of zinc-deficient rats was significantly higher than that of zinc-supplemented pair-fed controls. The contents of lipid and nitrogen appeared to be unaffected by zinc deficiency. The amount of hydroxyproline was increased in zinc-deficient rats as compared with zinc-supplemented controls. Further studies indicate that zinc-deficient rats had a significant increase of hydroxyproline content in insoluble fraction with a concomitant decrease of hydroxyproline in both neutral soluble fractions. Table 1 shows that the radioactivities recovered as soluble and insoluble fractions were reduced in the skin of zinc-deficient rats. Treatment with insulin appears to have a greater effect than that with growth hormone in [^{14}C]-2-glycine incorporation into skin collagens regardless of zinc status. There was a significant increase in the formation of labeled collagens by insulin-injected, zinc-deficient and zinc-supplemented rats as compared with

Table 1. Incorporation of [^{14}C]glycine-2 into skin collagen

Type of diet	Hormone treatment	Hours after isotope injection	Skin collagens[1] (DPM/100μg hydroxyproline)	
			Insoluble	Soluble
Zn supplemented		24	267 ± 28^a	993 ± 79^a
Zn deficient		24	50 ± 18	412 ± 141
Zn supplemented		4	136 ± 18^a	1043 ± 99^a
Zn deficient		4	49 ± 30	582 ± 303
Zn supplemented	Growth hormone[2]	4	205 ± 74^a	1356 ± 375^a
Zn deficient	Growth hormone	4	75 ± 23	500 ± 104
Zn supplemented	Insulin[3]	4	236 ± 73^b	1848 ± 787^b
Zn deficient	Insulin	4	155 ± 56	1025 ± 261

[1]Values are mean ± s.d.; superscript a indicates $p < 0.01$ and superscript b indicates $p < 0.05$.

[2]Each received 0.1 mg of NIH GH B_{14} subcutaneously daily for the last 6 days of a 14-day experiment.

[3]Each received 0.5 U of Iletin (insulin injection, Lilly and Co., Indianapolis, Ind.)

their respective control animals without hormone treatment. This report confirms and extends the early observations indicating that zinc-deficient animals had a decreased incorporation of amino acids into skin proteins, collagen in particular. Whether the impairment of collagen synthesis as a result of dietary zinc deficiency is related with the requirement of zinc in skin DNA synthesis is not clear. Findings of a zinc-containing DNA polymerase [Slater *et al.*, Biochem. Biophys. Res. Commun. 44: 37 (1971)] suggest a possible relationship between zinc and nucleic acids. In zinc-deficient rats the suppression of skin DNA synthesis from [^{3}H] thymidine has been reported [Stephen and Hsu, J. Nutr. 103: 548 (1973)]. The incorporation of labeled precursors of DNA into DNA could result from either direct stimulation of its synthesis or the alteration in the pool size of the precursor by the zinc. Nevertheless, the decreased incorporation of thymidine into skin DNA along with the finding that zinc-deficient rats had a reduced labeling index in skin cells supports the idea that zinc directly stimulates skin DNA synthesis. Lieberman *et al.* [J. Biol. Chem. 238: 3955 (1963)] have already shown that zinc-deficient mammalian cells cultured *in vitro* have decreased activities of DNA polymerase and thymidine kinase compared with control cells. The action of zinc on the synthesis of these enzymes in rat skin remains to be studied.

DISCUSSION

Sandstead (Grand Forks). Could you tell us a little bit about the characteristics of the collagen which is found under these conditions?

Hsu. We have recently made an electron microscopy study and our pathologists have noted several changes in the zinc-deficient skin.

Strain (Cleveland). Are you going to extend your studies to the alimentary tract in the course of time?

Hsu. No, I plan to do some separation of the skin into the dermal or epidermal layers, and plan to study these separately.

Santos-Ruiz (Madrid). Do you have any information about the urinary excretion of hydroxyproline under the conditions of your experiments?

Hsu. We consistently find that as early as one week, the zinc-deficient animals excreted more total hydroxyproline (not free, but total hydroxyproline).

Quarterman (Aberdeen). Have you determined the effect of feeding excess zinc on these changes in skin?

Hsu. No, I haven't tried.

PARALLELISM BETWEEN SULFUR AND SELENIUM AMINO ACIDS IN PROTEIN SYNTHESIS IN THE SKIN OF ZINC-DEFICIENT RATS

K. P. McCONNELL, JENG M. HSU, J. L. HERRMAN, and W. L. ANTHONY

Veterans Administration Hospital and University of Louisville School of Medicine, Louisville, Kentucky, and Veterans Administration Hospital, Baltimore, Maryland

Studies in our laboratory and elsewhere have led to the assumption that, in protein synthesis and other biochemical pathways, selenium metabolism occurs via seleno-analogs of the relevant sulfur-containing metabolites, methionine (Met) and cystine (Cyst) in the case of protein biosynthesis. Probably the most interesting information in these studies is the similarity between Met and selenomethionine (Se Met), because much attention has been directed towards the presence of selenium in proteins as a starting point for defining the biochemical role of this element [Rotruck *et al.*, Fed. Proc. 31: 2684 (1972)]. Using a cell-free protein-synthesizing system, we have observed in both eucaryotes and procaryotes that selenomethionyl-tRNA participates in polypeptide synthesis to about the same extent as methionyl-tRNA, and there is competitive inhibition between Met and Se Met [McConnell *et al.*, Proc. Soc. Exp. Biol. Med. 140: 6381 (1972); FEBS Lett. 24: 604 (1972); Hoffman *et al.*, Biochim. Biophys. Acta 199: 531 (1970)]. Similarity between Met and Se Met has been demonstrated in yet another way. It has been established that the condition for active transport of Se Met across the intestine of the hamster is nearly identical with that for active transport of Met [McConnell and Cho, Am. J. Physiol. 208: 1191 (1965), *ibid.* 213: 150 (1967)]. It was suggested that Se Met and Met utilized the same transport system.

Since it is known that uptake of various amino acids, which include L-Met and L-Cyst in the skin and hair of zinc-deficient rats (ZnD), is significantly lower than the zinc-supplemented controls (ZnS) [Hsu *et al.*, J.

Table 1. Twenty-four-hour incorporation of L-[^{75}Se]selenomethionine and L-[^{75}Se]selenocystine into skin proteins of the rat

Type of diet	Specific activity[1] (epm/mg protein)			
	Met[2]	Cyst[3]	Se Met	Se Cyst
Zn supplemented (pair-fed)	654 ± 170	1112 ± 295	103 ± 33	139 ± 23
Zn deficient	220 ± 84[a]	306 ± 156[a]	28 ± 14[a]	81 ± 17[a]

[1]Values are mean ± s.e. Five to six rats in each group were on the experimental diet for 15 days. Superscript [a] indicates that the difference between Zn-supplemented (pair-fed) and Zn-deficient rats is statistically significant ($p < 0.05$).
[2]L-[^{35}S], methionine [J. Nutr. 102: 118 (1972)].
[3]L-^{35}S cysteine [J. Nutr. 101: 445 (1971)].

Nutr. 101: 445 (1971), *ibid.* 102: 1181 (1972)], studies were made to investigate the parallelism between sulfur (Met and Cyst) and selenium (Se Met and Se Cyst) amino acids in protein synthesis in the skin of ZnD rats. As shown in Table 1, similar results to those with Met and Cyst were found with Se Met and Se Cyst. At 24 hr the specific activity (cpm/mg protein) present in the skin of the ZnD rats was 25% and 58% that of the pair-fed controls, after i.m. injection of L- ^{75}Se Met and L- ^{75}Se Cyst, respectively. No significant differences were observed in the ^{75}Se content of the liver, kidneys, pancreas, and muscle between the two groups which is consistent also with the sulfur amino acids. There was no significant difference in the nonradioactive selenium and zinc in the tissues of the ZnD and ZnS rats. An explanation of the impairment of skin protein synthesis in the ZnD rats of the selenium as well as sulfur amino acids may be found in studies of DNA synthesis in the skin [Stephan and Hsu, J. Nutr. 103: 548 (1972)]. Skin DNA synthesis was observed using labeled [^{3}H] thymidine as affected by zinc deficiency. Suppression of DNA formation was clearly demonstrated in rats receiving a ZnD diet. The data support the idea that zinc directly stimulates skin DNA synthesis. Results reported here clearly establish a parallelism between sulfur and selenium amino acids in protein synthesis and strongly suggest that zinc plays a role directly or indirectly in the metabolism of Se Met and Se Cyst in protein synthesis in the skin.

DISCUSSION

Frost (Schenectady). It has been reported that selenium from selenite is taken up in DNA and in transfer RNA. Would you comment on this?

McConnell. In some experiments that we'll probably report a little later, we've found that in *Escherichia coli,* a procaryotic cell, selenium does go into the nucleic acid bases. At the present time we are working on the mammalian system and we have no reason to believe that it is different. This may complicate the picture a little more.

THE EFFECT OF SUPPLEMENTARY LYSOZYME ON ZINC DEFICIENCY IN THE RAT

R. W. LUECKE, BETTY V. BALTZER, and D. L. WHITENACK

Michigan State University, East Lansing, Michigan

Published reports of experimentally produced zinc deficiency in the rat indicates a wide range in the severity of the skin lesions produced. In general, two types of diet have been used to produce zinc deficiency, one a demineralized protein, usually casein or soybean protein, and the other a diet containing egg white or egg albumin as the protein source.

We previously reported [Fed. Proc. 30: 2502 (1971)] that, when a zinc-low diet containing demineralized casein was fed to weanling rats, the external manifestations of zinc deficiency were much more severe than when biotin-fortified egg white was used as the protein source, even though the zinc content of both diets was similar. Although Chu *et al.* [Nutr. Rep. Int. 1: 11 (1970)] indicated that supplementary biotin partially alleviated the severe dermatitis produced by feeding a zinc-low diet, we were unable to obtain any improvement by including biotin at a level of 4 mg/kg diet. Pallauf and Kirchgessner [Int. J. Vit. Nutr. Res. 42: 555 (1972)] were also unable to find any relationship between biotin and zinc.

In our experience feeding the zinc-deficient (0.6 ppm Zn) diet containing demineralized casein to young rats produced severe skin lesions including alopecia, a serious exudation, swelling, and bleeding of the paws. There was alopecia and an exudation on the ventral abdomen and lower jaws of these rats. Some of them showed a kangaroo-like posture as described by others. Histopathological examination of the skin of zinc-deficient rats fed the demineralized casein diet indicated a severe inflammatory reaction on the feet and abdomen. These inflammatory lesions were characterized by a severe neutrophilic response in the dermis with desquamation of the epithelium. On the surface of the skin and between the keratotic layers, there were numerous bacteria and a few yeast cells that were always much more numerous on the rats fed the demineralized casein diet. It seemed clear that the rats fed the demineralized casein diet showed evidence of a much greater number of microorganisms on the skin surface and in the parakeratotic layers of the epithelium of the tongue and esophagus. It should be pointed out, however, that all of these pathologic symptoms could be prevented or cured by zinc supplementation of the diet.

An investigation of the possible causes of the differences in severity of skin lesions produced by feeding the two protein sources was begun. Increasing the vitamin fortification of the diet as well as the addition of L-carnitine, L-inositol, para-aminobenzoic acid, and ascorbic acid to the vitamin mixture was without effect. Increases in the levels of copper, manganese, and cobalt of the trace-mineral mixture was also without effect. In addition, it was felt that while there were differences in amino acid composition between the two protein sources, it seemed unlikely that the higher level of sulfur-containing amino acids in the egg-white diet could account for the differences observed. However, the possibility existed that the differences observed between the two protein sources could not be attributed to any nutrient deficiency, and that an alternative explanation might lie in the fact that the egg-white product used contained lysozyme. This enzyme, functioning as an antibacterial agent in the diet, might prevent the proliferation of bacteria on the skin, mouth, and esophagus. Furthermore, casein is known to be devoid of this enzyme. Subsequently, assay of the spray-dried egg white used in the diet confirmed the presence of lysozyme and, in addition, the enzyme was found in the feces of rats fed the egg-white diets, but not in the feces of rats fed the casein diet. Lysozyme is known to cause lysis of the cell wall of certain bacteria which contain specific polysaccharides (N-acetylglucosamine and N-acetylmuramic acid).

Several experiments were conducted in which a zinc-low egg-white diet was compared with a zinc-low demineralized casein diet, both alone and supplemented with crystalline egg-white lysozyme at levels of 50, 100, 250, and 500 mg/kg. In addition, the effect of another antibacterial agent, chlortetracycline, was also used as a supplement to the demineralized casein diet. The results of these studies revealed that the addition of the higher levels of lysozyme not only delayed the onset of the skin lesions, but also reduced their severity, at least for the four-week experimental period. The addition of chlortetracycline had much the same effect as the lysozyme. In no case was growth improved by any of the supplements, which was to be expected since the diet was very low in zinc. However, since Smith *et al.* [J. Nutr. 102: 711 (1972)] have found that the presence of certain microorganisms may increase the dietary zinc requirement, it is postulated that, in our experiments, the presence of the antibiotics may have altered the dietary zinc requirement by reducing the total body burden of certain microorganisms. [Michigan Agricultural Experiment Station Journal Article No. 6406.]

DISCUSSION

Chesters (Aberdeen). I should like to say that I think this is fascinating and it does explain a lot of the reported differences between symptoms of zinc deficiency in the literature, especially if you take into account the possible effect of the method of management of the animals, whether they are SPS, conventional, or germfree.

Elmes (Belfast). I have used an egg-white diet, but a slightly different source of egg white than the Rowett people. I think that my rats had a deficiency just about half way between your lots. They didn't have severe skin lesions, but they did have swollen red paws and a bit of prophyrin excretion around their noses.

Luecke. We do observe the swollen paws at times. The spray-dried product we use is processed by spray drying at about 135°F, which certainly does not destroy lysozyme.

THE EFFECTS OF ZINC DEFICIENCY OR EXCESS ON THE ADRENALS AND THE THYMUS IN THE RAT

J. QUARTERMAN

Rowett Research Institute, Bucksburn, Aberdeen, U.K.

Changes in tissue Zn distribution and metabolism have been detected following tissue trauma [Fell *et al.,* Lancet 1: 280 (1973)] and decreases of plasma Zn during immunization [Svitsun, Mikroelem. Med. 2: 101 (1971)], infection [Pekarek and Beisel, Appl. Microbiol. 18: 482 (1969)] and steroid therapy [Flynn *et al.,* Lancet 1: 789 (1973)]. Oral Zn administration is believed to improve wound healing [Pories *et al.,* Lancet 1: 121 (1967)] and Zn deficiency to delay it [Flynn *et al., loc. cit.*]. There is very little known, however, about the effects of Zn deficiency or supplementation on metabolic processes which may be related to wound healing or the resistance to infection.

The adrenal glands of Zn-deficient rats are heavier and the thymus glands lighter than those of pair-fed control rats given Zn (Table 1). The adrenal glands of Zn-deficient rats always contain more cholesterol than those of controls, and the amount of cholesterol lost from the glands within 3 hr of an intraperitoneal dose of ACTH is much greater in Zn-deficient animals than in controls [Quarterman, Proc. Nutr. Soc. 31: 74A (1972)].

The changes in adrenal cholesterol occur within 6–10 days of feeding the deficient diets, but a significant decrease in the weight of the thymus does not occur until the rats have been receiving such diets for 30 days. The Zn content of the adrenals and the thymus is little changed by deficiency, but, because of the decreased weight of the thymus, there is a constant or even an increased concentration of Zn in this tissue in Zn-deficient rats. The protein and DNA concentrations in the thymus are unchanged; therefore, the loss in weight is probably due to a decreased number of cells.

If the adrenal hypertrophy and increased cholesterol concentration were associated with an increased output of corticosterone, this could provide an explanation for the involution of the thymus. Plasma 11-hydroxysteroids

Table 1. Effect of Zn deficiency and excess on the adrenals and the thymus[1]

	Wt adrenals (mg/100 g b.w.)	Wt thymus (mg/100 g b.w.)	Thymus Zn (μg/100 g b.w.)	Thymus DNA (μg/mg w.w.)
Experiment I				
Zn deficient	33.5 ± 1.2^{c}	146.4 ± 12.7^{b}	3.3 ± 0.5	3.32 ± 0.08
Control	26.9 ± 0.7	215.1 ± 14.5	3.5 ± 0.3	3.57 ± 0.14
Experiment II				
Zn supplemented	24.7 ± 0.8^{b}	392.3 ± 10.0^{a}	7.6 ± 0.7	3.29 ± 0.10
Control	20.5 ± 0.6	358.4 ± 10.9	6.5 ± 0.6	3.34 ± 0.09

[1]Values significantly different from controls are indicated by the following superscripts: $^{a}P < 0.05$; $^{b}P < 0.01$; $^{c}P < 0.001$.

have been measured spectrofluorimetrically in anesthetized rats with or without prior ACTH treatment in 15 experiments. The plasma steroid concentrations of Zn-deficient rats were significantly higher than of controls in five experiments and never less than those of controls.

In addition to the above experiments with Zn-deficient rats, some rats were given a large excess of Zn (about 10 mg $ZnSO_4$ per day in the drinking water). The Zn dosing produced increases in weight of both adrenals and thymus and a decrease in plasma 11-hydroxysteroid concentration, but changes in adrenal cholesterol similar to those seen in Zn deficiency. We have found that thymus weight is also affected by the level of daily food intake, and this variable must be controlled in experiments with Zn. Other workers [Smythe *et al.*, Lancet 2: 939 (1971); Schonland *et al.*, Lancet 2: 435 (1972)] have shown that protein-calorie malnutrition is associated with poor wound healing, elevated plasma steroid levels, and thymolymphatic deficiency.

Thus Zn deficiency and excess have produced opposite effects on thymus weight and plasma steroids, and it may be that the observed effects of Zn on healing and disease are mediated by its effects on steroid metabolism. It is anomalous that both deficiency and excess have produced similar effects on adrenal weight and cholesterol concentration.

Other lymphoid tissue may be affected in the same way as the thymus. Dreosti and co-workers [Proc. Soc. Exp. Biol. Med. 128: 169 (1968)] have reported a decrease in lymphocyte numbers which occurs when rats have been eating a Zn-deficient diet for three weeks, but we have found no change in the weight of Peyer's patches after 30–40 days on a deficient diet.

DISCUSSION

Hoekstra (Madison). As you probably know, neonatal thymectomized rodents develop a wasting syndrome and often rough hair coats similar to a zinc-deficiency syndrome. Do you think that the lack of thymus function is an important part of the zinc-deficiency syndrome?

Quarterman. I like to think so, though I haven't any more evidence than I have produced here. The timing of the decrease in thymus weight coincides with the timing of the decrease of plasma lymphocyte numbers which L. Hurley reported a few years ago. In our case, it took about three weeks for plasma lymphocyte numbers to decrease significantly. The small thymus weight and the small number of lymphocytes which Dr. Hurley has mentioned correlate quite well with recent observations in the medical literature in relation to kwashiorkor. Kwashiorkor children have low thymus weight, they have increased susceptibility to infection, high plasma steroids, poor wound healing, and low blood complement levels. I hope to be able to demonstrate that there is a good analogy between these two states.

ZINC DEFICIENCY DURING THE CRITICAL PERIOD FOR BRAIN GROWTH

H. H. SANDSTEAD, Y. Y. AL-UBAIDI, E. HALAS, and G. FOSMIRE

Human Nutrition Laboratory, U.S. Department of Agriculture, Agricultural Research Service, and the University of North Dakota, Grand Forks, North Dakota

Zinc deficiency in suckling rats retards brain maturation. At 12 days of age the incorporation of thymidine into DNA and of sulfur into the TCA precipitable fraction of the brain are decreased as is the total lipid concentration [Sandstead *et al.*, Pediat. Res. 6: 119 (1972)]. Accordingly, the effect of neonatal zinc deficiency on brain polysomes and on subsequent behavioral development of the nutritionally rehabilitated rat were examined.

The methods used for raising the rats in this study have been described previously [Sandstead *et al.*, *loc. cit.*]. The polysome profile of the 0.5% deoxycholate-treated brain postmitochondrial supernatant was assayed at wavelength 260 mμ after centrifugation on a 16–48% w/v continuous sucrose gradient. Acquisition of an elevated Tolman-Honzig maze was assessed from 50 to 75 days of age. The findings in the zinc-deprived group were compared with data from pups nursed by pair-fed or *ad libitum* fed control dams.

Figure 1 is a representative sucrose-density polysome profile which shows the effect of zinc deficiency at five days of age on the brain. The polysome portion of the gradient is decreased, while the lighter fractions are increased in the postmitochondrial supernatant of brain from the zinc-deficient pup, compared with material from the pair-fed and *ad libitum* control pups. It is evident that nursing on the pair-fed dam had an effect on polysomes which was intermediate between that on the zinc-deficient and *ad libitum* control animals.

Figure 2 illustrates the growth curves, maze acquisition, and the running time of the animals subsequent to nutritional rehabilitation from weaning at 22 days of age. At 50–75 days of age rats exposed to zinc deficiency from birth to weaning performed significantly ($P < 0.01$) less well than either group of control animals.

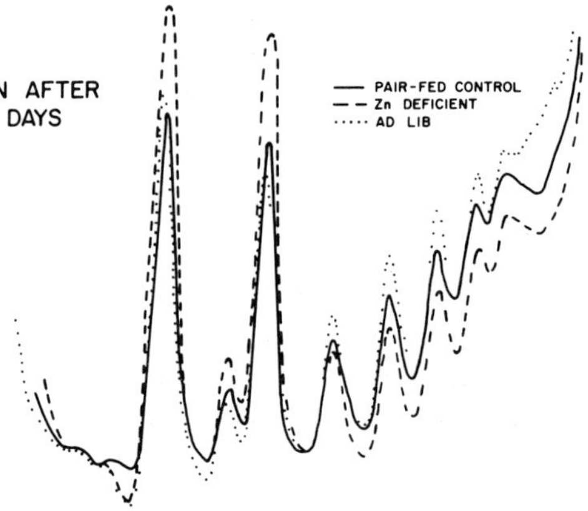

Fig. 1. Polysome profiles from Zn-deficient and control animals.

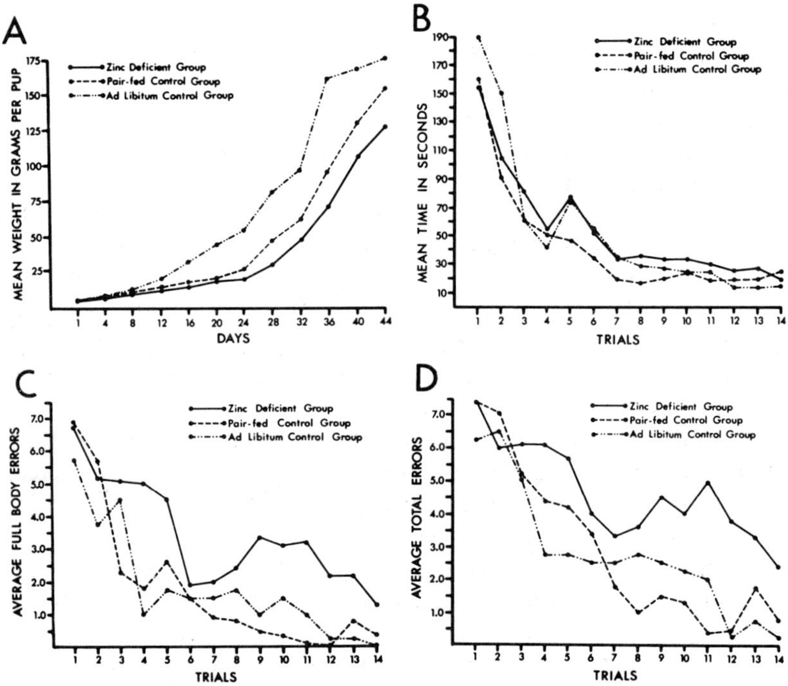

Fig. 2. Growth curves, maze acquisition and running time of animals subsequent to nutritional rehabilitation.

Zinc deficiency during the neonatal period appears to have retarded behavioral development. Changes were present in the polysome profile which may imply that abnormalities in protein synthesis occurred. Earlier studies had indicated that DNA synthesis in the brain was also impaired by zinc deficiency [Sandstead *et al., loc. cit.*]. Correlative with these biochemical findings in the zinc-deficient pups was their decreased maze acquisition following nutritional rehabilitation. In contrast, rats exposed to starvation as a consequence of being nursed by pair-fed dams acquired the maze as well as pups which had been nursed by *ad libitum* fed dams. These findings do not necessarily indicate that the rats which had been exposed to zinc deficiency were less intelligent; they may simply reflect changes in motivation or emotionality. In any case, they suggest that further investigation of the relationship between behavioral development and zinc nutriture is warranted. [This work was supported in part by Cooperative Agreement 1-12-14-100-11,178 (61) between the Agricultural Research Service and the University of North Dakota.]

DISCUSSION

Frost (Schenectady). How do these changes tie in with the changes observed by others working with protein inadequacy? Is this the same sort of thing that they measure?

Sandstead. There are a number of things that can be measured. The effects are similar to those one would find in protein malnutrition, and might be found if you made the rats deficient in many of the essential nutrients during the critical period of brain growth. The reason for picking zinc is that in human protein-calorie malnutrition, many of the individuals are also zinc deficient, and if zinc is essential for nitrogen utilization we would expect that there may be a mixed lesion present there. The whole story may not be just protein.

Karsarskis (Madison). You showed that all animals had the same running time as they progressed through the learning maze and yet the animals that were zinc deficient had a greater number of errors. Is this a common pattern to see in a general deprivation study of this type?

Sandstead. It depends when you deprive the animal. Sometimes you find that deficient animals will run faster because one of the effects of deficiency is simply to make them more emotional or to change their motivation. One cannot say that this study necessarily means we have affected learning; we simply have affected other aspects of behavior. Learning studies in experimental animals are extremely difficult to interpret, and it's dangerous to say that one has proven that learning is affected because there are so many aspects to behavior that influence the performance of an animal.

Chesters (Aberdeen). You showed that there was less lipid in the zinc-deficient animal. Have you determined the nature of the lipids which are involved in this?

Sandstead. Yes, we did a fatty acid pattern, and there was no difference in zinc-deficient and normal animals. I might say incidentally, Dr. Quarterman, that the thymus glands of the animals are significantly atrophied at 12 days of age if you produce the deficiency this way.

MANGANESE, ZINC, AND GENES IN OTOLITH DEVELOPMENT

LAWRENCE C. ERWAY and NICHOLAS A. PURICHIA

University of Cincinnati, Cincinnati, Ohio, and Marian College, Indianapolis, Indiana

Manganese deficiency during pregnancy in mice produced congenital otolith defects [J. Nutr. 100: 643 (1970)], thus explaining most of the behavioral defects observed by numerous investigators in manganese-deficient chicks, rats, and guinea pigs. The otolith defects in manganese-deficient mice were similar to those observed by Lyon in pallid mutant mice [J. Embryol. Exp. Morphol. 3: 230 (1955)]. Manganese supplementation of pallid mice prior to day 15 of gestation prevented the otolith defect [Genetics 67: 97 (1971)].

Histochemical evidence and $^{35}SO_4$-incorporation studies [Teratology 8: 257 (1973)] indicate that manganese is essential for the biosynthesis of the sulfated mucopolysaccharides of which the otolith membrane is composed. The requirement for manganese (days 11–15 *in utero*) preceded the appearance (days 14–15) and successive increase of a mass of otoconial crystals (days 15–17). The mucopolysaccharide matrix appears to be a prerequisite for the intra-otoconial matrix as observed by Lim [Ann. Oto. Rhino. Laryngol. 82: 23 (1973)], as well as for the otolith membrane.

The genetic basis for involvement of manganese in otolith development is a complex one. However, six pigment mutant genes in four species are now known to produce an associated otolith defect. Our most recent studies [J. Hered. 64: 111 (1973)] in pastel mink also demonstrate that manganese supplementation prevents an otolith defect, as well as the associated 'screw-neck' phenomenon. All of these pigment mutant genes interfere with the development of melanocytes within the inner ear. Moreover, the effect of the pallid gene on $^{35}SO_4$-incorporation is, in contrast to manganese-deficient mice, highly localized to the otolith matrix but not to the surrounding periotic cartilage. We believe that there is a causal relationship between the presence of pigment cells in the inner ear and availability of manganese for otolith development.

Nevertheless, manganese is not the only trace element required for formation of otoconia. Zinc deficiency also caused otolith defects when the pregnant mice were fed the diet beginning as late as day 16 of gestation [Develop. Biol. 27: 395 (1972)]. Moreover, dichlorophenamide caused otolith defects as late as day 18. This drug is a specific inhibitor of carbonic anhydrase, of which zinc is the metallic cofactor. Dichlorophenamide and other related sulfonamides may chelate zinc and thus interfere with normal, physiological conditions of the endolymphatic fluid of the inner ear, causing changes in otoconia.

In an attempt to understand the role of zinc in otolith formation, we determined, after the method of Nelson *et al.* [Phytochemistry 8: 2305 (1969)], carbonic anhydrase (CA) activity in single adult mouse ears and in pooled samples of 10 ears of fetal (days 14–18) and of newborn mice [unpublished]. The earliest detectable amounts of CA were observed at day 16, but it increased steadily through day 18 (~620 EU/10 fetal ears) and in newborn mice (~1200 EU/10 ears). The CA activity in the intact ear (~3500 EU) and for the isolated cochlea (~11,500 EU) was comparable to that observed for cat ears by Erulkar and Maren [Nature, Lond. 189: 459 (1961)]. Addition of dichlorophenamide to the incubation medium inhibited all CA activity. Marmo [Acta Embryol. Morphol. Exp. 9: 118 (1966)] demonstrated similar relationships between dichlorophenamide, carbonic anhydrase, and otolith formation in chick embryos.

These results indicate that adequate amounts of zinc are required for the formation and/or maintenance of the calcium carbonate (calcite) otoconia. Therefore we also determined the rate of incorporation of ^{45}Ca into the unossified, fetal mouse ear. The pregnant mice were injected with a single dose of 20 μCi of $^{45}CaCl_2$ on days 14–18. Fetuses were removed from 1 to 8 hr later, and single ears were dissected and prepared for liquid scintillation. The rate of incorporation was correlated with the more subjective evaluation of otolith formation, reaching the maximum (~900 dpm/fetal ear) by day 17.

These results suggest that formation of otoconia is a particularly sensitive bioassay for levels of manganese and zinc. Moreover, the mouse ear is especially easy to prepare and score for otoliths. Other chelating agents and nutritional conditions may lead to these defects and to related subtle but significant behavioral defects which we are now investigating. [Supported in part by USPHS grant AM 14613.]

DISCUSSION

Hill (London). This is a nice specific site to study; however, I'd like to know if other parts of the body being calcified or mineralized are also affected in these genetically abnormal mice. How much extra manganese do you need to overcome this genetic abnormality?

Erway. We had to go up to a dietary level of 1500–2000 ppm manganese; 1000 ppm was not enough. We have not yet done any work on how much was absorbed, or how much gets to the ear. These are things yet to be worked out. Zinc or magnesium will not supplement in the case of mutant animals. It seems to be specifically manganese.

Author Index

Figures in italics refer to contributed papers.

Subject Index